BASIC HEAT TRANSFER

Third Edition

A. F. Mills
Professor of Mechanical and Aerospace Engineering, Emeritus
The University of California at Los Angeles, Los Angeles, CA

C. F. M. Coimbra
Professor of Mechanical and Aerospace Engineering
The University of California at San Diego, La Jolla, CA

Temporal Publishing, LLC – San Diego, CA 92130

Library of Congress Cataloging-in-Publication Data

Mills, A. F. and Coimbra, C. F. M.
    Basic Heat Transfer 3/E by Anthony F. Mills & Carlos F. M. Coimbra
        p. cm.
    Includes bibliographical references and index.
    ISBN 978-0-9963053-1-0
CIP data available.

Address inquiries or comments to:
contact@temporalpublishing.com   www.temporalpublishing.com

Printed in the United States of America

1 0 9 8 7 6 5
ISBN 978-0-9963053-1-0

To Brigid
For your patience and understanding.

To Kaori
For your loving support.

PREFACE

For the third edition of *Basic Heat Transfer* Anthony Mills is joined by Carlos Coimbra as a co-author. Professor Coimbra brings to this venture the perspective and skills of a younger generation of heat transfer educators and his own special expertise in areas of heat transfer research. The second editions of the texts by Anthony Mills included *Heat Transfer*, *Basic Heat and Mass Transfer* and *Mass Transfer*, in order to provide texts suitable for a variety of courses and instructor's needs. In bringing out a third edition of this material we are adding a new title *Basic Heat Transfer* in response to requests for a text suitable for a junior or senior level course that covers heat transfer only. This goal was accomplished by essentially removing the chapter on mass transfer from *Basic Heat and Mass Transfer*, which reduced the text length from 1001 pages to 828 pages. We have designated *Basic Heat Transfer* as a third edition to avoid possible confusion since it is based on the third edition updated material of the textbook series. A third edition of the more advanced text *Heat Transfer* will be released later.

Fifteen years after the second edition was published, a new edition of these materials is perhaps overdue, but in a mature field such as heat transfer, it is not at all clear what topics should be introduced, and then what topics should be removed to retain an acceptable length for an introductory text. As a result, our main motivation in publishing a third edition has been a different consideration.

Our concern was the excessive prices of college textbooks, which in recent years have destroyed the established role played by these texts in the education of engineering students. Traditionally, students bought a required textbook, became familiar with it in taking the course, and then retained the book as a tool for subsequent courses and an engineering career. Nowadays the pattern is for a student to sell the textbooks back to the university bookstore at the end of the course in order to obtain funds for buying textbooks for the next term. Alternatively, electronic versions of portions of the text are used during the course, or course readers containing selected material from the text may be used. It is particularly frustrating to instructors of subsequent design and laboratory courses to find that the students no longer have appropriate textbooks. Also, the traditional role formerly played by textbooks as professional manuals for engineering practice has been significantly affected. Basic methodology and data are more easily and reliably obtained from a familiar text than from an internet search.

In an attempt to mitigate these problems and improve the experience of our engineering students we decided to retain creative and publishing rights over the content of this book for this and future editions. A company called Temporal Publishing LLC was created to publish quality engineering textbooks at more reasonable prices.[1] This entailed first converting the previous edition to LaTeX, which we could then modify efficiently. Since the conversion proved to be a major project in itself, our objective with this third edition is rather modest. We have focused on corrections, clarifications, minor updates and the production of a dedicated companion website.[2] We envisage this website to be an integral part of the project and hope to make it a really useful adjunct to the text, for both students and instructors. The website contains links to the dedicated software HT that automates most of the calculations done in the text, instructor aides (such as complete solutions manual for adoptees of the text, additional examples and exercises, presentations, etc.) and a compilation of answers to odd-numbered exercises to assist self-study by students. We will be continuously adding new technical content to the website while we work on future editions of the textbook. Also, given our closer association with the print-on-demand process, it will be easy for the authors to implement small improvements in subsequent printings of this edition. We certainly welcome input and suggestions from users to improve our product.

In preparing this new edition we have had valuable assistance from:

Marius Andronie

Kuang Chao

Kaori Yoshida Coimbra

We would like to dedicate the collaborative effort of bringing a new edition of *Basic Heat Transfer* to the memory of Prof. Donald K. Edwards, our teacher.

A. F. Mills
Santa Barbara, CA
amills@ucla.edu

C. F. M. Coimbra
La Jolla, CA
ccoimbra@ucsd.edu

---

[1] Books can be ordered directly at discounted prices at www.temporalpublishing.com

[2] www.temporalpublishing.com/bht-students

PREFACE TO THE SECOND EDITION

Basic Heat and Mass Transfer has been written for undergraduate students in mechanical engineering programs. Apart from the usual lower-division mathematics and science courses, the preparation required of the student includes introductory courses in fluid mechanics and thermodynamics, and preferably the usual junior-level engineering mathematics course. The ordering of the material and the pace at which it is presented have been carefully chosen so that the beginning student can proceed from the most elementary concepts to those that are more difficult. As a result, the book should prove to be quite versatile. It can be used as the text for an introductory course during the junior or senior year, although the coverage is sufficiently comprehensive for use as a reference work in undergraduate laboratory and design courses, and by the practicing engineer.

Throughout the text, the emphasis is on engineering calculations, and each topic is developed to a point that will provide students with the tools needed to practice the art of design. The worked examples not only illustrate the use of relevant equations but also teach modeling as both an art and science. A supporting feature of *Basic Heat and Mass Transfer* is the fully integrated software available from the author's website[3]. The software is intended to serve primarily as a tool for the student, both at college and after graduation in engineering practice. The programs are designed to reduce the effort required to obtain reliable numerical results and thereby increase the efficiency and effectiveness of the engineer. I have found the impact of the software on the educational process to be encouraging. It is now possible to assign more meaningful and interesting problems, because the students need not get bogged down in lengthy calculations. Parametric studies, which are the essence of engineering design, are relatively easily performed. Of course, computer programs are not a substitute for a proper understanding. The instructor is free to choose the extent to

---

[3] http://www.mae.ucla.edu/people/faculty/anthony-mills

which the software is used by students because of the unique exact correspondence between the software and the text material. My practice has been to initially require students to perform various hand calculations, using the computer to give immediate feedback. For example, they do not have to wait a week or two until homework is returned to find that a calculated convective heat transfer coefficient was incorrect because a property table was misread.

The extent to which engineering design should be introduced in a heat transfer course is a controversial subject. It is my experience that students can be best introduced to design methodology through an increased focus on equipment such as heat and mass exchangers: *Basic Heat and Mass Transfer* presents more extensive coverage of exchanger design than do comparable texts. In the context of such equipment one can conveniently introduce topics such as synthesis, parametric studies, tradeoffs, optimization, economics, and material or health constraints. The computer programs HEX2 and CTOWER assist the student to explore the consequences of changing the many parameters involved in the design process. If an appropriate selection of this material is taught, I am confident that Accreditation Board for Engineering and Technology guidelines for design content will be met. More important, I believe that engineering undergraduates are well served by being exposed to this material, even if it means studying somewhat less heat transfer science.

More than 300 new exercises have been added for this edition. They fall into two categories: (1) relatively straightforward exercises designed to help students understand fundamental concepts, and (2) exercises that introduce new technology and that have a practical flavor. The latter play a very important role in motivating students; considerable care has been taken to ensure that they are realistic in terms of parameter values and focus on significant aspects of real engineering problems. The practical exercises are first steps in the engineering design process and many have substantial design content. Since environmental considerations have required the phasing out of CFC refrigerants, R-12 and R-113 property data, worked examples and exercises, have been replaced with corresponding material for R-22 and R-134a.

Basic Heat and Mass Transfer complements *Heat Transfer*, which is published concurrently. *Basic Heat and Mass Transfer* was developed by omitting some of the more advanced heat transfer material from *Heat Transfer* and adding a chapter on mass transfer. As a result, *Basic Heat and Mass Transfer* contains the following chapters and appendixes:

1. Introduction and Elementary Heat Transfer

2. Steady One-Dimensional Heat Conduction

3. Multidimensional and Unsteady Conduction

4. Convection Fundamentals and Correlations

5. Convection Analysis

6. Thermal Radiation

7. Condensation, Evaporation, and Boiling

8. Heat Exchangers

9. Mass Transfer

A. Property Data

B. Units, Conversion Factors, and Mathematics

C. Charts

In a first course, the focus is always on the key topics of conduction, convection, radiation, and heat exchangers. Particular care has been taken to order the material on these topics from simpler to more difficult concepts. In Chapter 2 one-dimensional conduction and fins are treated before deriving the general partial differential heat conduction equation in Chapter 3. In Chapter 4 the student is taught how to use convection correlations before encountering the partial differential equations governing momentum and energy conservation in Chapter 5. In Chapter 6 radiation properties are introduced on a total energy basis and the shape factor is introduced as a geometrical concept to allow engineering problem solving before having to deal with the directional and spectral aspects of radiation. Also, wherever possible, advanced topics are located at the ends of chapters, and thus can be easily omitted in a first course.

Chapter 1 is a brief but self-contained introduction to heat transfer. Students are given an overview of the subject and some material needed in subsequent chapters. Interesting and relevant engineering problems can then be introduced at the earliest opportunity, thereby motivating student interest. All the exercises can be solved without accessing the property data in Appendix A.

Chapters 2 and 3 present a relatively conventional treatment of heat conduction, though the outdated and approximate Heissler and Grober charts are replaced by exact charts and the computer program COND2. The treatment of finite-difference numerical methods for conduction has been kept concise and is based on finite-volume energy balances. Students are encouraged to solve the difference equations by writing their own computer programs, or by using standard mathematics software such as Mathcad or MATLAB.

In keeping with the overall philosophy of the book, the objective of Chapter 4 is to develop the students' ability to calculate convective heat transfer coefficients. The physics of convection is explained in a brief introduction, and the heat transfer coefficient is defined. Dimensional analysis using the Buckingham pi theorem is used to introduce the required dimensional groups and to allow a discussion of the importance of laboratory experiments. A large number of correlation formulas follow; instructors can discuss selected geometrical configurations as class time allows, and students can use the associated computer program CONV to reliably calculate heat transfer coefficients and skin friction coefficients or pressure drop for a wide range of configurations. Being able to do parametric studies with a wide variety of correlations enhances the students' understanding more than can be accomplished by hand calculations. Design alternatives can also be explored using CONV.

Analysis of convection is deferred to Chapter 5: simple laminar flows are considered, and high-speed flows are treated first in Section 5.2, since an understanding of the recovery temperature concept enhances the students' problem-solving capabilities. Mixing length turbulence models are briefly discussed, and the chapter closes with a development of the general conservation equations.

Chapter 6 focuses on thermal radiation. Radiation properties are initially defined on a total energy basis, and the shape factor is introduced as a simple geometrical concept. This approach allows students to immediately begin solving engineering radiation exchange problems. Only subsequently need they tackle the more difficult directional and spectral aspects of radiation. For gas radiation, the ubiquitous Hottel charts have been replaced by the more accurate methods developed by Edwards; the accompanying computer program RAD3 makes their use particularly simple.

The treatment of condensation and evaporation heat transfer in Chapter 7 has novel features, while the treatment of pool boiling is quite conventional. Heatpipes are dealt with in some detail, enabling students to calculate the wicking limit and to analyze the performance of simple gas-controlled heatpipes.

Chapter 8 expands the presentation of the thermal analysis of heat exchangers beyond the customary and includes the calculation of exchanger pressure drop, thermal-hydraulic design, heat transfer surface selection for compact heat exchangers, and economic analysis leading to the calculation of the benefit-cost differential associated with heat recovery operations. The computer program HEX2 serves to introduce students to computer-aided design of heat exchangers.

Chapter 9 is an introduction to mass transfer. The focus is on diffusion in a stationary medium and low mass-transfer rate convection. As was the case with heat convection in Chapter 4, mass convection is introduced using dimensional analysis and the Buckingham pi theorem. Of particular importance to mechanical engineers is simultaneous heat and mass transfer, and this topic is given detailed consideration with a focus on problems involving water evaporation into air.

The author and publisher appreciate the-efforts of all those who provided input that helped develop and improve the text. We remain dedicated to further refining the text in future editions, and encourage you to contact us with any suggestions or comments you might have.

A. F. Mills
amills@ucla.edu

Bill Stenquist
Executive Editor
william_stenquist@prenhall.com

ACKNOWLEDGEMENTS TO THE FIRST AND SECOND EDITIONS

Reviewers commissioned for the first edition, published by Richard D. Irwin, Inc., provided helpful feedback. The author would like to thank the following for their contributions to the first edition.

Martin Crawford, University of Alabama—Birmingham

Lea Der Chen, University of Iowa

Prakash R. Damshala, University of Tennessee—Chattanooga

Tom Diller, Virginia Polytechnic Institute and State University

Abraham Engeda, Michigan State University

Glenn Gebert, Utah State University

Clark E. Hermance, University of Vermont

Harold R. Jacobs, Pennsylvania State University—University Park

John H. Lienhard V, Massachusetts Institute of Technology

Jennifer Linderman, University of Michigan—Ann Arbor

Vincent P. Mano, Tufts University

Robert J. Ribando, University of Virginia

Jamal Seyed-Yagoobi, Texas A&M University—College Station

The publisher would also like to acknowledge the excellent editorial efforts on the first edition. Elizabeth Jones was the sponsoring editor, and Kelley Butcher was the senior developmental editor.

Some of the material in *Basic Heat and Mass Transfer*, in the form of examples and exercises, has been adapted from an earlier text by my former colleagues at UCLA, D. K. Edwards and V. E. Denny (*Transfer Processes* 1/e, Holt, Rinehart & Winston, 1973; 2/e Hemisphere-McGraw-Hill, 1979). I have also drawn on material in radiation heat transfer from a more recent text by D. K. Edwards (*Radiation Heat Transfer Notes*, Hemisphere, 1981). I gratefully acknowledge the contributions of these gentlemen, both to this book and to my professional career. The late D. N. Bennion provided a chemical engineering perspective to some of the material on mass exchangers. The computer software was ably written by Baek Youn, Hae-Jin Choi, and Benjamin Tan. I would also like to thank former students S. W. Hiebert, R. Tsai, B. Cowan, E. Myhre, B. H. Chang, D. C. Weatherly, A. Gopinath, J. I. Rodriguez, B. P. Dooher, M. A. Friedman, and C. Yuen.

In preparing the second edition, I have had useful input from a number of people, including Professor F. Forster, University of Washington; Professor N. Shamsundar, University of Houston; Professor S. Kim, Kukmin University; and Professor A. Lavine, UCLA. Students who have helped include P. Hwang, M. Tari, B. Tan, J. Sigler, M. Fabbri, F. Chao, and A. Na-Nakornpanom.

My special thanks to the secretarial staff at UCLA and the University of Auckland: Phyllis Gilbert, Joy Wallace, and Julie Austin provided enthusiastic and expert typing of the manuscript. Mrs. Gilbert also provided expert typing of the solutions manual.

NOTES TO THE INSTRUCTOR AND STUDENT

These notes have been prepared to assist the instructor and student and should be read before the text is used. Topics covered include conventions for artwork and mathematics, the format for example problems, organization of the exercises, comments on the thermophysical property data in Appendix A, and a guide for use of the accompanying computer software.

ARTWORK

Conventions used in the figures are as follows.

- $\rightarrow$ Conduction or convection heat flow
- $\rightsquigarrow$ Radiation heat flow
- $\dashrightarrow$ Fluid flow

MATHEMATICAL SYMBOLS

Symbols that may need clarification are as follows.

- $\simeq$ Nearly equal
- $\sim$ Of the same order of magnitude
- $\big|_x$ All quantities in the term to the left of the bar are evaluated at x

EXAMPLES

Use of standard format for presenting the solutions of engineering problems is a good practice. The format used for the examples in *Basic Heat Transfer*, which is but one possible approach, is as follows.

Problem statement

Solution

Given:

Required:

Assumptions: 1.
                2. etc.

Sketch (when appropriate)
Analysis (diagrams when appropriate)
Properties evaluation
Calculations
Results (tables or graphs when appropriate)

Comments

1.
2. etc.

It is always assumed that the problem statement precedes the solution (as in the text) or that it is readily available (as in the *Solutions Manual*). Thus, the *Given* and *Required* statements are concise and focus on the essential features of the problem. Under *Assumptions*, the main assumptions required to solve the problem are listed; when appropriate, they are discussed further in the body of the solution. A sketch of the physical system is included when the geometry requires clarification; also, expected temperature and concentration profiles are given when appropriate. (Schematics that simply repeat the information in the problem statements are used sparingly. We know that many instructors always require a schematic. Our view is that students need to develop an appreciation of when a figure or graph is necessary, because artwork is usually an expensive component of engineering reports. For example, we see little use for a schematic that shows a 10 m length of straight 2 cm–O.D. tube.) The analysis may consist simply of listing some formulas from the text, or it may require setting up a differential equation and its solution. Strictly speaking, a property should not be evaluated until its need is identified by the analysis. However, in routine calculations, such as evaluation of convective heat transfer coefficients, it is often convenient to list all the property values taken from an Appendix A table in one place. The calculations then follow with results listed, tabulated, or graphed as appropriate. Under *Comments*, the significance of the results can be discussed, the validity of assumptions further evaluated, or the broader implications of the problem noted.

In presenting calculations for the examples in *Basic Heat Transfer*, we have rounded off results at each stage of the calculation. If additional figures are retained for the complete calculations, discrepancies in the last figure will be observed. Since many of the example calculations are quite lengthy, we believe our policy will facilitate checking a particular calculation step of concern. As is common practice, we have generally given results to more significant figures than is justified, so that these results can be conveniently used in further calculations. It is safe to say that no engineering heat transfer calculation will be accurate to within 1%, and that most experienced engineers will be pleased with results accurate to within 10% or 20%. Thus, preoccupation with a third or fourth significant figure is misplaced (unless required to prevent error magnification in operations such as subtraction). Fundamental constants are rounded off to no more than five significant figures.

EXERCISES

The diskette logo next to an exercise statement indicates that it can be solved using the *Heat Transfer* software, and that the sample solution provided to the instructor has been prepared accordingly. There are many additional exercises that can be solved using the software but that do not have the logo designation. These exercises are intended to give the student practice in hand calculations, and thus the sample solutions were also prepared manually.

The exercises have been ordered to correspond with the order in which the material is presented in the text, rather than in some increasing degree of difficulty. Since the range of difficulty of the exercises is considerable, the instructor is urged to give students guidance in selecting exercises for self-study. Answers to all exercises are listed in the *Solutions Manual* provided to instructors. Odd- and even-numbered exercises are listed separately; answers to odd-numbered exercises are available to students on the book website.

PROPERTY DATA

A considerable quantity of property data has been assembled in Appendix A. Key sources are given as references or are listed in the bibliography. Since *Basic Heat Transfer* is a textbook, our primary objective in preparing Appendix A was to provide the student with a wide range of data in an easily used form. Whenever possible, we have used the most accurate data that we could obtain, but accuracy was not always the primary concern. For example, the need to have consistent data over a wide range of temperature often dictated the choice of source. All the tables are in SI units, with temperature in kelvins. The computer program UNITS can be used for conversions to other systems of units. Appendix A should serve most needs of the student, as well as of the practicing engineer, for doing routine calculations. If a heat transfer research project requires accurate and reliable thermophysical property data, the prudent researcher should carefully check relevant primary data sources.

SOFTWARE

The HT software has a menu that describes the content of each program. The programs are also described at appropriate locations in the text. The input format and program use are demonstrated in example problems in the text. Use of the text index is recommended for locating the program descriptions and examples. There is a one-to-one correspondence between the text and the software. In principle, all numbers generated by the software can be calculated manually from formulas, graphs, and data given in the text. Small discrepancies may be seen when interpolation in graphs or property tables is required, since some of the data are stored in the software as polynomial curve fits.

The software facilitates self-study by the student. Practice hand calculations can be immediately checked using the software. When programs such as CONV, PHASE, and BOIL are used, properties evaluation and intermediate calculation steps can also be checked when the final results do not agree.

Since there is a large thermophysical property database stored in the software package, the programs can also be conveniently used to evaluate these properties for other purposes. For example, in CONV both the wall and fluid temperatures can be set equal to the desired temperature to obtain property values required for convection calculations. We can even go one step further when evaluating a convective heat transfer coefficient from a new correlation not contained in CONV: if a corresponding item is chosen, the values of relevant dimensionless groups can also be obtained from CONV, further simplifying the calculations.

Presently the HT software is available in both Windows and DOS versions. The latter can be used on both Mac OS X and Windows platforms with DOS emulators. Some examples in the text show sample inputs from the DOS version of the software. For the Windows version the inputs are essentially the same.

CONTENTS

**B   UNITS,
CONVERSION FACTORS,
AND MATHEMATICS   789**

INTRODUCTION AND ELEMENTARY HEAT TRANSFER

CONTENTS

1.1   INTRODUCTION

The process of heat transfer is familiar to us all. On a cold day we put on more clothing to reduce heat transfer from our warm body to cold surroundings. To make a cup of coffee we may plug in a kettle, inside which heat is transferred from an electrical resistance element to the water, heating the water until it boils. The engineering discipline of **heat transfer** is concerned with methods of calculating **rates** of heat transfer. These methods are used by engineers to design components and systems in which heat transfer occurs. Heat transfer considerations are important in almost all areas of technology. Traditionally, however, the discipline that has been most concerned with heat transfer is mechanical engineering because of the importance of heat transfer in energy conversion systems, from coal-fired power plants to solar water heaters.

Many *thermal design* problems require reducing heat transfer rates by providing suitable *insulation*. The insulation of buildings in extreme climates is a familiar example, but there are many others. The space shuttle has thermal tiles to insulate the vehicle from high-temperature air behind the bow shock wave during reentry into the atmosphere. Cryostats, which maintain the cryogenic temperatures required for the use of superconductors, must be effectively insulated to reduce the cooling load on the refrigeration system. Often, the only way to ensure protection from severe heating is to provide a fluid flow as a heat "sink". Nozzles of liquid-fueled rocket motors are cooled by pumping the cold fuel through passages in the nozzle wall before injection into the combustion chamber. A critical component in a fusion reactor is the "first wall" of the containment vessel, which must withstand intense heating from the hot plasma. Such walls may be cooled by a flow of helium gas or liquid lithium.

A common thermal design problem is the transfer of heat from one fluid to another. Devices for this purpose are called *heat exchangers*. A familiar example is the automobile radiator, in which heat is transferred from the hot engine coolant to cold air blowing through the radiator core. Heat exchangers of many different types are required for power production and by the process industries. A power plant, whether the fuel be fossil or nuclear, has a *boiler* in which water is evaporated to produce steam to drive the turbines, and a *condenser* in which the steam is condensed to provide a low back pressure on the turbines and for water recovery. The condenser patented by James Watt in 1769 more than doubled the efficiency of steam engines then being used and set the Industrial Revolution in motion. The common vapor cycle refrigeration or air-conditioning system has an *evaporator* where heat is absorbed at low temperature and a *condenser* where heat is rejected at a higher temperature. On a domestic refrigerator, the condenser is usually in the form of a tube coil with cooling *fins* to assist transfer of heat to the surroundings. An oil refinery has a great variety of heat transfer equipment, including rectification columns and thermal crackers. Many heat exchangers are used to transfer heat from one process stream to another, to reduce the total energy consumption by the refinery.

Often the design problem is one of *thermal control*, that is, maintaining the operating temperature of temperature-sensitive components within a specified range.

Cooling of all kinds of electronic gear is an example of thermal control. The development of faster computers is now severely constrained by the difficulty of controlling the temperature of very small components, which dissipate large amounts of heat. Thermal control of temperature-sensitive components in a communications satellite orbiting the Earth is a particularly difficult problem. Transistors and diodes must not overheat, batteries must not freeze, telescope optics must not lose alignment due to thermal expansion, and photographs must be processed at the proper temperature to ensure high resolution. Thermal control of space stations present even greater problems, since reliable life-support systems are also necessary.

From the foregoing examples, it is clear that heat transfer involves a great variety of physical phenomena and engineering systems. The phenomena must first be understood and quantified before a methodology for the thermal design of an engineering system can be developed. Chapter 1 is an overview of the subject and introduces key topics at an elementary level. In Section 1.2, the distinction between the subjects of heat transfer and thermodynamics is explained. The first law of thermodynamics is reviewed, and closed- and open-system forms required for heat transfer analysis are developed. Section 1.3 introduces the three important modes of heat transfer: **heat conduction**, **thermal radiation**, and **heat convection**. Some formulas are developed that allow elementary heat transfer calculations to be made. In practical engineering problems, these modes of heat transfer usually occur simultaneously. Thus, in Section 1.4, the analysis of heat transfer by combined modes is introduced. Engineers are concerned with the changes heat transfer processes effect in engineering systems and, in Section 1.5, an example is given in which the first law is applied to a simple model closed system to determine the temperature response of the system with time. Finally, in Section 1.6, the International System of units (SI) is reviewed, and the units policy that is followed in the text is discussed.

1.2   HEAT TRANSFER AND ITS RELATION TO THERMODYNAMICS

When a hot object is placed in cold surroundings, it cools: the object loses internal energy, while the surroundings gain internal energy. We commonly describe this interaction as a *transfer of heat* from the object to the surrounding region. Since the caloric theory of heat has been long discredited, we do not imagine a "heat substance" flowing from the object to the surroundings. Rather, we understand that internal energy has been transferred by complex interactions on an atomic or subatomic scale. Nevertheless, it remains common practice to describe these interactions as transfer, transport, or flow, of heat. The engineering discipline of heat transfer is concerned with calculation of the rate at which heat flows within a medium, across an interface, or from one surface to another, as well as with the calculation of associated temperatures.

It is important to understand the essential difference between the engineering discipline of heat transfer and what is commonly called thermodynamics. Classical thermodynamics deals with systems in equilibrium. Its methodology may be used to calculate the energy required to change a system from one equilibrium state to another, but it cannot be used to calculate the rate at which the change may occur. For example, if a 1 kg ingot of iron is quenched from 1000°C to 100°C in an oil bath, thermodynamics tells us that the loss in internal energy of the ingot is mass (1 kg) × specific heat capacity (~450 J/kg K) × temperature change (900 K), or approximately 405 kJ. But thermodynamics cannot tell us how long we will have to wait for the temperature to drop to 100°C. The time depends on the temperature of the oil bath, physical properties of the oil, motion of the oil, and other factors. An appropriate heat transfer analysis will consider all of these.

Analysis of heat transfer processes does require using some thermodynamics concepts. In particular, the **first law of thermodynamics** is used, generally in particularly simple forms since work effects can often be ignored. The first law is a statement of the *principle of conservation of energy*, which is a basic law of physics. This principle can be formulated in many ways by excluding forms of energy that are irrelevant to the problem under consideration, or by simply redefining what is meant by energy. In heat transfer, it is common practice to refer to the first law as the *energy conservation principle* or simply as an *energy* or *heat balance* when no work is done. However, as in thermodynamics, it is essential that the correct form of the first law be used. The student must be able to define an appropriate system, recognize whether the system is *open* or *closed*, and decide whether a steady state can be assumed. Some simple forms of the energy conservation principle, which find frequent use in this text, follow.

A closed system containing a fixed mass of a solid is shown in Fig. 1.1. The system has a volume $V [\text{m}^3]$, and the solid has a density $\rho [\text{kg/m}^3]$. There is *net* heat transfer into the system at a rate of $\dot{Q} [\text{J/s or W}]$, and heat may be generated within the solid, for example, by nuclear fission or by an electrical current, at a rate $\dot{Q}_v [\text{W}]$. Solids may be taken to be incompressible, so no work is done by or on the system. The principle of conservation of energy requires that over a time interval $\Delta t [\text{s}]$,

$$\begin{array}{ccc} \text{Change in internal energy} & & \text{Net heat transferred} & & \text{Heat generated} \\ \text{within the system} & = & \text{into the system} & + & \text{within the system} \end{array}$$

$$\Delta U = \dot{Q}\Delta t + \dot{Q}_v \Delta t \tag{1.1}$$

Dividing by Δt and letting Δt go to zero gives

$$\frac{dU}{dt} = \dot{Q} + \dot{Q}_v$$

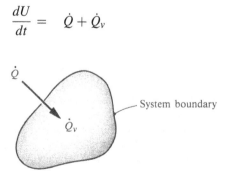

Figure 1.1   Application of the energy conservation principle to a closed system.

The system contains a fixed mass (ρV); thus, we can write $dU = \rho V du$, where u is the specific internal energy [J/kg]. Also, for an incompressible solid, $du = c_v dT$, where c_v is the constant-volume specific heat[1] [J/kg K], and T [K] is temperature. Since the solid has been taken to be incompressible, the constant-volume and constant-pressure specific heats are equal, so we simply write $du = cdT$ to obtain

$$\rho V c \frac{dT}{dt} = \dot{Q} + \dot{Q}_v \tag{1.2}$$

Equation (1.2) is a special form of the first law of thermodynamics that will be used often in this text. It is written on a *rate* basis; that is, it gives the rate of change of temperature with time. For some purposes, however, it will prove convenient to return to Eq. (1.1) as a statement of the first law.

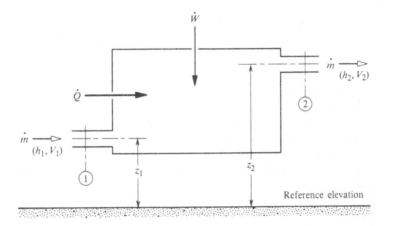

Figure 1.2 Application of the energy conservation principle to a steady-flow open system.

Figure 1.2 shows an *open* system (or *control volume*), for which a useful form of the first law is the **steady-flow energy equation**. It is used widely in the thermodynamic analysis of equipment such as turbines and compressors. Then

$$\dot{m}\Delta\left(h + \frac{V^2}{2} + gz\right) = \dot{Q} + \dot{W} \tag{1.3}$$

where $\dot{m}$ [kg/s] is the mass flow rate, h [J/kg] is the specific enthalpy, V [m/s] is velocity, g [m/s^2] is the gravitational acceleration, z is elevation [m], $\dot{Q}$ [W] is the net rate of heat transfer, as before, and $\dot{W}$ [W] is the rate at which external (shaft) work is done on the system.[2] Notice that the sign convention here is that external work done *on* the system is positive; the opposite sign convention is also widely used. The symbol ΔX means $X_{\text{out}} - X_{\text{in}}$, or the change in X. Equation (1.3) applies to a pure

---

[1] The terms *specific heat capacity* and *specific heat* are equivalent and interchangeable in the heat transfer literature.
[2] Equation (1.3) has been written as if h, V, and z are uniform in the streams crossing the control volume boundary. Often such an assumption can be made; if not, an integration across each stream is required to give appropriate average values.

substance when conditions within the system, such as temperature and velocity, are unchanging over some appropriate time interval. Heat generation within the system has not been included. In many types of heat transfer equipment, no external work is done, and changes in kinetic and potential energy are negligible; Eq. (1.3) then reduces to

$$\dot{m}\Delta h = \dot{Q} \tag{1.4}$$

The specific enthalpy h is related to the specific internal energy u as

$$h = u + Pv \tag{1.5}$$

where P [N/m^2 or Pa] is pressure, and v is specific volume [m^3/kg]. Two limit forms of Δh are useful. If the fluid enters the system at state 1 and leaves at state 2:

1. For ideal gases with $Pv = RT$,

$$\Delta h = \int_{T_1}^{T_2} c_p dT \tag{1.6a}$$

where R [J/kg K] is the gas constant and c_p [J/kg K] is the constant-pressure specific heat.

2. For incompressible liquids with $\rho = 1/v = $ constant

$$\Delta h = \int_{T_1}^{T_2} c\, dT + \frac{P_2 - P_1}{\rho} \tag{1.6b}$$

where $c = c_v = c_p$. The second term in Eq. (1.6b) is often negligible as will be assumed throughout this text.

Equation (1.4) is the usual starting point for the heat transfer analysis of steady-state open systems.

The *second law of thermodynamics* tells us that if two objects at temperatures T_1 and T_2 are connected, and if $T_1 > T_2$, then heat will flow spontaneously and irreversibly from object 1 to object 2. Also, there is an entropy increase associated with this heat flow. As T_2 approaches T_1, the process approaches a reversible process, but simultaneously the rate of heat transfer approaches zero, so the process is of little practical interest. All heat transfer processes encountered in engineering are irreversible and generate entropy. With the increasing realization that energy supplies should be conserved, efficient use of available energy is becoming an important consideration in thermal design. Thus, the engineer should be aware of the irreversible processes occurring in the system under development and understand that the optimal design may be one that minimizes entropy generation due to heat transfer and fluid flow. Most often, however, energy conservation is simply a consideration in the overall economic evaluation of the design. Usually there is an important trade-off between energy costs associated with the operation of the system and the capital costs required to construct the equipment.

1.3   MODES OF HEAT TRANSFER

In thermodynamics, *heat* is defined as energy transfer due to temperature gradients or differences. Consistent with this viewpoint, thermodynamics recognizes only two modes of heat transfer: *conduction* and *radiation*. For example, heat transfer across a steel pipe wall is by conduction, whereas heat transfer from the Sun to the Earth or to a spacecraft is by thermal radiation. These modes of heat transfer occur on a molecular or subatomic scale. In air at normal pressure, conduction is by molecules that travel a very short distance ($\sim 0.065 \mu m$) before colliding with another molecule and exchanging energy. On the other hand, radiation is by photons, which travel almost unimpeded through the air from one surface to another. Thus, an important distinction between conduction and radiation is that the energy carriers for conduction have a short *mean free path,* whereas for radiation the carriers have a long mean free path. However, in air at the very low pressures characteristic of high-vacuum equipment, the mean free path of molecules can be much longer than the equipment dimensions, so the molecules travel unimpeded from one surface to another. Then heat transfer by molecules is governed by laws analogous to those for radiation.

A fluid, by virtue of its mass and velocity, can transport momentum. In addition, by virtue of its temperature, it can transport energy. Strictly speaking, *convection* is the transport of energy by bulk motion of a medium (a moving solid can also convect energy in this sense). In the steady-flow energy equation, Eq. (1.3), convection of internal energy is contained in the term $\dot{m}\Delta h$, which is on the left-hand side of the equation, and heat transfer by conduction and radiation is on the right-hand side, as $\dot{Q}$. However, it is common engineering practice to use the term *convection* more broadly and describe heat transfer from a surface to a moving fluid also as convection, or *convective heat transfer,* even though conduction and radiation play a dominant role close to the surface, where the fluid is stationary. In this sense, convection is usually regarded as a distinct mode of heat transfer. Examples of convective heat transfer include heat transfer from the radiator of an automobile or to the skin of a hypersonic vehicle. Convection is often associated with a change of phase, for example, when water boils in a kettle or when steam condenses in a power plant condenser. Owing to the complexity of such processes, boiling and condensation are often regarded as distinct heat transfer processes.

The hot water home heating system shown in Fig. 1.3 illustrates the modes of heat transfer. Hot water from the furnace in the basement flows along pipes to radiators located in individual rooms. Transport of energy by the hot water from the basement is true convection as defined above; we do not call this a heat transfer process. Inside the radiators, there is convective heat transfer from the hot water to the radiator shell, conduction across the radiator shell, and both convective and radiative heat transfer from the hot outer surface of the radiator shell into the room. The convection is *natural* convection: the heated air adjacent to the radiator surface rises due to its buoyancy, and cooler air flows in to take its place. The radiators are heat exchangers. Although commonly used, the term *radiator* is misleading since heat transfer

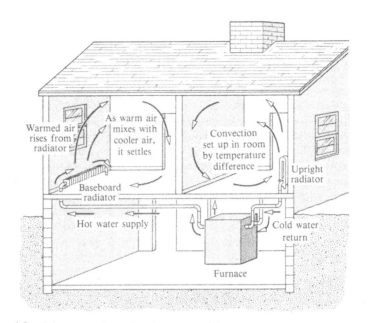

Figure 1.3   A hot-water home heating system illustrating the modes of heat transfer.

from the shell surface can be predominantly by convection rather than by radiation (see Exercise 1–20). Heaters that transfer heat predominantly by radiation are, for example, electrical resistance wire units.

Each of the three important subject areas of heat transfer will now be introduced: conduction, in Section 1.3.1; radiation, in Section 1.3.2; and convection, in Section 1.3.3.

1.3.1 Heat Conduction

On a microscopic level, the physical mechanisms of conduction are complex, encompassing such varied phenomena as molecular collisions in gases, lattice vibrations in crystals, and flow of free electrons in metals. However, if at all possible, the engineer avoids considering processes at the microscopic level, preferring to use *phenomenological laws,* at a macroscopic level. The phenomenological law governing heat conduction was proposed by the French mathematical physicist J. B. Fourier in 1822. This law will be introduced here by considering the simple problem of one-dimensional heat flow across a plane wall—for example, a layer of insulation.[3] Figure 1.4 shows a plane wall of surface area A and thickness L, with its face at $x = 0$ maintained at temperature T_1 and the face at $x = L$ maintained at T_2. The heat flow $\dot{Q}$ through the wall is in the direction of decreasing temperature: if

---

[3] In thermodynamics, the term *insulated* is often used to refer to a *perfectly* insulated (zero-heat-flow or adiabatic) surface. In practice, insulation is used to *reduce* heat flow and seldom can be regarded as perfect.

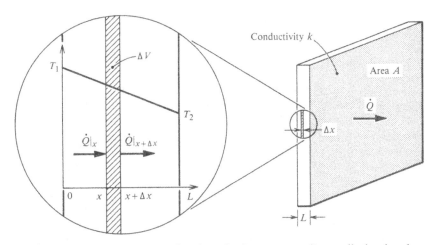

Figure 1.4 Steady one-dimensional conduction across a plane wall, showing the application of the energy conservation principle to an elemental volume Δx thick.

$T_1 > T_2$, $\dot{Q}$ is in the positive x direction.[4] The phenomenological law governing this heat flow is **Fourier's law of heat conduction,** which states that in a homogeneous substance, the local heat flux is proportional to the negative of the local temperature gradient:

$$\frac{\dot{Q}}{A} = q \quad \text{and} \quad q \propto -\frac{dT}{dx} \tag{1.7}$$

where q is the heat flux, or heat flow per unit area perpendicular to the flow direction [W/m^2], T is the local temperature [K or °C], and x is the coordinate in the flow direction [m]. When dT/dx is negative, the minus sign in Eq. (1.7) gives a positive q in the positive x direction. Introducing a constant of proportionality k,

$$q = -k\frac{dT}{dx} \tag{1.8}$$

where k is the **thermal conductivity** of the substance and, by inspection of the equation, must have units [W/m K]. Notice that temperature can be given in kelvins or degrees Celsius in Eq. (1.8): the temperature gradient does not depend on which of these units is used since one kelvin equals one degree Celsius (1 K = 1°C). Thus, the units of thermal conductivity could also be written [W/m°C], but this is not the recommended practice when using the SI system of units. The magnitude of the thermal conductivity k for a given substance very much depends on its microscopic structure and also tends to vary somewhat with temperature; Table 1.1 gives some selected values of k.

---

[4] Notice that this $\dot{Q}$ is the heat flow in the x direction, whereas in the first law, Eqs. (1.1)–(1.4), $\dot{Q} = \dot{Q}_{in} - \dot{Q}_{out}$ is the net heat transfer into the whole system. In linking thermodynamics to heat transfer, some ambiguity in notation arises when common practice in both subjects is followed.

Table 1.1   Selected values of thermal conductivity at 300 K ($\sim 25°C$).

Material	k W/m K
Copper	386
Aluminum	204
Brass (70% Cu, 30% Zn)	111
Mild steel	64
Stainless steel, 18–8	15
Mercury	8.4
Concrete	1.4
Pyrex glass	1.09
Water	0.611
Neoprene rubber	0.19
Engine oil, SAE 50	0.145
White pine, perpendicular to grain	0.10
Polyvinyl chloride (PVC)	0.092
Freon 12	0.071
Cork	0.043
Fiberglass (medium density)	0.038
Polystyrene	0.028
Air	0.027

Note: Appendix A contains more comprehensive data.

Figure 1.4 shows an elemental volume ΔV located between x and $x + \Delta x$; ΔV is a closed system, and the energy conservation principle in the form of Eq. (1.2) applies. If we consider a steady state, then temperatures are unchanging in time and $dT/dt = 0$; also, if there is no heat generated within the volume, $\dot{Q}_v = 0$. Then Eq. (1.2) states that the net heat flow into the system is zero. Because the same amount of heat is flowing into ΔV across the face at x, and out of ΔV across the face at $x + \Delta x$,

$$\dot{Q}|_x = \dot{Q}|_{x+\Delta x}$$

Since the rate of heat transfer is constant for all x, we simplify the notation by dropping the $|_x$ and $|_{x+\Delta x}$ subscripts (see the footnote on page 9), and write

$$\dot{Q} = \text{Constant}$$

But from Fourier's law, Eq. (1.8),

$$\dot{Q} = qA = -kA\frac{dT}{dx}$$

The variables are separable: rearranging and integrating across the wall,

$$\frac{\dot{Q}}{A}\int_0^L dx = -\int_{T_1}^{T_2} k\,dT$$

where $\dot{Q}$ and A have been taken outside the integral signs since both are constants. If the small variation of k with temperature is ignored for the present we obtain

$$\dot{Q} = \frac{kA}{L}(T_1 - T_2) = \frac{T_1 - T_2}{L/kA} \tag{1.9}$$

Comparison of Eq. (1.9) with Ohm's law, $I = E/R$, suggests that $\Delta T = T_1 - T_2$ can be viewed as a driving potential for flow of heat, analogous to voltage being the driving potential for current. Then $R \equiv L/kA$ can be viewed as a **thermal resistance** analogous to electrical resistance.

If we have a composite wall of two slabs of material, as shown in Fig. 1.5, the heat flow through each layer is the same:

$$\dot{Q} = \frac{T_1 - T_2}{L_A/k_A A} = \frac{T_2 - T_3}{L_B/k_B A}$$

Rearranging

$$\dot{Q}\left(\frac{L_A}{k_A A}\right) = T_1 - T_2$$

$$\dot{Q}\left(\frac{L_B}{k_B A}\right) = T_2 - T_3$$

Adding eliminates the interface temperature T_2:

$$\dot{Q}\left(\frac{L_A}{k_A A} + \frac{L_B}{k_B A}\right) = T_1 - T_3$$

or

$$\dot{Q} = \frac{T_1 - T_3}{L_A/k_A A + L_B/k_B A} = \frac{\Delta T}{R_A + R_B} \tag{1.10a}$$

Using the electrical resistance analogy, we would view the problem as two resistances in series forming a **thermal circuit**, and immediately write

$$\dot{Q} = \frac{\Delta T}{R_A + R_B} \tag{1.10b}$$

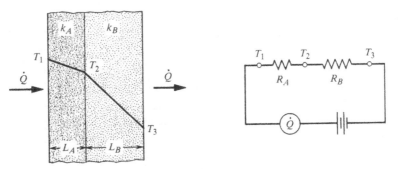

Figure 1.5 The temperature distribution for steady conduction across a composite plane wall and the corresponding thermal circuit.

EXAMPLE 1.1   Heat Transfer through Insulation

A refrigerated container is in the form of a cube with 2 m sides and has 5 mm-thick aluminum walls insulated with a 10 cm layer of cork. During steady operation, the temperatures on the inner and outer surfaces of the container are measured to be $-5°C$ and $20°C$, respectively. Determine the cooling load on the refrigerator.

Solution

Given: Aluminum container insulated with 10 cm—thick cork.

Required: Rate of heat gain.

Assumptions: 1. Steady state.
2. One-dimensional heat conduction (ignore corner effects).

Equation (1.10) applies:

$$\dot{Q} = \frac{\Delta T}{R_A + R_B} \quad \text{where } R = \frac{L}{kA}$$

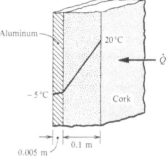

Let subscripts A and B denote the aluminum wall and cork insulation, respectively. Table 1.1 gives $k_A = 204$ W/m K, $k_B = 0.043$ W/m K. We suspect that the thermal resistance of the aluminum wall is negligible, but we will calculate it anyway. For one side of area $A = 4\,\text{m}^2$, the thermal resistances are

$$R_A = \frac{L_A}{k_A A} = \frac{(0.005 \text{ m})}{(204 \text{ W/m K})(4 \text{ m}^2)} = 6.13 \times 10^{-6} \text{ K/W}$$

$$R_B = \frac{L_B}{k_B A} = \frac{(0.10 \text{ m})}{(0.043 \text{ W/m K})(4 \text{ m}^2)} = 0.581 \text{ K/W}$$

Since R_A is five orders of magnitude less than R_B, it can be ignored. The heat flow for a temperature difference of $T_1 - T_2 = 20 - (-5) = 25$ K, is

$$\dot{Q} = \frac{\Delta T}{R_B} = \frac{25 \text{ K}}{0.581 \text{ K/W}} = 43.0 \text{ W}$$

For six sides, the total cooling load on the refrigerator is $6.0 \times 43.0 = 258$ W.

· **Comments**

1. In the future, when it is obvious that a resistance in a series network is negligible, it can be ignored from the outset (no effort should be expended to obtain data for its calculation).

2. The assumption of one-dimensional conduction is good because the 0.1 m insulation thickness is small compared to the 2 m-long sides of the cube.

3. Notice that the temperature difference $T_1 - T_2$ is expressed in kelvins, even though T_1 and T_2 were given in degrees Celsius.

4. We have assumed perfect thermal contact between the aluminum and cork; that is, there is no thermal resistance associated with the interface between the two materials (see Section 2.2.2).

1.3.2 Thermal Radiation

All matter and space contains electromagnetic radiation. A particle, or *quantum*, of electromagnetic energy is a photon, and heat transfer by radiation can be viewed either in terms of electromagnetic waves or in terms of photons. The flux of radiant energy incident on a surface is its **irradiation,** G [W/m²]; the energy flux leaving a surface due to emission and reflection of electromagnetic radiation is its **radiosity,** J [W/m²]. A **black surface** (or **blackbody**) is defined as a surface that absorbs all incident radiation, reflecting none. As a consequence, all of the radiation leaving a black surface is emitted by the surface and is given by the **Stefan-Boltzmann law** as

$$J = E_b = \sigma T^4 \tag{1.11}$$

where E_b is the **blackbody emissive power**, T is absolute temperature [K], and σ is the Stefan-Boltzmann constant ($\simeq 5.67 \times 10^{-8}$ W/m²K⁴). Table 1.2 shows how $E_b = \sigma T^4$ increases rapidly with temperature.

Table 1.2 Blackbody emissive power σT^4 at various temperatures.

Surface Temperature K	Blackbody Emissive Power W/m²
300 (room temperature)	459
1000 (cherry-red hot)	56,700
3000 (lamp filament)	4,590,000
5760 (Sun temperature)	62,400,000

Figure 1.6 shows a convex black object of surface area A_1 in an evacuated black isothermal enclosure at temperature T_2. At equilibrium, the object is also at temperature T_2, and the radiant energy incident on the object must equal the radiant energy leaving from the object:

$$G_1 A_1 = J_1 A_1 = \sigma T_2^4 A_1$$

Hence

$$G_1 = \sigma T_2^4 \tag{1.12}$$

and is uniform over the area. If the temperature of the object is now raised to T_1, its radiosity becomes σT_1^4 while its irradiation remains σT_2^4 (because the enclosure reflects no radiation). Then the net radiant heat flux through the surface, q_1, is the

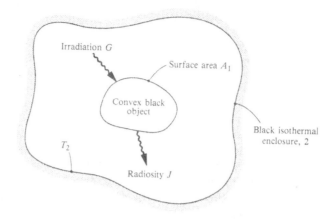

Figure 1.6   A convex black object (surface 1) in a black isothermal enclosure (surface 2).

radiosity minus the irradiation:

$$q_1 = J_1 - G_1 \tag{1.13}$$

or

$$q_1 = \sigma T_1^4 - \sigma T_2^4 \tag{1.14}$$

where the sign convention is such that a net flux away from the surface is positive. Equation (1.14) is also valid for two large black surfaces facing each other, as shown in Fig. 1.7.

The blackbody is an ideal surface. Real surfaces absorb less radiation than do black surfaces. The fraction of incident radiation absorbed is called the **absorptance** (or absorptivity), α. A widely used model of a real surface is the **gray surface,** which is defined as a surface for which α is a constant, irrespective of the nature of the incident radiation. The fraction of incident radiation reflected is the **reflectance** (or reflectivity), ρ. If the object is opaque, that is, not transparent to electromagnetic radiation, then

$$\rho = 1 - \alpha \tag{1.15}$$

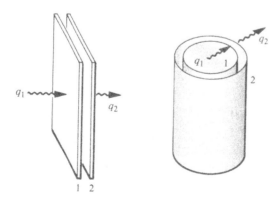

Figure 1.7   Examples of two large surfaces facing each other.

Table 1.3 Selected approximate values of emittance, ε (total hemispherical values at normal temperatures).

Surface	Emittance, ε
Aluminum alloy, unoxidized	0.035
Black anodized aluminum	0.80
Chromium plating	0.16
Stainless steel, type 312, lightly oxidized	0.30
Inconel X, oxidized	0.72
Black enamel paint	0.78
White acrylic paint	0.90
Asphalt	0.88
Concrete	0.90
Soil	0.94
Pyrex glass	0.80

Note: More comprehensive data are given in Appendix A. Emittance is very dependent on surface finish; thus, values obtained from various sources may differ significantly.

Real surfaces also emit less radiation than do black surfaces. The fraction of the blackbody emissive power σT^4 emitted is called the **emittance** (or emissivity), ε.[5] A gray surface also has a constant value of ε, independent of its temperature, and, as will be shown in Chapter 6, the emittance and absorptance of a gray surface are equal:

$$\varepsilon = \alpha \quad \text{(gray surface)} \tag{1.16}$$

Table 1.3 shows some typical values of ε at normal temperatures. Bright metal surfaces tend to have low values, whereas oxidized or painted surfaces tend to have high values. Values of α and ρ can also be obtained from Table 1.3 by using Eqs. (1.15) and (1.16).

If heat is transferred by radiation between two gray surfaces of finite size, as shown in Fig. 1.8, the rate of heat flow will depend on temperatures T_1 and T_2 and emittances ε_1 and ε_2, as well as the geometry. Clearly, some of the radiation leaving surface 1 will not be intercepted by surface 2, and vice versa. Determining the rate of heat flow is usually quite difficult. In general, we may write

$$\dot{Q}_{12} = A_1 \mathscr{F}_{12}\left(\sigma T_1^4 - \sigma T_2^4\right) \tag{1.17}$$

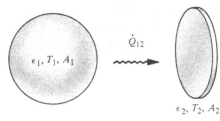

Figure 1.8 Radiation heat transfer between two finite gray surfaces.

---

[5] Both the endings *-ance* and *-ivity* are commonly used for radiation properties. In this text, *-ance* will be used for surface radiation properties. In Chapter 6, *-ivity* will be used for gas radiation properties.

where $\dot{Q}_{12}$ is the net radiant energy interchange (heat transfer) from surface 1 to surface 2, and $\mathscr{F}_{12}$ is a **transfer factor,** which depends on emittances and geometry. For the special case of surface 1 surrounded by surface 2, where either area A_1 is small compared to area A_2, or surface 2 is nearly black, $\mathscr{F}_{12} \simeq \varepsilon_1$ and Eq. (1.17) becomes

$$\dot{Q}_{12} = \varepsilon_1 A_1 (\sigma T_1^4 - \sigma T_2^4) \tag{1.18}$$

Equation (1.18) will be derived in Chapter 6. It is an important result and is often used for quick engineering estimates.

The T^4 dependence of radiant heat transfer complicates engineering calculations. When T_1 and T_2 are not too different, it is convenient to linearize Eq. (1.18) by factoring the term $(\sigma T_1^4 - \sigma T_2^4)$ to obtain

$$\dot{Q}_{12} = \varepsilon_1 A_1 \sigma (T_1^2 + T_2^2)(T_1 + T_2)(T_1 - T_2)$$

$$\simeq \varepsilon_1 A_1 \sigma (4T_m^3)(T_1 - T_2)$$

for $T_1 \simeq T_2$, where T_m is the mean of T_1 and T_2. This result can be written more concisely as

$$\dot{Q}_{12} \simeq A_1 h_r (T_1 - T_2) \tag{1.19}$$

where $h_r = 4\varepsilon_1 \sigma T_m^3$ is called the **radiation heat transfer coefficient** [W/m^2 K]. At 25°C (= 298 K),

$$h_r = (4)\varepsilon_1 (5.67 \times 10^{-8} \text{ W/m}^2 \text{K}^4)(298 \text{ K})^3$$

or

$$h_r \simeq 6\varepsilon_1 \text{ W/m}^2 \text{ K}$$

This result can be easily remembered: The radiation heat transfer coefficient at room temperature is about six times the surface emittance. For $T_1 = 320$ K and $T_2 = 300$ K, the error incurred in using the approximation of Eq. (1.19) is only 0.1%; for $T_1 = 400$ K and $T_2 = 300$ K, the error is 2%.

EXAMPLE 1.2   Heat Loss from a Transistor

An electronic package for an experiment in outer space contains a transistor capsule, which is approximately spherical in shape with a 2 cm diameter. It is contained in an evacuated case with nearly black walls at 30°C. The only significant path for heat loss from the capsule is radiation to the case walls. If the transistor dissipates 300 mW, what will the capsule temperature be if it is (i) bright aluminum and (ii) black anodized aluminum?

Solution

Given: 2 cm-diameter transistor capsule dissipating 300 mW.

Required: Capsule temperature for (i) bright aluminum and (ii) black anodized aluminum.

Assumptions: Model as a small gray body in large, nearly black surroundings.

Equation (1.18) is applicable with

$$\dot{Q}_{12} = 300 \text{ mW}$$
$$T_2 = 30°C = 303 \text{ K}$$

and T_1 is the unknown.

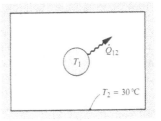

$$\dot{Q}_{12} = \varepsilon_1 A_1 (\sigma T_1^4 - \sigma T_2^4)$$

$$0.3 \text{ W} = (\varepsilon_1)(\pi)(0.02 \text{ m})^2 [\sigma T_1^4 - (5.67 \times 10^{-8} \text{ W/m}^2 \text{ K}^4)(303 \text{ K})^4]$$

Solving,

$$\sigma T_1^4 = 478 + \frac{239}{\varepsilon_1}$$

(i) For bright aluminum ($\varepsilon = 0.035$ from Table 1.3),

$$\sigma T_1^4 = 478 + 6828 = 7306 \text{ W/m}^2$$
$$T_1 = 599 \text{ K } (326°C)$$

(ii) For black anodized aluminum ($\varepsilon = 0.80$ from Table 1.3),

$$\sigma T_1^4 = 478 + 298 = 776 \text{ W/m}^2$$
$$T_1 = 342 \text{ K}(69°C)$$

Comments

1. The anodized aluminum gives a satisfactory operating temperature, but a bright aluminum capsule could not be used since 326°C is far in excess of allowable operating temperatures for semiconductor devices.

2. Note the use of kelvins for temperature in this radiation heat transfer calculation.

1.3.3 Heat Convection

As already explained, *convection* or *convective heat transfer* is the term used to describe heat transfer from a surface to a moving fluid, as shown in Fig. 1.9. The surface may be the inside of a pipe, the skin of a hypersonic aircraft, or a water-air interface in a cooling tower. The flow may be *forced*, as in the case of a liquid pumped through

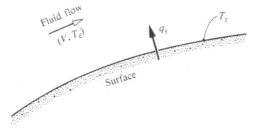

Figure 1.9 Schematic of convective heat transfer to a fluid at temperature T_e flowing at velocity V past a surface at temperature T_s.

the pipe or air on the flight vehicle propelled through the atmosphere. On the other hand, the flow could be *natural* (or *free*), driven by buoyancy forces arising from a density difference, as in the case of a natural-draft cooling tower. Either type of flow can be *internal*, such as the pipe flow, or *external*, such as flow over the vehicle. Also, both forced and natural flows can be either *laminar* or *turbulent*, with laminar flows being predominant at lower velocities, for smaller sizes, and for more viscous fluids. Flow in a pipe may become turbulent when the dimensionless group called the **Reynolds number**, $Re_D = VD/\nu$, exceeds about 2300, where V is the velocity [m/s], D is the pipe diameter [m], and ν is the kinematic viscosity of the fluid [m²/s]. Heat transfer rates tend to be much higher in turbulent flows than in laminar flows, owing to the vigorous mixing of the fluid. Figure 1.10 shows some commonly encountered flows.

The rate of heat transfer by convection is usually a complicated function of surface geometry and temperature, the fluid temperature and velocity, and fluid thermophysical properties. In an external forced flow, the rate of heat transfer is approximately proportional to the difference between the surface temperature T_s and the temperature of the free stream fluid T_e. The constant of proportionality is called the **convective heat transfer coefficient** h_c:

$$q_s = h_c \Delta T \tag{1.20}$$

where $\Delta T = T_s - T_e$, q_s is the heat flux from the surface into the fluid [W/m²], and h_c has units [W/m² K]. Equation (1.20) is often called *Newton's law of cooling* but is a definition of h_c rather than a true physical law. For natural convection, the situation is more complicated. If the flow is laminar, q_s varies as $\Delta T^{5/4}$; if the flow is turbulent, it varies as $\Delta T^{4/3}$. However, we still find it convenient to define a heat transfer coefficient by Eq. (1.20); then h_c varies as $\Delta T^{1/4}$ for laminar flows and as $\Delta T^{1/3}$ for turbulent ones.

An important practical problem is convective heat transfer to a fluid flowing in a tube, as may be found in heat exchangers for heating or cooling liquids, in condensers, and in various kinds of boilers. In using Eq. (1.20) for internal flows, $\Delta T = T_s - T_b$, where T_b is a properly averaged fluid temperature called the **bulk temperature** or mixed mean temperature and is defined in Chapter 4. Here it is sufficient to note that enthalpy in the steady-flow energy equation, Eq. (1.4), is also the bulk value, and T_b is the corresponding temperature. If the pipe has a uniform wall temperature T_s along its length, and the flow is laminar ($Re_D \lesssim 2300$), then sufficiently far from the pipe entrance, the heat transfer coefficient is given by the exact relation

$$h_c = 3.66 \frac{k}{D} \tag{1.21}$$

where k is the fluid thermal conductivity and D is the pipe diameter. Notice that the heat transfer coefficient is directly proportional to thermal conductivity, inversely proportional to pipe diameter, and—perhaps surprisingly —independent of flow velocity. On the other hand, for fully turbulent flow ($Re_D \gtrsim 10{,}000$), h_c is given

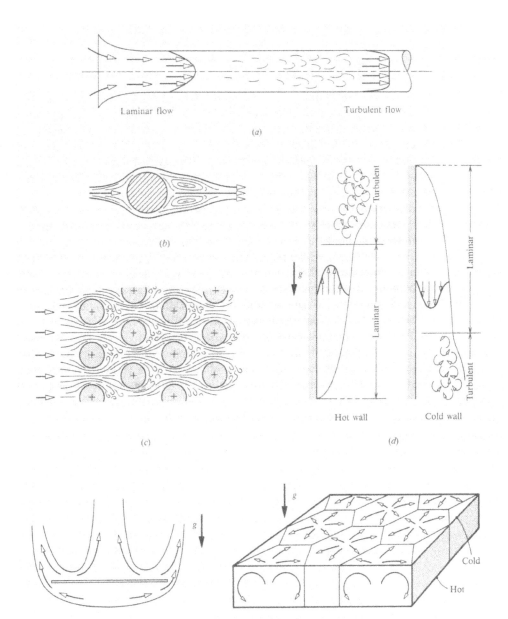

Figure 1.10   Some commonly encountered flows, *(a)* Forced flow in a pipe, $Re_D \simeq 50,000$. The flow is initially laminar because of the "bell-mouth" entrance but becomes turbulent downstream, *(b)* Laminar forced flow over a cylinder, $Re_D \simeq 25$. *(c)* Forced flow through a tube bank as found in a shell-and-tube heat exchanger, *(d)* Laminar and turbulent natural convection boundary layers on vertical walls, *(e)* Laminar natural convection about a heated horizontal plate, *(f)* Cellular natural convection in a horizontal enclosed fluid layer.

approximately by the following, rather complicated correlation of experimental data:

$$h_c = 0.023 \frac{V^{0.8} k^{0.6} (\rho c_p)^{0.4}}{D^{0.2} \nu^{0.4}}$$ (1.22)

In contrast to laminar flow, h_c is now strongly dependent on velocity, V, but only weakly dependent on diameter. In addition to thermal conductivity, other fluid properties involved are the kinematic viscosity, ν; density, ρ; and constant-pressure specific heat, c_p. In Chapter 4 we will see how Eq. (1.22) can be rearranged in a more compact form by introducing appropriate dimensionless groups. Equations (1.21) and (1.22) are only valid at some distance from the pipe entrance and indicate that the heat transfer coefficient is then independent of position along the pipe. Near the pipe entrance, heat transfer coefficients tend to be higher, due to the generation of large-scale vortices by upstream bends or sharp corners and the effect of suddenly heating the fluid.

Figure 1.11 shows a natural convection flow on a heated vertical surface, as well as a schematic of the associated variation of h_c along the surface. Transition from a laminar to a turbulent boundary layer is shown. In gases, the location of the transition is determined by a critical value of a dimensionless group called the **Grashof number**. The Grashof number is defined as $\mathrm{Gr}_x = (\beta \Delta T) g x^3 / \nu^2$, where $\Delta T = T_s - T_e$, g is the gravitational acceleration [m/s^2], x is the distance from the bottom of the surface where the boundary layer starts, and β is the volumetric coefficient of expansion, which for an ideal gas is simply $1/T$, where T is absolute temperature [K]. On a vertical wall, transition occurs at $\mathrm{Gr}_x \simeq 10^9$. For air, at normal temperatures, experiments show that the heat transfer coefficient for natural convection on a vertical wall can be approximated by the following formulas:

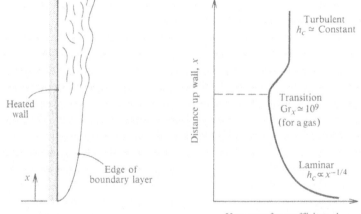

Figure 1.11   A natural-convection boundary layer on a vertical wall, showing the variation of local heat transfer coefficient. For gases, transition from a laminar to turbulent flow occurs at a Grashof number of approximately 10^9; hence $x_{tr} \simeq [10^9 \nu^2 / \beta \Delta T g]^{1/3}$.

Laminar flow: $h_c = 1.07(\Delta T/x)^{1/4}$ W/m²K $10^4 < Gr_x < 10^9$ **(1.23a)**

Turbulent flow: $h_c = 1.3(\Delta T)^{1/3}$ W/m²K $10^9 < Gr_x < 10^{12}$ **(1.23b)**

Since these are dimensional equations, it is necessary to specify the units of h_c, ΔT, and x, which are [W/m² K], [K], and [m], respectively. Notice that h_c varies as $x^{-1/4}$ in the laminar region but is independent of x in the turbulent region.

Usually the engineer requires the total heat transfer from a surface and is not too interested in the actual variation of heat flux along the surface. For this purpose, it is convenient to define an average heat transfer coefficient $\bar{h}_c$ for an *isothermal* surface of area A by the relation

$$\dot{Q} = \bar{h}_c A(T_s - T_e) \tag{1.24}$$

so that the total heat transfer rate, $\dot{Q}$, can be obtained easily. The relation between $\bar{h}_c$ and h_c is obtained as follows: For flow over a surface of width W and length L, as shown in Fig. 1.12,

$$d\dot{Q} = h_c(T_s - T_e)W\,dx$$

$$\dot{Q} = \int_0^L h_c(T_s - T_e)W\,dx$$

or

$$\dot{Q} = \left(\frac{1}{A}\int_0^A h_c\,dA\right)A(T_s - T_e), \quad \text{where } A = WL, dA = W\,dx \tag{1.25}$$

if $(T_s - T_e)$ is independent of x. Since T_e is usually constant, this condition requires an isothermal wall. Thus, comparing Eqs. (1.24) and (1.25),

$$\bar{h}_c = \frac{1}{A}\int_0^A h_c\,dA \tag{1.26}$$

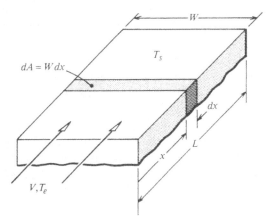

Figure 1.12   An isothermal surface used to define the average convective heat transfer coefficient $\bar{h}_c$.

Table 1.4   Orders of magnitude of average convective heat transfer coefficients.

Flow and Fluid	$\bar{h}_c$ W/m^2 K
Free convection, air	3–25
Free convection, water	15–1000
Forced convection, air	10–200
Forced convection, water	50–10,000
Forced convection, liquid sodium	10,000–100,000
Condensing steam	5000–50,000
Boiling water	3000–100,000

The surface may not be isothermal; for example, the surface may be electrically heated to give a uniform flux q_s along the surface. In this case, defining an average heat transfer coefficient is more difficult and will be dealt with in Chapter 4. Table 1.4 gives some order-of-magnitude values of average heat transfer coefficients for various situations. In general, high heat transfer coefficients are associated with high fluid thermal conductivities, high flow velocities, and small surfaces. The high heat transfer coefficients shown for boiling water and condensing steam are due to another cause: as we will see in Chapter 7, a large enthalpy of phase change (latent heat) is a contributing factor.

The complexity of most situations involving convective heat transfer precludes exact analysis, and *correlations* of experimental data must be used in engineering practice. For a particular situation, a number of correlations from various sources might be available, for example, from research laboratories in different countries. Also, as time goes by, older correlations may be superseded by newer correlations based on more accurate or more extensive experimental data. Heat transfer coefficients calculated from various available correlations usually do not differ by more than about 20%, but in more complex situations, much larger discrepancies may be encountered. Such is the nature of engineering calculations of convective heat transfer, in contrast to the more exact nature of the analysis of heat conduction or of elementary mechanics, for example.

EXAMPLE 1.3   Heat Loss through Glass Doors

The living room of a ski chalet has a pair of glass doors 2.3 m high and 4.0 m wide. On a cold morning, the air in the room is at 10°C, and frost partially covers the inner surface of the glass. Estimate the convective heat loss to the doors. Would you expect to see the frost form initially near the top or the bottom of the doors? Take $\nu = 14 \times 10^{-6}$ m^2/s for the air.

Solution

Given: Glass doors, width $W = 4$ m, height $L = 2.3$ m.

Required: Estimate of convective heat loss to the doors.

Assumptions: 1. Inner surface isothermal at $T_s \simeq 0°C$.
2. The laminar to turbulent flow transition occurs at $Gr_x \simeq 10^9$.

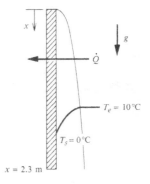

Equation (1.24) will be used to estimate the heat loss. The inner surface will be at approximately 0°C since it is only partially covered with frost. If it were warmer, frost couldn't form; and if it were much colder, frost would cover the glass completely. There is a natural convection flow down the door since $T_e = 10°$ C is greater than $T_s = 0°$ C. Transition from a laminar boundary layer to a turbulent boundary layer occurs when the Grashof number is about 10^9. For transition at $x = x_{tr}$,

$$Gr = 10^9 = \frac{(\beta \Delta T) g x_{tr}^3}{v^2}; \quad \beta = 1/T \text{ for an ideal gas}$$

$$x_{tr} = \left[\frac{10^9 v^2}{(\Delta T / T) g}\right]^{1/3} = \left[\frac{(10^9)(14 \times 10^{-6} \text{ m}^2/\text{s})^2}{(10/278)(9.81 \text{ m/s}^2)}\right]^{1/3} = 0.82 \text{ m}$$

where the average of T_s and T_e has been used to evaluate β. The transition is seen to take place about one third of the way down the door.

We find the average heat transfer coefficient, $\bar{h}_c$, by substituting Eqs. (1.23 a,b) in Eq. (1.26):

$$\bar{h}_c = \frac{1}{A}\int_0^A h_c \, dA; \quad A = WL, \quad dA = W \, dx$$

$$= \frac{1}{L}\int_0^L h_c \, dx$$

$$= \frac{1}{L}\left[\int_0^{x_{tr}} 1.07(\Delta T / x)^{1/4} \, dx + \int_{x_{tr}}^L 1.3(\Delta T)^{1/3} \, dx\right]$$

$$= (1/L)[(1.07)(4/3)\Delta T^{1/4} x_{tr}^{3/4} + (1.3)(\Delta T)^{1/3}(L - x_{tr})]$$

$$= (1/2.3)[(1.07)(4/3)(10)^{1/4}(0.82)^{3/4} + (1.3)(10)^{1/3}(2.3 - 0.82)]$$

$$= (1/2.3)[2.19 + 4.15]$$

$$= 2.75 \text{ W/m}^2\text{K}$$

Then, from Eq. (1.24), the total heat loss to the door is

$$\dot{Q} = \bar{h}_c A \Delta T = (2.75 \text{ W/m}^2 \text{ K})(2.3 \times 4.0 \text{ m}^2)(10 \text{ K}) = 253 \text{ W}$$

Comments

1. The local heat transfer coefficient is larger near the top of the door, so that the relatively warm room air will tend to cause the glass there to be at a higher temperature than further down the door. Thus, frost should initially form near the bottom of the door.

2. In addition, interior surfaces in the room will lose heat by radiation through the glass doors.

1.4   COMBINED MODES OF HEAT TRANSFER

Heat transfer problems encountered by the design engineer almost always involve more than one mode of heat transfer occurring simultaneously. For example, consider the nighttime heat loss through the roof of the house shown in Fig. 1.3. Heat is transferred to the ceiling by convection from the warm room air, and by radiation from the walls, furniture, and occupants. Heat transfer across the ceiling and its insulation is by conduction, across the attic crawlspace by convection and radiation, and across the roof tile by conduction. Finally, the heat is transferred by convection to the cold ambient air, and by radiation to the nighttime sky. To consider realistic engineering problems, it is necessary at the outset to develop the theory required to handle *combined modes* of heat transfer.

1.4.1 Thermal Circuits

The electrical circuit analogy for conduction through a composite wall was introduced in Section 1.3.1. We now extend this concept to include convection and radiation as well. Figure 1.13 shows a two-layer composite wall of cross-sectional area A with the layers A and B having thickness and conductivity L_A, k_A and L_B, k_B, respectively. Heat is transferred from a hot fluid at temperature T_i to the inside of the wall with a convective heat transfer coefficient $h_{c,i}$, and away from the outside of the wall to a cold fluid at temperature T_o with heat transfer coefficient $h_{c,o}$.

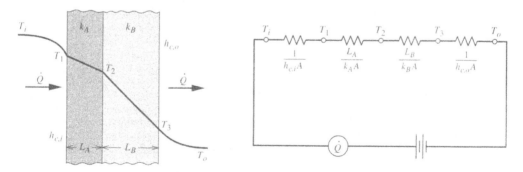

Figure 1.13   The temperature distribution for steady heat transfer across a composite plane wall, and the corresponding thermal circuit.

Newton's law of cooling, Eq. (1.20), can be rewritten as

$$\dot{Q} = \frac{\Delta T}{1/h_c A} \tag{1.27}$$

with $1/h_c A$ identified as a convective thermal resistance. At steady state, the heat flow through the wall is constant. Referring to Fig. 1.13 for the intermediate temperatures,

$$\dot{Q} = \frac{T_i - T_1}{1/h_{c,i}A} = \frac{T_1 - T_2}{L_A/k_A A} = \frac{T_2 - T_3}{L_B/k_B A} = \frac{T_3 - T_o}{1/h_{c,o}A} \tag{1.28}$$

Equation (1.28) is the basis of the thermal circuit shown in Fig. 1.13. The total resistance is the sum of four resistances in series. If we define the **overall heat transfer coefficient** U by the relation

$$\dot{Q} = UA(T_i - T_o) \tag{1.29}$$

then $1/UA$ is an overall resistance given by

$$\frac{1}{UA} = \frac{1}{h_{c,i}A} + \frac{L_A}{k_A A} + \frac{L_B}{k_B A} + \frac{1}{h_{c,o}A} \tag{1.30a}$$

or, since the cross-sectional area A is constant for a plane wall,

$$\frac{1}{U} = \frac{1}{h_{c,i}} + \frac{L_A}{k_A} + \frac{L_B}{k_B} + \frac{1}{h_{c,o}} \tag{1.30b}$$

Equation (1.29) is simple and convenient for use in engineering calculations. Typical values of U [W/m^2 K] vary over a wide range for different types of walls and convective flows.

Figure 1.14 shows a wall whose outer surface loses heat by both convection and radiation. For simplicity, assume that the fluid is at the same temperature as the surrounding surfaces, T_o. Using the approximate linearized Eq. (1.19),

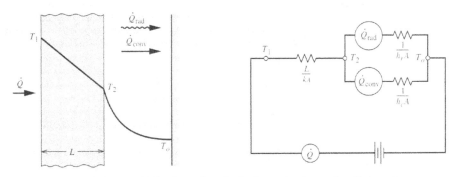

Figure 1.14 A wall that loses heat by both conduction and radiation; the thermal circuit shows resistances in parallel.

$$\dot{Q}_{rad} = \frac{\Delta T}{1/h_r A} \tag{1.31}$$

with $1/h_r A$ identified as a radiative thermal resistance. We now have two resistances in parallel, as shown in Fig. 1.14. The sum of the resistances is

$$\sum R = \frac{L}{kA} + \frac{1}{h_c A + h_r A}$$

so

$$\frac{1}{UA} = \frac{L}{kA} + \frac{1}{(h_c + h_r)A} \tag{1.32}$$

so that the convective and radiative heat transfer coefficients can simply be added. However, often the fluid and surrounding temperatures are not the same, or the simple linearized representation of radiative transfer [Eq. (1.19)] is invalid, so the thermal circuit is then more complex. When appropriate, we will write $h = h_c + h_r$ to account for combined convection and radiation.[6]

EXAMPLE 1.4   Heat Loss through a Composite Wall

The walls of a sparsely furnished single-room cabin in a forest consist of two layers of pine wood, each 2 cm thick, sandwiching 5 cm of fiberglass insulation. The cabin interior is maintained at 20°C when the ambient air temperature is 2°C. If the interior and exterior convective heat transfer coefficients are 3 and 6 W/m^2 K, respectively, and the exterior surface is finished with a white acrylic paint, estimate the heat flux through the wall.

Solution

Given: Pine wood cabin wall insulated with 5 cm of fiberglass.

Required: Estimate of heat loss through wall.

Assumptions: 1. Forest trees and shrubs are at the ambient air temperature, $T_e = 2°C$.
2. Radiation transfer inside cabin is negligible since inner surfaces of walls, roof, and floor are at approximately the same temperature.

From Eq. (1.29), the heat flux through the wall is

$$q = \frac{\dot{Q}}{A} = U(T_i - T_o)$$

From Eqs. (1.30) and (1.32), the overall heat transfer coefficient is given by

$$\frac{1}{U} = \frac{1}{h_{c,i}} + \frac{L_A}{k_A} + \frac{L_B}{k_B} + \frac{L_C}{k_C} + \frac{1}{(h_{c,o} + h_{r,o})}$$

---

[6] Notice that the notation used for this combined heat transfer coefficient, h, is the same as that used for enthalpy. The student must be careful not to confuse these two quantities. Other notation is also in common use, for example. α for the heat transfer coefficient and i for enthalpy.

The thermal conductivities of pine wood, perpendicular to the grain, and of fiberglass are given in Table 1.1 as 0.10 and 0.038 W/m K, respectively. The exterior radiation heat transfer coefficient is given by Eq. (1.19) as

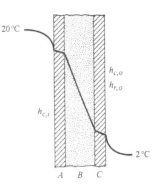

$$h_{r,o} = 4\varepsilon\sigma T_m^3$$

where $\varepsilon = 0.9$ for white acrylic paint, from Table 1.3, and $T_m \simeq 2°C = 275$ K (since we expect the exterior resistance to be small). Thus,

$$h_{ro} = 4(0.9)(5.67 \times 10^{-8} \text{ W/m}^2 \text{ K}^4)(275\text{K})^3$$

$$= 4.2 \text{ W/m}^2 \text{ K}$$

$$\frac{1}{U} = \frac{1}{3} + \frac{0.02}{0.10} + \frac{0.05}{0.038} + \frac{0.02}{0.10} + \frac{1}{6+4.2}$$

$$= 0.333 + 0.200 + 1.316 + 0.200 + 0.098$$

$$= 2.15 \text{ (W/m}^2 \text{ K)}^{-1}$$

$$U = 0.466 \text{ W/m}^2 \text{ K}$$

Then the heat flux $q = U(T_i - T_o) = 0.466(20 - 2) = 8.38$ W/m^2.
   The thermal circuit is shown below.

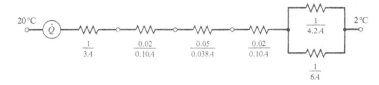

Comments

1. The outside resistance is seen to be $0.098/2.15 \simeq 5\%$ of the total resistance; hence, the outside wall of the cabin is only about 1 K above the ambient air, and our assumption of $T_m = 275$ K for the evaluation of $h_{r,o}$ is adequate.

2. The dominant resistance is that of the fiberglass insulation; therefore, an accurate calculation of q depends mainly on having accurate values for the fiberglass thickness and thermal conductivity. Poor data or poor assumptions for the other resistances have little impact on the result.

1.4.2 Surface Energy Balances

Section 1.4.1 assumed that the energy flow $\dot{Q}$ across the wall surfaces is continuous. In fact, we used a procedure commonly called a *surface energy balance*, which is used in various ways. Some examples follow. Figure 1.15 shows an opaque solid that is losing heat by convection and radiation to its surroundings. Two imaginary surfaces are located on each side of the real solid-fluid interface: an *s*-surface in

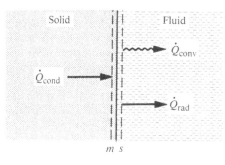

Figure 1.15   Schematic of a surface energy balance, showing the m- and s-surface in the solid and fluid, respectively.

the fluid just adjacent to the interface, and an m-surface in the solid located such that all radiation is emitted or absorbed between the m-surface and the interface. Thus, energy is transferred across the m-surface by conduction only. (The choice of s and m to designate these surfaces follows an established practice. In particular, the use of the s prefix is consistent with the use of the subscript s to denote a surface temperature T_s, in convection analysis.) The first law as applied to the closed system located between m- and s-surfaces requires that $\sum \dot{Q} = 0$; thus,

$$\dot{Q}_{cond} - \dot{Q}_{conv} - \dot{Q}_{rad} = 0 \tag{1.33}$$

or, for a unit area,

$$q_{cond} - q_{conv} - q_{rad} = 0 \tag{1.34}$$

where the sign convention for the fluxes is shown in Fig. 1.15. If the solid is isothermal, Eq. (1.33) reduces to

$$\dot{Q}_{conv} + \dot{Q}_{rad} = 0 \tag{1.35}$$

which is a simple energy balance on the solid. Notice that these surface energy balances remain valid for unsteady conditions, in which temperatures change with time, provided the mass contained between the s- and m-surfaces is negligible and cannot store energy.

EXAMPLE 1.5   Air Temperature Measurement

A machine operator in a workshop complains that the air-heating system is not keeping the air at the required minimum temperature of 20°C. To support his claim, he shows that a mercury-in-glass thermometer suspended from a roof truss reads only 17°C. The roof and walls of the workshop are made of corrugated iron and are not insulated; when the thermometer is held against the wall, it reads only 5°C. If the average convective heat transfer coefficient for the suspended thermometer is estimated to be $10 \text{ W/m}^2 \text{ K}$, what is the true air temperature?

Solution

Given: Thermometer reading a temperature of 17°C.

Required: True air temperature.

Assumptions: Thermometer can be modeled as a small gray body in large, nearly black surroundings at 5°C.

Let T_t be the thermometer reading, T_e the air temperature, and T_w the wall temperature. Equation (1.35) applies,

$$\dot{Q}_{conv} + \dot{Q}_{rad} = 0$$

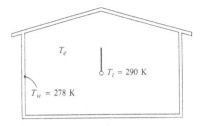

since at steady state there is no conduction within the thermometer. Substituting from Eqs. (1.24) and (1.18),

$$\bar{h}_c A(T_t - T_e) + \varepsilon \sigma A(T_t^4 - T_w^4) = 0$$

From Table 1.3, $\varepsilon = 0.8$ for pyrex glass. Canceling A,

$$10(290 - T_e) + (0.8)(5.67)(2.90^4 - 2.78^4) = 0$$

Solving,

$$T_e = 295 \text{ K} \simeq 22°C$$

Comments

1. Since $T_e > 20°C$, the air-heating system appears to be working satisfactorily.

2. Our model assumes that the thermometer is completely surrounded by a surface at 5°C: actually, the thermometer also receives radiation from machines, workers, and other sources at temperatures higher than 5°C, so that our calculated value of $T_e = 22°C$ is somewhat high.

1.5   TRANSIENT THERMAL RESPONSE

The heat transfer problems described in Examples 1.1 through 1.5 were *steady-state* problems; that is, temperatures were not changing in time. In Example 1.2, the transistor temperature was steady with the resistance $(I^2 R)$ heating balanced by the radiation heat loss. *Unsteady-state* or *transient* problems occur when temperatures change with time. Such problems are often encountered in engineering practice, and the engineer may be required to predict the temperature-time response of a system involved in a heat transfer process. If the system, or a component of the system, can be assumed to have a spatially uniform temperature, analysis involves a relatively simple application of the energy conservation principle, as will now be demonstrated.

1.5.1 The Lumped Thermal Capacity Model

If a system undergoing a transient thermal response to a heat transfer process has a nearly uniform temperature, we may ignore small differences of temperature within the system. Changes in internal energy of the system can then be specified in terms of changes of the assumed uniform (or average) temperature of the system. This approximation is called the **lumped thermal capacity** model.[7] The system might

---

[7] The term *capacitance* is also used, in analogy to an equivalent electrical circuit.

be a small solid component of high thermal conductivity that loses heat slowly to its surroundings via a large external thermal resistance. Since the thermal resistance to conduction in the solid is small compared to the external resistance, the assumption of a uniform temperature is justified. Alternatively, the system might be a well-stirred liquid in an insulated tank losing heat to its surroundings, in which case it is the mixing of the liquid by the stirrer that ensures a nearly uniform temperature. In either case, once we have assumed uniformity of temperature, we have no further need for details of the heat transfer within the system—that is, of the conduction in the solid component or the convection in the stirred liquid. Instead, the heat transfer process of concern is the interaction of the system with the surroundings, which might be by conduction, radiation, or convection.

Governing Equation and Initial Condition

For purposes of analysis, consider a metal forging removed from a furnace at temperature T_0 and suddenly immersed in an oil bath at temperature T_e, as shown in Fig. 1.16. The forging is a closed system, so the energy conservation principle in the form of Eq. (1.2) applies. Heat is transferred out of the system by convection. Using Eq. (1.24) the rate of heat transfer is $\bar{h}_c A(T - T_e)$, where $\bar{h}_c$ is the heat transfer coefficient averaged over the forging surface area A, and T is the forging temperature. There is no heat generated within the forging, so that $\dot{Q}_v = 0$. Substituting in Eq. (1.2):

$$\rho V c \frac{dT}{dt} = -\bar{h}_c A(T - T_e)$$

$$\frac{dT}{dt} = -\frac{\bar{h}_c A}{\rho V c}(T - T_e) \tag{1.36}$$

which is a first-order ordinary differential equation for the forging temperature, T, as a function of time, t. One initial condition is required:

$$t = 0: \quad T = T_0 \tag{1.37}$$

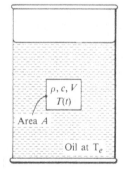

ρ, c, V
$T(t)$

Area A

Oil at T_e

Figure 1.16   A forging immersed in an oil bath for quenching.

Solution for the Temperature Response

A simple analytical solution can be obtained provided we assume that the bath is large, so T_e is independent of time, and that $\bar{h}_c A/\rho V c$ is approximated by a constant value independent of temperature. The variables in Eq. (1.36) can then be separated:

$$\frac{dT}{T-T_e} = -\frac{\bar{h}_c A}{\rho V c} dt$$

Writing $dT = d(T-T_e)$, since T_e is constant, and integrating with $T = T_0$ at $t = 0$,

$$\int_{T_0}^{T} \frac{d(T-T_e)}{T-T_e} = -\frac{\bar{h}_c A}{\rho V c} \int_{0}^{t} dt$$

$$\ln \frac{T-T_e}{T_0-T_e} = -\frac{\bar{h}_c A}{\rho V c} t$$

$$\frac{T-T_e}{T_0-T_e} = e^{-(\bar{h}_c A/\rho V c)t} = e^{-t/t_c} \tag{1.38}$$

where $t_c = \rho V c/\bar{h}_c A$ [s] is called the **time constant** of the process. When $t = t_c$, the temperature difference $(T-T_e)$ has dropped to be 36.8% of the initial difference $(T_0 - T_e)$. Our result, Eq. (1.38), is a relation between two dimensionless parameters: a dimensionless temperature, $T^* = (T-T_e)/(T_0-T_e)$, which varies from 1 to 0; and a dimensionless time, $t^* = t/t_c = \bar{h}_c At/\rho V c$, which varies from 0 to ∞. Equation (1.38) can be written simply as

$$T^* = e^{-t^*} \tag{1.39}$$

and a graph of T^* versus t^* is a single curve, as illustrated in Fig. 1.17.

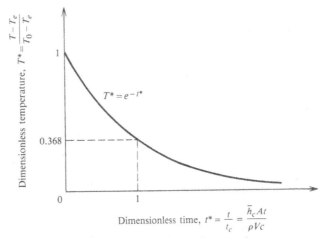

Figure 1.17   Lumped thermal capacity capacity temperature response in terms of dimensionless variables T^* and t^*.

Methods introduced in Chapter 2 can be used to deduce directly from Eqs. (1.36) and (1.37) that T^* must be a function of t^* alone [i.e., $T^* = f(t^*)$] without solving the equation. Of course, the solution also gives us the form of the function. Thus, the various parameters, $\overline{h}_c$, c, ρ, and so on, only affect the temperature response in the combination t^*, and not independently. If both $\overline{h}_c$ and c are doubled, the temperature at time t is unchanged. This dimensionless parameter t^* is a dimensionless group in the same sense as the Reynolds number, but it does not have a commonly used name.

Validity of the Model

We would expect our assumption of negligible temperature gradients within the system to be valid when the internal resistance to heat transfer is small compared with the external resistance. If L is some appropriate characteristic length of a solid body, for example, V/A (which for a plate is half its thickness), then

$$\frac{\text{Internal conduction resistance}}{\text{External convection resistance}} \simeq \frac{L/k_s A}{1/\overline{h}_c A} = \frac{\overline{h}_c L}{k_s} \simeq \frac{\overline{h}_c V}{k_s A} \qquad \textbf{(1.40)}$$

where k_s is the thermal conductivity of the solid material. The quantity $\overline{h}_c L/k_s$ [W/m^2 K][m]/[W/m K] is a dimensionless group called a **Biot number**, Bi.[8] More exact analyses of transient thermal response of solids indicate that, for bodies resembling a plate, cylinder, or sphere, $\text{Bi}_{LTC} = \overline{h}_c V/k_s A < 0.1$ ensures that the temperature given by the lumped thermal capacity (LTC) model will not differ from the exact volume averaged value by more than 5%, and that our assumption of uniform temperature is adequate. Nonetheless, the choice of both the length scale L and the threshold (e.g., $\text{Bi}_{LTC} < 0.1$) used to determine the validity of the lumped thermal capacity model should be done carefully if accurate calculations are critical (see Chapter 3). If the heat transfer is by radiation, the convective heat transfer coefficient in Eq. (1.40) can be replaced by the approximate radiation heat transfer coefficient h_r defined in Eq. (1.19).

In the case of the well-stirred liquid in an insulated tank, it will be necessary to evaluate the ratio

$$\frac{\text{Internal convection resistance}}{\text{External resistance}} \simeq \frac{1/h_{c,i}A}{1/UA} = \frac{U}{h_{c,i}} \qquad \textbf{(1.41)}$$

where U is the overall heat transfer coefficient, for heat transfer from the inner surface of the tank, across the tank wall and insulation, and into the surroundings. If this ratio is small relative to unity, the assumption of a uniform temperature in the liquid is justified.

The approximation or model used in the preceding analysis is called a lumped thermal capacity approximation since the thermal capacity is associated with a single temperature. There is an electrical analogy to the lumped thermal capacity model, owing to the mathematical equivalence of Eq. (1.36) to the equation governing the voltage in the simple resistance-capacitance electrical circuit shown

---

[8] To avoid confusion with the Biot number used in Chapter 3, we will denote the Biot number based on $L = V/A$ as Bi_{LTC} for use with the lumped thermal capacity model.

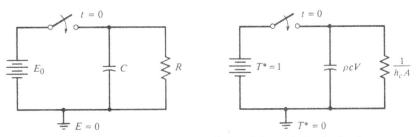

Figure 1.18   Equivalent electrical and thermal circuits for the lumped thermal capacity model of temperature response.

in Fig. 1.18,

$$\frac{dE}{dt} = -\frac{E}{RC} \tag{1.42}$$

with the initial condition $E = E_0$ at $t = 0$ if the capacitor is initially charged to a voltage E_0. The solution is identical in form to Eq. (1.38),

$$\frac{E}{E_0} = e^{-t/RC}$$

and the time constant is RC, the product of the resistance and capacitance [or $C/(1/R)$, the ratio of capacitance to conductance, to be exactly analogous to Eq. (1.38)].

EXAMPLE 1.6   Quenching of a Steel Plate

A steel plate 1 cm thick is taken from a furnace at 600°C and quenched in a bath of oil at 30°C. If the heat transfer coefficient is estimated to be 400 W/m² K, how long will it take for the plate to cool to 100°C? Take k, ρ, and c for the steel as 50 W/m K, 7800 kg/m³ and 450 J/kg K, respectively.

Solution

Given: Steel plate quenched in an oil bath.

Required: Time to cool from 600°C to 100°C.

Assumptions: Lumped thermal capacity model valid.

First the Biot number will be checked to see if the lumped thermal capacity approximation is valid. For a plate of width W, height H, and thickness L,

$$\frac{V}{A} \simeq \frac{WHL}{2WH} = \frac{L}{2}$$

where the surface area of the edges has been neglected.

$$\mathrm{Bi}_{LTC} = \frac{\bar{h}_c(L/2)}{k_s}$$

$$= \frac{(400\,\mathrm{W/m^2\,K})(0.005\,\mathrm{m})}{50\,\mathrm{W/m\ K}}$$

$$= 0.04 < 0.1$$

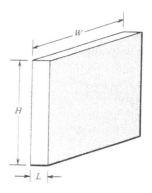

so the lumped thermal capacity model is applicable. The time constant t_c is

$$t_c = \frac{\rho V c}{\bar{h}_c A} = \frac{\rho(L/2)c}{\bar{h}_c} = \frac{(7800\,\mathrm{kg/m^3})(0.005\,\mathrm{m})(450\,\mathrm{J/kg\ K})}{(400\,\mathrm{W/m^2\ K})} = 43.9\,\mathrm{s}$$

Substituting $T_e = 30°\mathrm{C}$, $T_0 = 600°\mathrm{C}$, $T = 100°\mathrm{C}$ in Eq. (1.38) gives

$$\frac{100 - 30}{600 - 30} = e^{-t/43.9}$$

Solving,

$$t = 92\,\mathrm{s}$$

Comments

The use of a constant value of $\bar{h}_c$ may be inappropriate for heat transfer by natural convection or radiation.

1.5.2 Combined Convection and Radiation

The analysis of Section 1.5.1 assumes that the heat transfer coefficient was constant during the cooling period. This assumption is adequate for forced convection but is less appropriate for natural convection, and when thermal radiation is significant. Equation (1.23) shows that the natural convection heat transfer coefficient $\bar{h}_c$ is proportional to $\Delta T^{1/4}$ for laminar flow and to $\Delta T^{1/3}$ for turbulent flow. The temperature difference $\Delta T = T - T_e$ decreases as the body cools, as does $\bar{h}_c$. Radiation heat transfer is proportional to $(T^4 - T_e^4)$ and hence cannot be represented exactly by Newton's law of cooling. We now extend our lumped thermal capacity analysis to allow both for a variable convective heat transfer coefficient and for situations where both convection and radiation are important.

Governing Equation and Initial Condition

Figure 1.19 shows a body that loses heat by both convection and radiation. For a small gray body in large, nearly black surroundings also at temperature T_e, the radiation heat transfer is obtained from Eq. (1.18) as $\dot{Q} = \varepsilon A \sigma(T^4 - T_e^4)$. As in

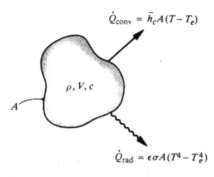

$$\dot{Q}_{conv} = \bar{h}_c A (T - T_e)$$

$$\rho, V, c$$

$$A$$

$$\dot{Q}_{rad} = \epsilon \sigma A (T^4 - T_e^4)$$

Figure 1.19 Schematic of a body losing heat by convection and radiation for a lumped thermal capacity model

Section 1.5.1, the energy conservation principle, Eq. (1.2), becomes

$$\rho V c \frac{dT}{dt} = -\bar{h}_c A (T - T_e) - \varepsilon A \sigma (T^4 - T_e^4)$$

or

$$\frac{dT}{dt} = -\frac{\bar{h}_c A}{\rho V c}(T - T_e) - \frac{\varepsilon A \sigma}{\rho V c}(T^4 - T_e^4) \tag{1.43}$$

The initial condition is again

$$t = 0 : \quad T = T_0 \tag{1.44}$$

This first-order ordinary differential equation has no convenient analytical solution even when the convective heat transfer coefficient $\bar{h}_c$ is constant as in forced convection. However, Eq. (1.43) can be solved easily using a numerical integration procedure. For this purpose, it can be rearranged as

$$\frac{dT}{dt} + \frac{hA}{\rho V c}(T - T_e) = 0 \tag{1.45}$$

$$h = \bar{h}_c + h_r = B(T - T_e)^n + \sigma \varepsilon (T^2 + T_e^2)(T + T_e) \tag{1.46}$$

where $(T^4 - T_e^4)$ has been factored, as was done in deriving Eq. (1.19). For forced convection, $n = 0$, $B = \bar{h}_c$; for laminar natural convection $n = 1/4$ and B is a constant [for example, for a plate of height L, Eq. (1.23a) gives $B = (4/3)(1.07)/L^{1/4}$]. Equation (1.46) defines a total heat transfer coefficient that accounts for both convection and radiation and changes continuously as the body cools. To put Eq. (1.45) in dimensionless form, we use the dimensionless variables introduced in Section 1.5.1:

$$T^* = \frac{T - T_e}{T_0 - T_e}, \qquad t^* = \frac{t}{t_c} \tag{1.47a,b}$$

The definition of the time constant t_c poses a problem since h is not a constant as before. We choose to define t_c in terms of the value of h at time $t = 0$, when the body temperature is T_0,

$$t_c = \frac{\rho V c}{h_0 A} = \frac{\rho V c}{[B(T_0 - T_e)^n + \sigma \varepsilon (T_0^2 + T_e^2)(T_0 + T_e)]A} \tag{1.48}$$

Equation (1.45) then becomes

$$\frac{dT^*}{dt^*} + \frac{h(T^*)}{h_0} T^* = 0 \tag{1.49}$$

with the initial condition

$$t^* = 0: \quad T^* = 1 \tag{1.50}$$

Computer Program LUMP

Numerical integration is appropriate for this problem. The computer program LUMP has been prepared accordingly. LUMP solves Eq. (1.49), that is, it obtains the temperature response of a body that loses heat by convection and/or radiation, based on the lumped thermal capacity model. The required input constant B is defined in Eq. (1.46). Any consistent system of units can be used. The output can be obtained either as graph or as numerical data.

EXAMPLE 1.7   Quenching of an Alloy Sphere

A materials processing experiment under microgravity conditions on the space station re-quires quenching in a forced flow of an inert gas. A 1 cm–diameter metal alloy sphere is removed from a furnace at 800°C and is to be cooled to 500°C by a flow of nitrogen gas at 25°C. Determine the effect of the convective heat transfer coefficient on cooling time for $10 < \bar{h}_c < 100$ W/m² K. Properties of the alloy include: $\rho = 14{,}000$ kg/m³; $c = 140$ J/kg K; $\varepsilon = 0.1$. The surrounds can be taken as nearly black at 25°C.

Solution

Given: A metal alloy sphere to be quenched.

Required: Effect of convective heat transfer coefficient on cooling time.

Assumptions: 1. Lumped thermal capacity model valid.
                  2. Constant convective heat transfer coefficient.

The computer code LUMP can be used to solve this problem.
The required inputs are:

T_0 and $T_e = 1073, 298$

$B = \bar{h}_c = 10$

$n = 0$

$\sigma = 5.67 \times 10^{-8}$

$\varepsilon = 0.1$

Final value of t^*: try $t^* = 1$

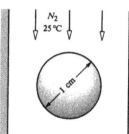

The required dimensionless temperature is

$$T^* = \frac{T - T_e}{T_0 - T_e} = \frac{773 - 298}{1073 - 298} = 0.613$$

and the code is used to obtain the corresponding dimensionless time t^*. For a sphere $V/A = (\pi D^3/6)(\pi D^2) = D/6$, so that the time constant is

$$t_c = \frac{\rho V c}{h_0 A} = \frac{\rho (D/6)c}{h_0} = \frac{(14,000)(0.01/6)(140)}{\bar{h}_c + (5.67 \times 10^{-8})(0.1)(1073^2 + 298^2)(1073 + 298)}$$

$$= \frac{3267}{\bar{h}_c + 9.64} \text{ s}$$

The actual time is $t = t^* t_c$. Results obtained using LUMP are tabulated below.

$\bar{h}_c$ W/m^2 K	t^*	t_c s	t s
10	0.59	166	98
20	0.54	110	61
30	0.53	82	43
50	0.52	55	29
100	0.51	30	15

Comments

1. Only two significant figures have been given since high accuracy is not warranted for the problem.

2. The heat transfer coefficient does not have a strong effect on t^*. Why?

3. For the lumped thermal capacity model to be valid, the Biot number, Bi_{LTC}, should be less than 0.1. The worst case is with $\bar{h}_c = 100$ W/m^2 K at time $t = 0$, giving $h_0 = 109.6$ W/m^2 K and $0.1 > (109.6)(0.01/6)/k_s$, that is, $k_s > 1.8$ W/m K, which certainly will be true for a metal alloy.

1.6   DIMENSIONS AND UNITS

Dimensions are physical properties that are measurable — for example, length, time, mass, and temperature. A system of *units* is used to give numerical values to dimensions. The system most widely used throughout the world in science and industry is the International System of units (SI), from the French name *Système International d' Unités*. This system was recommended at the General Conference on Weights and Measures of the International Academy of Sciences in 1960 and was adopted by the U.S. National Bureau of Standards in 1964.

In the United States, the transition from the older English system of units to the SI system has been slow and is not complete. The SI system is used in science education, in government contracts, by engineering professional societies, and by many industries. However, engineers in some more mature industries still prefer to use English units, and, of course, commerce and trade in the United States remains dominated by the English system. We buy pounds of vegetables, quarts of milk, drive miles to work, and say that it is a hot day when the temperature exceeds 80°F. (Wine is now sold in 750 ml bottles, though, which is a modest step forward!)

In this text, we will use the SI system, with which the student has become familiar from physics courses. For convenience, this system is summarized in the tables of Appendix B. Base and supplementary units, such as length, time, and plane angle, are given in Table B.1a; and derived units, such as force and energy, are given in Table B.1b. Recognized non-SI units (e.g., hour, bar) that are acceptable for use with the SI system are listed in Table B.1c. Multiples of SI units (e.g., kilo, micro) are defined in Table B.1d. Accordingly, the property data given in the tables of Appendix A are in SI units. The student should review this material and is urged to be careful when writing down units. For example, notice that the unit of temperature is a kelvin (not Kelvin) and has the symbol K (not °K). Likewise, the unit of power is the watt (not Watt). The symbol for a kilogram is kg (not KG). An issue that often confuses the student is the correct use of Celsius temperature. Celsius temperature is defined as $(T - 273.15)$ where T is in kelvins. However, the unit "degree Celsius" is equal to the unit "kelvin" ($1°C = 1$ K).

Notwithstanding the wide acceptance of the SI system of units, there remains a need to communicate with those engineers (or lawyers!) who are still using English units. Also, component dimensions, or data for physical properties, may be available only in English or cgs units. For example, most pipes and tubes used in the United States conform to standard sizes originally specified in English units. A 1 inch nominal-size tube has an outside diameter of 1 in. For convenience, selected dimensions of U.S. commercial standard pipes and tubes are given in SI units in Appendix A as Tables A.14a and A.14b, respectively. The engineer must be able to convert dimensions from one system of units to another. Table B.2 in Appendix B gives the conversion factors required for most heat transfer applications. The program UNITS is based on Table B.2 and contains all the conversion factors in the table. With the input of a quantity in one system of units, the output is the same quantity in the alternative units listed in Table B.2. It is recommended that the student or engineer perform all problem solving using the SI system so as to efficiently use the Appendix A property data and the computer software. If a problem is stated in English units, the data should be converted to SI units using UNITS; if a customer requires results in units other than SI, UNITS will give the required values.

1.7   CLOSURE

Chapter 1 had two main objectives:

1. To introduce the three important modes of heat transfer, namely, conduction, radiation, and convection.
2. To demonstrate how the first law of thermodynamics is applied to an engineering system to obtain the consequences of a heat transfer process.

For each mode of heat transfer, some working equations were developed, which, though simple, allow heat transfer calculations to be made for a wide variety of problems. Equations (1.9), (1.18), and (1.20) are probably the most frequently used equations for thermal design. An electric circuit analogy was shown to be a useful aid for problem solving when more than one mode of heat transfer is involved. In applying the first law to engineering systems, a closed system was considered, and the variation of temperature with time was determined for a solid of high conductivity or a well-stirred fluid. (An example of an open system is a heat exchanger, and Chapter 8 will show how the first law is applied to such systems.)

The student should be familiar with some of the Chapter 1 concepts from previous physics, thermodynamics, and fluid mechanics courses. A review of texts for such courses is appropriate at this time. Many new concepts were introduced, however, which will take a little time and effort to master. Fortunately, the mathematics in this chapter is simple, involving only algebra, calculus, and the simplest first-order differential equation, and should present no difficulties to the student. After successfully completing a selection of the following exercises, the student will be well equipped to tackle subsequent chapters.

A feature of this text is an emphasis on real engineering problems as examples and exercises. Thus, Chapter 1 has somewhat greater scope and detail than the introductory chapters found in most similar texts. With the additional material, more realistic problems can be treated, both in Chapter 1 and in subsequent chapters. In particular, conduction problems in Chapters 2 and 3 have more realistic convection and radiation boundary conditions. Throughout the text are exercises that require application of the first law to engineering systems, for it is always the consequences of a heat transfer process that motivate the engineer's concern with the subject.

A computer program accompanies Chapter 1. The program UNITS is a simple units conversion tool that allows unit conversions to be made quickly and reliably.

EXERCISES

Note to the student: Exercises 1-1 through 1-3 are included to provide a review of some concepts of mathematics and thermodynamics that are especially relevant to heat transfer. In addition, students are urged to keep their mathematics and thermodynamic texts in easy reach while studying heat and mass transfer, in order to review pertinent topics as the occasion arises. Too often, students tend to compartmentalize their learning experience, with each subject terminated by an end-of-semester examination. Continuous review of more elementary subjects, as the student proceeds through the degree program, is essential to the mastery of more advanced subjects.

1-1. Solve the following ordinary differential equations:

(i) $\dfrac{dy}{dx} + \beta y = 0$

(ii) $\dfrac{dy}{dx} + \beta y + \alpha = 0$

(iii) $\dfrac{d^2 y}{dx^2} - \lambda^2 y = 0$

(iv) $\dfrac{d^2 y}{dx^2} + \lambda^2 y = 0$

(v) $\dfrac{d^2 y}{dx^2} - \lambda^2 y + \alpha = 0$

where α, β, and λ are constants.

1-2. A low-pressure heat exchanger transfers heat between two helium streams, each with a flow rate of $\dot{m} = 5 \times 10^{-3}$ kg/s. In a performance test the cold stream enters at a pressure of 1000 Pa and a temperature of 50 K, and exits at 730 Pa and 350 K.

(i) If the flow cross-sectional area for the cold stream is 0.019 m^2, calculate the inlet and outlet velocities.
(ii) If the exchanger can be assumed to be perfectly insulated, determine the rate of heat transfer in the exchanger. For helium, $c_p = 5200$ J/kg K.

1-3. A shell-and-tube condenser for an ocean thermal energy conversion and fresh water plant is tested with a water feed rate to the tubes of 4000 kg/s. The water inlet and outlet conditions are measured to be $P_1 = 129$ kPa, $T_1 = 280$ K; and $P_2 = 108$ kPa, $T_2 = 285$ K.

(i) Calculate the rate of heat transfer to the water.
(ii) If saturated steam condenses in the shell at 1482 Pa, calculate the steam condensation rate.

For the feed water, take $\rho = 1000$ kg/m^3, $c_v = 4192$ J/kg K. (Steam tables are given as Table A.12a in Appendix A.)

1-4. A Pyrex glass vessel has a 5 mm-thick wall and is protected with a 1 cm-thick layer of neoprene rubber. If the inner and outer surface temperatures are 40°C and 20°C, respectively, and the total surface area of the vessel is 400 cm^2, calculate the rate of heat loss from the vessel. Also calculate the temperature of the interface between the glass and the rubber, and carefully sketch the temperature profile through the composite wall.

1-5. In the United States, insulations are often specified in terms of their thermal resistance in $[Btu/hr\ ft^2\ °F]^{-1}$, called the "R" value.

    (i) What is the R value of a 10 cm-thick layer of fiberglass insulation?
    (ii) How thick a layer of cork is required to give an R value of 18?
    (iii) What is the R value of a 2 cm-thick board of white pine?

1-6. A picnic icebox is 40 cm long, is 20 cm high and deep, and is insulated with 2 cm-thick polystyrene foam insulation. If the ambient air temperature is 30°C, estimate how much ice will melt in 8 hours. Use an enthalpy of melting for water of 335 kJ/kg.

1-7. A composite wall has a 6 cm layer of fiberglass insulation sandwiched between 2 cm–thick white pine boards. If the inner and outer surface temperatures are 20°C and 0°C, respectively, calculate the heat flow per unit area across the wall. Also calculate the wood-fiberglass interface temperatures, and accurately draw the temperature profile through the wall.

1-8. A freezer is 1 m wide and deep and 2 m high, and must operate at -10°C when the ambient air is at 30°C. What thickness of polystyrene is required if the load on the refrigeration unit should not exceed 200 W? Assume that the outer surface of the insulation is approximately at the ambient air temperature and that the base of the freezer is perfectly insulated.

1-9. A very effective insulation can be made from multiple layers of thin aluminized plastic film separated by rayon mesh and evacuated to a very low pressure ($\sim 10^{-5}$ torr). Such "superinsulation" can be used for insulating storage tanks holding cryogenic liquids. On a space station, a 1 m–O.D. spherical tank contains saturated nitrogen at 1 atm pressure. What thickness of a superinsulation having an effective thermal conductivity of 9×10^{-6} W/m K is required to have a boil-off rate of less than 2 mg/s when the ambient temperature is 250 K? The boiling point of nitrogen is 77.4 K, and its enthalpy of vaporization is 0.200×10^6 J/kg.

1-10. A blackbody radiates to a surrounding black enclosure. If the body is maintained at 100 K above the enclosure temperature, calculate the net radiative heat flux leaving the body when the enclosure is at 80 K, 300 K, 1000 K, and 5000 K.

1-11. An astronaut is at work in the service bay of a space shuttle and is surrounded by walls that are at -100°C. The outer surface of her space suit has an area of 3 m^2 and is aluminized with an emittance of 0.05. Calculate her rate of heat loss when the suit's outer temperature is 0°C. Express your answer in watts and kcal/hr.

1-12. An electronic device is contained in a cylinder 10 cm in diameter and 30 cm long. It operates inside an unpressurized module of an orbiting space station. The device dissipates 60 W, and its temperature must not exceed 80°C when the module walls are at −80°C. What value of emittance should be specified for the surface coating of the cylinder?

1-13. A high-vacuum chamber has its walls cooled to −190°C by liquid nitrogen. A sensor in the chamber has a surface area of 10 cm^2 and must be maintained at a temperature of 25°C. Plot a graph of the power required versus emittance of the sensor surface.

1-14. A semiconductor laser is attached to a diamond heat spreader on the top of a 1 cm copper cube heat sink. The assembly is located in an evacuated Dewar flask. The average surface temperature of the heat sink is 80 K when the inner surface of the Dewar flask is at 100 K. Estimate the parasitic heat gain by the sink due to radiation heat transfer. The emittance of the copper is 0.08 and Dewar flask inner surface is almost black.

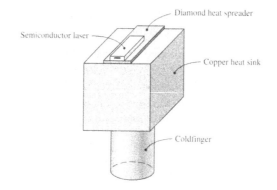

1-15. Consider a 3 m length of tube with a 1.26 cm inside diameter. Determine the convective heat transfer coefficient when

   (i) water flows at 2 m/s.
   (ii) oil (SAE 50) flows at 2 m/s.
   (iii) air at atmospheric pressure flows at 20 m/s.

Thermophysical property data at 300 K are as follows:

	ρ kg/m^3	ν m^2/s	k W/m K	c_p J/kg K
Water	996	0.87×10^{-6}	0.611	4178
SAE 50 oil	883	570×10^{-6}	0.145	1900
Air at 1 atm	1.177	15.7×10^{-6}	0.0267	1005

1-16. Consider flow of water at 300 K in a long pipe of 1 cm inside diameter. Plot a graph of the heat transfer coefficient versus velocity over the range 0.01 to 100 m/s. Repeat for air at 1 atm and 300 K. Use the property values given in Exercise 1–15.

1-17. A 1 m–high vertical wall is maintained at 310 K, when the surrounding air is at 1 atm and 290 K. Plot the local heat transfer coefficient as a function of location up the wall. Take $v = 15.7 \times 10^{-6}$ m²/s for air. Also calculate the convective heat loss per meter width of wall.

1-18. A 2 m–high vertical surface is maintained at 15°C when exposed to stagnant air at 1 atm and 25°C. Plot a graph showing the variation of the local heat transfer coefficient, and calculate the convective heat transfer for a 3 m width of wall. Take $v = 15.0 \times 10^{-6}$ m²/s for air.

1-19. A thermistor is used to measure the temperature of an air stream leaving an air heater. It is located in a 30 cm square duct and records a temperature of 42.6°C when the walls of the duct are at 38.1°C. What is the true temperature of the air? The thermistor can be modeled as a 3 mm-diameter sphere of emittance 0.7. The convective heat transfer coefficient from the air stream to the thermistor is estimated to be 31 W/m² K.

1-20. A room heater is in the form of a thin vertical panel 1 m long and 0.7 m high, with air allowed to circulate freely on both sides. If its rating is 800 W, what will the average panel surface temperature be when the room air temperature is 20°C? The emittance of the surface is 0.85. Take $v_{air} = 17.5 \times 10^{-6}$ m²/s.

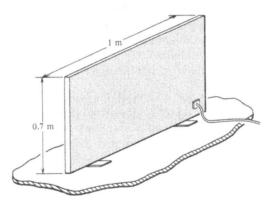

1-21. An electric water heater has a diameter of 1 m and a height of 2 m. It is insulated with 6 cm of medium-density fiberglass, and the outside heat transfer coefficient is estimated to be 8 W/m² K. If the water is maintained at 65°C and the ambient temperature is 20°C, determine

   (i) the rate of heat loss.
   (ii) the monthly cost attributed to heat loss if electricity costs 8 cents/kilowatt hour.

1-22. A 1 cm–diameter sphere is maintained at 60°C in an enclosure with walls at 35°C through which air at 40°C circulates. If the convective heat transfer coefficient is 11 W/m^2 K, estimate the rate of heat loss from the sphere when its emittance is

(i) 0.05.
(ii) 0.85.

1-23. Estimate the heating load for a building in a cold climate when the outside temperature is -10°C and the air inside is maintained at 20°C. The 350 m^2 of walls and ceiling are a composite of 1 cm-thick wallboard ($k = 0.2$ W/m K), 10 cm of vermiculite insulation ($k = 0.06$ W/m K), and 3 cm of wood ($k = 0.15$ W/m K). Take the inside and outside heat transfer coefficients as 7 and 35 W/m^2 K, respectively.

1-24. If a 2.5×10 m shaded wall in the building of Exercise 1–23 is replaced by a window, compare the heat loss through the wall if it is

(i) 0.3 cm-thick glass ($k = 0.88$ W/m K).
(ii) double-glazed with a 0.6 cm air gap between two 0.3 cm-thick glass panes.
(iii) the original wall.

1-25. Rework Exercise 1–20 for a panel 0.7 m high and 1.5 m long that is rated at 1 kW.

1-26. Saturated steam at 150°C flows through a 15 cm-O.D., uninsulated steam pipe ($\varepsilon = 0.8$). In order to reduce the amount of steam condensed, the pipe is painted with aluminum paint ($\varepsilon = 0.14$). Determine the reduction in the amount of steam condensed in kg/day for a 20 m length of pipe. Take the outside convective heat transfer coefficient to be 4 W/m^2 K, and surroundings at 20°C. Also, calculate the annual savings if the cost of thermal energy is 4 cents/kWh.

1-27. A 2 cm–square cross section, 10 cm-long bar consists of a 1 cm-thick copper layer and a 1 cm–thick epoxy composite layer. Compare the thermal resistances for heat flow perpendicular and parallel to the two layers. In both cases, assume that the two sides of the slab are isothermal. Take $k = 400$ W/m K for the copper and $k = 0.4$ W/m K for the epoxy composite.

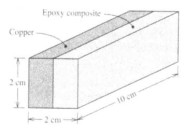

1-28. Suprathane, manufactured by Rubicon Chemicals Inc., is a wall insulation consisting of a sandwich of urethane foam covered with a protective layer on either side. Each protective layer is a composite of aluminum foil over kraft paper interlaced with high-strength glass fiber. For a $1^1/_4$ in-thick sandwich, the R value is 13.2 [Btu/hr ft^2 °F]$^{-1}$. When used in conjunction with a conventional $3^1/_2$ in–thick mineral wool blanket, a wall with a combined R value of 22.7 is claimed.

(i) What is the thermal resistance per unit area of the sandwich in SI units?
(ii) If inside and outside heat transfer coefficients are estimated to be 7 W/m^2 K and 20 W/m^2 K, respectively, what is the rate of heat loss through a 5 m–long, 3 m–high wall when the inside temperature is 20°C and the outside temperature is −20°C?

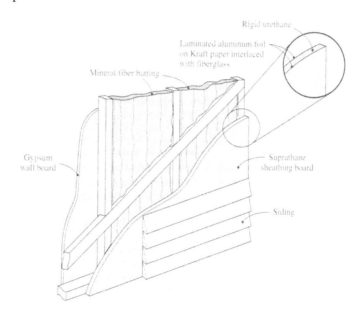

1-29. A polystyrene ice chest has exterior dimensions 30 cm × 20 cm × 15 cm deep, and a 3 cm wall thickness. It is filled with a mixture of ice cubes and water, and initially the ice is 70% by mass of the mixture. The ambient air is at 30°C, and the outside convective plus radiative heat transfer coefficient is estimated to be 10 W/m^2 K. If the heat gain through the chest base is negligible, determine the time required for the ice to melt. Take $\rho = 1000$ kg/m^3 for the mixture, and an enthalpy of melting of 335 J/kg K.

1-30. A kitchen oven has a maximum operating temperature of 280°C. Determine the thickness of fiberglass insulation required to ensure that the outside surfaces do not exceed 40°C when the kitchen air temperature is 25°C. The inside and outside heat transfer coefficients can be taken as 40 W/m^2 K and 15 W/m^2 K, respectively, and the conductivity of the fiberglass insulation as 0.07 W/m K.

1-31. A furnace wall is to operate with inner and outer surface temperatures of 1500 K and 320 K, respectively. Insulating bricks measuring 20 cm × 10 cm × 8 cm are available in two kinds at the same price. Type A has a thermal conductivity of 2.0 W/m K and a maximum allowable temperature of 1600 K. Type B has a thermal conductivity of 1.0 W/m K and a maximum allowable temperature of 1000 K. Determine how the bricks should be arranged so as not to exceed a heat flow per unit area of 1000 W/m^2, and minimize the cost of the walls.

1-32. A furnace wall has 0.3 m–thick inner layer of fire-clay brick ($k = 1.7$ W/m K), a 0.2 m-thick layer of kaolin brick ($k = 0.12$ W/m K), and a 0.1 m–thick outer layer of face brick ($k = 1.3$ W/m K). The furnace gases are at 1400 K, and the ambient air is at 310 K. The inside and outside heat transfer coefficients are 100 and 15 W/m^2 K, respectively.

(i) Determine the heat loss through a 4 m-high, 8 m-long wall.
(ii) It is later decided that the face brick temperature should not exceed 360 K. Can this constraint be met by increasing the thickness of the kaolin brick layer?

1-33. In cold climates, weather reports usually give both the actual air temperature and the "wind-chill" temperature, which can be interpreted as follows. At the prevailing wind speed there is a rate of heat loss per unit area q_s from a clothed person for an air temperature T_e. The wind-chill temperature T_{wc} is the air temperature that will give the same rate of heat loss on a calm day. Estimate the wind-chill temperature on a day when the air temperature is $-10°C$ and the wind speed is 10m/s, giving a convective heat transfer coefficient of 50 W/m^2 K. A radiation heat transfer coefficient of 5 W/m^2 K can be used, and under calm conditions the convective heat transfer coefficient can be taken to be 5.0 W/m^2 K. Assume a 3 mm layer of skin ($k = 0.35$ W/m K), clothing equivalent to 8 mm-thick wool ($k = 0.05$ W/m K), and a temperature of 35°C below the skin. Also calculate the skin outer temperature.

1-34. The cross section of a 20 cm-thick, 3 m-wide, and 1 m-high composite wall is shown. The conductivities of materials A, B, and C are 1.0, 0.1, and 0.05 W/m K, respectively. One side is exposed to air at 295 K with a heat transfer coefficient of 4 W/m^2 K, and the other side is exposed to air at 260 K with a heat transfer coefficient of 16 W/m^2 K. Estimate the heat flow through the wall. Carefully discuss any assumptions you need to make.

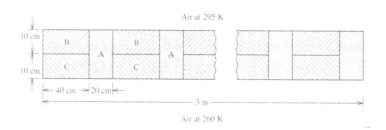

1-35. A mercury-in-glass thermometer used to measure the air temperature in an enclosure reads 15°C. The enclosure walls are all at 0°C. Estimate the true air temperature if the convective heat transfer coefficient for the thermometer bulb is estimated to be 12 W/m^2 K.

1-36. A tent is pitched on a mountain in an exposed location. The tent walls are opaque to thermal radiation. On a clear night the outside air temperature is $-1°C$, and the effective temperature of the sky as a black radiation sink is $-60°C$. The convective heat transfer coefficient between the tent and the ambient air can be taken to be 8 W/m^2 K. If the temperature of the outer surface of a sleeping bag on the tent floor is measured to be 10°C, estimate the heat loss from the bag in W/m^2,

   (i) if the emittance of the tent material is 0.7.
   (ii) if the outer surface of the tent is aluminized to give an emittance of 0.2.

   For the sleeping bag, take an emittance of 0.8 and a convective heat transfer coefficient of 4 W/m^2 K. Assume that the ambient air circulates through the tent.

1-37. A natural convection heat transfer coefficient meter is intended for situations where the air temperature T_e is known but the surrounding surfaces are at an unknown temperature T_w. The two sensors that make up the meter each have a surface area of 1 cm^2, one has a surface coating of emittance $\varepsilon_1 = 0.9$, and the other has an emittance of $\varepsilon_2 = 0.1$. The rear surface of the sensors is well insulated. When $T_e = 300$ K and the test surface is at 320 K, the power inputs required to maintain the sensor surfaces at 320 K are $\dot{Q}_1 = 21.7$ mW and $\dot{Q}_2 = 8.28$ mW. Determine the heat transfer coefficient at the meter location.

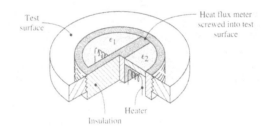

1-38. The horizontal roof of a building is surfaced with black tar paper of emittance 0.96. On a clear, still night the air temperature is 5°C, and the effective temperature of the sky as a black radiation sink is $-60°C$. The underside of the roof is well insulated.

   (i) Estimate the roof surface temperature for a convective heat transfer coefficient of 5 W/m^2 K.
   (ii) If the wind starts blowing, giving a convective heat transfer coefficient of 20 W/m^2 K, what is the new roof temperature?
   (iii) Repeat the preceding calculations for aluminum roofing of emittance 0.15.

1-39. A chemical reactor has a 5 mm-thick mild steel wall and is lined inside with a 2 mm-thick layer of polyvinylchloride. The contents are at 80°C, and the ambient air is at 20°C. The inside thermal resistance is negligible ($h_{c,i}$ very large), and the outside heat transfer coefficient for combined convection and radiation is 7 W/m² K.

(i) Draw the thermal circuit.
(ii) Plot a graph of the temperature profile through the wall.
(iii) Calculate the rate of heat loss for a surface area of 10 m².

1-40. To prevent misting of the windscreen of an automobile, recirculated warm air at 37°C is blown over the inner surface. The windscreen glass ($k = 1.0$ W/m K) is 4 mm thick, and the ambient temperature is 5°C. The outside and inside heat transfer coefficients are 70 and 35 W/m² K, respectively.

(i) Determine the temperature of the inside surface of the glass.
(ii) If the air inside the automobile is at 20°C, 1 atm, and 80% relative humidity, will misting occur? (Refer to your thermodynamics text for the principles of psychrometry.)

1-41. The horizontal roof of a building is coated with tar of emittance 0.94. On a cloudy, still night the air temperature is 5°C, and the convective heat transfer coefficient between the air and the roof is estimated to be 4 W/m² K.

(i) If the effective temperature of the sky as a black radiation sink is −10°C, determine the roof temperature. Assume that the under surface of the roof is well insulated.
(ii) If a wind starts blowing, resulting in a convective heat transfer coefficient of 12 W/m² K, what is the new roof temperature?
(iii) Repeat the preceding calculations for aluminum roofing of emittance 0.15.

1-42. An alloy cylinder 3 cm in diameter and 2 m high is removed from an oven at 200°C and stood on its end to cool in air at 20°C. Give a rough estimate of the time for the cylinder to cool to 100°C. For the alloy, take $\rho = 8600$ kg/m³, $c = 340$ J/kg K, $k = 110$ W/m K, and $\varepsilon = 0.74$; for air, take $\nu = 23 \times 10^{-6}$ m²/s.

1-43. A thermometer is used to check the temperature in a freezer that is set to operate at −5°C. If the thermometer initially reads 25°C, how long will it take for the reading to be within 1°C of the true temperature? Model the thermometer bulb as a 4 mm–diameter mercury sphere surrounded by a 2 mm–thick shell of glass. For mercury, take $\rho = 13{,}530$ kg/m³, $c = 140$ J/kg K; and for glass $\rho = 2640$ kg/m³, $c = 800$ J/kg K. Use a heat transfer coefficient of 15 W/m² K.

1-44. A thermocouple junction bead is modeled as a 1 mm-diameter lead sphere ($\rho = 11{,}340$ kg/m³, $c = 129$ J/kg K) and is initially at a room temperature of 20°C. If the thermocouple is suddenly immersed in ice water to serve as a reference junction, what will be the error in indicated temperature corresponding to 1, 2, and 3 times the time constant of the thermocouple? If the heat transfer coefficient is calculated to be 2140 W/m² K, what are the corresponding times?

1-45. A hot-water cylinder contains 150 liters of water. It is insulated, and its outer surface has an area of 3.5 m^2. It is located in an area where the ambient air is 25°C, and the overall heat transfer coefficient between the water and the surroundings is 1.0 W/m^2 K, based on outer surface area. If there is a power failure, how long will it take the water to cool from 65°C to 40°C? Take the density of water as 980 kg/m^3 and its specific heat as 4180 J/kg K.

1-46. An aluminum plate 10 cm square and 1 cm thick is immersed in a chemical bath at 50°C for cleaning. On removal, the plate is shiny bright and is allowed to cool in a vertical position in still air at 20°C. Estimate how long the plate will take to cool to 30°C by

   (i) assuming a constant heat transfer coefficient evaluated at the average ΔT of 20 K.
   (ii) allowing exactly for the $\Delta T^{1/4}$ dependence of h_c given by Eq. (1.23a).

For air, take $\nu = 16.5 \times 10^{-6}$ m^2/s, and for aluminum, take $k = 204$ W/m K, $\rho = 2710$ kg/m^3, $c = 896$ J/kg K.

1-47. A 2 cm–diameter copper sphere with a thermocouple at its center is suddenly immersed in liquid nitrogen contained in a Dewar flask. The temperature response is determined using a digital data acquisition system that records the temperature every 0.05 s. The maximum rate of temperature change dT/dt is found to occur when $T = 92.5$ K, with a value of 19.8 K/s.

   (i) Using the lumped thermal capacity model, determine the corresponding heat transfer coefficient.
   (ii) Check the Biot number to ensure that the model is valid.
   (iii) The fact that the cooling rate is a maximum toward the end of the cool-down period is unusual; what must be the reason?

The saturation temperature of nitrogen at 1 atm pressure is 77.4 K. Take $\rho = 8930$ kg/m^3, $c = 235$ J/kg K, and $k = 450$ W/m K for copper at 92.5 K.

1-48. A 1 cm–diameter alloy sphere is to be heated in a furnace maintained at 1000°C. If the initial temperature of the sphere is 25°C, calculate the time required for the sphere to reach 800°C

   (i) if the gas in the furnace is circulated to give a convective heat transfer coefficient of 100 W/m^2 K.
   (ii) if there is no forced convection, and the free-convection heat transfer coefficient is given by $\bar{h}_c \sim 5\Delta T^{1/4}$ W/m^2 K for ΔT in kelvins.

Properties of the alloy include $\rho = 4900$ kg/m^3, $c = 400$ J/kg K, and $\varepsilon = 0.45$.

1-49. A material sample, in the form of a 1 cm–diameter cylinder 10 cm long, is removed from a boiling water bath at 100°C and allowed to cool in air at 20°C. If the free-convection heat transfer coefficient can be approximated as $\bar{h}_c = 3.6\Delta T^{1/4}$ W/m^2 K for ΔT in kelvins, estimate the time required for the sample to cool to 25°C. For the sample properties take $\rho = 2260$ kg/m^3, $c = 830$ J/kg K, and $\varepsilon = 0.77$.

1-50. Two small blackened spheres of identical size—one of aluminum, the other of an unknown alloy of high conductivity—are suspended by thin wires inside a large cavity in a block of melting ice. It is found that it takes 4.8 minutes for the temperature of the aluminum sphere to drop from 3°C to 1°C, and 9.6 minutes for the alloy sphere to undergo the same change. If the specific gravities of the aluminum and alloy are 2.7 and 5.4, respectively, and the specific heat of the aluminum is 900 J/kg K, what is the specific heat of the alloy?

1-51. A mercury-in-glass thermometer is to be used to measure the temperature of a high-velocity air stream. If the air temperature increases linearly with time, $T_e = \alpha t +$ constant, perform an analysis to determine the error in the thermometer reading due to its thermal "inertia." Evaluate the error if the inside diameter of the mercury reservoir is 3 mm, its length is 1 cm, and the glass wall thickness is 0.5 mm, when the heat transfer coefficient is 60 W/m^2 K and the air temperature increases at a rate of

(i) 1°C per minute.
(ii) 1°C per second.

Property values for mercury are $\rho = 13{,}530$ kg/m^3, $c = 140$ J/kg K; for glass $\rho = 2640$ kg/m^3, $c = 800$ J/kg K.

1-52. Under high-vacuum conditions in the space shuttle service bay, radiation is the only significant mode of heat transfer. Obtain an analytical solution for a lumped thermal capacity model thermal response. Also, identify a dimensionless group analogous to the Biot number that can be used to determine if the model is valid. *(Hint:* A table of standard integrals found in mathematics handbooks may be of assistance.)

1-53. A thermocouple is immersed in an air stream whose temperature varies sinusoidally about an average value with angular frequency ω. The thermocouple is small enough for the Biot number to be less than 0.1, but the convective heat transfer coefficient is high enough for radiation heat transfer to be negligible compared to convection.

(i) Set up the differential equation governing the temperature of the thermocouple.
(ii) Solve the differential equation to obtain the amplitude and phase lag of the thermocouple temperature response.
(iii) The thermocouple can be modeled as a 2 mm-diameter lead sphere ($\rho = 11{,}340$ kg/m^3, $c = 129$ J/kg K). If the air temperature varies as $T = 320 + 10 \sin t$, for T in kelvins and t in seconds, calculate the amplitude and phase lag of the thermocouple for heat transfer coefficients of 30 and 100 W/m^2 K.

1-54. A system consists of a body in which heat is continuously generated at a rate $\dot{Q}_v$, while heat is lost from the body to its surroundings by convection. Using the lumped thermal capacity model, derive the differential equation governing the temperature response of the body. If the body is at temperature T_o when time $t = 0$, solve the differential equation to obtain $T(t)$. Also determine the steady-state temperature.

1-55. Electronic components are often mounted with good heat conduction paths to a finned aluminum base plate, which is exposed to a stream of cooling air from a fan. The sum of the mass times specific heat products for a base plate and components is 5000 J/K, and the effective heat transfer coefficient times surface area product is 10 W/K. The initial temperature of the plate and the cooling air temperature are 295 K when 300 W of power are switched on. Find the plate temperature after 10 minutes.

1-56. A reactor vessel's contents are initially at 290 K when a reactant is added, leading to an exothermic chemical reaction that releases heat at a rate of 4×10^5 W/m^3. The volume and exterior surface area of the vessel are 0.008 m^3 and 0.24 m^2, respectively, and the overall heat transfer coefficient between the vessel contents and the ambient air at 300 K is 5 W/m^2 K. If the reactants are well stirred, estimate their temperature after

   (i) 1 minute.
   (ii) 10 minutes.

Take $\rho = 1200$ kg/m^3 and $c = 3000$ J/kg K for the reactants.

1-57. A carbon steel butane tank weighs 4.0 kg (empty) and has a surface area of 0.22 m^2. When full it contains 2 kg of liquified gas. Butane gas is drawn off to a burner at a rate of 0.05 kg/h through a pressure-reducing valve. If the ambient temperature is 55°C, estimate the steady temperature of the tank and the time taken for 80% of the temperature drop to occur. Take the sum of the convective and radiative heat transfer coefficients from the tank to the surroundings as 5 W/m^2 K. Property values for butane are $c = 2390$ J/kg K and $h_{fg} = 3.86 \times 10^5$ J/kg; for the steel $c = 434$ J/kg K.

1-58. A 2.5 m–diameter, 3.5 m–high milk storage tank is located in a dairy factory in Onehunga, New Zealand, where the ambient temperature is 30°C. The tank has walls of stainless steel 2 mm thick and is insulated with a 7.5 cm–thick layer of polyurethane foam. The tank is filled with milk at 4°C and is continuously stirred by an impeller driven by an electric motor that consumes 400 W of power. What will the milk temperature be after 24 hours? For the milk, take $\rho = 1034$ kg/m^3, $c = 3894$ J/kg K; for the insulation, $k = 0.026$ W/m K; and for the outside heat transfer coefficient, $h = 5$ W/m^2 K. The impeller motor efficiency can be taken as 0.75.

1-59. Referring to Exercise 1-14, the laser dissipates heat at a rate of 1.5 W. Since the laser's action deteriorates above 100 K, it is sometimes necessary to operate the laser discontinuously. If a magnesium heat sink is initially at 50 K, estimate the time required for it to reach 100 K. A "cold finger" removes heat at a rate $\dot{Q}_c = 0.02(T - 50)$ W, where T is the block temperature in kelvins. Also determine the block equilibrium temperature. For the magnesium, take $\rho = 1750$ kg/m^3, $k = 250$ W/m K, $c = 450$ J/kg K. Ignore the heat capacity of the laser and the diamond spreader and parasitic heat gains from the Dewar flask.

1-60. A 3.5 cm–O.D., 2.75 mm–wall-thickness copper tube is used in a test rig for the measurement of convective heat transfer from a cylinder in a cross-flow of fluid. The tube is fitted with an internal electric heater. A 5 mm-square, 0.1 mm–thick heat flux meter is attached to the surface and measures both the local surface heat flux q_s and surface temperature T_s (see Exercise 1-74). In a series of tests to determine the heat transfer coefficient at the stagnation line, the cylinder is placed in a wind tunnel and the air speed varied incrementally over the desired range. How long will the experimenter have to wait after the fan speed is changed before taking data? The heat transfer coefficient is expected to be about 100 W/m^2 K. Take $\rho = 8950$ kg/m^3, $c = 385$ J/kg K for the copper.

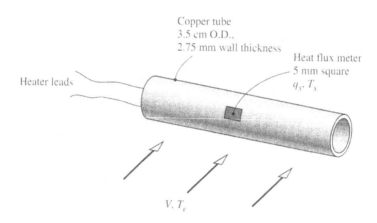

1-61. In a materials-processing experiment on a space station, a 1 cm–diameter sphere of alloy is to be cooled from 600 K to 400 K. The sphere is suspended in a test chamber by three jets of nitrogen at 300 K. The convective heat transfer coefficient between the jets and the sphere is estimated to be 180 W/m^2 K. Calculate the time required for the cooling process and the minimum quenching rate. Take the alloy density to be $\rho = 14,000$ kg/m^3, specific heat $c = 140$ J/kg K, and thermal conductivity $k = 240$ W/m K. Since the emittance of the alloy is very small, the radiation contribution to heat loss can be ignored.

1-62. Constant delivery of low-vapor-pressure reactive gases is required for semiconductor fabrication. In one process, tungsten fluoride WF$_6$ (normal boiling point 17°C) is supplied from a 80 cm–diameter spherical tank containing liquid WF$_6$ under pressure. The tank is located in surroundings at 21°C. After connecting a full tank to the gas delivery system, supply at a rate of 2500 sccm (standard cubic centimeters per minute) commences. In order to supply the required heat of vaporization, the liquid WF$_6$ temperature drops until a steady state is reached, for which the heat transferred into the tank from the surroundings balances the heat of vaporization required.

(i) Estimate the steady-state temperature of the liquid WF_6.
(ii) Estimate how long it will take for the liquid to approach within 1°C of its steady value.

Property values for liquid WF_6 include $\rho = 3440$ kg/m^3, $c_p = 1000$ J/kg K, $h_{fg} = 25.7 \times 10^3$ kJ/kmol, and for steel, $c = 434$ J/kg K. The weight of the empty tank is 30 kg, and the heat transfer coefficient for convection and radiation to the tank is $h = 8$ W/m^2 K.

1-63. A 83 mm–high Styrofoam cup has 1.5 mm-thick walls and is filled with 180 ml coffee at 80°C and covered with a lid. The outside diameter of the cup varies from 45 mm at its base to 73 mm at its top. The ambient air is at 24°C, and the combined convective and radiative heat transfer coefficient for the outside of the cup is estimated to be 10 W/m K.

(i) Determine the initial rate of heat loss through the side walls of the cup and the corresponding temperature of the outer surface.
(ii) Estimate the time for the coffee to cool to 60°C if the average of the heat fluxes through the lid and base are taken to be equal to the flux through the side walls.

Comment on the significance of your answer to actual cooling rates experienced at the morning coffee break. Take $k = 0.035$ W/m K for the Styrofoam, and $\rho = 985$ kg/m^3, $c_p = 4180$ J/kg K for the coffee.

1-64. Derive conversion factors for the following units conversions.

(i) Enthalpy of vaporization, Btu/lb to J/kg
(ii) Specific heat (capacity), Btu/lb °F to J/kg K
(iii) Density, lb/ft^3 to kg/m^3
(iv) Dynamic viscosity, lb/ft hr to kg/m s
(v) Kinematic viscosity, ft^2/hr to m^2/s
(vi) Thermal conductivity, Btu/hr ft °F to W/m K
(vii) Heat flux, Btu/hr ft^2 to W/m^2

1-65. In the United States, gas and liquid flow rates are commonly expressed in cubic feet per minute (CFM) and gallons per minute (GPM), respectively.

(i) For air at 1 atm and 300 K ($\rho = 1.177$ kg/m^3), prepare a table showing flow rates in m^3/s and kg/s corresponding to 1, 10, 100, 1000, and 10,000 CFM.
(ii) For water at 300 K ($\rho = 996$ kg/m^3), prepare a table showing flow rates in m^3/s and kg/s corresponding to 1, 10, 100, 1000, and 10,000 GPM.

1-66. In January 1989 the barometric pressure reached 31.84 inches of mercury at Northway, Alaska, a record for North America. On the other hand, a typical barometric pressure for Denver, Colorado, is 24.4 inches of mercury.

   (i) What are these pressures in mbar and pascals?

   (ii) At what temperature does water boil at these pressures?

1-67. Specify the following in the English system of units (Btu, hr, ft, °F or °R):

   (i) The Stefan-Boltzmann constant

   (ii) The radiation heat transfer coefficient at 25°C

   (iii) The free-convection formulas given by Eqs. (1.23a) and (1.23b).

1-68. Convert the problem statement of Example 1.3 to the English system of units, work the problem in English units, and convert your answer back into SI units.

1-69. Convert the problem statement of Example 1.4 to the English system of units, work the problem in English units, and convert your answers back into SI units.

1-70. Convert the problem statement of Example 1.5 to the English system of units, work the problem in English units, and convert your answers back into SI units.

1-71. Convert the problem statement of Example 1.6 to the English system of units, work the problem in English units, and convert your answers back into SI units.

1-72. Check the dimensions of Eq. (1.22) in

   (i) SI units.

   (ii) English units.

1-73. Asbestos is no longer used for insulating steam lines in power plants because it is a proven carcinogen. Substitutes include calcium silicate, ceramic fiber, and mineral wool. The thermal conductivity of these insulations is temperature-dependent, and appropriate data are required for design calculations. A simple quadratic curve fit $k(T) = A_0 + A_1 + A_2 T^2$ is usually adequate. Utility engineers in the United States tend to use English units, with k expressed in Btu in/hr ft^2 °F and temperature in degrees Fahrenheit. The table shows some data in current use.

Brand Name	Material	Company	A_0	A_1	A_2
Thermo-12	Calcium silicate	Manville	0.34	5.0×10^{-5}	2.5×10^{-7}
Kaowool	Ceramic fiber	Babcock and Wilcox	0.23	2.245×10^{-4}	3.75×10^{-7}
Epitherm-1200	Mineral wool	Fibrex	0.2809	-5.0×10^{-5}	1.0×10^{-6}
Kaylo	Calcium silicate	Owens/Corning	0.43	-1.25×10^{-4}	6.25×10^{-7}

   (i) Obtain quadratic curve fits for T in kelvins and k in W/m K.

   (ii) Prepare a graph of k versus T for these four insulations over the temperature range 300 K < T < 900 K.

(Hint: First obtain the appropriate curve fit for T in degrees Rankine, then convert to SI units.)

1-74. Miniature heat flux meters are being increasingly used to measure convective heat transfer coefficients. The cross section of a typical design is shown. Notice that the total thickness is approximately 0.1 mm, which is small enough not to disturb the flow in many applications. The two thermopiles (assemblies of thermocouples in series to multiply the signal) are located on each side of a 1 mil (2.54×10^{-2} mm)–thick Kapton film in order to measure the conduction heat transferred across the film. A separate thermocouple measures the temperature underneath the film. The manufacturer calibrates the meter by subjecting it to a known heat flux, and a calibration constant is supplied with the meter.

(i) The manufacturer specifies the calibration constant as 0.350 $\mu V/$(Btu/hr ft^2). Determine this constant in SI units.

(ii) The thermocouple does not measure the true surface temperature. Determine the correction required, named $(T_s - T_{tc})$, for heat fluxes of 100, 1000, and 10,000 W/m^2.

Take $k = 0.245$ W/m K for the Kapton.

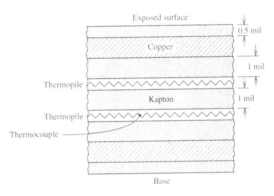

1-75. The thermal resistance per unit area of clothing is often expressed in the unit *clo*, where 1 clo = 0.88 ft^2 °F hr/Btu.

(i) What is 1 clo in SI units?

(ii) If a wool sweater is 2 mm thick and has a thermal conductivity of 0.05 W/m K, what is its thermal resistance in clos?

(iii) A cotton shirt has a thermal resistance of 0.5 clo. If the inner and outer surfaces are at 31°C and 28°C, respectively, what is the rate of heat loss per unit area?

STEADY ONE-DIMENSIONAL HEAT CONDUCTION

CONTENTS

2.1   INTRODUCTION

In this chapter we analyze problems involving steady one-dimensional heat conduction. By **steady** we mean that temperatures are constant with time; as a result, the heat flow is also constant with time. By **one-dimensional** we mean that temperature is a function of a single "dimension" or spatial coordinate. One-dimensional conduction can occur in a number of geometrical shapes. In Section 1.3.1, one-dimensional conduction across a plane wall was examined, with temperature as a function of Cartesian coordinate x only; that is, $T = T(x)$. Conduction in cylinders or spheres is one-dimensional when temperature is a function of only the radial coordinate r and does not vary with polar angle and axial distance, in the case of the cylinder, or with polar or azimuthal angles, in the case of the sphere; that is, $T = T(r)$. Analysis of steady one-dimensional heat conduction problems involves the solution of very simple ordinary differential equations to give algebraic formulas for the temperature variation and heat flow. Thus, if at all possible, engineers like to approximate, or *model*, a practical heat conduction problem as steady and one-dimensional, even though temperatures might vary slowly with time or vary a little in a second coordinate direction.

A wide range of practical heat transfer problems involve steady one-dimensional heat conduction. Examples include most heat insulation problems, such as the refrigerated container of Example 1.1, the prediction of temperatures in a nuclear reactor fuel rod, and the design of cooling fins for electronic gear. Often, complex systems involving two- or three-dimensional conduction can be divided into subsystems in which the conduction is one-dimensional. Cooling of integrated circuit components can often be satisfactorily analyzed in this manner.

In Section 2.2, Fourier's law of heat conduction is briefly revisited. The physical mechanisms of heat conduction are discussed, and the applicability of Fourier's law at the interface between two solids is examined. Conduction across plane walls has already been treated in Section 1.3.1; thus, in Section 2.3, we restrict our attention to conduction across cylindrical and spherical shells and include the effect of **heat generation** within the solid. Section 2.4 deals with the class of problems known as **fin** problems, including familiar cooling fins and an interesting variety of mathematically similar problems.

There is a common methodology to the analyses in Chapter 2. Each analysis begins with the application of the first law, Eq. (1.2), to a closed-system volume element and introduction of Fourier's law, to obtain the governing differential equation. This equation is then integrated to give the temperature distribution, with the constants of integration found from appropriate boundary conditions. Finally, the heat flow is obtained using Fourier's law.

2.2   FOURIER'S LAW OF HEAT CONDUCTION

Fourier's law of heat conduction was introduced in Section 1.3.1. A general statement of this law is: The conduction heat flux in a specified direction equals the negative of the product of the medium thermal conductivity and the temperature derivative in that direction. In Chapter 2, we are concerned with one-dimensional conduction. In Cartesian coordinates, with temperature varying in the x direction only,

$$q = -k\frac{dT}{dx} \tag{2.1}$$

Recall from Section 1.3.1 that the negative sign ensures that the heat flux q is positive in the positive x direction. In cylindrical or spherical coordinates, with temperature varying in the r direction only,

$$q = -k\frac{dT}{dr} \tag{2.2}$$

Equation (2.2) is the form of Fourier's law required for Section 2.3.

2.2.1 Thermal Conductivity

Table 1.1 gave a brief list of thermal conductivities to illustrate typical values for gases, liquids, and solids. Appendix A gives more complete tabulated data. The relevant tables are:

Table A.1   Solid metals

Table A.2   Solid dielectrics (nonmetals)

Table A.3   Insulators and building materials

Table A.4   Solids at cryogenic temperatures

Table A.7   Gases

Table A.8   Dielectric liquids

Table A.9   Liquid metals

Table A.13   Liquid solutions

Additional data may be found in the literature, for example, References [1] through [4]. In the case of commercial products, such as insulations, data can be obtained from the manufacturer.

The engineer needs conductivity data to solve heat conduction problems (as was seen in Chapter 1) and is usually not too concerned about the actual physical mechanism of heat conduction. However, conductivity data for a given substance are often sparse or nonexistent, and then a knowledge of the physics of heat conduction is useful to interpolate or extrapolate what data are available. Unfortunately, the physical mechanisms of conduction are many and complicated, and it is possible to develop simple theoretical models for gases and pure metals only. A brief account of some of the more important aspects of the conduction mechanisms follows.

Gases

The kinetic theory model gives a reliable basis for determining the thermal conductivity of a gas. Molecules are in a state of random motion. When collisions occur, there is an exchange of energy that results in heat being conducted down a temperature gradient from a hot region to a cold region. The simplest form of kinetic theory gives

$$k = \frac{1}{3} c \mathcal{N} \ell_t \left(m c_v + \frac{1}{2} \kappa_B \right) \text{ [W/m K]} \tag{2.3}$$

where c (for *celerity*) is the average molecular speed [m/s], $\mathcal{N}$ is the number of molecules per unit volume [m^{-3}], ℓ_t is the transport mean free path [m], m is the mass of the molecule [kg], c_v is the constant-volume specific heat [J/kgK], and κ_B is the Boltzmann constant equal to 1.38065×10^{-23} J/K. The value of κ_B can be obtained as the ratio between the universal gas constant $\mathcal{R}$ (8314.46 J/kmol K) and Avogadro's number $\mathcal{A}$ (6.02214×10^{26} molecules/kmol). The product $c \mathcal{N} \ell_t$ is virtually independent of pressure in the vicinity of atmospheric pressure; thus, conductivity is also independent of pressure. Table A.7 gives data for a nominal pressure at 1 atm but can be used for pressures down to about 1 torr (1/760 of an atmosphere, or 133.3 Pa). Notice also that the average molecular speed is higher and the mean free path is longer for small molecules; thus, conductivities of gases such as hydrogen and helium are much greater than those of xenon and the refrigerant R-22.

Dielectric Liquids

In liquids (excluding liquid metals such as mercury at normal temperatures and sodium at high temperatures), the molecules are relatively closely packed, and heat conduction occurs primarily by longitudinal vibrations, similar to the propagation of sound. The structure of liquids is not well understood at present, and there are no good theoretical formulas for conductivity.

Pure Metals and Alloys

The primary mechanism of heat conduction is the movement of free electrons. A smaller contribution is due to the transfer of atomic motions by lattice vibrations, or waves; however, this contribution is unimportant except at cryogenic temperatures. In alloys, the movement of free electrons is restricted, and thermal conductivity decreases markedly as alloying elements are added. The lattice wave contribution then becomes more important but is difficult to predict owing to the variable effects of heat treatment and cold working.

Dielectric Solids

Heat conduction is almost entirely due to atomic motions being transferred by lattice waves; hence, thermal conductivity is very dependent on the crystalline structure of the material. Many building materials and insulators have anomalously low values of conductivity because of their porous nature. The conductivity is an effective value for the porous medium and is low due to air or a gas filling the interstices and pores, through which heat is transferred rather poorly by conduction and radiation. Expanded plastic insulations, such as polystyrene, have a large molecule refrigerant gas filling the pores to reduce the conductivity.

In general, thermal conductivity is temperature-dependent. Fortunately, the variation in typical engineering problems is small, and it suffices to use an appropriate average value. An exception is solids at cryogenic temperatures, as an examination of Table A.4 will show. Another exception is gases in the vicinity of the critical point. Special care must be taken in such situations.

2.2.2 Contact Resistance

In Section 1.3.1, heat conduction through a composite wall was analyzed. Figure 2.1a shows the interface between two layers of a composite wall, with the surface of each layer assumed to be perfectly smooth. Two mathematical surfaces, the u- and s-surfaces, are located on each side of and infinitely close to the real interface, as shown. The first law of thermodynamics applied to the closed system located between the u- and s-surfaces requires that

$$\dot{Q}|_u = \dot{Q}|_s \tag{2.4}$$

since no energy can be stored in the infinitesimal amount of material in the system. Considering a unit area and introducing Fourier's law gives

$$-k_A \frac{dT}{dx}\bigg|_u = -k_B \frac{dT}{dx}\bigg|_s \tag{2.5}$$

Also, since the distance between the u- and s-surfaces is negligible, thermodynamic equilibrium requires

$$T_u = T_s \tag{2.6}$$

as shown on the temperature profile (for $k_A > k_B$). For perfectly smooth surfaces, there is no thermal resistance at the interface.

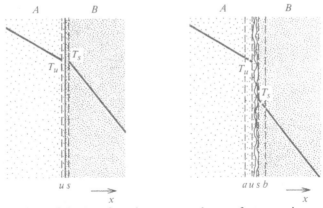

Figure 2.1 Interfaces between two layers of a composite wall. (a) Smooth surfaces. (b) Rough surfaces.

Figure 2.1b shows a more realistic situation, in which each surface has some degree of roughness. The solid materials are in contact at relatively few places, and the gaps may contain a fluid or, in some applications, a vacuum. The heat flow in the interface region is complicated: the conduction is three-dimensional as the heat tends to "squeeze" through the contact areas, and there are parallel paths of conduction and radiation through the gaps. The u- and s-surfaces are located just on either side of a somewhat arbitrarily defined interface location. In addition, a- and b-surfaces are located just far enough from the interface for the heat conduction to be one-dimensional. No temperature profile is shown between the a- and b-surfaces since no unique profile $T(x)$ exists there; instead, the temperature profiles are extrapolated from the bulk material to the interface as shown, thereby defining the temperatures and temperature gradients at the u- and s-surfaces. As was the case for the perfectly smooth surfaces, the first law requires

$$-k_A \frac{dT}{dx}\bigg|_u = -k_B \frac{dT}{dx}\bigg|_s \tag{2.7}$$

but now there is no continuity of temperature at the interface; that is, $T_u \neq T_s$. The thermal resistance to heat flow at the interface is called the **contact resistance** and is usually expressed in terms of an **interfacial conductance** h_i [W/m^2 K], defined in an analogous manner to Newton's law of cooling, namely,

$$\dot{Q} = h_i A (T_u - T_s) \tag{2.8}$$

or

$$-k_A \frac{dT}{dx}\bigg|_u = h_i(T_u - T_s) = -k_B \frac{dT}{dx}\bigg|_s \tag{2.9}$$

Figure 2.2 shows the contact resistance added to the thermal circuit of Fig. 1.5.

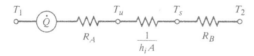

Figure 2.2   A contact resistance in a thermal circuit.

There is always a contact resistance to conduction across real solid-solid interfaces. The contact resistance can be the dominant thermal resistance when high-conductivity metals are involved — for example, in aircraft construction, where aluminum alloys are used extensively. The contact resistance depends on the pressure with which contact is maintained, with a marked decrease once the yield point of one of the materials is reached. Data for contact resistances are, unfortunately, sparse and unreliable. Table 2.1 does, however, show some representative values. Additional data can be found in the literature [5, 6, 7].

Table 2.1   Typical interfacial conductances (at moderate pressure and usual finishes, unless otherwise stated).

Interface	h_i W/m^2 K
Ceramic-ceramic	500–3000
Ceramic-metals	1500–8500
Graphite-metals	3000–6000
Stainless steel-stainless steel	1700–3700
Aluminum-aluminum	2200–12,000
Stainless steel-aluminum	3000–4500
Copper-copper	10,000–25,000
Rough aluminum-aluminum (vacuum conditions)	~150
Iron-aluminum	4000–40,000

2.3   CONDUCTION ACROSS CYLINDRICAL AND SPHERICAL SHELLS

Steady one-dimensional conduction in cylinders or spheres requires that temperature be a function of only the radial coordinate r. The analysis of steady heat flow across a plane wall in Section 1.3.1 was particularly simple because the flow area A did not change in the flow direction. In the case of a cylindrical or spherical shell, the area for heat flow changes in the direction of heat flow. For a cylindrical shell of length L, the area for heat flow is $A = 2\pi r L$; for a spherical shell, it is $A = 4\pi r^2$. In both cases, A increases with increasing r.

2.3.1 Conduction across a Cylindrical Shell

Figure 2.3 shows a cylindrical shell of length L, with inner radius r_1, and outer radius r_2. The inner surface is maintained at temperature T_1 and the outer surface is maintained at temperature T_2. An elemental control volume is located between radii r and $r + \Delta r$. If temperatures are unchanging in time and $\dot{Q}_v = 0$, the energy conservation principle, Eq. (1.2), requires that the heat flow across the face at r equal that at the face $r + \Delta r$

$$\dot{Q}|_r = \dot{Q}|_{r+\Delta r}$$

that is,

$$\dot{Q} = \text{Constant, independent of } r$$

Using Fourier's law in the form of Eq. (2.2)

$$\dot{Q} = Aq = 2\pi r L \left(-k \frac{dT}{dr} \right)$$

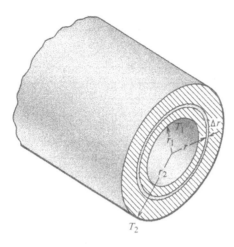

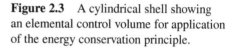

Figure 2.3   A cylindrical shell showing an elemental control volume for application of the energy conservation principle.

Dividing by $2\pi kL$ and assuming that the conductivity k is independent of temperature gives

$$\frac{\dot{Q}}{2\pi kL} = -r\frac{dT}{dr} = \text{Constant} = C_1 \tag{2.10}$$

which is a first-order ordinary differential equation for $T(r)$ and can be integrated easily:

$$\frac{dT}{dr} = -\frac{C_1}{r}$$

$$T = -C_1 \ln r + C_2 \tag{2.11}$$

Two boundary conditions are required to evaluate the two constants; these are

$$r = r_1; \quad T = T_1 \tag{2.12a}$$

$$r = r_2; \quad T = T_2 \tag{2.12b}$$

Substituting in Eq. (2.11) gives

$$T_1 = -C_1 \ln r_1 + C_2$$

$$T_2 = -C_1 \ln r_2 + C_2$$

which are two algebraic equations for the unknowns C_1 and C_2. Subtracting the second equation from the first:

$$T_1 - T_2 = -C_1 \ln r_1 + C_1 \ln r_2 = C_1 \ln(r_2/r_1)$$

or

$$C_1 = \frac{T_1 - T_2}{\ln(r_2/r_1)}$$

Using either of the two equations then gives

$$C_2 = T_1 + \frac{T_1 - T_2}{\ln(r_2/r_1)} \ln r_1$$

Substituting back in Eq. (2.11) and rearranging gives the temperature distribution as

$$\frac{T_1 - T}{T_1 - T_2} = \frac{\ln(r/r_1)}{\ln(r_2/r_1)} \tag{2.13}$$

which is a logarithmic variation, in contrast to the linear variation found for the plane wall in Section 1.3.1. The heat flow is found from Eq. (2.10) as $\dot{Q} = 2\pi k L C_1$, or

$$\dot{Q} = \frac{2\pi k L (T_1 - T_2)}{\ln(r_2/r_1)} \tag{2.14}$$

Equation (2.14) is again in the form of Ohm's law, and the thermal resistance of the cylindrical shell is

$$R = \frac{\ln(r_2/r_1)}{2\pi k L} \tag{2.15}$$

When $r_2 = r_1 + \delta$ and $\delta/r_1 \ll 1$, Eq. (2.15) reduces to the resistance of a slab, $\delta/2\pi r_1 k L = \delta/kA$.

It is now possible to treat composite cylindrical shells with convection and radiation from either side without any further analysis. Figure 2.4 shows the cross section of an insulated pipe of length L, through which flows superheated steam and which

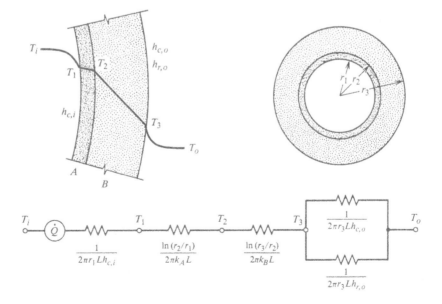

Figure 2.4 An insulated steam pipe showing the temperature distribution and thermal circuit.

loses heat by convection and radiation to its surroundings. The thermal circuit is also shown, with Eq. (2.15) used for the conductive resistances of the pipe and insulation. Notice that in contrast to the plane wall case, the area for heat flow is different on each side of the composite wall: on the inside it is $2\pi r_1 L$, and on the outside it is $2\pi r_3 L$. Again we define an *overall heat transfer coefficient* by Eq. (1.29):

$$\dot{Q} = UA(T_i - T_o) = \frac{T_i - T_o}{1/UA} \tag{2.16}$$

Then, summing the resistances in the thermal network,

$$\frac{1}{UA} = \frac{1}{2\pi r_1 L h_{c,i}} + \frac{\ln(r_2/r_1)}{2\pi k_A L} + \frac{\ln(r_3/r_2)}{2\pi k_B L} + \frac{1}{2\pi r_3 L(h_{c,o} + h_{r,o})} \tag{2.17}$$

The area A need not be specified since all we need is the UA product. However, often a value of U will be quoted based on either the inside or outside area; then the appropriate area must be used in Eqs. (2.16) and (2.17).

EXAMPLE 2.1   Heat Loss from an Insulated Steam Pipe

A mild steel steam pipe has an outside diameter of 15 cm and a wall thickness of 0.7 cm. It is insulated with a 5.3 cm-thick layer of 85% magnesia insulation. Superheated steam at 500 K flows through the pipe, and the inside heat transfer coefficient is 35 W/m²K. Heat lost by convection and radiation to surroundings at 300 K, and the sum of outside convection and radiation coefficients is estimated to be 8 W/m² K. Find the rate of heat loss for a 20 m length of pipe.

Solution

Given: Steam pipe with 85% magnesia insulation.

Required: Heat loss for 20 m length if $h_o = 8$ W/m² K.

Assumptions: Steady one-dimensional heat flow.

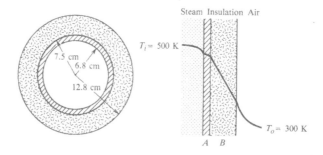

Equation (2.16) applies, with Eq. (2.17) used to obtain the UA product,

$$\dot{Q} = UA(T_i - T_o)$$

$$\frac{1}{UA} = \frac{1}{2\pi L}\left(\frac{1}{r_1 h_{c,i}} + \frac{\ln(r_2/r_1)}{k_A} + \frac{\ln(r_3/r_2)}{k_B} + \frac{1}{r_3 h_o}\right)$$

Tables A.1b and A.3 in Appendix A give the variation of conductivity with temperature for 1010 steel and magnesia, respectively. As a first step, we guess that the steel is close to the steam temperature (500 K), and since most of the temperature drop will be across the magnesia insulation, its average temperature will be about $(500 + 300)/2 = 400$ K. The corresponding conductivity values are $k_A = 54$ W/m K and $k_B = 0.073$ W/m K.

$$\frac{1}{UA} = \frac{1}{(2)(\pi)(20)}\left(\frac{1}{(0.068)(35)} + \frac{\ln(0.075/0.068)}{54} + \frac{\ln(0.128/0.075)}{0.073} + \frac{1}{(0.128)(8)}\right)$$

$$= \frac{1}{125.7}(0.42 + 0.002 + 7.32 + 0.98)$$

$$UA = 14.4\,\text{W/K}$$

$$\dot{Q} = UA\Delta T = (14.4)(500 - 300) = 2880\,\text{W}$$

Since the resistance of the steel wall is negligible, we do not need to check our guess for its conductivity. For the magnesia insulation, we estimate its average temperature by examining the relevant segment of the thermal circuit. For convenience, the thermal resistance of the insulation is split in half to estimate an average temperature $\overline{T}$:

$$\overline{T} - T_o = \dot{Q}\left[\left(\frac{1}{2}\right)\frac{\ln(r_3/r_2)}{2\pi L k_B} + \frac{1}{2\pi L r_3 h_o}\right]$$

$$\overline{T} - 300 = (2880)\left[\left(\frac{1}{2}\right)\frac{7.32}{125.7} + \frac{0.98}{125.7}\right]$$

$$= 106\,\text{K}$$

$$\overline{T} = 406\,\text{K}$$

A look at Table A.3 shows that our guess of 400 K introduced an error of less than 1%, so there is no need to calculate a new value of $\dot{Q}$ using an improved k value.

Comments

1. After one has gained some experience with this type of calculation, the problem can be simplified by ignoring the small resistance of the steel pipe.

2. In practice, the outside heat transfer coefficient varies somewhat around the circumference of the insulation, and the conduction is not truly one-dimensional. For an engineering calculation, we simply use an average value for h_o.

2.3.2 Critical Thickness of Insulation on a Cylinder

The insulation on the large steam pipe in Example 2.1 was installed because it reduced the heat loss. However, adding a layer of insulation to a cylinder does not necessarily reduce the heat loss. When the outer radius of the insulation r_o is small, there is the possibility that the added thermal resistance of the insulation is less than the reduction of the outside resistance $1/2\pi r_o L h_o$ due to the larger value of the area for convective and radiative heat transfer, $2\pi r_o L$. This phenomenon is often used for cooling electronic components that must dissipate $I^2 R$ heating. Figure 2.5 shows a resistor with an insulation sheath of inner radius r_i and outer radius r_o. Since the resistor usually has a relatively high thermal conductivity, we will assume it is isothermal at temperature T_i. The ambient air temperature is T_e, and the outside heat transfer coefficient is h_o. There are two resistances in series; denoting the total resistance as R the heat flow is

$$\dot{Q} = \frac{T_i - T_e}{R} = \frac{T_i - T_e}{\ln(r_o/r_i)/2\pi L k + 1/2\pi L r_o h_o} \tag{2.18}$$

$\dot{Q}$ will have a maximum value when the total resistance R has a minimum value; differentiating R with respect to r_o:

$$\frac{dR}{dr_o} = \frac{1}{2\pi L} \left(\frac{1}{r_o k} - \frac{1}{r_o^2 h_o} \right)$$

which equals zero when the outer radius of the insulation equals the **critical radius**,

$$r_o = r_{\mathrm{cr}} = \frac{k}{h_o} \tag{2.19}$$

To check whether r_{cr} gives a minimum resistance, we differentiate again and evaluate at $r_o = r_{\mathrm{cr}}$:

$$\frac{d^2 R}{dr_o^2} = \frac{1}{2\pi L} \left(-\frac{1}{r_o^2 k} + \frac{2}{r_o^3 h_o} \right)$$

and

$$\left. \frac{d^2 R}{dr^2} \right|_{r_o = r_{\mathrm{cr}}} = \frac{1}{2\pi L} \left(-\frac{h_o^2}{k^3} + \frac{2h_o^2}{k^3} \right) = \frac{h_o^2}{2\pi L k^3} > 0$$

as required.

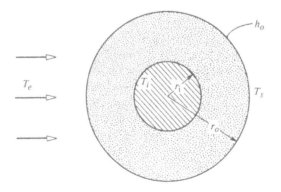

Figure 2.5   An electric resistor with an insulation sheath.

Equation (2.19) is an often-used formula for the critical radius, but it is only an approximate estimate since the heat transfer coefficient was assumed to be independent of r_o. In general, we can write $h_o = \alpha r_o^{-n}$ where, for example, $n = 1/2$ for laminar forced convection. Thus, the total resistance R is more correctly written as

$$R = \frac{\ln(r_0/r_i)}{2\pi Lk} + \frac{1}{2\pi L\alpha r_o^{1-n}}$$

hence,

$$\frac{dR}{dr_o} = \frac{1}{2\pi L}\left(\frac{1}{r_o k} + \frac{(n-1)r_o^{n-2}}{\alpha}\right)$$

which equals zero when

$$r_o = r_{cr} = \left(\frac{\alpha}{(1-n)k}\right)^{1/(n-1)} \tag{2.20}$$

For $n = 1/2$, $r_{cr} = (k/2\alpha)^2$. For natural convection, the situation is more complex: not only do we have $n \simeq 1/4$, but Newton's law of cooling is invalid, with $h_o \propto \Delta T^{1/4}$. In the case of radiation, h_r can be taken to be independent of r_o, but $h_r \propto T_s^3$ [for small $(T_s - T_e)$]. Exercises 2–30 and 2–31 examine these situations. Fortunately, from a practical standpoint, a precise value of r_{cr} is not needed. Because $\dot{Q}$ is a maximum at r_{cr}, the heat loss is not sensitive to the precise value of r when r is in the vicinity of r_{cr}.

EXAMPLE 2.2   Cooling of an Electrical Resistor

A 0.5 W, 1.5 MΩ graphite resistor has a diameter of 1 mm and is 20 mm long; it has a thin glass sheath and is encapsulated in micanite (crushed mica bonded by a phenolic resin). The micanite serves both as additional electrical insulation and to increase the heat loss. It can be assumed that 50% of the I^2R heating is dissipated by combined convection and radiation from the outer surface of the micanite to surroundings at 300 K with $h_o = 16$ W/m^2 K; the remainder is conducted through copper leads to a circuit board. If the conductivity of micanite is 0.1 W/m K, what radius will give the maximum cooling effect, and what is the corresponding resistor temperature?

Solution

Given: Cylindrical graphite resistor encapsulated in micanite.

Required: Critical radius of micanite insulation, and the resistor temperature.

Assumptions: 1. The resistance of the glass sheath is negligible.
2. The outside heat transfer coefficient h_o is constant.
3. The resistor temperature is uniform.

The critical radius is given by Eq. (2.19):

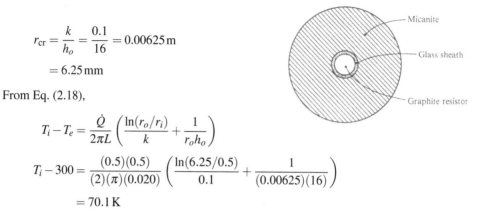

- Micanite
- Glass sheath
- Graphite resistor

$$r_{\text{cr}} = \frac{k}{h_o} = \frac{0.1}{16} = 0.00625\,\text{m}$$

$$= 6.25\,\text{mm}$$

From Eq. (2.18),

$$T_i - T_e = \frac{\dot{Q}}{2\pi L}\left(\frac{\ln(r_o/r_i)}{k} + \frac{1}{r_o h_o}\right)$$

$$T_i - 300 = \frac{(0.5)(0.5)}{(2)(\pi)(0.020)}\left(\frac{\ln(6.25/0.5)}{0.1} + \frac{1}{(0.00625)(16)}\right)$$

$$= 70.1\,\text{K}$$

Hence, $T_i = 300 + 70.1 = 370.1\,\text{K}$.

Comments

1. To obtain a more accurate result, additional data are required, particularly for the variation of h_o with radius.

2. Section 2.3.4 shows how to check the validity of assumption 3.

3. In general, h_o (and hence r_{cr}) vary around the circumference of the insulation. For engineering purposes, we ignore this complication and use an average value for h_o.

2.3.3 Conduction across a Spherical Shell

Figure 2.6 shows a spherical shell of inner radius r_1 and outer radius r_2. The inner surface is maintained at temperature T_1 and the outer surface at T_2. An elemental control volume is located between radii r and $r + \Delta r$. As was shown for the cylindrical shell in Section 2.3.1, energy conservation applied to the control volume requires that the heat flow $\dot{Q}$ be constant, independent of r if the temperatures are unchanging in time and $\dot{Q}_v = 0$. Using Fourier's law, Eq. (2.2),

$$\dot{Q} = Aq = 4\pi r^2\left(-k\frac{dT}{dr}\right)$$

Dividing by $4\pi k$ and assuming that the conductivity k is independent of temperature gives

$$\frac{\dot{Q}}{4\pi k} = -r^2\frac{dT}{dr} = \text{Constant} = C_1 \qquad\qquad (2.21)$$

$$\frac{dT}{dr} = -\frac{C_1}{r^2}$$

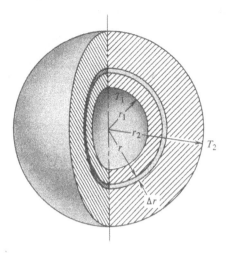

Figure 2.6   A spherical shell showing an elemental control volume for application of the energy conservation principle.

Integrating,

$$T = \frac{C_1}{r} + C_2 \qquad\qquad (2.22)$$

The boundary conditions required to evaluate the two constants are

$$r = r_1: \quad T = T_1 \qquad\qquad (2.23a)$$

$$r = r_2: \quad T = T_2 \qquad\qquad (2.23b)$$

Substituting in Eq. (2.22) and solving for C_1 and C_2 gives

$$C_1 = \frac{T_1 - T_2}{1/r_1 - 1/r_2}; \quad C_2 = T_1 - \frac{T_1 - T_2}{1 - r_1/r_2}$$

Then substituting back in Eq. (2.22) and rearranging gives the temperature distribution as

$$\frac{T_1 - T}{T_1 - T_2} = \frac{1/r_1 - 1/r}{1/r_1 - 1/r_2} \qquad\qquad (2.24)$$

The heat flow is found from Eq. (2.21) as $\dot{Q} = 4\pi k C_1$, or

$$\dot{Q} = \frac{4\pi k (T_1 - T_2)}{1/r_1 - 1/r_2} \qquad\qquad (2.25a)$$

Equation (2.25a) can be used to build up thermal circuits for composite spherical shells, as was done for composite cylindrical shells in Section 2.3.1. The thermal resistance of a spherical shell is

$$R = \frac{1/r_1 - 1/r_2}{4\pi k} \qquad\qquad (2.25b)$$

EXAMPLE 2.3   Determination of Thermal Conductivity

To measure the effective thermal conductivity of an opaque honeycomb material for an aircraft wall, a spherical shell of inner radius 26 cm and outer radius 34 cm was constructed and a 100 W electric light bulb placed in the center. At steady state, the temperatures of the inner and outer surfaces were measured to be 339 and 311 K respectively. What is the effective conductivity of the material?

Solution

Given: Spherical shell containing a 100 W heat source.

Required: Thermal conductivity of shell material.

Assumptions: 1. Steady state.
2. Spherical symmetry, $T = T(r)$.

Eq. (2.25a) applies, with $\dot{Q}$, T_1, T_2 known and k the unknown:

$$100 = \frac{4\pi k(339 - 311)}{1/0.26 - 1/0.34}$$

Solving, $k = 0.257$ W/m K.

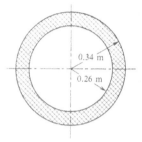

Comments

   The large thermal resistance of the honeycomb results in a relatively large temperature difference across it, which is easy to measure accurately. The same method would not be practical for determining the conductivity of a metal shell.

2.3.4 Conduction with Internal Heat Generation

In some situations, the thermal behavior of a body is affected by internally generated or absorbed thermal energy. The most common example is I^2R heating associated with the flow of electrical current I in an electrical resistance R. Other examples include fission reactions in the fuel rods of a nuclear reactor, absorption of radiation in a microwave oven, and emission of radiation by a flame. We will use the symbol $\dot{Q}_v'''$ [W/m³] for the heat generation rate per unit volume.[1] As an example, consider internal heat generation in a solid cylinder of outer radius r_1 as might occur in an electrical wire or a nuclear fuel rod. Figure 2.7 shows an elemental volume located between radii r and $r + \Delta r$. Applying the energy conservation principle, Eq. (1.2),

---

[1] The triple prime indicates "per unit volume" (per length dimension cubed). In Section 1.2, the symbol $\dot{Q}_v$ [W] was used for the heat generated within a system and is related to $\dot{Q}_v'''$ [W/m³] as $\dot{Q}_v = \int_V \dot{Q}_v''' \, dV$.

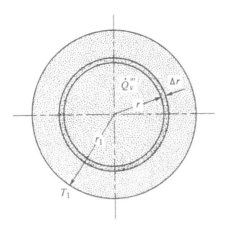

Figure 2.7   A solid cylindrical rod with internal heat generation.

requires that

$$\dot{Q}|_r - \dot{Q}|_{r+\Delta r} + \dot{Q}_v''' 2\pi r L \Delta r = 0$$

if temperatures are steady. Dividing by Δr and rearranging gives

$$\frac{\dot{Q}|_{r+\Delta r} - \dot{Q}|_r}{\Delta r} = \dot{Q}_v''' 2\pi r L$$

which for $\Delta r \to 0$ becomes

$$\frac{d\dot{Q}}{dr} = 2\pi r L \dot{Q}_v'''$$

Introducing Fourier's law, $\dot{Q} = Aq = 2\pi r L[-k(dT/dr)]$, and assuming k constant gives

$$\frac{d}{dr}\left(r\frac{dT}{dr}\right) = -\frac{\dot{Q}_v'''}{k}r \qquad\qquad (2.26)$$

which is a second-order linear ordinary differential equation for $T(r)$.

Two boundary conditions are required; the first comes from symmetry:

$$r = 0: \quad \frac{dT}{dr} = 0 \qquad\qquad (2.27a)$$

To obtain a result of some generality, we will take as the second boundary condition a specified temperature on the outer surface of the cylinder:

$$r = r_1; \quad T = T_1 \qquad\qquad (2.27b)$$

In a typical engineering problem, T_1 might not be specified; however, we shall see that the result will be in a form suitable for problem solving. Integrating Eq. (2.26) once gives

$$r\frac{dT}{dr} = -\frac{1}{2}\frac{\dot{Q}_v'''}{k}r^2 + C_1$$

or

$$\frac{dT}{dr} = -\frac{1}{2}\frac{\dot{Q}_v'''}{k}r + \frac{C_1}{r}$$

Applying the first boundary condition, Eq. (2.27a),

$$0 = 0 + \frac{C_1}{0}, \quad \text{or } C_1 = 0$$

Integrating again,

$$T = -\frac{1}{4}\frac{\dot{Q}_v'''}{k}r^2 + C_2$$

Applying the second boundary condition, Eq. (2.27b) allows C_2 to be evaluated:

$$T_1 = -\frac{1}{4}\frac{\dot{Q}_v'''}{k}r_1^2 + C_2$$

Substituting back gives the desired temperature distribution, $T(r)$:

$$T - T_1 = \frac{1}{4}\frac{\dot{Q}_v'''}{k}(r_1^2 - r^2) \tag{2.28}$$

The maximum temperature is at the centerline of the cylinder. Setting $r = 0$ in Eq. (2.28) gives

$$T_{\max} - T_1 = \frac{1}{4}\frac{\dot{Q}_v''' r_1^2}{k} \tag{2.29}$$

The use of this result is illustrated in the following example.

EXAMPLE 2.4   Temperature Distribution in a Nuclear Reactor Fuel Rod

Uranium oxide fuel is contained inside 0.825 cm–I.D., 0.970 cm–O.D. Zircaloy-4 tubes. The tubes have a 1.75 cm pitch in a square array. The power averaged over the volume including the space between the fuel rods is 152.4 W/cm³. At a specific location along the bundle the coolant water is at 400 K and the convective heat transfer coefficient h_c is 1.0×10^4 W/m² K. If the interfacial conductance between the fuel and the tube, h_i, is 6000 W/m²K, determine the maximum temperature in the fuel rods.

Solution

Given: Nuclear reactor fuel rod.

Required: Maximum rod temperature at location where the cool water is at $T_e = 400$ K.

Assumptions: Steady one-dimensional heat flow.

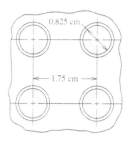

We cannot immediately use Eq. (2.29) to obtain $T_{\max}$ because the surface temperature of the fuel rod is unknown. We proceed as follows: first we calculate $\dot{Q}_v'''$ in the fuel itself,

(*a*)

$$\dot{Q}_v''' = 152.4 \frac{\text{Volume of array}}{\text{Volume of fuel}}$$

$$= 152.4 \frac{(1.75)^2}{(\pi/4)(0.825)^2}$$

$$= 873 \text{ W/cm}^3 = 8.73 \times 10^8 \text{ W/m}^3$$

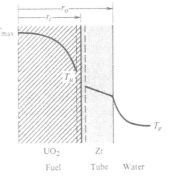

Next we find the temperature of the outer surface of the fuel rod. For unit length of rod, the heat flow across the Zircaloy tube wall is $\dot{Q} = \dot{Q}_v'''(\pi D^2/4)(1)$:

$$\dot{Q} = (8.73 \times 10^8)(\pi/4)(0.825 \times 10^{-2})^2(1)$$

$$= 46{,}700 \text{ W/m}$$

From the thermal circuit, as shown,

$$T_u = T_e + \dot{Q} \sum R$$

$$= 400 + 46{,}700 \left[\frac{1}{(2\pi)(0.00413)(6000)} + \frac{\ln(0.485/0.413)}{2\pi k_{Zr}} + \frac{1}{(2\pi)(0.00485)(10^4)} \right]$$

$$= 400 + 46{,}700(0.00642 + 0.0256/k_{Zr} + 0.00328)$$

As a guess, we take the mean temperature of the tube to be 600 K; from Table A.1b, the conductivity of Zircaloy-4 is 17.2 W/m K, and

$$T_u = 400 + 46{,}700(0.00642 + 0.00149 + 0.00328) = 923 \text{ K}$$

Now Eq. (2.29) can be used to obtain T_{max}. If we guess a mean temperature of 1500 K for the uranium oxide, Table A.2 gives $k_{UO_2} = 2.6$ W/m K, and

$$T_{max} = T_u + \frac{1}{4} \frac{\dot{Q}_v''' r_i^2}{k_{UO_2}} = 923 + \frac{(8.73 \times 10^8)(0.00413)^2}{(4)(2.6)} = 2355 \text{ K}$$

To check if our guessed mean temperatures are appropriate, we first determine the mean temperature of the tube. From the thermal circuit,

$$\overline{T}_{\text{tube}} \simeq 400 + (923 - 400) \frac{(0.00328 + 0.00149/2)}{(0.00642 + 0.00149 + 0.00328)} = 588 \text{ K}$$

which is close enough to our guess of 600 K. The mean temperature of the fuel rod is

$$\overline{T}_{UO_2} \simeq \frac{923 + 2355}{2} = 1639 \text{ K}$$

and at this temperature, $k_{UO_2} = 2.5$ W/m K. The new value of T_{max} is

$$T_{max} = 923 + \frac{(8.73 \times 10^8)(0.00413)^2}{(4)(2.5)} = 2412 \simeq 2400 \text{ K}$$

Comments

1. Since the k-value of UO_2 is given to only two significant figures, no further iteration is warranted.

2. Notice that the conductivity of Zircaloy-4 is lower than that of pure zirconium.

3. The largest thermal resistance in the circuit is at the fuel-cladding interface. The accuracy of the result depends primarily on our ability to obtain a reliable value of h_i. In fact, it could be argued that the second iteration for T_{max} was unwarranted due to uncertainty in the value of h_i.

4. Notice that we cannot extend the thermal circuit into the rod because $\dot{Q}$ is not constant when there is internal heat generation.

2.4   FINS

Heat transfer from a system can be increased by extending the surface area through the addition of fins. Fins are used when the convective heat transfer coefficient h_c is low, as is often the case for gases such as air, particularly under natural-convection conditions. Common examples are the cooling fins on electronics components, on the cylinders of air-cooled motorcycles and lawnmowers, and on the condenser tubes of a home refrigerator. Figure 2.8 shows a variety of fin configurations. A careful examination of an automobile radiator will show how it is designed to provide a large exterior surface.

Fins are added to increase the $h_c A$ product and hence decrease the convective thermal resistance $1/h_c A$. But the added area is not as efficient as the original surface area since there must be a temperature gradient along the fin to conduct the heat. Thus, for cooling, the average temperature difference $(T_s - T_e)$ is lower on a finned surface compared with the unfinned surface, and an appropriate thermal resistance for a fin is $1/h_c A \eta_f$, where A is the surface area of the fin and η_f is the *efficiency* of the fin $(0 < \eta_f < 1)$. For short fins of high thermal conductivity, η_f is large, but as the fin length increases, η_f decreases. Our objective here is to analyze heat flow in a fin to determine the temperature variation along the fin and, hence, to evaluate its efficiency η_f. Because fins are thin in one direction, it can be assumed that the temperature variation in this direction is negligible; this key assumption allows the conduction along the fin to be treated as if it were one-dimensional, which greatly simplifies the analysis.

2.4.1 The Pin Fin

Simple *pin fins*, such as those used to cool electronic components, will be analyzed to develop the essential concepts of fin theory. The first law is used to derive the governing differential equation, which, when solved subject to appropriate boundary conditions, gives the temperature distribution along the fin. The heat loss from the fin is then obtained and put in dimensionless form as the fin efficiency.

Figure 2.8 Some heat sinks incorporating fins for cooling of standard packages for integrated circuits. (Photograph courtesy of EG&G Wakefield Engineering, Wakefield, Mass.)

Governing Equation and Boundary Conditions

Consider the pin fin shown in Fig. 2.9. The cross-sectional area is $A_c = \pi R^2$ where R is the radius of the pin, and the perimeter $\mathscr{P} = 2\pi R$. Both A_c and R are uniform, that is, they do not vary along the fin in the x direction. The energy conservation principle, Eq. (1.2), is applied to an element of the fin located between x and $x + \Delta x$. Heat can enter and leave the element by conduction along the fin and can also be lost by convection from the surface of the element to the ambient fluid at temperature T_e. The surface area of the element is $\mathscr{P}\Delta x$; thus,

$$qA_c\big|_x - qA_c\big|_{x+\Delta x} - h_c\mathscr{P}\Delta x(T - T_e) = 0$$

Dividing by Δx and letting $\Delta x \to 0$ gives

$$-\frac{d}{dx}(qA_c) - h_c\mathscr{P}(T - T_e) = 0 \tag{2.30}$$

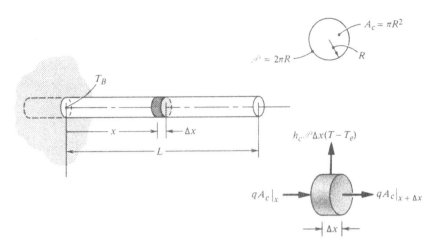

Figure 2.9   A pin fin showing the coordinate system, and an energy balance on a fin element.

For the pin fin, A_c is independent of x; using Fourier's law $q = -k \, dT/dx$ with k constant gives

$$kA_c \frac{d^2 T}{dx^2} - h_c \mathscr{P}(T - T_e) = 0 \qquad (2.31)$$

which is a second-order ordinary differential equation for $T = T(x)$. Notice that modeling of the conduction along the fin as one-dimensional has caused the convective heat loss from the sides of the fin to appear in the differential equation, in contrast to the problems dealt with in Section 2.3, where convection became involved as a boundary condition.

Next, boundary conditions for Eq. (2.31) must be specified. Since we wish to examine the performance of the fin itself, it is appropriate to take its base temperature as known; that is,

$$T|_{x=0} = T_B \qquad (2.32)$$

At the other end, the fin loses heat by Newton's law of cooling:

$$-A_c k \frac{dT}{dx}\bigg|_{x=L} = A_c h_c (T|_{x=L} - T_e) \qquad (2.33a)$$

where the convective heat transfer coefficient here is, in general, different from the one for the sides of the fin because the geometry is different. However, because the area of the end, A_c, is small compared to the side area, $\mathscr{P}L$, the heat loss from the end is correspondingly small and usually can be ignored. Then Eq. (2.33a) becomes

$$\frac{dT}{dx}\bigg|_{x=L} \simeq 0 \qquad (2.33b)$$

and this boundary condition is simpler to use than Eq. (2.33a). An even simpler result can be obtained if the temperature distribution along the fin is assumed identical to that for an infinitely long fin, for which the appropriate boundary condition is

$$\lim_{x \to \infty} T = T_e \tag{2.33c}$$

Figure 2.10 illustrates these boundary conditions.

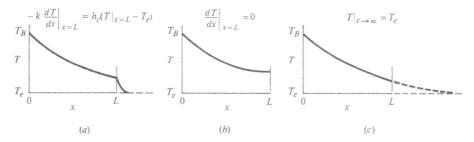

Figure 2.10   Three tip boundary conditions for the pin fin analysis. (a) Heat loss by convection, (b) Insulated tip. (c) Infinitely long fin.

Temperature Distribution

We will use Eq. (2.33b) for the second boundary condition as a compromise between accuracy and simplicity of the result. For mathematical convenience, let $\theta = T - T_e$ and $\beta^2 = h_c \mathscr{P}/kA_c$; then Eq. (2.31) becomes

$$\frac{d^2\theta}{dx^2} - \beta^2\theta = 0 \tag{2.34}$$

For β a constant, Eq. (2.34) has the solution

$$\theta = C_1 e^{\beta x} + C_2 e^{-\beta x}$$

or

$$\theta = B_1 \sinh\beta x + B_2 \cosh\beta x$$

The second form proves more convenient; thus, we have

$$T - T_e = B_1 \sinh\beta x + B_2 \cosh\beta x \tag{2.35}$$

Using the two boundary conditions, Eqs. (2.32) and (2.33b) give two algebraic equations for the unknown constants B_1 and B_2,

$$T_B - T_e = B_1 \sinh(0) + B_2 \cosh(0); \quad B_2 = T_B - T_e$$

$$\left.\frac{dT}{dx}\right|_{x=L} = \beta B_1 \cosh\beta L + \beta B_2 \sinh\beta L = 0; \quad B_1 = -B_2 \tanh\beta L$$

Substituting B_1 and B_2 in Eq. (2.35) and rearranging gives the temperature distribution as

$$\frac{T - T_e}{T_B - T_e} = \frac{\cosh \beta (L - x)}{\cosh \beta L}, \quad \text{where } \beta = \left(\frac{h_c \mathscr{P}}{k A_c} \right)^{1/2} \tag{2.36}$$

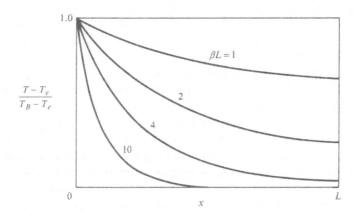

Figure 2.11   Fin temperature distributions calculated from Eq. (2.36).

Figure 2.11 shows a plot of Eq. (2.36). When β is small—for example, if the fin is made of aluminum and has a high thermal conductivity—the temperature T does not drop much below the base temperature T_B. For large β, T approaches the fluid temperature at the tip of the fin.[2]

Heat Loss

The heat dissipated from the fin can be found by integrating the heat loss over the side surface of the fin (there is no heat loss from the fin tip):

$$\dot{Q} = \int_0^L h_c \mathscr{P} (T - T_e) dx \tag{2.37}$$

with T obtained from Eq. (2.36). Substituting gives

$$\dot{Q} = \frac{h_c \mathscr{P} (T_B - T_e)}{\cosh \beta L} \int_0^L \cosh \beta (L - x) dx$$

To simplify the integration, let $\xi = \beta (L - x)$; then $dx = -d\xi / \beta$ and

---

[2] A_c in $\beta = (h_c \mathscr{P} / k A_c)^{1/2}$ is the cross-sectional area of the fin. The subscript c denotes "cross section" and not "convection" as in the heat transfer coefficient h_c. The area for convective heat loss is the surface area of the fin, $\mathscr{P} L$.

$$\dot{Q} = \frac{(h_c \mathscr{P}/\beta)(T_B - T_e)}{\cosh \beta L} \left[-\int_{\beta L}^{0} \cosh \xi \, d\xi \right]$$

$$= \frac{h_c \mathscr{P}}{\beta}(T_B - T_e) \left[-\frac{\sinh 0 - \sinh \beta L}{\cosh \beta L} \right]$$

$$= \frac{h_c \mathscr{P}}{\beta}(T_B - T_e) \tanh \beta L \qquad\qquad (2.38)$$

A less obvious alternative, but usually a more convenient way to find the heat dissipation, is to apply Fourier's law at the base of the fin:

$$\dot{Q} = -kA_c \frac{dT}{dx}\bigg|_{x=0} \qquad\qquad (2.39)$$

Substituting from Eq. (2.36),

$$\dot{Q} = -kA_c(T_B - T_e)\frac{[(d/dx)\cosh \beta(L-x)]_{x=0}}{\cosh \beta L}$$

$$= -kA_c(T_B - T_e)\frac{[-\beta \sinh \beta(L-x)]_{x=0}}{\cosh \beta L}$$

$$= kA_c \beta(T_B - T_e)\tanh \beta L \qquad\qquad (2.40)$$

Since $\beta^2 = h_c\mathscr{P}/kA_c$, Eqs. (2.38) and (2.40) give the same result, which is to be expected since there is no heat loss from the end of the fin.

Fin Efficiency

Let us now put Eq. (2.38) in **dimensionless** form by dividing through by $h_c\mathscr{P}L(T_B - T_e)$:

$$\frac{\dot{Q}}{h_c\mathscr{P}L(T_B - T_e)} = \frac{1}{\beta L}\tanh \beta L \qquad\qquad (2.41)$$

The dimensions of the left-hand side of this equation are $[W]/[W/m^2\ K][m][m][K] = 1$, as desired. The right-hand side must also be dimensionless since β has dimensions $[m^{-1}]$ and the group βL has dimensions $[m^{-1}][m] = 1$. (Of course, βL must be dimensionless to be the argument of the tanh function.) Now $h_c\mathscr{P}L(T_B - T_e)$ is the rate at which heat would be dissipated if the entire fin surface were at the base temperature T_B; in reality, there is a decrease in temperature along the fin, and the actual heat loss is less. Thus, the left-hand side of Eq. (2.41) can be viewed as the ratio of the actual heat loss to the maximum possible and is termed the **fin efficiency**, η_f. The right-hand side is a function of the dimensionless parameter βL only; we will set $\beta L = \chi$ as a *fin parameter*, and then Eq. (2.41) can be written in the compact form

$$\eta_f = \frac{\tanh \chi}{\chi} \qquad\qquad (2.42)$$

When χ is small, η_f is near unity; when χ is larger than about 4, $\tanh \chi \simeq 1$ and $\eta_f \simeq 1/\chi$. Since $\chi = \beta L = (h_c \mathscr{P} L^2 / k A_c)^{1/2}$, a small value of χ corresponds to relatively short, thick fins of high thermal conductivity, whereas large values of χ correspond to relatively long, thin fins of poor thermal conductivity. When χ is small, T does not fall much below T_B, and the fin is an efficient dissipator of heat. However, it is most important to understand that a thick fin with an efficiency of nearly 100% usually is not optimal from the viewpoint of heat transferred per unit weight or unit cost. The concept of fin efficiency refers only to the ability of the fin to transfer heat per unit area of exposed surface. Figure 2.12 shows a plot of Eq. (2.42). Use of dimensionless parameters has allowed the heat dissipation to be given by a single curve: different curves are not required for fins of various materials or lengths or for different values of the heat transfer coefficient. Likewise, storage of this information in a computer software package is efficient.

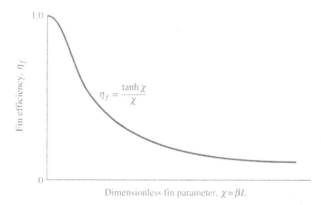

Figure 2.12   Efficiency of a pin fin as given in Eq. (2.42).

Straight Rectangular Fins

Although the pin fin shown in Fig. 2.9 was used for the purposes of this analysis, the results apply to any fin with a cross-sectional area A_c and perimeter $\mathscr{P}$ constant along the fin. The straight rectangular fin shown in Fig. 2.13 has a width W and thickness $2t$. The cross-sectional area is $A_c = 2tW$, and the perimeter is $\mathscr{P} = 2(W + 2t)$. For $W \gg t$, the ratio $\mathscr{P}/A_c$ is simply $1/t$, and $\beta = (h_c/kt)^{1/2}$.

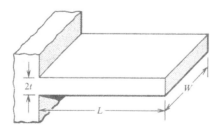

Figure 2.13   A straight rectangular fin.

Computer Program FIN1

The program FIN1 calculates the temperature distribution, fin efficiency, and base heat flow of straight rectangular fins. There are three options for the tip boundary condition: (1) infinitely long fin, (2) insulated, and (3) convective heat loss. The analysis for option 2 was given above; the analyses for options 1 and 3 are given as Exercises 2–58 and 2–59, respectively. For all three options, η_f is defined in terms of an isothermal fin heat loss of $\dot{Q} = h_c \mathscr{P} L(T_B - T_e)$. Use of FIN1 is illustrated in the example that follows.

EXAMPLE 2.5 Fins to Cool a Transistor

An array of eight aluminum alloy fins, each 3 mm wide, 0.4 mm thick, and 40 mm long, is used to cool a transistor. When the base is at 340 K and the ambient air is at 300 K, how much power do they dissipate if the combined convection and radiation heat transfer coefficient is estimated to be 8 W/m² K? The alloy has a conductivity of 175 W/m K.

Solution

Given: Aluminum fins to cool a transistor.

Required: Power dissipated by 8 fins.

Assumptions: 1. Heat transfer coefficient constant along fin.
2. Heat loss from fin tip negligible.

For one fin,

$$A_c = (0.003)(0.0004) = 1.2 \times 10^{-6}\,\mathrm{m^2}$$

$$\mathscr{P} = 2(0.003 + 0.0004) = 6.8 \times 10^{-3}\,\mathrm{m}$$

$$\beta^2 = \frac{h\mathscr{P}}{kA_c}$$

$$= \frac{(8.0\ \mathrm{W/m^2\ K})(6.8 \times 10^{-3}\ \mathrm{m})}{(175\ \mathrm{W/m\ K})(1.2 \times 10^{-6}\ \mathrm{m^2})}$$

$$= 259\ \mathrm{m^{-2}}$$

$$\beta = 16.1\ \mathrm{m^{-1}}$$

$$\chi = \beta L = (16.1\ \mathrm{m^{-1}})(0.040\ \mathrm{m}) = 0.644$$

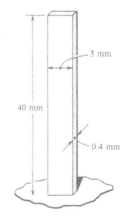

Substituting in Eq. (2.42)

$$\eta_f = \frac{1}{0.644}\tanh(0.644) = \frac{1}{0.644}\frac{e^{2(0.644)} - 1}{e^{2(0.644)} + 1} = 0.881$$

The side surface area of one fin is $\mathscr{P}L = (6.8 \times 10^{-3})(0.040) = 2.72 \times 10^{-4}\ \mathrm{m^2}$. If each fin were 100% efficient, it would be dissipate

$$h(\mathscr{P}L)(T_B - T_e) = (8)(2.72 \times 10^{-4})(340 - 300) = 8.70 \times 10^{-2}\ \mathrm{W}$$

Since the fins are only 88.1% efficient,

$$\dot{Q} = (0.881)(8.70 \times 10^{-2}) = 7.67 \times 10^{-2} \text{ W}$$

For 8 fins, $\dot{Q}_{total} = (8)(7.67 \times 10^{-2}) = 0.613$ W.

Solution using FIN1

The required input is:

Boundary condition = 2
Half-thickness, length, and width = 0.0002, 0.040, 0.003
Thermal conductivity = 175
Heat transfer coefficient = 8
Base temperature and ambient temperature = 340, 300
x-range for plot = 0.0, 0.04

FIN1 gives the output:

$$\eta_f = 0.881$$

$$\dot{Q} = 7.67 \times 10^{-2} \text{ (watts)}$$

Comments

1. Any consistent system of units can be used with FIN1. Since SI units were used here, the heat flow is in watts.

2. Notice the use of $h = h_c + h_r$ to account for radiation.

2.4.2 Fin Resistance and Surface Efficiency

It is useful to have an expression for the **thermal resistance** of a pin fin for use in thermal circuits. Equation (2.38) can be rewritten as

$$\dot{Q} = \frac{T_B - T_e}{1/[(h_c \mathscr{P}/\beta)\tanh \beta L]} \tag{2.43}$$

Thus, the thermal resistance of a pin fin is

$$R_{\text{fin}} = \frac{1}{(h_c \mathscr{P}/\beta)\tanh \beta L} = \frac{1}{h_c \mathscr{P} L \eta_f} \tag{2.44}$$

Notice that this thermal resistance accounts for both conduction along the fin and convection into the fluid. There are two parallel paths for heat loss from a finned surface—one through the fins and one through the area between the fins, as shown in Fig. 2.14. The respective conductances are thus additive; however, quite often the heat loss through the area between the fins is negligible.

The *total surface efficiency* η_t of a surface with fins of fin efficiency η_f is obtained by adding the unfinned portion of the surface area at 100% efficiency to the surface area of the fins at efficiency η_f:

$$A\eta_t = (A - A_f) + \eta_f A_f \tag{2.45}$$

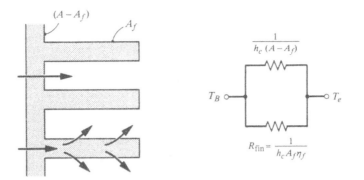

Figure 2.14   A finned surface showing the parallel paths for heat loss.

where A_f is the surface area of the fins and A is the total heat transfer surface area, including the fins and exposed tube or other surface. Solving for η_t,

$$\eta_t = 1 - \frac{A_f}{A}(1 - \eta_f) \qquad\qquad (2.46)$$

The corresponding thermal resistance of the finned surface is then

$$R = \frac{1}{h_c A \eta_t} \qquad\qquad (2.47)$$

Design calculations for the finned surfaces used in heat exchangers, such as automobile radiators, are conveniently made using Eq. (2.47).

2.4.3 Other Fin Type Analyses

The key feature of the fin analysis presented in Section 2.4.1 was that the thinness of the fin allowed us to ignore the temperature variation across the fin and, hence, to account for the convective loss from the surface directly in the differential equation for $T(x)$. The same assumption is valid for *extended surfaces* unrelated to cooling fins, and the results obtained in Section 2.4.1 are directly or indirectly applicable to these surfaces.

Sometimes it is quite obvious that the situation is similar to that for a cooling fin. For example, Fig. 2.15 shows a thermocouple installation used to measure the temperature of a hot air stream. The thermocouple junction is at a lower temperature than the air since the conduction heat flow along the thermocouple wires to the colder wall must be balanced by convection from the air. The temperature variation along

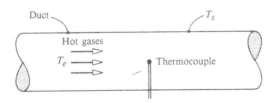

Figure 2.15   A thermocouple immersed in a fluid system.

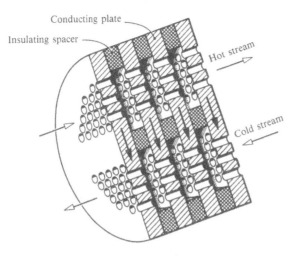

Figure 2.16   An element of a perforated-plate heat exchanger showing heat conduction along the plates.

the thermocouple is identical to that for a pin fin, so Eq. (2.36), with appropriate choices for the kA_c product, can be used to determine the error expected in the thermocouple reading.

Sometimes it is not obvious that the situation resembles that for a cooling fin, yet the assumption of negligible temperature variation in the thin direction of a wire or plate gives a differential equation similar to Eq. (2.30). The perforated plates in the heat exchanger shown in Fig. 2.16 can be treated as fins since the temperature variation across the plates is small compared to the temperature variation along the plates between the hot and cold streams. The copper conductors on the circuit board shown in Fig. 2.17 can be treated as fins, as can the circuit board between the conductors. The examples that follow relate to Figs. 2.15 and 2.16, and Exercise 2-88 is based on Fig. 2.17.

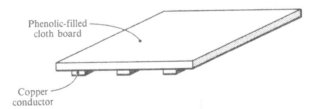

Figure 2.17   Copper conductors on a circuit board.

EXAMPLE 2.6   Error in Thermocouple Readings

Duplex thermocouple leads have both wires embedded in polyvinyl electrical insulation. One available size has wires of 0.25 mm diameter in insulation with an outside perimeter of 1.5 mm and is to be used in the situation depicted in Fig. 2.15. The air temperature is 350 K, and the wall temperature is 300 K. What length of immersion is required for the error in the thermocouple reading to be 0.1 K when the heat transfer coefficient on the perimeter is approximately 30 W/m^2 K? The wires are (i) copper and constantan (type T), (ii) iron and constantan (type J), and (iii) chromel and alumel (type K).

Solution

Given: Duplex thermocouple leads for types T, J, and K thermocouples.

Required: Length of immersion for a specified error.

Assumptions: Temperature variation across lead is small compared to the variation along the lead.

This is a fin-type problem since the temperature variation across the lead is small compared to the 50 K variation along the lead. The effective β^2 is

$$\beta^2 = \frac{h_c \mathscr{P}}{\sum k A_c}$$

where $h_c = 30$ W/m^2 K, $\mathscr{P} = 1.5 \times 10^{-3}$ m, and $\sum k A_c$ must be evaluated for the thermal resistances of the two wires and the insulation in *parallel*. For each wire, $A_c = (\pi/4)(0.25 \times 10^{-3})^2 = 4.91 \times 10^{-8}$ m^2, and for the insulation $A_c \simeq 10 \times 10^{-8}$ m^2. Thermal conductivity and kA_c values are given in the following table.

Wire Material	k W/m K	kA_c W m/K
Copper	385	19×10^{-6}
Constantan (55% Cu, 45% Ni)	23	1.1×10^{-6}
Iron	73	3.6×10^{-6}
Chromel-P (90% Ni, 10% Cr)	17	0.83×10^{-6}
Alumel (95% Ni, 2% Mn, 2% Al)	48	2.36×10^{-6}
Insulation	0.1	0.01×10^{-6}

The contribution of the insulation to $\sum k A_c$ is seen to be negligible; hence its precise shape or composition is unimportant. The temperature of the thermocouple junction (located at $x = L$) is given by Eq. (2.36) as

$$\frac{T_L - T_e}{T_B - T_e} = \frac{-0.1}{300 - 350} = \frac{1}{\cosh \beta L}$$

Solving,

$$\cosh \beta L = 500$$

$$\beta L = 6.91 \text{ from a calculator, or use } \cosh x = (1/2)(e^x + e^{-x})$$

$$L = 6.91/\beta$$

Evaluating β for each thermocouple pair gives the following results:

Thermocouple Type	$\sum kA_c$ W m/K	β m^{-1}	L cm
T	20.1×10^{-6}	47.3	14.6
J	4.7×10^{-6}	97.8	7.1
K	3.2×10^{-6}	119	5.8

Comments

1. Type T thermocouples are to be avoided when conduction along the wires may cause a significant error.

2. There are other criteria for choosing thermocouple pairs, including operating temperature range, emf output, and corrosion resistance.

3. Type T thermocouples are widely used because the component wires are relatively free from inhomogeneities; hence, calibration charts are reliable.

EXAMPLE 2.7   A Perforated Plate Heat Exchanger

Perforated-plate heat exchangers are used in cryogenic refrigeration systems. The fluids flow through high-conductivity perforated plates separated by insulating spacers. Heat is transferred from the hot stream to the cold stream by conduction along the plates, as shown in Fig. 2.16. A counterflow helium to helium unit has 0.5 mm-thick rectangular aluminum plates with 0.9 mm-diameter holes in a square array of pitch 1.3 mm. The plate length exposed to each stream is 20 mm, and the plate width is 80 mm. The spacer is 4 mm wide and 0.86 mm thick. If the heat transfer coefficient is 400 W/m^2 K for both streams, calculate the overall heat transfer coefficient. Take $k = 200$ W/m K for the aluminum.

Solution

Given: A perforated-plate heat exchanger.

Required: Overall heat transfer coefficient U.

Assumptions: 1. The plates are thin enough for a fin-type analysis to be valid.
2. Heat flow along the plates is one-dimensional.

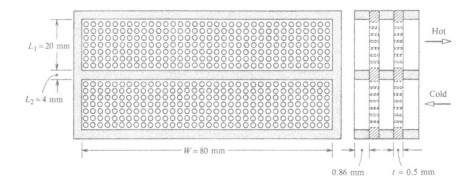

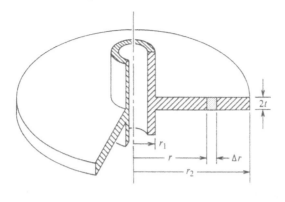

Figure 2.18 An annular fin of uniform thickness.

Governing Equation and Boundary Conditions

Figure 2.18 shows an annular fin of uniform thickness $2t$, as might be found on the outside of a tube. Such fins have extensive application in liquid-gas heat exchangers such as air-cooled evaporators for refrigeration systems. The energy conservation principle, Eq. (1.2), applied to a differential element between radii r and $r + \Delta r$ requires that

$$q(2\pi r)(2t)|_r - q(2\pi r)(2t)|_{r+\Delta r} - h_c(2)2\pi r\Delta r(T - T_e) = 0$$

Dividing through by $4\pi\Delta r$ and letting $\Delta r \to 0$.

$$-\frac{d}{dr}(rtq) - h_c r(T - T_e) = 0$$

Substituting Fourier's law, $q = -k(dT/dr)$, and dividing by (tk) gives

$$\frac{d}{dr}\left(r\frac{dT}{dr}\right) - \frac{h_c r}{tk}(T - T_e) = 0 \tag{2.48}$$

Note that this equation also could have been obtained directly from Eq. (2.30) by substituting $A_c = 4\pi rt$ and $\mathcal{P} = 4\pi r$, since A_c and $\mathcal{P}$ had not yet been given constant values. Equation (2.48) can be rearranged as

$$r^2\frac{d^2T}{dr^2} + r\frac{dT}{dr} - \beta^2 r^2(T - T_e) = 0$$

where $\beta^2 = h_c/tk$; then introducing new variables $z = \beta r$ and $\theta = (T - T_e)/(T_B - T_e)$ gives

$$z^2\frac{d^2\theta}{dz^2} + z\frac{d\theta}{dz} - z^2\theta = 0 \tag{2.49}$$

Suitable boundary conditions are as used for the pin fin in Section 2.4.1, that is, a specified base temperature and zero heat flow through the tip of the fin,

$$r = r_1: \quad T = T_B$$

$$r = r_2: \quad \frac{dT}{dr} = 0$$

or

$$z = z_1 = \beta r_1: \quad \theta = 1 \tag{2.50a}$$

$$z = z_2 = \beta r_2: \quad \frac{d\theta}{dz} = 0 \tag{2.50b}$$

Temperature Distribution

Equation (2.49) is a modified Bessel's equation of zero order and has the solution

$$\theta = C_1 I_0(z) + C_2 K_0(z) \tag{2.51}$$

where I_0 and K_0 are zero-order modified Bessel functions of the first and second kinds, respectively. Properties of Bessel functions are given in Appendix B. Applying the boundary conditions and using the differentiation formulas in Appendix B,

$$1 = C_1 I_0(z_1) + C_2 K_0(z_1)$$

$$0 = C_1 I_1(z_2) - C_2 K_1(z_2)$$

since $dI_0/dz = I_1$, $dK_0/dz = -K_1$, where I_1 and K_1 are first-order modified Bessel functions. Solving for C_1 and C_2,

$$C_1 = \frac{K_1(z_2)}{F(z_1, z_2)}, \quad C_2 = \frac{I_1(z_2)}{F(z_1, z_2)}$$

where

$$F(z_1, z_2) \equiv I_0(z_1) K_1(z_2) + I_1(z_2) K_0(z_1)$$

Substitution in Eq. (2.51) gives the temperature distribution along the fin.

Heat Loss and Efficiency

Next we obtain the heat dissipated by the fin and its efficiency. The heat flow through the base of the fin is

$$\dot{Q} = -kA_c \frac{dT}{dr}\bigg|_{r=r_1} = -k(2\pi r_1)(2t)(T_B - T_e)\beta \frac{d\theta}{dz}\bigg|_{z=z_1}$$

since $dT = (T_B - T_e)d\theta$ and $dr = dz/\beta$. Differentiating Eq. (2.51) gives

$$\frac{d\theta}{dz}\bigg|_{z=z_1} = C_1 I_1(z_1) - C_2 K_1(z_1)$$

and hence

$$\dot{Q} = k(4\pi r_1 t)(T_B - T_e)\beta[C_2 K_1(\beta r_1) - C_1 I_1(\beta r_1)]$$

The maximum possible heat loss is from an isothermal fin and is simply the product of the heat transfer coefficient, surface area, and temperature difference: $(h_c)(2)(\pi r_2^2 - \pi r_1^2)(T_B - T_e)$. The fin efficiency η_f is the ratio of the actual heat loss to that for an isothermal fin and can be rearranged as

$$\eta_f = \frac{(2r_1/\beta)}{(r_2^2 - r_1^2)} \frac{K_1(\beta r_1)I_1(\beta r_2) - I_1(\beta r_1)K_1(\beta r_2)}{K_0(\beta r_1)I_1(\beta r_2) + I_0(\beta r_1)K_1(\beta r_2)} \tag{2.52}$$

Equation (2.52) can be evaluated using the tables of Bessel functions given in Appendix B.

Other Fin Profiles: Computer Program FIN2

A variety of fin profiles are used in practice. Table 2.2 gives the efficiency of a selection of straight fins, annular fins, and spines. To facilitate the calculation of heat flow and fin mass, the surface area per unit width S' and profile area A_p are given for straight fins, and the surface area S and volume V are given for annular fins and spines. The efficiencies for items 1, 5, 6, and 7 were obtained using the boundary condition of zero heat flow through the tip. For thick rectangular fins, a simple approximate rule to account for heat loss from the tip is to add half the fin thickness to the fin length L for the straight fin and to the outer radius r_2 for the annular fin.

The computer program FIN2 calculates the efficiency, base heat flow, and mass for the 10 fin profiles listed in Table 2.2. For straight fins, the heat flow and mass are per unit width of fin. Its use is illustrated in Examples 2.8 and 2.9.

Cooling Fin Design

The proper design of cooling fins is an optimization problem: usually the objective is to minimize the amount of material in the fins in order to minimize either weight or cost. Exercises 2–71 and 2–113 show how optimal dimensions can be found for a given fin shape, and Exercises 2–105, 2–107, 2–120, and 2–125 illustrate that there is an optimal fin shape. The engineer is also free to choose the fraction of area covered by fin "footprints." This is a more difficult problem because as the fins are moved closer to each other, the value of the heat transfer coefficient h_c changes in a complicated way. There is always the question of whether fins should be provided at all. Exercise 2–104 shows that when the heat transfer coefficient is large, adding fins can actually reduce the heat loss. The conduction resistance in the fin can exceed the decrease in convective resistance due to the increased surface area. A useful rule is not to use fins unless $k/h_c t > 5$.

Table 2.2 Fins of various shapes: Efficiency, surface area per unit width $(S')^a$ and profile area (A_p) for straight fins; surface area $(S)^b$ and volume (V) for annular fins and spines: $\beta = (h_c/kt)^{1/2}$.

Straight Fins

1. Rectangular
$y = t$

$$\eta_f = \frac{1}{\beta L}\tanh\beta L \qquad S' = 2L, \qquad A_p = 2tL$$

2. Parabolic
$y = t(1-x/L)^{1/2}$

$$\eta_f = \frac{1}{\beta L}\frac{I_{2/3}\left(\frac{4}{3}\beta L\right)}{I_{-1/3}\left(\frac{4}{3}\beta L\right)}$$

$$S' = tB + (t^2/2L)\ln(2L/t + B)$$
$$B = \sqrt{1+4L^2/t^2}, \qquad A_p = \tfrac{4}{3}tL$$

3. Triangular
$y = t(1-x/L)$

$$\eta_f = \frac{1}{\beta L}\frac{I_1(2\beta L)}{I_0(2\beta L)} \qquad S' = 2\sqrt{t^2+L^2}$$
$$A_p = tL$$

4. Parabolic
$y = t(1-x/L)^2$

$$\eta_f = \frac{2}{\sqrt{4(\beta L)^2+1}+1} \qquad S' = LB + (L^2/2t)\ln(2t/L + B)$$
$$B = \sqrt{1+4t^2/L^2}, \qquad A_p = \tfrac{2}{3}tL$$

Annular Fins

5. Rectangular
$y = t$

$$\eta_f = \frac{(2r_1/\beta)}{(r_2^2-r_1^2)}\frac{K_1(\beta r_1)I_1(\beta r_2) - I_1(\beta r_1)K_1(\beta r_2)}{K_0(\beta r_1)I_1(\beta r_2) + I_0(\beta r_1)K_1(\beta r_2)}$$

$$S = 2\pi(r_2^2-r_1^2), \qquad V = 2\pi(r_2^2-r_1^2)t$$

6. Hyperbolic $y = t(r_1/r)$ 	$\eta_f = -\dfrac{2r_1/\beta}{(r_2+r_1)}\dfrac{I_{2/3}\left(\frac{2}{3}\beta r_1\right)L_{-2/3}\left(\frac{2}{3}\beta r_2\sqrt{r_2/r_1}\right) - I_{-2/3}\left(\frac{2}{3}\beta r_2\sqrt{r_2/r_1}\right)L_{-2/3}\left(\frac{2}{3}\beta r_1\right)}{I_{1/3}\left(\frac{2}{3}\beta r_1\right)I_{2/3}\left(\frac{2}{3}\beta r_2\sqrt{r_2/r_1}\right) - I_{-2/3}\left(\frac{2}{3}\beta r_2\sqrt{r_2/r_1}\right)L_{-1/3}\left(\frac{2}{3}\beta r_1\right)}$ $S = 2\pi r_1\left\{C - B + (t/2)\ln\dfrac{(C-t)(B+t)}{(C+t)(B-t)}\right\}$ $\qquad B = \sqrt{r_1^2 + t^2}$ $V = 4\pi t r_1(r_2 - r_1)$ $\qquad\qquad\qquad\qquad\qquad C = \sqrt{(r_2^2/r_1)^2 + t^2}$
Spines (Circular Cross Section)		
7. Pin $y = t$ 	$\eta_f = \dfrac{1}{\sqrt{2}\beta L}\tanh(\sqrt{2}\beta L),\qquad S = 2\pi t L,\qquad V = \pi t^2 L$
8. Parabolic $y = t(1 - x/L)^{1/2}$ 	$\eta_f = \dfrac{2}{\left(\frac{4}{3}\sqrt{2}\beta L\right)}\dfrac{I_1\left(\frac{4}{3}\sqrt{2}\beta L\right)}{I_0\left(\frac{4}{3}\sqrt{2}\beta L\right)},\qquad\begin{array}{l} S = (t^4\pi/6L^2)\{(4L^2/t^2+1)^{3/2} - 1\} \\[4pt] V = (\pi/2)t^2 L \end{array}$
9. Triangular $y = t(1 - x/L)$ 	$\eta_f = \dfrac{4}{(2\sqrt{2}\beta L)}\dfrac{I_2(2\sqrt{2}\beta L)}{I_1(2\sqrt{2}\beta L)},\qquad S = \pi t\sqrt{L^2 + t^2},\qquad V = (\pi/3)t^2 L$
10. Parabolic $y = t(1 - x/L)^2$ 	$\eta_f = \dfrac{2}{\sqrt{8/9(\beta L)^2 + 1} + 1},\qquad\qquad V = (\pi/5)t^2 L$ $S = (\pi L^3/16t)\{AB - (L/4t)\ln[(4tB/L) + A]\}$ $A = 1 + (8t^2/L^2),\qquad B = \sqrt{1 + (4t^2/L^2)}$

[a,b]Note that the formulas for η_f were derived assuming $dS' = dx$ (straight fins) and $dS = \mathscr{P}dx$ (spines). Use of the exact expressions for S' and S when calculating $\dot{Q}$ from η_f gives a result in better agreement with exact numerical solutions for two-dimensional conduction in short fins (for long fins the difference is negligible).

EXAMPLE 2.8   Cooling Fin for a Transistor

An aluminum annular fin is used to cool a transistor. The inner and outer radii are 5 mm and 20 mm, respectively, and the thickness is 0.2 mm. Calculate its efficiency and the heat dissipated when its base is at 380 K, the ambient air temperature is 300 K, and the estimated heat transfer coefficient is 8.2 W/m^{2}K. Take the conductivity of aluminum as 205 W/m K.

Solution

Given: Aluminum annular fin.

Required: Efficiency and heat dissipated.

Assumptions: 1. Heat transfer coefficient constant over the fin surface.
2. Heat loss from tip negligible.

For an annular fin, $\beta^2 = h_c/kt$:

$$\beta^2 = \frac{(8.2)}{(205)(0.1 \times 10^{-3})} = 400\,\text{m}^{-2}$$

$$\beta = 20\,\text{m}^{-1}$$

The fin effectiveness is given by Eq. (2.52):

$$\eta_f = \frac{(2r_1/\beta)}{(r_2^2 - r_1^2)} \frac{K_1(\beta r_1)I_1(\beta r_2) - I_1(\beta r_1)K_1(\beta r_2)}{K_0(\beta r_1)I_1(\beta r_2) + I_0(\beta r_1)K_1(\beta r_2)}$$

$$\beta r_1 = (20)(0.005) = 0.1; \quad \beta r_2 = (20)(0.020) = 0.4$$

From Appendix B, Table B.3b, the required values of Bessel functions are:

βr	I_0	I_1	K_0	K_1
0.1	1.0025	0.0501	2.4271	9.8538
0.4		0.2040		2.1843

Substituting in Eq. (2.52),

$$\eta_f = \frac{(2)(0.005)/(20)}{(0.020^2 - 0.005^2)} \frac{(9.8538)(0.2040) - (0.0501)(2.1843)}{(2.4271)(0.2040) + (1.0025)(2.1843)} = 0.944$$

The heat dissipation is the efficiency times the dissipation for an isothermal fin:

$$\dot{Q} = \eta_f(h_c)(2)(\pi)(r_2^2 - r_1^2)(T_B - T_e)$$

$$= (0.944)(8.2)(2)(\pi)(0.020^2 - 0.005^2)(380 - 300)$$

$$= 1.46\,\text{W}$$

Solution using FIN2

The required input in SI units is:

> Item number = 5
> Thermal conductivity and density of the fin = 205, 2700
> Heat transfer coefficient = 8.2
> Base temperature and ambient temperature = 380, 300
> $t = 0.0001$
> r_1 and $r_2 = 0.005, 0.020$

FIN2 gives the following output:

> Fin efficiency = 0.944
> Base heat flow = 1.459 (watts)
> Mass of fin = 6.362×10^{-4} (kilograms)

Comments

1. Notice that Table B.3b gives e^{-x} times $I_0(x)$ and $I_1(x)$ to simplify the tabulation.

2. The high efficiency suggests that the thickness of such fins is determined by rigidity rather than by heat transfer considerations.

3. Any consistent system of units can be used in FIN2. Since SI units were used here, the base heat flow is in watts, and the mass of the fin is in kilograms.

EXAMPLE 2.9 Heat Loss from a Parabolic Fin

A straight Duralumin fin has a parabolic profile $y = t(1 - x/L)^2$, with $t = 3$ mm and $L = 20$ mm. Determine the heat dissipation by the fin when its base temperature is 500 K and it is exposed to fluid at 300 K with a heat transfer coefficient of 2800 W/m^2 K. Also calculate the fin mass.

Solution

Given: Straight fin with a parabolic profile.

Required: Heat dissipation, mass.

Assumptions: Constant heat transfer coefficient over fin surface.

From Table A.1b, the conductivity of Duralumin at a guessed average fin temperature of $(1/2)(500 + 300) = 400$ K is 187 W/m K. Using Table 2.2, item 4,

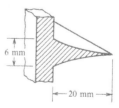

$$\beta = \left(\frac{h_c}{kt}\right)^{1/2} = \left[\frac{2800}{(187)(0.003)}\right]^{1/2} = 70.65\,\mathrm{m}^{-1}$$

$$\beta L = (70.65)(0.02) = 1.413$$

$$\eta_f = \frac{2}{[4(\beta L)^2 + 1]^{1/2} + 1} = \frac{2}{[4(1.413)^2 + 1]^{1/2} + 1} = 0.500$$

$$B = \left[1 + 4\left(\frac{t}{L}\right)^2\right]^{1/2} = \left[1 + 4\left(\frac{0.003}{0.02}\right)^2\right]^{1/2} = 1.044$$

$$S' = LB + \left(\frac{L^2}{2t}\right)\ln\left(\frac{2t}{L} + B\right)$$

$$= (0.02)(1.044) + \frac{0.02^2}{2 \times 0.003}\ln\left(\frac{2 \times 0.003}{0.02} + 1.044\right)$$

$$= 0.0406\,\text{m}$$

For a unit width of fin,

$$\dot{Q} = h_c S'(T_B - T_e)\eta_f = (2800)(0.0406)(500 - 300)(0.500) = 11,370\ \text{W/m}$$

From Table A.1a the density of Duralumin is $2770\,\text{kg/m}^3$; thus,

$$\text{Fin mass} = A_p\rho = \frac{2}{3}tL\rho = \left(\frac{2}{3}\right)(0.003)(0.02)(2770) = 0.1108\,\text{kg/m}$$

Solution using FIN2

The required input in SI units is:

Item number $= 4$
Thermal conductivity and density of the fin $= 187, 2770$
Heat transfer coefficient $= 2800$
Base temperature and ambient temperature $= 500, 300$
$t = 0.003$
$L = 0.02$

FIN2 gives the following output:

Fin efficiency $= 0.500$
Base heat flow $= 11,370$ (watts/meter)
Mass of fin $= 0.1108$ (kilograms/meter)

Comments

Exercise 2–113 shows that this fin profile gives the maximum heat loss for a given weight of any profile.

2.4.5 The Similarity Principle and Dimensional Analysis

To conclude our analysis of fins we use the pin fin problem of Section 2.4.1 to illustrate the **similarity principle** and **dimensional analysis**. These are important concepts used in the analysis of more complex heat transfer problems. Equation (2.42) showed that the fin performance could be expressed as a relation between just two dimensionless parameters: the fin efficiency, $\eta_f = \dot{Q}/h_c \mathscr{P} L (T_B - T_e)$, and a fin parameter $\chi = (h_c \mathscr{P} L^2 / k A_c)^{1/2}$. The similarity principle for this problem is simply the statement that η_f is a function of χ only. Thus, for example, if $\mathscr{P}$ and A_c are both doubled, η_f remains the same. We say that all pin fins with the same value of χ are *similar*, even though their sizes, materials, or heat transfer coefficients may be quite different.

The dimensionless groups relevant to a given problem are required for use of the similarity principle. We can deduce these dimensionless groups without actually solving the governing equations, as was done in Section 2.4.1. For this purpose, we use *dimensional analysis*, for which a number of methods are available. The pin fin problem will be used to demonstrate a method that requires a transformation of variables to make the governing equation and boundary conditions dimensionless. The first step is to choose dimensionless forms of the independent variable x and the dependent variable T. For x, an obvious choice is $\xi = x/L$, where L is the length of the fin; ξ then varies from zero to unity as x varies from zero to L. For T we will choose $\theta = (T - T_e)/(T_B - T_e)$; θ has a value of unity at the fin base and will approach zero at the tip of an infinitely long fin. Next, we transform the problem statement into the new variables. The rules of the transformation are

$$x = L\xi \qquad T = (T_B - T_e)\theta + T_e$$

$$dx = L d\xi \qquad dT = (T_B - T_e)d\theta$$

and Eq. (2.31) becomes

$$kA_c \frac{(T_B - T_e)}{L^2} \frac{d^2\theta}{d\xi^2} - h_c \mathscr{P}(T_B - T_e)\theta = 0$$

or

$$\frac{kA_c}{L^2} \frac{d^2\theta}{d\xi^2} - h_c \mathscr{P}\theta = 0$$

The boundary condition Eq. (2.32) becomes

$$\xi = 0: \quad \theta = 1 \tag{2.53a}$$

and the boundary condition Eq. (2.33b) becomes

$$\xi = 1: \quad \frac{(T_B - T_e)}{L} \frac{d\theta}{d\xi} = 0$$

or

$$\frac{d\theta}{d\xi} = 0 \tag{2.53b}$$

The differential equation is now put in dimensionless form. Dividing by kA_c/L^2,

$$\frac{d^2\theta}{d\xi^2} - \frac{h_c \mathscr{P} L^2}{kA_c}\theta = 0$$

or

$$\frac{d^2\theta}{d\xi^2} - \chi^2\theta = 0; \quad \chi = \beta L, \quad \beta^2 = \frac{h_c \mathscr{P}}{kA_c} \tag{2.54}$$

and we see that the fin parameter χ appears quite naturally in the dimensionless form of the governing equation. The boundary conditions are already dimensionless. Equation (2.54) is a differential equation for θ as a function of ξ and contains one dimensionless parameter, χ; the boundary conditions contain no further parameters. Thus, the solution must be of the form

$$\theta = \theta(\xi, \chi)$$

We now transform Eq. (2.37) for the rate of heat dissipation:

$$\dot{Q} = \int_0^L h_c \mathscr{P}(T - T_e)dx$$

$$= h_c \mathscr{P}L(T_B - T_e)\int_0^1 \theta d\xi$$

or

$$\frac{\dot{Q}}{h_c \mathscr{P}L(T_B - T_e)} = \eta_f = \int_0^1 \theta d\xi \tag{2.55}$$

For this simple case, the fin efficiency appears quite naturally as the dimensionless form of the heat dissipation rate. Although $\theta = (\xi, \chi)$, the definite integral in Eq. (2.55) is not a function of ξ, so that

$$\eta_f = \eta_f(\chi) \tag{2.56}$$

which corresponds to the analytical solution, Eq. (2.42).

More complex heat transfer problems are often governed by differential equations that are difficult or impossible to solve analytically. Use of the above procedure allows the most concise form of the solution to be determined, which can be used as a basis for correlating experimental data or the results of numerical solutions. Also, when properly used, dimensional analysis facilitates the estimation of errors incurred in making simplifying assumptions. Such an approach is especially important for the analysis of convective heat transfer, as is shown in advanced texts. As a rather simple example of error estimation, consider the pin fin problem when the tip heat loss is not neglected. The appropriate second boundary condition is then Eq. (2.33a), which transforms into

$$\xi = 1: \quad -k\frac{(T_B - T_e)}{L}\frac{d\theta}{d\xi} = h_c(T_B - T_e)\theta$$

or

$$\frac{d\theta}{d\xi} = -\text{Bi}\theta \tag{2.57}$$

The Biot number $\text{Bi} = h_c L/k$ was discussed in Section 1.5.1. This boundary condition introduces a second parameter into the problem; the temperature distribution must now be of the form

$$\theta = \theta(\xi, \chi, \text{Bi}) \tag{2.58}$$

and the fin efficiency is a function of both χ and Bi. Using physical intuition, we would expect the tip heat loss to be significant only when the tip temperature is slightly less than that of the base, that is, when the fin parameter χ is small. Then the ratio of the heat loss from the tip to the heat loss from the sides is on the order of the area ratio $A_c/\mathscr{P}L$. At first sight, this area ratio may appear to be a new dimensionless parameter, but it is simply Bi/χ^2. The analytical solution for this case is given as Exercise 2–69. It confirms that the tip loss is significant only for small χ, in which case the fractional tip loss is approximately Bi/χ^2.

2.5   CLOSURE

In this chapter, the first law of thermodynamics and Fourier's law of heat conduction were used to solve a variety of steady one-dimensional heat conduction problems. Simple analytical results were obtained for conduction through cylindrical and spherical shells, and these can be used to build up thermal circuits for more complicated problems. The concept of a critical radius of insulation for a cylinder was introduced, which showed that insulation should not be added to a small cylinder (or sphere) for the purpose of reducing heat loss. The temperature distribution in a cylinder with internal heat generation was obtained, and the result was applied to calculation of the maximum temperature in a nuclear reactor fuel rod.

Much of the chapter dealt with the very important subject of extended surfaces, or fins. Cooling fins are widely used to reduce convective heat transfer resistance, particularly when gases are involved. The fin efficiency of a pin fin and annular fin were obtained by analysis. The result for an annular fin is obtained in terms of Bessel functions, which may be new to the student. Use of these functions is similar to use of the familiar trigonometric functions, and Appendix B conveniently specifies the required differentiation rules and provides tables. The efficiencies for eight additional fin profiles are given in Table 2.2 for engineering use. The key assumption in the analysis of cooling fins was that the temperature variation across the fin can be ignored. The same assumption is valid for a variety of other extended surfaces—for example, thermocouples and copper conductors. In some cases, the results of the cooling fin analyses can be used directly; in others, a new but similar analysis is required. The discussion of fins concluded by using the pin fin problem to demonstrate the use of dimensional analysis and the principle of similarity, both very important concepts that will be used throughout this text.

Two computer programs were introduced in Chapter 2. FIN1 is primarily an instructional aid and allows the student to explore the effect of fin parameters and boundary conditions on the temperature profile along a rectangular fin and on the fin efficiency. FIN2 is an engineering tool that gives the fin efficiency and fin mass for 10 different fin profiles, including straight fins, annular fins, and spines.

REFERENCES

1. Touloukian, Y. S., and Ho, C. Y., eds., *Thermophysical Properties of Matter. Vol. I, Thermal Conductivity of Metallic Solids; Vol. 2, Thermal Conductivity of Nonmetallic Solids*, Plenum Press, New York (1972).

2. Vargaftik, N. B., *Tables of Thermophysical Properties of Liquids and Gases*, 2nd ed., Hemisphere Publishing Corp., Washington, D.C. (1975).

3. Desai, D. P., Chu, T. K., Bogaard, R. H., Ackerman, M. W., and Ho, C. Y, *CINDAS Special Report. Part I: Thermophysical Properties of Carbon Steels; Part II; Thermophysical Properties of Low Chromium Steels; Part III: Thermophysical Properties of Nickel Steels; Part IV: Thermophysical Properties of Stainless Steels*, Purdue University, West Lafayette, Ind., September (1976).

4. American Society of Heating, Refrigerating and Air Conditioning Engineers, *ASHRAE Handbook of Fundamentals*, ASHRAE, New York (1981).

5. Salerno, L. J., and Kittel, P. "Thermal Contact Resistance," *NASA TM 110429*, (1997).

6. Jacobs, G., and Todreas, N. "Thermal contact conductance in reactor fuel elements," *Nuclear Science and Engineering*, 50, 283–290 (1973).

7. Madhusudana, C.V., Fletcher, L.S., and Peterson, G.P., "Thermal conductance of cylindrical joints – a critical review,"*Journal of Thermophysics and Heat Transfer*, 4, 204–211 (1990).

8. Jakob, M., *Heat Transfer*, vol. 1, John Wiley & Sons, New York (1949).

9. Zhang, B. X., and Chung. B. T. E, "A multi-objective fuzzy optimization for a convective fin," Paper No. 3-NT-34, *Proceedings of the Tenth International Heat Transfer Conference*, Brighton UK, ed. Hewitt, G. F., Taylor and Francis, Bristol Pa. (1994).

10. Shamsundar, N., University of Houston, private communication.

EXERCISES

2-1. The thermal conductivity of a solid may often be assumed to vary linearly with temperature, $k = k_0[1 + a(T - T_0)]$, where $k = k_0$ at a reference temperature T_0 and a is a constant coefficient. Consider a solid slab, $0 < x < L$, with the face at $x = 0$ maintained at temperature T_1. Determine the temperature profile across the slab, $T = T(x)$, in terms of T_1 and the heat flux q. Sketch profiles for zero, positive, and negative coefficients.

2-2. A typical polystyrene coffee cup has a wall thickness of 2 mm. We are all familiar with the insulation properties of such a cup from the time required for coffee to cool. To obtain the same insulating effect, how thick a layer of each of the following materials is required?

   (i) polyvinylchloride
   (ii) paper
   (iii) oak wood
   (iv) stainless steel
   (v) brass
   (vi) copper
   (vii) diamond (type IIb)

2-3. A long, flat slab is made of n pairs of square bars of different thermal conductivities, k_A and k_B. Determine the effective thermal conductivity of the slab

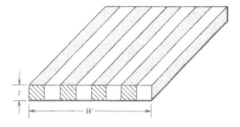

   (i) across the width.
   (ii) through the thickness.

2-4. Use Eq. (2.3) to estimate the thermal conductivity of

   (i) air at 300 K and 1 atm pressure.
   (ii) air at 400 K and 1 atm pressure.
   (iii) air at 300 K and 0.01 atm pressure.
   (iv) helium at 300 K and 1 atm pressure.

Compare your results to the data given in Table A.7, and discuss the effects of temperature and pressure. The average molecular speed c and transport mean free path ℓ_t may be calculated from the following formulas:

$$c = \left(\frac{8\kappa_B T}{\pi m}\right)^{1/2}; \quad \ell_t = v\left(\frac{9\pi m}{8\kappa_B T}\right)^{1/2}$$

2-5. An aluminum/aluminum interface may have an interfacial conductance in the range 150–12,000 W/m² K (the lower-limit value corresponds to vacuum conditions). Consider conduction from an aluminum channel to an aluminum plate, as shown. The channel is 3 mm thick, and the plate is 1 mm thick. Plot a graph of the ratio of interfacial thermal resistance to total thermal resistance

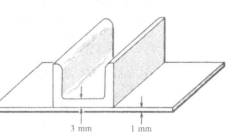

3 mm   1 mm

(channel, interface, plate) for this range of interfacial conductances. Take $k = 200$ W/m K for the aluminum.

2-6. A test rig for the measurement of interfacial conductance is used to determine the effect of surface anodization treatment on the interfacial conductance for aluminum-aluminum contact. The specimens themselves form the heat flux meters because they are each fitted with a pair of thermocouples as shown. The heat flux is determined from the measured temperature gradient and a known thermal conductivity of the aluminum of 185 W/m K as

$$q_1 = \frac{k(T_1 - T_2)}{L_1}; \quad q_2 = \frac{k(T_3 - T_4)}{L_2}; \quad q = \frac{1}{2}(q_1 + q_2)$$

and the interfacial temperatures obtained by linear extrapolation.

$$T_5 = \frac{1}{2}(T_1 + T_2) - \left[\frac{L_3 + \dfrac{L_1}{2}}{L_1} \right](T_1 - T_2);$$

$$T_6 = \frac{1}{2}(T_3 + T_4) + \left[\frac{L_4 + \dfrac{L_2}{2}}{L_2} \right](T_3 - T_4)$$

The interfacial conductance is then obtained as

$$h_i = \frac{q}{(T_5 - T_6)}$$

The main possible sources of error in the values of h_i so determined are due to uncertainty in temperature measurement and thermocouple locations. Since only temperature differences are involved, the absolute uncertainty in the individual temperature measurements is not of concern; rather it is the relative uncertainties. Previous calibrations of similar type and grade thermocouples indicate that the relative uncertainties are $\pm 0.2°C$. Also, the technician who drilled the thermocouple holes and installed the thermocouples estimates an uncertainty of ± 0.5 mm in the thermocouple junction locations. At a particular pressure, the temperatures recorded are $T_1 = 338.7$ K, $T_2 = 328.7$ K, $T_3 = 305.3$ K, $T_4 = 295.0$ K.

(i) Assuming that the uncertainties in temperature measurement and thermocouple location can be treated as random errors, estimate the uncertainty in the interfacial conductance.

(ii) In reality, the uncertainties in temperature and location are bias errors (for example, the location of a thermocouple does not vary from test to test). Thus it is more appropriate to determine bounds on the possible error in h_i, by considering best and worst cases. Determine these bounds.

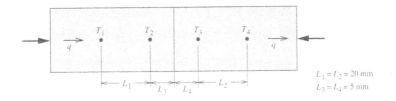

$L_1 = L_2 = 20$ mm
$L_3 = L_4 = 5$ mm

2-7. Consider steady one-dimensional conduction through a plane wall. The figure shows the resulting temperature profiles for three materials. Explain how the thermal conductivity k must vary with temperature to account for the behavior of these profiles for each material. Support your explanations with appropriate equations.

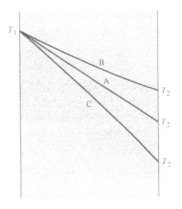

2-8. A 1 cm-thick, AISI 302 stainless steel wall is insulated with a 5 cm-thick layer of fiberglass (28 kg/m^3 density) on its outside surface. In a test with ambient air at 25°C, the temperature of the surface of the insulation is measured to be 40°C. The outside convective heat transfer coefficient is estimated to be 6 W/m^2 K, and the emittance of the insulation can be taken as 0.8.

(i) Draw the thermal circuit.
(ii) Determine the temperature of the inner surface of the stainless steel.

2-9. In order to prevent fogging, the 3 mm-thick rear window of an automobile has a transparent film electrical heater bonded to the inside of the glass. During a test, 200 W are dissipated in a 0.567 m^2 area of window when the inside and outside air temperatures are 22°C and 1°C, and the inside and outside heat transfer coefficients are 9 W/m^2 K and 22 W/m^2 K, respectively. Determine the temperature of the inside surface of the window.

2-10. A 2 cm-thick composite plate has electric heating wires arranged in a grid in its centerplane. On one side there is air at 20°C, and on the other side there is air at 100°C. If the heat transfer coefficient on both sides is 40 W/m^2 K, what is the maximum allowable rate of heat generation per unit area if the composite temperature should not exceed 300°C? Take $k = 0.45$ W/m K for the composite material.

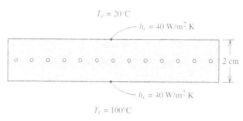

2-11. A horizontal fireclay brick partition, 20 cm thick, is covered by a 5 cm-thick layer of medium-density ($\rho = 28$ kg/m³) fiberglass insulation. The under surface of the brick is maintained at 540 K, and the upper surface of the fiberglass is maintained at 300 K. Calculate the heat flow per unit area. Evaluate the k values of each layer at the average temperature of the layer.

2-12. A composite wall consists of a 1 mm-thick stainless steel plate, 2 cm of 4-ply laminated asbestos paper, and 2 cm of 8-ply laminated asbestos paper. The stainless steel surface is maintained at 380 K while the other side loses heat to ambient air at 300 K, with a combined convection and radiation heat transfer coefficient of 5 W/m² K. Estimate the heat flow per unit area. Evaluate the k values of the asbestos layers at suitable average temperatures.

2-13. Estimate the heat flow through the composite wall shown for $k_A = 1$ W/m K, $k_B = 0.3$ W/m K, $k_C = 0.6$ W/m K, $k_D = 10$ W/m K. Assume

  (i) surfaces of constant x are isothermal.
  (ii) no heat flow transverse to the x-direction.

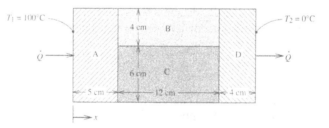

2-14. A 2 m-long cylindrical chemical reactor has an inside diameter of 5 cm, has a 1 cm-thick stainless steel shell, and is insulated on the outside by a layer of medium-density ($\rho = 28$ kg/m³) fiberglass 5 cm thick. The ambient air is at 25°C, and the surroundings can be assumed large and black. The convective heat transfer coefficient between the insulation and air is 6 W/m² K, and the emittance of the insulation is 0.8. At steady state the outer surface of the insulation is measured at 32°C. Draw the thermal circuit and determine the temperature of the inner surface of the stainless steel shell. Take $k = 16$ W/m K for the stainless steel.

2-15. A 4 cm-O.D., 2 mm-wall-thickness stainless steel tube is insulated with a 5 cm-thick layer of cork. Chilled milk flows through the tube. At a given location the milk temperature is 5°C when the ambient temperature is 25°C. If the inside and outside heat transfer coefficients are estimated to be 50 and 5 W/m² K, respectively, calculate the rate of heat gain per meter length of tube.

2-16. Saturated steam at 200°C flows through an AISI 1010 steel tube with an outer diameter of 10 cm and a 4 mm wall thickness. It is proposed to add a 5 cm-thick layer of 85% magnesia insulation. Compare the heat loss from the insulated tube to that from the bare tube when the ambient air temperature is 20°C. Take outside heat transfer coefficients of 6 and 5 W/m² K for the bare and insulated tubes, respectively.

2-17. A hollow cylinder, of inner and outer diameters 3 and 5 cm, respectively, has an inner surface temperature of 400 K. The outer surface temperature is 326 K when exposed to fluid at 300 K with an outside heat transfer coefficient of 27 W/m² K. What is the thermal conductivity of the cylinder?

2-18. Superheated steam at 500 K flows in Schedule 40 steel pipe of nominal size 6 in. Determine the effect of adding magnesia insulation to the pipe as a function of insulation thickness and outside heat transfer coefficient. Assume an inside heat transfer coefficient of 7000 W/m² K and surroundings at 300 K. Prepare a graph of heat loss per unit length as a function of insulation thickness, with the outside heat transfer coefficient ($10 < h_o < 100$ W/m² K) as a parameter.

2-19. A thermal conductivity cell consists of concentric thin-walled copper tubes with an electrical heater inside the inner tube and is used to measure the conductivity of granular materials. The inner and outer radii of the annular gap are 2 and 4 cm. In a particular test the electrical power to the heater was 10.6 W per meter length, and the inner and outer tube temperatures were measured to be 321.4 K and 312.7 K, respectively. Calculate the thermal conductivity of the sample.

2-20. A thick-walled cylindrical tube has inner and outer diameters of 2 cm and 5 cm, respectively. The tube is evacuated and contains a high-temperature radiation source along its axis giving a net radiant heat flux into the inner surface of the tube of 10^5 W/m². The outer surface of the tube is convectively cooled by a coolant at 300 K with a convective heat transfer coefficient of 120 W/m² K. If the conductivity of the tube material is 2.2 W/m K, determine the temperature distribution $T(r)$ in the tube wall. Also determine the inner-surface temperature.

2-21. Modern fossil-fueled boilers in power plants produce steam at about 550°C and 200 bar. The steam flows along steam lines to turbines for the generation of electrical power. These steam lines are insulated to reduce heat loss to the ambient air and to prevent skin burns due to accidental contact. Until the mid 1970s, relatively inexpensive asbestos products were used as insulators because of their low thermal conductivity and their resistance to wear and corrosion. However, asbestos proved to be a carcinogen and substitutes are required. Apart from a low thermal conductivity, other desirable properties include workability, high strength, and reusability. The table shows data for the thermal conductivity of candidate insulations at 600°F taken from the catalogs of U.S. companies.

Brand Name	Material	Company	k Btu in/hr ft ^{2}°F	Configuration
Thermo-12	Calcium silicate	Manville	0.46	Cylindrical block
Durablanket	Ceramic fiber	Standard Oil	0.48	Blanket
Paroc-1200	Mineral wool	Partek	0.52	Cylindrical block
Kaylo	Calcium silicate	Owens/Corning	0.58	Cylindrical block
CeraWool	Ceramic fiber	Manville	0.60	Blanket
Calsilite	Calcium silicate	Calsilite	0.64	Cylindrical block
Fiberglass	Fibrous glass	Owens/Corning	0.66	Cylindrical block
Micro-lok	Fibrous alass	Manville	0.72	Cylindrical block

(i) Make a table of the k-values in SI units.

(ii) Compare these k-values with other insulations in Table A.3. In particular, make comparisons for magnesia pipe insulation, insulating brick, and fiberglass batts used for home insulation. Comment on your observations.

(iii) Typically, a 40 cm-O.D. steam line will be insulated with a 13 cm-thick layer of insulation. The temperature outside the insulation should be low enough not to burn bare skin on contact. For a steam temperature of 560°C, give a rough estimate of energy saved by insulating a 100 m length of pipe, and its value if the cost to produce thermal energy is 3 cents/kWh. Assume outside heat transfer coefficients for the insulated and bare pipes of 10 and 50 W/m^2 K, respectively.

2-22. Calcium silicate has replaced asbestos as the preferred insulation for steam lines in power plants. Consider a 40 cm-O.D., 4 cm-wall-thickness steel steam line insulated with a 12 cm thickness of calcium silicate. The insulation is protected from damage by an aluminum sheet lagging that is 2.5 mm thick. The steam temperature is 565°C and ambient air temperature in the power plant is 26°C. The inside convective resistance is negligible, the outside convective heat transfer coefficient can be taken as 11 W/m^2 K, and the emittance of the aluminum is 0.1. Calculate the rate of heat loss per meter. The thermal conductivity of the pipe wall is 40 W/m K, and the table gives values for Calsilite calcium silicate insulation blocks.

T, K	500	600	700	800	900
k, W/m K	0.074	0.096	0.142	0.211	0.303

2-23. Repeat Exercise 2–22 for a ceramic fiber insulation, for which the following thermal conductivity data for Manville's CeraWool insulation blanket can be used.

T, K	400	500	600	700	800	900
k, W/m K	0.048	0.069	0.089	0.105	0.120	0.132

2-24. Use of multilayer insulations can be cost-effective and reduce energy losses for steam lines in power plants. Most of the organic binders commonly used in insulations such as fiberglass or mineral wool can withstand temperatures up to 180°C, so these insulations can be used as an outer layer surrounding a stronger, high-temperature resistant inner layer. Repeat Exercise 2–22 for an insulation consisting of 9 cm of ceramic fiber and 3 cm of fiberglass. Thermal conductivity data for the ceramic fiber can be found in Exercise 2–23, and is presented in the table for a candidate fiberglass insulation.

T, K	300	350	400	450
k, W/m K	0.035	0.048	0.063	0.080

2-25. Insulated steam lines in power plants are lagged to protect the insulation from damage and to prevent absorption of moisture into the insulation. Canvas cloth or thin aluminum sheet is currently used. The aluminum is typically 1.5 mm thick, whereas the cloth may be 1.5 to 3 mm thick. Use of aluminum, with its low emittance, reduces the heat loss but increases the surface temperature by 5 to 15°C. To avoid skin burns, it is recommended that maximum safe lagging surface temperatures are 40°C for metallic jackets and 60°C for canvas jackets. A 36 cm–O.D., 7 cm-wall-thickness steel steam line is insulated with a 10 cm thickness of calcium silicate. The steam temperature is 560°C, and the ambient air temperature is 26°C. The inside convective resistance is negligible and the outside convective heat transfer coefficient can be taken as 10 W/m^2 K.

   (i) Calculate the rate of heat loss per meter and surface temperature for a 1.5 mm-thick aluminum sheet lagging.

  (ii) Calculate the rate of heat loss per meter and surface temperature for a 3 mm-thick canvas lagging. Take $\varepsilon = 0.9$ for the canvas.

 (iii) If the aluminum is painted with an oil-based paint with an emittance of 0.9, calculate the increase in heat loss and reduction in surface temperature.

Use $\varepsilon = 0.1$ for the aluminum, $k = 0.04$ W/m K for canvas, and the data for k of Calsilite insulation from Exercise 2–22.

2-26. Liquid oxygen at 90 K flows inside a 3 cm–O.D., 2 mm-wall-thickness, AISI 303 stainless steel tube. The tube is insulated with 3 cm-thick fiberglass insulation of thermal conductivity 0.021 W/m K. Will atmospheric water vapor condense on the outside of the insulation when the ambient air temperature is 300 K and the dewpoint temperature is 285 K? Take the outside convective and radiative heat transfer coefficients as 5 W/m^2 K and 4 W/m^2 K, respectively.

2-27. Refrigerant-12 at −35°C flows in a copper tube of 8 mm outer diameter and 1 mm wall thickness. It has been suggested that a foam insulation with an aluminized outer surface be used to reduce heat leakage to the refrigerant. If the inside and outside heat transfer coefficients are taken as 300 and 5 W/m² K, respectively, and the surrounding air is at 20°C, plot a graph of heat leakage per meter versus insulation thickness. Should the insulation be used? Take $k = 0.035$ W/m K for the insulation.

2-28. A 1 in Schedule 10 copper pipe carries 10 GPM of brine at $-5°C$. The ambient air is at 20°C and has a dewpoint of 10°C. How thick a layer of insulation of thermal conductivity 0.2 W/m K is required to prevent condensation on the outside of the insulation? Take the outside heat transfer coefficient as 11.0 W/m² K.

2-29. A 1.5 mm-diameter wire is to be insulated with a plastic material of thermal conductivity 0.37 W/m K. It is found by experiment that a given current will heat the bare wire to 40°C when the ambient air temperature is 20°C. Plot the wire temperature and the surface temperature of the insulation as a function of insulation thickness for the same current. The emittance of the insulation is 0.9. and that of the bare wire is 0.07. The convective heat transfer coefficient can be calculated from an approximate relation for a horizontal cylinder in air at normal temperature, $\bar{h}_c = 1.3(\Delta T/D)^{1/4}$ W/m² K, for ΔT in kelvins and D in meters.

2-30. A 6 mm-O.D. tube is to be insulated with an insulation of thermal conductivity 0.08 W/m K and a very low surface emittance. Heat loss is by natural convection, for which the heat transfer coefficient can be taken as $\bar{h}_c = 1.3(\Delta T/D)^{1/4}$ W/m² K for $\Delta T = T_s - T_e$ in kelvins and the diameter D in meters. Determine the critical radius of the insulation and the corresponding heat loss for a tube surface temperature of 350 K, and an ambient temperature of 300 K.

2-31. A 1 mm-diameter resistor has a sheath of thermal conductivity $k = 0.12$ W/m K and is located in an evacuated enclosure. Determine the radius of the sheath that maximizes the heat loss from the resistor when it is maintained at 450 K and the enclosure is at 300 K. The surface emittance of the sheath is 0.85.

2-32. A 2 mm-diameter resistor, for an electronic component on a space station, is to have a sheath of thermal conductivity 0.1 W/m K. It is cooled by forced convection with $\bar{h}_c \simeq 1.1D^{-1/2}$ W/m² K, for diameter D in meters, and by radiation with $q_{rad} = \sigma\varepsilon(T_s^4 - T_e^4)$. Determine the radius of the sheath that maximizes the heat loss when the resistor is at 400 K and the surroundings are at 300 K. Take the value of the surface emittance ε as

(i) 0.9.
(ii) 0.5.

2-33. A 2 mm-diameter electrical wire has a 1 mm-thick electrical insulation with a thermal conductivity of 0.12 W/m K. The combined convection and radiation heat transfer coefficient on the outside of the insulation is 12 W/m² K.

(i) Would increasing the thickness of the insulation to 3 mm increase or decrease the heat transfer?

(ii) Would the presence of a contact resistance between the wire and insulation of 5×10^{-4} [W/m^2 K]$^{-1}$ affect your conclusion?

2-34. An experimental boiling water reactor is spherical in shape and operates with a water temperature of 420 K. The shell is made from nickel alloy steel ($k = 21$ W/m K) and has an inside radius of 0.7 m with a wall thickness of 7 cm. The reactor is surrounded by a layer of concrete 20 cm thick. If the outside heat transfer coefficient is 8 W/m^2 K and the ambient air is at 300 K, what are the temperatures of the internal and external surfaces of the concrete? Also, if the reactor operates at a power level of 30 kW, what fraction of the power generated is lost by heat transfer through the shell? The resistance to heat flow from the water to the shell can be taken to be negligible.

2-35. (i) Derive an expression for the relation between heat loss and temperature difference across the inner and outer surfaces of a hollow sphere, the conductivity of which varies with temperature in manner given by $k = k_0[1 + a(T - T_0)]$, where T_0 is a reference temperature.

   (ii) Find the corresponding result for a hollow cylinder.

   (iii) Compare the expressions derived for parts (i) and (ii) for the special case of the outside radius becoming infinite. Explain the different values obtained.

2-36. A 5 cm-high stainless steel truncated cone has a base diameter of 10 cm and a top diameter of 5 cm. The sides are insulated, and the base and top temperatures are 100°C and 50°C, respectively Assuming one-dimensional heat flow, estimate the heat flow. Take $k = 15$ W/m K for the stainless steel.

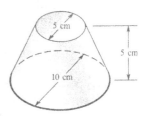

2-37. The concept of a critical radius for maximum heat loss developed for a cylinder in Section 2.3.2 also applies to a sphere. Derive expressions for critical radius for the following conditions:

   (i) The outside heat transfer coefficient has a constant value h_o.

   (ii) The outside heat transfer coefficient is proportional to $r_0^{-1/2}$.

2-38. A 1 m-diameter liquid oxygen (LOX) tank is insulated with a 10 cm-thick blanket of fiberglass insulation having a thermal conductivity of 0.022 W/m K. The tank is vented to the atmosphere. Determine the boil-off rate if the ambient air is at 310 K and the outside convective and radiative heat transfer coefficients are 3 W/m^2 K and 2 W/m^2 K, respectively. The boiling point of oxygen is 90 K, and its enthalpy of vaporization is 0.213×10^6 J/kg.

2-39. Consider steady conduction across a cylindrical or spherical shell. It is sometimes convenient to express the heat flow as

$$\dot{Q} = \frac{kA_{\text{eff}}(T_1 - T_2)}{r_2 - r_1}$$

that is, in the same form as for a plane slab where A_{eff} is an effective cross-sectional area.

   (i) Determine A_{eff} for a cylindrical shell.
   (ii) Repeat for a spherical shell.
   (iii) If the arithmetic mean area $A_m = \frac{1}{2}(A_1 + A_2)$ is used instead of A_{eff}, determine the error in heat flow for $r_2/r_1 = 1.5, 3$, and 5.

2-40. A spherical metal tank has a 2.5 m outside diameter and is insulated with a 0.5 m-thick cork layer. The tank contains liquefied gas at $-60°C$ and the ambient air is at 20°C. The inside heat transfer coefficient can be assumed to be large, and the combined convection and radiation outside heat transfer coefficient is estimated to be 8 W/m^2 K. Atmospheric water vapor diffuses into the cork, and a layer of ice forms adjacent to the tank wall. Determine the thickness of the layer. Assume that the cork thermal conductivity of 0.06 W/m K is unaffected by the ice and water, but comment on the validity of this assumption.

2-41. A thin-wall, spherical stainless steel vessel has an outside diameter of 40 cm and contains biochemical reactants maintained at 160°C. The tank is located in a laboratory where the air is maintained at 20°C. A layer of insulation is to be added to prevent workers from being burned by accidental contact with the vessel. The insulation chosen has a thermal conductivity of 0.1 W/m K. How thick should the layer be if the threshold for a skin burn on a nonmetallic surface can be taken as 55°C? Also calculate the heat loss. The combined convective and radiative heat transfer coefficients on the outside of the insulation is estimated to be 9 W/m^2 K.

2-42.   (i) Heat is generated uniformly in a plate 2L thick at a rate $\dot{Q}_v'''$ W/m^3. If the surfaces are maintained at temperature T_s, determine the temperature distribution across the plate.
   (ii) A 1 cm-thick stainless steel plate is heated by an electric current giving an I^2R internal heat generation of 1×10^6 W/m^3. The plate is cooled by an air stream at temperature $T_e = 300$ K, and the air velocity is adjusted to give a maximum plate temperature of 360 K. What is the average convective heat transfer coefficient?

2-43. An electrical current of 15 A flows in an 18 gage copper wire (1.02 mm diameter). If the wire has an electrical resistance of 0.0209 Ω/m, calculate

   (i) the rate of heat generation per meter length of wire.
   (ii) the rate of heat generation per unit volume of copper.
   (iii) the heat flux across the wire surface at steady state.

2-44. An electrical cable has a 2 mm-diameter copper wire encased in a 4 mm-thick insulator of conductivity 0.2 W/m K. The cable is located in still air at 25°C. and the convective heat transfer coefficient can be approximated as $h_c = 1.3(\Delta T/D)^{1/4}$ W/m^2 K. If the temperature limit for the insulator is 150°C, determine the maximum I^2R losses that can be allowed in the wire.

(i) Ignore thermal radiation.

(ii) Account for thermal radiation if the emittance of the insulator is 0.8.

2-45. Determine the allowable current in a 10 gage (2.59 mm diameter) copper wire that is insulated with a 1 cm-O.D. layer of rubber. The outside heat transfer coefficient is 20 W/m^2 K, and the ambient air is at 310 K. The allowable maximum temperature of the rubber is 380 K. Take $k = 0.15$ W/m K for the rubber and an electrical resistancethe rubber and an electrical resistance of 0.00328 Ω/m for the copper wire.

2-46. An explosive is to be stored in large slabs of thickness $2L$ clad on both sides with a protective sheath. The rate at which heat is generated within the explosive is temperature-dependent and can be approximated by the linear relation $\dot{Q}_v''' = a + b(T - T_e)$, where T_e is the prevailing ambient air temperature. If the overall heat transfer coefficient between the slab surface and the ambient air is U, show that the condition for an explosion is $L = (k/b)^{1/2} \tan^{-1}[U/(kb)^{1/2}]$. Determine the slab thickness if $k = 0.9$ W/m K, $U = 0.20$ W/m^2 K, $a = 60$ W/m^3, $b = 6.0$ W/m^3 K.

2-47. On the flight of Apollo 12, plutonium oxide ($Pu^{238}O_2^{16}$) was used to generate electrical power. Heat was generated uniformly through the loss of kinetic uniformly through the loss of kinetic energy from alpha particles emitted by the Pu^{238}. Consider a sphere of plutonium oxide of 3 cm diameter covered with thermo-electric elements for converting heat to electricity. The physical properties of these elements (tellurides) and heat rejection considerations suggest that the surface of the sphere be at 200°C. On the other hand, the ceramic nature of the plutonium oxide allows a maximum temperature of 1750°C. With these constraints, determine

(i) the maximum allowable volumetric heating rate.

(ii) the electrical power generated, assuming a thermal efficiency of 4%.

Take $k_{PuO_2} = 4$ W/m K.

2-48. In a laboratory experiment, a long, 2 cm-diameter, cylinder of fissionable material is encased in a 1 cm-thick graphite shell. The unit is immersed in a coolant at 330 K, and the convective heat transfer coefficient on the graphite surface is estimated to be 1000 W/m^2 K. If heat is generated uniformly within the fissionable material at a rate of 100 MW/m^3, determine the temperature at the center-line of the cylinder. Allow for an interfacial conductance between the material and the graphite shell of 3000 W/m^2 K. and take the thermal conductivities of the material and graphite as 4.1 W/m K and 50 W/m K, respectively.

2-49. A slab of semitransparent material of thickness L has one surface irradiated by a radiant energy flux G [W/m^2] from a high-temperature source. The rate of absorption of the radiation decays exponentially into the material: the resulting volumetric heating can be expressed as $\dot{Q}_v''' = \kappa G e^{-\kappa x}$ [W/m^3] where κ [m^{-1}] is the *absorption coefficient*. If the front and back surfaces are maintained at temperatures T_1 and T_2, respectively, with $T_1 > T_2$, derive an expression for the temperature profile $T(x)$. Also determine the location of the maximum temperature, and show how it depends on the problem parameters.

2-50. A slab of Li_2O ceramic of height 0.5 m and thickness 1 cm is part of a blanket of an experimental fusion reactor. The purpose of the blanket is to produce tritium through neutron interaction with lithium. The volumetric nuclear heating in the Li_2O can be assumed uniform at 10 MW/m^3. The slab has a 1 mm-thick cladding of 304 stainless steel, and is cooled by water at 30°C on each side, with a convective heat transfer coefficient of 290 W/m^2 K. Determine the maximum temperature in the ceramic if its conductivity is 3 W/m K.

2-51. Heat is generated at a rate $\dot{Q}_v'''$ in a large slab of thickness 2L. The side surface lose heat by convection to a liquid at temperature T_e. Obtain the steady-state temperature distributions for the following cases:

    (i) $\dot{Q}_v'''$ is constant.

    (ii) $\dot{Q}_v''' = \dot{Q}_{v0}'''[1 - (x/L)^2]$, with x measured from the centerplane.

    (iii) $\dot{Q}_v''' = a + b(T - T_e)$.

2-52. Heat is generated at a rate $\dot{Q}_v'''$ in a long solid cylinder of radius R. The cylinder has a thin metal sheath and is immersed in a liquid at temperature T_e. Heat transfer from the cylinder surface to the liquid can be characterized by an overall heat transfer coefficient U. Obtain the steady-state temperature distributions for the following cases:

    (i) $\dot{Q}_v'''$ is constant.

    (ii) $\dot{Q}_v''' = \dot{Q}_{v0}'''[1 - (r/R)^2]$.

    (iii) $\dot{Q}_v''' = a + b(T - T_e)$.

2-53. A 5 kW electric heater using Nichrome wire is to be designed to heat air to 400 K. The maximum allowable wire temperature is 1500 K, and a minimum heat transfer coefficient of 600 W/m^2 K is expected. A variable voltage power supply up to 130 V is available. Determine the length of 1.0 mm-diameter wire required. Also check the current and voltage. Take the electrical resistivity of Nichrome wire as 100 $\mu\Omega$ cm and its thermal conductivity as 30 W/m K.

2-54. An electric heater consists of a thin ribbon of metal and is used to boil a dielectric liquid. The liquid temperature T_e is uniform at its boiling point, and the heat transfer coefficient on the ribbon can be assumed to be uniform as well. A resistance measurement allows the average temperature of the ribbon $\overline{T}$ to be determined. Obtain an expression for $\overline{T} - T_e$ in terms of the ribbon dimensions (width W, thickness $2t$), the ribbon thermal conductivity k, the heat transfer coefficient h_c, and the ribbon electrical conductivity $\sigma[\Omega^{-1}\ m^{-1}]$.

2-55. Radioactive wastes are stored in a spherical type 316 stainless steel tank of inner diameter 1 m and 1 cm wall thickness. Heat is generated uniformly in the wastes at a rate of 3×10^4 W/m^3. The outer surface of the tank is cooled by air at 300 K with a heat transfer coefficient of 100 W/m^2 K. Determine the maximum temperature in the tank. Take the thermal conductivity of the wastes as 2.0 W/m K.

2-56. A 5 mm × 2 mm × 1 mm-thick semiconductor laser is mounted on a 1 cm cube copper heat sink and enclosed in a Dewar flask. The laser dissipates 2 W, and a cryogenic refrigeration system maintains the copper block at a nearly uniform temperature of 90 K. Estimate the top surface temperature of the laser chip for the following models of the dissipation process:

(i) The energy is dissipated in a 10 μm-thick layer underneath the top surface of the laser.

(ii) The energy is dissipated in a 10 μm-thick layer at the midplane of the chip.

(iii) The energy is dissipated uniformly through the chip.

Take $k = 170$ W/m K for the chip, and neglect parasitic heat gains from the Dewar flask.

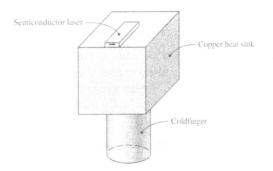

2-57. Heat is generated uniformly in a 8 cm-thick slab at a rate of 450 kW/m^3. One face of the slab is insulated and the other is cooled by water at 20°C, giving a heat transfer coefficient of 800 W/m^2 K. If the conductivity of the slab is 12.0 W/m K, determine the maximum temperature in the slab.

2-58. Show that the temperature distribution along an infinitely long pin fin is given by

$$\frac{T - T_e}{T_B - T_e} = e^{-\beta x}$$

Also find the heat dissipated by determining the base heat flow. Compare this result to Eq. (2.40) for an insulated tip and discuss.

2-59. Show that the temperature distribution along a short pin fin, for which heat loss from the tip cannot be neglected, is

$$\frac{T - T_e}{T_B - T_e} = \frac{\cosh \beta (L - x) + (h_c/\beta k) \sinh \beta (L - x)}{\cosh \beta L + (h_c/\beta k) \sinh \beta L}$$

where the heat transfer coefficient h_c is the same on the tip and sides.

2-60. A copper tube has a 2 cm inside diameter and a wall thickness of 1.5 mm. Over the tube is an aluminum sleeve of 1.5 mm thickness having 100 pin fins per centimeter length. The pin fins are 1.5 mm in diameter and are 4 cm long. The fluid inside the tube is at 100°C, and the inside heat transfer coefficient is 5000 W/m^2 K. The fluid outside the tube is at 250°C, and the heat transfer coefficient on the outer surface is 7 W/m^2 K. Calculate the heat transfer per meter length of tube. Take $k = 204$ W/m K for the aluminum.

2-61. A gas turbine rotor has 54 AISI 302 stainless steel blades of dimensions $L = 6$ cm, $A_c = 4 \times 10^{-4}$ m^2, and $\mathscr{P} = 0.1$ m. When the gas stream is at 900°C, the temperature at the root of the blades is measured to be 500°C. If the convective heat transfer coefficient is estimated to be 440 W/m^2 K, calculate the heat load on the rotor internal cooling system.

2-62. Aluminum alloy straight rectangular fins for cooling a semiconductor device are 1 cm long and 1 mm thick. Investigate the effect of choice of tip boundary condition on heat loss as a function of convective heat transfer coefficient. Use $k = 175$ W/m K for the alloy and a range of h_c values from 10 to 200 W/m^2 K.

2-63. Inconel-X-750 straight rectangular fins are to be used in an application where the fins are 2 mm thick and the convective heat transfer coefficient is 300 W/m^2 K. Investigate the effect of tip boundary condition on estimated heat loss for fin length L varying from 6 mm to 20 mm. Take $T_B = 800$ K, $T_e = 300$ K, $k = 18.8$ W/m K.

2-64. A straight rectangular fin has a constant incident radiation heat flux q_{rad} W/m^2 on one side from a high-temperature source and loses heat by convection from both sides. If $L = 10$ cm, half-thickness $t = 5$ mm, $k = 30$ W/m K, $q_{rad} = 30,000$ W/m^2, and $h_c = 100$ W/m^2 K, determine the base temperature to give tip temperature of 400 K when the ambient fluid is at 300 K.

2-65. (i) Repeat the pin fin analysis of Section 2.4.1 using the exponential rather than the hyperbolic function form of the solution to the governing differential equation.
(ii) As for case (i), but locate the origin for x at the fin tip.
(iii) As in the text using hyperbolic functions, but locate the origin for x at the fin tip.
Comment on the ease of solution using these different approaches.

2-66. In many situations the convective heat transfer coefficient on the tip of a fin is different from that on the sides, owing to a different flow geometry. Determine the temperature distribution, heat loss, and fin efficiency for such a fin. Denote the side and tip heat transfer coefficients as h_{cs} and h_{ct}, respectively.

2-67. Calculation of heat transfer coefficients on finned surfaces is difficult because the flow patterns are often complex and the resulting heat transfer coefficients are not constant (as assumed in conventional fin analyses). Experimentally determined effective average values of h_c are thus useful. A test rig maintains the fin base temperatures at 100°C in an air flow at 20°C. Thermocouples are installed to allow measurement of the fin tip temperatures. In a test the rectangular aluminum fins are 30 mm long and 0.3 mm thick, and are at a pitch of 3 mm. For a particular air velocity, the tip temperatures are measured to be 61.2°C. Assuming a constant heat transfer coefficient, determine its value. Take $k = 180$ W/m K for the aluminum.

2-68. A 16 mm-square chip has sixteen 2 mm-diameter, 15 mm-long aluminum pin fins in an aligned array at a pitch of 4 mm. A fan blows 25°C air through the array, giving a heat transfer coefficient of 110 W/m² K. If the chip is not to exceed a 75°C operating temperature, what is the allowable power rating of the chip?

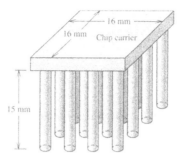

2-69. If the heat loss from the tip of a pin fin is not neglected, show that the fin efficiency is given by

$$\eta_f = \frac{\sinh \chi / \chi + \zeta \cosh \chi}{(1 + \zeta)(\cosh \chi + \zeta \chi \sinh \chi)}$$

where $\chi = \beta L = (h_c \mathscr{P}/k A_c)^{1/2} L$ and $\zeta = A_c / \mathscr{P} L$. By comparing the heat loss with that given by Eq. (2.40), show that tip loss is important only for small values of χ and is then of order ζ. Was this condition met in Example 2.5?

2-70. A heat sink assembly capable of mounting 36 power transistors may be idealized as a 15 cm cube containing four rows of 24 aluminum fins per row, each fin being

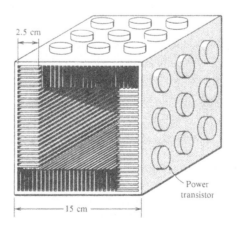

15 cm wide, 2.5 cm high, and 2 mm thick. A fan is an integral part of the assembly and blows air at a velocity that gives a heat transfer coefficient of 50 W/m^2 K. If the manufacturer's transistor temperature limit is 360 K, specify the allowable power dissipation per transistor. The mean air temperature is 310 K. If the rise in air temperature is limited to 10 K, specify the required capacity of the fan in m^3/min.

2-71. The dimensions of a straight rectangular fin can be optimized to give maximum heat transfer for a given mass. If the fin has thickness $2t$ and length L, the mass per unit width is ρA_p, where $A_p = 2tL$ is the profile area. For negligible tip heat loss, show that the heat flow $\dot{Q}$ is a maximum when

$$\tanh \chi = \frac{3\chi}{\cosh^2 \chi}$$

where $\chi = \beta A_p / 2t$ and $\beta^2 = h_c / kt$. Hence show that the optimal dimensions are

$$\frac{L}{t} = 1.419(k/h_c t)^{1/2}$$

Heat transfer coefficients of 150 W/m^2 K are typical for air-cooled reciprocating aircraft engines. What is the optimal length of 1 mm-thick rectangular fins if made from

(i) mild steel?
(ii) aluminum?

2-72. Some cooling fins lose heat predominantly by radiation. Show that the solution for the temperature distribution along a long pin fin, which loses heat by radiation only, can be given as an integral that can be evaluated numerically. Also find the heat loss from a pin fin of 1 cm diameter, 10 cm long, when the base temperature is 1000 K and the surroundings are black at 300 K. The thermal conductivity of the fin material is 10 W/m K. Assume a constant transfer factor $\mathscr{F}$ along the fin, with a value of 0.8.

2-73. A long gas turbine blade receives heat from combustion gases by convection and radiation. If emission from the blade can be neglected ($T_s \ll T_e$), determine the temperature distribution along the blade. Assume

(i) the blade tip is insulated.
(ii) the heat transfer coefficient on the tip equals that on the blade sides.

The cross-sectional area of the blade may be taken to be constant.

2-74. A 60 cm-long, 3 cm-diameter AISI 1010 steel rod is welded to a furnace wall and passes through 20 cm of insulation before emerging into the surrounding air. The furnace wall is at 300°C, and the air temperature is 20°C. Estimate the temperature of the bar tip if the

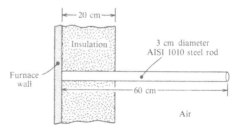

heat transfer coefficient between the rod and air is taken to be 13 W/m^2 K.

2-75. Jakob [8] suggests that Eq. (2.40) for the heat loss from a fin can be corrected to account for heat loss from the tip by adding to the length L the "tip size" $(A_c/\mathscr{P})$. Select a number of test cases and use FIN1 to check the validity of this rule.

2-76. A rectangular fin has a length of 2 cm, a width of 4 cm. and a thickness of 1 mm. It has a base temperature of 120°C and is exposed to air at 20°C with a convective heat transfer coefficient of 20 W/m^2 K. Determine the fin efficiency, heat loss and tip temperature for each of the three tip boundary conditions given by Eqs. (2.33a,b,c). Take the fin material as

  (i) aluminum, $k = 220$ W/m K.
  (ii) stainless steel, $k = 15$ W/m K.

2-77. A test technique for measuring the thermal conductivity of copper-nickel alloys is based on the measurement of the tip temperature of pin fins made from the alloys. The standard fin dimensions are a diameter of 5 mm and a length of 20 cm. The test fin and a reference brass fin ($k = 111$ W/m K) are mounted on a copper base plate in a wind tunnel. The test data include $T_B = 100°C$, $T_e = 20°C$, and tip temperatures of 64.2°C and 49.7°C for the brass and test alloy fins, respectively.

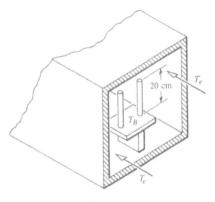

  (i) Determine the conductivity of the test alloy.
  (ii) If the conductivity should be known to ± 1.0 W/m K, how accurately should the tip temperatures be measured?

2-78. Exercise 2-71 requires an analytical proof that the dimensions of a straight rectangular fin that result in a maximum heat transfer for a given weight are $(L/t) = 1.419(k/h_c t)^{1/2}$. For an aluminum alloy fin ($k = 175$ W/m K) and a convective heat transfer coefficient of 200 W/m^2 K, use FIN1 to check this result using a 2 mm-thick fin as a base case.

2-79. The term *fin effectiveness* is used in two ways in the heat transfer literature. It can be a synonym for fin efficiency, and it may be defined as the ratio of the fin heat transfer rate to the rate that would exist without the fin, denoted ε_f. An AISI 302 stainless steel ($k = 15$ W/m K) straight rectangular fin is 1 cm wide, is 2 mm thick, and is cooled by an air flow giving a convective heat transfer coefficient on the sides and tip of 25 W/m^2 K. Calculate $\dot{Q}$ as a function of fin length, and hence prepare a graph or table of ε_f versus L. Comment on the significance of this result to the design of such fins.

2-80.

(i) A straight rectangular fin has a conductivity of 40 W/m K, and the heat transfer coefficient on its surface is 10 W/m² K. The fin is required to dissipate 100 W/m when the base and ambient temperatures differ by 40 K. Using the result of Exercise 2-71, determine the length and half-thickness of the fin of minimum mass.

(ii) B. X. Zhang and B.T.F. Chung [9] have demonstrated the use of fuzzy set theory for the optimization of a straight rectangular fin. The aim was to obtain a fin profile of a smaller aspect ratio ($L/2t$) than the optimal one, in order to simplify manufacture, allowing the specified heat load to be met within a specified tolerance. For the data in part (i), they obtain a reduction in aspect ratio from 20 to 8.5, but with a penalty of 33% increase in mass. FIN2 allows such studies to be performed by simple parameter variation. But decreasing L below the optimum value and adjusting t to give the desired heat load, determine a range of aspect ratios and corresponding fin masses that might be useful.

2-81. Electronic components are attached to a 10 cm-square, 2 mm-thick aluminum plate, and the backface is cooled by a flow of air. The backface has rectangular aluminum fins 25 mm long, 0.3 mm thick, at a pitch of 3 mm. If the cooling air is at 20°C and the heat transfer coefficient on the fins is 30 W/m² K, what is the allowable heat dissipation rate if the plate temperature should not exceed 70°C. Take $k = 180$ W/m K for the aluminum.

2-82. A 1 cm-wide, 2 cm-long semiconductor device dissipates 5 W and is mounted on a 2 cm-square, 2 mm-thick aluminum plate via a 0.1 mm-thick diamond wafer that acts as a heat spreader. The underside of the plate is fitted with ten 15 mm-long, 0.2 mm-thick rectangular aluminum fins at a pitch of 2 mm. Cooling air at 25°C flows through the fin array and gives a heat transfer coefficient of 28 W/m² K. Estimate the temperature of the base of the semiconductor. Take $k = 175$ W/m K for the aluminum.

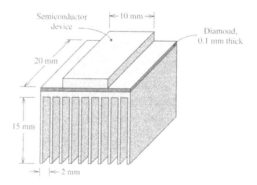

2-83. In testing a prototype unit, a finned surface is found not to perform according a design specifications. The fins are press-fitted onto rectangular ducts, and a contact resistance between the wall and the fin bases is suspected. The duct and fin are aluminum ($k = 185$ W/m K), and the fins have a rectangular profile 20 mm long and 0.3 mm thick, and are at a pitch of 3 mm. The duct wall is 1 mm thick. Thermocouples were then installed on the inside wall surface of the duct and on selected fin tips. In a particular test the following data apply: $T_{iw} = 100.0°C$, $T_{tip} = 41.6°C$, $T_e = 23.2°C$, and $h_c = 21$ W/m^2 K. Estimate the magnitude of the interfacial conductance (if significant).

2-84. A mercury-in-glass thermometer is to be used to measure the temperature of a hot gas flowing in a duct. To protect the thermometer, a pocket is made from a 7 mm-diameter, 0.7 mm-wall-thickness stainless steel tube, with one end sealed and the other welded to the duct wall. The small gap between the thermometer and the pocket wall is filled with oil to ensure good thermal contact and the thermometer bulb is in contact with the sealed end. The gas stream is at 320°C and the duct wall is at 240°C. How long should the pocket be for the error in the thermometer reading to be less than 2°C? Take the thermal conductivity of the stainless steel as 15 W/m K, and the convective heat transfer coefficient on the outside of the pocket as 30 W/m^{2}K.

2-85. A pool of liquid is heated by an immersed long, thin wire of length L and diameter d, through which an electrical current I is passed. The end of the wire at $x = 0$ is at temperature T_1 and at $x = L, T_2$. Assuming the convective heat transfer coefficient h_c to be constant along the wire, determine the x-location at which the maximum wire temperature occurs.

2-86. Consider a heat barrier consisting of a 2 mm-thick brass plate to which 3 mm copper tubing is soldered. The tubes are spaced 10 cm apart. Cooling water passed through the tubes keeps them at approximately 315 K. The underside of the brass wall is insulated with a 1.5 cm-thick asbestos layer, which in turn contacts a hot wall at 600 K.

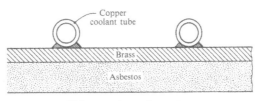

Assuming that the heat transfer from the cold side of the brass plate is negligible, estimate the temperature of the hottest spot on the brass wall. Take $k = 0.16$ W/m K for the asbestos and 111 W/m K for the brass.

2-87. A pressure transducer is connected to a high-temperature furnace by a copper tube "pigtail" of 3 mm outer diameter and 0.5 mm wall thickness. If the furnace operates at 1000 K and the transducer must not exceed 340 K, how long should the tube be? Take the

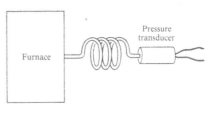

ambient temperature as 300 K and assume a heat transfer coefficient of 30 W/m² K.

2-88. An encapsulated semiconductor chip is connected by a brass lead 5 mm long and with a 1.25 × 0.25 mm cross section, to the end of a copper conductor on a circuit board, as shown in Fig. 2.17. The board is 10 cm wide and 1.5 mm thick and has a conductivity of 0.2 W/m K. The conductors are 2 mm wide and 0.75 mm thick and are spaced at 12 mm intervals along the board. If 60% of the power generated by the chip is to be dissipated by the board, what is the allowable rating for the chip if its temperature the chip if its temperature is not to exceed 350 K? The average cooling air temperature is 310 K, and the convective heat transfer coefficient between the air and the board is estimated to be 5 W/m² K. Take $k = 386$ W/m K for the copper.

2-89. A copper-constantan (45% Ni) thermocouple is constructed from 24 gage (0.510 mm diameter) wire and protrudes into a steam chamber. The steam is at 320 K, and the chamber wall is at 300 K. The wires are bare and well separated.

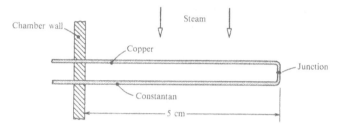

If the length of protrusion is 5 cm, calculate the error due to conduction along the wires for a heat transfer coefficient of 100 W/m² K.

2-90. An electrical current is passed through a horizontal copper rod 2 mm in diameter and 30 cm long, located in an air stream at 20°C. If the ends of the rod are also maintained at 20°C and the convective heat transfer coefficient is estimated to be 30 W/m² K, determine the maximum current that can be passed if the midpoint temperature is not to exceed 50°C.

(i) Ignore thermal radiation.
(ii) Include the effect of thermal radiation.

For the copper rod, take $k = 386$ W/m K, $\varepsilon = 0.8$, and electrical resistivity of 1.72×10^{-8} Ω m.

2-91. A stainless steel tube of 1 cm outer diameter and 1 mm wall thickness receives a uniform radiative heat flux of 100 W/cm² from a high-temperature plasma over 180° of its outer circumference; the remaining 180° is insulated. Water at 300 K flows through the tube, and the inside heat transfer coefficient is 1000 W/m² K.

(i) Determine the wall temperature distribution around the tube.

(ii) If there is also a volumetric heat source of 50 W/cm³ in the tube wall due to neutron absorption, find the new temperature distribution.

Take $k = 20.0$ W/m K for the stainless steel.

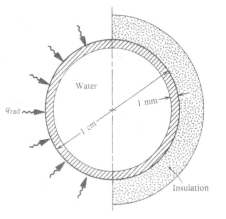

2-92. The absorber of a simple flat-plate solar collector with no coverplate consists of a 2 mm-thick aluminum plate with 6 mm-diameter aluminum water tubes spaced at a pitch of 10 cm, as shown.

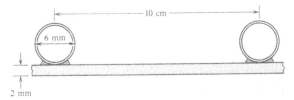

On a clear summer day near the ocean, the air temperature is 20°C, and a steady wind is blowing. The solar radiation absorbed by the plate is calculated to be 680 W/m², and the convective heat transfer coefficient is estimated to be 14 W/m² K. If water at 20°C enters the collector at 7×10^{-3} kg/s per meter width of collector, and the collector is 3 m long, estimate the outlet water temperature. For the aluminum take $k = 200$ W/m K and $\varepsilon = 0.20$. (*Hint:* Evaluate the heat lost by reradiation using Eq. (1.19) with a constant value of h_r corresponding to a guessed average plate temperature. Then apply the steady-flow equation to an elemental length of the collector, and so derive a differential equation governing the water temperature increase along the collector.)

2-93. The end of a soldering iron consists of a 4 mm-diameter copper rod, 5 cm long. If the tip must operate at 350°C when the ambient air temperature is 20°C, determine the base temperature and heat flow. The heat transfer coefficient from the rod to the air is estimated to be about 10 W/m² K. Take $k = 386$ W/m K for the copper.

2-94. A skin panel for an actively cooled hypersonic aircraft has square passages through which supercritical hydrogen flows before being used as fuel in a scramjet engine. Possible dimensions are shown for a panel made from Inconel-X-750 nickel alloy. If the heating load is 100 kW/m^2 and the inside convective heat transfer coefficient is 6000 W/m^2 K, estimate the maximum skin temperature at a location along the panel where the bulk coolant temperature is 100 K.

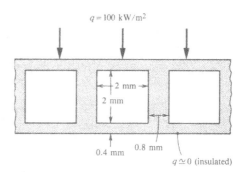

2-95. An exhaust stack thermocouple is inserted into a 20 cm-long well made of 5 mm-O.D., 0.7 mm-wall-thickness AISI 316 stainless steel tube and is held in good thermal contact with the sealed end by a spring-loaded plug. The convective heat transfer coefficient for the exhaust gases flowing across the tubing is estimated to be 65 W/m^2 K. If the thermocouple reading is 221°C when the stack walls are at 178°C, determine the gas temperature.

2-96. A 2 cm-O.D. stainless steel tube with a 1 mm wall thickness receives a radiative heat flux from a high-temperature gas distributed as $q = q_0 \cos \phi$ over 180° of its circumference, as shown. The remaining 180° is insulated. Water at 320 K flows inside the tube, and the inside heat transfer coefficient is 800 W/m^2 K. Determine the wall temperature distribution around the tube for $q_0 = 105$ W/m^2. Take $k = 18.0$ W/m K for the stainless steel.

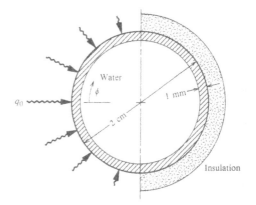

2-97. The convective heat transfer coefficient around a cylinder held perpendicular to a flow varies in a complicated manner. A test cylinder to investigate this behavior consists of a 0.001 in-thick, 12.7 mm-wide stainless steel heater ribbon (cut from shim stock) wound around a 2 cm-O.D., 2 mm-wall-thickness Teflon tube. A single thermocouple is located just underneath the ribbon and measures the local ribbon temperature $T_s(\theta)$.

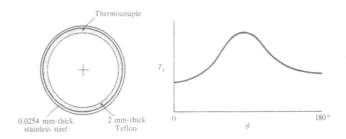

The cylinder is installed in a wind tunnel, and a second thermocouple is used to measure the ambient air temperature T_e. The power input to the heater is metered, from which the electrical heat generation per unit area $\dot{Q}/A$ can be calculated (A is the surface area of one side of the ribbon).

As a first approximation, the local heat transfer coefficient $h_c(\theta)$ can be obtained from

$$h_c(\theta) = \frac{\dot{Q}/A}{T_s(\theta) - T_e}$$

Hence, by rotating the cylinder with the power held constant, the variation of h_c can be obtained from the variation of T_s: where $T_s(\theta)$ is low, $h_c(\theta)$ is high, and vice versa. A typical variation of $T_s(\theta)$ is shown in the graph. A problem with this technique is that conduction around the circumference of the tube causes the local heat flux $q_s(\theta)$ to not exactly equal $\dot{Q}/A$.

(i) Derive a formula for $h_c(\theta)$ that approximately accounts for circumferential conduction.

(ii) The following table gives values of $T_s(\theta)$ in a sector where circumferential conduction effects are expected to be large. Use these values together with $\dot{Q}/A = 5900$ W/m^2 and $T_e = 25°$C to estimate the conduction effect at $\theta = 110°$.

Angle (degrees)	T_s (°C)
100	65.9
110	65.7
120	64.4

(iii) Comment on the design of the cylinder. Would a 3 mm-thick brass tube, directly heated by an electric current, be a suitable alternative?

Use $k = 15$ W/m K for the stainless steel and 0.38 W/m K for Teflon.

2-98. A 4 mm-diameter, 25 cm-long aluminum alloy rod has an electric heater wound over the central 5 cm length. The outside of the heater is well insulated. The two 10 cm-long exposed portions of rod are cooled by an air stream at 300 K giving an average convective heat transfer coefficient of 50 W/m^2 K. If the power input to the heater is 10 W, determine the temperature at the ends of the rod. Take $k = 190$ W/m K for the aluminum alloy.

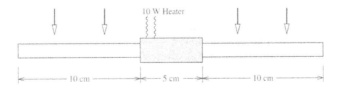

2-99. An electrical current is passed through a 1 mm–diameter, 20 cm–long copper wire located in an air flow at 290 K. If the ends of the wire are maintained at 300 K, determine the maximum current that can be passed if the midpoint temperature is not to exceed 400 K. The convective heat transfer coefficient is estimated to be 20 W/m^2 K. For the copper wire, take $k = 386$ W/m K, $\varepsilon = 0.8$. and an electrical resistance of $2.2 \times 10^{-2} \Omega$/m.

2-100.   A 3 cm–O.D., 1.25 mm–wall-thickness copper tube spans a 20 cm–wide wind tunnel. The 10 cm-long midsection is heated by a heater tape attached to the inner surface of the tube. A test is performed with 25 W power input and an airflow at 300 K, giving an average heat transfer coefficient of 145 W/m^2 K on the outside surface of the cylinder.

(i) Use a fin-type analysis to find the temperature distribution along the tube.
(ii) Determine the temperature difference between the midplane and the end of the heated section.

Take $k = 401$ W/m K for the copper. Assume a negligible heat loss out of the ends of the tube.

2-101.   The absorber of a flat-plate solar collector with no coverplate consists of a 4 mm–thick aluminum plate with 8 mm–O.D. aluminum water tubes spaced at a pitch of 12 cm, as shown. During a test the air temperature is 22°C, and a steady wind gives an estimated convective heat transfer coefficient of 15 W/m^2 K. The solar radiation absorbed by the plate is estimated to be 750 W/m^2. At a location along the collector where the water is 40°C, calculate the temperature midway between two tubes, and the rate of heat transfer to the water (per meter). Assume an emittance of 0.2 in order to calculate the radiation emitted by the plate. Take $k = 200$ W/m K for the aluminum.

2-102.   Two air flows are separated by a 2 mm-thick plastic wall. A 20.2 cm-long, 2 cm-diameter aluminum rod transfers heat from one flow to the other as shown. The hot air flow is at 70°C, and the convective heat transfer coefficient to the rod is 48 W/m^2 K; the cold air flow is at 20°C and is at a lower velocity, giving a heat transfer coefficient of only 24 W/m^2 K. Determine the rate of heat transfer and the temperature of the midsection of the rod. Take $k = 190$ W/m K for the aluminum.

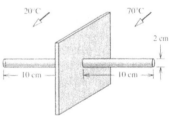

2-103.   A space radiator is made of 0.3 mm-thick aluminum plate with heatpipes at a pitch of 8 cm. The heatpipes reject heat at 330 K. The back of the radiator is insulated, and the front sees outer space at 0 K. If the aluminum surface is hard-anodized to give an emittance of 0.8, determine the fin effectiveness of the radiator and the rate of heat rejection per unit area. Take $k = 200$ W/m K for the aluminum. (*Hint:* We do not expect the plate temperature to vary more than a few kelvins: assume q_{rad} is constant at an average value to obtain an approximate analytical solution.)

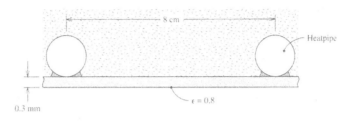

2-104. It is proposed to redesign a cast-iron channel to have fins in the streamwise direction of approximately triangular cross section, with height, base width, and pitch all of 2 cm. Determine the effect of adding the fins on the surface thermal resistance if the heat transfer coefficient on both the finned and unfinned surfaces is

(i) 1000 W/m^2 K.
(ii) 8000 W/m^2 K.

2-105. Referring to Example 2.9, show that straight rectangular and triangular fins, with the same base thickness and mass per unit width as the parabolic fin, dissipate less heat.

2-106. A straight Duralumin fin has a parabolic profile $y = t(1 - x/L)^2$ with $t = 3$ mm and $L = 10$ mm. Determine the heat dissipation by the fin when the base temperature is 400 K and it is exposed to fluid at 300 K with a heat transfer coefficient of 40 W/m^2 K. Also calculate the fin mass.

2-107. (i) An aluminum pin fin has a diameter of 4 mm and a length of 20 mm. Calculate the heat dissipated when the base temperature is 600 K, the fluid temperature is 400 K, and the heat transfer coefficient is 100 W/m^2 K. Take $k = 180$ W/m K.
(ii) Compare the heat dissipation obtained above with that which would be obtained with parabolic and triangular spines (items 8, 9, 10 of Table 2.2) of the same base area and equal mass.

2-108. A perforated-plate heat exchanger has a cross section as shown, with $D_1 = 5.00$ cm, $D_2 = 5.30$ cm, and D_3 chosen to give equal flow areas for each stream. The plates are 1 mm-thick aluminum with 1.5 mm-diameter holes taking up 30% of the plate area. If the convective heat transfer coefficient is 300 W/m^2 K, determine the overall heat transfer coefficient. Take $k = 190$ W/m K for the aluminum.

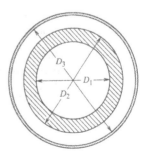

2-109. A steel heat exchanger tube of 2 cm outer diameter is fitted with a steel spiral annular fin of 5 cm outer diameter and 0.2 mm thickness, wound at a pitch of 3 mm. If the outside heat transfer coefficient is 20 W/m^2 K on the bare tube and 15 W/m^2 K on the finned tube, determine the reduction in the outside thermal resistance achieved by adding the fins. Take $k_{steel} = 42$ W/m K.

2-110. A thin metal disk is insulated on one side and exposed to a jet of hot air at temperature T_e on the other. The periphery at $r = R$ is maintained at a uniform temperature T_R. If the convective heat transfer coefficient h_c can be taken to be constant over the disk, obtain an expression for the temperature at the center of the disk.

2-111. Referring to comment 2 of Example 2.7, obtain the temperature distribution along a fin for which the temperature difference $[T(x) - T_e]$ is constant. Assume the tip is insulated. If the fin efficiency is defined as $\eta_f = \dot{Q}/h_c \mathscr{P} L (T_B - \overline{T}_e)$. show that $\eta_f = 1/[(\chi^2/3) + 1]$. Calculate the value of η_f appropriate to Example 2.7.

2-112. A transistor has a cylindrical cap of radius R and height L.

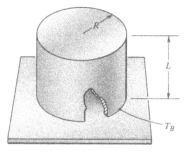

(i) Show that the heat dissipated is

$$\dot{Q} = 2\pi k R t (T_B - T_e)\beta \left[\frac{I_0(\beta R)\sinh\beta L + I_1(\beta R)\cosh\beta L}{I_0(\beta R)\cosh\beta L + I_1(\beta R)\sinh\beta L} \right]$$

where the metal thickness is t, $\beta = (h_c/kt)^{1/2}$, and the heat transfer coefficient on the sides and top is assumed to be the same.

(ii) The cap is fabricated from steel ($k = 50$ W/m K) of thickness 0.3 mm and has a height of 9 mm and a diameter of 8 mm. Air at 310 K blows over the cap to give a heat transfer coefficient of 25 W/m^2 K. What is the base temperature for 400 mW dissipation? Compare your answer to the manufacturer's allowable limit of 370 K.

2-113. The straight fin with a parabolic profile $y = t(1 - x/L)^2$ is of particular interest since it can give the maximum heat loss for a given weight of any profile.

(i) By substituting appropriate values of A_c and $\mathscr{P}$ in Eq. (2.30), show that the governing differential equation is

$$z^2\frac{d^2T}{dz^2} + 2z\frac{dT}{dz} - \beta^2 L^2(T - T_e) = 0; \quad \beta = (h_c/kt)^{1/2}$$

where $z = L - x$. This is an *Euler* equation.

(ii) Show that a solution of the differential equation that satisfies the boundary conditions $T = T_B$ at $x = 0$ and $T = T_e$ at $x = L$ is

$$\frac{T - T_e}{T_B - T_e} = \left(1 - \frac{x}{L}\right)^p; \quad p = -\frac{1}{2} + \frac{1}{2}(1 + 4\beta^2 L^2)^{1/2}$$

(iii) If $p = 1$, the temperature profile is linear. Since q_x is then constant, this fin proves to be the fin of least material and hence gives the maximum heat dissipation for a given weight. Show that the efficiency of such a fin is 50% (see Example 2.9).

2-114. An annular fin of uniform thickness has an inner radius of 2 cm, an outer radius of 4 cm, and a thickness of 2 mm. The material is steel with $k = 60$ W/m K. It is cooled by air at 20°C, giving a convective heat transfer coefficient of 24 W/m² K. When the fin base is at 110°C, determine the rate at which heat is dissipated by the fin.

2-115. A wall has its surface maintained at 180°C and is in contact with a fluid at 80°C. Find the percent increase in the heat dissipation if triangular fins are added to the surface. The fins are 6 mm thick at the base, are 30 mm long, and are spaced at a pitch of 15 mm. Assume that the heat transfer coefficient is 20 W/m² K for both the plain and finned surfaces, and that the fin material thermal conductivity is 50 W/m K.

2-116. A thin metal disk is insulated on one side, and the other side is exposed to a high-temperature radiation source and convective cooling. Obtain an expression for the difference between the temperature at the center and at the outer edge. Ignore reradiation.

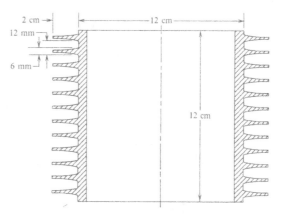

2-117. Cylinders of air-cooled internal combustion engines are provided with cooling fins owing to the large amount of heat that must be dissipated. A two-stroke motorcycle engine has a cast aluminum alloy 195 cylinder of height 12 cm and outside diameter 12 cm, with hyperbolic fins of base width 6 mm, pitch 12 mm, and length 20 mm. In a test simulating a road speed of 90 km/h, the fin base temperatures are measured to average 485 K for ambient air at 300 K. If the heat transfer coefficient is estimated to be 60 W/m^2 K, determine the heat loss from the cylinder. Under these conditions, how does the efficiency and mass of the hyperbolic fin compare with the rectangular fin of the same base and length?

2-118. So-called "compact" heat exchanger cores often consist of finned passages between parallel plates. A particularly simple configuration has square passages with the effective fin length equal to half the plate spacing L. In a particular application with $L = 5$ mm, a convective heat transfer coefficient of 160 W/m^2 K is expected. If 95% efficient fins are desired, how thick should they be if the core is constructed from

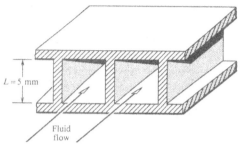

(i) an aluminum alloy with $k = 180$ W/m K?
(ii) mild steel with $k = 64$ W/m K?
(iii) a plastic with $k = 0.33$ W/m K?

Discuss the significance of your results to the design of such cores.

2-119. A steel heat exchanger tube of 2 cm outer diameter is fitted with a steel spiral annular fin of 4 cm outer diameter, thickness 0.4 mm, and wound at a pitch of 3 mm. If the outside heat transfer coefficient is 20 W/m^2 K on the original bare tube and 15 W/m^2 K on the finned tube, determine the reduction in the outside thermal resistance achieved by adding the fins. Take $k_{steel} = 42$ W/m K.

2-120. The fin efficiency concept is not useful when optimal proportions of fins are sought that maximize the heat loss for a given mass. Shamsundar [10] suggests that the fin heat loss be nondimensionalized as

$$\dot{Q}^+ = \frac{\dot{Q}}{2kW(T_B - T_e)[A_p/2(k/h_c)^2]^{1/3}}$$

where $A_p = 2CtL$ is the profile area with $C = 1, 1/2$, and $1/3$ for rectangular, triangular, and parabolic profiles, respectively. For a given mass, A_p is constant, and maximizing $\dot{Q}^+$ gives the corresponding maximum heat loss.

(i) Show that the heat loss from a straight rectangular fin with an insulated tip is

$$\dot{Q}^+ = \frac{\tanh \chi}{\chi^{1/3}}$$

where $\chi = \beta L$ for $\beta = (h_c/kt)^{1/2}$. Plot $\dot{Q}^+$ versus χ for $0 < \chi < 5$. Show that $\dot{Q}^+$ has a maximum value of 0.791 at $\chi = 1.419$, for which the corresponding fin efficiency is $\eta_f = 0.627$.

(ii) Using Table 2.2, item 3, show that for a straight triangular fin,

$$\dot{Q}^+ = \left(\frac{2}{\chi}\right)^{1/3} \frac{I_1(2\chi)}{I_0(2\chi)}$$

Also plot $\dot{Q}^+$ versus χ, and show that $\dot{Q}^+$ has a maximum value of 0.895 at $\chi = 1.309$, for which $\eta_f = 0.594$.

(iii) Using Table 2.2, item 4, show that for a straight parabolic fin,

$$\dot{Q}^+ = \frac{2(3\chi^2)^{1/3}}{1 + (1 + 4\chi^2)^{1/2}}$$

Also plot $\dot{Q}^+$ versus χ, and show that $\dot{Q}^+$ has a maximum value of 0.909 at $\chi = 1.414$, for which $\eta_f = 0.5$.

(iv) Compare these results for rectangular, triangular, and parabolic fins.

2-121. For a given base thickness and weight, will a triangular fin always give a higher heat flow than a rectangular fin? To explore this question consider an aluminum fin with $k = 180$ W/m K, $\rho = 2770$ kg/m^3, a base thickness of 3 mm, and a heat transfer coefficient of 50 W/m^2 K. Consider fin lengths up to 25 cm, and use FIN2 to produce a graph of heat loss versus fin mass for both profiles. Discuss the implications of this result for the design of finned surfaces. Plot numerical values for unit temperature difference and unit fin width.

2-122. An air-cooled R-22 condenser has 10 mm-O.D., 9 mm-I.D. aluminum tubes with 50 mm-O.D. annular rectangular aluminum fins of thickness 0.2 mm at a pitch of 2 mm. The heat transfer coefficient inside the tubes is 800 W/m^2 K, and on the fins it is 20 W/m^2 K. If the R-22 is at 320 K and the air is at 300 K, calculate the heat transfer per unit length of tube. Take $k = 180$ W/m K for the aluminum.

2-123. The fin model of Section 2.4 assumes a uniform temperature across the fin and that the base temperature is equal to the wall temperature. An investigation of the validity of these assumptions is planned using numerical simulation computer software (see Section 3.5). A straight rectangular fin of half thickness $t = 1$ mm and length $L = 10$ mm is of interest. However, to avoid some numerical error problems it is advisable to numerically simulate a larger fin of length, say, 10 m. The fin conductivity is $k = 100$ W/m K, and the heat transfer coefficients of interest are 40, 1000, and 25,000 W/m^2 K. If the same conductivity is used in the computer program, what values of t and h_c should be used to obtain a correct simulation of the 10 mm-long fin? For the tip boundary condition, use Eq. (2.33a), that is, a convective heat loss.

2-124. A 2 cm-long aluminum alloy pin fin has a reduction in diameter from 2 mm to 1 mm at a location 8 mm from its base. The fin is attached to a wall at 200°C and is exposed to a gas flow at 20°C, giving heat transfer coefficients of 42 and 60 W/m² K on the larger and smaller diameter portions, respectively. Calculate the heat loss from the fin and the temperatures at the contraction and tip. Take $k = 185$ W/m K for the alloy.

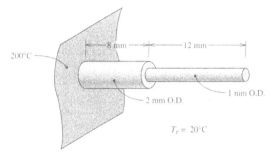

2-125.   Optimization of finned surfaces is usually a rather complicated task, with some subtle issues involved. Exercise 2-121 is a simple demonstration that for a given weight a triangular fin will give a higher heat flow than a rectangular fin. For this purpose an aluminum fin with a 3 mm base thickness, $k = 180$ W/m K, and $h_c = 50$ W/m² K was considered. A rectangular fin 6 cm long gives a heat flow of 4.947 W for W = 1 m, $T_B - T_e = 1$ K, and has a mass of 0.499 kg/m. A triangular fin that gives the same heat flow proves to have a mass of 0.288 kg/m, i.e., 57.7% of the mass of the rectangular fin. Although this is a useful result, we should look deeper into the problem and recognize that neither of these fins has been optimized to give a minimum mass for the desired heat flow. Exercise 2-120 gives formulas for optimal dimensions of rectangular and triangular fins. Use these results to determine optimal rectangular and triangular fins that dissipate 4.947 W, and discuss.

2-126.   Low-weight, high-performance perforated-plate heat exchangers are required for space vehicle cryogenic refrigeration systems. Current technology uses copper or aluminum plates with hole diameters of 1-2 mm (see Example 2.7). Smaller hole sizes give a larger heat transfer area, and new manufacturing methods developed for MEMS application (microelectromechanical systems) allow hole diameters down to 10 μm to be used. However, performance is then limited by the fin efficiency of the plates, and hence by the conductivity of the plate material. Use of single crystal silicon becomes attractive owing to its high thermal conductivity and low density (compared with copper). Consider a redesign of the unit described in Example 2.7. Holes of 90 μm diameter are to be arranged in a square array of pitch 130 μW. The convective heat transfer coefficient is 800 W/m² K at the nominal cold side operating temperature of 50 K.

(i) Calculate the fin efficiency of a copper plate ($k = 890$ W/m K at 50 K).
(ii) Repeat for single crystal silicon ($k = 2500$ W/m K at 50 K).

Also compare the UA products with the value obtained in Example 2.7.

MULTIDIMENSIONAL AND UNSTEADY CONDUCTION

CONTENTS

3.1 INTRODUCTION

The analyses of steady one-dimensional heat conduction in Chapter 2 were relatively simple but nevertheless gave results that are widely used by engineers for the design of thermal systems. In general, however, heat conduction can be **unsteady**; that is, temperatures change with time. An example is heat flow through the cylinder wall of an automobile engine. Also, heat conduction can be **multidimensional**; that is, temperatures vary significantly in more than one coordinate direction. An example is heat loss from a hot oil line buried underneath the ground. Heat conduction can also be simultaneously unsteady and multidimensional, for example, when a rectangular block forging is quenched.

In Chapter 2, each new analysis commenced with the application of the first law to an elemental volume to yield the governing differential equation. In Chapter 3, our approach will be different. We will derive a partial differential equation that governs the temperature distribution in a solid under very general conditions; we will then start each analysis by choosing the form of this equation appropriate to the problem under consideration. This **general heat conduction equation** is derived in Section 3.2, where boundary and initial conditions as well as solution methods are discussed. Solution methods are broadly divided into two groups: (1) classical mathematical methods, and (2) numerical methods. Classical mathematical methods are demonstrated in Section 3.3 for multidimensional steady conduction, and in Section 3.4 for unsteady conduction. In particular, the method of separation of variables is used, which leads to the need to construct Fourier series expansions. Results of the analyses are presented in the form of formulas and charts that find extensive use in engineering practice. Numerical methods commonly used to solve the heat conduction equation include the finite-difference method, the finite-volume method, and the finite-element method. In Section 3.5, use of the finite-difference method is demonstrated for steady two-dimensional conduction and unsteady one-dimensional conduction.

The classical mathematical methods used to solve the heat conduction equation might at first appear intimidating to the student. However, these methods rely on concepts normally studied in freshman- and sophomore-level mathematics courses, such as partial differentiation, integration, and second-order ordinary differential equations. Sufficient detail is given in the analyses for the student to proceed step by step without having to refer to a text on advanced engineering mathematics. Those students who have already had a junior- or senior-level engineering mathematics course should find the mathematics straightforward (and even perhaps old-fashioned!).

3.2 THE HEAT CONDUCTION EQUATION

In this section, the energy conservation principle and Fourier's law of heat conduction are used to derive various forms of the differential equation governing the temperature distribution in a stationary medium. The types of boundary and initial conditions encountered in practical problems are then discussed and classified. Finally, various methods available for solving the equation are introduced.

3.2.1 Fourier's Law as a Vector Equation

Chapter 2 used one-dimensional forms of Fourier's law of conduction. In general, the temperature in a body may vary in all three coordinate directions, which requires a more general form of Fourier's law. For simplicity, we will restrict our attention to *isotropic* media, for which the conductivity is the same in all directions. Most materials are isotropic. Exceptions include timber, which has different values of thermal conductivity k along and perpendicular to the grain, and pyrolytic graphite, for which the value of k can vary by an order of magnitude in different directions. For an isotropic medium, Fourier's law in terms of Cartesian coordinates is

$$q_x = -k\frac{\partial T}{\partial x}; \quad q_y = -k\frac{\partial T}{\partial y}; \quad q_z = -k\frac{\partial T}{\partial z} \tag{3.1}$$

where q_x is the component of the heat flux in the x direction, $\partial T/\partial x$ is the *partial derivative* of $T(x,y,z,t)$ with respect to x, and so on. As indicated in Fig. 3.1, Eq. (3.1) can be written more compactly in vector form as

$$\mathbf{q} = -k\nabla T \tag{3.2}$$

where $\mathbf{q}$ is the conduction heat flux vector, and ∇T is the gradient of the scalar temperature field. In Cartesian coordinates,

$$\mathbf{q} = \mathbf{i}q_x + \mathbf{j}q_y + \mathbf{k}q_z$$

$$\nabla T = \mathbf{i}\frac{\partial T}{\partial x} + \mathbf{j}\frac{\partial T}{\partial y} + \mathbf{k}\frac{\partial T}{\partial z}$$

where $\mathbf{i}$, $\mathbf{j}$, and $\mathbf{k}$ are the unit vectors in the x, y, and z directions, respectively.

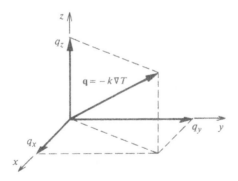

Figure 3.1 Vector representation of Fourier's law of heat conduction.

3.2.2 Derivation of the Heat Conduction Equation

The general heat conduction equation will first be derived in a Cartesian coordinate system. Subsequently, the result will be written in cylindrical and spherical coordinates.

Derivation in Cartesian Coordinates

Figure 3.2 depicts an elemental volume Δx by Δy by Δz located in a solid. The energy conservation principle, Eq. (1.2), applied to the elemental volume as a closed system gives

$$\rho(\Delta x \Delta y \Delta z)c\frac{\partial T}{\partial t} = \dot{Q} + \dot{Q}_v \tag{3.3}$$

where the time derivative is a partial derivative, since T is also a function of the spatial coordinates x, y, and z.

The term $\dot{Q}$ represents heat transfer across the volume boundaries by conduction. The rate of heat inflow across the face at x is

$$q_x|_x \Delta y \Delta z$$

and the outflow across the face at $x + \Delta x$ is

$$q_x|_{x+\Delta x} \Delta y \Delta z$$

The net inflow in the x direction is then

$$(q_x|_x - q_x|_{x+\Delta x})\Delta y \Delta z$$

The outflow heat flux can be expanded in a Taylor series as

$$q_x|_{x+\Delta x} = q_x|_x + \frac{\partial q_x}{\partial x}\Delta x + \text{Higher-order terms}$$

Substituting and dropping the higher-order terms gives the net inflow in the x direction as

$$-\frac{\partial q_x}{\partial x}\Delta x \Delta y \Delta z$$

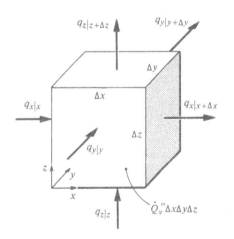

Figure 3.2   Three-dimensional Cartesian elemental volume in a solid for derivation of the heat conduction equation.

Similar terms arise from conduction in the y and z directions. Thus, the net heat transfer into the volume by conduction is

$$\left(-\frac{\partial q_x}{\partial x} - \frac{\partial q_y}{\partial y} - \frac{\partial q_z}{\partial z} \right) \Delta x \Delta y \Delta z$$

The rate of generation of thermal energy within the volume $\dot{Q}_v$ is simply

$$\dot{Q}_v''' \Delta x \Delta y \Delta z$$

where $\dot{Q}_v'''$ [W/m^3] is the rate of internal or volumetric heat generation introduced in Section 2.3.4. Substituting in Eq. (3.3) and dividing by $\Delta x \Delta y \Delta z$ gives

$$\rho c \frac{\partial T}{\partial t} = -\left(\frac{\partial q_x}{\partial x} + \frac{\partial q_y}{\partial y} + \frac{\partial q_z}{\partial z} \right) + \dot{Q}_v'''$$

Introducing Fourier's law, Eq. (3.1), for q_x, q_y, and q_z,

$$\rho c \frac{\partial T}{\partial t} = \frac{\partial}{\partial x}\left(k\frac{\partial T}{\partial x} \right) + \frac{\partial}{\partial y}\left(k\frac{\partial T}{\partial y} \right) + \frac{\partial}{\partial z}\left(k\frac{\partial T}{\partial z} \right) + \dot{Q}_v''' \tag{3.4}$$

Notice that the thermal conductivity k has been left inside the derivatives since, in general, k is a function of temperature. However, we usually simplify heat conduction analysis by taking k to be independent of temperature; k is then also independent of position, and Eq. (3.4) becomes

$$\rho c \frac{\partial T}{\partial t} = k\left(\frac{\partial^2 T}{\partial x^2} + \frac{\partial^2 T}{\partial y^2} + \frac{\partial^2 T}{\partial z^2} \right) + \dot{Q}_v''' \tag{3.5}$$

When there is no internal heat generation, $\dot{Q}_v''' = 0$, and Eq. (3.5) reduces to

$$\frac{\partial T}{\partial t} = \alpha\left(\frac{\partial^2 T}{\partial x^2} + \frac{\partial^2 T}{\partial y^2} + \frac{\partial^2 T}{\partial z^2} \right) \tag{3.6}$$

where $\alpha = k/\rho c$ [m^2/s] is a thermophysical property of the material called the **thermal diffusivity**. Table 3.1 gives selected values of the thermal diffusivity. Additional data are given in Appendix A, as are values for k, ρ, and c, from which α can be calculated. Equation (3.6) is called **Fourier's equation** (or the *heat* or *diffusion equation*) and governs the temperature distribution $T(x,y,z,t)$ in a solid. The relevance of the thermal diffusivity can be seen in Fourier's equation: when there is no internal heat generation, it is the only physical property that influences temperature changes in the solid. The thermal diffusivity is the ratio of thermal conductivity to a volumetric heat capacity: the larger α, the faster temperature changes will propagate through the solid.

For a timewise steady state, $\partial/\partial t = 0$, Eq. (3.5) reduces to *Poisson's equation:*

$$\frac{\partial^2 T}{\partial x^2} + \frac{\partial^2 T}{\partial y^2} + \frac{\partial^2 T}{\partial z^2} = -\frac{\dot{Q}_v'''}{k} \tag{3.7}$$

Table 3.1 Selected values of thermal diffusivity at 300 K ($\sim$25°C). (Other values may be calculated from the data given in Appendix A.)

Material	α m²/s $\times 10^6$
Copper	112
Aluminum	84
Brass, 70% Cu, 30% Zn	34.2
Air at 1 atm pressure	22.5
Mild steel	18.8
Mercury	4.43
Stainless steel, 18-8	3.88
Fiberglass (medium density)	1.6
Concrete	0.75
Pyrex glass	0.51
Cork	0.16
Water	0.147
Engine oil, SAE 50	0.086
Neoprene rubber	0.079
White pine, perpendicular to grain	0.071
Refrigerant R-12	0.056
Polyvinylchloride (PVC)	0.051

Note: This table should be read as $\alpha \times 10^6$ m²/s = listed value; for example, for copper $\alpha = 112 \times 10^{-6}$ m²/s.

Finally, for a steady state and no internal heat generation, $\partial/\partial t = 0$, $\dot{Q}_v''' = 0$, so

$$\frac{\partial^2 T}{\partial x^2} + \frac{\partial^2 T}{\partial y^2} + \frac{\partial^2 T}{\partial z^2} = 0 \tag{3.8}$$

which is **Laplace's equation**.

The Fourier, Poisson, and Laplace equations are *partial differential equations* and have been thoroughly studied by mathematicians [1,2,3]. They are important because each is the governing equation for many different physical phenomena in fields as diverse as heat conduction, mass diffusion, electrostatics, and fluid mechanics. Solutions of Laplace's equation are called *potential* or *harmonic* functions.

Other Coordinate Systems

The solution of partial differential equations is simpler when boundary conditions are specified on *coordinate surfaces*, for example, $x =$ Constant in the Cartesian coordinate system. Thus, for conduction problems in cylindrical or spherical bodies, the Cartesian coordinate system is inappropriate. Such problems require heat conduction equations in terms of the cylindrical and spherical coordinate systems shown in Fig. 3.3. We can proceed in a number of ways. The most direct approach is

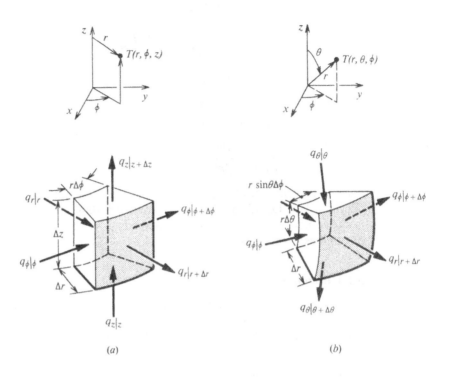

Figure 3.3 (a) Cylindrical coordinate system, r, ϕ, z, and an elemental volume. (b) Spherical coordinate system, r, θ, ϕ, and an elemental volume.

to repeat our derivation using appropriate volume elements, as shown in Fig. 3.3. Alternatively, we can transform the equations already derived in Cartesian coordinates to cylindrical or spherical coordinates; Exercises 3–1, 3–2, and 3–4 illustrate these procedures. To present the results in compact form, we introduce the *del-squared* or *Laplacian operator*. In Cartesian coordinates,

$$\nabla^2 = \frac{\partial^2}{\partial x^2} + \frac{\partial^2}{\partial y^2} + \frac{\partial^2}{\partial z^2}$$

and Eq. (3.5) becomes

$$\rho c \frac{\partial T}{\partial t} = k \nabla^2 T + \dot{Q}_v'''$$ (3.9)

Writing the heat conduction equation in any coordinate system then simply requires the proper expression for ∇^2. For cylindrical coordinates r, ϕ, z,

$$\nabla^2 = \frac{1}{r}\frac{\partial}{\partial r}\left(r\frac{\partial}{\partial r}\right) + \frac{1}{r^2}\frac{\partial^2}{\partial \phi^2} + \frac{\partial^2}{\partial z^2}$$ (3.10)

For spherical coordinates r, θ, ϕ,

$$\nabla^2 = \frac{1}{r^2}\frac{\partial}{\partial r}\left(r^2\frac{\partial}{\partial r}\right) + \frac{1}{r^2\sin\theta}\frac{\partial}{\partial\theta}\left(\sin\theta\frac{\partial}{\partial\theta}\right) + \frac{1}{r^2\sin^2\theta}\frac{\partial^2}{\partial\phi^2} \tag{3.11}$$

Two additional useful relations are

$$\frac{1}{r}\frac{\partial}{\partial r}\left(r\frac{\partial}{\partial r}\right) = \frac{\partial^2}{\partial r^2} + \frac{1}{r}\frac{\partial}{\partial r} \tag{3.12a}$$

$$\frac{1}{r^2}\frac{\partial}{\partial r}\left(r^2\frac{\partial}{\partial r}\right) = \frac{\partial^2}{\partial r^2} + \frac{2}{r}\frac{\partial}{\partial r} \tag{3.12b}$$

As a final comment regarding the heat conduction equation, it is noted that Fourier's law, Eq. (3.2), is a vector equation; thus, the heat conduction equation is most efficiently derived using the methods of vector calculus. Such a derivation is required as Exercise 3–3.

3.2.3 Boundary and Initial Conditions

In solving heat conduction problems in Chapters 1 and 2, boundary and initial conditions were used to evaluate integration constants. We now classify the types of boundary and initial conditions required to solve heat conduction problems.

Boundary Conditions

In Chapters 1 and 2, it was shown how practical heat conduction problems involve adjacent regions that may be quite different. For example, in Example 2.4, a uranium oxide nuclear fuel rod is enclosed in a Zircaloy-4 sheath and cooled by flowing water. Heat is generated within the fuel and flows by conduction to the fuel-sheath interface, by conduction across the sheath to the sheath-water interface, and by convection into the water. To analyze such problems, it is necessary to specify thermal conditions at solid-solid and solid-liquid interfaces. In general, it is required that both the heat flux and the temperature be continuous across an interface (although when a contact resistance model is used, as described in Section 2.2.2, the effect is to have a discontinuity in temperature). Thus, the solutions of the heat conduction equation in each region are *coupled*.

When analyzing more difficult heat transfer problems, we often find it convenient to uncouple the regions and consider each region independently. The boundary condition is then simply one of specified temperature. Considering the coordinate surface $x = L$ and referring to Fig. 3.4a,

$$T|_{x=L} = T_s \tag{3.13}$$

which is called a **first-kind** or **Dirichlet** boundary condition.

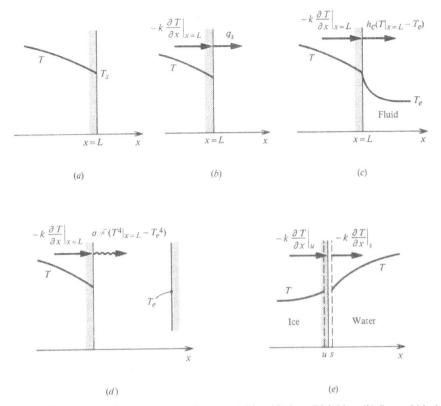

Figure 3.4 Types of boundary conditions. (*a*) First kind, or Dirichlet. (*b*) Second kind, or Neumann. (*c*) Third kind, or mixed. (*d*) Fourth kind, or radiation. (*e*) Phase change.

Sometimes it is convenient to apply a boundary condition at a surface where the heat flux is known. Referring to Fig. 3.4*b*,

$$-k\frac{\partial T}{\partial x}\bigg|_{x=L} = q_s \tag{3.14}$$

which is called a **second-kind** or **Neumann** boundary condition. Often the known heat flux is zero, such as at a plane of symmetry, or approximately zero, such as when the adjoining region is a good insulator.

Another commonly encountered situation is one in which the adjacent region is a fluid and we wish to describe heat transfer to the fluid using Newton's law of cooling. Referring to Fig. 3.4*c*,

$$-k\frac{\partial T}{\partial x}\bigg|_{x=L} = h_c(T|_{x=L} - T_e) \tag{3.15}$$

which is the **third-kind** or **mixed** boundary condition. Notice that Eq. (3.15) involves both the value of the dependent variable and that of its derivative at the boundary. Similarly, if there is heat loss by thermal radiation into adjacent surroundings (see Fig. 3.4d), using Eq. (1.17),

$$-k \frac{\partial T}{\partial x}\bigg|_{x=L} = \sigma \mathscr{F} \left(T|_{x=L}^4 - T_e^4 \right) \tag{3.16}$$

which is a **fourth-kind** or **radiation** boundary condition.[1] (Note, however, that some texts refer to the third-kind boundary condition as a radiation boundary condition, a practice with historical precedent but preferably avoided.)

There are other, more complex boundary conditions encountered in practice. The contact resistance described in Section 2.2.2 is one example. Another example is when there is a change of phase at the interface, such as ice melting or steam condensing. Figure 3.4e represents the surface of a block of ice melting in warm water. The s- and u-surfaces are on either side and infinitesimally close to the actual water-ice interface. Thus, the s-surface is in water, and the u-surface is in solid ice. Only the temperature is the same at the s- and u-surfaces; in general, the physical properties and temperature gradients are different. If the ice is melting at a rate per unit area $\dot{m}''$ [kg/m^2 s] and the interface is imagined to be fixed in space, then ice flows toward the interface, and water flows away at the melting rate. Application of the steady-flow energy equation to the control volume bounded by the u- and s-surfaces gives

$$\dot{m}''(h_s - h_u) = -k \frac{\partial T}{\partial x}\bigg|_u - \left(-k \frac{\partial T}{\partial x}\bigg|_s \right) \tag{3.17}$$

for $\dot{m}''$ positive. Equation (3.17) can be rearranged as

$$k \frac{\partial T}{\partial x}\bigg|_s = k \frac{\partial T}{\partial x}\bigg|_u + \dot{m}'' h_{\text{fs}} \tag{3.18}$$

where $h_{\text{fs}} = h_s - h_u$ is the enthalpy of fusion of the ice. Equation (3.18) can be interpreted as stating that the heat conducted from the warm water to the interface must balance both the heat conducted away from the interface into the cold ice and the heat required to melt the ice.

Initial Conditions

Transient heat conduction problems usually require specification of an **initial condition**, which simply means that the temperature throughout the region must be known at some instant in time before its subsequent variation with time can be

---

[1] This simple form is strictly valid only when the surroundings are isothermal and have a uniform emittance. More general situations are treated in Chapter 6.

determined. An exception is when the temperature varies periodically, for example, conduction in a spacecraft orbiting the earth. Then a periodic condition must be imposed on the solution.

3.2.4 Solution Methods

During the 19th century, considerable progress was made in developing mathematical methods for solving the various forms of the heat conduction equation. The first major contribution was by J. Fourier. His book, published in 1822 [4], developed the use of the method of **separation of variables**, which leads to the need to express an arbitrary function in a **Fourier series expansion**. Subsequently, **transform methods**—particularly the use of the *Laplace transform*—as well as other methods of classical mathematics were widely used. The treatise on heat conduction by Carslaw and Jaeger [5] contains an extensive compilation of solutions obtained using classical mathematical methods. Use of these methods usually requires that (1) the bounding surfaces be of relatively simple shape, (2) the boundary conditions be of simple mathematical form, and (3) the thermophysical properties be constant. Notwithstanding these limitations, many analytical results have wide engineering utility. Analytical solutions are useful benchmarks for checking the accuracy of numerical methods of solution. Also, the exercise of obtaining an analytical solution gives valuable insight into the essential features of heat transfer by conduction.

More recently, numerical methods, including **finite-difference** and **finite-element** methods, have been developed that allow solutions to be easily obtained for problems involving unusual shapes, complicated boundary conditions, and variable thermophysical properties. An early example was the application of the numerical *relaxation* method to steady heat conduction by H. Emmons in 1943 [6]. However, with the advent of the modern high-speed computer in the 1960s, numerical methods have been greatly improved. The wide availability of the personal computer in the 1980s has led to the marketing of versatile computer programs. These can be used to solve a great variety of heat conduction problems without requiring the user to have detailed knowledge of the numerical methods involved.

Some other methods have been used to obtain solutions to heat conduction problems. Two graphical methods, the *flux plotting* method and the *Schmidt plot*, are described in some heat transfer texts. The first is used for steady two-dimensional conduction and involves the free-hand sketching of isotherms and lines of heat flow; the latter is used for transient conduction and is the graphical equivalent to a finite-difference numerical method. A number of *analog* methods have also been used. Since Laplace's equation also governs electrical potential fields, two-dimensional steady conduction problems have been solved by making voltage and current measurements in appropriate shapes cut out of graphite-coated paper of high electrical resistance [7]. Before digital computers were developed, the analog computer or "differential analyzer" was used to solve heat conduction problems [8].

Methods of mathematical analysis are demonstrated in Sections 3.3 and 3.4, and finite-difference methods are dealt with in Section 3.5.

3.3   MULTIDIMENSIONAL STEADY CONDUCTION

One-dimensional steady conduction was dealt with in Chapter 2. Although the simple analytical results obtained are very useful, they have obvious limitations. Often the heat flow is multidimensional, that is, in two or three directions. For example, consider the furnace shown in Fig. 3.5. If the insulation is thin compared to the furnace dimensions, the assumption of one-dimensional heat flow is adequate (see Example 1.1). High-temperature furnaces, however, require thick insulation to reduce heat loss; in these furnaces, the heat flow through the edges is two-dimensional, and through the corners it is three-dimensional. Multidimensional steady conduction with no internal heat generation is governed by Laplace's equation. The classical approach to solving Laplace's equation is the *separation of variables* method; Section 3.3.1 uses a simple two-dimensional problem to demonstrate this approach. Often we are concerned with conduction between two isothermal surfaces, all other surfaces present being adiabatic. A **conduction shape factor** can be defined for such configurations, and a compilation of useful shape factors is given in Section 3.3.3.

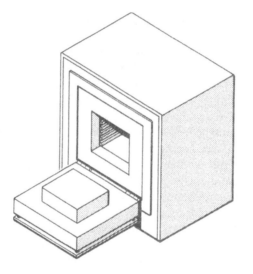

Figure 3.5   An insulated furnace.

3.3.1 Steady Conduction in a Rectangular Plate

The relatively simple problem of two-dimensional steady heat conduction in a rectangular plate will be used to demonstrate the method of separation of variables for solving Laplace's equation.

The Governing Equation and Boundary Conditions

Figure 3.6 depicts a thin rectangular plate with negligible heat loss from its surface. Temperature variations across the plate in the z direction are assumed to be zero ($\partial^2 T / \partial z^2 = 0$), and the thermal conductivity is assumed to be constant.

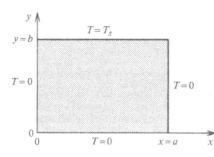

Figure 3.6 Boundary conditions for two-dimensional steady conduction in a rectangular plate.

The temperature distribution $T(x,y)$ is then governed by the two-dimensional form of Laplace's equation:

$$\frac{\partial^2 T}{\partial x^2} + \frac{\partial^2 T}{\partial y^2} = 0 \qquad (3.19)$$

To demonstrate the method of solution, we will first consider a model problem and later demonstrate how our solution can also be used to construct solutions for practical situations. Thus, we choose the boundary conditions shown in Fig. 3.6, namely,

$$x = 0, \quad 0 < y < b: \quad T = 0$$
$$y = 0, \quad 0 < x < a: \quad T = 0 \qquad (3.20a)$$
$$x = a, \quad 0 < y < b: \quad T = 0$$
$$y = b, \quad 0 < x < a: \quad T = T_s \qquad (3.20b)$$

The zero values for the boundary conditions, Eqs. (3.20a), will simplify the problem without loss of generality. Such boundary conditions are *homogeneous*. In general, a differential equation or boundary condition is said to be homogeneous if a constant times a solution is also a solution, that is, if T_1 is a solution, CT_1 is also a solution. The boundary condition Eq. (3.20b) looks simple, but we shall see that it actually makes solution of the problem quite difficult.

Solution for the Temperature Distribution

To use the method of separation of variables, we first assume that the function $T(x,y)$ can be expressed as the product of a function of x only, $X(x)$, and a function of y only, $Y(y)$:

$$T(x,y) = X(x)Y(y) \qquad (3.21)$$

Substitution in Eq. (3.19) gives

$$Y\frac{d^2 X}{dx^2} + X\frac{d^2 Y}{dy^2} = 0$$

where total derivatives have replaced the partial derivatives, since X is a function of independent variable x only, and Y is a function of independent variable y only.

Rearranging,

$$-\frac{1}{X}\frac{d^2X}{dx^2} = \frac{1}{Y}\frac{d^2Y}{dy^2} \tag{3.22}$$

Since each side of Eq. (3.22) is a function of a single independent variable, the equality can hold only if both sides are equal to a constant, which can be positive, negative, or zero. Using hindsight, the constant will be chosen to be a positive number λ^2. The reason for this choice will be discussed later. Two ordinary differential equations are thus obtained,

$$\frac{d^2X}{dx^2} + \lambda^2 X = 0 \qquad \frac{d^2Y}{dy^2} - \lambda^2 Y = 0$$

which have the solutions

$$X = B\cos\lambda x + C\sin\lambda x \qquad Y = De^{-\lambda y} + Ee^{\lambda y}$$

Substituting in Eq. (3.21) gives a tentative solution for $T(x,y)$:

$$T(x,y) = (B\cos\lambda x + C\sin\lambda x)(De^{-\lambda y} + Ee^{\lambda y}) \tag{3.23}$$

Applying boundary conditions Eqs. (3.20a) and taking care to ensure that the x- and y-dependences remain,

$$x = 0: \quad B(De^{-\lambda y} + Ee^{\lambda y}) = 0; \quad \text{thus, } B = 0$$

$$y = 0: \quad C\sin\lambda x(D + E) = 0; \quad \text{thus, } E = -D$$

$$x = a: \quad CD\sin\lambda a(e^{-\lambda y} - e^{\lambda y}) = -2CD\sin\lambda a\sinh\lambda y = 0$$

This requires that $\sin\lambda a = 0$, which has the roots $\lambda_n = n\pi/a$, for $n = 0, 1, 2, 3, \ldots$. These values of λ are called the **eigenvalues** or *characteristic values* of the problem. There is a distinct solution for each eigenvalue, each with its own constant. Writing the constant $-2CD$ for the nth solution as A_n,

$$T_n(x,y) = A_n\sin\frac{n\pi x}{a}\sinh\frac{n\pi y}{a}; \quad n = 0, 1, 2, 3, \ldots \tag{3.24}$$

Equation (3.19) is a *linear* differential equation, so its general solution is a sum of the series of solutions given by Eq. (3.24):

$$T(x,y) = \sum_{n=1}^{\infty} A_n\sin\frac{n\pi x}{a}\sinh\frac{n\pi y}{a} \tag{3.25}$$

where the solution for $n = 0$ has been deleted since $\sinh 0 = 0$. We now apply the last boundary condition, Eq. (3.20b), which requires that at $y = b$,

$$T|_{y=b} = T_s = \sum_{n=1}^{\infty} A_n\sin\frac{n\pi x}{a}\sinh\frac{n\pi b}{a} \tag{3.26}$$

that is, the constant T_s must be expressed in terms of an infinite series of sine functions, or a *Fourier series*.

Construction of a Fourier Series Expansion

In general, the temperature distribution along the boundary $y = b$ will be some arbitrary function $f(x)$, and the required expansion is then

$$f(x) = \sum_{n=1}^{\infty} C_n \sin \frac{n\pi x}{a}; \quad C_n = A_n \sinh \frac{n\pi b}{a} \tag{3.27}$$

The constants C_n are determined as follows. Multiply Eq. (3.27) by $\sin n\pi x/a$, and integrate term by term from $x = 0$ to $x = a$:

$$\int_0^a f(x) \sin \frac{n\pi x}{a} dx = \int_0^a C_1 \sin \frac{\pi x}{a} \sin \frac{n\pi x}{a} dx + \cdots + \int_0^a C_n \sin^2 \left(\frac{n\pi x}{a} \right) dx$$

$$+ \cdots + \int_0^a C_m \sin \frac{m\pi x}{a} \sin \frac{n\pi x}{a} dx + \cdots \tag{3.28}$$

Using standard integral tables, we find:

$$\int_0^a \sin \frac{n\pi x}{a} \sin \frac{m\pi x}{a} dx = \left[\frac{\sin(n\pi x/a - m\pi x/a)}{2(n\pi/a - m\pi/a)} - \frac{\sin(n\pi x/a + m\pi x/a)}{2(n\pi/a + m\pi/a)} \right]_0^a$$

$$= 0 \text{ for } n \neq m$$

$$\int_0^a \sin^2 \left(\frac{n\pi x}{a} \right) dx = \frac{a}{2n\pi} \left[\frac{n\pi x}{a} - \frac{1}{2} \sin \frac{2n\pi x}{a} \right]_0^a = \frac{a}{2}$$

Thus, only the nth term on the right-hand side of Eq. (3.28) remains, and solving for C_n gives

$$C_n = \frac{2}{a} \int_0^a f(x) \sin \frac{n\pi x}{a} dx \tag{3.29}$$

The set of sine functions $\sin \pi x/a, \sin 2\pi x/a, \ldots, \sin n\pi x/a, \ldots$ is said to be **orthogonal** over the interval $0 \leq x \leq a$, because the integral of $\sin m\pi x/a \, \sin n\pi x/a$ is zero if $m \neq n$. As shown in texts on applied mathematics [2, 3], if the function $f(x)$ is piecewise continuous, it can always be expressed in terms of a uniformly converging series of orthogonal functions. The cosine function is also orthogonal over an appropriate interval, as are many other functions, including Bessel functions and Legendre polynomials.

Temperature Distribution for $f(x) = T_s$

For the function $f(x)$ equal to a constant value T_s, the integral in Eq. (3.29) can be evaluated analytically:

$$C_n = \frac{2}{a} \left[-\frac{T_s a}{n\pi} \cos \frac{n\pi x}{a} \right]_0^a = T_s \frac{2}{n\pi} [1 - (-1)^n]$$

and from Eq. (3.27),

$$A_n = \frac{C_n}{\sinh(n\pi b/a)} = T_s \frac{2[1-(-1)^n)]}{n\pi \sinh(n\pi b/a)}$$

Substituting into Eq. (3.25) gives the desired temperature distribution for steady conduction in a rectangular plate:

$$T(x,y) = T_s \sum_{n=1}^{\infty} \frac{2[1-(-1)^n]}{n\pi \sinh(n\pi b/a)} \sin\frac{n\pi x}{a} \sinh\frac{n\pi y}{a} \tag{3.30}$$

Lines of constant temperature, or *isotherms*, are shown in Fig. 3.7. The temperature discontinuities at the top corners are physically unrealistic since an infinite heat flow is implied. In fact, the heat flow across the plate edge at $y = b$, evaluated from Eq. (3.30) using Fourier's law, is infinite. Thus, the isotherms in the vicinity of these corners correspond to a mathematical problem only; in real physical problems, there might be a very marked variation in temperature along the edges near these corners, but there cannot be an actual discontinuity.

Since the variables in the partial differential equation, Eq. (3.19), did separate, and because a solution could be found to satisfy the boundary conditions, the method of solution has been successful. If a negative constant is chosen for Eq. (3.22), the boundary conditions cannot be satisfied. Use of a negative constant just reverses the roles of the independent variables x and y, and the negative constant is appropriate if the temperature T_s is specified on the edge $x = a$.

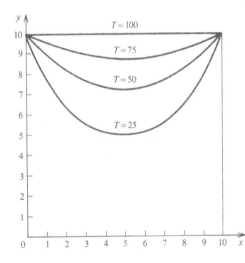

Figure 3.7   Isotherms for conduction in a rectangular plate obtained from Eq. (3.30); $a = b = 10, T_s = 100$.

Generalization Using the Principle of Superposition

Laplace's equation is a linear differential equation. A useful consequence of this property is that the solution of a problem with complicated boundary conditions can be constructed by adding solutions for problems having simpler boundary conditions.

To illustrate the procedure, consider a problem where the plate temperature is specified along two edges. If the boundary conditions Eqs. (3.20) are replaced by those shown in Fig. 3.8,

$$x=0, \quad 0<y<b: \quad T=0$$
$$y=0, \quad 0<x<a: \quad T=0$$
$$x=a, \quad 0<y<b: \quad T=f_1(y)$$
$$y=b, \quad 0<x<a: \quad T=f_2(x)$$

the *superposition principle* can be used as follows. Let $T(x,y) = T_1(x,y) + T_2(x,y)$, where T_1 and T_2 satisfy

$$\frac{\partial^2 T_1}{\partial x^2} + \frac{\partial^2 T_1}{\partial y^2} = 0 \qquad\qquad \frac{\partial^2 T_2}{\partial x^2} + \frac{\partial^2 T_2}{\partial y^2} = 0$$
$$x=0, y=0, y=b: T_1=0 \qquad x=0, y=0, x=a: T_2=0$$
$$x=a: T_1=f_1(y) \qquad\qquad y=b: T_2=f_2(x)$$

Addition of the equations and boundary conditions shows that $T = T_1 + T_2$ satisfies the original problem. Clearly, this approach can be extended to a problem where the temperature is specified on three or all four sides.

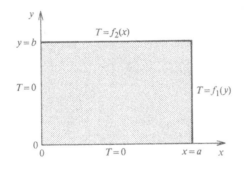

Figure 3.8   Rectangular plate with nonhomogeneous boundary conditions specified on two edges.

EXAMPLE 3.1   Heat Flow across a Neoprene Rubber Pad

A long neoprene rubber pad of width $a = 2$ cm and height $b = 4$ cm is a component of a spacecraft structure. Its sides and bottom are bonded to a metal channel at temperature $T_e = 20°$ C, and the temperature distribution along the top can be approximated as a simple sine curve, $T = T_e + \Delta T_m \sin(\pi x/a)$, where $\Delta T_m = 80$ K. Determine the heat flow across the pad per meter length.

Solution

Given: Long rubber pad with a rectangular cross section. rpm.

Required: Heat flow across pad for given boundary conditions.

Assumptions: Two-dimensional, steady conduction.

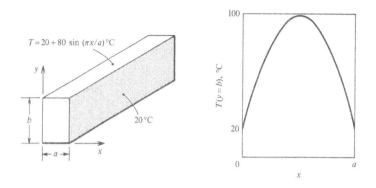

Conservation of energy requires that at steady state, the heat flow across the top surface equal the heat flow out across the sides and bottom; thus,

$$\dot{Q} = (1) \int_0^a q_{y|_{y=b}}\, dx = \int_0^a -k \frac{\partial T}{\partial y}\bigg|_{y=b} dx$$

The sides and bottom of the pad can be taken to have a temperature of $20°C$. thus, T is the solution of Laplace's equation that satisfies the boundary conditions

$$x = 0, x = a, y = 0 : T = T_e$$

$$y = b : T = T_e + \Delta T_m \sin \frac{\pi x}{a}$$

Let $T_1 = T - T_e$; then T_1 satisfies the boundary conditions

$$x = 0, x = a, y = 0 : T_1 = 0$$

$$y = b : T_1 = \Delta T_m \sin \frac{\pi x}{a}$$

Equation (3.25) satisfies the homogeneous boundary conditions. Applying the boundary condition at $y = b$ gives

$$T_1|_{y=b} = \Delta T_m \sin \frac{\pi x}{a} = \sum_{n=1}^{\infty} A_n \sin \frac{n\pi x}{a} \sinh \frac{n\pi b}{a}$$

which can be satisfied if

$$A_1 = \frac{\Delta T_m}{\sinh(\pi b/a)}; \quad A_2 = A_3 = \cdots = 0$$

That is, only the first term of the infinite series is required. Hence,

$$T - T_e = T_1 = \Delta T_m \sin \frac{\pi x}{a} \frac{\sinh(\pi y/a)}{\sinh(\pi b/a)}$$

$$\frac{\partial T}{\partial y}\bigg|_{y=b} = \Delta T_m \frac{(\pi/a)\sin(\pi x/a)}{\tanh(\pi b/a)}$$

$$\dot{Q} = \int_0^a \left. -k\frac{\partial T}{\partial y} \right|_{y=b} dx = \frac{\Delta T_m k}{\tanh(\pi b/a)}(\cos\pi - \cos 0) = -\frac{2\Delta T_m k}{\tanh(\pi b/a)} \text{ W/m}$$

For $\Delta T_m = 80$ K, $k = 0.19$ W/m K (Table A.2), and $b/a = 2.0$, the heat flow is

$$\dot{Q} = -\frac{(2)(80)(0.19)}{\tanh 2\pi} = -\frac{30.4}{1.0000} = -30.4 \text{ W/m}$$

Comments

Notice that in the limit of large b/a, $\tanh\pi b/a \to 1$ and $\dot{Q} = -2\Delta T_m k$ is independent of b/a; even for a square pad, $b/a = 1$, $\tanh\pi = 0.996$. In the opposite limit of $b/a \to 0$, we can use the expansion $\tanh x = x - x^3/3 + \cdots$ to obtain

$$\dot{Q} = -\frac{2\Delta T_m ka}{\pi b}$$

which is the result for one-dimensional heat flow across a thin wall. For a 1 m length of pad,

$$\dot{Q} = \int_0^a q\,dx = -\int_0^a \frac{k}{b}\left(T_e + \Delta T_m \sin\frac{\pi x}{a} - T_e\right)dx$$

$$= -\frac{k\Delta T_m}{b}\int_0^a \sin\frac{\pi x}{a}dx = -\frac{2\Delta T_m ka}{\pi b}$$

which agrees with the result obtained above.

3.3.2 Steady Conduction in a Rectangular Block

Consider a rectangular block with boundary conditions that cause the temperature to vary in all three coordinate directions, x, y, and z. For an assumed constant thermal conductivity, the temperature distribution $T(x,y,z)$ is governed by the three-dimensional form of Laplace's equation:

$$\frac{\partial^2 T}{\partial x^2} + \frac{\partial^2 T}{\partial y^2} + \frac{\partial^2 T}{\partial z^2} = 0 \tag{3.31}$$

Again the method of separation of variables can be used and $T(x,y,z)$ taken to have the form

$$T(x,y,z) = X(x)Y(y)Z(z)$$

The analysis proceeds as for the rectangular plate in Section 3.3.1, but the mathematics is very cumbersome, even for the simplest of boundary conditions. In practice, one cannot expect a simple behavior of the boundary conditions over a three-dimensional object, and the numerical methods of solution described in Section 3.5 are more appropriate than analytical methods.

Table 3.2   Shape factors for steady-state conduction for use in Eq. (3.32), $\dot{Q} = kS\Delta T$; $\Delta T = T_1 - T_2$. (See also the bibliography for Chapter 3.)

Configuration	Shape Factor
1. Plane wall 	$\dfrac{A}{L}$
2. Concentric cylinders $L \gg r_2$	$\dfrac{2\pi L}{\ln(r_2/r_1)}$ Note there is no steady-state solution for $r_2 \to \infty$, i.e., for a cylinder in an infinite medium.
3. Concentric spheres 	$(a)\quad \dfrac{4\pi}{1/r_1 - 1/r_2}$ $(b)\quad 4\pi r_1 \quad$ for $r_2 \to \infty$
4. Eccentric cylinders $L \gg r_2$	$\dfrac{2\pi L}{\cosh^{-1}\left(\dfrac{r_2^2 + r_1^2 - e^2}{2r_1 r_2}\right)}$
5. Concentric square cylinders $L \gg a$	$\dfrac{2\pi L}{0.93\ln(a/b) - 0.0502}\quad$ for $\dfrac{a}{b} > 1.4$ $\dfrac{2\pi L}{0.785\ln(a/b)}\quad$ for $\dfrac{a}{b} < 1.4$
6. Concentric circular and square cylinders 	$\dfrac{2\pi L}{\ln(0.54a/r)}\quad a > 2r$

Table 3.2   (Concluded)

Configuration	Shape Factor
7. Buried sphere	$\dfrac{4\pi r_1}{1 - r_1/2h}$
	For $h \to \infty$, the result for item 3(b) is recovered
Medium at infinity also at T_2	
8. Buried cylinder	$\dfrac{2\pi L}{\cosh^{-1}(h/r_1)}$
	$\dfrac{2\pi L}{\ln(2h/r_1)}$ for $h > 3r_1$
	For $h/r_1 \to \infty$, $S \to 0$ since steady flow is impossible
Medium at infinity also at T_2 $L \gg r_1$	
9. Buried rectangular beam	$2.756L\left[\ln\left(1 + \dfrac{h}{a}\right)\right]^{-0.59}\left(\dfrac{h}{b}\right)^{-0.078}$
Medium at infinity also at T_2 $L \gg h, a, b$	
10. The edge of adjoining walls	$0.54W$ for $W > L/5$
	(W is the inner edge)
11. The corner of three adjoining walls	$0.15L$ for $W > L/5$
12. Disk area on the adiabatic surface of a semi-infinite solid	$4r$
Medium at infinity at T_2	

3.3.3 Conduction Shape Factors

Many multidimensional conduction problems involve heat flow between two surfaces, each of uniform temperature, with any other surfaces present being adiabatic. The conduction *shape factor*, S, is defined such that the heat flow between the surfaces, $\dot{Q}$, is

$$\dot{Q} = kS\Delta T \tag{3.32}$$

where k is the thermal conductivity and ΔT is the difference in surface temperatures; S is seen to have the dimensions of length. The results we have already obtained for one-dimensional conduction can also be expressed in terms of the shape factor. For example, a plane slab of area A and thickness L has $S = A/L$ from Eq. (1.9). Table 3.2 lists shape factors for various configurations.

Some points to note when using Table 3.2 are:

1. There is no internal heat generation: $\dot{Q}_v''' = 0$.
2. The thermal conductivity, k, is constant.
3. The two surfaces should be isothermal. If these temperatures are not prescribed, but are intermediate temperatures in a series thermal circuit, the isothermal condition may not be satisfied. The surfaces will generally be isothermal when the component in question has the dominant thermal resistance. Example 3.3 illustrates this point.
4. Special care must be taken with the configurations involving an infinite medium. For example, in item 7, not only the plane surface but also the medium at infinity must be at temperature T_2.
5. Item 8 is often used incorrectly for calculating heat loss or gain from buried pipelines. It is essential that the deep soil be at the same temperature as the surface, which is a condition seldom met in reality. Also, the buried pipeline problem often involves transient conduction.
6. The shape factors given in items 10 and 11 were developed by the physicist I. Langmuir and coworkers in 1913 for calculating the heat loss from furnaces. Example 3.2 illustrates their use.

EXAMPLE 3.2   Heat Loss from a Laboratory Furnace

A small laboratory furnace is in the form of a cube and is insulated with a 10 cm layer of fiberglass insulation, with an inside edge 30 cm long. If the only significant resistance to heat flow across the furnace wall is this insulation, determine the power required for steady operation at a temperature of 600 K when the outer casing temperature is 350 K. The thermal conductivity of the fiberglass insulation at the mean temperature of 475 K is approximately 0.11 W/m K.

Solution

Given: Insulated laboratory furnace.

Required: Power required for operation at 600 K.

Assumptions: 1. Steady state.
  2. The outer convective resistance is negligible.

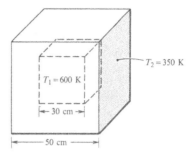

The shape factors given as items 1, 10, and 11 of Table 3.2 should be used, assuming independent parallel paths for heat flow through the 6 sides, 12 edges, and 8 corners of the enclosure. Thus, if L is the insulation thickness and W the inside edge,

$$\dot{Q} = k\Delta T S$$

$$= k(T_1 - T_2)[6W^2/L + (12)(0.54)W + (8)(0.15)L]$$

$$= (0.11)(600 - 350)[(6)(0.3)^2/(0.1) + (12)(0.54)(0.3) + (8)(0.15)(0.1)]$$

$$= (0.11)(250)[5.40 + 1.94 + 0.12]$$

$$= 205 \text{ W}$$

Comments

If an effective area for heat flow A_{eff} is defined by the equation for one-dimensional heat flow,

$$\dot{Q} = \frac{kA_{\text{eff}}(T_1 - T_2)}{L}$$

then $A_{\text{eff}} = 6W^2 + (12)(0.54)WL + (8)(0.15)L^2 = 0.746 \text{ m}^2$. Notice that A_{eff} is significantly less than either the arithmetic average of the inner and outer areas, 1.02 m^2, or the value midway through the wall, 0.96 m^2.

EXAMPLE 3.3   Heat Loss from a Buried Oil Line

An oil pipeline has an outside diameter of 30 cm and is buried with its centerline 1 m below ground level in damp soil. The line is 5000 m long, and the oil flows at 2.5 kg/s. If the inlet temperature of the oil is 120°C and the ground level soil is at 23°C, estimate the oil outlet temperature and the heat loss (i) for an uninsulated pipe, and (ii) if the pipe is insulated with a 15 cm layer of insulation with conductivity $k = 0.03$ W/m K. Take the soil thermal conductivity as 1.5 W/m K and the oil specific heat as 2000 J/kg K.

Solution

Given: Buried oil pipeline.

Required: Oil outlet temperature and heat loss if (i) uninsulated, (ii) insulated.

Assumptions: 1. Steady state.
2. Deep ground temperature same as surface temperature.
3. Negligible resistance of the pipe wall and for convection from the oil.
4. Isothermal surface exposed to soil.

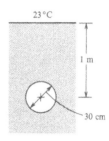

(i) The temperature difference between the oil and the ground surface, $\Delta T = T - T_s$, decreases continuously in the flow direction, so how do we use Eq. (3.32)? We need to derive a differential equation governing the change in the oil temperature. Consider an element of pipe Δx long, as shown in the figure. Application of the steady-flow energy equation, Eq. (1.4), gives

$$\dot{m}\Delta h = \Delta \dot{Q}$$

$$\dot{m}c_p(T|_{x+\Delta x} - T|_x) = k\Delta S(T_s - T)$$

where $\Delta S = 2\pi\Delta x/\ln(2h/r_1)$ from Table 3.2, item 8, since $h/r_1 = 1.0/0.15 = 6.67 > 3$. Substituting for ΔS and dividing by Δx,

$$\dot{m}c_p\left(\frac{T|_{x+\Delta x} - T|_x}{\Delta x}\right) = \frac{2\pi k}{\ln(2h/r_1)}(T_s - T)$$

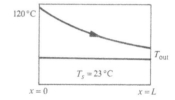

Letting $\Delta x \to 0$ and rearranging gives the desired differential equation for $T(x)$,

$$\frac{dT}{dx} - \frac{2\pi k}{\dot{m}c_p\ln(2h/r_1)}(T_s - T) = 0$$

Integrating with $T = T_{in}$ at $x = 0$,

$$T - T_s = (T_{in} - T_s)e^{-[2\pi k/\dot{m}c_p\ln(2h/r_1)]x}$$

In particular, the oil outlet temperature is obtained for $x = L$,

$$T_{out} - T_s = (T_{in} - T_s)e^{-2\pi kL/\dot{m}c_p\ln(2h/r_1)}$$

$$\frac{2\pi kL}{\dot{m}c_p\ln(2h/r_1)} = \frac{(2)(\pi)(1.5)(5000)}{(2.5)(2000)\ln(2 \times 1.0/0.15)} = 3.64$$

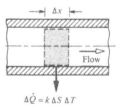

$$T_{out} - 23 = (120 - 23)e^{-3.64} = 2.546$$

$$T_{out} = 2.55 + 23 = 25.5°C$$

The heat loss is equal to the enthalpy given up by the oil,

$$\dot{Q} = \dot{m}c_p(T_{in} - T_{out}) = (2.5)(2000)(120 - 25.5) = 472\,\text{kW}$$

(ii) This problem cannot be solved exactly using the shape factor concept, since, when the insulation is added to the pipe, the outer surface of the insulation will not be isothermal.

However, to get some idea of the effect of the insulation, we will assume an isothermal surface. Then for two resistances in series,

$$\dot{m}\Delta h = \Delta\dot{Q}; \quad \Delta\dot{Q} = \frac{T - T_s}{\dfrac{\ln(r_2/r_1)}{2\pi k_{ins}\Delta x} + \dfrac{1}{k_{soil}\Delta S}}$$

$$\frac{h}{r_2} = \frac{1.0}{0.3} = 3.33; \quad \text{hence } \Delta S = 2\pi\Delta x / \ln(2h/r_2)$$

Proceeding as in part (i),

$$T_{out} - T_s = (T_{in} - T_s)\exp\left(-\frac{1}{\dot{m}c_p\left[\dfrac{\ln(r_2/r_1)}{2\pi k_{ins}L} + \dfrac{\ln(2h/r_2)}{2\pi k_{soil}L}\right]}\right)$$

$$\frac{\ln(r_2/r_1)}{2\pi k_{ins}L} = \frac{\ln(0.30/0.15)}{(2\pi)(0.03)(5000)} = 7.35 \times 10^{-4}\,(\text{W/K})^{-1}$$

$$\frac{\ln(2h/r_2)}{2\pi k_{soil}L} = \frac{\ln(2 \times 1.0/0.3)}{(2\pi)(1.5)(5000)} = 0.403 \times 10^{-4}\,(\text{W/K})^{-1}$$

$$T_{out} = 23 + (120 - 23)e^{-1/[(2.5)(2000)(7.35+0.403)\times 10^{-4}]}$$

$$T_{out} = 23 + (120 - 23)e^{-0.258} = 98.0°\text{C}$$

$$\dot{Q} = (2.5)(2000)(120 - 98.0) = 110\text{ kW}$$

Comments

The pipeline is located on a tropical island where the annual ground temperature variation is relatively small, so a steady-state analysis is reasonably valid. Problems concerning the freezing of buried water pipes often require a transient analysis because of ground temperature variations (see Exercise 3–39).

3.4   UNSTEADY CONDUCTION

In unsteady or *transient* conduction, temperature is a function of both time and spatial coordinates. In the absence of internal heat generation, the temperature response of a body is governed by Fourier's equation. Again the method of separation of variables is useful, and examples of its use are given in Sections 3.4.1 and 3.4.3. The method, however, fails under certain circumstances—for example, when the medium extends to infinity. Then possible methods include the use of *Laplace transforms* or a **similarity** transformation of the partial differential equation into an ordinary differential equation. The latter method is demonstrated in Section 3.4.2. Analytical results of unsteady conduction tend to be complicated and awkward to use. Thus, where possible, approximate solutions of adequate accuracy will be indicated. Often the results are conveniently presented in graphical form for rapid engineering calculations.

3.4.1 The Slab with Negligible Surface Resistance

Figure 3.9 shows a slab $2L$ thick. It is initially at a uniform temperature T_0, and at time $t = 0$ the surfaces at $x = +L$ and $x = -L$ are suddenly lowered to temperature T_s. Such a situation is encountered in practice when a poorly conducting slab is suddenly immersed in a liquid for which the convective heat transfer coefficient is very large, that is, under conditions where the convective resistance to heat transfer is negligible. Note that this situation is the opposite limit to that considered in Section 1.5, for which the lumped thermal capacity model was applicable. In that case, the Biot number $\mathrm{Bi}_{LTC} = \bar{h}_c V / k_s A$ had to be small; for this case, the Biot number must be large. In addition to its practical utility, the solution of this problem allows a simple demonstration of some important features of transient conduction.

The Governing Equation and Conditions

With no internal heat generation and an assumed constant thermal conductivity, Fourier's equation, Eq. (3.6), applies. Since the temperature does not vary in the y and z directions,

$$\frac{\partial T}{\partial t} = \alpha \frac{\partial^2 T}{\partial x^2} \tag{3.33}$$

which is the governing differential equation for $T(x,t)$ and must be solved subject to the initial condition

$$t = 0: \quad T = T_0 \tag{3.34a}$$

The symmetry of the problem allows the differential equation to be solved in the region $0 \leq x \leq L$, for which the appropriate boundary conditions are

$$x = 0: \quad \frac{\partial T}{\partial x} = 0 \tag{3.34b}$$

$$x = L: \quad T = T_s \tag{3.34c}$$

Equation (3.34b) follows from the symmetry of the temperature profile about the plane $x = 0$, or equivalently, from the condition that there can be no heat flow across this plane.

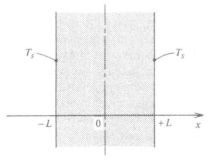

Figure 3.9   Coordinate system for the analysis of unsteady conduction in an infinite slab.

Dimensional Analysis

Before an attempt is made to solve Eq. (3.33), it is helpful to perform a dimensional analysis. We begin by constructing suitable dimensionless independent and dependent variables. Suitable choices for x and T are $\eta = x/L$ and $\theta = (T - T_s)/(T_0 - T_s)$, since η and θ will both vary between zero and unity. However, there is no obvious time scale for the problem that can be used to make t dimensionless, so we will let the differential equation itself indicate an appropriate choice. Transforming Eq. (3.33) with

$$x = L\eta; \quad dx = L\,d\eta$$

$$T = (T_0 - T_s)\theta + T_s; \qquad dT = (T_0 - T_s)d\theta$$

gives

$$(T_0 - T_s)\frac{\partial\theta}{\partial t} = \frac{\alpha(T_0 - T_s)}{L^2}\frac{\partial^2\theta}{\partial\eta^2}$$

or

$$\frac{\partial\theta}{\partial(\alpha t/L^2)} = \frac{\partial^2\theta}{\partial\eta^2}$$

Thus, we choose $\zeta = t/(L^2/\alpha)$ as dimensionless time, and the result is

$$\frac{\partial\theta}{\partial\zeta} = \frac{\partial^2\theta}{\partial\eta^2} \tag{3.35}$$

The initial and boundary conditions transform into

$$\zeta = 0: \quad \theta = 1 \tag{3.36a}$$

$$\eta = 0: \quad \frac{\partial\theta}{\partial\eta} = 0 \tag{3.36b}$$

$$\eta = 1: \quad \theta = 0 \tag{3.36c}$$

There are no parameters in the transformed statement of the problem, so the solution is simply $\theta(\zeta, \eta)$. The dimensionless time variable ζ is also commonly called the **Fourier number**, Fo $= \alpha t/L^2$. Notice that the Fourier number can also be written as Fo $= t/t_c$; $t_c = L^2/\alpha$ is a characteristic time (time constant) for this conduction problem. The behavior of the solution will depend on the value of t relative to t_c, that is, on whether Fo is much smaller than unity, of order unity, or much larger than unity. The relation $\theta = \theta(\zeta, \eta)$ is a statement of the similarity principle for this problem, a concept introduced in Section 2.4.5. We now know that the solution for the dimensionless temperature θ will be a function of dimensionless time ζ and dimensionless position η only.

Solution for the Temperature Response

As in Section 3.3.1, for steady conduction in a rectangular plate, the method of separation of variables will be used to solve the partial differential equation. We assume that the function $\theta(\zeta, \eta)$ can be expressed as the product of a function of ζ

only, $Z(\zeta)$, and a function of η only, $H(\eta)$

$$\theta(\zeta, \eta) = Z(\zeta)H(\eta) \tag{3.37}$$

Substitution in Eq. (3.35) yields

$$H\frac{dZ}{d\zeta} = Z\frac{d^2H}{d\eta^2}$$

where total derivatives have replaced partial derivatives, since Z is a function of independent variable ζ only, and H is a function of independent variable η only. Rearranging gives

$$\frac{1}{Z}\frac{dZ}{d\zeta} = \frac{1}{H}\frac{d^2H}{d\eta^2} \tag{3.38}$$

Since each side of Eq. (3.38) is a function of a single independent variable, the equality can hold only if both sides are equal to a constant. To satisfy the boundary conditions, this constant must be a negative number, which will be written $-\lambda^2$. Two ordinary differential equations are obtained:

$$\frac{dZ}{d\zeta} + \lambda^2 Z = 0 \qquad \frac{d^2H}{d\eta^2} + \lambda^2 H = 0$$

which have the solutions

$$Z = C_1 e^{-\lambda^2 \zeta} \qquad H = C_2 \cos \lambda\eta + C_3 \sin \lambda\eta$$

Hence,

$$\theta(\zeta, \eta) = e^{-\lambda^2 \zeta}(A\cos \lambda\eta + B\sin \lambda\eta)$$

where $A = C_1 C_2$ and $B = C_1 C_3$. Applying boundary condition Eq. (3.36b),

$$\frac{\partial \theta}{\partial \eta}\Big|_{\eta=0} = e^{-\lambda^2 \zeta}(-A\lambda \sin \lambda\eta + B\lambda \cos \lambda\eta)_{\eta=0} = 0$$

which requires that $B = 0$. Next, applying boundary condition Eq. (3.36c) gives

$$\theta|_{\eta=1} = Ae^{-\lambda^2 \zeta} \cos \lambda\eta|_{\eta=1} = 0$$

which requires that $\cos \lambda = 0$, or $\lambda_n = (n+1/2)\pi$, $n = 0, 1, 2, 3, \ldots$ The λ_n are the eigenvalues for this problem, and the solution corresponding to the nth eigenvalue may be written as

$$\theta_n(\zeta, \eta) = A_n e^{-(n+1/2)^2\pi^2\zeta} \cos\left(n+\frac{1}{2}\right)\pi\eta \tag{3.39}$$

The general solution to Eq. (3.35) is the sum of the series of solutions given by Eq. (3.39)

$$\theta(\zeta, \eta) = \sum_{n=0}^{\infty} A_n e^{-(n+1/2)^2\pi^2\zeta} \cos\left(n+\frac{1}{2}\right)\pi\eta \tag{3.40}$$

The constants A_n are determined from the initial condition, Eq. (3.36a):

$$\theta|_{\zeta=0} = \sum_{n=0}^{\infty} A_n \cos\left(n+\frac{1}{2}\right)\pi\eta = 1$$

That is, a constant must be expressed in terms of an infinite series of cosine functions, which again is a Fourier series expansion. More generally, $\theta|_{\zeta=0} = f(\eta)$ is an arbitrary function of η, and the required expansion is then

$$f(\eta) = \sum_{n=0}^{\infty} A_n \cos\left(n+\frac{1}{2}\right)\pi\eta \tag{3.41}$$

Determination of the constants A_n follows the procedure demonstrated in Section 3.3.1. The result is

$$A_n = 2\int_0^1 f(\eta)\cos\left(n+\frac{1}{2}\right)\pi\eta\, d\eta \tag{3.42}$$

For $f(\eta)=1$, $A_n = \dfrac{2}{(n+1/2)\pi}\left[\sin\left(n+\frac{1}{2}\right)\pi\eta\right]_0^1 = \dfrac{2(-1)^n}{(n+1/2)\pi}$

Substituting in Eq. (3.40) gives the solution for the temperature distribution,

$$\theta(\zeta,\eta) = \sum_{n=0}^{\infty} \frac{2(-1)^n}{(n+1/2)\pi} e^{-(n+1/2)^2\pi^2\zeta} \cos\left(n+\frac{1}{2}\right)\pi\eta \tag{3.43}$$

or, in terms of temperature $T(x,t)$, Fourier number $\mathrm{Fo} = \alpha t/L^2$, and x/L,

$$\frac{T-T_s}{T_0-T_s} = \sum_{n=0}^{\infty} \frac{2(-1)^n}{(n+1/2)\pi} e^{-(n+1/2)^2\pi^2\mathrm{Fo}} \cos\left(n+\frac{1}{2}\right)\pi\frac{x}{L} \tag{3.44}$$

which is plotted in Fig. 3.10.

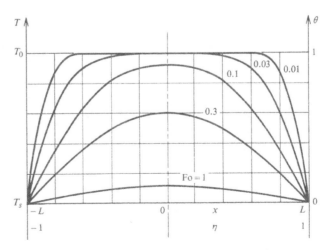

Figure 3.10   Temperature response in a slab with negligible surface resistance, calculated from Eq. (3.44).

The Surface Heat Flux

Of particular interest is the rate of heat transfer out of the slab. The surface heat flux at $x = L$ is obtained from Fourier's law and Eq. (3.44) as

$$q_s = -k\frac{\partial T}{\partial x}\bigg|_{x=L}$$

$$= -k(T_0 - T_s) \sum_{n=0}^{\infty} \frac{2(-1)^n}{(n+1/2)\pi} e^{-(n+1/2)^2\pi^2 \text{Fo}} \frac{\partial}{\partial x}\left[\cos\left(n+\frac{1}{2}\right)\pi\frac{x}{L}\right]_{x=L}$$

$$= -k(T_0 - T_s) \sum_{n=0}^{\infty} \frac{2(-1)^n}{(n+1/2)\pi} e^{-(n+1/2)^2\pi^2 \text{Fo}} \left[\frac{-(n+1/2)\pi(-1)^n}{L}\right]$$

$$= \frac{2k(T_0 - T_s)}{L} \sum_{n=0}^{\infty} e^{-(n+1/2)^2\pi^2 \text{Fo}} \tag{3.45}$$

or

$$q_s = \frac{2k(T_0 - T_s)}{L}\left[e^{-(\pi/2)^2\text{Fo}} + e^{-(3\pi/2)^2\text{Fo}} + e^{-(5\pi/2)^2\text{Fo}} + e^{-(7\pi/2)^2\text{Fo}} + \cdots\right]$$

Table 3.3 gives the first four terms in the series for values of the Fourier number $\text{Fo} = \alpha t/L^2$ equal to 0.01, 0.1, 0.2, 0.3, and 1.0. It can be seen that the series converges rapidly unless the Fourier number is very small. For $\text{Fo} \gtrsim 0.2$, only the first term in the series need be retained, and

$$q_s \simeq \frac{2k(T_0 - T_s)}{L} e^{-(\pi/2)^2\text{Fo}} \tag{3.46}$$

with an error of less than 2%. Since $\text{Fo} = \alpha t/L^2$, Eq. (3.46) is a solution valid for *long times*.

For small values of the Fourier number (that is, soon after the slab is immersed in the liquid), the series converges slowly, and sufficient terms must be retained for an accurate result. However, in the limit $\text{Fo} \to 0$, there is the simple mathematical result:

Table 3.3   The first four terms of the series of Eq. (3.45) for the surface heat flux of a slab with negligible surface resistance.

Fo	$e^{-(\pi/2)^2\text{Fo}}$	$e^{-(3\pi/2)^2\text{Fo}}$	$e^{-(5\pi/2)^2\text{Fo}}$	$e^{-(7\pi/2)^2\text{Fo}}$
0.01	0.9756	0.8009	0.5396	0.2985
0.1	0.7813	0.1085	0.0021	$\sim 10^{-5}$
0.2	0.6105	0.0118	$\sim 10^{-6}$	
0.3	0.4770	0.0013		
1.0	0.0848	$\sim 10^{-10}$		

$$\sum_{n=0}^{\infty} e^{-(n+1/2)^2 \pi^2 \text{Fo}} = \frac{1}{2\pi^{1/2} \text{Fo}^{1/2}}; \quad \text{Fo} \to 0$$

Substituting in Eq. (3.45) gives

$$q_s \simeq \frac{k(T_0 - T_s)}{(\pi \alpha t)^{1/2}} \tag{3.47}$$

which can be used for $\text{Fo} \lesssim 0.05$. Equation (3.47) is thus a valid solution for *short times*. Figure 3.10 shows that for sufficiently short times, the temperature changes in a thin region near the surface only, while the temperature of the slab interior remains unaffected. The heat conduction process is confined to this thin region, and the thickness of the slab is of no consequence: notice that L does not appear in Eq. (3.47). It is this fact that motivates the analysis of the semi-infinite solid in Section 3.4.2.

Nonsymmetrical Boundary Conditions

In our analysis, we specified the same temperature on both sides of the slab. Symmetry allowed the problem to be solved in the half slab, $0 \le x \le L$; upon transformation, the boundary condition at $\eta = 1$ was $\theta = 0$, that is, homogeneous. If we now specify a temperature $T_{s'}$ at $x = 0$ and T_s at $x = L$, as shown in Fig. 3.11, how do we proceed? It is not possible to define a dimensionless temperature θ such that $\theta = 0$ on both boundaries, and without homogeneous boundary conditions, it is not possible to obtain an eigenvalue problem as before. To get around this hurdle, we reduce the problem to the **superposition** of a steady and a transient problem. Let $T(x,t) = T_1(x) + T_2(x,t)$ such that

$$\frac{d^2 T_1}{dx^2} = 0 \qquad\qquad \frac{\partial T_2}{\partial t} = \alpha \frac{\partial^2 T_2}{\partial x^2}$$
$$x = 0: \quad T_1 = T_{s'} \qquad\qquad x = 0, L: \quad T_2 = 0$$
$$x = L: \quad T_1 = T_s \qquad\qquad t = 0: \quad T_2 = T_0 - T_1$$

$$T_1 = T_{s'} + (T_s - T_{s'})\frac{x}{L} \qquad T_2 = \sum_{n=1}^{\infty} B_n e^{-n^2 \pi^2 \text{Fo}} \sin\left(\frac{n\pi x}{L}\right)$$

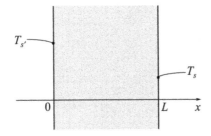

Figure 3.11 Schematic for the analysis of unsteady conduction in a slab with unequal surface temperatures.

Following the solution procedure for Eq.(3.38) the constants B_n must be determined from

$$T_2\,(t=0) = T_0 - T_{s'} - (T_s - T_{s'})\frac{x}{L} = \sum_{n=1}^{\infty} B_n \sin\left(\frac{n\pi x}{L}\right) \tag{3.48}$$

Figure 3.12 illustrates this superposition technique.

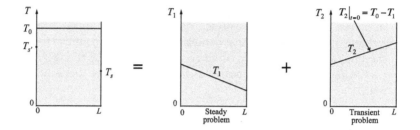

Figure 3.12   Schematic of superposition of solutions for a slab with unequal surface temperatures.

EXAMPLE 3.4   Transient Steam Condensation on a Concrete Wall

In a postulated nuclear reactor accident scenario, a concrete wall 20 cm thick at an initial temperature of 20°C is suddenly exposed on both sides to pure steam at atmospheric pressure. If the thermal resistance of the condensate flowing down the wall is negligible, estimate the rate of steam condensation on 160 m² wall area after (i) 10 s, (ii) 10 min, and (iii) 3 h.

Solution

Given: Concrete wall suddenly exposed to steam.

Required: Rate of steam condensation at various times.

Assumptions: 1. Thermal resistance of condensate negligible.

2. Pure steam, T_{sat} (1 atm) = 100°C.

The first step is to calculate the time constant for the wall, $t_c = L^2/\alpha$. From Table A.3, we take $\alpha = 0.75 \times 10^{-6}$ m²/s for concrete. Then

$$t_c = \frac{L^2}{\alpha} = \frac{(0.1)^2}{0.75 \times 10^{-6}} = 1.33 \times 10^4 \text{s (3.7 h)}$$

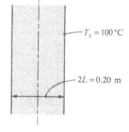

(i) $t = 10$ s:

Fo $= t/t_c = 10/1.33 \times 10^4 = 7.5 \times 10^{-4} < 0.05$. Thus, Eq. (3.47) applies. The condensation rate is $\dot{m} = q_s A / h_{fg}$ where h_{fg} is the enthalpy of vaporization of steam at 100°C. From steam tables (for example, Table A. 12a), $h_{fg} = 2.257 \times 10^6$ J/kg. Also required is the conductivity of concrete, which from Table A.3 is 1.4 W/m K.

$$\dot{m} = \frac{q_s A}{h_{fg}} = \frac{k(T_s - T_0)A}{(\pi \alpha t)^{1/2} h_{fg}} = \frac{(1.4)(100 - 20)(160)}{[\pi(0.75 \times 10^{-6})(10)]^{1/2}(2.257 \times 10^6)} = 1.64 \text{kg/s}$$

(ii) $t = 10$ min:

Fo $= t/t_c = (10)(60)/1.33 \times 10^4 = 0.0451 < 0.05$. Thus, Eq. (3.47) remains valid. Since q_s and, hence, $\dot{m}$ decrease like $t^{-1/2}$, the condensation rate is now

$$\dot{m} = 1.64 \left(\frac{600}{10} \right)^{-1/2} = 0.212 \text{kg/s}$$

(iii) $t = 3$ h:

Fo $= t/t_c = (3)(3600)/1.33 \times 10^4 = 0.812 > 0.2$. Equation (3.46) now applies, and the condensation rate is

$$\dot{m} = \frac{2k(T_s - T_0)A}{L h_{fg}} e^{-(\pi/2)^2 \text{Fo}} = \frac{(2)(1.4)(100 - 20)(160)}{(0.10)(2.257 \times 10^6)} e^{-(\pi/2)^2(0.812)} = 0.0214 \text{ kg/s}$$

Comments

These estimates may be regarded as upper limits: in practice the steam is likely to contain some noncondensables, such as air, and condense at a temperature lower than 100°C. The problem is then one involving simultaneous heat and mass transfer at high mass transfer rates, and is beyond the scope of this text.

3.4.2 The Semi-Infinite Solid

In Section 3.4.1, we saw that for sufficiently short times, temperature changes did not penetrate far enough into the slab for the thickness of the slab to have any effect on the heat conduction process. Such situations are encountered in practice. One method of case-hardening tool steel involves rapid quenching from a high temperature for a short time. Only the metal close to the surface is rapidly cooled and hardened; the interior cools slowly after the quenching process and remains ductile. Thus, during the quenching process, the interior temperature remains unchanged, and the precise thickness and shape of the tool is irrelevant. The conduction process is confined to a thin region near the surface into which temperature changes have penetrated. Thus, it is useful to have formulas giving the temperature distribution and heat flow for various kinds of boundary conditions that are applicable to these *penetration* problems.

The Governing Equation and Conditions

An appropriate model for penetration problems is transient conduction in a semi-infinite solid, as shown in Fig. 3.13. If we assume constant thermal conductivity, no internal heat generation, and negligible temperature variations in the y and z directions, Eq. (3.6) again reduces to

$$\frac{\partial T}{\partial t} = \alpha \frac{\partial^2 T}{\partial x^2} \tag{3.49}$$

If the solid is initially at a uniform temperature T_0, the appropriate initial condition is

$$t = 0: \quad T = T_0 \tag{3.50a}$$

The left face of the solid is suddenly raised to temperature T_s at time zero and held at that value. The two required boundary conditions are then

$$x = 0: \quad T = T_s \tag{3.50b}$$

$$x \to \infty: \quad T \to T_0 \tag{3.50c}$$

It is the second boundary condition, Eq. (3.50c), that makes this problem different from the slab problem of Section 3.4.1.

Solution for the Temperature Distribution

It might at first appear that the *separation of variables* solution method can be used once again. As in the slab analysis of Section 3.4.1, the variables are separable in the differential equation, Eq. (3.49). However, a necessary requirement for completing the solution is that the boundary conditions of the eigenvalue problem be specified on *coordinate surfaces*, and $x \to \infty$ is not a coordinate surface of the Cartesian coordinate system. Instead, we proceed as follows.

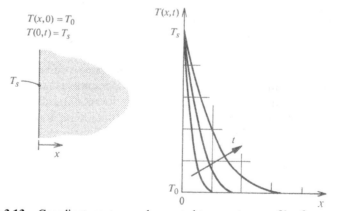

Figure 3.13 Coordinate system and expected temperature profiles for transient conduction in a semi-infinite solid with a step change in surface temperature.

We first normalize the dependent variable $T(x,t)$ for algebraic convenience. Defining $\theta = (T - T_0)/(T_s - T_0)$, the problem statement transforms into

$$\frac{\partial \theta}{\partial t} = \alpha \frac{\partial^2 \theta}{\partial x^2} \tag{3.51}$$

$$t = 0: \quad \theta = 0 \tag{3.52a}$$

$$x = 0: \quad \theta = 1 \tag{3.52b}$$

$$x \to \infty: \quad \theta \to 0 \tag{3.52c}$$

This mathematical problem can be reduced to the solution of an ordinary differential equation for θ. The independent variable in this equation is called a **similarity variable** and is an appropriate combination of x and t. One way to discover the combination is based on physical reasoning and dimensional considerations. Figure 3.13 shows the expected temperature profiles as a function of time as the heat penetrates into the solid. We can arbitrarily define some penetration depth δ, for example, the location where $\theta = 0.01$ or 0.001. Then δ is clearly a function of time t and thermal diffusivity α: the larger α, the deeper the penetration at a given time. Notice that the problem statement, Eqs. (3.51) and (3.52), contains no other quantities on which δ can depend. Time has units [s], and thermal diffusivity has units [m²/s]. The only combination of these two variables that will give the required units of [m] for the penetration depth is $(\alpha t)^{1/2}$. For the temperature profile to be a function of a single variable, distances into the solid must be *scaled* to this penetration depth. Thus, we will choose

$$\eta = \frac{x}{(4\alpha t)^{1/2}} \tag{3.53}$$

where the factor of $4^{1/2}$ has been inserted for future algebraic convenience. Eq. (3.51) is transformed by careful use of the rules of partial differentiation. The required differential operators are

$$\frac{\partial \theta}{\partial t} = \frac{d\theta}{d\eta} \frac{\partial \eta}{\partial t}\bigg|_x = \frac{d\theta}{d\eta}\left(-\frac{x}{2t(4\alpha t)^{1/2}}\right)$$

$$\frac{\partial \theta}{\partial x} = \frac{d\theta}{d\eta} \frac{\partial \eta}{\partial x}\bigg|_t = \frac{d\theta}{d\eta}\left(\frac{1}{(4\alpha t)^{1/2}}\right)$$

$$\frac{\partial^2 \theta}{\partial x^2} = \frac{1}{(4\alpha t)^{1/2}} \frac{d^2\theta}{d\eta^2} \frac{\partial \eta}{\partial x}\bigg|_t = \frac{1}{(4\alpha t)} \frac{d^2\theta}{d\eta^2}$$

Notice that we have written the derivatives of θ with respect to η as total derivatives since we have assumed that θ is a function of η alone. Substituting in Eq. (3.51) gives

$$-\frac{x}{2t(4\alpha t)^{1/2}} \frac{d\theta}{d\eta} = \frac{\alpha}{(4\alpha t)} \frac{d^2\theta}{d\eta^2}$$

which simplifies to

$$-2\eta \frac{d\theta}{d\eta} = \frac{d^2\theta}{d\eta^2} \tag{3.54}$$

Equation (3.54) is a second-order linear ordinary differential equation with a non-constant coefficient, which requires two boundary conditions. Transforming Eqs. (3.52) gives

$$\eta = 0: \quad \theta = 1, \quad \text{since } \eta = 0 \text{ when } x = 0 \tag{3.55a}$$

$$\eta \to \infty: \quad \theta = 0, \quad \text{since } \eta \to \infty \text{ when } x \to \infty, \text{ or } t \to 0 \tag{3.55b}$$

Since both the transformed equation and conditions do not depend on x or t, the *similarity transformation* has been successful. Let $d\theta/d\eta = p$; then Eq. (3.54) becomes the first-order equation

$$-2\eta p = \frac{dp}{d\eta}$$

or

$$\frac{dp}{p} = -2\eta \, d\eta = -d\eta^2$$

Integrating once,

$$p = \frac{d\theta}{d\eta} = C_1 e^{-\eta^2}$$

Integrating again,

$$\theta = C_1 \int_0^\eta e^{-u^2} du + C_2 \tag{3.56}$$

where u is a dummy variable for the integration. The integration constants are evaluated from the boundary conditions, Eq. (3.55). From Eq. (3.55a), $\theta = 1$ at $\eta = 0$; hence, $C_2 = 1$. From Eq. (3.55b), $\theta \to 0$ as $\eta \to \infty$,

$$0 = C_1 \int_0^\infty e^{-u^2} du + 1$$

The definite integral is given by standard integral tables as $\pi^{1/2}/2$; hence,

$$0 = C_1 \frac{\pi^{1/2}}{2} + 1 \quad \text{or} \quad C_1 = -\frac{2}{\pi^{1/2}}$$

Substituting back in Eq. (3.56) gives the temperature distribution as

$$\theta = \frac{T - T_0}{T_s - T_0} = 1 - \frac{2}{\pi^{1/2}} \int_0^\eta e^{-u^2} du \tag{3.57}$$

The function $(2/\pi^{1/2}) \int_0^\eta e^{-u^2} du$ is called the **error function**, erf η. For our purposes, it is more convenient to use the **complementary error function**, erfc $\eta = 1 - $ erf η, which is tabulated as Table B.4 in Appendix B. Then

$$\frac{T - T_0}{T_s - T_0} = \text{erfc} \frac{x}{(4\alpha t)^{1/2}} \tag{3.58}$$

Equation (3.58) is plotted as Fig. 3.14. Since the temperature profiles at any time fall on a single curve when plotted as $\theta(\eta)$, they are said to be *self-similar*, which is why η is called the *similarity variable* for the problem.

The Surface Heat Flux

The heat flux at the surface is found from Eq. (3.58) using Fourier's law and the chain rule of differentiation,

$$
\begin{aligned}
q_s &= -k \frac{\partial T}{\partial x}\bigg|_{x=0} \\
&= k(T_s - T_0) \frac{\partial}{\partial x} \left[1 - \frac{2}{\pi^{1/2}} \int_0^\eta e^{-u^2} du \right]_{x=0} \\
&= k(T_s - T_0) \left[-\frac{2}{\pi^{1/2}} e^{-\eta^2} \frac{1}{(4\alpha t)^{1/2}} \right]_{\eta=0} \\
q_s &= \frac{k(T_s - T_0)}{(\pi\alpha t)^{1/2}}
\end{aligned}
\tag{3.59}
$$

which is identical to the short-time solution for the slab problem, Eq. (3.47).

Solutions of the semi-infinite solid problem for other kinds of boundary conditions are also useful. A selection follows.

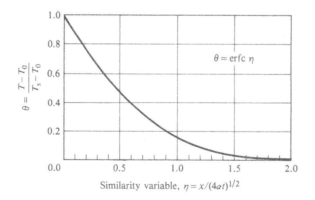

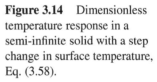

Figure 3.14 Dimensionless temperature response in a semi-infinite solid with a step change in surface temperature, Eq. (3.58).

Constant Surface Heat Flux

If at time $t = 0$ the surface is suddenly exposed to a constant heat flux q_s — for example, by radiation from a high-temperature source—the resulting temperature response is

$$T - T_0 = \frac{q_s}{k} \left[\left(\frac{4\alpha t}{\pi} \right)^{1/2} e^{-x^2/4\alpha t} - x \operatorname{erfc} \frac{x}{(4\alpha t)^{1/2}} \right] \tag{3.60}$$

This temperature response is shown in Fig. 3.15.

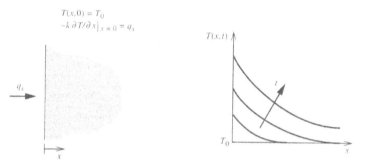

Figure 3.15   Temperature response in a semi-infinite solid exposed to a constant surface heat flux, Eq. (3.60).

Convective Heat Transfer to the Surface

If at time $t = 0$ the surface is suddenly exposed to a fluid at temperature T_e, with a convective heat transfer coefficient h_c, the resulting temperature response is

$$\frac{T - T_0}{T_e - T_0} = \operatorname{erfc} \frac{x}{(4\alpha t)^{1/2}} - e^{h_c x/k + (h_c/k)^2 \alpha t} \operatorname{erfc} \left(\frac{x}{(4\alpha t)^{1/2}} + \frac{h_c}{k} (\alpha t)^{1/2} \right) \tag{3.61}$$

as shown in Fig. 3.16.

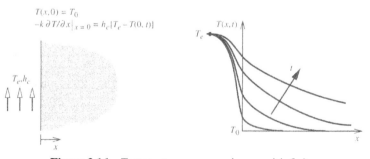

Figure 3.16   Temperature response in a semi-infinite solid suddenly exposed to a fluid, Eq. (3.61).

Surface Energy Pulse

If an amount of energy E per unit area is released instantaneously on the surface at $t = 0$ (e.g., if the surface is exposed to an energy pulse from a laser), and none of this energy is lost from the surface, the resulting temperature response is

$$T - T_0 = \frac{E}{\rho c (\pi \alpha t)^{1/2}} e^{-x^2/4\alpha t} \tag{3.62}$$

as shown in Fig. 3.17.

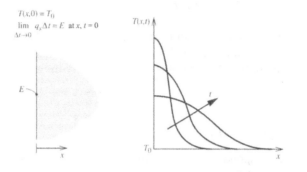

Figure 3.17 Temperature response in a semi-infinite solid after an instantaneous release of energy on the surface, Eq. (3.62).

Periodic Surface Temperature Variation

The surface temperature varies periodically as $(T_s - T_0) = (T_s^* - T_0)\sin \omega t$, as shown in Fig. 3.18. The resulting periodic (long-time) temperature response is

$$\frac{T - T_0}{T_s^* - T_0} = e^{-x(\omega/2\alpha)^{1/2}} \sin[\omega t - x(\omega/2\alpha)^{1/2}] \tag{3.63}$$

Notice how the amplitude of the temperature variation decays into the solid exponentially, while a phase lag $x(\omega/2\alpha)^{1/2}$ develops.

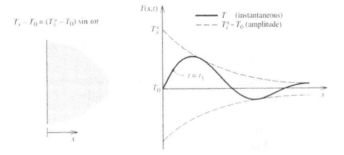

Figure 3.18 Periodic surface temperature variation for a semi-infinite solid: instantaneous temperature profile at $t = t_1$, Eq. (3.63).

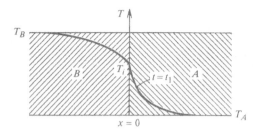

Figure 3.19   Contact of two semi-infinite solids: instantaneous temperature profile at $t = t_1$ for $k_B > k_A$.

Contact of Two Semi-Infinite Solids

Now consider two semi-infinite solids, A and B, of different materials with uniform temperatures T_A and T_B, brought together at time $t = 0$, as shown in Fig. 3.19. The solution of this problem shows that the interface temperature T_i is constant in time and is given by

$$\frac{T_A - T_i}{T_i - T_B} = \frac{k_B}{k_A}\left(\frac{\alpha_A}{\alpha_B}\right)^{1/2} = \left(\frac{(k\rho c)_B}{(k\rho c)_A}\right)^{1/2} \tag{3.64}$$

with corresponding erfc function temperature distributions in each solid. For example, if solid A is a carbon steel with $k = 48$ W/m K and $\alpha = 13.3 \times 10^{-6}$ m²/s at $100°$C, and solid B is neoprene rubber with $k = 0.19$ W/m K and $\alpha = 0.079 \times 10^{-6}$ m²/s at $0°$C, the interface temperature T_i is calculated to be $95.1°$C. This is much closer to the initial temperature of the steel than to that of the rubber. Equation (3.64) shows why a high-conductivity material at room temperature feels colder to the touch than does a low-conductivity material at the same temperature.

The instantaneous heat flux at the surface $q_s(t)$ can be found from Eqs. (3.61) through (3.63) by applying Fourier's law. The derivation of Eq. (3.63) is given as Exercise 3–37. The other solutions are best obtained using Laplace transforms [5, 9].

Computer Program COND1

COND1 calculates the thermal response of a semi-infinite solid initially at temperature T_0. There is a choice of five boundary conditions imposed at time $t = 0$:

1. The surface temperature is changed to T_s.
2. A heat flux q_s is imposed on the surface.
3. The surface is exposed to a fluid at temperature T_e, with a convective heat transfer coefficient h_c.
4. An amount of energy E is released instantaneously at the surface.
5. The surface temperature varies periodically as $T_s - T_0 = (T_s^* - T_0)\sin \omega t$.

Plotting options include $T(x), T_s(t)$, or $q_s(t)$, which can be chosen appropriately. The analysis for boundary condition 1 was given in Section 3.4.2. The temperature responses for boundary conditions 2 through 5 are given as Eqs. (3.60) through (3.63), respectively.

EXAMPLE 3.5   Cooling of a Concrete Slab

A thick concrete slab initially at 400 K is sprayed with a large quantity of water at 300 K. How long will the location 5 cm below the surface take to cool to 320 K?

Solution

Given: Hot concrete slab sprayed with water at 300 K.

Required: Rate of cooling 5 cm below surface.

Assumptions: 1. The rate of spraying is sufficient to maintain the surface at 300 K.
                      2. The slab can be treated as a semi-infinite solid.

Equation (3.58) applies and can be written as

$$\theta = \operatorname{erfc}\eta \quad \text{where } \theta = \frac{T - T_0}{T_s - T_0}, \eta = \frac{x}{(4\alpha t)^{1/2}}$$

$$\theta = \frac{320 - 400}{300 - 400} = 0.8$$

Thus, $0.8 = \operatorname{erfc}\eta$; from Table B.4, $\eta = 0.179$.

$$t = \frac{x^2}{4\alpha\eta^2}$$

From Table A.3, α for concrete is 0.75×10^{-6} m²/s. Thus, the required time is

$$t = \frac{(0.05)^2}{(4)(0.75 \times 10^{-6})(0.179)^2} = 2.60 \times 10^4 \text{ s} \simeq 7\,\text{h}$$

Comments

1. A temperature penetration depth δ_t may be defined as the location where the tangent to the temperature profile at $x = 0$ intercepts the line $T = 400$ K, as shown in the figure. The temperature gradient at $x = 0$ is found by differentiating Eq. (3.58):

$$-\left.\frac{\partial T}{\partial x}\right|_{x=0} = \frac{T_s - T_0}{(\pi\alpha t)^{1/2}} \equiv \frac{T_s - T_0}{\delta_t}$$

Hence, $\delta_t = 1.772(\alpha t)^{1/2} = 1.772(0.75 \times 10^{-6} \times 2.60 \times 10^4)^{1/2} = 0.247$ m.

2. Check t using COND1.

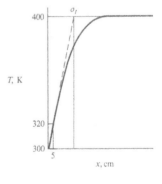

EXAMPLE 3.6   Radiative Heating of a Firewall

A 15 cm-thick concrete firewall has a black silicone paint surface. The wall is suddenly exposed to a radiant heat source that can be approximated as a blackbody at 1000 K. How long will it take for the surface to reach 500 K if the initial temperature of the wall is 300 K?

Solution

Given: Concrete firewall exposed to radiant heat source.

Required: Temperature response of surface.

1. Wall can be modeled as a semi-infinite solid.
2. Negligible temperature drop across the paint layer.
3. Negligible heat capacity of the paint layer.
4. An absorptance $\alpha = 0.9$ for high-heat black silicone paint (Table A.5a); negligible radiation emitted by the wall.

Equation (3.60) evaluated at $x = 0$ is

$$T_s - T_0 = \frac{q_s}{k}\left(\frac{4\alpha t}{\pi}\right)^{1/2}$$

or

$$t = \frac{\pi}{\alpha}\left[\frac{(T_s - T_0)k}{2q_s}\right]^2$$

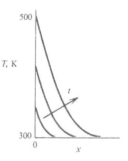

From Table A.3, the required concrete properties are $k = 1.4$ W/m K and $\alpha = 0.75 \times 10^{-6}$ m²/s.

Since the surface temperature is low compared to the radiation source temperature, we can neglect radiation emitted by the surface, so

$$q_s \simeq \alpha\sigma T^4 = (0.9)(5.67 \times 10^{-8})(1000)^4 = 51.0 \times 10^3 \text{ W/m}^2$$

$$t = \frac{\pi}{0.75 \times 10^{-6}}\left[\frac{(500 - 300)(1.4)}{(2)(51.0 \times 10^3)}\right]^2 = 31.6\,\text{s}$$

Comments

1. At most, $\varepsilon\sigma T_s^4 = (0.9)(5.67 \times 10^{-8})(500^4) = 3.19 \times 10^3$ W/m², which is only 6% of q_s and is justifiably neglected in making this engineering estimate.

2. Check to see if the temperature drop and heat capacity of the paint layer are negligible. Choose an appropriate thickness and properties.

3. To check whether the assumption of a semi-infinite solid is valid, we estimate a penetration depth δ_t:

$$\delta_t \simeq \frac{T_s - T_0}{-(\partial T/\partial x)_{x=0}}$$

$$= \frac{T_s - T_0}{q_s/k}$$

$$= \frac{500 - 300}{(51.0 \times 10^3)/(1.4)}$$

$$= 0.0055 \text{ m (5.5 mm), which is small.}$$

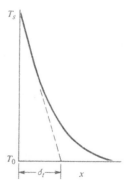

Solution Using COND1

The required input in SI units is:

Boundary condition $= 2$
Plot option $= 2$ (T_s versus t)
t-range for plot $= 0, 60$
Thermal conductivity $= 1.4$
Thermal diffusivity $= 0.75 \times 10^{-6}$
Initial temperature $T_0 = 300$
Surface heat flux $q_s = 51.0 \times 10^3$

From the graph of T_s versus t, $t \simeq 32$ (seconds) when $T_s = 500$ (kelvins).

Comments

1. A more accurate answer can be obtained from COND1 by adjusting the t-range.

2. Any consistent units can be used in COND1. With SI units, temperature may be in kelvins or degrees Celsius.

EXAMPLE 3.7   Thermal Response of Soil

On a tropical island, a large refrigerated shed has been operating at 5°C for a number of years. It is then put out of service; a wood floor is removed, and ambient air at 27°C is allowed to circulate freely through the shed. How long will it take for the ground 1 m below the surface to reach 15°C? Assume a convective heat transfer coefficient of 3.0 W/m² K, and use thermal properties of a wet soil ($k = 2.6$ W/m K, $\alpha = 0.45 \times 10^{-6}$ m²/s).

Solution

Given: Ground initially at 5°C, exposed to air at 27°C.

Required: Temperature response 1 m below surface.

Assumptions: 1. Semi-infinite solid model valid.
2. The initial temperature is uniform (i.e., 5°C for an appreciable distance below the surface).

Equation (3.61) applies:

$$\frac{T - T_0}{T_e - T_0} = \text{erfc} \, \frac{x}{(4\alpha t)^{1/2}} - e^{h_c x/k + (h_c/k)^2 \alpha t} \text{erfc} \left(\frac{x}{(4\alpha t)^{1/2}} + \frac{h_c}{k} (\alpha t)^{1/2} \right)$$

$$\frac{T - T_0}{T_e - T_0} = \frac{15 - 5}{27 - 5} = 0.4545; \quad \frac{h_c}{k} = \frac{3.0}{2.6} = 1.154 \, \text{m}^{-1}, \quad x = 1 \, \text{m}$$

An iterative solution is required, but how do we make a reasonable first guess for t? A lower limit is obtained if we use only the first term of Eq. (3.61), which corresponds to $h_c \to \infty$,

$T_s = T_e$.

$$0.4545 = \text{erfc } \eta$$

From Table B.4, $\eta = 0.53$.

$$0.53 = \frac{x}{(4\alpha t)^{1/2}} = \frac{1}{(4 \times 0.45 \times 10^{-6} t)^{1/2}}; \quad t = 2.0 \times 10^6 \text{ s}$$

The actual time will be greater than 2.0×10^6 s; taking $t = 4 \times 10^6$ s as a first guess, the following table summarizes the results:

Time, t $s \times 10^{-6}$	$\dfrac{T - T_0}{T_e - T_0}$
4	0.368
5	0.415
6	0.452
6.05	0.4540
6.06	0.4545

Hence, $t = 6.06 \times 10^6$ s ($\sim$ 70 days)

Comments

1. If the solution is done by hand, care must be taken to evaluate the erfc function accurately. COND1 will perform the required calculations rapidly and reliably.

2. The long time required suggests that problems involving conduction into the ground are almost always *transient* problems.

EXAMPLE 3.8   Temperature Fluctuations in a Diesel Engine Cylinder Wall

A thermocouple is installed in the 5 mm-thick cylinder wall of a stationary diesel engine, 1 mm below the inner surface. In a particular test, the engine operates at 1000 rpm, and the thermocouple reading is found to have a mean value of 322°C and an amplitude of 0.79°C. If the temperature variation can be assumed to be approximately sinusoidal, estimate the amplitude and phase difference of the inner-surface temperature variation. Take $\alpha = 12.0 \times 10^{-6}$ m^2/s and $k = 40$ W/m K for the carbon steel wall.

Solution

Given: Thermocouple installed in diesel engine cylinder wall.

Required: Amplitude and phase difference of inner-surface temperature.

Assumptions: Model wall as a semi-infinite solid.

At first it would appear that the analysis of Section 3.4.2 does not apply to this problem. The cylinder wall is not very thick, and since the outer surface is cooled, there will be a temperature gradient through the wall. However, if the temperature wave is damped out in a very short distance from the surface, Eq. (3.63) can be used to estimate the amplitude decay and phase lag:

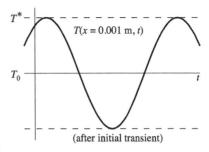

(after initial transient)

$$\frac{T - T_0}{T_s^* - T_0} = e^{-x(\omega/2\alpha)^{1/2}} \sin[\omega t - x(\omega/2\alpha)^{1/2}]$$

$$\omega = 2\pi \left(\frac{1000}{60}\right) = 104.7 \text{ rad/s}; \quad (\omega/2\alpha)^{1/2} = \left(\frac{104.7}{(2)(12.0 \times 10^{-6})}\right)^{1/2} = 2089 \text{ m}^{-1}$$

If $(T^* - T_0)$ is the amplitude of the temperature variation,

$$\frac{T^* - T_0}{T_s^* - T_0} = e^{-x(\omega/2\alpha)^{1/2}} = e^{-(0.001)(2089)} = 0.124$$

Thus,

$$T_s^* - T_0 = \frac{T^* - T_0}{0.124} = \frac{0.79}{0.124} = 6.38°C$$

The phase lag is $x(\omega/2\alpha)^{1/2} = 2.09 \text{ rad} = 120$ degrees.

Comments

Use COND1 to examine some spatial and temporal temperature profiles.

3.4.3 Convective Cooling of Slabs, Cylinders, and Spheres

We now consider the more general problem of transient conduction in three common shapes: the infinite slab, the infinite cylinder, and the sphere, with surface cooling (or heating) by convection. The slab problem will be analyzed first, and the results will be generalized to the cylinder and sphere.

Analysis for the Slab

In Section 3.4.1, we considered the temperature response $T(x,t)$ of a slab suddenly immersed in a fluid under conditions where the convective heat transfer resistance is negligible, that is, $Bi = h_c L/k$ is large. On the other hand, the lumped thermal capacity model of Section 1.5 applies when the conduction resistance in the slab is negligible, that is, the Biot number is small. We now consider the general case where the convection and conduction resistances are of comparable magnitudes, giving a Biot number of order unity. The slab and coordinate system are shown in Fig. 3.20.

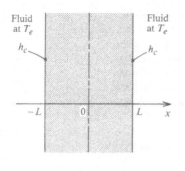

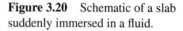

Figure 3.20 Schematic of a slab suddenly immersed in a fluid.

We again define $\eta = x/L$ and $\zeta = \alpha t/L^2$, but now we define the dimensionless temperature θ in terms of the initial slab temperature T_0 and fluid temperature T_e as $\theta = (T - T_e)/(T_0 - T_e)$. The analysis proceeds as in Section 3.4.1 to obtain

$$\theta(\zeta, \eta) = e^{-\lambda^2 \zeta}(A \cos \lambda \eta + B \sin \lambda \eta)$$

The boundary condition, Eq. (3.36b), is as before; namely, $\partial \theta/\partial \eta|_{\eta=0} = 0$ so $B = 0$ and

$$\theta(\zeta, \eta) = A e^{-\lambda^2 \zeta} \cos \lambda \eta \qquad (3.65)$$

However, the second boundary condition is now obtained from the requirement that the heat conduction at the surface of the solid equal the heat convection into the fluid:

$$-k \frac{\partial T}{\partial x}\bigg|_{x=L} = h_c(T|_{x=L} - T_e)$$

which transforms into

$$-\frac{k}{L} \frac{\partial \theta}{\partial \eta}\bigg|_{\eta=1} = h_c \theta|_{\eta=1}$$

or

$$-\frac{\partial \theta}{\partial \eta}\bigg|_{\eta=1} = \text{Bi } \theta|_{\eta=1} \qquad (3.66)$$

We see how a Biot number $\text{Bi} = h_c L/k$, where L is the slab half-width, occurs naturally when the convective boundary condition is put in dimensionless form. Substituting Eq. (3.65) into Eq. (3.66) gives

$$A e^{-\lambda^2 \zeta} \lambda \sin \lambda = \text{Bi } A e^{-\lambda^2 \zeta} \cos \lambda$$

or

$$\cot \lambda = \frac{\lambda}{\text{Bi}} \qquad (3.67)$$

which is a transcendental equation with an infinite number of roots or eigenvalues.

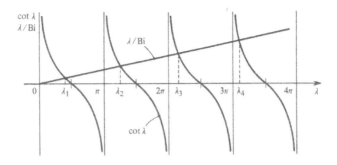

Figure 3.21 Graphical solution of the transcendental equation, Eq. (3.67): $\cot \lambda = \lambda/\text{Bi}$.

Figure 3.21 shows a plot of Eq. (3.67); for a given value of the Biot number, the eigenvalues $\lambda_n (n = 1, 2, 3, \ldots)$ can be calculated. To each of the eigenvalues corresponds a solution with eigenfunction $\cos \lambda_n \eta$ and arbitrary constant A_n. Their sum is the general solution:

$$\theta(\zeta, \eta) = \sum_{n=1}^{\infty} A_n e^{-\lambda_n^2 \zeta} \cos \lambda_n \eta \tag{3.68}$$

The constants A_n are evaluated from the initial condition, Eq. (3.36a):

$$\theta(0, \eta) = \sum_{n=1}^{\infty} A_n \cos \lambda_n \eta = 1$$

Thus, a Fourier series expansion in terms of the eigenfunctions $\cos \lambda_n \eta$ is required. The details are required as Exercise 3–56, and the result is

$$A_n = \frac{2 \sin \lambda_n}{\lambda_n + \sin \lambda_n \cos \lambda_n} \tag{3.69}$$

The solution in terms of temperature, Fourier number, and x/L is

$$\frac{T - T_e}{T_0 - T_e} = \sum_{n=1}^{\infty} e^{-\lambda_n^2 \text{Fo}} \frac{2 \sin \lambda_n}{\lambda_n + \sin \lambda_n \cos \lambda_n} \cos \lambda_n \frac{x}{L} \tag{3.70}$$

Figure 3.22 shows a plot of Eq. (3.70) for Bi = 3.0. Notice how the tangents to the temperature curves at the surface all intersect at a common point, the location of which is given by boundary condition, Eq. (3.66).

In addition to the temperature distribution, it is often useful to know the **fractional energy loss** Φ, which is the actual energy loss in time t divided by the total loss in cooling completely to the ambient temperature. An energy balance on unit area of the half slab gives

$$\Phi = \frac{\int_0^t q_s dt}{\rho c L (T_0 - T_e)} = \frac{\rho c L (T_0 - \overline{T})}{\rho c L (T_0 - T_e)} = 1 - \frac{\overline{T} - T_e}{T_0 - T_e}$$

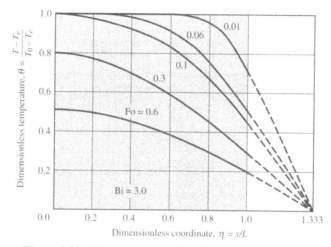

Figure 3.22 Temperature profiles for convective cooling of a slab calculated from Eq. (3.70); Bi = 3.0.

where $\overline{T}$ is the volume-averaged temperature. Evaluating $\overline{T}$ from Eq. (3.70) gives

$$\Phi = 1 - \sum_{n=1}^{\infty} e^{-\lambda_n^2 Fo} \frac{2\sin\lambda_n}{\lambda_n + \sin\lambda_n\cos\lambda_n} \frac{\sin\lambda_n}{\lambda_n} \tag{3.71}$$

Figure 3.23 shows a plot of Eq. (3.71) for various values of the Biot number.

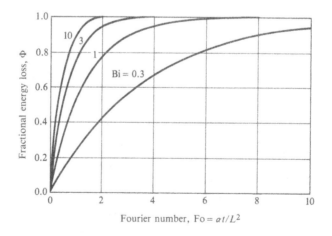

Figure 3.23 Fractional energy loss Φ as a function of Fourier number $\alpha t/L^2$ for convective cooling of a slab calculated from Eq. (3.71). Biot number Bi = 0.3, 1, 3, and 10.

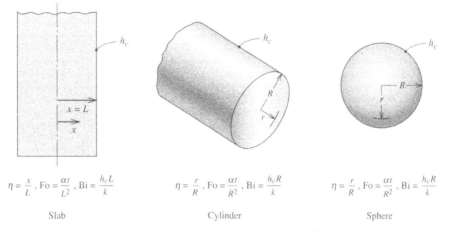

$$\eta = \frac{x}{L}, \text{Fo} = \frac{\alpha t}{L^2}, \text{Bi} = \frac{h_c L}{k} \qquad \eta = \frac{r}{R}, \text{Fo} = \frac{\alpha t}{R^2}, \text{Bi} = \frac{h_c R}{k} \qquad \eta = \frac{r}{R}, \text{Fo} = \frac{\alpha t}{R^2}, \text{Bi} = \frac{h_c R}{k}$$

Slab   Cylinder   Sphere

Figure 3.24 Schematic for the generalized solution of the temperature response of a convectively cooled slab, cylinder, or sphere.

Generalized Form of the Solution

It is convenient to have a general form of the solution applicable to a slab, cylinder, or sphere. Referring to Fig. 3.24,

$$\theta = \sum_{n=1}^{\infty} A_n e^{-\lambda_n^2 \text{Fo}} f_n(\lambda_n \eta) \tag{3.72}$$

$$\Phi = 1 - \bar{\theta} = 1 - \sum_{n=1}^{\infty} A_n e^{-\lambda_n^2 \text{Fo}} B_n \tag{3.73}$$

with the eigenvalues given by

Slab: $\text{Bi} \cos \lambda - \lambda \sin \lambda = 0$ (3.74a)

Cylinder: $\lambda J_1(\lambda) - \text{Bi} \, J_0(\lambda) = 0$ (3.74b)

Sphere: $\lambda \cos \lambda + (\text{Bi} - 1) \sin \lambda = 0$ (3.74c)

For the slab, $\eta = x/L$, $\text{Fo} = \alpha t/L^2$, and $\text{Bi} = h_c L/k$, as before; for the cylinder and sphere, $\eta = r/R$, $\text{Fo} = \alpha t/R^2$, and $\text{Bi} = h_c R/k$. Table 3.4 gives A_n, f_n, and B_n. In Eq. (3.74b) J_0 and J_1 are Bessel functions of the first kind, of orders 0 and 1, respectively. These functions are defined and tabulated in Appendix B.

Notice that the characteristic lengths used to define the Biot number are the slab half-width, L, and cylinder or sphere radius, R. Recall that in the lumped thermal capacity analysis of Section 1.5, we defined the characteristic length as volume/area, V/A. For a slab, $V/A = L$, so the Biot numbers are the same. However, for the cylinder and sphere, $V/A = R/2$ and $R/3$, respectively, and the Biot number definitions are different.

Table 3.4 The constants A_n and B_n and the function f_n for the transient thermal response of slabs, cylinders, and spheres.

Geometry	$A_n(\lambda_n)$	$B_n(\lambda_n)$	$f_n(\lambda_n\eta)$
Slab	$2\dfrac{\sin\lambda_n}{\lambda_n + \sin\lambda_n\cos\lambda_n}$	$\dfrac{\sin\lambda_n}{\lambda_n}$	$\cos\left(\lambda_n\dfrac{x}{L}\right)$
Cylinder	$2\dfrac{J_1\lambda_n}{\lambda_n[J_0^2(\lambda_n) + J_1^2(\lambda_n)]}$	$2\dfrac{J_1(\lambda_n)}{\lambda_n}$	$J_0\left(\lambda_n\dfrac{r}{R}\right)$
Sphere	$2\dfrac{\sin\lambda_n - \lambda_n\cos\lambda_n}{\lambda_n - \sin\lambda_n\cos\lambda_n}$	$3\dfrac{\sin\lambda_n - \lambda_n\cos\lambda_n}{\lambda_n^3}$	$\dfrac{\sin[\lambda_n(r/R)]}{\lambda_n(r/R)}$

Computer Program COND2

The generalized form of the solution described above is implemented in COND2. The program computes the eigenvalues from Eqs. (3.74) using Newton's method. Up to 40 eigenvalues are calculated in order to meet a specified accuracy of 10^{-4} in the dimensionless temperature θ. For very short times, more than 40 eigenvalues are required to obtain the desired accuracy. Thus, for Fourier number Fo $< 10^{-3}$, COND1 should be used, since the semi-infinite solid model is quite appropriate for such short times. The output can be obtained as numerical data, a plot of the temperature profile, $\theta(\eta)$, or a plot of the fractional energy loss as a function of time, Φ (Fo).

Approximate Solutions for Long Times

In Section 3.4.1, it was shown that the series solution converged rapidly for long times, and for Fo > 0.2, only the first term of the series need be retained for 2% accuracy. Usually we are most interested in the temperature at the center of the body ($x = 0$ or $r = 0$), where the response is slowest. Denoting the dimensionless center temperature as $\theta_c = (T_c - T_e)/(T_0 - T_e)$ and retaining only the first term of Eq. (3.72) gives

$$\theta_c = A_1 e^{-\lambda_1^2 \text{Fo}}, \quad \text{Fo} > 0.2 \tag{3.75}$$

since $f_1(\lambda_1\eta) = 1$ for $\eta = 0$. When only the first term of the series is retained, the *shape* of the temperature distribution is unchanging with time. Thus, the temperature at any location is simply related to the center temperature as

$$\theta = \theta_c f_1(\lambda_1\eta), \quad \text{Fo} > 0.2 \tag{3.76}$$

Similarly, retaining only one term in Eq. (3.73), the fractional energy loss is

$$\Phi = 1 - B_1\theta_c, \quad \text{Fo} > 0.2 \tag{3.77}$$

Table 3.5 gives values of λ_1^2, A_1, and B_1 as a function of Biot number.

Table 3.5 Constants in the one-term approximation for convective cooling of slabs, cylinders, and spheres.

Slab Bi	λ_1^2	A_1	B_1	Bi	λ_1^2	A_1	B_1
0.02	0.01989	1.0033	0.9967	2	1.160	1.180	0.8176
0.04	0.03948	1.0066	0.9934	4	1.600	1.229	0.7540
0.06	0.05881	1.0098	0.9902	6	1.821	1.248	0.7229
0.08	0.07790	1.0130	0.9871	8	1.954	1.257	0.7047
0.10	0.09678	1.016	0.9839	10	2.042	1.262	0.6928
0.2	0.1873	1.031	0.9691	20	2.238	1.270	0.6665
0.4	0.3519	1.058	0.9424	30	2.311	1.272	0.6570
0.6	0.4972	1.081	0.9192	40	2.349	1.272	0.6521
0.8	0.6257	1.102	0.8989	50	2.371	1.273	0.6490
1.0	0.7401	1.119	0.8811	100	2.419	1.273	0.6429
				∞	2.467	1.273	0.6366

Cylinder Bi	λ_1^2	A_1	B_1	Bi	λ_1^2	A_1	B_1
0.02	0.03980	1.0051	0.9950	2	2.558	1.338	0.7125
0.04	0.07919	1.010	0.9896	4	3.641	1.470	0.6088
0.06	0.1182	1.015	0.9844	6	4.198	1.526	0.5589
0.08	0.1568	1.020	0.9804	8	4.531	1.553	0.5306
0.10	0.1951	1.025	0.9749	10	4.750	1.568	0.5125
0.2	0.3807	1.049	0.9526	20	5.235	1.593	0.4736
0.4	0.7552	1.094	0.9112	30	5.411	1.598	0.4598
0.6	1.037	1.135	0.8753	40	5.501	1.600	0.4527
0.8	1.320	1.173	0.8430	50	5.556	1.601	0.4485
1.0	1.577	1.208	0.8147	100	5.669	1.602	0.4401
				∞	5.784	1.602	0.4317

Sphere Bi	λ_1^2	A_1	B_1	Bi	λ_1^2	A_1	B_1
0.02	0.05978	1.0060	0.9940	2	4.116	1.419	0.6445
0.04	0.1190	1.012	0.9881	4	6.030	1.720	0.5133
0.06	0.1778	1.018	0.9823	6	7.042	1.834	0.4516
0.08	0.2362	1.024	0.9766	8	7.647	1.892	0.4170
0.10	0.2941	1.030	0.9710	10	8.045	1.925	0.3952
0.2	0.5765	1.059	0.9435	20	8.914	1.978	0.3500
0.4	1.108	1.116	0.8935	30	9.225	1.990	0.3346
0.6	1.599	1.171	0.8490	40	9.383	1.994	0.3269
0.8	2.051	1.224	0.8094	50	9.479	1.996	0.3223
1.0	2.467	1.273	0.7740	100	9.673	1.999	0.3131
				∞	9.869	2.000	0.3040

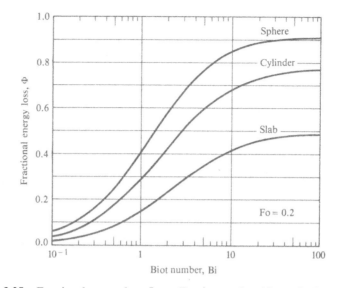

Figure 3.25 Fractional energy loss Φ at a Fourier number (dimensionless time) of Fo = 0.2 for the slab, cylinder, and sphere: effect of Biot number Bi. The one-term approximation is valid above each curve.

The single-term approximation is useful because it is valid during much of the cooling period. Figure 3.25 shows the fractional energy loss Φ at Fo = 0.2 as a function of Bi for the three shapes. The fraction of the cooling period over which the approximation is valid increases as the Biot number and surface-area/volume ratio (sphere-cylinder-slab) decrease. The single-term approximation is also used for problems in which the fluid temperature T_e changes slowly with time. It is then useful to define an *interior heat transfer coefficient* for conduction into the body for use in a thermal-circuit representation of the complete system. This concept is developed and used in Exercises 3–81 through 3–84.

Temperature Response Charts

Graphs of the various solutions for convective cooling, known as *temperature response charts*, were indispensable to engineers until handheld calculators and personal computers became standard tools. Charts for the series solution were given by Gurney and Lurie [10] as early as 1923, but better examples are now available. Perhaps the best widely available selection of charts is found in the *Handbook of Heat Transfer Fundamentals* [11]. In addition to charts for the convective cooling problem discussed here, Reference [11] gives charts for numerous other cooling and heating problems. For example, there are charts for radiation heating, conduction with internal heat generation, and composite solids. These charts were first published by Schneider in 1963 [12].

Heisler published charts for the one-term approximate solution in 1947 [13]. Heisler-type charts are found in many textbooks and are widely used. However, such charts have three major limitations:

1. The charts are invalid for Fo < 0.2.

2. The charts are often difficult to read accurately for Fo $\lesssim$ 1.

3. The cooling process is nearly complete over a large region of the charts. For example, for a sphere with Bi $\gtrsim$ 1, it is 90% complete for Fo $\gtrsim$ 1.

Appendix C contains temperature response charts for convective cooling. Two types of charts for the slab, cylinder, and sphere are given. Figure C.1 gives the temperature response of the body center ($x = 0$ or $r = 0$), and Fig. C.2 gives the fractional energy loss Φ as a function of time. Both figures are based on the complete solution and are thus valid for all values of Fourier number. The curves for Bi = 1000 in Fig. C.1 correspond to negligible convective resistance and, hence, to a prescribed surface temperature $T_s = T_e$.

Problem-Solving Strategy

Our solutions for convective cooling (or heating) of slabs, cylinders, and spheres can be organized in the form of a problem-solving strategy. The basic problem is one where the heat transfer coefficient is known, and the temperature or fractional heat loss must be calculated at a given time. Then we may proceed as follows:

1. Calculate the Fourier number. If Fo < 0.05, the semi-infinite solid solution, Eq. (3.61), applies. A handheld calculator or COND1 can be used.

2. If 0.05 < Fo < 0.2, the complete series solution, Eqs. (3.72) and (3.73), applies. COND2 or the temperature response charts in Appendix C should be used.

3. If Fo > 0.2, the long-time approximate solution, Eqs. (3.75) through (3.77), applies. COND2 can be used if available. Otherwise, the temperature response charts or a handheld calculator will suffice.

The simple lumped thermal capacity (LTC) model of Section 1.5 can be used in some situations:

4. Calculate Bi_{LTC}.[2] If $Bi_{LTC} < 0.1$, the LTC model can be used to calculate the fractional heat loss. However, if temperatures are required, Bi_{LTC} should be significantly less than 0.1 to ensure an accurate result near the surface.

Sometimes a problem may be posed in such a manner that the foregoing procedure cannot be followed exactly (as will be the case in Examples 3.9 and 3.10). Also, as in all engineering problem solving, the required accuracy of a particular calculation should be viewed in the context of the complete problem. It is of little value to obtain θ or Φ to even two-figure accuracy when the model is a poor simulation of the real

---

[2] $Bi_{LTC} = h_c(V/A)/k$; $V/A = L$ for an infinite slab, $R/2$ for an infinite cylinder, and $R/3$ for a sphere.

engineering problem. The major source of error here is in the specification of the heat transfer coefficient. Not only is it difficult to specify a value of h_c with less than 10% error, but in many cooling problems, h_c is not a constant as assumed by the model. For example, when the cooling is by natural convection, Eq. (1.23) shows that h_c is proportional to $(T_s - T_e)$ to the 1/4 power for laminar flow, and to the 1/3 power for turbulent flow. Thus, h_c must be estimated at some average temperature difference.

EXAMPLE 3.9   Annealing of Steel Plate

When steel plates are thinned by rolling, periodic reheating is required. Plain carbon steel plate 8 cm thick, initially at 440°C, is to be reheated to a minimum temperature of 520°C in a furnace maintained at 600°C. If the sum of the convective and radiative heat transfer coefficients is estimated to be 200 W/m² K, how long will the reheating take? Take $k = 40$ W/m K and $\alpha = 8.0 \times 10^{-6}$ m²/s for the steel.

Solution

Given: Steel plate, thickness $2L = 8$ cm.

Required: Temperature response of the center of the plate.

Assumptions: The heat transfer coefficient is constant at 200 W/m² K.

We first calculate the time constant for the heating process:

$$t_c = \frac{L^2}{\alpha} = \frac{(0.04)^2}{8 \times 10^{-6}} = 200\text{s}$$

which tells us the order of magnitude of the time required, namely, a few minutes (not seconds and not hours). Next we calculate the Biot number:

$$\text{Bi}_{LTC} = \frac{hL}{k} = \frac{(200)(0.040)}{40} = 0.2 \quad (= \text{Bi})$$

Since $\text{Bi}_{LTC} = \text{Bi} > 0.1$, the lumped thermal capacity model should not be used. Since we cannot calculate the Fourier number yet (it is the answer to the problem), the complete series solution will be used. The minimum temperature is at the center of the plate, and the desired value of θ is

$$\theta_c = \frac{T_c - T_e}{T_0 - T_e} = \frac{520 - 600}{440 - 600} = 0.50$$

Using the temperature response chart, Fig. C.1a in Appendix C,

$$\text{Fo} = 3.9; \quad t = t_c \, \text{Fo} = (200)(3.9) = 780 \text{ s} \ (13 \text{ min})$$

Since Fo > 0.2, we could have used the one-term approximation, Eq. (3.75):

$$\theta_c = 0.5 = A_1 e^{-\lambda_1^2 Fo}$$

From Table 3.5 for Bi = 0.2, $\lambda_1^2 = 0.1873$ and $A_1 = 1.031$. Solving, Fo = 3.86.

Solution Using COND2

The required inputs are:

Geometry $= 1$ (slab)
Bi $= 0.2$
Output option $= 2$ (θ vs. η plot)
Fo $=$ (must guess and iterate)
η range $= 0, 1$

A few iterations will give Fo $= 3.9$ for $\theta_c = 0.50$.

EXAMPLE 3.10   A Pebble Bed Air Heater

A pebble bed for storing thermal energy in a solar heating system has pebbles that can be approximated as 6 cm-diameter spheres. The bed is initially at 350 K before cold air at 280 K is admitted to the bed. If the heat transfer coefficient is 80 W/m² K, how long will it take the pebbles at the inlet of the bed to lose 90% of their available energy? Take $k = 1.6$ W/m K and $\alpha = 0.7 \times 10^{-6}$ m²/s for the pebbles.

Solution

Given: Hot pebbles suddenly exposed to a cold air stream.

Required: Time for pebbles at inlet to lose 90% of their available energy.

Assumptions: Pebbles are spherical.

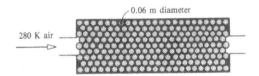

We first calculate the time constant for the cooling process:

$$t_c = \frac{R^2}{\alpha} = \frac{(0.03)^2}{0.7 \times 10^{-6}} = 1286 \text{ s } (21 \text{ min})$$

Next we calculate the lumped thermal capacity model Biot number:

$$\text{Bi}_{LTC} = \frac{h_c(R/3)}{k} = \frac{(80)(0.03/3)}{1.6} = 0.5$$

Since $\text{Bi}_{LTC} > 0.1$, the lumped thermal capacity method cannot be used. Although we suspect that Fo > 0.2 for $\Phi = 0.9$, we can conveniently use the complete series solution given as the temperature response chart, Fig. C.2c of Appendix C. For Bi $= h_c R/k = (3)(0.5) = 1.5$ and $\Phi = 0.9$, the chart gives Bi² Fo $= 1.5$. Thus,

$$\text{Fo} = 1.5/\text{Bi}^2 = 1.5/(1.5)^2 = 0.67$$

$$t = t_c \text{ Fo} = (1286)(0.67) = 860 \text{ s } (\sim 14 \text{ min})$$

Since Fo > 0.2, we could have used the one-term approximate solution as follows. From Eq. (3.77), $\Phi = 1 - B_1\theta_c$. For Bi $= 1.5$, interpolation in Table 3.5 gives $B_1 = 0.70$. Thus,

$$0.9 = 1 - (0.70)\theta_c \quad \text{or} \quad \theta_c = 0.143$$

By Eq. (3.75), $\theta_c = A_1 e^{-\lambda_1^2 \text{Fo}}$. For Bi $= 1.5$, interpolation in Table 3.5 gives $\lambda_1^2 = 3.33$, $A_1 = 1.38$.

$$0.143 = (1.38)e^{-(3.33)\text{Fo}} \quad \text{or} \quad \text{Fo} = 0.68$$

which agrees well with the chart solution.

Solution Using COND2

The required input is:

Geometry $= 3$ (sphere)
Bi $= 1.5$
Output option $= 1$ (Numerical data), or 3 (Φ vs. Fo plot)
Fo range $= 0, 1$

Fo $= 0.67$ for $\Phi = 0.9$.

Comments

The relatively short time of 14 min does not mean that warm air cannot be obtained for a long period. The bed is a regenerative heat exchanger: a temperature "wave" passes slowly through the bed, and the useful operating time is the time taken for this wave to break through the outlet end of the bed.

3.4.4 Product Solutions for Multidimensional Unsteady Conduction

Consider a long rectangular bar, with sides $2L_1$ and $2L_2$ wide, that is initially at temperature T_0 and suddenly immersed in a fluid at temperature T_e. The heat transfer coefficients on the sides are h_{c1} and h_{c2}. The task is to determine the temperature distribution $T(x,y,t)$. Again, a dimensionless temperature $\theta = (T - T_e)/(T_0 - T_e)$ is defined. The governing equation and appropriate initial and boundary conditions in a coordinate system such that $-L_1 \leq x \leq L_1, -L_2 \leq y \leq L_2$ are

$$\frac{\partial\theta}{\partial t} = \alpha\left(\frac{\partial^2\theta}{\partial x^2} + \frac{\partial^2\theta}{\partial y^2}\right) \tag{3.78}$$

$$t = 0: \quad \theta = 1$$

$$x = 0: \quad \frac{\partial\theta}{\partial x} = 0; \quad y = 0: \quad \frac{\partial\theta}{\partial y} = 0$$

$$x = L_1: \quad -k\frac{\partial\theta}{\partial x} = h_{c1}\theta; \quad y = L_2: \quad -k\frac{\partial\theta}{\partial y} = h_{c2}\theta$$

If the *separation of variables* method is to be used, one might assume a product solution of the form

$$\theta(t,x,y) = \mathscr{T}(t)X(x)Y(y)$$

where the functions $\mathscr{T}(t)$, $X(x)$, and $Y(y)$ are to be determined as before. However, it will now be shown that the solution can be expressed as the product of known solutions for the infinite slab. Consider two slabs of thickness $2L_1$ and $2L_2$, for which the dimensionless temperature governing equations and boundary conditions are

$$\theta_1 = \frac{T_1 - T_e}{T_0 - T_e} \qquad \theta_2 = \frac{T_2 - T_e}{T_0 - T_e}$$

$$\frac{\partial \theta_1}{\partial t} = \alpha \frac{\partial^2 \theta_1}{\partial x^2} \qquad \frac{\partial \theta_2}{\partial t} = \alpha \frac{\partial^2 \theta_2}{\partial y^2} \qquad \text{(3.79a,b)}$$

$$t = 0: \quad \theta_1 = 1 \qquad\qquad t = 0: \quad \theta_2 = 1$$

$$x = 0: \quad \frac{\partial \theta_1}{\partial x} = 0 \qquad\qquad y = 0: \quad \frac{\partial \theta_2}{\partial y} = 0$$

$$x = L_1: \quad -k\frac{\partial \theta_1}{\partial x} = h_{c1}\theta_1 \qquad y = L_2: \quad -k\frac{\partial \theta_2}{\partial y} = h_{c2}\theta_2$$

The product of the solutions of these two problems satisfies the original problem. Let

$$\theta(t,x,y) = \theta_1(t,x)\theta_2(t,y)$$

Then

$$\frac{\partial^2 \theta}{\partial x^2} = \theta_2 \frac{\partial^2 \theta_1}{\partial x^2}; \qquad \frac{\partial^2 \theta}{\partial y^2} = \theta_1 \frac{\partial^2 \theta_2}{\partial y^2} \qquad \text{(3.80a,b)}$$

$$\frac{\partial \theta}{\partial t} = \theta_1 \frac{\partial \theta_2}{\partial t} + \theta_2 \frac{\partial \theta_1}{\partial t} \qquad \text{(3.81)}$$

Substituting Eqs. (3.79a,b) into Eq. (3.81),

$$\frac{\partial \theta}{\partial t} = \alpha \left(\theta_1 \frac{\partial^2 \theta_2}{\partial y^2} + \theta_2 \frac{\partial^2 \theta_1}{\partial x^2} \right)$$

Then substituting from Eqs. (3.80a,b) gives

$$\frac{\partial \theta}{\partial t} = \alpha \left(\frac{\partial^2 \theta}{\partial x^2} + \frac{\partial^2 \theta}{\partial y^2} \right)$$

which is the original differential equation, Eq. (3.78). Also, the initial and boundary conditions become

$$t = 0: \quad \theta(0, x, y) = \theta_1(0, x)\theta_2(0, y) = (1)(1) = 1$$

$$x = 0: \quad \frac{\partial \theta}{\partial x} = \theta_2 \frac{\partial \theta_1}{\partial x} = \theta_2 \times 0 = 0$$

$$y = 0: \quad \frac{\partial \theta}{\partial y} = \theta_1 \frac{\partial \theta_2}{\partial y} = \theta_1 \times 0 = 0$$

$$x = L_1: \quad -k \frac{\partial \theta}{\partial x} = \theta_2 \left(-k \frac{\partial \theta_1}{\partial x}\right) = \theta_2(h_{c1}\theta_1) = h_{c1}\theta$$

$$y = L_2: \quad -k \frac{\partial \theta}{\partial y} = \theta_1 \left(-k \frac{\partial \theta_2}{\partial y}\right) = \theta_1(h_{c2}\theta_2) = h_{c2}\theta$$

which are the original conditions. Figure 3.26 shows a schematic of the product solution.

Shapes amenable to product solutions of this type are shown in Table 3.6. Langston [14] has recently shown how the product rule can be applied to obtain the fractional energy loss. If the shape is formed by the intersection of two bodies, for example, the short cylinder of item 4 in Table 3.6, the fractional energy loss is

$$\Phi = \Phi_1 + \Phi_2(1 - \Phi_1) = \Phi_1 + \Phi_2 - \Phi_1\Phi_2 \qquad (3.82a)$$

where subscripts 1 and 2 refer to the infinite slab and infinite cylinder, respectively. If the shape is formed by the intersection of three bodies, for example, the rectangular block of item 6 in Table 3.6, then

$$\Phi = \Phi_1 + \Phi_2(1 - \Phi_1) + \Phi_3(1 - \Phi_1)(1 - \Phi_2) \qquad (3.82b)$$

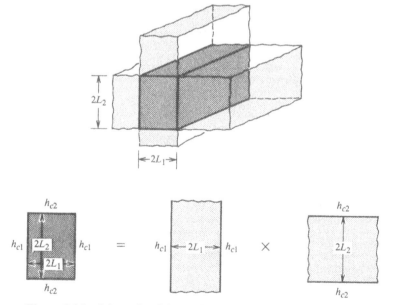

Figure 3.26   Schematic of the product solution procedure for the temperature response of a convectively cooled long rectangular bar.

Table 3.6   Shapes amenable to product solutions. *Caution:* The dimensionless temperatures S, P, and C are evaluated at the same value of actual time (not Fourier number).

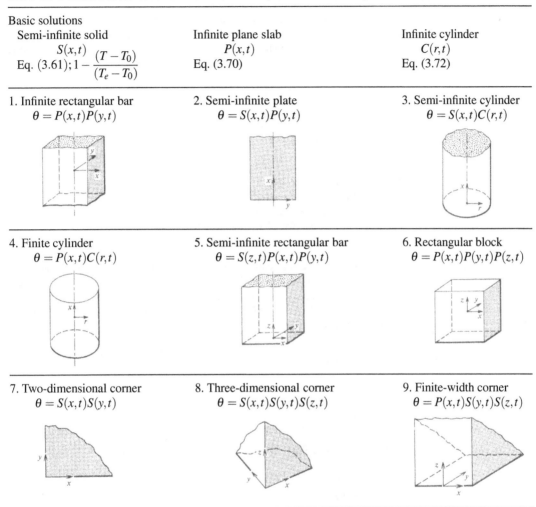

Basic solutions

Semi-infinite solid	Infinite plane slab	Infinite cylinder
$S(x,t)$	$P(x,t)$	$C(r,t)$
Eq. (3.61); $1 - \dfrac{(T - T_0)}{(T_e - T_0)}$	Eq. (3.70)	Eq. (3.72)

1. Infinite rectangular bar	2. Semi-infinite plate	3. Semi-infinite cylinder
$\theta = P(x,t)P(y,t)$	$\theta = S(x,t)P(y,t)$	$\theta = S(x,t)C(r,t)$

4. Finite cylinder	5. Semi-infinite rectangular bar	6. Rectangular block
$\theta = P(x,t)C(r,t)$	$\theta = S(z,t)P(x,t)P(y,t)$	$\theta = P(x,t)P(y,t)P(z,t)$

7. Two-dimensional corner	8. Three-dimensional corner	9. Finite-width corner
$\theta = S(x,t)S(y,t)$	$\theta = S(x,t)S(y,t)S(z,t)$	$\theta = P(x,t)S(y,t)S(z,t)$

Notice that these product solutions are not applicable when

1. The initial temperature of the body is nonuniform.

2. The fluid temperature T_e is not the same on all sides of the body.

3. The surface boundary conditions are of the second kind.

The requirement that the initial temperature of the body be uniform is, of course, the specified initial condition for the solutions given by Eqs. (3.72) and (3.73). However, the beginning student often overlooks this point in problem solving.

EXAMPLE 3.11   Sterilization of a Can of Vegetables

A can of vegetables 10 cm in diameter and 8 cm high is to be sterilized by immersion in saturated steam at 105°C. If the initial temperature is 40°C, what will be the minimum temperature in the can after 80 minutes? Also, calculate the total heat transfer to the can in this period.

Solution

Given: Can of vegetables to be sterilized in steam.

Required: (i) Minimum temperature in can after 80 minutes, and
(ii) total heat transfer.

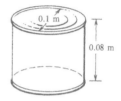

Assumptions: **1.** The heat transfer coefficient for condensing steam is very large; hence, Bi → ∞.
2. No circulation inside can.
3. Thermal diffusivity of contents approximates that of water.

(i) Since Bi → ∞, the lumped thermal capacity model cannot be used. Item 4 of Table 3.6 applies, with the minimum temperature at the center of the can. The thermal diffusivity is evaluated at a guessed average temperature of 360 K; using the data in Table A.8 of Appendix A, $k = 0.676$ W/m K, $\rho = 967$ kg/m^3, $c = 4200$ J/kg K; hence, $\alpha = k/\rho c = (0.676)/(967)(4200) = 0.166 \times 10^{-6}$ m^2/s. The two Fourier numbers are:

$$\text{Slab with 4 cm half-width:} \qquad \text{Fo}_1 = \frac{\alpha t}{L^2} = \frac{(0.166 \times 10^{-6})(4800)}{(0.04)^2} = 0.498$$

$$\text{Infinite cylinder of 5 cm radius:} \quad \text{Fo}_2 = \frac{\alpha t}{R^2} = \frac{(0.166 \times 10^{-6})(4800)}{(0.05)^2} = 0.319$$

Then $\theta_c = P(0,t)C(0,t)$, where $P(0,t)$ and $C(0,t)$ can be obtained from Fig. C.1a and b of Appendix C or by using COND2. Using the charts with Bi = 1000, $P(0,t) = 0.38$ and $C(0,t) = 0.27$.

$$\theta_c = \frac{T_c - T_e}{T_0 - T_e} = P(0,t)C(0,t)$$

$$\frac{T_c - 105}{40 - 105} = (0.38)(0.27); \qquad \text{solving, } T_c = 98.3°\text{C}$$

(ii) The total heat transfer is related to the fractional energy gain as

$$Q = \Phi \rho c V (T_e - T_0)$$

where V is the volume of the can and $\Phi = \Phi_1 + \Phi_2 - \Phi_1\Phi_2$ from Eq. (3.82a). Using COND2 or Fig. C.2a and b with Bi = 50, $\Phi_1 = 0.75$, $\Phi_2 = 0.88$. Thus,

$$\Phi = 0.75 + 0.88 - (0.75)(0.88) = 0.97$$

$$Q = (0.97)(967)(4200)(\pi)(0.05)^2(0.08)(105 - 40) = 161 \text{ kJ}$$

Comments

Although the heat transfer coefficient for condensing steam is large (see Chapter 7), it is not infinite. Using the curves for the largest values of Bi available in the charts gives satisfactory estimates for this problem.

3.5   NUMERICAL SOLUTION METHODS

Although many simple steady-state and transient heat conduction problems can be solved analytically, solutions for more complex problems are best obtained numerically. Numerical solution methods are particularly useful when the shape of the solid is irregular, when thermal properties are temperature- or position-dependent, and when boundary conditions are nonlinear. Numerical methods commonly used include the *finite-difference method*, the *finite-volume method*, and the *finite-element method*. *Boundary-element* and other meshless numerical methods are popular as research tools but not for commercial codes. The finite-difference method was the first numerical method to be used extensively for heat conduction. It remains a popular method, not because it is superior to other methods for heat conduction, but because it is easier to implement and is also one of the most widely used numerical solution methods for heat convection problems.

The first step in a finite-difference solution procedure is to discretize the spatial and time coordinates to form a *mesh of nodes*. Next, finite-difference approximations are made to the derivatives appearing in the heat conduction equation to convert the *differential* equation to an algebraic *difference* equation. Alternatively, the algebraic equation can be constructed by applying the energy conservation principle directly to a volume element surrounding the node, resulting in a finite-volume formulation. The finite-volume method makes use of the divergence theorem to reduce volume integrals to surface integrals that are then discretized numerically to produce algebraic coefficients for each surface of the volume. In steady-state problems, a set of linear algebraic equations is obtained for both methods with as many unknowns as the number of nodes in the mesh. These equations can be solved by matrix inversion or by iteration. For transient conduction, temperatures at the current time step may be found directly using values at the preceding time step. In some formulations, iteration may be required, since values at the current time step are also involved. In the first applications of finite-difference methods to heat conduction, the calculations were done by hand, limiting consideration to a coarse mesh with relatively few nodes. Because the accuracy of a finite-difference approximation increases with number of nodes, these solutions were inadequate.

The availability of mainframe digital computers in the late 1950s completely changed the picture, and by the 1960s, engineers could use as fine a grid as was necessary to meet their requirements. The 1970s saw increased use of the programmable calculator to obtain finite-difference solutions to heat conduction problems, which meant that an engineer would write a computer program for the specific problem under consideration. In the 1980s and 1990s, powerful personal computers have become available. Furthermore, there is no longer the need, nor is it cost-effective, for engineers to write their own computer programs to implement a numerical solution method. There are many standard computer programs available for this purpose. Some examples are PHOENICS, SINDA, ANSYS, COMSOL and OpenFOAM.

Thus, in this section, finite-difference methods are presented with the modest objective of giving the student an appreciation of the essential ideas involved, so that the available computer programs can be used intelligently. Any serious endeavor to develop engineering tools based on computational solution procedures should be preceded by an appropriate course in numerical analysis.

Finite-element methods are widely used in structural mechanics and are also applicable to heat conduction problems. An object is divided into discrete spatial regions called *finite elements*. The most common two-dimensional element is the triangle, and the most common three-dimensional element is the tetrahedron. The finite-element method allows the heat conduction equation to be satisfied in an average sense over the finite element; thus, the elements can be much larger than the control volumes used in finite-difference methods. The use of triangles or tetrahedrons for elements allows the approximation of complex and irregularly shaped objects. Application of the method leads to a set of algebraic equations, which are solved by matrix inversion or iteration. Compared to finite-difference methods, the formulation of these equations is considerably more involved and requires more effort, as does writing a computer program to implement the procedure. However, once written, finite-element computer programs tend to be more versatile than their finite-difference counterparts. Choice of which method to use is perhaps dictated by the objective rather than by the intrinsic virtues of the method. For example, calculation of temperature variations in solids is often required for the purpose of determining thermal stresses. Since the finite-element method is preeminent for stress calculations, many standard computer codes (e.g., COMSOL) use the finite-element method to calculate both temperatures and stresses in one package.

Numerical solution methods can be also readily implemented in modern numerical or symbolic computing environments such as Python, MATLAB or Mathematica.

3.5.1 A Finite-Difference Method for Two-Dimensional Steady Conduction

Consider two-dimensional steady conduction with volumetric heat generation and constant thermal properties. Using Cartesian coordinates, Fig. 3.27 shows a finite control volume Δx by Δy of unit depth surrounding node (m,n) within the solid. The m and n indices denote x and y locations, respectively, in a uniform mesh of node points. For convenience, we will use compass directions to denote the faces of the element (N, S, E, and W). The energy conservation principle, Eq. (1.2), applied to the finite control volume reduces to

$$0 = \dot{Q} + \dot{Q}_v$$

The heat transfer $\dot{Q}$ is by conduction across the four faces of the element; thus,

$$0 = \dot{Q}_x|_W + \dot{Q}_y|_S - \dot{Q}_x|_E - \dot{Q}_y|_N + \Delta\dot{Q}_v \tag{3.83}$$

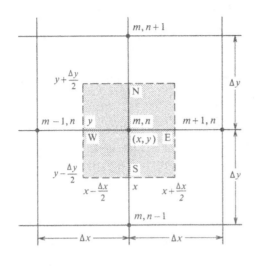

Figure 3.27 Finite control volume Δx by Δy by 1 surrounding node (m,n) at the location (x,y) used to derive the difference equation for two-dimensional steady conduction.

since the sign convention in Eq. (1.2) requires heat transfer into the system to be positive. The heat conduction across the left-hand face of the element is

$$\dot{Q}_x|_W = -k\frac{\partial T}{\partial x}\bigg|_W \Delta y \cdot 1$$

To approximate the derivative of $T(x,y)$ we will assume a linear temperature gradient between the nodes $(m-1,\,n)$ and $(m,\,n)$; then

$$\dot{Q}_x|_W = -k\frac{T_{m,n}-T_{m-1,n}}{\Delta x}\Delta y$$

which is seen to be positive for $T_{m-1,n} > T_{m,n}$. Similarly, for the right-hand, bottom, and top faces,

$$\dot{Q}_x|_E = -k\frac{T_{m+1,n}-T_{m,n}}{\Delta x}\Delta y$$

$$\dot{Q}_y|_S = -k\frac{T_{m,n}-T_{m,n-1}}{\Delta y}\Delta x$$

$$\dot{Q}_y|_N = -k\frac{T_{m,n+1}-T_{m,n}}{\Delta y}\Delta x$$

The internal heat generation is simply $\dot{Q}_v'''$ times the volume of the element:

$$\Delta\dot{Q}_v = \dot{Q}_v''' \Delta x \Delta y \cdot 1$$

Substituting in Eq. (3.83), multiplying by $\Delta x / k \Delta y$, and rearranging,

$$2(1+\beta)T_{m,n} = T_{m-1,n} + T_{m+1,n} + \beta(T_{m,n-1} + T_{m,n+1}) + \frac{\dot{Q}_v'''}{k}\Delta x^2 \qquad (3.84)$$

where $\beta = (\Delta x / \Delta y)^2$ is a geometric mesh factor. For a square mesh $\Delta x = \Delta y, \beta = 1$, and no internal heat generation,

$$4T_{m,n} = T_{m,n+1} + T_{m+1,n} + T_{m,n-1} + T_{m-1,n} \qquad (3.85)$$

which simply states that the temperature at each node is the arithmetic average of the temperatures at the four nearest neighboring nodes.

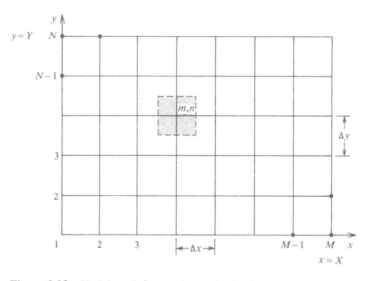

Figure 3.28 Nodal mesh for steady conduction in a rectangular plate.

Boundary Conditions

Figure 3.28 shows a complete mesh for a rectangular plate of dimensions X, Y. There are M node points in the x direction and N node points in the y direction. If the boundary condition is one of prescribed temperature, the temperatures at the boundary nodes are known. If the boundary condition is one of prescribed heat flux or convection, then the finite-difference form of the condition is obtained by making an energy balance on a finite control volume adjacent to the boundary. For example, consider the convection boundary condition shown in Fig. 3.29, and assume $\dot{Q}_v''' = 0$ for simplicity,

$$\begin{array}{c}\text{Net heat conduction} \\ \text{into the volume}\end{array} + \begin{array}{c}\text{Heat convection across} \\ \text{face at } x = 0\end{array} = 0$$

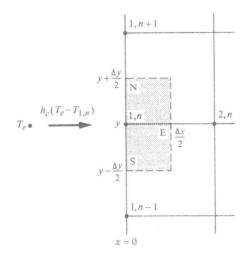

Figure 3.29 Finite control volume used to derive the difference equation for a convective boundary condition at $x = 0$: two-dimensional steady conduction.

As before,

$$\dot{Q}_x|_E = -k\frac{T_{2,n} - T_{1,n}}{\Delta x}\Delta y$$

$$\dot{Q}_y|_S = -k\frac{T_{1,n} - T_{1,n-1}}{\Delta y}\frac{\Delta x}{2}$$

$$\dot{Q}_y|_N = -k\frac{T_{1,n+1} - T_{1,n}}{\Delta y}\frac{\Delta x}{2}$$

and

$$\dot{Q}_x|_0 = h_c(T_e - T_{1,n})\Delta y$$

Substituting in the energy balance and taking $\Delta x = \Delta y$,

$$T_{2,n} + \frac{1}{2}(T_{1,n-1} + T_{1,n+1}) - 2T_{1,n} + \frac{h_c\Delta x}{k}(T_e - T_{1,n}) = 0$$

A mesh Biot number is defined as $\mathrm{Bi} = h_c\Delta x/k$; then solving for $T_{1,n}$ gives

$$T_{1,n} = \frac{1}{2 + \mathrm{Bi}}\left[T_{2,n} + \frac{1}{2}(T_{1,n-1} + T_{1,n+1}) + \mathrm{Bi}T_e\right] \qquad (3.86)$$

Table 3.7 gives similar results for a variety of boundary conditions, including interior and exterior corners. In this table, a simplified node numbering scheme is used for clarity.

Table 3.7   Finite-difference approximations for steady-state conduction, square mesh.

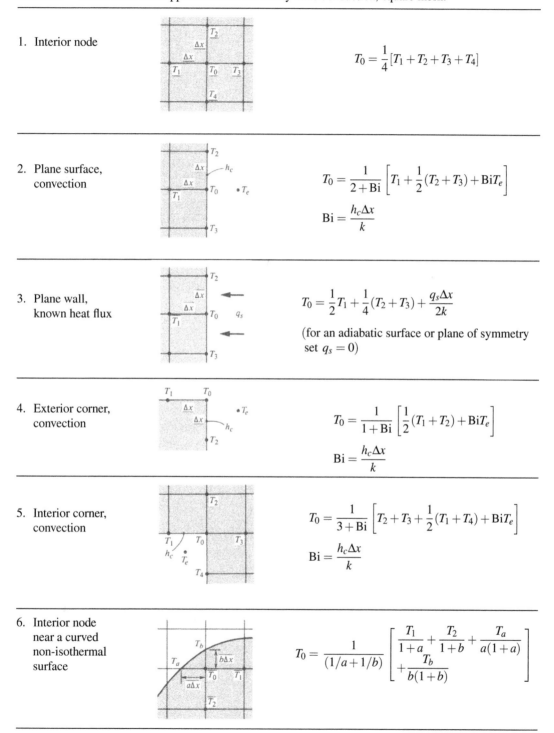

1. Interior node

$$T_0 = \frac{1}{4}[T_1 + T_2 + T_3 + T_4]$$

2. Plane surface, convection

$$T_0 = \frac{1}{2 + \mathrm{Bi}}\left[T_1 + \frac{1}{2}(T_2 + T_3) + \mathrm{Bi}T_e\right]$$

$$\mathrm{Bi} = \frac{h_c \Delta x}{k}$$

3. Plane wall, known heat flux

$$T_0 = \frac{1}{2}T_1 + \frac{1}{4}(T_2 + T_3) + \frac{q_s \Delta x}{2k}$$

(for an adiabatic surface or plane of symmetry set $q_s = 0$)

4. Exterior corner, convection

$$T_0 = \frac{1}{1 + \mathrm{Bi}}\left[\frac{1}{2}(T_1 + T_2) + \mathrm{Bi}T_e\right]$$

$$\mathrm{Bi} = \frac{h_c \Delta x}{k}$$

5. Interior corner, convection

$$T_0 = \frac{1}{3 + \mathrm{Bi}}\left[T_2 + T_3 + \frac{1}{2}(T_1 + T_4) + \mathrm{Bi}T_e\right]$$

$$\mathrm{Bi} = \frac{h_c \Delta x}{k}$$

6. Interior node near a curved non-isothermal surface

$$T_0 = \frac{1}{(1/a + 1/b)}\left[\frac{T_1}{1 + a} + \frac{T_2}{1 + b} + \frac{T_a}{a(1 + a)} + \frac{T_b}{b(1 + b)}\right]$$

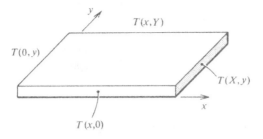

Figure 3.30 Schematic of rectangular plate with prescribed edge temperatures.

Solution Procedures

Consider the simplest case of a rectangular plate with prescribed boundary temperatures, as shown in Fig. 3.30. Referring to Fig. 3.28, the plate has dimensions X and Y, with M and N nodes in the x and y directions, respectively; then $\Delta x = X/(M-1), \Delta y = Y/(N-1)$. Equation (3.85) is a set of $(M-2) \times (N-2)$ linear algebraic equations in the $(M-2) \times (N-2)$ unknown interior nodal temperatures $T_{m,n}$. If a coarse mesh is chosen so that there are relatively few nodes, *matrix inversion* or *Gaussian elimination* can be used to solve the equation set. In writing a computer program to effect the solution, all that is required is to call on a standard subroutine. However, the accuracy of a finite-difference method will increase as the mesh size is reduced (provided round-off error in the numerical computations is not introduced). Thus, typically many nodes will be used, perhaps 100 or more. Then matrix inversion is uneconomical, and more sophisticated elimination methods should be used to take advantage of the sparseness of the matrix generated by the simple finite-difference approximations described here. Such methods are widely used in practice. Alternatively, iterative solution methods can be used. A simple and easily programmed iterative method is *Gauss-Seidel iteration*, which proceeds as follows:

1. A reasonable initial guess $T^0_{m,n}$ is made for each of the unknown interior nodal temperatures. This is iteration zero.

2. New values $T^1_{m,n}$ are calculated by applying Eq. (3.85) to each node sequentially. Initial T^0 values or, if available, new T^1 values are substituted in the right-hand side of the equation.

3. The mesh is swept repeatedly until, at iteration k, the temperatures at each node are seen to change by less than a prescribed small amount,

$$\left| 1 - \frac{T^k_{m,n}}{T^{k-1}_{m,n}} \right| < \varepsilon \tag{3.87}$$

The finite-difference solution is then said to have *converged* to the exact solution of the difference equation.

The system of linear algebraic equations, Eq. (3.85), is *diagonally dominant*; that is, when written in matrix form, the largest elements on each row of the coefficient matrix are on the main diagonal. The Gauss-Seidel procedure applied to such a system always converges uniformly to the proper solution and never becomes unstable; however, for a large number of equations, the convergence can be slow. Many methods have been devised to obtain faster convergence, for example, the *alternating-direction implicit* and *successive over-relaxation methods*. Such methods are described in the references listed in the bibliography at the end of the text.

Surface Heat Flux

Once the temperature field is obtained, it is sometimes necessary to determine the heat flux at a boundary surface. We make an energy balance on a finite control volume adjacent to the boundary, as shown in Fig. 3.31, and proceed as in the derivation of the boundary condition, Eq. (3.86).

$$\frac{\text{Net heat conduction}}{\text{into the volume}} + \frac{\text{Heat flux across}}{\text{face at } x = 0} = 0$$

As before,

$$\dot{Q}_x|_E = -k\frac{T_{2,n} - T_{1,n}}{\Delta x}\Delta y$$

$$\dot{Q}_y|_S = -k\frac{T_{1,n} - T_{1,n-1}}{\Delta y}\frac{\Delta x}{2}$$

$$\dot{Q}_y|_N = -k\frac{T_{1,n+1} - T_{1,n}}{\Delta y}\frac{\Delta x}{2}$$

and

$$\dot{Q}_x|_0 = q_s\Delta y$$

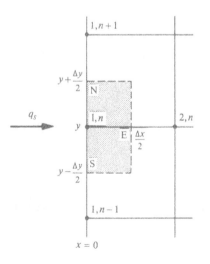

Figure 3.31 Finite control volume used to derive a difference equation for the surface heat flux: steady two-dimensional conduction.

Substituting in the energy balance, taking $\Delta x = \Delta y$, and solving for q_s gives

$$q_s = \frac{k}{\Delta x}\left[2T_{1,n} - T_{2,n} - \frac{1}{2}(T_{1,n-1} + T_{1,n+1})\right] \tag{3.88}$$

Equation (3.88) is simply item 3 of Table 3.7 rearranged to give q_s. Equation (3.88) is quite general; however, for a convective boundary condition, the surface heat can be just as easily calculated from Newton's law of cooling as $q_s = h_c(T_e - T_{1,n})$.

EXAMPLE 3.12   Steady Conduction in a Square Plate

An 8×8 cm square plate has one edge maintained at $100°C$; the other three edges are maintained at $0°C$. Use the finite-difference method to determine the temperature distribution in the plate. Compare the result with the exact solution given in Section 3.3.1.

Solution

Given: Square plate with edge temperatures prescribed.

Required: Steady-state temperature distribution using the finite-difference method.

Assumptions: Temperatures are constant across the thickness of the plate to give a two-dimensional problem.

The figure shows the mesh and the prescribed temperatures along the edges. There are nine interior nodes, but symmetry about the centerline results in only six unknown temperatures, which are labeled $T_1, T_2, \ldots, T_6$ as shown. From Eq. (3.85), these temperatures are given by

$$T_1 = \frac{1}{4}(T_3 + T_2 + 0 + 100)$$

$$T_2 = \frac{1}{4}(T_4 + T_1 + T_1 + 100)$$

$$T_3 = \frac{1}{4}(T_5 + T_4 + 0 + T_1)$$

$$T_4 = \frac{1}{4}(T_6 + T_3 + T_3 + T_2)$$

$$T_5 = \frac{1}{4}(0 + T_6 + 0 + T_3)$$

$$T_6 = \frac{1}{4}(0 + T_5 + T_5 + T_4)$$

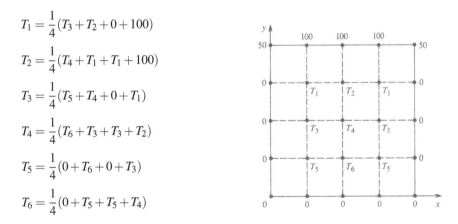

For an initial guess, a linear variation in y is assumed, and the preceding equations are evaluated in order, using latest available values. For example, when T_2 is evaluated, the new value of T_1 just obtained is used. The process is repeated until the convergence is satisfactory. The following table shows 16 iterations. Also shown is the exact solution obtained by evaluating Eq. (3.30).

Temperature	T_1	T_2	T_3	T_4	T_5	T_6
Initial guess, °C	75	75	50	50	25	25
Iteration level:						
$k=1$	56.25	65.62	32.81	39.06	14.45	16.99
$k=2$	49.61	59.57	25.78	32.03	10.69	13.35
$k=4$	44.61	54.43	20.51	26.76	8.02	10.70
$k=8$	42.97	52.79	18.86	25.11	7.20	9.88
$k=16$	42.86	52.68	18.75	25.00	7.14	9.82
Exact solution	43.20	54.05	18.20	25.00	6.80	9.54
Percent error	0.80	2.54	3.01	0.00	5.08	2.93

Comments

1. Notice that the center temperature, T_4, is the average of the edge temperatures.

2. At the corners where the temperature is discontinuous, the average value of 50°C can be assigned. Assigning the average value of 50°C at the corners ensures a zero net heat flow into the corner control volume; however these values are not used in the calculations.

3.5.2 Finite-Difference Methods for One-Dimensional Unsteady Conduction

Consider one-dimensional unsteady conduction with no internal heat generation and constant properties. Figure 3.32 shows a finite control volume $\Delta x \cdot 1 \cdot 1$ surrounding node m at location x in the solid. Figure 3.33 shows the mesh where the time coordinate is discretized in steps Δt and the index i denotes time. The energy conservation principle, Eq. (1.1), is applied to the finite control volume over a time interval Δt, from time step i to step $(i+1)$:

$$\Delta U = \dot{Q}\Delta t \tag{3.89}$$

The increase in internal energy from time step i to step $(i+1)$ is

$$\Delta U = \rho c(\Delta x \cdot 1 \cdot 1)(T_m^{i+1} - T_m^i)$$

The conduction over the time interval Δt is taken as

$$\dot{Q}_x|_W \Delta t = -k\frac{T_m^i - T_{m-1}^i}{\Delta x}(1 \cdot 1)\Delta t$$

$$\dot{Q}_x|_E \Delta t = -k\frac{T_{m+1}^i - T_m^i}{\Delta x}(1 \cdot 1)\Delta t$$

where the fluxes have been evaluated at time step i. Substituting in Eq. (3.89),

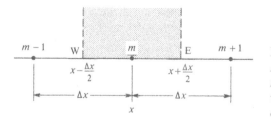

Figure 3.32 Finite control volume Δx by 1 by 1 surrounding node m at location x used to derive the difference equation for one-dimensional unsteady conduction.

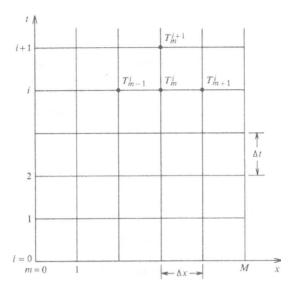

Figure 3.33 Nodal mesh for one-dimensional unsteady conduction.

dividing by Δt, and rearranging gives

$$T_m^{i+1} = \text{Fo}(T_{m-1}^i + T_{m+l}^i) + (1 - 2\,\text{Fo})T_m^i \tag{3.90}$$

where $\text{Fo} = \alpha\Delta t/\Delta x^2$ is the mesh Fourier number. Equation (3.90) is an explicit relation for T_m^{i+1}, the temperature of node m at the $(i+1)th$ time step, in terms of temperatures at the previous time step. In this *explicit* method, no iteration is required. Using given initial temperatures at time $t = 0, T_m^0$, the new temperatures, T_m^1, can be calculated. Temperatures at the boundary nodes 0 and M are fixed by the specified boundary conditions. The process is then repeated, marching forward in time.

Although simple, the explicit method has a major drawback: the allowable size of time step is limited by stability requirements. To avoid divergent oscillations in the solution, the coefficient for T_m^i in Eq. (3.90) must not be negative. That is,

$$\text{Fo} \leq \frac{1}{2} \tag{3.91}$$

Since the spatial discretization step Δx is usually chosen to give a desired spatial resolution of the temperature profile, Eq. (3.91) sets a limit on the time step:

$$\Delta t \leq \frac{\Delta x^2}{2\alpha} \tag{3.92}$$

If the boundary condition is other than that of prescribed temperature, the associated stability requirement is more stringent than Eq. (3.92). Consider the convective boundary condition illustrated in Fig. 3.34. An energy balance requires that, during the time step Δt,

$$\begin{matrix} \text{Increase in internal} \\ \text{energy within volume} \end{matrix} = \begin{matrix} \text{Net conduction} \\ \text{into volume} \end{matrix} + \begin{matrix} \text{Convection across} \\ \text{boundary} \end{matrix}$$

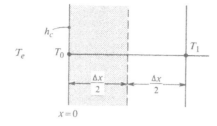

Figure 3.34   Finite control volume used to derive the difference equation for a convective boundary condition at $x = 0$: one-dimensional unsteady conduction.

or

$$\rho c \frac{\Delta x}{2}(1 \cdot 1)(T_0^{i+1} - T_0^i) = k(1 \cdot 1)\frac{T_1^i - T_0^i}{\Delta x}\Delta t + h_c(1 \cdot 1)(T_e^i - T_0^i)\Delta t$$

Rearranging and solving for the new nodal temperature, T_0^{i+1},

$$T_0^{i+1} = 2\,\text{Fo}\,(T_1^i + \text{Bi}T_e^i) + (1 - 2\text{Fo} - 2\text{FoBi})T_0^i \tag{3.93}$$

where again $\text{Bi} = h_c \Delta x / k$. For stability, the coefficient of T_0^i must not be negative:

$$1 - 2\,\text{Fo} - 2\,\text{Fo Bi} \geq 0$$

or

$$\text{Fo} \leq \frac{1}{2(1 + \text{Bi})} \tag{3.94}$$

Since the Biot number is positive, Eq. (3.94) is always a more stringent stability condition than Eq. (3.91).

The stability criterion limits Δt to no more than $(\Delta x)^2 / 2\alpha$, and less if the mesh Biot number is not small. Thus, if we wish to improve accuracy by halving the mesh size Δx, the time step Δt must be divided by four. For accurate solutions using a fine spatial mesh size, alternative methods are available. The *implicit* method evaluates the conduction fluxes at time step $(i + 1)$ rather than at step i:

$$\dot{Q}_x|_W \Delta t = -k\frac{T_m^{i+1} - T_{m-1}^{i+1}}{\Delta x}(1 \cdot 1)\Delta t$$

$$\dot{Q}_x|_E \Delta t = -k\frac{T_{m+1}^{i+1} - T_m^{i+1}}{\Delta x}(1 \cdot 1)\Delta t$$

The new form of Eq. (3.90) is

$$T_m^{i+1} = \frac{\text{Fo}\,(T_{m+1}^{i+1} + T_{m-1}^{i+1}) + T_m^i}{1 + 2\text{Fo}} \tag{3.95}$$

There are three unknown temperatures in each nodal equation. The set of algebraic equations is *tridiagonal*; that is, when written in matrix form, all the elements of the coefficient matrix are zero except for those that are on or to either side of the main diagonal. To advance the solution through each time step, Gauss-Seidel iteration works well.[3] At time step $(i + 1)$, the nodal equations are swept repeatedly until

---

[3] Direct methods, such as successive substitution, may also be used when the equation set is tridiagonal.

Table 3.8 Finite-difference approximations for one-dimensional unsteady conduction.

Item	Configuration	Explicit Form and Stability Criterion	Implicit Form
1.	Interior node	$T_0^{i+1} = \text{Fo}(T_1^i + T_2^i) + (1 - 2\text{Fo})T_0^i$ $\text{Fo} = \dfrac{\alpha \Delta t}{\Delta x^2} \leq \dfrac{1}{2}$	$T_0^{i+1} = \dfrac{\text{Fo}(T_1^{i+1} + T_2^{i+1}) + T_0^i}{1 + 2\text{Fo}}$
2.	Convection at surface	$T_0^{i+1} = 2\text{Fo}(T_1^i + \text{Bi}T_e^i) + (1 - 2\text{Fo} - 2\text{Fo}\text{Bi})T_0^i$ $\text{Fo} \leq \dfrac{1}{2(1 + \text{Bi})}$; $\text{Bi} = \dfrac{h_c \Delta x}{k}$	$T_0^{i+1} = \dfrac{2\text{Fo}(T_1^{i+1} + \text{Bi}T_e^{i+1}) + T_0^i}{1 + 2\text{Fo} + 2\text{Fo}\,\text{Bi}}$
3.	Known surface heat flux	$T_0^{i+1} = 2\text{Fo}\left(T_1^i + \dfrac{q_s^i \Delta x}{k}\right) + (1 - 2\text{Fo})T_0^i$ $\text{Fo} \leq \dfrac{1}{2}$	$T_0^{i+1} = \dfrac{2\text{Fo}[T_1^{i+1} + (q_s^{i+1}\Delta x/k)] + T_0^i}{1 + 2\text{Fo}}$
4.	Adiabatic surface or plane of symmetry	$T_0^{i+1} = 2\text{Fo}T_1^i + (1 - 2\text{Fo})T_0^i$ $\text{Fo} \leq \dfrac{1}{2}$	$T_0^{i+1} = \dfrac{2\text{Fo}T_1^{i+1} + T_0^i}{1 + 2\text{Fo}}$

Table 3.9 Finite-difference approximations and stability criteria for two-dimensional unsteady conduction, square mesh. For an adiabatic surface or plane of symmetry, set $q_s = 0$ in item 3.

| 1. Interior node | | Explicit: $T_0^{i+1} = \text{Fo}(T_1^i + T_2^i + T_3^i + T_4^i) + (1 - 4\text{Fo})T_0^i; \qquad \text{Fo} \leq \dfrac{1}{4}$ |
| | | Implicit: $T_0^{i+1} = \dfrac{\text{Fo}(T_1^{i+1} + T_2^{i+1} + T_3^{i+1} + T_4^{i+1}) + T_0^i}{1 + 4\text{Fo}}$ |

| 2. Plane surface, convection | | Explicit: $T_0^{i+1} = 2\text{Fo}\left[T_1^i + \dfrac{1}{2}(T_2^i + T_3^i) + \text{Bi}T_e^i\right] + (1 - 4\text{Fo} - 2\text{FoBi})T_0^i; \qquad \text{Fo} \leq \dfrac{1}{2(2 + \text{Bi})}$ |
| | | Implicit: $T_0^{i+1} = \dfrac{2\text{Fo}[T_1^{i+1} + (1/2)(T_2^{i+1} + T_3^{i+1}) + \text{Bi}T_e^{i+1}] + T_0^i}{1 + 2\text{Fo}(2 + \text{Bi})}$ |

| 3. Plane surface, known heat flux | | Explicit: $T_0^{i+1} = 2\text{Fo}\left[T_1^i + \dfrac{1}{2}(T_2^i + T_3^i) + \dfrac{q_s^i \Delta x}{k}\right] + (1 - 4\text{Fo})T_0^i; \qquad \text{Fo} \leq \dfrac{1}{4}$ |
| | | Implicit: $T_0^{i+1} = \dfrac{2\text{Fo}[T_1^{i+1} + (1/2)(T_2^{i+1} + T_3^{i+1}) + (q_s^{i+1} \Delta x/k)] + T_0^i}{1 + 4\text{Fo}}$ |

| 4. Exterior corner, convection | | Explicit: $T_0^{i+1} = 2\text{Fo}(T_1^i + T_2^i + 2\text{Bi}T_e^i) + (1 - 4\text{Fo} - 4\text{FoBi})T_0^i; \qquad \text{Fo} \leq \dfrac{1}{4(1 + \text{Bi})}$ |
| | | Implicit: $T_0^{i+1} = \dfrac{2\text{Fo}(T_1^{i+1} + T_2^{i+1} + 2\text{Bi}T_e^{i+1}) + T_0^i}{1 + 4\text{Fo}(1 + \text{Bi})}$ |

| 5. Interior corner, convection | | Explicit: $T_0^{i+1} = \dfrac{4}{3}\text{Fo}\left[\dfrac{1}{2}(T_1^i + T_4^i) + T_2^i + T_3^i + \text{Bi}T_e^i\right] + \left(1 - 4\text{Fo} - \dfrac{4}{3}\text{FoBi}\right)T_0^i; \qquad \text{Fo} \leq \dfrac{3}{4(3 + \text{Bi})}$ |
| | | Implicit: $T_0^{i+1} = \dfrac{(4/3)\text{Fo}[(1/2)(T_1^{i+1} + T_4^{i+1}) + T_2^{i+1} + T_3^{i+1} + \text{Bi}T_e^{i+1}] + T_0^i}{1 + 4\text{Fo}[1 + (1/3)\text{Bi}]}$ |

convergence to sufficient accuracy is obtained. The implicit method is unconditionally stable, and the choice of the time step size Δt is dictated by accuracy rather than stability considerations. As mentioned in Section 3.5.1, there are iteration schemes that give faster convergence than Gauss-Seidel iteration, and these may be found in numerical methods texts. When the boundary condition is other than that of prescribed temperature, an energy balance must be used at the boundary control volumes, as was shown for the explicit method, but with spatial derivatives evaluated at time step $(i + 1)$. Table 3.8 lists both explicit and implicit forms for a variety of boundary conditions. Table 3.9 lists corresponding results for two-dimensional unsteady conduction.

In addition to the explicit and implicit methods, a third method often used is the *Crank-Nicolson* method. Whereas the explicit method evaluates conduction fluxes at the old time step i, and the implicit method uses the new time step $(i + 1)$, the Crank-Nicolson method uses an average of the values at time steps i and $(i + 1)$. The nodal equation is then more complicated (see Exercise 3–117). For a given mesh size, the Crank-Nicolson method gives more accurate results than either the explicit or implicit methods. Although oscillations can occur, they never become unstable.

EXAMPLE 3.13 Convective Heating of a Resin Slab

An 8 cm-thick slab of resin is to be cured under an array of air jets at 100°C, as shown in the accompanying sketch. If the initial temperature of the resin is 20°C, determine the temperature of the back face after one hour. Take the heat transfer coefficient as 40 W/m^2 K, and for the resin $\rho = 2600$ kg/m^3, $c = 800$ J/kg K, $k = 1.0$ W/m K.

Solution

Given: Slab, convectively heated on one face.

Required: Back face temperature after one hour.

Assumptions: 1. The back face is well insulated.
2. The heat transfer coefficient h_c is uniform over the surface, and k, ρ, and c are constant.
3. Edge losses are negligible.

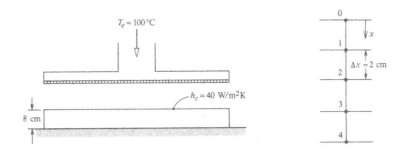

The explicit finite-difference method will be used. Let $\Delta x = 2$ cm; then the mesh size-based Biot number is

$$\text{Bi} = \frac{h_c \Delta x}{k} = \frac{(40)(0.02)}{1.0} = 0.8$$

The stability criterion, Eq. (3.94), is

$$\text{Fo} \le \frac{1}{2(1+\text{Bi})} = \frac{1}{2(1+0.8)} = 0.277$$

Choose Fo $= 0.25$, so that the time step is

$$\Delta t = \frac{\text{Fo}(\Delta x)^2}{\alpha} = \frac{(0.25)(0.02)^2}{1.0/(2600 \times 800)} = 208\,\text{s}\,(3.47\text{min})$$

Equation (3.90) for the interior nodes becomes

$$T_m^{i+1} = [1 - (2)(0.25)]T_m^i + 0.25(T_{m-1}^i + T_{m+1}^i)$$

$$= 0.5T_m^i + 0.25(T_{m-1}^i + T_{m+1}^i)$$

and Eq. (3.93) for the surface node is

$$T_0^{i+1} = (2)(0.25)(T_1^i + 0.8T_e) + [1 - 2(0.25) - 2(0.25)(0.8)]T_0^i$$

$$= 0.5(T_1^i + 0.8T_e) + 0.1T_0^i$$

To obtain an appropriate equation for node 4 at the adiabatic surface, we simply set Bi $= 0$ in Eq. (3.93) to obtain

$$T_4^{i+1} = (2)(0.25)T_3^i + [1 - (2)(0.25)]T_4^i = 0.5(T_3^i + T_4^i)$$

The initial condition is $T = 20°C$; thus, the temperatures at the first time step are

$$T_0^1 = 0.5[20 + (0.8)(100)] + 0.1(20) = 52$$

$$T_1^1 = 0.5(20) + 0.25(20 + 20) \qquad = 20$$

$$T_2^1 = 0.5(20) + 0.25(20 + 20) \qquad = 20$$

$$T_3^1 = 0.5(20) + 0.25(20 + 20) \qquad = 20$$

$$T_4^1 = 0.5(20 + 20) \qquad\qquad\quad = 20$$

and at the second time step,

$$T_0^2 = 0.5[20 + (0.8)(100)] + 0.1(52) = 55.2$$

$$T_1^2 = 0.5(20) + 0.25(52 + 20) \qquad = 28.0$$

$$T_2^2 = 0.5(20) + 0.25(20 + 20) \qquad = 20$$

$$T_3^2 = 0.5(20) + 0.25(20 + 20) \qquad = 20$$

$$T_4^2 = 0.5(20 + 20) \qquad\qquad\quad = 20$$

and so on. The results for 20 time steps obtained using a programmable hand calculator are given in the following table. Also shown is the surface temperature T_0, calculated using computer program COND2, which is essentially exact.

Time step	Time min	T_0 °C	T_1 °C	T_2 °C	T_3 °C	T_4 °C	T_0 (exact) °C
0	0.00	20.00	20.00	20.00	20.00	20.00	20.00
1	3.47	52.00	20.00	20.00	20.00	20.00	46.32
2	6.93	55.20	28.00	20.00	20.00	20.00	53.30
3	10.40	59.52	32.80	22.00	20.00	20.00	57.68
4	13.87	62.35	36.78	24.20	20.50	20.00	60.85
5	17.33	64.63	40.03	26.42	21.30	20.25	63.32
6	20.80	66.48	42.78	28.54	22.32	20.78	65.33
7	24.27	68.04	45.14	30.54	23.49	21.55	67.01
8	27.73	69.37	47.22	32.43	24.77	22.52	68.45
9	31.20	70.55	49.06	34.21	26.12	23.64	69.71
10	34.67	71.58	50.72	35.90	27.52	24.88	70.82
11	38.13	72.52	52.23	37.51	28.96	26.20	71.81
12	41.60	73.37	53.62	39.05	30.41	27.58	72.72
13	45.07	74.15	54.92	40.53	31.86	28.99	73.54
14	48.53	74.87	56.13	41.96	33.31	30.43	74.31
15	52.00	75.55	57.27	43.34	34.75	31.87	75.02
16	55.47	76.19	58.36	44.68	36.18	33.31	75.69
17	58.93	76.80	59.40	45.97	37.59	34.75	76.32
18	62.40	77.38	60.39	47.23	38.97	36.17	76.92
19	65.87	77.93	61.35	48.46	40.34	37.57	77.50
20	69.33	78.47	62.27	49.65	41.67	38.95	78.05

Comments

Notice that, for this crude calculation using $\Delta x = 2$ cm, the stability criterion is not a significant limitation on the time step. However, if Δx were reduced to, say, 0.5 cm to improve spatial resolution of the temperature profile, Δt becomes only 13 s. Even this time step poses no problem to present-day personal computers.

EXAMPLE 3.14   Quenching of a Slab with Nucleate Boiling

A 4 cm-thick slab of steel initially at 500 K is immersed in a water bath at 310 K and 1 atm. Under these conditions, *nucleate boiling* (see Chapter 7) occurs on the slab surface; the heat transfer coefficient is very large and is strongly dependent on temperature difference. An appropriate empirical equation for h_c under these conditions is $h_c = 140(T - T_{sat})^2$ W/m^2 K, where T_{sat} is the saturation temperature (boiling point). Determine the temperature profile across the slab for a period of 30 s. For the steel, take $k = 54$ W/m K, $\alpha = 1.5 \times 10^{-5}$ m^2/s.

Solution

Given: Slab immersed in water; nucleate boiling on surface.

Required: Temperature profiles.

Assumptions: Edge effects negligible to give a one-dimensional problem.

The implicit finite-difference method will be used for this problem, and the results will be obtained using a computer. Since the problem is symmetrical about the center plane of the slab, node $M = 21$ is located on the center plane as shown. Choosing a time step of 1 s, the mesh Fourier number is

$$\text{Fo} = \frac{\alpha \Delta t}{\Delta x^2} = \frac{(1.5 \times 10^{-5})(1)}{(0.02/20)^2} = 15$$

Temperatures at the interior nodes $m = 2, \ldots, M-1$ are given by Eq. (3.95):

$$T_m^{i+1} = \frac{1}{1+(2)(15)} \left[15 \left(T_{m+1}^{i+1} + T_{m-1}^{i+1} \right) + T_m^i \right]$$

$$= \frac{1}{31} \left[15 \left(T_{m+1}^{i+1} + T_{m-1}^{i+1} \right) + T_m^i \right]$$

The temperature at node M is given in Table 3.8, item 4:

$$T_M^{i+1} = \frac{1}{1+(2)(15)} \left[(2)(15)T_{M-1}^{i+1} + T_M^i \right]$$

$$T_{21}^{i+1} = \frac{1}{31} \left[30 T_{20}^{i+1} + T_{21}^i \right]$$

The temperature at the surface node $m = 1$ is given in Table 3.8, item 2, with T_e replaced by T_{sat}:

$$T_1^{i+1} = \frac{2(15)(T_2^{i+1} + \text{Bi}\, T_{sat}) + T_1^i}{1 + 2(15) + (2)(15)\text{Bi}}$$

The Biot number is not a constant in this problem and, when the implicit formulation is used, must be evaluated at the current time step:

$$B_i = \frac{h_c \Delta x}{k} = \frac{140(T_1^{i+1} - T_{sat})^2 (0.02/20)}{54} = 2.59 \times 10^{-3} (T_1^{i+1} - T_{sat})^2$$

Thus,

$$T_1^{i+1} = \frac{30(T_2^{i+1} + 2.59 \times 10^{-3} (T_1^{i+1} - T_{sat})^2 T_{sat}) + T_1^i}{31 + 7.78 \times 10^{-2} (T_1^{i+1} - T_{sat})^2}$$

Since T_1^{i+1} appears on both sides of this equation, it should be solved for by iteration. A flow diagram for a simple program based on Gauss-Seidel and Newton iteration follows. Sample results are given in the accompanying table.

Node	1	6	11	16	21
Location x, cm	0	0.5	1.0	1.5	2.0
Nodal temperatures, K:					
$t = 0$ s	500.0	500.0	500.0	500.0	500.0
1	394.9	471.0	492.0	497.6	498.8
2	390.8	451.2	481.2	492.8	495.6
5	387.6	425.8	455.1	472.4	478.0
10	385.4	409.4	429.3	442.4	447.0
20	382.5	393.3	402.3	408.2	410.2
30	380.5	385.6	389.8	392.6	393.5

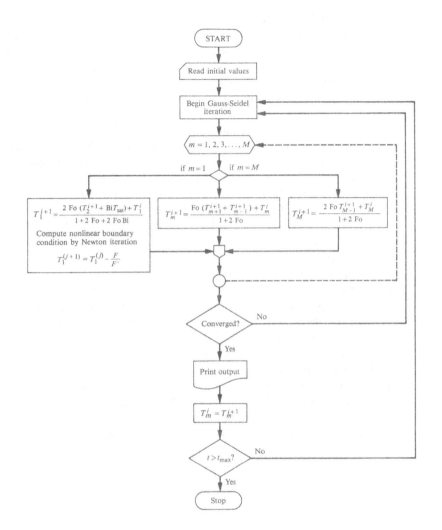

Comments

1. The effect of Δx and Δt on the accuracy of the solution should be explored.

2. The task of writing a computer program is given as Exercise 3–119.

3.6   CLOSURE

The temperature distribution in a solid is governed by the general heat conduction equation. This partial differential equation can be solved using classical mathematical methods or using numerical methods. In either case, considerable effort is required to obtain the solution for a particular problem.

The use of the classical separation-of-variables method was demonstrated for both steady multidimensional conduction and unsteady one-dimensional conduction. The method of superposition of solutions was used to build up the solution of a problem with complicated boundary conditions from solutions for simple boundary conditions. A useful product rule allows the temperature response of a number of shapes of finite dimensions to be obtained as a product of the responses of simpler shapes with infinite dimensions. For example, the response for a finite-length cylinder is obtained from the response of an infinite cylinder and an infinite slab. The conduction shape factor is convenient for calculating two-dimensional heat conduction between two isothermal surfaces. Solutions for conduction into a semi-infinite solid are always applicable for times short enough for the penetration of the thermal response to be small compared to the body dimensions. The computer program COND1 is a useful tool for such calculations. Determining the temperature response of convectively cooled (or heated) slabs, cylinders, and spheres is made simple by the computer program COND2 and by the availability of results in graphical form as temperature response charts.

When the shape of a solid is irregular, or when boundary conditions are complex, solutions to the heat conduction equation are best obtained numerically. Standard computer programs for this purpose are widely available, and such programs should be used for any serious thermal design activity. Thus, numerical methods were not presented in great detail. Only the finite-difference method was considered, and then only with the objective of conveying the essential ideas involved. Actually, the heat conduction equation is one of the easiest equations to solve numerically, and even the simplest methods yield satisfactory results. At this level, the ideas involved are almost intuitive. However, any serious effort to develop versatile and efficient computer programs to solve the heat conduction equation should be preceded by an appropriate course in numerical analysis, so that questions concerning stability, rate of convergence, and accuracy are properly handled.

REFERENCES

1. Haberman, R., *Applied Partial Differential Equations with Fourier Series and Boundary Value Problems*, 5th ed., Pearson, New York (2012).

2. Boyce, W. E., and DiPrima, R. C, *Elementary Differential Equations*, 10th ed., John Wiley & Sons, New York (2013).

3. Kreyszig, E., *Advanced Engineering Mathematics*, 10th ed., John Wiley & Sons, New York (2011).

4. Fourier, J., *Théorie Analytique de la Chaleur*, Firmin Didot Père et Fils, Paris (1822).

5. Carslaw, H. S., and Jaeger, J. C, *Conduction of Heat in Solids*, 2nd ed., Oxford Science Publications (1986).

6. Emmons, H. W, "The numerical solution of heat conduction problems," *Trans. ASME*, 65, 607–615 (1943).

7. Kayan, C. R, "An electrical geometrical analogue for complex heat flow," *Trans. ASME*, 67, 713–718 (1945).

8. Karplus, W. J., and Soroka, W. W., *Analog Methods: Computation and Simulation*, 2nd ed., McGraw-Hill, New York (1959).

9. Myers, G. E., *Analytical Methods in Conduction Heat Transfer*, AMCHT Publications, Madison, WI (1987).

10. Gurney, H. P., and Lurie, J., "Charts for estimating temperature distributions in heating and cooling solid shapes," *Ind. Eng. Chem.*, 15, 1170–1172 (1923).

11. Rohsenow, W. M., Hartnett, J. P., and Ganic, E. N., eds., *Handbook of Heat Transfer Fundamentals*, 2nd ed., McGraw-Hill, New York (1985).

12. Schneider, P. J., *Temperature Response Charts*, John Wiley & Sons, New York (1963).

13. Heisler, M. P., "Temperature charts for induction and constant temperature heating," *Trans. ASME*, 69, 227–236 (1947).

14. Langston, L. S., "Heat transfer from multidimensional objects using one-dimensional solutions for heat loss," *Int. J. Heat Mass Transfer*, 25, 149–150 (1982).

15. Mitra, K., Kumar S., Vedavarz, A., and Moallemi, M. K., "Experimental evidence of hyperbolic heat conduction in processed meat," *J. Heat Transfer*, 117, 568–573 (1995).

EXERCISES

3-1. Derive the general heat conduction equation in cylindrical coordinates by applying the first law to the volume element shown in Fig. 3.3a.

3-2. Derive the general heat conduction equation in spherical coordinates by applying the first law to the volume element shown in Fig. 3.3b.

3-3. Use the methods of vector calculus to derive the general heat conduction equation. (*Hint:* Apply the first law to a volume V with surface S, and use the Gauss divergence theorem to convert the surface integral of heat flow across S to a volume integral over V.)

3-4. The cylindrical and spherical coordinate systems are examples of *orthogonal curvilinear coordinates*. In general, we can denote these coordinates by u_1, u_2, u_3, which are defined by specifying the Cartesian coordinates x, y, z as

$$x = x(u_1, u_2, u_3)$$

$$y = y(u_1, u_2, u_3)$$

$$z = z(u_1, u_2, u_3)$$

A coordinate system is orthogonal when the three families of surfaces $u_1 = $ Const, $u_2 = $ Const, $u_3 = $ Const are orthogonal to one another. The figure shows an elemental parallelepiped whose faces coincide with planes u_1 or u_2 or $u_3 = $ Const, with edge lengths $h_1 du_1, h_2 du_2, h_3 du_3$ where h_1, h_2, h_3 are called the *metric coefficients*. The length of a diagonal is given by

$$ds^2 = h_1^2 du_1^2 + h_2^2 du_2^2 + h_3^2 du_3^2$$

In terms of these coordinates, the components of the temperature gradient are

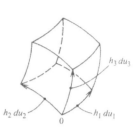

$$\frac{1}{h_1}\frac{\partial T}{\partial u_1}, \quad \frac{1}{h_2}\frac{\partial T}{\partial u_2}, \quad \frac{1}{h_3}\frac{\partial T}{\partial u_3}$$

The divergence of the heat flux is

$$\nabla \cdot \mathbf{q} = \frac{1}{h_1 h_2 h_3}\left[\frac{\partial}{\partial u_1}(h_2 h_3 q_1) + \frac{\partial}{\partial u_2}(h_3 h_1 q_2) + \frac{\partial}{\partial u_3}(h_1 h_2 q_3)\right]$$

and $\nabla^2 T$ is

$$\nabla^2 T = \frac{1}{h_1 h_2 h_3}\left[\frac{\partial}{\partial u_1}\left(\frac{h_2 h_3}{h_1}\frac{\partial T}{\partial u_1}\right) + \frac{\partial}{\partial u_2}\left(\frac{h_3 h_1}{h_2}\frac{\partial T}{\partial u_2}\right) + \frac{\partial}{\partial u_3}\left(\frac{h_1 h_2}{h_3}\frac{\partial T}{\partial u_3}\right)\right]$$

(i) Identify the metric coefficients for the cylindrical coordinate system, and hence write down $\nabla^2 T$ in cylindrical coordinates.

(ii) Repeat for the spherical coordinate system.

3-5. One face of a block of ice is observed to recede at a rate of 0.22 mm/min. If the ice has been melting for some time, calculate the temperature gradient in the water adjacent to the ice surface. The density and enthalpy of fusion of ice at 0°C are 910 kg/m³ and 0.335×10^6 J/kg, respectively.

3-6. A thin rectangular plate, $0 \le x \le a, 0 \le y \le b$, with negligible heat loss from its sides, has a linear temperature variation along the edge at $y = b$ given by $T = 20 + 100(x/a)$°C. The other three edges are maintained at 20°C. Determine the temperature distribution $T(x, y)$.

3-7. A thin rectangular plate $0 \le x \le a, 0 \le y \le b$ has the following temperature distribution around its boundary:

$$x = 0, \quad 0 < y < b: \ T = 300\text{K}$$

$$x = a, \quad 0 < y < b: \ T = 300 + 100\sin(\pi y/b)\text{K}$$

$$y = 0, \quad 0 < x < a: \ T = 300\text{K}$$

$$y = b, \quad 0 < x < a: \ T = 300 + 200\sin(\pi x/a)\text{K}$$

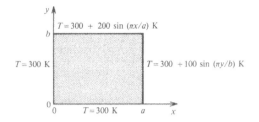

(i) Determine the steady-state temperature distribution if there is negligible heat loss from the sides.

(ii) If $a = b$, determine the center temperature.

3-8. A thin, 20 cm–square plate, with negligible heat loss from its sides, has three edges maintained at 20°C, and a fourth edge has a temperature distribution given by $T = 20(1 + \sin \pi x/L)$°C, where $L = 20$ cm, and x [cm] is measured from one corner.

(i) Determine the temperature at the midpoint of the plate.

(ii) If the plate is 3 mm–thick stainless steel with $k = 16$ W/m K, determine the rate of heat supply required to maintain a steady state.

3-9. A thin rectangular plate, $0 \le x \le a, 0 \le y \le b$, has the following boundary conditions:

$$x = 0, \quad 0 < y < b : T = 300\text{K}$$

$$x = a, \quad 0 < y < b : T = 300\text{K}$$

$$y = 0, \quad 0 < x < a : q_y = 0 \text{ (insulated)}$$

$$y = b, \quad 0 < x < a : T = 300 + 300\sin(\pi x/a)$$

(i) Determine the temperature distribution in the plate if it has negligible heat loss from its surface.

(ii) If $a = b$, find the center temperature.

3-10. A thin rectangular plate, $0 \le x \le a, 0 \le y \le b$, has the following temperature distribution around its boundary:

$$x = 0, \quad 0 < y < b : T = 300\text{K}$$

$$x = a, \quad 0 < y < b : T = 400\text{K}$$

$$y = 0, \quad 0 < x < a : T = 320\text{K}$$

$$y = b, \quad 0 < x < a : T = 380\text{K}$$

Determine the temperature distribution in the plate if it has negligible heat loss from its surface.

3-11. A thin rectangular plate, $0 \le x \le a, 0 \le y \le b$, with negligible heat loss from its sides, has the following boundary conditions:

$$x = 0, \quad 0 < y < b : \ T = 300\text{K}$$

$$x = a, \quad 0 < y < b : \ T = 400\text{K}$$

$$y = 0, \quad 0 < x < a : \ q_y = 0 \ (\text{insulated})$$

$$y = b, \quad 0 < x < a : \ T = 500\text{K}$$

(i) Determine the steady-state temperature distribution.
(ii) If $a = b$, determine the center temperature.

3-12. An 8×8 cm square plate with negligible heat loss from its sides has boundary conditions as indicated. Obtain an analytical solution for the temperature distribution, and evaluate the center temperature.

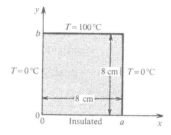

3-13. A thin rectangular plate $0 \le x \le a, 0 \le y \le b$, has a constant heat flux q_s through the edge at $y = b$, and all other edges are isothermal at temperature T_1. Determine the temperature distribution and also the heat flow through the edge at $y = 0$ for $a = b$.

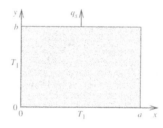

3-14. A long rectangular bar $0 \le x \le a, -b \le y \le b$ with $a = 2b$, and $a, b \ll L$, the bar length, is heated at $x = 0$ with a uniform heat flux and is insulated at $x = a$ and $y = 0$. The side at $y = b$ loses heat by convection to a fluid at temperature T_e. Determine the temperature distribution $T(x,y)$.

3-15. A long square bar has three sides maintained at temperature T_1 and the fourth loses heat by convection to surroundings at temperature T_e. Determine the temperature distribution over the bar cross section.

3-16. Determine the temperature distribution in a thick, long rectangular fin by formulating and solving the appropriate two-dimensional conduction problem. (Use Eq. (2.33c) as the tip boundary condition.) Also determine the base heat flow and compare it to the result for a long thin fin, namely, $\dot{Q} = 2Wtk(T_B - T_e)\beta$.

3-17. A long rectangular bar $0 \le x \le a, 0 \le y \le b$ and $a, b \ll L$, the bar length, is heated with a uniform heat flux q_s on the surface $y = b$. The other surfaces are cooled by steam condensing at temperature T_{sat}. Determine the temperature distribution $T(x,y)$ and the maximum temperature.

3-18. (i) Show that the heat dissipated by a straight rectangular fin of thickness $2t$, allowing for two-dimensional heat conduction, with boundary conditions $T = T_B$ at $x = 0$ and $\partial T / \partial x = 0$ at $x = L$, is

$$\dot{Q}_{2D} = 8kW(T_B - T_e) \sum_{n \text{ odd}} \tanh \frac{n\pi t}{2L} \bigg/ \left[n\pi \left(\frac{n\pi t}{2\mathrm{Bi}L} \tanh \frac{n\pi t}{2L} + 1 \right) \right]; \quad \mathrm{Bi} = \frac{h_c t}{k}$$

(ii) Write a computer program to evaluate Q_{2D}/Q_{1D}, with Q_{1D} given by Eq. (2.40). Explore the error incurred by using the one-dimensional fin model as a function of Bi and t/L.

3-19. In a natural convection test rig, a glass partition 1 cm thick and 4 cm high separates two copper plates. The lower plate is maintained at 340 K, while the upper plate is maintained at 300 K. The upper plate is maintained at 300 K. The air temperature on each side of the partition is 320 K, and the heat transfer coefficient is 6.0 W/m² K.

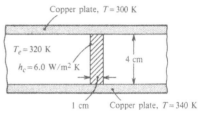

Copper plate, $T = 300$ K
$T_e = 320$ K
$h_c = 6.0$ W/m² K
4 cm
1 cm
Copper plate, $T = 340$ K

Determine the heat flow at the bottom and the top of the partition

(i) assuming one-dimensional conduction.
(ii) assuming two-dimensional conduction.

Take $k = 0.78$ W/m K for the glass. You will need the roots of Eq. (3.67), which are given in the following table for an appropriate range of Biot number.

Bi	λ_1	λ_2	λ_3	λ_4
0.01	0.100	3.145	6.285	9.426
0.02	0.141	3.148	6.286	9.427
0.05	0.222	3.157	6.291	9.430
0.1	0.311	3.173	6.299	9.435
0.2	0.433	3.204	6.315	9.446
0.5	0.653	3.292	6.362	9.477
1.0	0.861	3.426	6.437	9.529

3-20. Radioactive wastes dissipating 500 W are temporarily stored in a 2 m–diameter spherical container, the center of which is buried 5 m below the ground in a location where the soil is relatively dry. If the ground-level temperature does not exceed 30°C at any time, estimate the maximum temperature the container might attain. Take k for the soil as 0.6 W/m K.

3-21. Exhaust gas from a furnace flows at 4.50 kg/s up a concrete chimney 26 m high, 90 cm square on the inside, with a wall thickness of 30 cm. The gas inlet temperature is 500 K, and the ambient air temperature is 300 K. Determine the gas outlet condition for wind conditions that give an outside heat transfer coefficient of approximately 12 W/m^2 K. Take the inside heat transfer coefficient as 15 W/m^2 K, the specific heat of the gas mixture as 1100 J/kg K, and the concrete conductivity as 1.13 W/m K.

3-22. A small kiln for firing ceramic products has inside dimensions 1 m wide, 1 m high, and 1.5 m deep. The walls are 50 cm thick and made of zirconia brick. At steady operating conditions the inside temperature is 700°C, and the ambient air temperature is 25°C. If the outside heat transfer coefficient is approximately 5 W/m^2 K, estimate the heat loss from the kiln. Take $k = 2.4$ W/m K for zirconia brick.

3-23. The heat flow between two concentric cylinders of radii r_1 and r at temperatures T_1 and T, respectively, is, from Eq. (2.14),

$$\dot{Q} = \frac{2\pi kL(T_1 - T)}{\ln(r/r_1)} \quad \text{or} \quad T = T_1 - \frac{\dot{Q}}{2\pi kL}\ln\frac{r}{r_1}$$

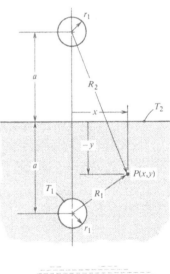

which can be interpreted as the temperature due to a line source of strength $\dot{Q}$ at $r = 0$, expressed in terms of the temperature T_1 at a reference radius r_1. Derive the shape factor S given by item 8 of Table 3.2 by superimposing the temperature fields of a line source $\dot{Q}$ below the ground and a line sink $-\dot{Q}$ above the ground, in a mirror image position as shown. Define an excess temperature $T - T_2$, where T_2 is the ground surface temperature, and show that the isotherms are concentric circles with origins at $x = 0$, $y = a(1+c)/(1-c)$, and radii $2c^{1/2}a/(1-c)$, where c is a constant parameter. Also show that the ground surface is an isotherm.

3-24. An 8 cm-O.D. pipe, 200 m long, is enclosed in 15 cm-square concrete and submerged in seawater at 10°C. Although originally used as a fresh water pipeline, it is under consideration as a temporary oil pipeline. If the oil flow rate is 0.6 kg/s and the inlet temperature is 120°C, estimate the oil outlet temperature. The oil specific heat can be taken as 2000 J/kg K.

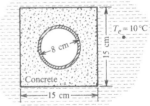

3-25. A 15 cm-O.D. pipe is buried with its centerline 1 m below the surface of the ground. An oil of specific gravity 0.8 and specific heat 1950 J/kg K flows in the pipe at 0.5 m^3/min. If the ground surface temperature is 25°C and the pipe wall temperature is 95°C, estimate the oil temperature drop, in kelvins per meter. Take $k = 1$ W/m K for the soil.

3-26. A buried insulated power cable has an outside diameter of 3 cm and is 1 m below the surface of the ground. What is the maximum allowable dissipation per unit length if the outer surface of the insulation must not exceed 350 K when the ground surface and the deep soil are at 300 K? Take $k = 1$ W/m K for the soil.

3-27. A 25 cm-diameter oil line is buried with its centerline 60 cm below the ground. Assuming a soil temperature of 10°C and a soil thermal conductivity of 0.8 W/m K, estimate the steady-state heat loss from the pipe (in W/m) when the oil is at 90°C.

   (i) Ignore the inside thermal resistance.
   (ii) Recalculate your result if the inside heat transfer coefficient is 300 W/m^2 K.

3-28. A 20 cm-O.D. steam pipe is buried 1.5 m below the ground surface in dry soil. Steam flows through the pipe at 1.2 kg/s. Assuming a ground surface temperature of 15°C and saturated steam at 1.1×10^5 Pa, estimate the steam condensation rate per 100 m of pipeline.

3-29. A small laboratory oven is cubical in shape with an inside edge 20 cm long. It is insulated with 6 cm of medium-density fiberglass. What power supply is required to maintain an interior temperature of 440 K when the ambient temperature is 20°C and the outside heat transfer coefficient is 7 W/m^2 K?

3-30. A 20 cm-O.D. pipe is buried 1 m below the surface of the ground. Hot water flows in the pipe at 500 gal/min. Assuming a pipe wall temperature of 70°C, estimate the length of pipe in which the water temperature decreases by 1°C when the ground surface is at 10°C. Take the soil thermal conductivity as 1 W/m K.

3-31. An electrical power cable has a 20 cm-O.D. sheath and is buried with its center 1.2 m below ground level. The trench is back-filled with Fire Valley thermal sand, which has a thermal conductivity of 1.1 W/m K dry, and 2.0 W/m K when saturated with water. Note that this latter value is an effective one: it is higher than the thermal conductivity of water (~ 0.6 W/m K). In the presence of a temperature gradient, water evaporates in hotter regions and condenses in colder regions, thereby transporting enthalpy of vaporization that augments the ordinary thermal conduction. If the allowable outside temperature of the sheath is 60°C, determine the maximum power dissipation per unit length for a ground surface temperature of 25°C.

3-32. A slab 2L thick, initially at a uniform temperature T_0, has its surfaces suddenly lowered to temperature T_s. Simultaneously a volumetric heat source $\dot{Q}_v'''$ is activated within the slab. Show how you would analyze this problem to determine the temperature response. (*Hint:* Try a superposition of a steady and a transient problem.)

3-33. Show why the constant for Eq. (3.38) cannot be positive or zero.

3-34. A thick slab of Pyrex glass, initially at 350 K, has the temperature on one surface suddenly dropped to 300 K. Calculate the time needed for the location 1 cm below the surface to decrease in temperature by 5 K.

3-35. A 4 cm-thick slab has an initial temperature of 100°C. At time $t = 0$ the temperature of the one surface is lowered to 50°C and the other to 0°C. Determine the heat flux out the colder side at $t = 2000$ s. Take $k = 0.35$ W/m K, $\alpha = 0.15 \times 10^{-6}$ m²/s for the slab material.

3-36. On Thanksgiving Day in St. Louis a sudden storm reduces the ambient air to -15°C. If the ground was at a uniform temperature of 20°C before the storm, what is the surface temperature after 4 hours, and to what depth below the surface will the freezing temperature have penetrated? Take $k = 1.2$ W/m K, $\alpha = 0.40 \times 10^{-6}$ m²/s, $h_c = 20$ W/m² K. Ignore any phase change effects.

3-37. Derive Eq. (3.63) by assuming a solution that has the functional form $T = C\exp(-px)\sin(\omega t - qx) + D$, and determine C, p, and q by substituting back into the governing equation.

3-38. To determine whether a new composite material can withstand thermal cycling without degrading, a thick slab is subjected to heating, which causes the surface temperature to vary in an approximate sinusoidal manner with an amplitude of 200°C and a period 50 s. If the thermal conductivity and diffusivity of the composite are 10 W/m K and 2.0×10^{-6} m²/s, respectively, determine the amplitude and phase lag of the temperature variation 1 cm below the surface.

3-39. Water pipes are to be buried in a geographical area that has a mean winter temperature of about 5°C but is subject to sudden drops in air temperature to about -10°C for a maximum duration of 48 hours. Estimate the minimum depth of the pipes needed to prevent freezing. Take $\alpha = 0.5 \times 10^{-6}$ m²/s for the soil.

3-40. A 5 mm-thick steel firewall is suddenly exposed to a radiant heat source that can be approximated as a blackbody at 1000 K. How long will it take for the surface to reach 500 K if the initial temperature of the wall is 300 K?

3-41. The *integral method* can be used to solve the problem of conduction in a semi-infinite solid subject to a step change in surface temperature. If it is assumed that temperature changes penetrate only to $x = \delta$, the first law applied to a control volume of unit cross-sectional area located between $x = 0$ and $x = \delta$ gives

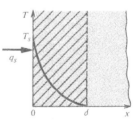

$$\dot{Q} = \frac{dU}{dt} \quad \text{or} \quad q_s - 0 = \frac{d}{dt}\int_0^\delta \rho u\, dx = \frac{d}{dt}\int_0^\delta \rho c(T - T_0)\, dx \tag{1}$$

where the initial temperature T_0 is chosen as the datum state for internal energy.

(i) Assume that the temperature profile is parabolic, $T = A + Bx + Cx^2$ with constants obtained from boundary conditions, and hence show that

$$\frac{T - T_0}{T_s - T_0} = \left(1 - \frac{x}{\delta}\right)^2; \quad q_s = \frac{2k}{\delta}(T_s - T_0) \tag{2a,b}$$

(ii) By substituting Eqs. (2) in Eq. (1), show that the thermal penetration δ satisfies the differential equation

$$\frac{d\delta}{dt} = \frac{6\alpha}{\delta} \tag{3}$$

(iii) Finally, show that the heat flow into the solid is given by

$$q_s = \frac{k(T_s - T_0)}{(3\alpha t)^{1/2}}$$

and compare this result with the exact solution, Eq. (3.59).

3-42. At a ski resort in the Sierra Nevada mountains, the winter ground-surface temperature typically varies between -10 and $12°C$ each day, while the in-depth soil temperature is $+1°C$. How far below the surface should a water pipe be buried to prevent freezing?

3-43. A thermocouple is installed in the cylinder wall of a two-stroke internal combustion engine 1.0 mm below the inner surface. In a particular test the engine operates at 2500 rpm, and the thermocouple reading is found to have a mean value of $290.0°C$ and an amplitude of $1.08°C$. If the temperature variation is assumed to be sinusoidal, estimate the amplitude of the cylinder wall surface temperature variation and the phase difference. Take $\alpha = 12.0 \times 10^{-6}$ m²/s for the carbon steel wall.

3-44. A thin, flat electrical resistance strip heater is sandwiched between a firebrick wall and a thick AISI 1010 steel plate, both initially at 300 K. The heater is switched on and generates heat at a rate of 50,000 W/m². Plot the temperature of the steel surface adjacent to the heater as a function of time for 10 seconds.

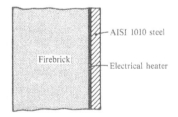

3-45. A long rod is well insulated along its sides. A heater varies the temperature of one end face sinusoidally from 100 to 200°C with a period of 90.9 s. Two thermocouples, 10 and 70 cm from the heated end, record temperatures. The phase difference between the peak temperatures is found to be 15.0 min. If the density of the rod material is 8300 kg/m^3 and its specific heat 470 J/kg K, what is its thermal conductivity?

3-46.   Estimate the depth below the ground at which the annual temperature variation will be 10% of that at the surface for

(i) dry soil, $k = 1.0$ W/m K.
(ii) wet soil, $k = 2.0$ W/m K.

Repeat for the diurnal temperature variation.

3-47.   A turbine component in the preburner section of a turbopump for a rocket motor is made from Inconel X-750. A "thermal barrier" coating of 0.3 mm-thick YSZ (yttrium-stabilized zirconia) is plasma-sprayed on the surface. This coating has a high reflectance and can significantly reduce the radiation heat loads on cooled surfaces. Upon start-up, the surfaces are initially at 30°C and are suddenly exposed to hot gases at 800°C, with an estimated effective heat transfer coefficient of 180 W/m^2 K. Plot the surface and interface temperature as a function of time for short times. Take $k = 5.0$ W/m K for YSZ.

3-48.   Associated with the annual variation of the weather, the average ground surface temperature at a given location will have a maximum in summer and a minimum in winter. If you dig a hole exactly six months after the maximum occurs, how deep will you have to dig to reach the peak of the resulting temperature response, and what will be the percent decay of the amplitude? Take $\alpha = 1.2 \times 10^{-6}$ m^2/s for the soil diffusivity.

3-49.   The temperature of the outer 2 mm of skin on the forearm can be taken to be 32°C when the ambient air is at 25°C. If your arm suddenly contacts a slab of aluminum at 100°C, what is the skin surface temperature until blood flow responds to the change in conditions? Repeat for slabs of 18-8 stainless steel, Pyrex glass, and Teflon. For skin tissue take $k = 0.37$ W/m K, $\alpha = 0.1 \times 10^{-6}$ m^2/s.

3-50.   The Fourier heat conduction equation has been used to obtain the temperature-time response during cooking of meat (see, for example, Exercise 3–69). However, recent experimental studies [15] have found processed bologna meat to exhibit an anomalous behavior. In one experiment, two identical samples at different temperatures were brought into contact with each other. One sample was refrigerated at 8.2°C, the other was at a room temperature of 23.1°C. Thermocouples were inserted at the interface and in the room temperature sample at a distance of 6.3 mm from the interface. The graph shows the measured temperature-time response. Compare this response with a predicted response based on the heat conduction equation. Measured properties of bologna meat are $k = 0.80 \pm 0.04$ W/m K, $\rho = 1230 \pm 10$ kg/m^3, $c_p = 4660 \pm 200$ J/kg K.

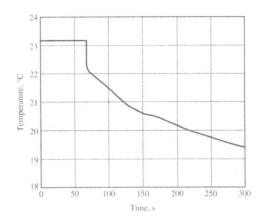

3-51. A method for experimentally determining convective heat transfer coefficients involves use of a plastic model coated with a liquid crystal paint. The liquid crystal is chosen to have a color transition at a convenient temperature. The model is initially isothermal at temperature T_0, and is suddenly exposed to the gas flow at temperature T_e. The time elapsed till the color transition is observed is recorded. Conduction into the plastic is modeled as conduction into a semi-infinite solid in order to relate the observed temperature response to the heat transfer coefficient. In a particular experiment, $T_0 = 20°C$, $T_e = 45°C$, and the color transition temperature is 38.0°C. If the time measured is 9.3 s, determine the heat transfer coefficient. Take $k = 0.24$ W/m K and $\alpha = 0.12 \times 10^{-6}$ m^2/s for the plastic.

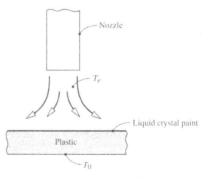

3-52. A proposed procedure for measuring convective heat transfer coefficients underneath an impinging jet uses the end of a copper rod as the heat transfer surface. The end of the rod is coated with a thin layer of material for which the melting temperature is precisely known. The surface is suddenly exposed to a hot gas flow and time to melting measured. By embedding the rod in a low-conductivity insulator, conduction out the sides of the rod can be neglected and the rod modeled as a semi-infinite solid. Proposed test conditions include an initial temperature $T_0 = 25°C$, a gas stream temperature of 300°C, and a coating melting point of 55.5°C.

  (i) If the time to melting is measured to be 400 s, determine the heat transfer coefficient. Also determine the length of rod required for the semi-infinite solid model to be adequate.

  (ii) Repeat the calculations for a type AISI 302 stainless steel rod. Does the use of stainless steel improve the experiment?

  (iii) Repeat the calculations for a Teflon rod and comment on the result.

Evaluate properties at 300 K.

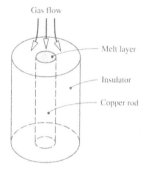

3-53. During vulcanization a tire carcass is heated by exposure to saturated steam at 150°C on both sides. If the carcass is 2 cm thick and has an initial temperature of 20°C, how long will it take for the centerplane to reach 130°C? Take $\alpha = 0.06 \times 10^{-6}$ m²/s for the rubber.

3-54. An AISI 1010 carbon steel slab 5 cm thick is insulated on one side and is initially at 700 K. To cool the slab, it is exposed to jets of water at 300 K which give a heat transfer coefficient of 2000 W/m² K. What is the temperature of the back face after

  (i) 5 minutes?

  (ii) 1 hour?

Repeat for a chrome brick slab.

3-55. A sphere of 1 cm diameter at 320 K is suddenly immersed in an air stream at 280 K. If the average heat transfer coefficient is 100 W/m² K, determine the time required for the sphere to lose 90% of its energy content. Consider four materials:

  (i) cork

  (ii) Teflon

  (iii) AISI 302 stainless steel

  (iv) pure aluminum

3-56. Derive Eq. (3.69); that is, show that the constants A_n in the solution for the temperature response of a slab with finite surface resistance are given by

$$A_n = \frac{2\sin\lambda_n}{\lambda_n + \sin\lambda_n \cos\lambda_n}$$

where λ_n are the roots of $\cot\lambda = \lambda/\text{Bi}$.

3-57. The nozzle of an experimental rocket motor is fabricated from 5 mm-thick alloy steel. The combustion gases are at 2100 K, and the effective heat transfer coefficient is 6000 W/m^2 K. If the nozzle is initially at 300 K and the maximum allowable operating temperature for the steel is specified as 1300 K, what is the allowable duration of firing? To obtain a conservative estimate, neglect the heat loss from the outer surface of the nozzle. Take $k = 43$ W/m K, $\alpha = 12.0 \times 10^{-6}$ m^2/s for the steel.

3-58. If the pebble bed air heater of Example 3.10 were packed with 6 cm iron spheres, what would be the result?

3-59. A 2 cm-thick ceramic slab has a thin metal sheath for protection. The contact resistance between the ceramic and metal gives an interfacial conductance of 1600 W/m^2 K. The slab is initially at 300 K and is suddenly exposed to a hot air stream at 1500 K with a convective heat transfer coefficient of 800 W/m^2 K. How long will it take for the center of the slab to reach 1200 K? The ceramic properties are

$$\rho = 2600 \text{ kg/m}^3$$
$$c = 1150 \text{ J/kg K}$$
$$k = 3.0 \text{ W/m K}$$

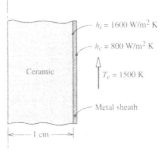

3-60. A 1 cm-thick slab of hard rubber, initially at 15°C, is heated by saturated steam at 140°C. As soon as the center temperature reaches 125°C, the slab is removed and allowed to cool in 25°C air until the center temperature falls to 35°C. How long does each stage take? Assume a very large heat transfer coefficient for the heating process, and a value of 8 W/m^2 K for the cooling process.

3-61. A 3 cm-diameter rod of a composite material is initially at 30°C. It is immersed into a chamber in which saturated steam at 120°C condenses on the rod and heats it. When the centerline temperature reaches 110°C, the rod is removed and allowed to cool in 20°C air until the center falls to 30°C. How long does each stage take? Assume a very large heat transfer coefficient for the heating process, and a value of 15 W/m^2 K for the cooling process. Properties of the composite include $\rho = 1500$ kg/m^3, $c = 1800$ J/kg K, $k = 1.2$ W/m K.

3-62. A slab $-L < x < L$ is initially at a temperature of $100 \cos(\pi x/2L)°C$. For times $t > 0$, the surfaces are insulated. Determine the temperature response $T(x,t)$ in the slab.

3-63. The lumped thermal capacity model is a valid approximation when spatial temperature variations are negligible within a component. A criterion for its use is $Bi_{LTC} = h_c(V/A)/k < 0.1$. A fellow student makes the suggestion that this model should be valid at long times, irrespective of Biot number, since toward the end of the cooling process, temperature variations within the component are small. Evaluate this suggestion for convective cooling of a slab by determining the rate of change of average temperature when it is one-tenth of its initial value, using both the lumped thermal capacity model and the one-term approximate solution for long times. Consider Biot numbers of 0.02, 0.1, 1.0, 10, and ∞. Explain your results.

3-64. A long bar of AISI 316 stainless steel has an 8 cm × 8 cm square cross section. It is hot-rolled at 620°C and then cooled by cold air jets at 30°C, giving a heat transfer coefficient of 400 W/m² K. Determine the time required for the center temperature to decrease to 130°C.

3-65. Find the time required to heat the center of a 5 × 10 × 20 cm clay brick from 550°C to 1500°C in a convective oven at 1700°C. Take the heat transfer coefficient to be 100 W/m² K, and $k = 1.7$ W/m K, $\alpha = 0.35 \times 10^{-6}$ m²/s for the brick.

3-66. A 2 cm-thick slab of cloth-filled thermoplastic resin, initially at 20°C, is placed in a press with walls maintained at 100°C by condensing steam. As soon as the center temperature reaches 85°C, the slab is removed and allowed to cool in 20°C air until the center temperature falls to 40°C. How long does each stage of the process take? Use $k = 0.6$ W/m K, $\rho = 1700$ kg/m³, and $c = 1200$ J/kg K. The heat transfer coefficient for the cooling process can be taken as 7 W/m² K.

3-67. A can of beer at 300 K is placed in a refrigerator that maintains an air temperature of 277 K. The can is 8 cm in diameter and 12 cm high. The outside heat transfer coefficient is 5 W/m² K. After 6 hours the beer is removed from the refrigerator and poured into a glass. Estimate the beer temperature.

3-68. A physicist working for your company has proposed the following simple procedure for measuring the thermal conductivity of soil. An electrically heated sphere 6 cm in diameter is to be buried 30 cm below the surface of a very large box of the soil. The sphere will be maintained at 30°C while the laboratory air-conditioner maintains the surface of the soil at 20°C. The power input to the heater

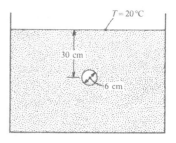

required to maintain a steady state will be measured and Table 3.2, item 7 used to calculate k. Evaluate the practicality of this procedure.

3-69. A 4 kg beef roast, roughly spherical in shape with a diameter of 20 cm, is removed from a 5°C refrigerator and placed in a 150°C oven. It is desired to raise the center temperature to 70°C. How long should the roast be cooked? Assume that the meat has the thermal properties of water and that the heat transfer coefficient is 12 W/m^2 K. If the roast lost 0.4 kg of its mass by evaporation, make a rough estimate of the enthalpy of vaporization absorbed and compare this to the sensible enthalpy change of the roast.

3-70. A blank for a telescope mirror is a 25 cm-diameter, 5 cm-thick disk of glass. The blank is at room temperature, 20°C, and is placed in an oven at 420°C for stress relieving. If the heat transfer coefficient is 12 W/m^2 K, how long will it be before the minimum temperature in the glass is 400°C? Take $k = 1.09$ W/m K, $\alpha = 0.51 \times 10^{-6}$ m^2/s for the glass.

3-71. A 15 cm-diameter, 30 cm-long 18-8 stainless steel billet at 20°C is placed in an oil bath at 300°C. The heat transfer coefficient may be taken to be 400 W/m^2 K. Determine the center temperature after 500 s have elapsed

  (i) using the lumped thermal capacity model.
  (ii) assuming the billet is long compared to its diameter.
 (iii) accounting for the finite length of the cylinder.

3-72. Plywood is to be cured between plates maintained at 105°C by condensing steam. How long will it take to cure a sheet 1 cm thick if the minimum temperature must reach 95°C and its initial temperature is 25°C?

3-73. Rework Exercise 3–65 for $h_c = 200$ W/m^2 K.

3-74. An egg, which may be modeled as a 4 cm-diameter sphere with the thermal properties of water, is initially at 5°C and is immersed in boiling water. Determine the temperature at the center of the egg after

  (i) 4 minutes.
  (ii) 7 minutes.

Take the outside heat transfer coefficient as 1200 W/m^2 K.

3-75. A large rectangular safe has a 10 cm-thick asbestos insulation. In a fire the outside of the insulation is estimated to be 800°C. How long must the fire last to destroy papers that char at 150°C contained in the safe? At the outbreak of the fire the safe was at 20°C. Take $\alpha = 0.4 \times 10^{-6}$ m^2/s for the asbestos.

3-76. A cylindrical AISI 302 stainless steel pin 2 cm in diameter and 10 cm high initially at 20°C is suddenly exposed to saturated steam at 1 atm pressure. A thermocouple is located on the centerline of the cylinder 1 cm from the top of the pin. What temperature will the thermocouple record after

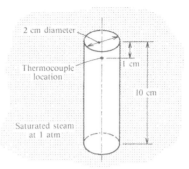

(i) 10 s?
(ii) 100 s?

3-77. Repeat Exercise 3–76 for a 2 cm-square pin.

3-78. A 20 cm-long, 25 cm-O.D. fused silica glass cylinder initially at 30°C is placed in a furnace at 800°C. If the heat transfer coefficient is estimated to be 20 W/m^2 K, prepare a graph of the center temperature as a function of time.

3-79. A water storage tank for a nuclear power plant in Tennessee is 10 m high and has a 13.5 m diameter. It is insulated on the top and sides with a 7.6 cm-thick insulation of $k = 0.05$ W/m K. The water is maintained at 22°C by four immersion heaters. If the power to the heaters is switched off, how long will it take the water temperature to drop to 16°C if the average ambient air temperature is 5°C? The tank is located in a sheltered position adjacent to some large buildings. (*Hint:* A rough estimate of the outside heat transfer coefficient will suffice.)

3-80. A 10 cm cube of Inconel-X alloy is removed from an oven at 810°C and quenched in coolant at 40°C, giving a heat transfer coefficient of 800 W/m^2 K. Calculate the time required for

(i) the center temperature to undergo 95% of its decrease to the equilibrium value.
(ii) a 95% fractional energy loss.

Compare the two results and comment.

3-81. An *interior heat transfer coefficient h* can be defined for heat conduction out of a convectively cooled slab of thickness 2L as

$$h = \frac{-k(\partial T/\partial x)|_{x=L}}{\overline{T} - T_s}$$

with a corresponding dimensionless **Nusselt number** Nu $= h(2L)/k$. Show that for Fo > 0.2, Nu has a constant value

$$Nu = \frac{2\lambda_1^2 \sin\lambda_1}{\sin\lambda_1 - \lambda_1 \cos\lambda_1}$$

which for Bi $\to \infty$ is $\pi^2/2 = 4.934$.

3-82. Repeat Exercise 3–81 for a cylinder.

3-83. Repeat Exercise 3–81 for a sphere.

3-84. A regenerative heat exchanger (see Section 8.2) has a matrix in the form of parallel plates of plastic 6 mm thick, at a spacing of b mm. If the convective heat transfer coefficient for the fluid flow is approximately $4k/b$, where k is the gas conductivity, estimate the overall heat transfer coefficient for heat transfer from the fluid into the plates. Tabulate your results for $b = 1, 2, \ldots, 10$ mm. Take $k = 0.15$ W/m K for the plastic and 0.10 W/m K for the fluid.

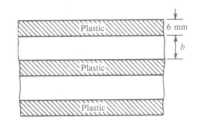

3-85. Transient heat conduction in citrus fruits is of concern to farmers who must develop strategies to prevent freezing during cold weather. A relatively simple procedure for determining the thermal diffusivity of a citrus fruit involves installing a thermocouple at the center of the fruit and determining the temperature response when the fruit is placed in a rapidly stirred water bath maintained at 0°C. The following table gives some typical data for a 6.8 cm-diameter grapefruit. Estimate the thermal diffusivity and compare your value to that of water at 10°C.

t, min	0	5	10	15	20	25	30	35	40	45
T_c,°C	20.0	20.0	19.6	18.0	15.8	13.0	10.8	8.4	7.2	5.8

3-86. (i) A slab of thickness L is initially at a uniform temperature T_0. The face at $x = 0$ is perfectly insulated. At time $t = 0$, a constant heat flux q_s is imposed on the face at $x = L$. Show that the temperature response is

$$\frac{T - T_0}{q_s L/k} = \frac{3x^2 - L^2}{6L^2} + \frac{\alpha t}{L^2} - \frac{2}{\pi^2}\sum_{n=1}^{\infty}\frac{(-1)^n}{n^2}e^{-(\alpha n^2 \pi^2/L^2)t}\cos\frac{n\pi x}{L}$$

(ii) If a negative heat flux equal to $-q_s$ is subsequently imposed at time $t = t_1$, show that the temperature response for $t > t_1$ is

$$\frac{T - T_0}{q_s L/k} = \frac{\alpha t_1}{L^2} - \frac{2}{\pi^2}\sum_{n=1}^{\infty}\frac{(-1)^n}{n^2}e^{-(\alpha n^2 \pi^2/L^2)t}\left[1 - e^{(\alpha n^2 \pi^2/L^2)t_1}\right]\cos\frac{n\pi x}{L}$$

(*Hint:* Use the principle of superposition for a linear differential equation.)

3-87.   (i) A slab of thickness L is initially at a uniform temperature T_0. The back face at $x = 0$ is perfectly insulated. At time $t = 0$, a laser deposits an amount of energy E per unit area on the face at $x = L$. Determine the temperature response. (*Hint:* Let $\theta(x,t)$ be the temperature response for a unit uniform surface heat flux obtained by setting $q_s = 1$ in the result of Exercise 3-86, part (i). Then, following part (ii) of Exercise 3–86, show that the required temperature response is

$$T - T_0 = E\frac{\partial\theta(x,t)}{\partial t}$$

by imposing a heat flux q_s for time Δt and then letting $\Delta t \to 0$ such that $q_s\Delta t = E$.)

(ii) Hence show that the time for the back-face temperature rise to equal half of its maximum rise is $t_{1/2} = 1.39L^2/\pi^2\alpha$.

3-88.   The analysis of Exercise 3-87 forms the basis of a technique for measuring the thermal diffusivity of thin samples. In an experiment on a 1 mm-thick laminate, a CO_2 gas laser is used to deposit approximately 0.2×10^4 J/m^2 of energy over a period of 200 nanoseconds. The back-face temperature is measured using a type K thermocouple, and the μV-time trace is as shown. Estimate the thermal diffusivity of the sample.

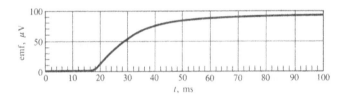

3-89.   Write a computer program to solve Example 3-12.

3-90.   Write a computer program to solve the problem of Example 3-12, but allow for a $N \times N$ mesh. Explore the effect of N on the accuracy and rate of convergence of the solution.

3-91.   Modify the computer program of Exercise 3-90 to solve Exercise 3-10 numerically. Take $a/b = 2$.

3-92.   An 8×8 cm square plate has one edge maintained at 100°C, whereas the other three edges are exposed to a fluid at 0°C with a heat transfer coefficient of 10 W/m^2 K. Modify the computer program of Exercise 3-90 to obtain the temperature distribution in the plate. Take $k = 1.0$ W/m K.

3-93.   A straight rectangular stainless steel fin is 2 mm thick and 2 cm long. It has a base temperature of 100°C and is exposed to an airflow at 20°C with a convective heat transfer coefficient of 300 W/m^2 K.

(i) Develop a finite-difference formulation of this steady one-dimensional heat conduction problem.

(ii) Using a node spacing of $\Delta x = 0.5$ cm, use Gauss-Seidel iteration to obtain temperatures T_2 through T_5 for five iterations.

Take $k = 15$ W/m K for the stainless steel. Neglect the tip heat loss.

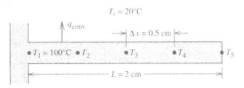

3-94. Write a computer program to solve Exercise 3-93. Explore the effect of node spacing on the accuracy of the solution.

3-95. A 2 mm-thick, 4 cm-long, straight rectangular composite material fin has a base temperature of 400 K and is located in a vacuum system. The fin has an emittance of 0.85 and sees nearly black vessel walls at 360 K.

  (i) Develop a finite-difference formulation of this steady one-dimensional conduction problem. Use a radiation heat transfer coefficient to account for the radiation heat transfer.
  (ii) Using a mesh size of $\Delta x = 1$ cm, obtain temperatures T_2 through T_5 for the first iteration.

Take $k = 4$ W/m K for the composite material. Neglect the tip heat loss.

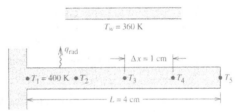

3-96. Write a computer program to solve Exercise 3-95. Explore the effect of mesh size on the accuracy of the solution.

3-97. Write a computer program to calculate the shape factor given as item 5 of Table 3.2. Take $a/b = 2.0$. Check the formulas given in the table.

3-98. Derive the finite-difference approximation formulas for steady conduction given as items 3, 4, and 5 of Table 3.7.

3-99. For steady conduction, derive the finite-difference approximation for an interior node near a curved nonisothermal surface. (See item 6 of Table 3.7.)

3-100. Write a computer program to solve for the temperature distribution and fin efficiency of a straight rectangular fin, allowing for two-dimensional conduction.

Compare your numerical results with the analytical formula given in Exercise 3-18 for $t = 0.5$ cm, $L = 3$ cm, $k = 1.0$ W/m K, $h_c = 4$ W/m^2 K.

3-101. Modify the finite-difference approximation formula for a convective boundary condition, given as Eq. (3.86), to include the effect of internal heat generation $\dot{Q}_v'''$.

3-102. A 4 cm-square plate has edge temperatures maintained at 60°C, 40°C, 20°C, and 0°C, as shown. The faces of the plate are well insulated. Use Gauss-Seidel iteration on a 1 cm-square mesh to solve for the steady-state temperature distribution. Give the interior node temperatures after the first and second iterations.

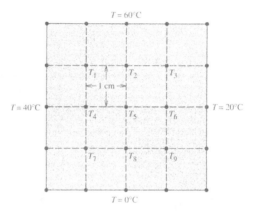

3-103. Write a computer program to solve Exercise 3-102, allowing for an arbitrary mesh size. Also attempt to calculate the heat flow out the edge at 0°C if the plate is 3 mm thick and has a thermal conductivity of 200 W/m K. Explore the effect of mesh size on the calculated heat flow by obtaining results for mesh sizes of 1 cm, 0.5 cm, 0.2 cm, and 0.1 cm: explain the anomalous behavior obtained.

3-104. An 8 cm × 4 cm plate has edge temperatures maintained at 40°C, 20°C, 20°C, and 0°C, as shown. The faces of the plate are well insulated. Use Gauss-Seidel iteration on a 2 cm × 1 cm mesh to solve for the steady-state temperature distribution. Give the interior node temperatures after the first and second iterations.

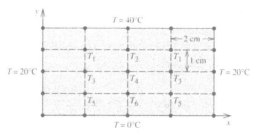

3-105. An 8 cm × 4 cm plate has edge temperatures maintained at 40°C, 20°C, 20°C, and 0°C, as shown. The faces are well insulated. Write a computer program to solve for the temperature distribution using Gauss-Seidel iteration. Allow for an arbitrary mesh size. If the plate is 1 cm thick and has a thermal conductivity of 200 W/m K, attempt to calculate the heat flow across the edge at 40°C. Obtain results for mesh sizes of 2 cm × 1 cm, 1 cm × 0.5 cm, and 0.25 cm × 0.125 cm. Explain the anomalous behavior of the heat flow. (*Hint:* Determine the corner temperatures by requiring that the net heat flow into a corner control volume is zero.)

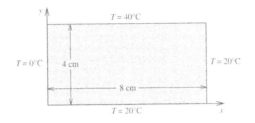

3-106. A long, 4 cm-square bar has opposite sides maintained at 100°C and 0°C, and the other two sides lose heat by convection to a fluid at 0°C. The conductivity of the bar material is 2 W/m K, and the convective heat transfer coefficient is 100 W/m² K. For the given mesh use Gauss-Seidel iteration to determine the temperatures T_1 through T_9 after the first and second iterations.

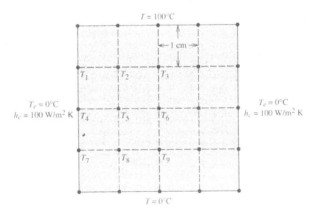

3-107. Write a computer program to solve Exercise 3-106, allowing for an arbitrary mesh size. Also determine the heat flow into the side at 100°C. Explore the effect of mesh size on the calculated heat flow for mesh sizes of 5 mm, 3 mm, and 1 mm.

3-108. A long, 8 cm-square bar is heated uniformly at 80 W/m² on one side; the opposite side is maintained at 0°C, and the other two sides lose heat by convection to a fluid at 0°C. The convective heat transfer coefficient is 20 W/m² K, and the thermal conductivity of the bar is 0.08 W/m K.

  (i) Derive the finite-difference approximation at the top corners for steady conduction using a square mesh.

  (ii) For the given mesh, use Gauss-Seidel iteration to obtain temperatures T_1 through T_{12} for two iterations.

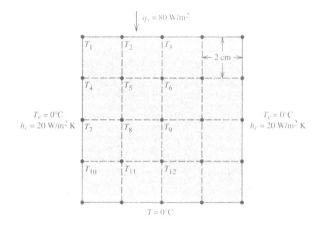

3-109. Write a computer program to solve Exercise 3-108, allowing for an arbitrary mesh size and any input heat transfer coefficient. Also calculate the heat flow out the side at 0°C for $h_c = 20$ W/m² K, and mesh sizes of 1 cm, 0.5 cm, and 0.2 cm.

3-110. A 3 cm-square bar has opposite surface temperatures maintained at 60°C and 0°C, and the other two surfaces are maintained at 30°C.

  (i) Use Gauss-Seidel iteration on a 1 cm-square mesh to solve for the steady-state temperature distribution. Iterate until the solution has converged within 0.01°C.

  (ii) If the conductivity of the bar is 1 W/m K, obtain the heat flow across each surface and show that energy is conserved.

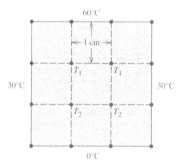

3-111. A 3 cm-square plate has its edge temperatures maintained at 30°C, 20°C, 10°C, and 0°C, as shown. The faces of the plate are well insulated. Use Gauss-Seidel iteration on a 1 cm-square mesh to obtain the steady-state temperature distribution. Give the interior node temperatures after the first four iterations.

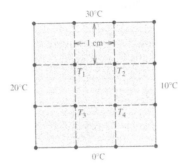

3-112. Repeat Exercise 3-111 using a direct method to solve the finite-difference equations.

3-113. A long, 3 cm-square bar has one side maintained at 50°C, the opposite side is well insulated, and the other two sides lose heat by convection to a fluid at 0°C. The conductivity of the bar material is 2 W/m K, and the convective heat transfer coefficient is 200 W/m² K.

  (i) Derive finite-difference approximations at the bottom corners for steady conduction.
  (ii) For the given mesh, use Gauss-Seidel iteration to obtain temperatures T_1 through T_6 for three iterations.
  (iii) After the third iteration, perform an energy balance on the bar.

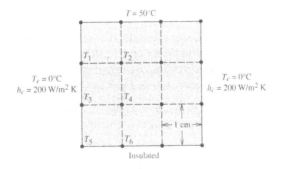

3-114. Equation (3.88) gives a finite-difference approximation for the surface heat flux valid for steady two-dimensional conduction. Extend this result to account for an internal heat generation $\dot{Q}_v'''$.

3-115.   A long, 4 cm-square bar has opposite faces maintained at 80°C and 0°C, respectively, and the other two lose heat by convection to a fluid at 0°C. An electric current flows in the bar and causes a resistance heating of 3×10^6 W/m³. The bar has a thermal conductivity of 15 W/m K, and the convective heat transfer coefficient is 1500 W/m² K.

   (i) Determine the finite-difference approximation for an interior node, and for a surface node on the convectively cooled faces.
   (ii) For the given mesh, use Gauss-Seidel iteration to determine the temperatures T_1 through T_9 after the first and second iterations.

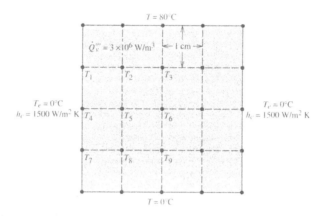

3-116.   Write a computer program to solve Exercise 3-115, allowing for an arbitrary mesh size and resistance heating rate $\dot{Q}_v'''$. Also calculate the heat loss to the fluid. Explore the effect of mesh size on the calculated heat flow for mesh sizes of 1 cm, 0.5 cm, and 0.2 cm. Also explore the effect of the magnitude of $\dot{Q}_v'''$ on the location of maximum temperature.

3-117.   The Crank-Nicolson method has been widely used for finite-difference numerical solution of unsteady heat conduction problems. Whereas the explicit method evaluates conduction fluxes at the old time step, and the implicit method uses the new time step, this method uses an average of the old and the new. Derive the nodal equations corresponding to Eqs. (3.90) and (3.95).

3-118.   Write a computer program to solve Example 3.13. Allow for an arbitrary mesh size Δx and time step Δt. Investigate the effect of Δx and Δt on the accuracy of the solution.

3-119.   Write a computer program to implement the flow diagram shown in Example 3.14. Rework the example for

   (i) $h_c = 280(T - T_{sat})^2$ W/m² K.
   (ii) $h_c = 600(T - T_{sat})^{1.8}$ W/m² K.

3-120. For one-dimensional unsteady conduction, derive the implicit form of the finite-difference formula for the convective boundary condition given as item 2 of Table 3.8.

3-121. For one-dimensional unsteady conduction, derive both the explicit and implicit forms of the finite-difference formula for a boundary where the surface heat flux is known (item 3 of Table 3.8). In the case of the explicit form, also derive the stability criterion.

3-122. For two-dimensional unsteady conduction, derive both the explicit and implicit forms of the finite-difference approximation for a convective boundary condition at a plane surface. For the explicit case, also derive the stability criterion. (See Table 3.9, item 2.)

3-123. For two-dimensional unsteady conduction, derive both the explicit and implicit forms of the finite-difference approximation for a known heat flux boundary condition at a plane surface. For the explicit case, also derive the stability criterion. (See Table 3.9, item 3.)

3-124. For two-dimensional unsteady conduction, derive both the explicit and implicit forms of the finite-difference approximation for a convective boundary condition at an exterior corner. For the explicit case, also derive the stability criterion. (See Table 3.9, item 4.)

3-125. For two-dimensional unsteady conduction, derive both the explicit and implicit forms of the finite-difference approximation for a convective boundary condition at an interior corner. For the explicit case, also derive the stability criterion. (See Table 3.9, item 5.)

3-126. A large slab of polyvinylchloride 1 cm thick is in contact with fluid on either side. On one side the heat transfer coefficient is 20 W/m^2 K, and on the other it is 40 W/m^2 K. Initially both fluids are at 20°C, and the system is in thermal equilibrium. Suddenly the fluid temperatures are raised to 100°C. Determine the time required for the center of the plate to reach 80°C.

3-127. A large slab of 5 cm-thick Pyrex glass is to be cooled slowly from an initial uniform temperature of 400°C by an array of air jets impinging on each side of the slab. The ducting of the air is such that on one side the air is at 20°C, whereas on the other side it is at 70°C. The convective heat transfer coefficient on both sides can be taken as 30 W/m^2 K. Determine the temperature at the center of the slab after 3 hours have elapsed.

3-128. A plate-glass slab with initial temperature T_0 is suddenly exposed to a gas and surroundings at temperature T_e. The slab loses heat by forced convection and radiation. The convective heat transfer coefficient can be assumed constant, and radiation can be calculated for a small gray body in large, nearly black surroundings.

  (i) Give the equations required to solve for the temperature response using the explicit finite-difference method. Discuss the appropriate stability criterion.

(ii) Obtain numerical results for two time steps with a mesh size of 1 cm, and a mesh Fourier number of 0.1. Numerical values of the parameters are $k = 1.75$ W/m K, $\alpha = 0.75 \times 10^{-6}$ m^2/s, slab thickness $2L = 8$ cm, $T_0 = 600°C$, $T_e = 0°C$, $h_c = 50$ W/m^2 K, $\varepsilon = 0.91$.

3-129. A large slab of 8 cm-thick soda lime glass is to be cooled slowly from an initial temperature of 300°C by an array of air jets impinging on each side of the plate. The air temperature is 30°C, and the convective heat transfer coefficient on both sides is 44 W/m^2 K. On the given mesh, use the explicit method to calculate the nodal temperatures for the first five time steps. Use a mesh Fourier number of 0.3 but show that it satisfies the appropriate stability criterion.

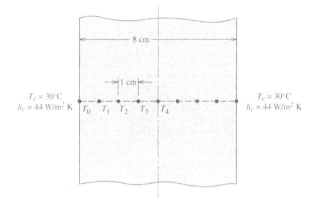

3-130. Write a computer program to solve Exercise 3-129, allowing for an arbitrary number of nodes. Obtain the temperature distribution at $t = 20$ minutes using 5, 10, 20, and 40 nodes. Compare your result to the exact solution obtained from COND2.

3-131. A large slab of 3 cm-thick Teflon at 20°C is exposed to a radiative heat flux from a high temperature heat source of 3500 W/m^2 on one side, and is cooled by an array of air jets on the other side. The jet air temperature is 10°C, and the convective heat transfer coefficient is 35 W/m^2 K. For the given mesh, calculate the nodal temperatures for the first four time steps using the explicit method. Use a Fourier number of 0.2 but show that it satisfies appropriate stability criteria. Neglect radiative heat transfer from the slab surfaces to the surroundings.

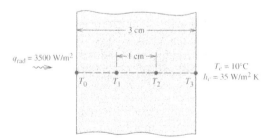

3-132. A 3 cm-thick AISI 302 stainless steel plate coated with black oxide is suddenly exposed to a solar radiation flux of 900 W/m^2, of which it absorbs 89%. The back side of the plate is insulated. The exposed side loses heat by convection to air at 20°C with a heat transfer coefficient of 6 W/m^2 K, and by reradiation to the surroundings also at 20°C. The emittance of the surface is 0.75. The initial temperature of the plate is 20°C.

  (i) Derive the explicit form and stability criterion for the finite-difference approximation of one-dimensional conduction at the surface node T_0.
  (ii) For the given mesh, calculate the nodal temperatures for the first four time steps by the explicit method. Use a Fourier number of 0.4 but show that it satisfies the appropriate stability criterion.

  (*Hint:* Use a radiation heat transfer coefficient to account for reradiation.)

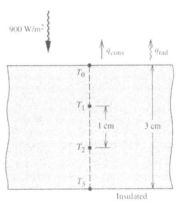

3-133. A type 316 stainless steel bar is 20 cm long, 3 cm wide, and 4 mm thick. Water at 20°C impinges on each side face of the bar, giving a convective heat transfer coefficient of 6500 W/m^2 K. The initial temperature of the bar is also 20°C, and at time $t = 0$ an electric current is passed through the bar, giving a volumetric heat generation of 86.7 MW/m^3.

  (i) Derive explicit finite-difference approximations for nodes T_0, T_1, and T_2.
  (ii) For the given mesh, calculate the nodal temperatures for the first four time steps by the explicit method. Use a Fourier number of 0.3 but show that it satisfies the appropriate stability criterion.

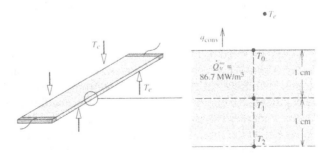

3-134.   A large slab of 4 cm-thick glass is to be cooled from an initial temperature of 400°C by an array of air jets impinging on each side of the slab. The air temperature is 20°C, and the convective heat transfer coefficient on both sides is estimated to be 30 W/m² K. The glass properties can be taken as $k = 1.5$ W/m K, $\alpha = 0.8 \times 10^{-6}$ m²/s. On the given mesh, use the explicit method to obtain the node temperatures for five time steps. Use a mesh Fourier number of 0.4 but show that it satisfies the appropriate stability criterion.

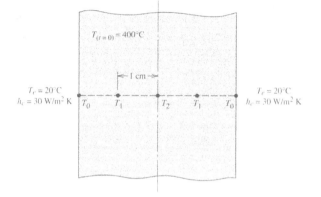

3-135.   Write a computer program (or use a spreadsheet) to implement the solution procedure of Exercise 3-134. Allow for an arbitrary mesh size and mesh Fourier number. Explore the effect of mesh size on the temperature profile at $t = 1000$ s. Also compare your results with the exact solution from COND2.

3-136.   Repeat Exercise 3-134 using the implicit method. Give the finite-difference approximation for the given mesh, and by Gauss-Seidel iteration perform five iterations at the first time step. Use Fo = 0.4.

3-137.   Write a computer program (or use a spreadsheet) to implement the solution procedure of Exercise 3-136. Allow for an arbitrary number of nodes and length of time step. Investigate the effect of number of nodes and size of time step on the temperature profile at $t = 1000$ s. Also compare your results with the analytical solution using COND2.

3-138.   Repeat Exercise 3-136 using a direct method to solve the finite-difference equations at the first time step.

3-139.   A large slab of 3 cm-thick Teflon at 20°C is suddenly exposed to a radiative heat flux of 3500 W/m² from a high temperature source on one side, and is insulated on the other. Reradiation and convection from the exposed side can be assumed to be negligible. Using the implicit method, give the finite-difference approximations for the 1 cm mesh shown, and perform five iterations at the first time step using Gauss-Seidel iteration. Also check the temperatures at the first time step by direct solution of the difference equations. Use Fo = 0.4.

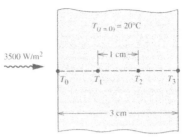

3-140. Write a computer program (or use a spreadsheet) to implement the solution procedure of Exercise 3-139. Allow for an arbitrary mesh size and length of time step. Investigate the effect of number of nodes and length of time step on the temperature profile at $t = 100$ s.

3-141. Repeat Exercise 3-132 using the implicit method. Give the finite-difference approximations for the given mesh, and use Gauss-Seidel iteration to perform three iterations at the first time step. Also check the temperatures at the first time step using a direct method.

3-142. Write a computer program (or use a spreadsheet) to implement the solution procedure of Exercise 3-132. Allow for an arbitrary mesh size and length of time step. Investigate the effect of mesh size and time step length on the temperature profile at $t = 1000$ s.

3-143. Repeat Exercise 3-133 using the implicit method. Give the finite-difference approximations for the specified mesh, and use Gauss-Seidel iteration to perform five iterations at the first time step. Also check the temperatures at the first time step using a direct method.

3-144. Write a computer program (or use a spreadsheet) to implement the solution procedure of Exercise 3-143. Allow for an arbitrary mesh size and length of time step. Investigate the effect of mesh size and time step length on the temperature profile at $t = 100$ s.

CONVECTION FUNDAMENTALS AND CORRELATIONS

CONTENTS

4.1 INTRODUCTION

In the preceding chapters, we have seen that engineering calculations of heat conduction are based on analysis. A model is formulated, and the resulting differential or algebraic equations are solved. The nature of engineering calculations for heat convection is different. The most frequent task an engineer performs is the estimation of heat transfer coefficients from correlations of experimental data, because the differential equations governing convection can be analytically solved only for the simplest of flows. The engineer must rely on experimental data for most of the great variety of flows encountered in practice.

Chapter 4 has been organized accordingly. In Section 4.2, the fundamentals of convection are discussed rather briefly, and the focus is on those concepts required by the engineer to use heat transfer coefficient correlations effectively. The subsequent sections present correlations for a wide range of commonly encountered flows. Analyses of convection are deferred to Chapter 5, because although these analyses may give insight into the nature of convection, their mastery is not essential to the use of heat transfer coefficient correlations.

Calculation of heat transfer coefficients using a handheld calculator can be tedious. The correlations are often complicated functions, and fluid properties must be evaluated by interpolating in tables. Also, the risk of making a careless error is significant. Thus, an ideal tool is a computer program in which the large database is efficiently stored and that executes the calculations rapidly and reliably. Section 4.8 describes the computer program CONV, which was developed for this purpose. CONV should prove indispensable and will probably be used more frequently than any other program in the software package. Before CONV is used for a particular problem, the relevant section of the text in Chapter 4 should be consulted to ensure that an appropriate correlation is used.

4.2 FUNDAMENTALS

This section presents the fundamentals of convection required to understand and use correlations of the convective heat transfer coefficient. First, the convective heat transfer coefficient is defined for three types of flow. In each case, the physics of the flow is described in detail to give the student an appreciation of the nature of convection. Next, dimensional analysis based on the Buckingham pi theorem is used to derive the dimensionless groups pertinent to convection, such as the Reynolds, Grashof, and Nusselt numbers. Experimental data for convective heat transfer coefficients are best correlated in terms of the pertinent dimensionless groups. Heat transfer to flow over a cylinder is used to demonstrate the approach. Finally, we address the awkward problem of how to evaluate temperature-dependent fluid properties in dimensionless groups. Experience has shown that unless this problem is tackled head-on at the outset, it tends to remain a source of confusion for the student.

4.2.1 The Convective Heat Transfer Coefficient

Fluid motion past a surface increases the rate of heat transfer between the surface and the fluid. We are all aware of how a brisk wind increases our discomfort on a cold day. A flowing fluid transports thermal energy by virtue of its motion, and it does so very effectively. Resulting convective heat transfer coefficients can be very large. To define the heat transfer coefficient, we consider three types of flow: (1) an external forced flow, (2) an internal natural-convection flow, and (3) an internal forced flow.

Forced Flow over a Cylinder

Figure 4.1 shows isotherms around a heated cylinder mounted transversely in a wind tunnel. As the air velocity increases, the isotherms on the windward side move closer together. At a sufficiently high velocity, the heated fluid is confined to a very thin **thermal boundary layer**, which has approximately the same thickness as the hydrodynamic boundary layer. The flow on the leeward side of the cylinder is more complex, and when flow separation occurs, large-scale vortices are shed from the cylinder. Fourier's law is not only valid in a stationary solid, but it also applies in a moving fluid. Thus, as the isotherms move closer together and the temperature gradient normal to the cylinder surface increases, so does the resulting heat conduction. Indeed, the fluid in contact with the cylinder is stationary due to viscous action, and heat can be transferred from the cylinder surface into the fluid

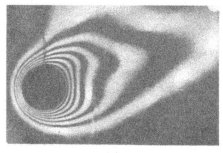

(a)

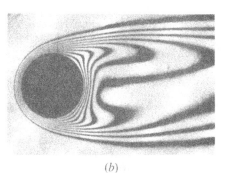

(b)

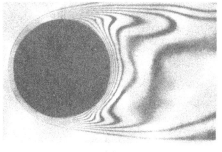

(c)

Figure 4.1 Flow over a heated cylinder: the effect of velocity on isotherms, (a) Re = 23, (b) Re = 120, (c) Re = 597. (Photograph by E. Soehngen, courtesy Professor J. P. Holman, Southern Methodist University, Dallas.)

by conduction only. Thus, the local convective heat transfer rate is given by

$$q_s = -k \frac{\partial T}{\partial y}\bigg|_{y=0} \tag{4.1}$$

where y is the coordinate direction normal to the surface and k is the fluid thermal conductivity. On the other hand, Newton's law of cooling, Eq. (1.20), defined the convective heat transfer coefficient by the relation

$$q_s = h_c(T_s - T_e) \tag{4.2}$$

where T_s is the surface temperature and T_e is the fluid temperature in the free stream. Combining Eqs. (4.1) and (4.2),

$$h_c = \frac{-k(\partial T/\partial y)|_{y=0}}{(T_s - T_e)} \tag{4.3}$$

The effect of increasing the fluid velocity is to steepen the temperature gradient at the surface, thereby increasing the heat transfer coefficient.

Natural Convection between Two Horizontal Plates

Figure 4.2 shows an internal natural-convection flow. A layer of fluid of thickness L is confined between two isothermal plates and heated from below. For small values of the difference between the plate temperatures $(T_H - T_C)$, the fluid is stationary, and the heat transfer through the layer is by conduction only. However, if $(T_H - T_C)$ is increased to a critical value, the fluid becomes unstable, and a cellular flow pattern is set up. The circulation of fluid in a cell convects warm fluid upward and cold fluid downward, and, to accommodate the higher rate of heat transfer across the layer, the isotherms adjacent to each plate move closer together. If $(T_H - T_C)$ is further increased, there are transitions to increasingly more complex flows until cellular flow is replaced by a chaotic turbulent motion. The core of the fluid layer is then almost isothermal, with the major temperature variation confined to a very thin *viscous sublayer* adjacent to each plate, where viscosity serves to damp the turbulence motions. The engineer is usually concerned with heat transfer from one plate to the other, as, for example, in a covered flat-plate solar collector, rather than from one of the plates into the fluid. Thus, it is usual to define the average heat transfer coefficient for such configurations in terms of $(T_H - T_C)$, that is,

$$\bar{h}_c = \frac{\dot{Q}/A}{T_H - T_C} \tag{4.4}$$

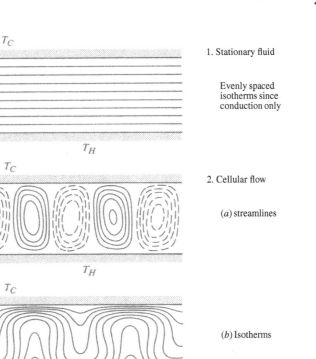

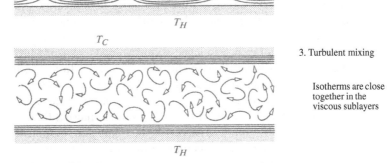

Figure 4.2 Natural-convection flow regimes for a layer of fluid between two horizontal isothermal plates spaced L apart. The temperature difference $(T_H - T_C)$ increases from regime (1) through regime (3). Cellular flow streamlines and isotherms courtesy Professor G. Mallinson, University of Auckland.

rather than in terms of some fluid temperature. An average heat transfer coefficient is appropriate since for cellular flow h_c is not constant over each plate. As before, the heat flow can be expressed in terms of conduction in the fluid adjacent to the plate surfaces at $y = 0$ and $y = L$:

$$\dot{Q} = \int_A q_s dA = \int_A -k \frac{\partial T}{\partial y}\bigg|_{y=0} dA = \int_A -k \frac{\partial T}{\partial y}\bigg|_{y=L} dA \tag{4.5}$$

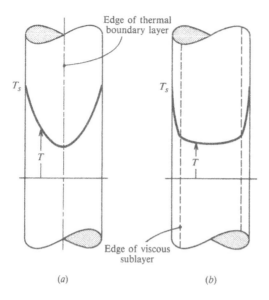

Figure 4.3   Temperature profiles for laminar and turbulent flow in a tube. (*a*) Laminar flow, (*b*) Turbulent flow.

Flow in a Tube

Figure 4.3*a* shows a temperature profile for laminar flow in a tube some distance from the entrance. In this case, the thermal boundary layer extends to the centerline of the tube, and its thickness is independent of velocity. As a result, the temperature gradient at the wall is also independent of velocity, as is the heat transfer coefficient (cf Eq. 1.21). Turbulent flow is shown in Fig. 4.3*b*. In this case, an increase in velocity produces a more vigorous turbulent mixing in the core of flow, and there is a resulting thinning of the viscous sublayer adjacent to the wall, with an increase in the heat transfer coefficient (cf Eq. 1.22). For an internal flow, such as flow in a tube or annulus, it is customary to define the heat transfer coefficient in terms of the *bulk* temperature, which is the temperature the fluid would attain at a given axial location if it were diverted into an adiabatic mixing chamber and thoroughly mixed, as shown in Fig. 4.4 for flow in a tube. The velocity profile $u(r)$ will be parabolic if the flow is laminar, or much flatter if the flow is turbulent. The mass flow rate $\dot{m}$ [kg/s] is obtained by integrating over the cross section,

$$\dot{m} = \int_0^R \rho u 2\pi r \, dr \qquad (4.6)$$

and, by mass conservation, is the same entering and leaving the chamber. Application of the steady-flow energy equation, Eq. (1.4), to the chamber requires simply that the rate at which enthalpy, h, enters and leaves the chamber be equal:

$$\int_0^R \rho u h 2\pi r \, dr = \int_0^R \rho u h_b 2\pi r \, dr$$

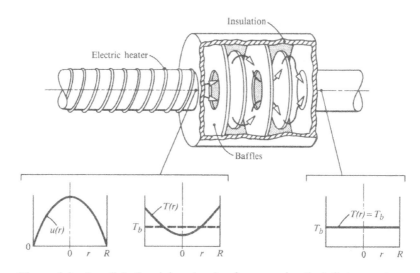

Figure 4.4 An adiabatic mixing chamber for measuring the bulk temperature.

where h_b is the bulk enthalpy. If we assume constant properties (ρ and c_p) and an enthalpy datum state of $h = 0$ at $T = 0$, we can write $h = c_p T$, and

$$\int_0^R \rho u c_p T 2\pi r \, dr = \int_0^R \rho u c_p T_b 2\pi r \, dr = c_p T_b \dot{m}$$

from Eq. (4.6), since T_b is not a function of r. Canceling c_p gives

$$T_b = \frac{\int_0^R \rho u T 2\pi r \, dr}{\dot{m}} \tag{4.7}$$

which allows the bulk temperature to be calculated if the velocity and temperature profiles across the flow are known. Now consider an element of tube Δx long, as shown in Fig. 4.5, and once again apply the steady-flow energy equation:

$$q_s 2\pi R \Delta x = \int_0^R \rho u c_p T 2\pi r \, dr \bigg|_{x+\Delta x} - \int_0^R \rho u c_p T 2\pi r \, dr \bigg|_x$$

where q_s is the wall heat flux. Using Eq. (4.7),

$$q_s 2\pi R \Delta x = \dot{m} c_p T_b \bigg|_{x+\Delta x} - \dot{m} c_p T_b \bigg|_x$$

Dividing by Δx and letting $\Delta x \to 0$,

$$q_s 2\pi R = \dot{m} c_p \frac{dT_b}{dx} \tag{4.8}$$

If we define the heat transfer coefficient in terms of bulk temperature, that is,

$$h_c = \frac{q_s}{T_s - T_b} \tag{4.9}$$

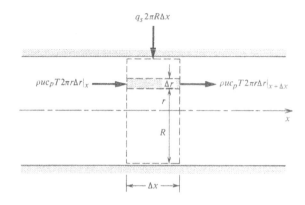

Figure 4.5   An energy balance on an element of tube Δx long.

then Eq. (4.8) becomes

$$h_c(T_s - T_b)2\pi R = \dot{m}c_p\frac{dT_b}{dx} \tag{4.10}$$

If the wall temperature is constant along the tube, there is a single dependent variable $T_b(x)$, and this equation can be integrated directly. It would be less convenient had h_c been defined in terms of, for example, the centerline temperature.

For a tube of length L and T_s constant, Eq. (4.10) can be integrated with $T_b = T_{b,in}$ at $x = 0$, and $T_b = T_{b,out}$ at $x = L$, to give

$$T_{b,out} = T_s - (T_s - T_{b,in})e^{-\bar{h}_c 2\pi RL/\dot{m}c_p} \tag{4.11}$$

where $\bar{h}_c$ is an average heat transfer coefficient defined by

$$\bar{h}_c = \frac{1}{L}\int_0^L h_c\, dx \tag{4.12}$$

T_s and $T_b(x)$ are plotted in Fig. 4.6. The definition of average heat transfer coef-

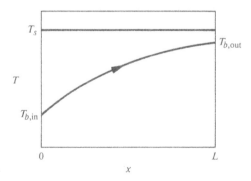

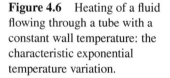

Figure 4.6   Heating of a fluid flowing through a tube with a constant wall temperature: the characteristic exponential temperature variation.

ficient is seen to be equivalent to Eq. (1.26) for an external flow on an isothermal wall. Equation (4.11) will prove useful for problem solving. When the exponent $\bar{h}_c 2\pi R L / \dot{m} c_p$ is much less than unity, the fluid temperature change is small, $T_{b,\text{out}} \simeq T_{b,\text{in}}$. On the other hand, when it is greater than about 3, the fluid outlet temperature approaches the wall temperature, $T_{b,\text{out}} \simeq T_s$.

Table 4.1 summarizes the three definitions of the convective heat transfer coefficient.

Table 4.1   Summary of defining equations for the convective heat transfer coefficient.

Flow Configuration	Heat Transfer Coefficient Defining Equation	Comments
External flow over a surface	$q_s = h_c(T_s - T_e)$ (4.2)	T_e is the free-stream temperature.
Natural flow in an enclosure	$\dfrac{\dot{Q}}{A} = \bar{h}_c(T_H - T_C)$ (4.4)	T_H and T_C are the hot and cold surface temperatures, respectively.
Internal flow in a duct	$q_s = h_c(T_s - T_b)$ (4.9)	T_b is the bulk temperature defined by Eq. (4.7).

4.2.2 Dimensional Analysis

Experimental data for convective heat transfer coefficients, as well as the results of analysis, can be conveniently and concisely organized as relationships between dimensionless groups of the pertinent variables. The Reynolds and Grashof numbers introduced in Chapter 1 are examples of such groups. The simplest method for selecting dimensionless groups appropriate to a given problem is dimensional analysis using the **Buckingham pi theorem** and the **method of indices**. It is assumed that the student has already had an introduction to fluid mechanics, in which the methodology of dimensional analysis is applied to viscous flows. The application of this methodology to convective heat transfer is somewhat more difficult, owing to the larger number of variables involved.

Before commencing the dimensional analysis of convective heat transfer, a short review of the dimensionless groups pertinent to simple viscous flows is appropriate. The dimensionless group characterizing a viscous flow is the **Reynolds number**, Re:

$$\text{Re} = \frac{\rho V L}{\mu} = \frac{V L}{\nu} \tag{4.13}$$

where L is a characteristic length of the configuration, and for external flows, V is usually the velocity of the undisturbed free stream. For internal flows, ρV [kg/m^{2}s] is taken as the *mass velocity* $G = \dot{m}/A_c$, where $\dot{m}$ is the mass flow rate and A_c is the cross-sectional area for flow. If the density can be assumed constant, then $G = \rho u_b$, where u_b is the *bulk velocity*. The SI units for dynamic viscosity μ are [kg/m s],[1] and the units of kinematic viscosity $\nu(= \mu/\rho)$ are [m²/s]. Momentum transfer to a wall is made dimensionless as the **skin friction coefficient**, C_f:

$$C_f = \frac{\tau_s}{(1/2)\rho V^2} \tag{4.14}$$

where τ_s [N/m²] is the wall shear stress. In pipe flows, the pressure gradient is made dimensionless as the **friction factor**, f, for incompressible flow:

$$f = \frac{\Delta P/L}{(1/2)\rho V^2/D} \tag{4.15}$$

where ΔP [N/m²] is the pressure drop over a length L of a pipe with diameter D. This friction factor is called the *Darcy* friction factor. Some texts use the *Fanning* friction factor, which is one fourth the Darcy value. *We will use the Darcy friction factor exclusively.* If the flow in the pipe is hydrodynamically fully developed, that is, if the velocity profile is unchanging with axial position, then f and C_f are simply related as

$$f = 4C_f \tag{4.16}$$

which can be derived from the force balance shown in Fig. 4.7. The pressure drop across an orifice can be made dimensionless as the **Euler number**, Eu:[2]

$$\text{Eu} = \frac{\Delta P}{G^2/\rho} = \frac{\Delta P}{\rho V^2} \tag{4.17}$$

To introduce the dimensionless groups of convective heat transfer, we will consider a number of sample flows.

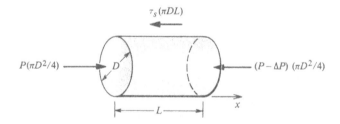

Figure 4.7 A force balance on a length of pipe, used to relate the friction factor to the skin friction coefficient.

---

[1] Strictly speaking, the SI units for dynamic viscosity are *pascal seconds* [Pa s]. However, the equivalent units [kg/m s] and [N s/m²] are more widely used in engineering practice. For dimensional analysis, the most convenient form is [kg/m s], and this form will be used throughout this text.
[2] In some texts the Euler number is defined as $\text{Eu} = \Delta P/(1/2)\rho V^2$. Thus, the definition of the Euler number should be checked when consulting relevant literature.

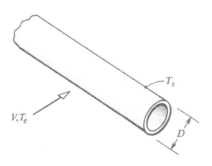

Figure 4.8 Schematic of forced flow across a cylinder.

Flow across a Cylinder

External forced flow across a long cylinder is shown in Fig. 4.8 and was discussed in detail in Section 4.2.1. The flow field is expected to depend on the upstream velocity V, diameter D, and the fluid density ρ and dynamic viscosity μ. The average wall heat flux is expected to depend on these variables, as well as on the temperature difference, $\Delta T = T_s - T_e$, and the fluid specific heat c_p and thermal conductivity k. Thus, the expected functional dependence of $\bar{q}_s$ is

$$\bar{q}_s = f(V, D, \rho, \mu, \Delta T, c_p, k)$$

The units of the variables may be written as follows:

$\bar{q}_s$	V	D	ρ	μ	ΔT	c_p	k
$\dfrac{W}{m^2}$	$\dfrac{m}{s}$	m	$\dfrac{kg}{m^3}$	$\dfrac{kg}{m\,s}$	K	$\dfrac{W\,s}{kg\,K}$	$\dfrac{W}{m\,K}$

Since we only use SI units in this text, we need not distinguish between the general concept of *dimensions* and a specific system of *units*. Notice that the units of c_p are written [W s/kg K] rather than [J/kg K] so as to utilize the set of *primary* dimensions or units kg, m, s, K, W. There are five primary dimensions and eight variables; thus, according to the Buckingham pi theorem, $(8 - 5) = 3$ independent dimensionless groups or products can be formed from the variables. The most familiar method of obtaining these groups is the *method of indices,* which is used as follows. We begin by writing any dimensionless group Π as a product of the variables, each raised to an unknown exponent,

$$\Pi = \bar{q}_s^a V^b D^c \rho^d \mu^e \, \Delta T^f c_p^g k^h$$

and substitute the appropriate dimensions:

$$\Pi = \left[\frac{W}{m^2}\right]^a \left[\frac{m}{s}\right]^b [m]^c \left[\frac{kg}{m^3}\right]^d \left[\frac{kg}{m\,s}\right]^e [K]^f \left[\frac{W\,s}{kg\,K}\right]^g \left[\frac{W}{m\,K}\right]^h$$

For Π to be dimensionless, the sum of exponents of each primary dimension must add to zero:

$$
\begin{aligned}
\text{kg}: & & d+e-g &= 0 \\
\text{m}: & & -2a+b+c-3d-e-h &= 0 \\
\text{s}: & & -b-e+g &= 0 \\
\text{K}: & & f-g-h &= 0 \\
\text{W}: & & a+g+h &= 0
\end{aligned}
$$

This constitutes a set of five linear algebraic equations in eight unknowns. A basic theorem of linear algebra states that the number of distinct solutions is equal to the number of unknowns minus the number of linearly independent equations. The number of linearly independent equations is given by the *rank* of the coefficient matrix, which for this problem can be shown to equal five. Thus, there are $8-5=3$ distinct solutions, as stated by the pi theorem.

To obtain these solutions, we are free to choose values for three exponents, and we let the equations give the values for the remaining five. Because the dimensionless groups initially obtained depend on our choice, we are faced with a dilemma. We might proceed as follows. Based on fluid mechanics experience, we know that the Reynolds number characterizes the flow field; hence, set $b=1, a=g=0$:

$$
\begin{aligned}
d+e &= 0 \\
1+c-3d-e-h &= 0 \\
-1-e &= 0 \\
f-h &= 0 \\
h &= 0
\end{aligned}
$$

Hence, $h=0, f=0, e=-1, d=1, c=1$, which gives the Reynolds number

$$
\Pi_1 = \frac{VD\rho}{\mu}
$$

Next, since we would like to have q_s in the dependent variable group, we set $a=1$ and we do not set $f=0$ so as to obtain the heat transfer coefficient $q_s/\Delta T$ in this group. Thus, somewhat arbitrarily, we set $b=g=0$:

$$
\begin{aligned}
d+e &= 0 \\
-2+c-3d-e-h &= 0 \\
e &= 0 \\
f-h &= 0 \\
1+h &= 0
\end{aligned}
$$

Hence, $h=-1, f=-1, e=0, d=0, c=1$, which gives

$$
\Pi_2 = \frac{\bar{q}_s D}{\Delta T k}
$$

which is the average **Nusselt number**, $\overline{\text{Nu}}$.

To obtain a third group, we recognize that fluid properties affect heat transport in the fluid and thus attempt to obtain a dimensionless group of properties by setting $g = 1, a = b = 0$:

$$d + e - 1 = 0$$
$$c - 3d - e - h = 0$$
$$-e + 1 = 0$$
$$f - 1 - h = 0$$
$$1 + h = 0$$

Hence, $h = -1, f = 0, e = 1, d = 0, c = 0$, which gives

$$\Pi_3 = \frac{c_p \mu}{k}$$

which is the **Prandtl number**, Pr. Thus, the result of the dimensionless analysis can be written in terms of three independent dimensionless groups as

$$\overline{\text{Nu}} = f(\text{Re}, \text{Pr}) \tag{4.18}$$

The ratio $\bar{q}_s / \Delta T$ is the convective heat transfer coefficient; thus, the Nusselt number may be viewed as a dimensionless heat transfer coefficient. In general,

$$\text{Nu} = \frac{h_c L}{k} \tag{4.19}$$

where the characteristic length L may be the distance along a flat plate, the diameter of a cylinder in cross-flow, or pipe diameter for flow in a pipe. The Prandtl number can be written as

$$\text{Pr} = \frac{c_p \mu}{k} = \frac{\nu}{k/\rho c_p} = \frac{\nu}{\alpha} \tag{4.20}$$

where it is seen that this fluid properties group is the ratio of the kinematic viscosity to thermal diffusivity. Indeed, the kinematic viscosity is more appropriately termed the *momentum diffusivity,* so that the Prandtl number is the ratio of momentum and thermal diffusivities. Table 4.2 gives values of the Prandtl number for various commonly encountered fluids. There are three broad groups: liquid metals, with $\text{Pr} \ll 1$; gases, with $\text{Pr} \sim 1$; and oils, with $\text{Pr} \gg 1$. The value for water ranges from about 1 to 10, depending on temperature. It will be seen that the convective heat transfer characteristics of a fluid are very much dependent on its Prandtl number.

In the dimensional analysis, we could have obtained an alternative Π_2 as follows. Again, $a = 1, f \neq 0$, but now we choose $c = h = 0$:

$$d + e - g = 0$$
$$-2 + b - 3d - e = 0$$
$$-b - e + g = 0$$
$$f - g = 0$$
$$1 + g = 0$$

Table 4.2 Prandtl number values for various fluids ($\mathrm{Pr} = c_p\mu/k = \nu/\alpha$).

Fluid	Temperature K	Prandtl Number
Sodium	1000	0.0038
Mercury	500	0.012
Lithium	700	0.031
Argon	400	0.67
Air	300	0.69
Saturated steam	373	0.98
Liquid ammonia	270	1.49
Water	460	0.98
	360	2.00
	275	12.9
Liquid refrigerant-12	250	4.6
Therminol 60	350	31.3
Ethylene glycol	300	151
SAE 50 oil	400	154
	300	6600

Hence, $g = -1, f = -1, b = -1, d = -1, e = 0$, which gives

$$\Pi_2' = \frac{\bar{q}_s}{\Delta T \rho c_p V}$$

which is the average **Stanton number,** $\overline{\mathrm{St}}$. Again, writing $\bar{q}_s/\Delta T = \bar{h}_c$, we have

$$\overline{\mathrm{St}} = \frac{\bar{h}_c}{\rho c_p V} = \frac{\bar{h}_c}{c_p G} \tag{4.21}$$

The Stanton number is an alternative dimensionless heat transfer coefficient and is related to the Nusselt number as $\mathrm{St} = \mathrm{Nu/RePr}$. Of course, other alternatives can be obtained by combining the preceding groups. For example, the **Péclet number,**

$$\mathrm{Pe} = \mathrm{RePr} = \frac{VL}{\alpha} \tag{4.22}$$

is often used for creeping external flow ($\mathrm{Re} \simeq 1$) and laminar internal flows.

High-Speed Flow

In the foregoing dimensional analysis, five primary dimensions were used. We will now rework the analysis using only four primary dimensions: kg, s, m, and K. Then, since power equals force times velocity,

$$[\mathrm{W}] = [\mathrm{N}] \left[\frac{\mathrm{m}}{\mathrm{s}}\right] = \left[\frac{\mathrm{kg\ m}}{\mathrm{s}^2}\right] \left[\frac{\mathrm{m}}{\mathrm{s}}\right] = \left[\frac{\mathrm{kg\ m}^2}{\mathrm{s}^3}\right]$$

and the units of the variables that involved watts are now

$\overline{q}_s$	c_p	k
$\dfrac{kg}{s^3}$	$\dfrac{m^2}{s^2\,K}$	$\dfrac{kg\,m}{s^3\,K}$

$$\Pi = \left[\frac{kg}{s^3}\right]^a \left[\frac{m}{s}\right]^b [m]^c \left[\frac{kg}{m^3}\right]^d \left[\frac{kg}{m\,s}\right]^e [K]^f \left[\frac{m^2}{s^2\,K}\right]^g \left[\frac{kg\,m}{s^3\,K}\right]^h$$

$$\text{kg}: \qquad a+d+e+h = 0$$
$$\text{s}: \qquad -3a-b-e-2g-3h = 0$$
$$\text{m}: \qquad b+c-3d-e+2g+h = 0$$
$$\text{K}: \qquad f-g-h = 0$$

According to the pi theorem, there are now $(8-4) = 4$ independent dimensionless groups. The Reynolds, Nusselt, and Prandtl numbers can be obtained as before, but now we must choose the fourth group. We could reason as follows. In eliminating the watt as a primary dimension by using the form power equals force times velocity, we have recognized that kinetic energy can be converted into thermal energy; thus, the fourth group might involve the kinetic energy of the flow. Choosing $b = 2, a = d = e = 0$,

$$h = 0$$
$$-2-2g-3h = 0$$
$$2+c+2g+h = 0$$
$$f-g-h = 0$$

Hence, $h = 0$, $g = -1, c = 0, f = -1$, which gives

$$\Pi_4 = \frac{V^2}{c_p \Delta T}$$

Since V^2 in the numerator of Π_4 is associated with kinetic energy of the fluid, we choose to introduce a factor of 1/2 and define the **Eckert number,** Ec, as

$$\text{Ec} = \frac{(1/2)V^2}{c_p \Delta T} \qquad\qquad\qquad (4.23)$$

This result appears satisfactory and perhaps could be used for correlating heat transfer data for high-speed flows. But experience tells us that the conversion of kinetic energy into thermal energy in an incompressible flow is due to the action of viscous stresses, and yet viscosity does not appear in the Eckert number. Indeed, this phenomenon is called *viscous dissipation*. So we try again and choose $b = 2, a = d = g = 0$:

$$e + h = 0$$

$$-2 - e - 3h = 0$$

$$2 + c - e + h = 0$$

$$f - h = 0$$

Hence, $h = -1, e = 1, f = -1, c = 0$, which gives

$$\Pi_4' = \frac{V^2 \mu}{k \Delta T}$$

This product is the **Brinkman number,** Br, and is most appropriate for characterizing viscous dissipation. It is easily seen that

$$\text{Br} = 2 \frac{(1/2)V^2}{c_p \Delta T} \cdot \frac{c_p \mu}{k} = 2\text{EcPr} \tag{4.24}$$

That is, the Eckert-Prandtl number product is equivalent to the Brinkman number. Thus, for a high-speed flow,

$$\overline{\text{Nu}} = f(\text{Re, Pr, Br})$$
$$= F(\text{Re, Pr, EcPr}) \tag{4.25}$$

Natural-Convection Flow on a Vertical Plate

A vertical plate of height L at temperature T_s is immersed in a fluid at temperature T_e, as shown in Fig. 4.9. Before postulating the functional dependence of the average heat flux, it is necessary to first examine the underlying physics. In natural convection, motion is caused by density variations. From Archimedes' buoyancy principle, the force per unit volume acting on the heated fluid adjacent to the wall is $(\rho_e - \rho)g$, where ρ is the local fluid density, ρ_e is the ambient fluid density, and g is the acceleration due to gravity. The buoyancy force per unit mass is then $(\rho_e - \rho)g/\rho$. If β [K^{-1}]

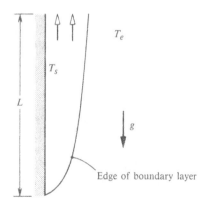

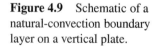

Figure 4.9 Schematic of a natural-convection boundary layer on a vertical plate.

is the volumetric coefficient of thermal expansion, then $\beta(T - T_e) \simeq (\rho_e - \rho)/\rho$. In terms of β, the buoyancy force per unit mass becomes $(T - T_e)\beta g$, and it is this force that causes the natural-convection motion.

We now let $\Delta T = T_s - T_e$ and suggest the following functional dependence for the average heat flux, $\bar{q}_s$:

$$\bar{q}_s = f(\Delta T, \beta, g, \rho, \mu, k, c_p, L)$$

$$\Pi = \bar{q}_s^a \Delta T^b \beta^c g^d \rho^e \mu^f k^g c_p^h L^i$$

which, in terms of the primary dimensions kg, m, s, K, and W, is

$$\Pi = \left[\frac{W}{m^2}\right]^a [K]^b [K^{-1}]^c \left[\frac{m}{s^2}\right]^d \left[\frac{kg}{m^3}\right]^e \left[\frac{kg}{m\,s}\right]^f \left[\frac{W}{m\,K}\right]^g \left[\frac{W\,s}{kg\,K}\right]^h [m]^i$$

Equating the sum of exponents of each primary dimension gives

$$
\begin{aligned}
\text{kg}: &\quad e + f - h = 0 \\
\text{m}: &\quad -2a + d - 3e - f - g + i = 0 \\
\text{s}: &\quad -2d - f + h = 0 \\
\text{K}: &\quad b - c - g - h = 0 \\
\text{W}: &\quad a + g + h = 0
\end{aligned}
$$

There are $(9 - 5) = 4$ independent dimensionless groups. At this point, we use experience gained in analyzing forced convection to immediately choose the Nusselt number $Nu = q_s L/k\Delta T$ as the dimensionless heat flux, and the Prandtl number $Pr = c_p \mu/k$ as a fluid properties group. Also, by inspection, $\Pi_3 = \beta\Delta T$ is obviously dimensionless and independent of Nu and Pr. Thus, there remains but one group to obtain, which we can ensure is independent from the other groups by choosing $d = 1$, $a = b = g = 0$:

$$
\begin{aligned}
e + f - h &= 0 \\
1 - 3e - f + i &= 0 \\
-2 - f + h &= 0 \\
-c - h &= 0 \\
h &= 0
\end{aligned}
$$

Hence, $c = 0, f = -2, e = 2, i = 3$, which gives

$$\Pi_4 = \frac{g\rho^2 L^3}{\mu^2} = \frac{gL^3}{\nu^2}$$

Thus,

$$\overline{Nu} = f\left(Pr, \beta\Delta T, \frac{gL^3}{\nu^2}\right) \tag{4.26}$$

The groups Π_3 and Π_4 do not have names because both theory and experiment show that natural convection depends on the product $\Pi_3\Pi_4$ rather than on each group independently.[3] This group is the familiar **Grashof number,** Gr:

$$\mathrm{Gr} = \frac{\beta \Delta T g L^3}{\nu^2} \tag{4.27}$$

Thus,

$$\overline{\mathrm{Nu}} = f(\mathrm{Gr}, \mathrm{Pr}) \tag{4.28}$$

In fact, there is a further simplification for either very high Prandtl-number fluids (e.g., oils) or very low Prandtl-number fluids (e.g., liquid metals). For $\mathrm{Pr} \gg 1$, the Nusselt number is found to depend on the product of Grashof and Prandtl numbers; this group is the **Rayleigh number,** Ra:

$$\mathrm{Ra} = \mathrm{GrPr} = \frac{\beta \Delta T g L^3}{\nu \alpha} \tag{4.29}$$

$$\overline{\mathrm{Nu}} = f(\mathrm{Ra}); \qquad \mathrm{Pr} \gg 1 \tag{4.30}$$

For $\mathrm{Pr} \ll 1$, the Nusselt number is found to depend on the product of Grashof number and Prandtl number squared; this group is the **Boussinesq number,** Bo:

$$\mathrm{Bo} = \mathrm{GrPr}^2 = \frac{\beta \Delta T g L^3}{\alpha^2} \tag{4.31}$$

$$\overline{\mathrm{Nu}} = f(\mathrm{Bo}); \qquad \mathrm{Pr} \ll 1 \tag{4.32}$$

Natural Convection in an Inclined Enclosure

Figure 4.10 shows an inclined enclosure of the kind found in flat-plate solar collectors. The flow regimes for a horizontal enclosure were discussed in Section 4.2.1. We let $\Delta T = T_H - T_C$ and suggest the following functional dependence for the

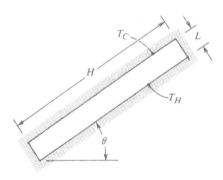

Figure 4.10   Schematic of an inclined enclosure.

---

[3] Since the buoyancy force per unit mass is $\beta g \Delta T$, one could argue that the product βg should be taken as a single variable; then the Grashof number is obtained directly.

average heat flux $\bar{q}_s$:

$$\bar{q}_s = f(\Delta T, \beta, g, \rho, \mu, k, c_p, L, H, \theta)$$

There are now 11 variables in 5 primary dimensions, giving $11 - 5 = 6$ dimensionless groups. Since we have already performed a dimensional analysis of natural convection, we can immediately write down the four groups just obtained,

$$\frac{\bar{q}_s L}{\Delta T k}, \quad \frac{c_p \mu}{k}, \quad \beta \Delta T, \quad \frac{gL^3}{\nu^2}$$

leaving two groups to be obtained. The **aspect ratio**, H/L, is obviously dimensionless, as is angle θ; these are independent of the four groups listed. Thus, the desired result is

$$\overline{\mathrm{Nu}} = f\left(\mathrm{Pr}, \beta \Delta T, \frac{gL^3}{\nu^2}, \frac{H}{L}, \theta\right) \qquad (4.33)$$

Notice that the characteristic length, in both the Nusselt number and the group gL^3/ν^2, was chosen to be L, the spacing of the plates forming the enclosure, and not H. This choice is obvious for the Nusselt number since heat is transferred across the enclosure. For the group gL^3/ν^2, the choice is also obvious for a horizontal enclosure ($\theta = 0$) because the buoyancy force acts in the vertical direction, so that L characterizes the flow length of the motion. For a vertical enclosure ($\theta = 90°$), the choice proves to be inappropriate for some flow regimes. *Whenever there is the possibility for ambiguity in the choice of appropriate length scale for dimensionless groups, the groups should be subscripted accordingly* (e.g., $\overline{\mathrm{Nu}}_L, \mathrm{Gr}_L, \mathrm{Re}_D$, etc.). Figure 4.11 illustrates this practice for natural convection in a rectangular enclosure.

In general, dimensionless transfer coefficients can be *local* values at a particular location on a surface, or *average* values over a surface. In the preceding discussion of dimensional analysis, only average values were considered. In the remainder of

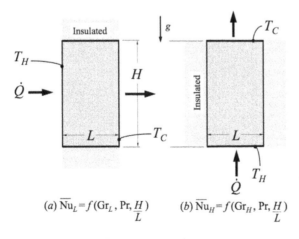

(a) $\overline{\mathrm{Nu}}_L = f(\mathrm{Gr}_L, \mathrm{Pr}, \frac{H}{L})$ (b) $\overline{\mathrm{Nu}}_H = f(\mathrm{Gr}_H, \mathrm{Pr}, \frac{H}{L})$

Figure 4.11 Characteristic lengths for natural convection in a rectangular enclosure.

the chapter, both local and average values will be of concern. The bar notation will be used to denote average coefficients, for example, $\overline{Nu}$ and $\overline{C}_f$; local values will simply be Nu, C_f, and so on.

The preceding dimensional analyses of forced and natural convection have served to introduce the dimensionless groups commonly used in the study of heat convection. Table 4.3 summarizes these groups. Although the application of this methodology of dimensional analysis to simple problems is quite straightforward, its successful application to heat convection problems requires some physical insight and experience. Dimensional analysis based on the Buckingham pi theorem will not yield useful

Table 4.3 Summary of the major dimensionless groups characterizing momentum transfer and convective heat transfer.

Group	Definition	Use
Skin friction coefficient	$C_f = \dfrac{\tau_s}{(1/2)\rho V^2}$	External flows
Friction factor	$f = \dfrac{\Delta P}{(L/D)(1/2)\rho V^2}$	Internal flows
Euler number	$Eu = \dfrac{\Delta P}{\rho V^2}$	Flows through orifices
Reynolds number	$Re = \dfrac{VL}{\nu}$	Forced flows
Nusselt number	$Nu = \dfrac{h_c L}{k}$	Forced and natural flows
Stanton number	$St = \dfrac{h_c}{\rho c_p V}$	Forced flows
Prandtl number	$Pr = \dfrac{c_p \mu}{k} = \dfrac{\nu}{\alpha}$	Forced and natural flows
Grashof number	$Gr = \dfrac{\beta \Delta T g L^3}{\nu^2}$	Natural flows with $Pr \sim 1$
Péclet number	$Pe = \dfrac{VL}{\alpha}$	Laminar internal flows, creeping external flows
Rayleigh number	$Ra = \dfrac{\beta \Delta T g L^3}{\nu \alpha}$	Natural flows with $Pr \gg 1$
Boussinesq number	$Bo = \dfrac{\beta \Delta T g L^3}{\alpha^2}$	Natural flows with $Pr \ll 1$
Brinkman number	$Br = \dfrac{V^2 \mu}{k \Delta T}$	Flows with viscous dissipation

results unless it is accompanied by careful thought. Nevertheless, it has proven to be a valuable tool in many areas of engineering. Exercises 4-4 through 4-7 are further examples of the application of the pi theorem to heat convection problems.

4.2.3 Correlation of Experimental Data

Most engineering calculations of convective heat transfer use heat transfer coefficients obtained from experimental data. Dimensional analysis has proven to be an invaluable tool for the efficient planning of experiments and organization of the resulting data. Once again we will use flow across a heated cylinder to illustrate the procedure. It is relatively simple to perform an experiment to obtain the average heat transfer coefficient in an air flow, as will now be described.

A 3 cm-diameter copper tube containing an electrical heater is shown in Fig. 4.12. Thermocouples are located as shown to measure the surface temperature T_s and the air temperature T_e. Use of a copper tube with a high thermal conductivity ensures an isothermal surface. The experiments are conducted in a conveniently available wind tunnel. In addition to temperatures, required measurements are the power input to the heater $\dot{Q}$, and air speed V. If A is the heated area of the cylinder, the average heat transfer coefficient is (if radiation and end losses are negligible)

$$\bar{h}_c = \frac{\dot{Q}/A}{T_s - T_e}$$

The air speed V is varied over the tunnel operating range while the power is adjusted to hold T_s constant, at about 10 K above the air temperature. In this manner, secondary effects due to variation of fluid properties are eliminated.

Sample data for h_c versus V are shown in Fig. 4.13. As presented in the figure, the data are of limited utility. The graph applies only to heat transfer from a 3 cm-diameter cylinder to air at 300 K, 1 atm. To extend the utility of the data, we use the result of our dimensional analysis, Eq. (4.18):

$$\overline{Nu} = f(Re, Pr)$$

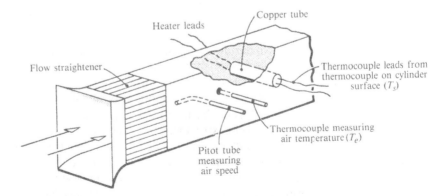

Figure 4.12 An experimental rig for investigating forced-convection heat transfer from a cylinder.

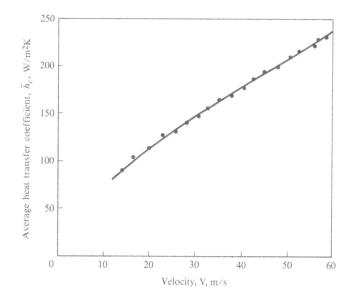

Figure 4.13 Average heat transfer coefficient versus velocity of air flow over a 3 cm-diameter cylinder: air at 300 K, 1 atm.

Table A.7 in Appendix A shows that the Prandtl number of air is constant over a wide range of temperature, equal to 0.69. Thus, we present the data as

$$\overline{\text{Nu}} = f(\text{Re}); \qquad \text{Pr} = 0.69 \qquad\qquad\qquad (4.34)$$

or

$$\frac{\overline{h}_c D}{k} = f\left(\frac{VD}{\nu}\right)$$

with the fluid properties k and ν evaluated at the average of T_s and T_e. The result is shown in Fig. 4.14. Since a simple power law of the form $\text{Nu} = C_1 \text{Re}^n$ would be a convenient correlation formula for the data, Fig. 4.14 is a log-log plot on which a power law will be a straight line with slope n and intercept C_1. A least squares linear regression analysis shows that $n = 0.63$ is a good fit. Figure 4.14 applies to any combination of cylinder diameter and air velocity, provided that the Reynolds number is in the indicated range of $1.5 \times 10^4 < \text{Re} < 10^5$. For example, it applies to a 12 cm-diameter cylinder provided the air speed does not exceed $(3/12)(60) = 15$ m/s to ensure that $\text{Re} < 10^5$ This important observation is a statement of the *similarity principle,* which is the basis of modeling. The engineer often exploits the similarity principle to perform experiments more conveniently. For example, a small model can be tested in a wind tunnel, provided the air speed is increased accordingly to give the desired Reynolds number range. Notice also that the Reynolds number range of the data in Fig. 4.14 can be extended by testing larger- and smaller-diameter cylinders. The similarity principle, which was introduced in Chapters 2 and 3 for heat conduction, is discussed further in Chapter 5.

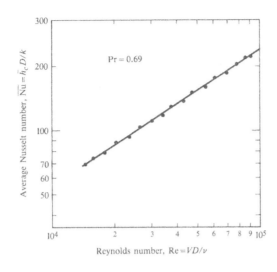

Figure 4.14 Average Nusselt number versus Reynolds number for flow over a cylinder: $Pr = 0.69$.

Figure 4.14 applies to any fluid whose Prandtl number is approximately 0.7. An examination of Table A.7 shows that this condition is met by many gases, including carbon dioxide, helium, and hydrogen. Notice that these gases can have values of k, μ, and c_p that are quite different from those for air, yet their Prandtl numbers are almost equal. To estimate heat transfer coefficients for fluids with Prandtl numbers not equal to 0.7, further experiments must be performed. Examination of Table 4.2 suggests experiments with a liquid metal such as mercury, cold water, and Therminol 60 to obtain a large Prandtl number range. However, experiments with mercury and other liquid metals are difficult to perform. Mercury is highly toxic, sodium ignites spontaneously in air, and there are materials compatibility problems. Liquid metals are not routinely used in experiments unless a specific application requiring a liquid metal is contemplated, for example, in a sodium-cooled nuclear reactor. Thus, we will restrict our attention to higher values of Prandtl number. Figure 4.15a shows the results of a series of experiments. It can be seen that for each fluid, the data are correlated well with straight lines of slope 0.63. Thus, we can now seek a power law correlation of the form

$$\overline{Nu} = C_2 Re^{0.63} Pr^m$$

Using the correlations for each fluid shown in Fig. 4.15a, $\log(\overline{Nu}/Re^{0.63})$ is plotted versus $\log Pr$ in Fig. 4.15b, where it is seen that $m = 0.36, C_2 = 0.19$. Thus, the recommended correlation is

$$\overline{Nu} = 0.19 Re^{0.63} Pr^{0.36} \tag{4.35}$$

which is valid for $10^4 \leq Re \leq 10^5, 0.69 < Pr < 31.3$. Modest extrapolation outside these Reynolds and Prandtl number ranges would be warranted. Of course, the preferred procedure to determine the constant C_2 and exponents n and m is a multivariable least squares fit of all the data points using a standard computer subroutine.

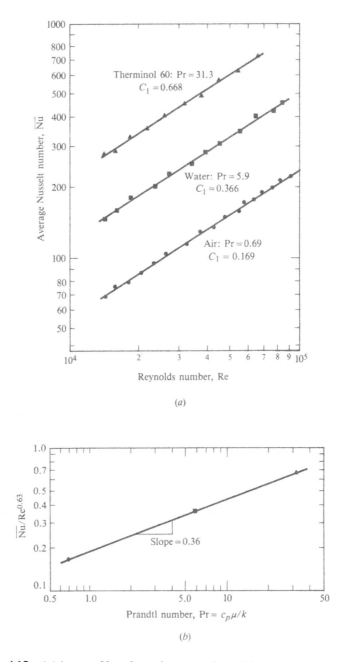

Figure 4.15 *(a)* Average Nusselt number versus Reynolds number for flow across a cylinder with three fluids: (1) air at 300 K, 1 bar; (2) water at 300 K; (3) Therminol 60 at 350 K. The constant C_1 is defined by the relation $\overline{Nu} = C_1 Re^{0.63}$. *(b)* The ratio $\overline{Nu}/Re^{0.63}$ versus Prandtl number for flow over cylinder; $1.5 \times 10^4 < Re < 10^5$.

As mentioned earlier, most engineering calculations of convective heat transfer use heat transfer coefficients obtained from experimental data. For some simple flows (e.g., laminar flow in tubes), we have exact analytical solutions of the governing differential equations. In the process of obtaining these solutions, the appropriate dimensionless groups arise quite naturally; thus, the final results are also presented in terms of the commonly used groups. Examples of such analyses are given in Chapter 5.

The use of numerical methods to solve the governing differential equations is rapidly increasing. This activity is often called *computational fluid dynamics (CFD)*, but in relation to convective heat transfer, the subject has broader scope than this name implies. The computer codes that execute the numerical computations yield results that, in many ways, resemble the data obtained from physical experiments. The term *numerical experiments* is a good description of the activity. As for physical experiments, the data are organized in terms of dimensionless groups whenever possible. However, since the numerical results are obtained by solving model governing differential equations, these equations can be used to determine the appropriate dimensionless groups, as was done for the pin fin problem in Section 2.4.5.

4.2.4 Evaluation of Fluid Properties

The fluid properties appearing in the dimensionless groups pertinent to convective heat transfer are density ρ [kg/m^3], dynamic viscosity μ [kg/m s] or kinematic viscosity ν [m^2/s], thermal conductivity k [W/m K], specific heat c_p [J/kg K], and the volume-expansion coefficient β [1/K]. Data for these properties for selected fluids are given in Tables A.7–A.10 of Appendix A and can be seen to be temperature-dependent to a greater or lesser extent. For liquids, ρ is essentially constant, whereas for gases, $\rho \propto T^{-1}$. For liquids such as water and oils, the viscosity varies markedly with temperature. For water, the viscosity more than doubles as the temperature changes from 330 to 290 K. For gases, the dynamic viscosity and conductivity increase with temperature, approximately as $\mu \propto T^{0.7}$, $k \propto T^{0.7}$. The specific heat does not vary much unless there is a change in molecular structure.

In elementary fluid mechanics, the problems the student faces, such as evaluation of the Reynolds number for calculating drag on an immersed object, usually involve an isothermal fluid; thus, the temperature at which the kinematic viscosity must be evaluated is obvious. But in convective heat transfer, there is always a temperature difference between the surface and the bulk or free-stream fluid, so the question arises as to what is the proper temperature to use when calculating fluid properties in dimensionless groups. There is no simple answer. Strictly speaking, the property variation itself is an additional problem parameter to be characterized by an additional dimensionless group. For example, the friction factor for flow in a pipe depends on the viscosity ratio μ_s/μ_b, where μ_s and μ_b are the viscosity values at the wall and bulk temperatures, respectively. Such **variable-property effects** can be determined only by careful experiment or detailed analysis. For engineering calculations, the effects of variable properties are usually approximately accounted for as follows.

External Flows

For external flows, all properties are evaluated at a *reference temperature T_r,* where

$$T_r = T_s - \alpha(T_s - T_e) \tag{4.36a}$$

Unless otherwise stated, the value of α should be taken as 1/2, that is, the reference temperature is the arithmetic mean of the surface and free-stream temperatures. This value of the reference temperature is also called the **mean film temperature,** perhaps because the boundary layer is sometimes imagined to be a thin film of stagnant fluid adjacent to the surface.

Internal Flows

For internal flows, the reference temperature approach can also be used, with T_r given by

$$T_r = T_s - \alpha(T_s - T_b) \tag{4.36b}$$

and with α taken as 1/2 unless otherwise stated. For liquids with an essentially constant density, the reference temperature approach is straightforward to use, but for gases with a variable density, the evaluation of Reynolds and Stanton numbers is awkward. In an internal flow, the mass velocity ρV is always $\dot{m}/A_c$, and if ρ varies across the duct, the separation of ρ and V so as to evaluate ρ at the reference temperature is best avoided. Thus, for internal flows, the *property ratio* or *temperature ratio* approach is usually used. For liquids, a viscosity or Prandtl number ratio is used since viscosity varies more than any other property. For gases, a temperature ratio is used since density, viscosity, and conductivity are all well-behaved functions of absolute temperature. We can write

$$\frac{f}{f_b} = \left(\frac{\text{Pr}_s}{\text{Pr}_b}\right)^m \quad \text{or} \quad \left(\frac{\mu_s}{\mu_b}\right)^m \quad \text{or} \quad \left(\frac{T_s}{T_b}\right)^m \tag{4.37}$$

$$\frac{\text{Nu}}{\text{Nu}_b} = \left(\frac{\text{Pr}_s}{\text{Pr}_b}\right)^n \quad \text{or} \quad \left(\frac{\mu_s}{\mu_b}\right)^n \quad \text{or} \quad \left(\frac{T_s}{T_b}\right)^n \tag{4.38}$$

where f_b and Nu_b, are evaluated using bulk properties.

A further complication for internal flows is that both the wall temperature T_s and the bulk temperature T_b may vary along the duct as heat is added or removed. In calculating local values of friction factor or Nusselt number, the temperatures at the location are used to evaluate properties. In calculating average values for a duct of length L, it is usually adequate to use arithmetic averages of the inlet and outlet temperatures.

In the sections that follow, specific instructions on the evaluation of fluid properties will be given for each correlation or group of correlations.

4.3   FORCED CONVECTION

Forced flows may be *internal* or *external*. In an internal flow, such as in a heat exchanger tube, the flow is forced by a fan if the fluid is a gas, or a pump if it is a liquid. An external flow over a model in a wind tunnel is forced by the tunnel fan. Alternatively, the surface may move through a stationary fluid; an example is the flight of a hypersonic vehicle. In Section 4.3, we consider only simple forced-convection flows over smooth surfaces. More complicated forced-convection flows are dealt with in Section 4.5, and the effect of surface roughness will be discussed in Section 4.7. Unless otherwise noted, the formulas presented are correlations of experimental data.

4.3.1 Forced Flow in Tubes and Ducts

Fully Developed Flow in Round Tubes (or Pipes)

For laminar flow sufficiently far from the entrance of a tube or pipe, where the flow is *hydrodynamically fully developed* and has the characteristic parabolic velocity profile of *Poiseuille* flow, some simple analytical results are available. The friction factor has a constant value

$$f = \frac{64}{\mathrm{Re}_D}; \qquad \mathrm{Re}_D = \frac{GD}{\mu} \tag{4.39}$$

where D is the tube diameter and G is the mass velocity $(G = \dot{m}/A_c)$. Note the use of the subscript D to indicate that the characteristic length in the Reynolds number is the tube diameter. If the wall temperature is uniform, for example, if steam is condensing on the outside of the tube wall, then sufficiently far downstream of where heating starts, the flow becomes *thermally fully developed*, the shape of the temperature profile is unchanging, and the Nusselt number has a constant value given by Eq. (1.21) rearranged into dimensionless form:

$$\mathrm{Nu}_D = 3.66 \tag{4.40}$$

If, on the other hand, the heat flux through the tube wall is uniform, for example, if the tube is wound with an electrical resistance wire at constant pitch, then

$$Nu_D = \frac{48}{11} = 4.364 \tag{4.41}$$

Equations (4.39) and (4.41) are derived in Chapter 5.

Transition to turbulence takes place at $\mathrm{Re}_D \simeq 2300$, although the turbulence becomes fully established only for $\mathrm{Re}_D > 10,000$. For hydrodynamically fully developed flow, the friction factor can be obtained from a Moody chart (given later in this chapter as Fig. 4.49) or, for a smooth wall, from Petukhov's formula[1]:

$$f = (0.790 \ln \mathrm{Re}_D - 1.64)^{-2}; \qquad 10^4 < \mathrm{Re}_D < 5 \times 10^6 \tag{4.42}$$

Alternatively, there is a less accurate power law formula:

$$f = 0.184\, \mathrm{Re}_D^{-0.2}; \qquad 4 \times 10^4 < \mathrm{Re}_D < 10^6 \tag{4.43}$$

In contrast to laminar flow, the effect of wall boundary condition (e.g., whether or not the wall is at a uniform temperature or the heat flux is uniform along the tube) is unimportant for turbulent flow of all fluids except low-Prandtl-number liquid metals. For thermally fully developed flow in a smooth tube with $\mathrm{Pr} > 0.5$, a simple power law formula is

$$\mathrm{Nu}_D = 0.023\, \mathrm{Re}_D^{0.8} \mathrm{Pr}^{0.4}; \qquad \mathrm{Re}_D > 10{,}000 \tag{4.44}$$

which is in fact Eq. (1.22) in dimensionless form.[4] If more accurate results are desired. Gnielinski's formula is recommended [2]:

$$\mathrm{Nu}_D = \frac{(f/8)(\mathrm{Re}_D - 1000)\mathrm{Pr}}{1 + 12.7(f/8)^{1/2}(\mathrm{Pr}^{2/3} - 1)}; \qquad 3000 < \mathrm{Re}_D < 10^6 \tag{4.45}$$

where the friction factor must be calculated from Eq. (4.42), for which the lower limit on Re_D can be ignored. Equation (4.45) agrees with most available experimental data within 20%: an example is shown in Fig. 4.16. Below $\mathrm{Re}_D \simeq 10{,}000$, the turbulence can be intermittent and the correlation is less reliable. For

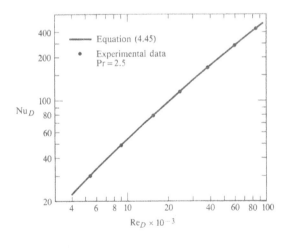

Figure 4.16   Turbulent flow in tubes: comparison of Eq. (4.45) with experimental data of Sparrow and Ohadi [3].

---

[4] There are many power law formulas similar to Eq. (4.44) but with different constants, Prandtl-number exponents, and schemes for accounting for variable properties. The *Dittus-Boelter, McAdams,* and *Colburn* formulas, all developed in the 1930s, are widely used but are often misquoted in the literature. Equation (4.44) should be viewed as the formula Colburn might have proposed if log-log slide rules or handheld calculators had been in common use at the time.

low-Prandtl-number liquid metals, Notter and Sleicher [4] recommend, for a uniform wall temperature,

$$\mathrm{Nu}_D = 4.8 + 0.0156\mathrm{Re}_D^{0.85}\mathrm{Pr}^{0.93}; \qquad 0.004 < \mathrm{Pr} < 0.01$$
$$10^4 < \mathrm{Re}_D < 10^6 \tag{4.46}$$

and for a uniform wall heat flux,

$$\mathrm{Nu}_D = 6.3 + 0.0167\mathrm{Re}_D^{0.85}\mathrm{Pr}^{0.93}; \qquad 0.004 < \mathrm{Pr} < 0.01$$
$$10^4 < \mathrm{Re}_D < 10^6 \tag{4.47}$$

Entrance Effects

Near the entrance of a tube, the friction and rate of heat transfer are generally higher than far downstream, where the velocity and temperature profiles are fully developed. A *hydrodynamic entrance length* L_{ef} can be defined as the distance required for the friction factor to decrease to within 5% of its fully developed value f_∞. If the flow is laminar, and if fluid enters the tube through a smooth, rounded entrance as shown in Fig. 4.17, the velocity profile is initially uniform, and analysis gives

$$\frac{L_{ef}(5\%)}{D} \simeq 0.05\mathrm{Re}_D \tag{4.48}$$

Similarly, a *thermal entrance length* L_{eh} can be defined as the distance required for the Nusselt number to decrease to within 5% of its fully developed value Nu_∞. If at $x = 0$ the flow is laminar and is already fully developed hydrodynamically (i.e., the velocity profile is parabolic), and heating commences with a uniform wall tempera-

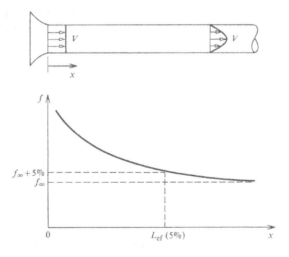

Figure 4.17 Laminar flow in a tube with smooth, rounded entrance: definition of the hydrodynamic entrance length.

ture, then analysis gives

$$\frac{L_{eh}\,(5\%)}{D} = 0.033 \mathrm{Re}_D \mathrm{Pr} \tag{4.49}$$

The corresponding average Nusselt number for a tube of length L is [5]

$$\overline{\mathrm{Nu}}_D = 3.66 + \frac{0.065(D/L)\mathrm{Re}_D\mathrm{Pr}}{1 + 0.04[(D/L)\mathrm{Re}_D\mathrm{Pr}]^{2/3}}; \qquad \mathrm{Re}_D \lesssim 2300 \tag{4.50}$$

which is seen to have the asymptote $\mathrm{Nu}_D = 3.66$ as $L/D \rightarrow \infty$. Equation (4.49) shows that thermal entrance lengths tend to be very short for low-Prandtl-number liquid metals but long for high-Prandtl-number oils. For example, Fig. 4.18 shows oil of $\mathrm{Pr} = 200$ at $\mathrm{Re}_D = 100$ being cooled in a heat exchanger, in which each tube is 100 diameters long. From Eq. (4.48), the hydrodynamic entrance length is 5 diameters, so an assumption of fully developed hydrodynamics throughout would be appropriate. However, the thermal entrance length from Eq. (4.49) is 660 diameters; therefore, the heat transfer is not fully developed. In fact, Eq. (4.50) gives $\overline{\mathrm{Nu}}_D = 9.15$, which is 2.5 times larger than the fully developed value.

Turbulent flows with simply defined hydrodynamics at a tube entrance are seldom encountered in engineering practice. Most often there is a sharp 90° edge, a bend, or an elbow, as shown in Fig. 4.19. The corresponding hydrodynamic entrance lengths vary from about $10 - 15$ diameters, when no large-scale eddies are present, to about $30 - 40$ diameters, when there are large-scale eddies. Similarly, the thermal entrance length depends on entrance configuration as well as Prandtl number. At usual Reynolds numbers, the thermal entrance length can be less than 5 diameters for high-Prandtl-number oils and low-Prandtl-number liquid metals, provided there are no large-scale eddies. For a fluid with Prandtl number of order unity, including gases and water at higher temperatures, the thermal entrance length varies between 15 and 40 diameters. Figure 4.20 shows some typical variations of the local heat transfer coefficient in the entrance region for some practical entrance configurations, based

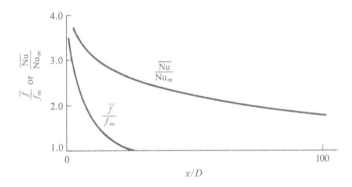

Figure 4.18 Laminar flow of oil in a heat exchanger tube: variation of friction factor and Nusselt number for $\mathrm{Re}_D = 100$, $\mathrm{Pr} = 200$, $L/D = 100$.

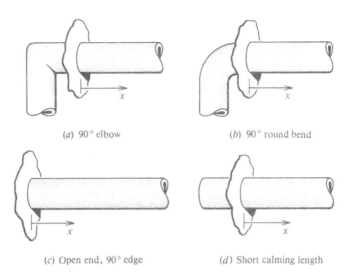

(a) 90° elbow   (b) 90° round bend

(c) Open end, 90° edge   (d) Short calming length

Figure 4.19 Various tube entrance configurations used in practice. (a) 90° elbow. (b) 90° round bend. (c) Open end, 90° edge. (d) Short calming length.

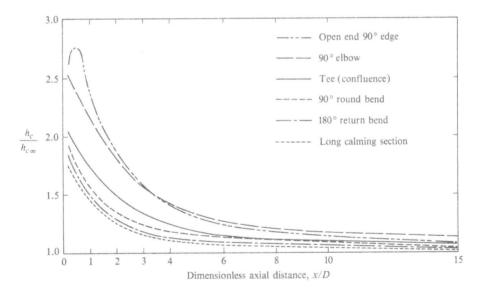

Figure 4.20 Entrance region heat transfer for turbulent flow of gases with various entrance configurations; $h_{c\infty}$ is the heat transfer coefficient far from the entrance, x is the distance from where heating commences, and D is the tube diameter [6].

Table 4.4 Effect of entrance configuration on average heat transfer for turbulent pipe flow: the ratio of average Nusselt number to the fully developed value, $\overline{\text{Nu}}/\text{Nu}_\infty$, for gas flow or Pr about unity [6].

Entrance	Pipe Length, in Diameters									
	2	4	6	8	10	20	40	80	160	320
Long calming section	1.49	1.34	1.26	1.21	1.17	1.10	1.06	1.03	1.01	1.01
Open end, 90° edge	2.36	1.95	1.73	1.60	1.54	1.32	1.18	1.09	1.05	1.02
90° elbow	2.15	1.86	1.68	1.57	1.49	1.32	1.18	1.09	1.05	1.02
Tee (confluence)	1.77	1.56	1.44	1.36	1.31	1.19	1.10	1.06	1.03	1.01
90° round bend	1.63	1.44	1.34	1.28	1.24	1.16	1.10	1.05	1.03	1.01
180° return bend	1.54	1.37	1.28	1.23	1.19	1.12	1.08	1.04	1.02	1.01

on experiments with air. Table 4.4 shows corresponding average Nusselt numbers for various tube lengths; the table applies to gases and other fluids with Prandtl numbers of about unity.

Use of average Nusselt numbers for internal flows requires some care. Consider first an isothermal wall. If the flow is laminar, Eq. (4.50) may be used to determine the average heat transfer coefficient in Eq. (4.12). If the flow is turbulent and Pr $\sim$ 1, Eq. (4.45) together with Table 4.4 can be used. For turbulent flows and Pr $\gg$ 1 or Pr $\ll$ 1, entrance effects can be neglected unless the tube is very short: then Eq. (4.45) can be used directly. Examples 4.1 and 4.2 illustrate the procedure. For a nonisothermal wall, such as that found in a two-stream heat exchanger, the problem is more difficult and will be discussed in Chapter 8.

Flow in Ducts of Various Cross Sections

Table 4.5 gives friction factors and Nusselt numbers for fully developed laminar flow in ducts of various cross sections. For noncircular ducts, the length scale in the Reynolds and Nusselt numbers is the **hydraulic diameter**, $D_h = 4A_c/\mathscr{P}$ where A_c is the cross-sectional area for flow and $\mathscr{P}$ is the wetted perimeter. (For a round tube, $D_h = D$ since $A_c = \pi D^2/4$ and $\mathscr{P} = \pi D$.)

Entrance lengths and the variation of Nusselt number in the thermal entrance length are similar to those for a round tube. For example, the average Nusselt number for flow between isothermal parallel plates of length L is [5]

$$\overline{\text{Nu}}_{D_h} = 7.54 + \frac{0.03(D_h/L)\text{Re}_{D_h}\text{Pr}}{1+0.016[(D_h/L)\text{Re}_{D_h}\text{Pr}]^{2/3}} \qquad \text{Re}_{D_h} \lesssim 2800 \qquad \textbf{(4.51)}$$

where the hydraulic diameter is simply twice the spacing of the plates. Transition Reynolds numbers are a little different from those for a round tube, with a value of 2800 being more appropriate for flow between parallel plates.

For turbulent flow, the correlations for a round tube can be used, with the diameter D replaced by the hydraulic diameter D_h. The same correlations can be used for all turbulent duct flows because the viscous sublayer around the perimeter of the duct

Table 4.5 Nusselt numbers and the product of friction factor times Reynolds number for fully developed laminar flow in ducts of various cross sections.

Cross Section	Nu_{D_h}		$f\mathrm{Re}_{D_h}$
	Constant Axial Wall Heat Flux	Constant Axial Wall Temperature	
Equilateral triangle	3.1	2.4	53
Circle	4.364	3.657	64
Square (1 × 1)	3.6	2.976	57
Rectangle (1 × 1.4)	3.8	3.1	59
1 × 2	4.1	3.4	62
1 × 3	4.8	4.0	69
1 × 4	5.3	4.4	73
1 × 8	6.5	5.6	82
∞	8.235	7.541	96
Heated / ∞ Insulated	5.385	4.861	96

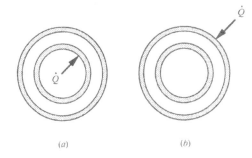

Figure 4.21   Schematic of annular ducts. (a) Heat transfer through the inner wall. (b) Heat transfer through the outer wall.

is very thin, and the velocity and temperature are nearly uniform across the core fluid. Since the viscous sublayer is the major resistance to momentum and heat transfer, the precise shape of the core fluid is not critical. For the annular ducts shown in Fig. 4.21, greater accuracy can be obtained if the Nusselt number given by Eq. (4.45) is multiplied by the correction factors recommended by Petukov and Roizen [7]. For heat transfer through the inner wall with the outer wall insulated, the factor is

$$0.86 \left(\frac{D_i}{D_o} \right)^{-0.16} \tag{4.52a}$$

where D_i and D_o are the inner and outer diameters, respectively. For the inner wall insulated and heat transfer through the outer wall, the factor is

$$1 - 0.14 \left(\frac{D_i}{D_o} \right)^{0.6} \tag{4.52b}$$

When Eq. (4.52) is used, the appropriate heat transfer area is that of the heated wall only.

Reynolds Analogy

The physical processes of momentum and heat transfer in turbulent flow are very similar. The turbulent eddy that transports momentum from the core fluid to the fluid near the wall also transports heat. For fluids with Prandtl number values close to unity, the resistance of the viscous sublayer to transfer of momentum is almost the same as the resistance to heat transfer. Thus, we would expect a simple relation between friction and heat transfer. Using Eqs. (4.43) and (4.44),

$$\frac{f}{8} = \frac{C_f}{2} = St Pr^{0.6} \tag{4.53}$$

For Pr = 1,

$$\frac{C_f}{2} = St$$

which is the famous *Reynolds analogy* between momentum and heat transfer, first proposed by O. Reynolds in 1874.

Variable-Property Effects

Recommended exponents m and n for property and temperature ratio corrections are given in Table 4.6. These data are for pipe flow but are also approximately valid for other duct shapes.

Table 4.6   Exponents for property and temperature ratio corrections for use in Eqs. (4.37) and (4.38): flow in tubes.

Type of Flow	Fluid	Wall Condition	m	n
Laminar	Liquids (μ_s/μ_b)	Heating	0.58	−0.11
		Cooling	0.50	−0.11
	Gases (T_s/T_b)	Heating and cooling	1	0
Turbulent	Liquids (μ_s/μ_b)	Heating	0.25	−0.11
		Cooling	0.25	−0.25
	Gases (T_s/T_b)	Heating	−0.2	−0.55
		Cooling	−0.1	0.0

EXAMPLE 4.1   Laminar Flow of Oil

SAE 50 oil flows at 0.007 kg/s through a 1 cm-I.D. tube, 1.5 m long. The wall temperature is 300 K, and the inlet oil bulk temperature is 377 K. Estimate the average heat transfer coefficient.

Solution

Given: SAE 50 oil flowing inside a tube.

Required: Average heat transfer coefficient.

Assumptions: The flow is hydrodynamically fully developed at the commencement of cooling.

The average Nusselt number is to be evaluated using the bulk temperature for fluid properties, with a subsequent viscosity ratio correction for variable property effects from Table 4.6. The bulk temperature of the oil decreases along the tube, but an arithmetic average of the inlet and outlet values is usually adequate. There remains a difficulty, however. We do not know the oil outlet temperature since it depends on the heat transfer coefficient we 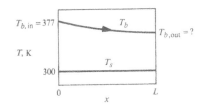 are attempting to calculate. We must guess and check later. Therefore, expecting an oil temperature drop of 10° to 20°C, we take the average of the inlet and outlet bulk oil temperatures to be 370 K. From Table A.8, $k = 0.137$ W/m K, $\mu = 1.89 \times 10^{-2}$ kg/m s, $c_p = 2200$ J/kg K, and Pr $= 300$.

We first calculate the tube cross-sectional area and the Reynolds number:

$$A_c = \pi \frac{D^2}{4} = \pi \left(\frac{0.01^2}{4} \right) = 7.854 \times 10^{-5} \, \text{m}^2$$

$$\text{Re}_D = \frac{(\dot{m}/A_c)D}{\mu} = \frac{(0.007/7.854 \times 10^{-5})(0.01)}{1.89 \times 10^{-2}} = 47.2$$

Since $\text{Re}_D < 2300$, the flow is laminar, and Eq. (4.50) applies:

$$\overline{\text{Nu}}_D = 3.66 + \frac{(0.065)(D/L)\text{Re}_D\text{Pr}}{1 + 0.04[(D/L)\text{Re}_D\text{Pr}]^{2/3}}$$

$$= 3.66 + \frac{(0.065)(0.01/1.5)(47.2)(300)}{1 + 0.04[(0.01/1.5)(47.2)(300)]^{2/3}} = 3.66 + 3.35 = 7.01$$

We now correct for variable properties. From Table 4.6, $n = -0.11$, and from Table A.8, $\mu_s = \mu(300\,\text{K}) = 50.3 \times 10^{-2}$ kg/m s:

$$\left(\frac{\mu_s}{\mu_b} \right)^{-0.11} = \left(\frac{50.3 \times 10^{-2}}{1.89 \times 10^{-2}} \right)^{-0.11} = 0.697$$

The corrected Nusselt number is $\overline{\text{Nu}}_D = (7.01)(0.697) = 4.89$. Thus,

$$\bar{h}_c = \left(\frac{k}{D} \right) \overline{\text{Nu}}_D = \left(\frac{0.137}{0.01} \right) 4.89 = 66.9 \, \text{W/m}^2 \, \text{K}$$

To check the guessed average bulk temperature, we must find the outlet oil temperature. Since the wall temperature T_s is constant, Eq. (4.11) applies.

$$T_{b,\text{out}} = T_s - (T_s - T_{b,\text{in}})e^{-\bar{h}_c 2\pi RL/\dot{m}c_p}$$

$$= 300 - (300 - 377)e^{-(66.9)(2\pi)(0.005)(1.5)/(0.0070)(2200)}$$

$$= 300 - (300 - 377)e^{-0.205} = 363 \, \text{K}$$

The guessed average temperature was thus appropriate, and no iteration is required.

Solution using CONV

The required input is:

Configuration number $= 2$ (tubes: laminar flow)
Fluid $= 2$ (SAE 50 engine oil)
$T_s = 300$
$T_b = 370$
$P = $ Any value for a liquid
$D = 0.01$
$L = 1.5$
$\dot{m} = 0.007$

The output is: .

SAE 50 oil properties at 370 K
Re = 47.2
$f = 7.00$
$\Delta P / L = 3303$ Pa/m
Nu = 4.89
$h_c = 67.0$ W/m^2 K

Comments

1. Input to CONV must be in SI units with temperatures in kelvins.

2. Notice that for a liquid, any value for pressure can be input into CONV.

3. Notice that the value of the exponent $\bar{h}_c 2\pi RL / \dot{m} c_p$ was 0.205, indicating a rather small change in the oil temperature along the tube. For the outlet oil temperature to approach the wall temperature, the value of the exponent should be greater than about 3 or 4.

EXAMPLE 4.2   Turbulent Flow of Air

Air flows at 0.11 kg/s through a 1 cm−wide, 0.5 m−high channel of a plate-type heat exchanger. The channel is 0.8 m long and its walls are at 600 K. If the pressure is 1 atm and the average of the inlet and outlet bulk air temperatures is estimated to be 400 K, determine the average heat transfer coefficient. The entrance has a 90° edge.

Solution

Given: Air flowing between parallel plates.

Required: Average heat transfer coefficient, $\bar{h}_c$.

Assumptions: The data for tube flow in Table 4.4 can be used to correct for entrance effects.

At 400 K, 1 atm, air properties from Table A.7 are: k = 0.0331 W/m K, Pr = 0.69, and $\mu = 22.5 \times 10^{-6}$ kg/m s. Thus,

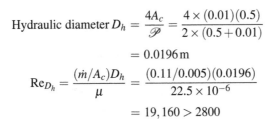
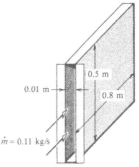

$$\text{Hydraulic diameter } D_h = \frac{4A_c}{\mathscr{P}} = \frac{4 \times (0.01)(0.5)}{2 \times (0.5 + 0.01)}$$

$$= 0.0196 \, \text{m}$$

$$\text{Re}_{D_h} = \frac{(\dot{m}/A_c)D_h}{\mu} = \frac{(0.11/0.005)(0.0196)}{22.5 \times 10^{-6}}$$

$$= 19,160 > 2800$$

The flow is turbulent, so that Eqs. (4.42) and (4.45) apply:

$$f = (0.790 \ln \mathrm{Re}_{D_h} - 1.64)^{-2} = (0.790 \ln 19{,}160 - 1.64)^{-2} = 0.0264$$

$$\mathrm{Nu}_{D_h} = \frac{(f/8)(\mathrm{Re}_{D_h} - 1000)\mathrm{Pr}}{1 + 12.7(f/8)^{1/2}(\mathrm{Pr}^{2/3} - 1)} = \frac{(0.0264/8)(19{,}160 - 1000)(0.69)}{1 + 12.7(0.0264/8)^{1/2}(0.69^{2/3} - 1)} = 49.2$$

We now correct for variable-property effects. Table 4.6 gives $n = -0.55$:

$$\left(\frac{T_s}{T_b}\right)^{-0.55} = \left(\frac{600}{400}\right)^{-0.55} = 0.800$$

$$\mathrm{Nu}_{D_h} = (49.2)(0.800) = 39.4$$

To correct for entrance effects, Table 4.4 applies approximately. For item 2 and $L/D_h = 41$,

$$\overline{\mathrm{Nu}}/\mathrm{Nu}_\infty = 1.18$$

$$\overline{\mathrm{Nu}}_{D_h} = (39.4)(1.18) = 46.5$$

$$\bar{h}_c = (k/D_h)\overline{\mathrm{Nu}}_{D_h} = (0.0331/0.0196)(46.5) = 78.5\,\mathrm{W/m^2\,K}$$

Comments

1. Notice that the friction factor f is *not* corrected for variable property effects before substitution in Eq. (4.45) to calculate Nu_{D_h}.

2. Use CONV to check Re_{D_h} and Nu_{D_h}; note that the friction factor given by CONV has been corrected for variable property effects.

4.3.2 External Forced Flows

In this section, various external forced flows are considered and correlations for skin friction and heat transfer given. The heat transfer correlations all apply to an *isothermal surface;* other wall boundary conditions are discussed at the end of the section.

Flow along a Flat Plate

Figure 4.22 shows a schematic of flow along a flat plate. A laminar boundary layer forms from the leading edge, and transition to turbulent flow usually occurs at a value of $\mathrm{Re}_x = u_e x/\nu$ where u_e is the free-stream velocity, in the range 50,000–500,000, for x measured from the leading edge. Higher values are associated with careful wind-tunnel tests, and lower values are more characteristic of practical situations where such factors as surface roughness and vibration are present. If the Reynolds number were based on an appropriate thickness of the boundary layer, the transition value would be of the same order as that given in Section 4.3.1 for pipe flow. Both the local shear stress τ_{sx} and the local heat transfer coefficient h_{cx} vary along the plate as shown, and a constant asymptotic value is never attained; that is, conditions are similar to the entrance region of internal duct flows.

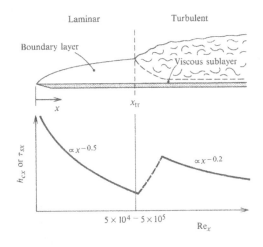

Figure 4.22   Forced flow along a flat plate, showing boundary layer growth and the resulting variation of the local shear stress and heat transfer coefficient.

For laminar flow, the local skin friction coefficient at a distance x from the leading edge is

$$C_{fx} = 0.664\mathrm{Re}_x^{-1/2}; \qquad \mathrm{Re}_x \gtrsim 10^3 \tag{4.54}$$

where the subscript x on the Reynolds number again indicates the length scale on which it is based. The average shear stress for a plate of length L can be obtained from the average skin friction coefficient,

$$\overline{C}_f = \frac{1}{L}\int_0^L C_{fx}dx = 1.328\mathrm{Re}_L^{-1/2}; \qquad 10^3 < \mathrm{Re}_L \lesssim 5\times10^5 \tag{4.55}$$

where the Reynolds number is now based on plate length, L. The local Nusselt number, $\mathrm{Nu}_x = h_{cx}x/k$, is given by

$$\mathrm{Nu}_x = 0.332\mathrm{Re}_x^{1/2}\mathrm{Pr}^{1/3}; \qquad \mathrm{Pr} > 0.5 \tag{4.56}$$

To obtain a correlation for the average heat transfer coefficient, we cannot simply average the Nusselt number, since it contains x. Instead, we must evaluate

$$\overline{h}_c = \frac{1}{L}\int_0^L h_{cx}\,dx = \frac{1}{L}\int_0^L \left(\frac{k}{x}\right)(0.332)\left(\frac{u_e x}{\nu}\right)^{1/2}\mathrm{Pr}^{1/3}dx$$

to obtain

$$\overline{\mathrm{Nu}} = \frac{\overline{h}_c L}{k} = 0.664\mathrm{Re}_L^{1/2}\mathrm{Pr}^{1/3}; \qquad \mathrm{Pr} > 0.5 \tag{4.57}$$

For low-Prandtl-number liquid metals, an appropriate expression is

$$\overline{\mathrm{Nu}} = 1.128\mathrm{Re}_L^{1/2}\mathrm{Pr}^{1/2}; \qquad \mathrm{Pr} \ll 1 \tag{4.58}$$

Equations (4.54) through (4.58) are all based on exact analysis and have been confirmed by experiment. The analysis is given in Chapter 5.

For turbulent flow, the local skin friction coefficient is given by simple power law expressions,

$$C_{fx} = 0.0592\text{Re}_x^{-1/5} \qquad 10^5 < \text{Re}_x < 10^7 \tag{4.59a}$$

$$C_{fx} = 0.026\text{Re}_x^{-1/7}; \qquad 10^6 < \text{Re}_x < 10^9 \tag{4.59b}$$

or, if greater accuracy is required, White's formula may be used [8]:

$$C_{fx} = \frac{0.455}{(\ln 0.06\text{Re}_x)^2}; \qquad 10^5 < \text{Re}_x < 10^9 \tag{4.60}$$

In these expressions, x is the distance measured from the *virtual origin* of the turbulent boundary layer, shown in Fig. 4.23; for flow along a flat plate, this can be taken to be the leading edge with sufficient accuracy for most engineering purposes. To determine the total drag, an average skin friction coefficient is required. If transition is assumed to occur abruptly at x_{tr}, the average shear stress on a plate of length L is

$$\overline{\tau}_s = \frac{1}{L}\left[\int_0^{x_{tr}} \tau_s(\text{laminar})dx + \int_{x_{tr}}^{L} \tau_s(\text{turbulent})dx\right]$$

Dividing by $(1/2)\rho u_e^2$,

$$\overline{C}_f = \frac{1}{L}\left[\int_0^{x_{tr}} C_{fx}(\text{laminar})dx + \int_{x_{tr}}^{L} C_{fx}(\text{turbulent})dx\right]$$

and substituting from Eqs. (4.54) and (4.59a),

$$\overline{C}_f = \frac{1}{L}\left[\int_0^{x_{tr}} 0.664\text{Re}_x^{-1/2}dx + \int_{x_{tr}}^{L} 0.0592\text{Re}_x^{-1/5}dx\right]$$

It is convenient to integrate with respect to Re_x rather than x:

$$\text{Re}_x = \frac{u_e x}{\nu}; \qquad d\text{Re}_x = \frac{u_e}{\nu}dx \qquad \text{or} \qquad dx = \frac{\nu}{u_e}d\text{Re}_x$$

$$\overline{C}_f = \frac{\nu}{u_e L}\left[\int_0^{\text{Re}_{tr}} 0.664\text{Re}_x^{-1/2}d\text{Re}_x + \int_{\text{Re}_{tr}}^{\text{Re}_L} 0.0592\text{Re}_x^{-1/5}d\text{Re}_x\right]$$

$$= \frac{1}{\text{Re}_L}\left[(2)(0.664)\text{Re}_{tr}^{1/2} + (5/4)(0.0592)\left(\text{Re}_L^{4/5} - \text{Re}_{tr}^{4/5}\right)\right]$$

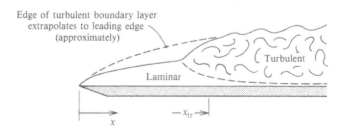

Figure 4.23 Virtual origin of a turbulent boundary layer on a flat plate.

which can be rearranged as

$$\overline{C}_f = 1.328 \mathrm{Re}_{\mathrm{tr}}^{-1/2}\left(\frac{\mathrm{Re}_{\mathrm{tr}}}{\mathrm{Re}_L}\right) + 0.0740 \mathrm{Re}_L^{-1/5}\left[1 - \left(\frac{\mathrm{Re}_{\mathrm{tr}}}{\mathrm{Re}_L}\right)^{4/5}\right] \qquad \textbf{(4.61)}$$

Equation (4.61) is accurate for $\mathrm{Re}_L < 10^7$. For higher Reynolds numbers, the integration can be extended using Eq. (4.59b). Alternatively, although Eq. (4.60) cannot be integrated analytically, it can be integrated numerically and the result curve-fitted. When used in place of Eq. (4.59), the resulting expression for $\overline{C}_f$ is

$$\overline{C}_f = 1.328 \mathrm{Re}_{\mathrm{tr}}^{-1/2}\left(\frac{\mathrm{Re}_{\mathrm{tr}}}{\mathrm{Re}_L}\right) + \frac{0.523}{(\ln 0.06\,\mathrm{Re}_L)^2}\left(1 - \frac{\mathrm{Re}_{\mathrm{tr}}}{\mathrm{Re}_L}\right) \qquad \textbf{(4.62)}$$

and is accurate for $\mathrm{Re}_L < 10^9$. Equation (4.62) is recommended for general use. The total viscous drag force on a plate of width W and length L is $F = \overline{C}_f(1/2)\rho u_e^2 WL$.

For heat transfer across a turbulent boundary layer, the local Nusselt number is given by White [9] as

$$\mathrm{Nu}_x = \frac{(C_{fx}/2)\mathrm{Re}_x\mathrm{Pr}}{1 + 12.7(C_{fx}/2)^{1/2}(\mathrm{Pr}^{2/3} - 1)} \qquad \textbf{(4.63)}$$

with C_{fx} given by Eq. (4.59a); this form is valid for $0.5 < \mathrm{Pr} < 2000$, $5 \times 10^5 < \mathrm{Re}_x < 10^7$. Alternatively, there is a simpler power law expression recommended by Whitaker [10],

$$\mathrm{Nu}_x = 0.029 \mathrm{Re}_x^{0.8}\mathrm{Pr}^{0.43} \qquad \textbf{(4.64)}$$

which is valid for $0.7 < \mathrm{Pr} < 400$, $5 \times 10^5 < \mathrm{Re}_x < 3 \times 10^7$. To determine the average Nusselt number, we first evaluate the average heat transfer coefficient:

$$\overline{h}_c = \frac{1}{L}\left[\int_0^{x_{\mathrm{tr}}} h_{cx}(\text{laminar})\,dx + \int_{x_{\mathrm{tr}}}^L h_{cx}(\text{turbulent})\,dx\right]$$

Substituting Eqs. (4.56) and (4.64),

$$\overline{h}_c = \frac{1}{L}\left[\int_0^{x_{\mathrm{tr}}} (k/x)0.332\mathrm{Re}_x^{1/2}\mathrm{Pr}^{1/3}\,dx + \int_{x_{\mathrm{tr}}}^L (k/x)0.029\mathrm{Re}_x^{0.8}\mathrm{Pr}^{0.43}\,dx\right]$$

from which $\overline{\mathrm{Nu}} = \overline{h}_c L/k$ is obtained as

$$\overline{\mathrm{Nu}} = 0.664\mathrm{Re}_{\mathrm{tr}}^{1/2}\mathrm{Pr}^{1/3} + 0.036\mathrm{Re}_L^{0.8}\mathrm{Pr}^{0.43}\left[1 - \left(\frac{\mathrm{Re}_{\mathrm{tr}}}{\mathrm{Re}_L}\right)^{0.8}\right] \qquad \textbf{(4.65)}$$

For the flat plate, there is an analogy between momentum and heat transfer for both laminar and turbulent flow For laminar flow using the relation $\mathrm{St}_x = \mathrm{Nu}_x/\mathrm{Re}_x\mathrm{Pr}$ and Eqs. (4.54) and (4.56),

$$\mathrm{St}_x = \left(\frac{C_{fx}}{2}\right)\mathrm{Pr}^{-2/3}; \qquad \mathrm{Pr} > 0.5 \qquad \textbf{(4.66)}$$

For turbulent flow, using Eqs. (4.59a) and (4.64),

$$St_x = \left(\frac{C_{fx}}{2}\right) Pr^{-0.57}; \qquad 0.7 < Pr < 400 \tag{4.67}$$

is a good approximation. Equation (4.63) itself can also be viewed as a form of the analogy.

Flow across a Cylinder

The flow pattern around a cylinder depends very much on the Reynolds number VD/ν, where V is the velocity of the undisturbed flow. Figure 4.24 gives the main flow regimes, showing in particular the shedding of vortices, separation of the boundary layer, and at higher Reynolds numbers, transition from a laminar to a

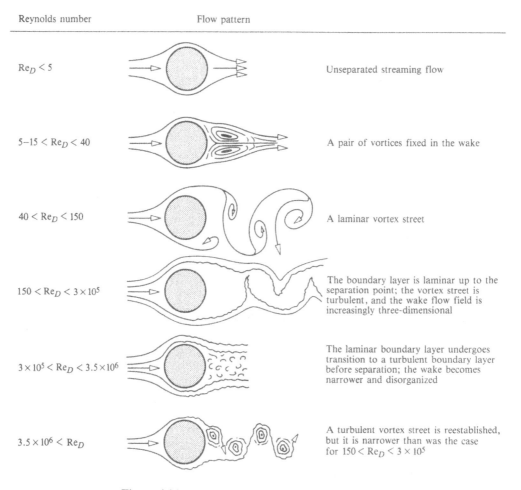

Reynolds number	Flow pattern	
$Re_D < 5$		Unseparated streaming flow
$5{-}15 < Re_D < 40$		A pair of vortices fixed in the wake
$40 < Re_D < 150$		A laminar vortex street
$150 < Re_D < 3 \times 10^5$		The boundary layer is laminar up to the separation point; the vortex street is turbulent, and the wake flow field is increasingly three-dimensional
$3 \times 10^5 < Re_D < 3.5 \times 10^6$		The laminar boundary layer undergoes transition to a turbulent boundary layer before separation; the wake becomes narrower and disorganized
$3.5 \times 10^6 < Re_D$		A turbulent vortex street is reestablished, but it is narrower than was the case for $150 < Re_D < 3 \times 10^5$

Figure 4.24   The main flow regimes for flow across a cylinder.

turbulent boundary layer before separation. The local skin friction coefficient varies in a complicated fashion around the cylinder. However, in practice, what is usually required is the total drag force F on the cylinder due to both friction, or *viscous drag,* and pressure imbalance, *or form drag.* The drag force can be obtained from the drag coefficient C_D:

$$C_D = \frac{F}{(1/2)\rho V^2 A_f} \tag{4.68}$$

where A_f is the area of the cylinder *normal* to the flow, that is, DL for a cylinder of diameter D and length L. Figure 4.25 is a graph of C_D versus Re_D. At low Reynolds numbers, viscous drag predominates, and a useful formula is

$$C_D = 1 + \frac{10}{Re_D^{2/3}}; \qquad 1 < Re_D < 10^4 \tag{4.69}$$

Above about $Re_D = 10^3$, the flow separates at $\theta \simeq 80°$, and the form drag dominates to give $C_D \simeq 1.2$, nearly independent of Reynolds number. At about $Re_D = 2 \times 10^5$, there is a transition from a laminar to a turbulent boundary layer, which has the effect of moving the separation point to the rear of the cylinder (up to $\theta = 130°$), thereby causing a significant decrease in the form drag. Figure 4.26 is a schematic showing laminar and turbulent boundary layer separation.

The corresponding variation of the local heat transfer coefficient around the cylinder is also very complicated, as might be expected. Figure 4.27 shows some experimental data. The Nusselt number is high on the front of the cylinder, where

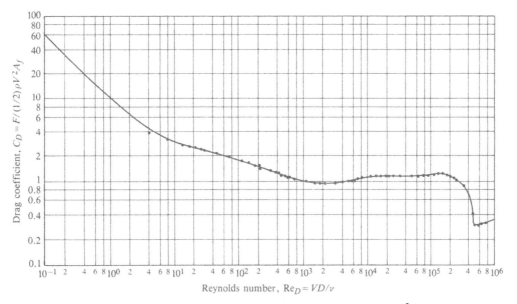

Figure 4.25 The drag coefficient $C_D = F/(1/2)\rho V^2 A_f$ for flow across a cylinder [11]. (Adapted with permission.)

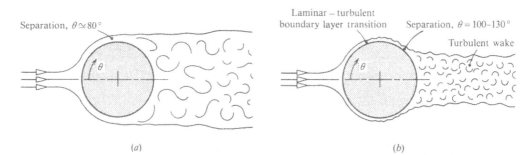

Figure 4.26   Schematic showing the difference between *(a)* laminar and *(b)* turbulent boundary layer separation for flow across a cylinder.

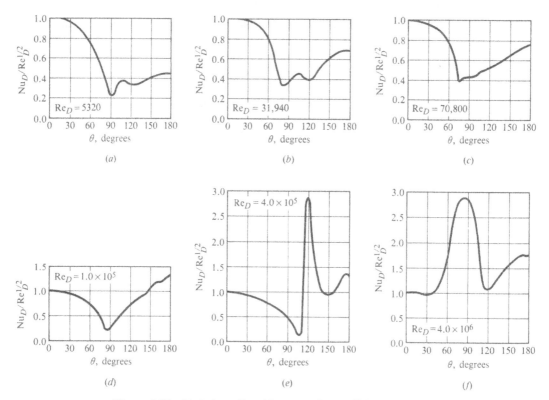

Figure 4.27   Variation of local heat transfer coefficient around an isothermal cylinder in a cross-flow of air [12, 13, 14]. Notice that Eq. (4.70) gives $Nu/Re_D^{1/2} \simeq 1.00$ at the forward stagnation line.

the boundary layer is thin, and decreases as the boundary layer grows in thickness around the cylinder. When the flow does not separate, the Nusselt number decreases steadily toward the rear of the cylinder. For $Re_D \gtrsim 10^3$, the flow separates at about $\theta = 80°$, and there is a Nusselt number minimum in the separated region. On the rear of the cylinder, there can be a local maximum at the point of reattachment of the flow; for example, Fig. 4.27b shows a local maximum at 105°. For $Re_D \gtrsim 2 \times 10^5$, there is a transition from laminar to turbulent flow in the boundary layer that delays separation. Figure 4.27e shows two Nusselt number minima, the first just before transition, at about $\theta = 110°$, and the second in the separated region, at about $\theta = 150°$. At $Re_D = 4.0 \times 10^6$, Fig. 4.27f shows the corresponding Nusselt number minima at 30° and 120°, respectively. The local Nusselt number at the forward stagnation line can be obtained by analysis [15]. For $Pr > 0.5$,

$$Nu_D = 1.15 Re_D^{1/2} Pr^{1/3} \tag{4.70}$$

The average Nusselt number for $Pr > 0.5$ is given by a rather complicated correlation suggested by Churchill and Bernstein [16]:

$$\overline{Nu}_D = 0.3 + \frac{0.62 Re_D^{1/2} Pr^{1/3}}{[1+(0.4/Pr)^{2/3}]^{1/4}}; \qquad Re_D < 10^4 \tag{4.71a}$$

$$\overline{Nu}_D = 0.3 + \frac{0.62 Re_D^{1/2} Pr^{1/3}}{[1+(0.4/Pr)^{2/3}]^{1/4}} \left[1+\left(\frac{Re_D}{282,000}\right)^{1/2} \right]; \tag{4.71b}$$

$$2 \times 10^4 < Re_D < 4 \times 10^5$$

$$\overline{Nu}_D = 0.3 + \frac{0.62 Re_D^{1/2} Pr^{1/3}}{[1+(0.4/Pr)^{2/3}]^{1/4}} \left[1+\left(\frac{Re_D}{282,000}\right)^{5/8} \right]^{4/5}; \tag{4.71c}$$

$$4 \times 10^5 < Re_D < 5 \times 10^6$$

However, for very low Reynolds numbers, Nakai and Okazaki [17] recommend

$$\overline{Nu}_D = \frac{1}{0.8237 - \ln(Re_D Pr)^{1/2}}; \qquad Re_D Pr < 0.2 \tag{4.72}$$

Flow over a Sphere

The flow pattern around a sphere is somewhat similar to that for a cylinder except that it does not exhibit the same regular eddy shedding phenomena. Figure 4.28 shows a graph of C_D versus Re_D and is very similar to Fig. 4.25 for a cylinder. For very low Reynolds-number *creeping* flows, *Stokes' law* is valid [18]:

$$C_D = \frac{24}{Re_D}; \qquad Re_D < 0.5 \tag{4.73}$$

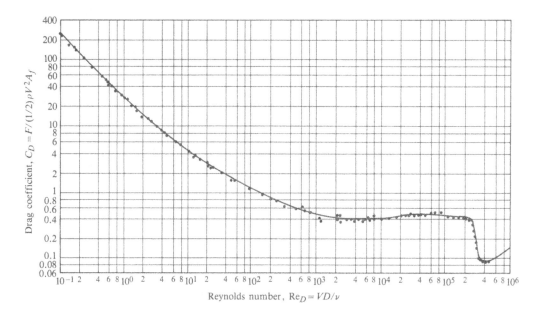

Figure 4.28 The drag coefficient $C_D = F/(1/2)\rho V^2 A_f$ for flow across a flow across a sphere [11]. (Adapted with permission.)

For somewhat higher Reynolds numbers, a useful correlation is

$$C_D \simeq \frac{24}{\mathrm{Re}_D}\left(1 + \frac{\mathrm{Re}_D^{2/3}}{6}\right); \qquad 2 < \mathrm{Re}_D < 500 \tag{4.74}$$

At still higher Reynolds numbers, *Newton's law* applies, and C_D is a constant, equal to approximately 0.44 in the range $500 < \mathrm{Re}_D < 2 \times 10^5$.

The local Nusselt number at the forward stagnation point can be obtained by analysis [15]. For $\mathrm{Pr} > 0.5$,

$$\mathrm{Nu}_D = 1.32\mathrm{Re}_D^{1/2}\mathrm{Pr}^{1/3} \tag{4.75}$$

Whitaker [10] recommends that the average Nusselt number for $0.7 < \mathrm{Pr} < 380$ be calculated from

$$\overline{\mathrm{Nu}}_D = 2 + (0.4\mathrm{Re}_D^{1/2} + 0.06\mathrm{Re}_D^{2/3})\mathrm{Pr}^{0.4}; \qquad 3.5 < \mathrm{Re}_D < 8 \times 10^4 \tag{4.76}$$

Notice that Eq. (4.76) has a lower limit of $\overline{\mathrm{Nu}}_D = 2$, which corresponds to conduction from a sphere into stationary infinite surrounds. Figure 4.29 illustrates this important point. From Section 2.3.3, the heat flow by conduction across a spherical shell is given by

$$\dot{Q} = \frac{4\pi k(T_1 - T_2)}{1/r_1 - 1/r_2} \tag{4.77}$$

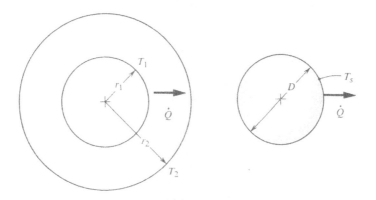

Figure 4.29   Schematic for conduction across a spherical shell, and from a sphere into stationary infinite surrounds.

which for $D = 2r_1$ and $r_2 \to \infty$ is

$$\dot{Q} = \frac{2k}{D}\pi D^2 (T_1 - T_2) = h_c A (T_s - T_e)$$

Hence, the heat transfer coefficient is

$$h_c = \frac{2k}{D} \quad \text{and} \quad \mathrm{Nu}_D = 2 \tag{4.78}$$

Variable Property Effects

All the correlations given in this section for external forced flows should be used with fluid properties evaluated at the mean film temperature. An exception is Eq. (4.76), for which better results are obtained if fluid properties are evaluated at the free stream temperature, with a viscosity ratio correction $(\mu_s/\mu_e)^n, n = -0.25$, applied to the convection contribution.

Other Wall Boundary Conditions

All the foregoing correlations for external flows are valid for an isothermal surface. Correlations are also available for a uniform wall heat flux, for which the local Nusselt numbers tend to be higher than their counterparts for an isothermal surface. For laminar flow along a uniformly heated flat plate, analysis gives the local Nusselt number [15, 19]

$$\mathrm{Nu}_x = 0.453\mathrm{Re}_x^{1/2}\mathrm{Pr}^{1/3}; \qquad \mathrm{Pr} > 0.5 \tag{4.79}$$

which is 36% higher than the values obtained from Eq. (4.56) for an isothermal surface. For turbulent gas flow along a uniformly heated flat plate, Kays and Crawford

[15] recommend for the local Nusselt number

$$\text{Nu}_x = 0.030\text{Re}_x^{0.8}\text{Pr}^{0.4}; \qquad 0.5 < \text{Pr} < 400, \tag{4.80}$$

$$5 \times 10^5 < \text{Re}_x < 5 \times 10^6$$

which is only 4% higher than the isothermal plate result, Eq. (4.64).

Appropriately averaged Nusselt numbers are far less sensitive to the wall boundary condition. For a nonisothermal wall, Eq. (1.26), the definition of an average heat transfer coefficient, used to derive Eq. (4.57), namely,

$$\bar{h}_c = \frac{1}{A}\int_0^A h_c \, dA \qquad \left(= \frac{1}{L}\int_0^L h_{cx} \, dx \quad \text{for a rectangular surface}\right)$$

is of little practical utility. When the temperature difference $(T_s - T_e)$ is not constant, heat transfer from the plate cannot be calculated from the simple formula used for isothermal surfaces, namely,

$$\dot{Q} = \bar{h}_c A(T_s - T_e) \tag{4.81}$$

which was introduced as Eq. (1.24). For the special case of a uniformly heated wall, the average heat transfer coefficient is preferably defined in terms of $(\overline{T_s - T_e})$ as

$$\bar{h}_c = \frac{q_s}{(\overline{T_s - T_e})}; \qquad \overline{T_s - T_e} = \frac{1}{L}\int_0^L (T_s - T_e)dx \tag{4.82}$$

Such a definition is particularly useful when resistance thermometry is used to measure the average temperature of the wall. Since $\text{Nu}_x = q_s x/k(T_s - T_e)$, Eq. (4.79) can be integrated to obtain

$$0.453k(V/v)^{1/2}\text{Pr}^{1/3}(\overline{T_s - T_e}) = \frac{q_s}{L}\int_0^L x^{1/2}dx = \frac{2}{3}q_s L^{1/2}$$

$$\overline{\text{Nu}_L} = \frac{q_s L}{k(\overline{T_s - T_e})} = 0.680\text{Re}_L^{1/2}\text{Pr}^{1/3} \tag{4.83}$$

which is only 2.3% higher than the isothermal wall result, Eq. (4.57). For turbulent flow, the difference is even smaller. As a general rule, the average heat transfer coefficient for external flows over uniformly heated surfaces defined by Eq. (4.82) can be taken to be equal to that for an isothermal surface [19]. For surfaces that are neither isothermal nor uniformly heated, the advanced texts on convective heat transfer listed in the bibliography at the end of the text should be consulted.

EXAMPLE 4.3   Heat Loss from a Hut Roof

A wind blows at 8 m/s over the 5 m square flat roof of a hut that is part of an Antarctic research station. If the ambient air is at approximately 250 K, estimate the average heat transfer coefficient. Use a transition Reynolds number of 10^5.

Solution

Given: Wind blowing over roof of a hut.

Required: Average heat transfer coefficient.

Assumptions: 1. Roof can be modeled as a flat
plate.
2. $\mathrm{Re}_{tr} = 10^5$.
3. Properties can be evaluated at
250 K since we do not expect
a large temperature difference
between the roof and air.

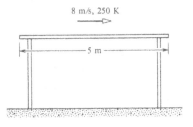

Equation (4.65) applies. From Table A.7, the required air properties at 250 K are $k = 0.0235$
W/m K, $v = 11.42 \times 10^{-6}$ m²/s, Pr = 0.69. The Reynolds number at the end of the roof is
$\mathrm{Re}_L = VL/v = (8)(5)/(11.42 \times 10^{-6}) = 3.50 \times 10^6$. Since $\mathrm{Re}_{tr} = 0.1 \times 10^6$, only the first
3% of the boundary layer is laminar; the remainder is turbulent. Using Eq. (4.65),

$$\overline{\mathrm{Nu}}_L = 0.664\mathrm{Re}_{tr}^{1/2}\mathrm{Pr}^{1/3} + 0.036\mathrm{Re}_L^{0.8}\mathrm{Pr}^{0.43}\left[1 - \left(\frac{\mathrm{Re}_{tr}}{\mathrm{Re}_L}\right)^{0.8}\right]$$

$$= 0.664(0.1 \times 10^6)^{1/2}(0.69)^{1/3} + 0.036(3.5 \times 10^6)^{0.8}(0.69)^{0.43}\left[1 - \left(\frac{0.1}{3.5}\right)^{0.8}\right]$$

$$= 186 + 4968 = 5150$$

$$\overline{h}_c = (k/L)\overline{\mathrm{Nu}}_L = (0.0235/5)5150 = 24.2 \ \mathrm{W/m^2 \ K}.$$

Comments

1. The precise location of transition in this problem is unimportant. The boundary layer
could have been taken as turbulent from the leading edge with little effect on the result.

2. Use CONV to check $\overline{h}_c$.

EXAMPLE 4.4   Cooling of a Molten Droplet of Aluminum

During arc welding of aluminum, molten metal droplets are ejected. Some of these droplets
are hot enough ($\gtrsim 2300$ K) to ignite and form sparks. Most droplets are ejected at lower
temperatures and simply cool down. If a particular molten droplet has a diameter of 0.5 mm,
an initial temperature of 1700 K, and an initial velocity of 1 m/s, estimate its initial rate of
cooling in air at 300 K. Liquid aluminum properties at 1700 K are estimated to be $\rho = 2100$
kg/m³, $c_p = 1100$ J/kg K, $\varepsilon = 0.20$.

Solution

Given: Molten aluminum droplet at 1700 K.

Required: Initial rate of cooling, dT/dt.

Assumptions: 1. Model as a small gray object in large surroundings to calculate q_{rad}.

2. The lumped thermal capacity model is applicable.

3. Negligible formation of solid oxide on the droplet surface.

Equation (4.76) will be used to calculate the average Nusselt number with air properties evaluated at the mean film temperature of $(1700 + 300)/2 = 1000$ K. From Table A.7, $k = 0.0672$ W/m K, $\nu = 117.3 \times 10^{-6}$ m²/s, Pr = 0.70. the Reynolds number is

$$Re_D = \frac{VD}{\nu} = \frac{(1)(0.0005)}{117.3 \times 10^{-6}} = 4.26$$

Substituting in Eq. (4.76),

$$\overline{Nu}_D = 2 + (0.4Re_D^{1/2} + 0.06Re_D^{2/3})Pr^{0.4}$$

$$= 2 + [0.4(4.26)^{1/2} + 0.06(4.26)^{2/3}](0.70)^{0.4} = 2.85$$

$$\overline{h}_c = \left(\frac{k}{D}\right)\overline{Nu}_D = \left(\frac{0.0672}{0.0005}\right)2.85 = 383 \text{ W/m}^2 \text{ K}$$

$$q_{conv} = \overline{h}_c(T_s - T_e) = (383)(1700 - 300) = 5.37 \times 10^5 \text{ W/m}^2$$

The radiation heat loss can be estimated using Eq. (1.18) for a small gray object in large surroundings:

$$q_{rad} = \sigma\varepsilon(T_s^4 - T_e^4) = (5.67 \times 10^{-8})(0.2)(1700^4 - 300^4) = 0.946 \times 10^5 \text{ W/m}^2$$

Taking $k \simeq 200$ W/m K for molten aluminum, the Biot number based on convective heat transfer from Eq. (1.40) is

$$Bi_{LTC} = \frac{\overline{h}_c(D/6)}{k_s} = \frac{(383)(0.0005/6)}{(200)} \simeq 1.6 \times 10^{-4} \ll 0.1$$

Thus, the lumped thermal capacity model is certainly valid. From Section 1.5.2,

$$\rho c V \frac{dT}{dt} = -(q_{conv} + q_{rad})A$$

Substituting $V = \pi D^3/6, A = \pi D^2$ gives

$$\frac{dT}{dt} = -\frac{6(q_{conv} + q_{rad})}{\rho c D} = -\frac{6(5.37 + 0.95)(10^5)}{(2100)(1100)(0.0005)} = -3280 \text{ K/s}$$

Comments

1. Such small droplets cool off very fast.

2. Although the temperature is high, the radiation contribution is relatively small. The reason is the low value of ε and the large value of h_c (due to the small size).

3. Use CONV to check the value of h_c.

4. Is natural convection significant for such a hot droplet? Section 4.4.3 will deal with mixed natural and forced convection.

5. Recall that a viscosity ratio correction for variable properties, $(\mu_s/\mu_e)^n$, $n = -0.25$, applied to the convection contribution should give a better result than use of the mean film temperature (CONV uses the mean film temperature).

6. Some solid oxide will form on the droplet surface. However, the heat of oxidation liberated will be small unless the droplet temperature is close to the ignition value of 2300 K.

4.4 NATURAL CONVECTION

A heated fluid rises. Density differences and the earth's gravitational field act to produce a **buoyancy force**, which drives the flow. Such flows are called *natural, free*, or *buoyant convection*. Whenever a fluid is heated or cooled in a gravitational field, there is the possibility of natural convection. Density differences can also be caused by composition gradients. For example, moist air rises mainly due to the lower density of the water vapor present in the mixture. A related phenomenon is flow induced by a centripetal acceleration field, as in rotating machinery. We will restrict our attention to pure fluids in the earth's gravity field. For a pure fluid, density gradients can be related to temperature gradients through the volumetric coefficient of expansion, β. For an ideal gas, $\beta = 1/T$, where T is *absolute* temperature. Data for β of liquids are given in Table A.10.

Natural convection flows can be either *external* or *internal*. External flows include flow up a heated wall and the plume rising above a power plant stack. Internal flows are found between the cover plate and absorbing surface of a solar collector and inside hollow insulating walls. Velocities associated with natural convection are relatively small, not much more than 2 m/s. Thus, natural-convection heat transfer coefficients tend to be much smaller than those for forced convection. For gases, these coefficients are of the order of only 5 W/m^2 K, and the engineer must be careful to always check if simultaneous radiation heat transfer is significant to the thermal design. Since there is no obvious characteristic velocity of a natural convection flow, the Reynolds number of forced convection does not play a role. It is replaced by the Grashof or Rayleigh number.

4.4.1 External Natural Flows

In this section, various *external* natural convection flows are considered, and correlations are given for heat transfer from isothermal surfaces. Other wall boundary conditions are discussed at the end of the section.

Flow on a Vertical Wall

Figure 4.30 shows a natural-convection boundary layer on a vertical plate. A laminar boundary layer forms at the lower end, and transition to a turbulent boundary layer occurs at a critical value of the Rayleigh number $\text{Ra}_x = \beta \Delta T g x^3 / \nu \alpha \simeq 10^9$.

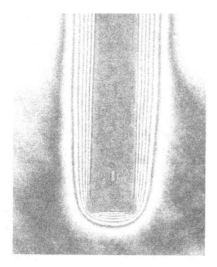

Figure 4.30   A thermal boundary layer on a heated vertical plate. The black lines are *interference fringes*, which are contours of constant density, and hence are isotherms. The spacing of the fringes is inversely proportional to the temperature gradient (Interferogram courtesy of Professor J. Gryzagoridis, University of Cape Town.)

Since $Ra_x = Gr_x Pr$, where the Grashof number $Gr_x = \beta \Delta T g x^3 / \nu^2$, for *gases* with $Pr \simeq 1$, transition can be said to occur at $Gr_x \simeq 10^9$ (see Section 1.3.3). Following Churchill and Usagi [21], we define a Prandtl number function Ψ as

$$\Psi = \left[1 + \left(\frac{0.492}{Pr} \right)^{9/16} \right]^{-16/9} \tag{4.84}$$

Churchill and Chu [22] correlated the average Nusselt number for laminar flow on a plate with a sharp leading edge and of height L as

$$\overline{Nu}_L = 0.68 + 0.670 (Ra_L \Psi)^{1/4}; \quad Ra_L \lesssim 10^9 \tag{4.85}$$

and for turbulent flow,

$$\overline{Nu}_L = 0.68 + 0.670 (Ra_L \Psi)^{1/4} (1 + 1.6 \times 10^{-8} Ra_L \Psi)^{1/12}; \quad 10^9 \lesssim Ra_L < 10^{12} \tag{4.86}$$

Notice that for a large Rayleigh number, Eq. (4.86) shows that the heat transfer coefficient is independent of plate height L. Equations (4.85) and (4.86) do not match exactly at $Ra_L \sim 10^9$.

Flow on a Horizontal Cylinder

Figure 4.31 is an interferogram showing isotherms around an isothermal heated cylinder in air. A plume of heated air is seen to rise above the cylinder, and the boundary layer is rather thick compared to the forced-flow case, which is typical. The flow is laminar for $Ra_D = Gr_D Pr \lesssim 10^9$. Churchill and Chu [23] correlated the average Nusselt number as

$$\overline{Nu}_D = 0.36 + \frac{0.518 Ra_D^{1/4}}{[1 + (0.559/Pr)^{9/16}]^{4/9}} : \quad 10^{-6} < Ra_D \lesssim 10^9 \tag{4.87}$$

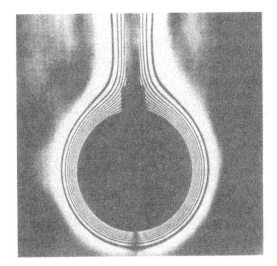

Figure 4.31 A thermal boundary layer on a heated horizontal cylinder. (Interferogram courtesy of Professor U. Grigull, and W. Hauf, Technische Universität, München.)

When $\mathrm{Ra}_D \gtrsim 10^9$, there is a transition from a laminar to a turbulent boundary layer, and the rate of increase of Nusselt number with Rayleigh number is greater. The recommended correlation then is

$$\overline{\mathrm{Nu}}_D = \left\{ 0.06 + 0.387 \left[\frac{\mathrm{Ra}_D}{[1 + (0.559/\mathrm{Pr})^{9/16}]^{16/9}} \right]^{1/6} \right\}^2 ; \quad \mathrm{Ra}_D \gtrsim 10^9 \quad \textbf{(4.88)}$$

Equations (4.87) and (4.88) do not match exactly at $\mathrm{Ra}_D \sim 10^9$.

Flow on a Sphere

For fluids with Prandtl number of order unity, which includes all gases, Yuge [24] gives the average Nusselt number as

$$\overline{\mathrm{Nu}}_D = 2 + 0.43\mathrm{Ra}_D^{1/4}; \quad 1 < \mathrm{Ra}_D < 10^5 \quad \textbf{(4.89)}$$

A more general formula valid for $\mathrm{Pr} > 0.5$ is due to Churchill [25]:

$$\overline{\mathrm{Nu}}_D = 2 + \frac{0.589\mathrm{Ra}_D^{1/4}}{[1 + (0.469/\mathrm{Pr})^{9/16}]^{4/9}}; \quad \mathrm{Ra}_D \lesssim 10^{11} \quad \textbf{(4.90)}$$

Objects of Arbitrary Shape

For a laminar natural convection boundary layer on an object of any shape, in fluids other than those for which $\mathrm{Pr} \ll 1$, Lienhard [26] suggests that the average Nusselt number is approximately

$$\overline{\mathrm{Nu}}_L = 0.52\mathrm{Ra}_L^{1/4} \quad \textbf{(4.91)}$$

where the characteristic length L is the length of the boundary layer, for example, $L = \pi D/2$ for a cylinder or sphere.

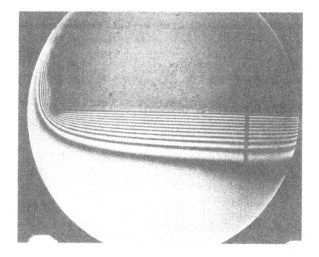

Figure 4.32   The thermal boundary layer below a 10 cm × 30 cm heated horizontal plate $Ra_L = 3.33 \times 10^6$, $Nu_L = 15.0$. The pin marks the plate center. Air rises from some distance below the plate until, in the vicinity of the plate, it flows sideways and then around the corner, carrying the heat away in a vertical plume. There are 10 isotherms visible: these are 3°C apart adjacent to the plate where the air is hot (and index of refraction is lower), decreasing to 2°C apart as the ambient air is approached. (Photograph courtesy of Professor D. K. Edwards, University of California, Irvine.)

Heated Horizontal Plate Facing Down, or Cooled Horizontal Plate Facing Up

Figure 4.32 shows an interferogram of the flow field underneath a heated horizontal plate. For a square plate of side length L, Kadambi and Drake [28] give the average Nusselt number as

$$\overline{Nu}_L = 0.82 Ra_L^{1/5}; \quad 10^5 < Ra_L < 10^{10} \tag{4.92}$$

and for an infinitely long strip of width L, Fugii and Imura [29] recommend

$$\overline{Nu}_L = 0.58 Ra_L^{1/5}; \quad 10^6 < Ra_L < 10^{11} \tag{4.93}$$

Alternatively, there is a more general correlation developed by Hatfield and Edwards [27] that applies to all aspect ratios and also allows for adiabatic extensions. If, as shown in Fig. 4.33, L is the length of the shorter side, W the length of the longer side, and L_a the length of adiabatic extensions on the shorter side, then

$$\overline{Nu}_L = 6.5 \left[1 + 0.38 \frac{L}{W}\right][(1 + X)^{0.39} - X^{0.39}]Ra_L^{0.13}$$

$$X = 13.5 Ra_L^{-0.16} + 2.2 \left(\frac{L_a}{L}\right)^{0.7} \tag{4.94}$$

which is based on experimental data for $10^6 < Ra_L < 10^{10}$, $0.7 < Pr < 4800$, and $0 < L_a/L < 0.2$.

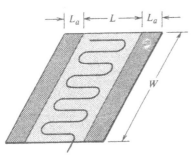

Figure 4.33 Schematic of a heated horizontal plate facing downward, showing adiabatic extensions on the shorter side.)

Heated Horizontal Plate Facing Up, or Cooled Horizontal Plate Facing Down

In contrast to the previous case, the flow for the configuration of Fig. 4.34 is unstable. McAdams [30] recommends

$$\overline{\mathrm{Nu}}_L = 0.54 \mathrm{Ra}_L^{1/4}; \quad 10^5 < \mathrm{Ra}_L < 2 \times 10^7 \tag{4.95}$$

where L is the side length of a square plate, or the length of the shorter side of a rectangular plate. For $\mathrm{Ra}_L > 10^7$, turbulent thermals rise irregularly above the plate and result in an average Nusselt number independent of plate size or shape [30]:

$$\overline{\mathrm{Nu}}_L = 0.14 \mathrm{Ra}_L^{1/3}; \quad 2 \times 10^7 < \mathrm{Ra}_L < 3 \times 10^{10} \tag{4.96}$$

Since the characteristic length L cancels, this result can also be written as

$$\frac{h_c (\nu \alpha / g)^{1/3}}{k} = 0.14 \left(\frac{\Delta \rho}{\rho} \right)^{1/3} \tag{4.97}$$

where the left-hand side can be viewed as a Nusselt number with a characteristic length $L = (\nu \alpha / g)^{1/3}$. These formulas are based on experimental data for air but may be used for any fluid with $\mathrm{Pr} > 0.5$.

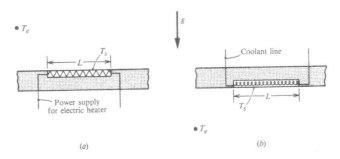

Figure 4.34 Schematic of (*a*) a heated plate facing upward, and (*b*) a cooled plate facing downward. The flow is unstable and the streamlines are unsteady.

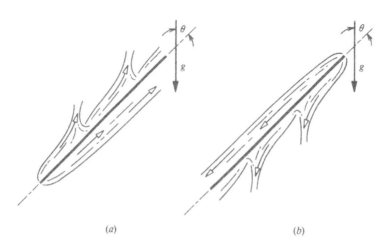

(a)   *(b)*

Figure 4.35   Flow patterns around (*a*) heated and (*b*) cooled inclined plates.

Inclined Plates

Figure 4.35 shows flow patterns around heated and cooled inclined plates. Provided θ is not too close to 90°, one would expect Eq. (4.85) for a vertical wall to apply to inclined walls, with g replaced by $g \cos \theta$. For the lower side of hot plates and the upper side of cold plates, the limit is $\theta \simeq 88°$ for $10^5 < \mathrm{Gr}_L < 10^{11}$; for the upper side of hot plates and the lower side of cold plates, the limit is $\theta \simeq 60°$.

Variable Property Effects

In the correlations of Section 4.4.1, all properties, including β, should be evaluated at the *mean film temperature.* Note that there are correlations for natural convection that require β to be evaluated at T_e, but this is not the case for the correlations presented here.

Other Wall Boundary Conditions

All the foregoing correlations are for an isothermal surface. Except for the laminar boundary layer on a vertical wall, there are few results available for nonisothermal surfaces. For a uniform wall heat flux, use of Eq. (4.82) and an isothermal wall correlation will give an engineering estimate of the average wall temperature. See also Exercise 4–86.

EXAMPLE 4.5   Heat Loss from a Solar Power Plant Central Receiver

The central receiver of a solar power plant is in the form of a cylinder 7 m in diameter and 13 m high. A tower supports the receiver high above the ground, and solar radiation is reflected to the receiver by rows of reflectors at ground level. If the operating temperature of its surface is 700 K, calculate the convective heat loss into still ambient air at 300 K. If the total irradiation of the collector is 20 MW, express the loss as a percentage of the irradiation.

Solution

Given: Vertical cylinder at 700 K.

Required: Convective heat loss into still ambient air.

Assumptions: 1. Curvature effects are negligible.
 2. The free-convection boundary layer commences at the
 bottom of the cylinder.

Depending on whether the flow is laminar or turbulent, either Eq. (4.85) or Eq. (4.86) should apply. Evaluate all properties at a mean film temperature of 500 K; $\beta = 1/T_r = 1/500, k = 0.0389$ W/m K, $\nu = 37.3 \times 10^{-6}\,\mathrm{m}^2/\mathrm{s}$, Pr $= 0.69$. The Rayleigh number is

$$\mathrm{Ra}_L = \mathrm{Gr}_L \mathrm{Pr} = \frac{\beta \Delta T g L^3}{\nu^2}\mathrm{Pr}$$

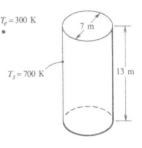

$T_e = 300$ K

$T_s = 700$ K

7 m

13 m

$$= \frac{(1/500)(400)(9.81)(13)^3(0.69)}{(37.3 \times 10^{-6})^2}$$

$$= 8.55 \times 10^{12}$$

This Rayleigh number is somewhat larger than the value of 10^{12} quoted as the upper limit of validity for Eq. (4.86). But $\bar{h}_c$ is almost independent of L for large values of Ra_L, so the result will not be too much in error. Using Eq. (4.86),

$$\Psi = \left[1 + \left(\frac{0.492}{\mathrm{Pr}}\right)^{9/16}\right]^{-16/9} = \left[1 + \left(\frac{0.492}{0.69}\right)^{9/16}\right]^{-16/9} = 0.343$$

$$\mathrm{Ra}_L \Psi = (8.55 \times 10^{12})(0.343) = 2.93 \times 10^{12}$$

$$\overline{\mathrm{Nu}}_L = 0.68 + 0.670(\mathrm{Ra}_L \Psi)^{1/4}(1 + 1.6 \times 10^{-8}\mathrm{Ra}_L \Psi)^{1/12}$$

$$= 0.68 + 0.670(2.93 \times 10^{12})^{1/4}[1 + (1.6 \times 10^8)(2.93 \times 10^{12})]^{1/12} = 2150$$

$$\bar{h}_c = (k/L)\overline{\mathrm{Nu}}_L = (0.0389/13)(2150) = 6.43 \text{ W/m}^2\text{ K}$$

$$\dot{Q} = \bar{h}_c A(T_s - T_e) = (6.43)(\pi)(7)(13)(700 - 300) = 7.35 \times 10^5 \text{ W} = 0.735 \text{ MW}$$

$$\frac{\dot{Q}_{\mathrm{loss}}}{\dot{Q}_{\mathrm{gross}}} = \frac{0.735}{20} = 3.7\%$$

Solution using CONV

The required input in SI units is:

Configuration number $= 9$ (vertical wall: turbulent flow)
Fluid number $= 21$
$T_s = 700$
$T_e = 300$
$P = 1.013 \times 10^5$
$L = 13$

The output is:

Air properties at 500 K, 1.013×10^5 Pa
$Gr = 1.241 \times 10^{13}$
$Ra = 8.57 \times 10^{12}$
$\overline{Nu} = 2150$
$\bar{h}_c = 6.43$ W/m^2 K

Comments

1. The thermal plume above the receiver carries away 0.735 MW of thermal energy. Can you propose a method for recovering some of this waste heat?

2. For $\beta \Delta T$ not much less than unity, Barrow and Sitharamarao [31] recommend that the Nusselt number be increased by a factor of $(1 + 0.6\beta\Delta T)^{1/4}$. For this problem, the correction is 10%.

EXAMPLE 4.6   Heat Loss from a Steam Pipe

A 30 cm–O.D. horizontal steam pipe has an outer surface temperature of 500 K and is located in still air at 300 K. Calculate the average heat transfer coefficient and the convective heat loss per meter length of pipe.

Solution

Given: Natural convection from a horizontal steam pipe.

Required: $\bar{h}_c$ and heat loss per meter.

Evaluate air properties at the mean fluid temperature of $(500 + 300)/2 = 400$ K. From Table A.7, $k = 0.0331$ W/m K, $v = 25.5 \times 10^{-6}$ m^2/s, Pr $= 0.69$, and $\beta = 1/400$ for an ideal gas. The Rayleigh number is

$$Ra_D = Gr_D Pr = \frac{\beta \Delta T g D^3}{v^2} Pr = \frac{(1/400)(200)(9.81)(0.3)^3(0.69)}{(25.5 \times 10^{-6})^2} = 1.41 \times 10^8$$

Equation (4.87) applies:

$$\overline{Nu}_D = 0.36 + \frac{0.518 Ra_D^{1/4}}{[1 + (0.559/Pr)^{9/16}]^{4/9}} = 0.36 + \frac{0.518(1.41 \times 10^8)^{1/4}}{[1 + (0.559/0.69)^{9/16}]^{4/9}} = 42.9$$

$$\bar{h}_c = \left(\frac{k}{D}\right)\overline{Nu}_D = \left(\frac{0.0331}{0.3}\right)42.9 = 4.73 \text{ W/m}^2 \text{ K}$$

$$\dot{Q} = h_c A(T_s - T_e) = (4.73)(\pi)(0.3)(1)(500 - 300) = 892 \text{ W/m}$$

We will check this result using the general correlation for laminar natural flows, Eq. (4.91).

For $L = \pi D/2 = 0.47$ m,

$$\overline{\mathrm{Nu}}_D = 0.52(\mathrm{Gr}_L\mathrm{Pr})^{1/4} = (0.52)\left(\frac{(1/400)(200)(9.81)(0.47)^3(0.69)}{(25.5 \times 10^{-6})^2}\right)^{1/4} = 79.3$$

$$\overline{h}_c = \left(\frac{k}{L}\right)\mathrm{Nu}_L = \left(\frac{0.0331}{0.47}\right)79.3 = 5.58 \text{ W/m}^2 \text{ K}$$

Comments

1. The more approximate Eq. (4.91) gives a value of $\overline{h}_c$ that is 18% higher than that from Eq. (4.87).
2. Use CONV to check $\overline{h}_c$.

4.4.2 Internal Natural Flows

Figure 4.36 shows a selection of enclosures in which natural convection is of engineering concern—for example, in flat-plate solar collectors, wall cavities, and window glazing. The horizontal layer heated from below was discussed in Section 4.2.1, where an appropriate definition of the convective heat transfer coefficient was shown to be

$$\overline{h}_c = \frac{\dot{Q}/A}{T_H - T_C}$$

The length parameter commonly used to define the Nusselt number is the plate spacing L. If the temperature difference $(T_H - T_C)$ is less than the critical value required

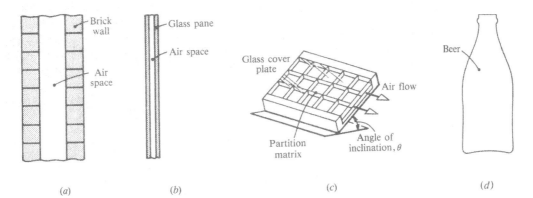

Figure 4.36 Enclosures. (a) A double wall with an air gap. (b) Double window glazing. (c) A flat-plate solar collector with a partition to suppress natural convection. (d) Sterilization of beer by condensing steam.

for the fluid to become unstable, then heat transfer across the layer is by conduction only, and from Eq. (1.9),

$$\dot{Q} = \frac{kA}{L}(T_H - T_C)$$

or

$$\bar{h}_c = \frac{k}{L}, \quad \overline{Nu}_L = 1$$

Thus, correlations for the Nusselt number always have a lower limit of $\overline{Nu}_L = 1$, corresponding to pure conduction. In Section 4.2.1, it was indicated that a horizontal layer heated from below becomes unstable at a critical value of $(T_H - T_C)$. In dimensionless form, the criterion for instability and the onset of cellular convection is a critical value of the Rayleigh number,

$$Ra_L = \frac{g\beta(T_H - T_C)L^3}{\nu\alpha} = 1708$$

As the temperature difference $(T_H - T_C)$ and, hence, the Rayleigh number increases, there are transitions to increasingly more complex flow patterns until finally the flow in the core is turbulent. In the case of a vertical layer of fluid contained between parallel plates maintained at different temperatures, circulation occurs for any $Ra_L > 0$; however, heat transfer is essentially by pure conduction for $Ra_L < 10^3$. As the Rayleigh number is increased, the circulating flow develops and cells are formed. At $Ra_L \simeq 10^4$ the flow changes to a boundary layer type with a boundary layer flowing upward on the hot wall and downward on the cold wall, while the fluid in the core region remains relatively stationary. At $Ra_L \simeq 10^5$, vertical rows of horizontal vortices develop in the core; and at $Ra_L \simeq 10^6$, the flow in the core finally becomes turbulent.

The marked changes in flow pattern with changes in Rayleigh number are characteristic of internal natural convection in all shapes of enclosures. Thus, it would be unreasonable to seek a single simple correlation formula valid over wide Rayleigh and Prandtl number ranges. Simple power law-type formulas are usually valid for small ranges of Ra; more general formulas are usually quite complex. Thus, only a few configurations will be considered here.

Heat transfer across thin air layers is of considerable engineering importance. Referring to Fig. 4.37, correlations recommended by Hollands and coworkers for aspect ratios of $H/L > 10$ are as follows.

1. $0 \leq \theta < 60°$ [32]:

$$\overline{Nu}_L = 1 + 1.44\left[1 - \frac{1708}{Ra_L\cos\theta}\right]\left\{1 - \frac{1708(\sin 1.8\theta)^{1.6}}{Ra_L\cos\theta}\right\} + \left[\left(\frac{Ra_L\cos\theta}{5830}\right)^{1/3} - 1\right]$$

(4.98)

where if either of the terms in square brackets is negative, it must be set equal to zero. Equation (4.98) is valid for $0 < Ra_L < 10^5$.

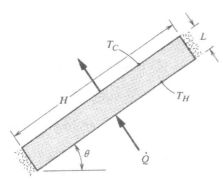

Figure 4.37 Schematic of a large-aspect-ratio inclined enclosure. The angle θ is measured from the horizontal.

2. $\theta = 60°$ [33]:

$$\overline{Nu}_{L60°} = \max\{Nu_1, Nu_2\} \tag{4.99}$$

where

$$Nu_1 = \left\{1 + \left[\frac{0.0936Ra_L^{0.314}}{1 + \{0.5/[1 + (Ra_L/3160)^{20.6}]^{0.1}\}}\right]^7\right\}^{1/7}$$

$$Nu_2 = \left(0.104 + \frac{0.175}{H/L}\right)Ra_L^{0.283}$$

and is valid for $0 < Ra_L < 10^7$.

3. $60° < \theta < 90°$ [33]:

$$\overline{Nu}_L = \left(\frac{90 - \theta}{30}\right)\overline{Nu}_{L60°} + \left(\frac{\theta - 60}{30}\right)\overline{Nu}_{L90°} \tag{4.100}$$

4. $\theta = 90°$ [33]:

$$\overline{Nu}_{L90°} = \max\{Nu_1, Nu_2, Nu_3\} \tag{4.101}$$

where

$$Nu_1 = 0.0605Ra_L^{1/3}$$

$$Nu_2 = \left\{1 + \left[\frac{0.104Ra_L^{0.293}}{1 + (6310/Ra_L)^{1.36}}\right]^3\right\}^{1/3}$$

$$Nu_3 = 0.242\left(\frac{Ra_L}{H/L}\right)^{0.272}$$

and is valid for $10^3 < Ra_L < 10^7$; for $Ra_L \leq 10^3$, $\overline{Nu}_{L90°} \simeq 1$.

For liquids of moderate Prandtl number, such as water, Eq. (4.98) can also be used for Ra $< 10^5$. For higher Rayleigh numbers, the Globe and Dropkin correlation [34] may be used for horizontal layers:

$$\overline{Nu}_L = 0.069Ra_L^{1/3}Pr^{0.074}; \quad 3 \times 10^5 < Ra_L < 7 \times 10^9 \tag{4.102}$$

Also, for horizontal layers of air, that is, $\theta = 0°$, the range of validity of Eq. (4.98) extends to $Ra_L = 10^8$.

Data are available in the literature for inclined layers of small aspect ratio but have not been correlated in a satisfactory manner.

In vertical cavities of small aspect ratio, with the horizontal surfaces insulated, as shown in Fig. 4.38, the following correlations due to Berkovsky and Polevikov [35] may be used for fluids of any Prandtl number.

1. $2 < H/L < 10$:

$$\overline{Nu}_L = 0.22 \left(\frac{Pr}{0.2 + Pr} Ra_L \right)^{0.28} \left(\frac{H}{L} \right)^{-1/4}; \quad Ra_L < 10^{10} \tag{4.103a}$$

2. $1 < H/L < 2$:

$$\overline{Nu}_L = 0.18 \left(\frac{Pr}{0.2 + Pr} Ra_L \right)^{0.29}; \quad 10^3 < \frac{Pr}{0.2 + Pr} Ra_L \tag{4.103b}$$

The flow and convective heat transfer in small-aspect-ratio cavities can depend on the temperature variation along the separating walls and, hence, on conduction in the walls and on radiation exchange within the cavity.

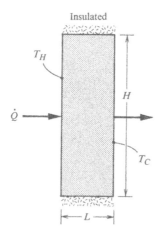

Figure 4.38   Schematic of a small-aspect-ratio vertical enclosure.

Concentric Cylinders and Spheres

Figure 4.39 shows isotherms for natural convection between concentric cylinders, with the inner cylinder heated and the outer cylinder cooled. The correlations recommended by Raithby and Hollands [36] for natural convection between concentric

Figure 4.39 Isotherms for natural convection between two concentric cylinders: $D_o/D_i = 3.9$, $L = 2.9$ cm, $\Delta T = 14.5$ K. (Interferogram courtesy of Professor U. Grigüll and W. Haüf, Technische Universität, München.)

cylinders and spheres are in the form of an effective thermal conductivity for use in the equations for conduction between concentric cylinders and spheres [Eqs. (2.14) and (2.25), respectively].

1. Concentric cylinders:

$$\frac{k_{\text{eff}}}{k} = 0.386 \left(\frac{\text{Pr}}{0.861 + \text{Pr}} \right)^{1/4} \text{Ra}_{\text{cyl}}^{1/4}; \quad 10^2 < \text{Ra}_{\text{cyl}} < 10^7 \qquad \textbf{(4.104)}$$

where

$$\text{Ra}_{\text{cyl}} = \frac{[\ln(D_o/D_i)]^4}{L^3 \left(D_i^{-3/5} + D_o^{-3/5} \right)^5} \text{Ra}_L$$

and $L = (D_o - D_i)/2$ is the gap width.

2. Concentric spheres:

$$\frac{k_{\text{eff}}}{k} = 0.74 \left(\frac{\text{Pr}}{0.861 + \text{Pr}} \right)^{1/4} \text{Ra}_{\text{sph}}^{1/4}; \quad 10^2 < \text{Ra}_{\text{sph}} < 10^4 \qquad \textbf{(4.105)}$$

where

$$\text{Ra}_{\text{sph}} = \frac{L}{(D_o D_i)^4} \frac{\text{Ra}_L}{\left(D_i^{-7/5} + D_o^{-7/5} \right)^5}$$

These equations are valid only when $k_{\text{eff}}/k \gtrsim 1$; otherwise, $k_{\text{eff}} = k$.

Variable Property Effects

All properties should be evaluated at a reference temperature $T_r = (1/2)(T_H + T_C)$.

EXAMPLE 4.7   Heat Loss through a Double Wall

A vertical double wall 3 m high has an air gap 10 cm thick. Estimate the convective heat transfer coefficient when the wall faces are at 305 K and 295 K, and the pressure is 1 atm. Compare the convective and radiative losses across the gap.

Solution

Given: Vertical air gap, 10 cm thick.

Required: Convective and radiative heat transfer across gap.

Assumptions: The facing surfaces are black.

Evaluate air properties at a mean fluid temperature of 300 K. From Table A.7, $k = 0.0267$ W/m K, $\rho = 1.177$ kg/m^3, $v = 15.7 \times 10^{-6}$ m^2/s, Pr $= 0.69$, and $\beta = 1/300$. The Rayleigh number and aspect ratio are

$$\mathrm{Ra}_L = \mathrm{Gr}_L \mathrm{Pr} = \frac{(\beta \Delta T)gL^3 \mathrm{Pr}}{v^2}$$

$$= \frac{(10/300)(9.81)(0.1)^3(0.69)}{(15.7 \times 10^{-6})^2}$$

$$= 9.15 \times 10^5$$

$$\frac{H}{L} = \frac{3}{0.1} = 30$$

Equation (4.101) applies:

$$\mathrm{Nu}_1 = 0.0605(9.15 \times 10^5)^{1/3} = 5.87$$

$$\mathrm{Nu}_2 = \left\{ 1 + \left[\frac{0.104(9.15 \times 10^5)^{0.293}}{1 + (6310/9.15 \times 10^5)^{1.36}} \right]^3 \right\}^{1/3} = 5.81$$

$$\mathrm{Nu}_3 = 0.242 \left(\frac{9.15 \times 10^5}{30} \right)^{0.272} = 4.01$$

$$\overline{\mathrm{Nu}}_L = \max\{5.87, 5.81, 4.01\} = 5.87$$

$$\bar{h}_c = \left(\frac{k}{L} \right) \overline{\mathrm{Nu}}_L = \left(\frac{0.0267}{0.1} \right) 5.87 = 1.57 \text{ W/m}^2 \text{ K}$$

$$q_{\mathrm{conv}} = \bar{h}_c \Delta T = (1.57)(305 - 295) = 15.7 \text{ W/m}^2$$

For black surfaces, the radiative transfer is given by Eq. (1.13):

$$q_{\mathrm{rad}} = \sigma T_1^4 - \sigma T_2^4 = (5.67 \times 10^{-8})(305^4 - 295^4) = 61.3 \text{ W/m}^2$$

$$q_{\mathrm{total}} = q_{\mathrm{conv}} + q_{\mathrm{rad}} = 15.7 + 61.3 = 77.0 \text{ W/m}^2$$

Comments

1. Use CONV to check $\bar{h}_c$.

2. Two possible methods of reducing the heat loss are as follows:

   (a) The gap could be filled with fiberglass insulation, with conductivity $k = 0.048$ W/m K, $q_{total} = (k/L)\Delta T = (0.048/0.1)(10) = 4.8$ W/m^2.

   (b) Since the radiation loss dominates, the walls could be lined with aluminum foil of $\varepsilon = 0.04$. In Chapter 6, the radiation transfer factor is shown to be $\mathscr{F}_{12} = \varepsilon/(2 - \varepsilon) = 0.0204$. Using Eq. (1.17), $q_{rad} = \mathscr{F}_{12}\sigma(T_1^4 - T_2^4) = 1.25$ W/m^2, $q_{total} = 15.7 + 1.25 = 17.0$ W/m^2.

   Economic considerations will have an impact on the final choice.

EXAMPLE 4.8   Convection Loss in a Flat-Plate Solar Collector

A flat-plate solar collector 2 m long and 1 m wide is inclined 60° to the horizontal. The cover plate is separated from the absorber plate by an air gap 2 cm thick. If the average temperatures of the cover and absorber plates are 305 K and 335 K, respectively, estimate the convective heat loss. Take the pressure as 1 atm.

Solution

Given: Horizontal air gap, 2 cm thick.

Required: Convective heat transfer.

Assumptions: Surfaces are isothermal.

Evaluate properties at a mean film temperature of 320 K. From Table A.7, $k = 0.0281$ W/m K, $v = 17.44 \times 10^{-6}$ m^2/s, Pr $= 0.69$, and $\beta = 1/T_r = 1/320$. The Rayleigh number and aspect ratio are

$$\mathrm{Ra}_L = \mathrm{Gr}_L\mathrm{Pr} = \frac{(\beta\Delta T)gL^3\mathrm{Pr}}{v^2}$$

$$= \frac{(30/320)(9.81)(0.02)^3(0.69)}{(17.44 \times 10^{-6})^2}$$

$$= 16{,}700$$

$$\frac{H}{L} = \frac{2}{0.02} = 100$$

Equation (4.99) applies:

$$\mathrm{Nu}_1 = \left\{1 + \left[\frac{(0.0936)(16{,}700)^{0.314}}{1 + \{0.5/[1 + (16{,}700/3160)^{20.6}]^{0.1}\}}\right]^7\right\}^{1/7} = 1.95$$

$$\text{Nu}_2 = \left(0.104 + \frac{0.175}{100}\right)(16{,}700)^{0.283} = 1.66$$

$$\overline{\text{Nu}}_{L60^\circ} = \max\{1.95, 1.66\} = 1.95$$

$$\bar{h}_c = \left(\frac{k}{L}\right)\overline{\text{Nu}} = \left(\frac{0.0281}{0.02}\right)1.95 = 2.74 \text{ W/m}^2 \text{ K}$$

$$\dot{Q} = h_c A \Delta T = (2.74)(2 \times 1)(30) = 164 \text{ W}$$

Comments

1. Use CONV to check $\bar{h}_c$.

2. It may be desirable to place a honeycomb structure between the plates to suppress natural convection.

4.4.3 Mixed Forced and Natural Flows

In treating forced flows in Section 4.3, it was implicitly assumed that buoyancy effects were negligible. But heat transfer requires a temperature gradient, and if this gradient causes the fluid to be unstable, there is the possibility of the buoyancy force contributing to the flow. Figure 4.40 shows some possible situations, where it is shown that the buoyancy force may *assist* or *oppose* the forced flow, or perhaps act perpendicularly to the forced flow. Such **mixed flows** are obviously very complex and cannot be dealt with in detail here. In this section, criteria for deciding whether forced or natural convection dominates will be discussed, and correlations and data for some very simple mixed flows will be presented.

External Flows

We have seen that the dimensionless group that characterizes a forced flow is the Reynolds number, whereas for a natural flow it is the Grashof or Rayleigh number. To gain some insight into the significance of the Grashof number, consider natural convection on a vertical isothermal wall, as shown in Fig. 4.41. If we follow a fluid element along the depicted streamline and ignore the viscous forces, then, as a rough approximation, the gain in kinetic energy of the fluid element must equal the work done by the buoyancy force:

$$\frac{1}{2}\rho V^2 \sim g\Delta\rho L$$

or,

$$V \sim \left(\frac{\Delta\rho}{\rho}gL\right)^{1/2} \tag{4.106}$$

where the factor of 1/2 can be ignored in making this *estimate* of the characteristic natural convection velocity V. A Reynolds number based on this velocity and

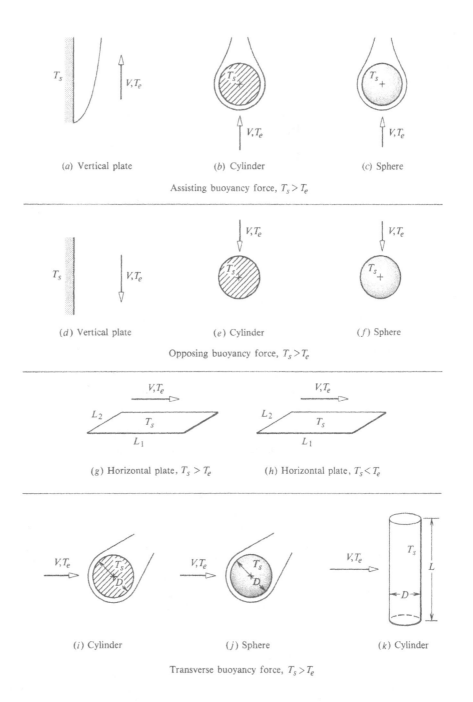

(a) Vertical plate   (b) Cylinder   (c) Sphere

Assisting buoyancy force, $T_s > T_e$

(d) Vertical plate   (e) Cylinder   (f) Sphere

Opposing buoyancy force, $T_s > T_e$

(g) Horizontal plate, $T_s > T_e$ (h) Horizontal plate, $T_s < T_e$

(i) Cylinder   (j) Sphere   (k) Cylinder

Transverse buoyancy force, $T_s > T_e$

Figure 4.40   Schematic of possible mixed forced- and natural-convection flows.

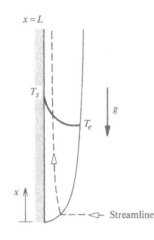

Figure 4.41 A streamline in a laminar natural convection boundary layer on a vertical isothermal wall.

plate length L is thus

$$Re = \frac{[(\Delta\rho/\rho)gL]^{1/2}L}{\nu} \tag{4.107}$$

and

$$Re^2 = \frac{(\Delta\rho/\rho)gL^3}{\nu^2} = Gr \tag{4.108}$$

So the Grashof number is simply the square of a Reynolds number based on an estimate of the free-convection velocity, V. The numerical value of V given by Eq. (4.106) will be somewhat larger than the maximum velocity in the natural-convection boundary layer but will be of the same order of magnitude. Since we have ignored viscous forces in estimating V, we would expect Eq. (4.106) to be less accurate for very viscous fluids, that is, if the Prandtl number is large. In the limit Pr $\to \infty$, the fluid motion is determined by a balance of buoyancy and viscous forces, and the characteristic velocity must be estimated accordingly.

To ascertain whether forced or natural convection dominates, or whether there is mixed convection, we must compare the velocities associated with each: for forced convection, the appropriate velocity is either the free-stream velocity for external flows, or the bulk velocity for pipe and duct flows. For fluids of Prandtl number of order unity or less, the velocity given by Eq. (4.106) is appropriate; equivalently, the Grashof number for natural convection can be compared to the square of the Reynolds number. For high-Prandtl-number fluids, the Grashof number divided by the cube root of the Prandtl number should be compared to the square of the Reynolds number. The classification in Table 4.7 has been prepared accordingly. Care must be taken to use appropriate length parameters. For example, in Fig. 4.40*b*, the proper parameter is the cylinder diameter D, and in Fig. 4.40*k*, it is D for forced convection and the tube length L for natural convection. In the case of the horizontal rectangular plate in Fig. 4.40*g*, the length parameter is L_1 for forced convection and L_2 for natural convection, if $L_2 < L_1$ (although if the flow is turbulent, L_2 is of no consequence).

Table 4.7 Criteria for mixed convection in external flows.

	$\text{Pr} \lesssim 1$	$\text{Pr} \to \infty$
Forced convection	$(\text{Gr/Re}^2) \ll 1$	$(\text{Gr/Pr}^{1/3}\text{Re}^2) \ll 1$
Mixed convection	$(\text{Gr/Re}^2) \simeq 1$	$(\text{Gr/Pr}^{1/3}\text{Re}^2) \simeq 1$
Natural convection	$(\text{Gr/Re}^2) \gg 1$	$(\text{Gr/Pr}^{1/3}\text{Re}^2) \gg 1$

Internal Flows

Mixed convection in duct flows is more complicated than in external flows. In vertical ducts, an aiding buoyancy force increases heat transfer in laminar flows but usually decreases heat transfer in turbulent flows. Similarly, an opposing buoyancy force decreases heat transfer in laminar flow but usually increases heat transfer in turbulent flow. Heating a fluid flowing in a horizontal pipe produces a secondary motion in which fluid circulates upward on the vertical walls and downward in the central region, with a resulting increase in heat transfer. Cooling reverses the direction of circulation and similarly increases the heat transfer.

Figure 4.42*a* and *b* indicates the mixed convection regime, defined as a flow for which the Nusselt number deviates by more than 10% from the pure forced-or natural-convection value. For horizontal flow, in Fig. 4.42*b*, the boundary between mixed and natural convection is omitted since it has yet to be satisfactorily established.

In situations where it is not obvious that forced or natural convection dominates, the criteria given here, namely, Table 4.7 for external flows and Fig. 4.42 for internal flows, must be used to ascertain whether special correlations for mixed convection are necessary.

Correlations

Some appropriate correlations for mixed convection in external flows recommended by Churchill [38] are as follows.

1. Laminar or turbulent boundary layer flows with an assisting buoyancy force on vertical plates, cylinders, or spheres (Fig. 4.40*a*, *b*, *c*):

$$(\overline{\text{Nu}} - \overline{\text{Nu}}_0)^3 = (\overline{\text{Nu}}_f - \overline{\text{Nu}}_0)^3 + (\overline{\text{Nu}}_n - \overline{\text{Nu}}_0)^3 \qquad \textbf{(4.109)}$$

where the subscripts f and n refer to forced and natural convection, respectively, and $\overline{\text{Nu}}_0 = 0, 0.3,$ and 2 for plates, cylinders, and spheres, respectively. Equation (4.109) is well established for laminar flows but is expected to be less reliable for turbulent flows. Also, it should be noted that a buoyancy force can suppress transition of a laminar forced-convection boundary layer.

2. Boundary layer flows with an opposing buoyancy force (Fig. 4.40*d*, *e*, *f*):

$$(\overline{\text{Nu}} - \overline{\text{Nu}}_0)^3 = |(\overline{\text{Nu}}_f - \overline{\text{Nu}}_0)^3 + (\overline{\text{Nu}}_n - \overline{\text{Nu}}_0)^3| \qquad \textbf{(4.110)}$$

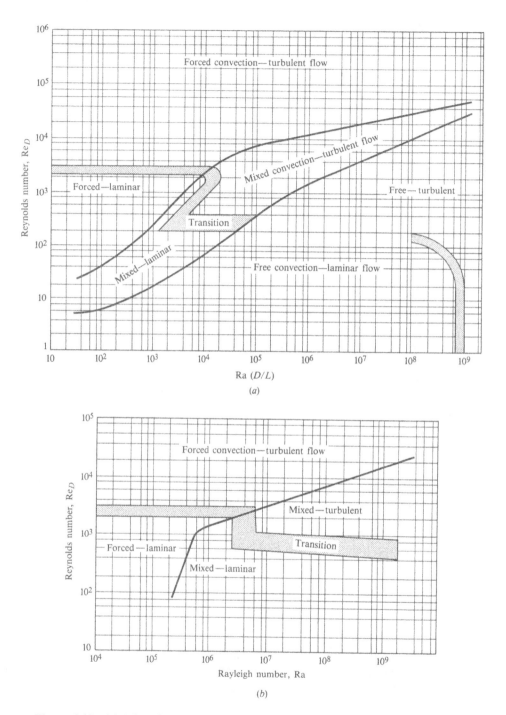

Figure 4.42 (a) Mixed forced- and natural-convection flow regimes in vertical tubes; $10^{-2} < \mathrm{Pr}D/L < 1$. (b) Mixed forced- and natural-convection flow regimes in horizontal tubes; $10^{-2} < \mathrm{Pr}D/L < 1$ [37]. (Adapted with permission.)

3. Forced boundary layer flow along a horizontal plate with a transverse buoyancy force (Fig. 4.40g, h):

$$\overline{\text{Nu}}^{7/2} = \overline{\text{Nu}}_f^{7/2} \pm \overline{\text{Nu}}_n^{7/2}; \quad L_1 < L_2 \tag{4.111}$$

where the positive sign is applicable to a heated plate facing up or a cooled plate facing down, and vice versa for the negative sign. Since separation of the boundary layer occurs before $\overline{\text{Nu}}_n$, exceeds $\overline{\text{Nu}}_f$ the limit of applicability of Eq. (4.111) is $\overline{\text{Nu}} > 0$.

4. Cross-flow over an immersed cylinder or sphere with a transverse buoyancy force (Fig. 4.40i, j):

$$(\overline{\text{Nu}} - \overline{\text{Nu}}_0)^4 = (\overline{\text{Nu}}_f - \overline{\text{Nu}}_0)^4 + (\overline{\text{Nu}}_n - \overline{\text{Nu}}_0)^4 \tag{4.112}$$

These combining rules for mixed convection, Eqs. (4.109) through (4.112), are clearly rather crude; however, they are adequate for most engineering purposes. Correlations for mixed convection in internal flows are more complicated; some useful results may be found in handbooks listed in the bibliography for Chapter 4.

EXAMPLE 4.9   Cooling of an Electronics Package

The sides of an electronics package are rectangular aluminum plates 70 cm wide and 40 cm high and act as a heat sink for the contents. The plate temperature must not exceed 340 K. If air at 300 K is blown upward over the plates at 1 m/s, estimate the power each plate dissipates by convective heat transfer.

Solution

Given: Mixed forced and natural convection on a vertical plate.

Required: Convective heat transfer rate.

Assumptions: The boundary layer starts at the bottom of the plate.

Evaluate air properties at a mean film temperature of 320 K: $\beta = 1/T_r = 1/320, k = 0.0281$ W/m K, $\nu = 17.44 \times 10^{-6}$ m²/s, and Pr = 0.69. We must first calculate average Nusselt numbers assuming there is only forced or only natural convection.

$$\text{Re}_L = \frac{(1)(0.4)}{17.44 \times 10^{-6}} = 2.29 \times 10^4 \quad \text{(laminar flow)}$$

Using Eq. (4.57),

$$\overline{\text{Nu}}_f = 0.664\text{Re}_L^{1/2}\text{Pr}^{1/3} = 0.664(2.29 \times 10^4)^{1/2}(0.69)^{1/3}$$

$$= 88.8$$

$$\text{Gr}_L = \frac{\beta\Delta T g L^3}{\nu^2} = \frac{(40/320)(9.81)(0.4)^3}{(17.44 \times 10^{-6})^2}$$

$$= 2.58 \times 10^8$$

$$\text{Ra}_L = \text{Gr}_L\text{Pr} = (2.58 \times 10^8)(0.69) = 1.780 \times 10^8 < 10^9 \quad \text{(laminar)}$$

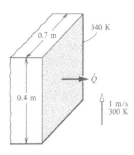

Using Eqs. (4.84) and (4.85),

$$\Psi = \left[1 + \left(\frac{0.492}{0.69}\right)^{9/16}\right]^{-16/9} = 0.3426$$

$$\overline{Nu}_n = 0.68 + 0.670(1.780 \times 10^8 \times 0.3426)^{1/4} = 59.9$$

For mixed convection, we use Eq. (4.109) with $Nu_0 = 0$:

$$\overline{Nu} = (\overline{Nu}_f^3 + \overline{Nu}_n^3)^{1/3} = (88.8^3 + 59.9^3)^{1/3} = 97.1$$

$$\overline{h}_c = \left(\frac{k}{L}\right)\overline{Nu} = \left(\frac{0.0281}{0.4}\right)97.1 = 6.82 \text{ W/m}^2 \text{ K}$$

$$\dot{Q} = \overline{h}_c A \Delta T = (6.82)(0.4 \times 0.7)(340 - 300) = 76.4 \text{ W}$$

Comments

1. Since the heat transfer coefficient is relatively low (~ 7 W/m^2 K), radiative heat transfer will play an important role, particularly if the aluminum surface is black-anodized to give a high emittance.

2. Use of vertical fins would markedly improve the convective heat transfer.

3. Use CONV to check $\overline{Nu}_f$ and $\overline{Nu}_n$.

EXAMPLE 4.10   Heat Loss from a Workshop Roof

A wind blows at 2 m/s over the 10 m square horizontal roof of a workshop on a sunny day. The ambient air temperature is 295 K, and the roof temperature is measured to be approximately 315 K. Estimate the convective heat loss from the roof. Take $Re_{tr} = 10^5$.

Solution

Given: Mixed forced and natural convection on a horizontal surface.

Required: Convective heat transfer rate.

Assumptions: The wind direction is normal to one edge of the roof.

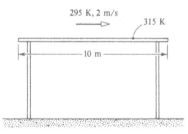

Evaluate fluid properties at a mean film temperature of 305 K: $\beta = 1/T_r = 1/305$, $k = 0.0270$ W/m K, $c_p = 1005$ J/kg K, $\nu = 16.1 \times 10^{-6}$ m^2/s, and Pr = 0.69. The Reynolds number is

$$Re_L = \frac{VL}{\nu} = \frac{(2)(10)}{16.1 \times 10^{-6}} = 1.242 \times 10^6$$

For forced convection, $\overline{Nu}$ is given by Eq. (4.65):

$$\overline{Nu}_f = 0.664Re_{tr}^{1/2}Pr^{1/3} + 0.036Re_L^{0.8}Pr^{0.43}\left[1 - \left(\frac{Re_{tr}}{Re_L}\right)^{0.8}\right]$$

$$= 0.664(10^5)^{1/2}(0.69)^{1/3} + 0.036(1.242 \times 10^6)^{0.8}(0.69)^{0.43}\left[1 - \left(\frac{0.1}{1.242}\right)^{0.8}\right]$$

$$= 186 + 1996 = 2182$$

The Grashof number is

$$Gr_L = \frac{\beta\Delta T g L^3}{\nu^2} = \frac{(20/305)(9.81)(10)^3}{(16.1 \times 10^{-6})^2} = 2.48 \times 10^{12}$$

Thus, Eq. (4.96) applies for natural convection:

$$\overline{Nu}_n = 0.14Ra_L^{1/3} = 0.14(Gr_L Pr)^{1/3} = 0.14[(2.48 \times 10^{12})(0.69)]^{1/3} = 1674$$

For mixed convection, Eq. (4.111) applies:

$$\overline{Nu}^{7/2} = \overline{Nu}_f^{7/2} + \overline{Nu}_n^{7/2} = (2182)^{7/2} + (1674)^{7/2} = 4.85 \times 10^{11} + 1.92 \times 10^{11}$$

$$= 6.77 \times 10^{11}$$

$$\overline{Nu} = 2400$$

$$\bar{h}_c = \left(\frac{k}{L}\right)\overline{Nu} = \left(\frac{0.0270}{10}\right)2400 = 6.48 \text{ W/m}^2 \text{ K}$$

$$\dot{Q} = \bar{h}_c A\Delta T = (6.48)(10 \times 10)(315 - 295) = 13.0 \text{ kW}$$

Comments

1. Use CONV to check $\overline{Nu}_f$ and $\overline{Nu}_n$.

2. Equation (4.96) has been used outside its stated range of applicability ($Ra_L < 3 \times 10^{10}$).

4.5   TUBE BANKS AND PACKED BEDS

Flow through tube banks (bundles) or packed beds in heat exchangers are forced flows and have characteristics of both internal and external flows. The flow is confined in a shell or duct to give an overall internal flow behavior, but the flow over a particular tube or packing piece has similarities to an external flow.

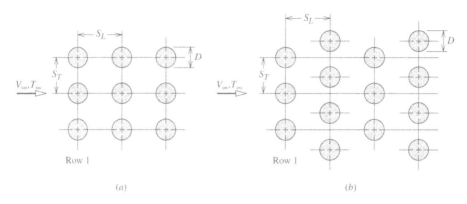

Figure 4.43   Tube bank configurations, (*a*) Aligned, (*b*) Staggered.

4.5.1 Flow through Tube Banks

Cross-flow over tube banks occurs in shell-and-tube heat exchangers, which are widely used as evaporators and condensers in power and refrigeration systems and for a great variety of applications in the process industries. Often the tubes are finned, but here attention will be restricted to smooth tubes. Both aligned and staggered tube banks are used, as shown in Fig. 4.43. The flow inside a tube bank is complex, involving boundary layer separation on each tube and interactions of the resulting wakes with adjacent wakes and downstream tubes. Figure 4.43 shows the geometric parameters of a tube bank, which include the tube diameter D and the transverse and longitudinal pitches, S_T and S_L, respectively. The number of rows of tubes transverse to the flow is N.

To calculate heat transfer, a procedure similar to that of Gnielinski et al. [39] is recommended. The heat transfer coefficient on the first row of tubes is higher than for a single tube in cross-flow, as the fluid must accelerate to pass through the spaces between adjacent tubes. However, if the Reynolds number is based on an average velocity in the space between two adjacent tubes, that is, defined by the relation

$$\frac{\overline{V}}{V_0} = \frac{S_T}{S_T - (\pi/4)D} \tag{4.113}$$

where V_0 is the velocity of the fluid in the empty cross section of the shell or duct, then Eq. (4.71) for a single tube in cross-flow applies.

The heat transfer coefficient increases from the first to about the fifth row of a tube bank. The average Nusselt number for a tube bank with 10 or more rows may be calculated from

$$\overline{\mathrm{Nu}}_D^{10+} = \Phi \overline{\mathrm{Nu}}_D^1 \tag{4.114}$$

where $\overline{\mathrm{Nu}}_D^1$ is the Nusselt number for the first row, and Φ is an *arrangement factor*. We define the dimensionless transverse pitch as $P_T = S_T/D$, the dimensionless longitudinal pitch as $P_L = S_L/D$, and a factor ψ as

$$\psi = 1 - \frac{\pi}{4P_T} \quad \text{if } P_L \geq 1 \tag{4.115a}$$

$$\psi = 1 - \frac{\pi}{4P_T P_L} \quad \text{if } P_L < 1 \tag{4.115b}$$

Then the arrangement factors are

$$\Phi_{\text{aligned}} = 1 + \frac{0.7}{\psi^{1.5}} \frac{S_L/S_T - 0.3}{(S_L/S_T + 0.7)^2} \tag{4.116}$$

$$\Phi_{\text{staggered}} = 1 + \frac{2}{3P_L} \tag{4.117}$$

For tube banks of fewer than 10 rows, a simple interpolation formula applies:

$$\overline{\text{Nu}}_D = \frac{1 + (N-1)\Phi}{N} \overline{\text{Nu}}_D^1 \tag{4.118}$$

Properties are to be evaluated at the average mean film temperature for gases; for liquids, properties are first evaluated at the average bulk temperature, $(T_{\text{in}} + T_{\text{out}})/2$, and then a Prandtl number correction factor is used with $n = -0.25$ for heating and $n = -0.11$ for cooling.

Zukauskus [40] recommends that the pressure drop in a tube bank be calculated as

$$\Delta P = N\chi \left(\frac{\rho V_{\text{max}}^2}{2} \right) f \tag{4.119}$$

where the friction factor f and correction factor χ are given in Fig. 4.44a and b. Notice that $\chi = 1$ for a square aligned configuration and for an equilateral-triangle staggered configuration. The velocity V_{max} is the maximum velocity in the tube bank. For an aligned bank, the maximum velocity occurs between adjacent tubes of a transverse row; hence,

$$\frac{V_{\text{max}}}{V_0} = \frac{S_T}{S_T - D} \tag{4.120a}$$

For a staggered bank, V_{max} may occur between adjacent tubes of a transverse row or of a diagonal row. Then

$$\frac{V_{\text{max}}}{V_0} = \max\left\{ \frac{S_T}{S_T - D}, \frac{S_T/2}{[S_L^2 + (S_T/2)^2]^{1/2} - D} \right\} \tag{4.120b}$$

Properties for the pressure drop calculation are to be evaluated at the average bulk temperature.

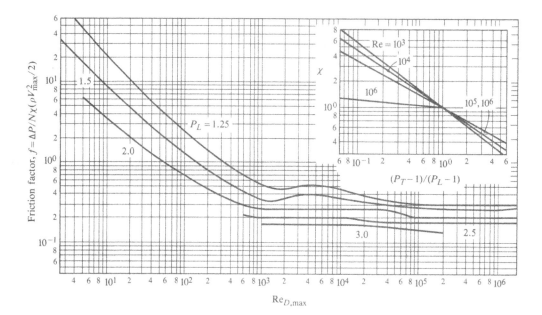

(a)

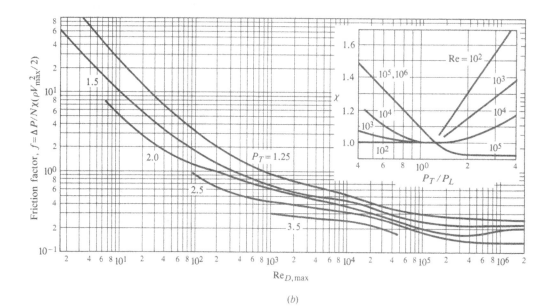

(b)

Figure 4.44   Friction factor f and correction factor χ in Eq. (4.119). (a) For an aligned tube bank. (b) For a staggered tube bank [40]. (Adapted with permission.)

EXAMPLE 4.11   A Tube Bank Air Heater

Air at 290 K and 1 atm flows at 2.0 kg/s along a duct of cross section 1 m × 0.4 m. The duct contains a tube bank of 1 m-long, 15 mm–O.D. tubes in a staggered arrangement, with both transverse and longitudinal pitches of 30 mm. If steam at atmospheric pressure is condensed in the tubes, how many rows of tubes are required to heat the air to 324 K? Also estimate the pressure drop.

Solution

Given: Air flowing through a bank of tubes, inside which steam condenses at 1 atm pressure.

Required: Number of tubes needed to heat air to 324 K, and the pressure drop.

Assumptions: Negligible condensing side and wall thermal resistances; hence, the outside wall temperature can be taken to be P_{sat} (1 atm) = 373 K.

To calculate heat transfer, the air properties are evaluated at the mean film temperature based on the tube wall temperature, and the average of the inlet and outlet bulk air temperatures:

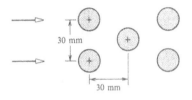

$$T_r = \frac{1}{2}\left(373 + \frac{290 + 324}{2}\right) = 340 \text{ K}$$

From Table A.7, $k = 0.0294$ W/m K, $v = 19.32 \times 10^{-6}$ m²/s, $\rho = 1.042$ kg/m³, Pr $= 0.69$, and $c_p = 1007$ J/kg K.

The empty duct velocity is $V_0 = \dot{m}/\rho A_c = (2.0)/(1.042)(0.4) = 4.80$ m/s. Using Eq. (4.113) for the first row of tubes,

$$\overline{V} = V_0 \frac{S_T}{S_T - (\pi/4)D} = (4.80)\frac{30}{30 - (\pi/4)15} = 7.90 \text{ m/s}$$

$$\text{Re}_D = \frac{\overline{V}D}{v} = \frac{(7.90)(0.015)}{19.32 \times 10^{-6}} = 6130$$

Using Eq. (4.71a),

$$\overline{\text{Nu}}_D^1 = 0.3 + \frac{0.62\text{Re}_D^{1/2}\text{Pr}^{1/3}}{[1 + (0.4/\text{Pr})^{2/3}]^{1/4}} = 0.3 + \frac{0.62(6130)^{1/2}(0.69)^{1/3}}{[1 + (0.4/0.69)^{2/3}]^{1/4}} = 37.9$$

We will assume that more than 10 tube rows are required and check later. Using Eqs. (4.117) and (4.114),

$$P_L = \frac{S_L}{D} = \frac{30}{15} = 2$$

$$\Phi_{\text{staggered}} = 1 + \frac{2}{3P_L} = 1 + \frac{2}{(3)(2)} = 1.333$$

$$\overline{\text{Nu}}_D^{10+} = \Phi\overline{\text{Nu}}_D^1 = (1.333)(37.9) = 50.5$$

$$\overline{h}_c = \frac{\overline{\text{Nu}}_D k}{D} = \frac{(50.5)(0.0294)}{0.015} = 99.0 \text{ W/m}^2\text{K}$$

Since the tube wall temperature is constant through the bank, Eq. (4.11) applies, with the tube wall area $2\pi RL$ replaced by the transfer area of the tube bank A,

$$T_{b,out} = T_s - (T_s - T_{b,in})e^{-\bar{h}_c A/\dot{m}c_p}$$

Substituting the known temperatures and solving,

$$324 = 373 - (373 - 290)e^{-\bar{h}_c A/\dot{m}c_p}; \quad \frac{\bar{h}_c A}{\dot{m}c_p} = 0.528$$

The required transfer area is then

$$A = \frac{0.528\dot{m}c_p}{\bar{h}_c} = \frac{0.528(2.0)(1007)}{99.0} = 10.74 \text{ m}^2$$

The number of tubes per row is the duct width divided by the transverse pitch:

$$\frac{W}{S_T} = \frac{0.4}{0.030} = 13.3 \simeq 13$$

Then

$$A = (\text{Number of rows})(\text{Number of tubes per row})(\text{Area of one tube})$$
$$= N(13)(\pi)(0.015)(1)$$
$$= 0.613N$$

Hence, $N = 10.74/0.613 = 17.5 \simeq 18$ rows (> 10).

The pressure drop is obtained from Eq. (4.119) and Fig. 4.44b. From Eq. (4.120b)

$$\frac{V_{max}}{V_0} = \max\left\{ \frac{S_T}{S_T - D}, \frac{S_T/2}{[S_L^2 + (S_T/2)^2]^{1/2} - D} \right\}$$

$$= \max\left\{ \frac{30}{30 - 15}, \frac{30/2}{[30^2 + (30/2)^2]^{1/2} - 15} \right\}$$

$$= \max\{2, 0.809\} = 2$$

To calculate pressure drop, properties are evaluated at the average bulk temperature:

$$T_r = \frac{1}{2}(290 + 324) = 307 \text{ K}$$

From Table A.7, $\rho = 1.151 \text{ kg/m}^3$, $v = 16.27 \times 10^{-6} \text{ m}^2/\text{s}$.

$$V_0 = \frac{\dot{m}}{\rho A} = \frac{2.0}{(1.151)(0.4)} = 4.34 \text{ m/s}$$

$$V_{max} = (2)(4.34) = 8.68 \text{ m/s}$$

$$Re_{D,max} = \frac{(8.68)(0.015)}{16.27 \times 10^{-6}} = 8000$$

$$P_L = P_T = 30/15 = 2.0, \quad P_T/P_L = 1.0$$

Using Fig. 4.44b $f = 0.38$, $\chi = 1.0$.

$$\Delta P = N\chi \left(\frac{\rho V_{max}^2}{2} \right) f = (18)(1.0) \left[\frac{(1.151)(8.68)^2}{2} \right] (0.38) = 297 \text{ Pa}$$

Comments

1. Use CONV to check $\bar{h}_c$ and ΔP.

2. In general, there may be an uncomfortably large discrepancy in ΔP given by CONV, because CONV contains an approximate curve fit to the data in Fig. 4.44. However, the original data are not too reliable, so greater precision is perhaps not warranted.

3. This air heater is termed a single-stream heat exchanger since only the air stream temperature changes in the unit (see Section 8.4).

4. The pressure drop is less than 1% of the total pressure; thus, its effect on density can be ignored. When the pressure drop is a significant fraction of the total pressure, an iterative calculation procedure is required.

5. Notice the different property evaluation schemes for heat transfer and pressure drop. However, the difference is of little practical significance owing to the large expected uncertainty in f values obtained from Fig. 4.44.

EXAMPLE 4.12   A Tube Bank Water Heater

A tube bank is 30 rows deep and has 15 mm-O.D. tubes in a staggered arrangement, with a transverse pitch of 24 mm and a longitudinal pitch of 15 mm. Steam at 2×10^5 Pa condenses inside the tubes. In a performance test, 304.0 K water enters the tube bank at 2 m/s and leaves at 316.0 K. Estimate the average heat transfer coefficient of the tube bank, and the pressure drop.

Solution

Given: Water flowing through a bank of tubes, inside which steam condenses at 2×10^5 Pa.

Required: Average heat transfer coefficient $\bar{h}_c$, and pressure drop ΔP.

Assumptions: The tube wall temperature is approximately constant through the bundle.

For a liquid flowing through a tube bank, properties are evaluated at the average of the inlet and outlet bulk temperatures, with a Prandtl number ratio correction applied later.

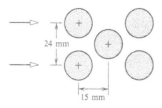

$$T_r = \frac{304 + 316}{2} = 310 \text{ K}$$

From Table A.8, water properties are $k = 0.628$ W/m K, $\rho = 993$ kg/m^3, $c_p = 4174$ J/kg K, $\nu = 0.70 \times 10^{-6}$ m^2/s, and Pr $= 4.6$. Using Eq. (4.113) for the first row of tubes,

$$\overline{V} = V_0 \frac{S_T}{S_T - (\pi/4)D} = (2)\frac{24}{24 - (\pi/4)(15)} = 3.93 \text{ m/s}$$

$$\text{Re}_D = \frac{\overline{V}D}{\nu} = \frac{(3.93)(0.015)}{0.70 \times 10^{-6}} = 84{,}200$$

Using Eq. (4.71b),

$$\overline{\text{Nu}}_D^1 = 0.3 + \frac{0.62\text{Re}_D^{1/2}\text{Pr}^{1/3}}{[1+(0.4/\text{Pr})^{2/3}]^{1/4}}\left[1+\left(\frac{\text{Re}_D}{282{,}000}\right)^{1/2}\right]$$

$$= 0.3 + \frac{0.62(84{,}200)^{1/2}(4.6)^{1/3}}{[1+(0.4/4.6)^{2/3}]^{1/4}}\left[1+\left(\frac{84{,}200}{282{,}000}\right)^{1/2}\right] = 443$$

$$P_L = \frac{S_L}{D} = \frac{15}{15} = 1.00$$

$$\Phi_{\text{staggered}} = 1 + \frac{2}{3P_L} = 1 + \frac{2}{(3)(1.0)} = 1.667$$

$$\overline{\text{Nu}}_D^{10+} = \Phi\overline{\text{Nu}}_D^1 = (1.667)(443) = 738$$

To obtain the Prandtl number ratio correction, $(\text{Pr}_s/\text{Pr}_b)^{-1/4}$, we need an estimate of the tube wall temperature. For this purpose, we assume a constant tube wall temperature through the bank, so that Eq. (4.11) applies, with the tube wall area $2\pi RL$ replaced by the transfer area of the tube bank A,

$$T_{b,\text{out}} = T_s - (T_s - T_{b,\text{in}})e^{-\overline{h}_cA/\dot{m}c_p} \tag{1}$$

The uncorrected value of $\overline{\text{Nu}}_D$ is used to estimate $\overline{h}_c$:

$$\overline{h}_c = \left(\frac{k}{D}\right)\overline{\text{Nu}}_D = \left(\frac{0.628}{0.015}\right)738 = 3.09 \times 10^4 \text{ W/m}^2 \text{ K}$$

$$A = (\text{Number of rows})(\text{Number of tubes per row})(\text{Area of one tube})$$

$$= (30)(1/0.024)(\pi)(0.015)(1) \quad \text{for 1 m}^2\text{of cross section}$$

$$= 58.9 \text{ m}^2$$

$$\dot{m} = \rho VA_c = (995)(2)(1) = 1990 \text{ kg/s} \quad (\text{using } \rho \text{ at 304 K})$$

$$\frac{\overline{h}_cA}{\dot{m}c_p} = \frac{(3.09 \times 10^4)(58.9)}{(1990)(4174)} = 0.219$$

Substituting in Eq. (1),

$$316 = T_s - (T_s - 304)e^{-0.219}$$

$$T_s = 364.9 \text{ K}$$

At 364.9 K, Table A.8 gives Pr = 1.93; hence,

$$(\text{Pr}_s/\text{Pr}_b)^{-1/4} = (1.93/4.6)^{-1/4} = 1.24$$

Thus, the corrected Nusselt number and heat transfer coefficients are

$$\overline{\mathrm{Nu}}_D = (1.24)(738) = 917$$

$$\overline{h}_c = (0.628/0.015)(917) = 3.84 \times 10^4 \ \mathrm{W/m^2 \ K}$$

Since the correction is quite large, a second iteration is perhaps justified.

$$\overline{h}_c A / \dot{m} c_p = 0.272; \quad T_s = 354; \quad \mathrm{Pr}_s = 2.24$$

$$(\mathrm{Pr}_s/\mathrm{Pr}_b)^{-1/4} = 1.20; \quad \overline{\mathrm{Nu}}_D = 886; \quad \overline{h}_c = 3.71 \times 10^4 \ \mathrm{W/m^2 \ K}$$

The pressure drop is obtained from Eqs. (4.119) and (4.120b) and Fig. 4.44b:

$$\frac{S_T}{S_T - D} = \frac{24}{24 - 15} = 2.67$$

$$\frac{S_T/2}{[S_L^2 + (S_T/2)^2]^{1/2} - D} = \frac{24/2}{[15^2 - (24/2)^2]^{1/2} - 15} = 2.85$$

$$V_{\max} = 2.85 V_0 = (2.85)(2) = 5.70 \ \mathrm{m/s}$$

$$\mathrm{Re}_{D,\max} = \frac{(5.70)(0.015)}{0.70 \times 10^{-6}} = 1.22 \times 10^5$$

$$P_L = 15/15 = 1; \quad P_T = 24/15 = 1.6; \quad P_T/P_L = 1.6$$

From Fig. 4.44b, $f = 0.21$, $\chi = 0.95$; thus,

$$\Delta P = N\chi \left(\frac{\rho V_{\max}^2}{2} \right) f = (30)(0.95) \left[\frac{(993)(5.70)^2}{2} \right] (0.21) = 0.97 \times 10^5 \ \mathrm{Pa}$$

Comments

1. In contrast to the previous example, the tube wall temperature could not be assumed equal to the saturation temperature of the steam: in this situation, the water-side resistance is comparable to the steam-side resistance, and the latter cannot be ignored.

2. Use CONV to check $\overline{h}_c$ and ΔP.

3. For a liquid, the properties used to calculate $V_{\max}$ and Re_D for heat transfer and pressure drop are the same (for a gas, they are different; see Example 4.11).

4.5.2 Flow through Packed Beds

Packed beds of solid particles, such as the pebble bed shown in Fig. 4.45, are often used as heat exchangers or for storage of thermal energy. The pebbles are heated by passing a hot fluid through the bed, and the energy stored in the pebbles is subsequently extracted by passing a cold fluid through the bed. The objective may be to simply transfer heat from one fluid stream to another: then the bed is termed a *regenerative* heat exchanger, or simply a *regenerator*, such exchangers are described in Section 8.2. Alternatively, the objective may be to store thermal energy for a spe-

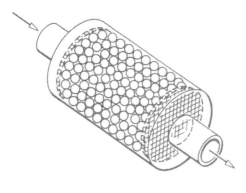

Figure 4.45   A packed pebble bed.

cific period of time. For example, in a solar home-heating system, the bed is heated during the day, when solar energy is available, and extracted during the night to heat the home. Packed beds are also very widely used as mass exchangers in process engineering practice, where a great variety of particle shapes are encountered.

The **void fraction** ε_v of a packed bed is defined as

$$\varepsilon_v = \frac{\text{Bed volume} - \text{Packing volume}}{\text{Bed volume}} \tag{4.121}$$

Values of ε_v between 0.3 and 0.5 are typical. The surface area for transfer depends on the particle size and shape and the void fraction. The **specific surface area** $a[m^{-1}]$ of a packed bed is the wetted or transfer area per unit volume of bed:

$$a = \frac{\text{Total surface area of particles}}{\text{Bed volume}} \tag{4.122}$$

If a bed consists of particles of volume V_p and surface area A_p, then

$$a = \frac{A_p}{V_p}(1 - \varepsilon_v) \tag{4.123}$$

For example, for a spherical particle of diameter d_p, $A_p/V_p = \pi d_p^2/(1/6)\pi d_p^3 = 6/d_p$, and $a = 6(1 - \varepsilon_v)/d_p$. In general, the specific surface area can be calculated if the geometry of the particles and the void fraction in the bed are known. The hydraulic diameter of the bed D_h is defined in an analogous manner as for duct flow:

$$D_h = \frac{\text{Volume of bed available for flow}}{\text{Wetted surface in bed}} = \frac{\text{Void volume/unit volume}}{\text{Wetted surface/unit volume}} = \frac{\varepsilon_v}{a} \tag{4.124}$$

Using Eq. (4.123),

$$D_h = \left(\frac{\varepsilon_v}{1 - \varepsilon_v}\right)\frac{V_p}{A_p} \tag{4.125}$$

We next define a characteristic length and velocity for flow through packing. For the characteristic length $\mathscr{L}$, we choose $6D_h$; then, if we also define the effective particle diameter as

$$d_p = 6\frac{V_p}{A_p} \tag{4.126}$$

which yields the actual diameter for a spherical particle, the characteristic length becomes

$$\mathscr{L} = d_p\left(\frac{\varepsilon_v}{1-\varepsilon_v}\right) \tag{4.127}$$

For the characteristic velocity, we choose the average velocity of the fluid flowing in the void space, $\mathscr{V}$. The **superficial velocity** in the bed V is defined as the velocity in the bed if no packing were present:

$$V = \frac{\dot{m}}{\rho A_c} \tag{4.128}$$

where $\dot{m}$ [kg/s] is the mass flow, ρ is the fluid density, and A_c is the cross-sectional area of the bed. The superficial velocity can also be viewed as the velocity immediately upstream or downstream of the packing. Since the average cross-sectional area available for flow is simply $\varepsilon_v A_c$, the characteristic velocity is

$$\mathscr{V} = \frac{\dot{m}}{\rho \varepsilon_v A_c} \tag{4.129}$$

We would expect correlations of pressure drop and heat transfer based on $\mathscr{L}$ and $\mathscr{V}$ to be valid for a variety of particle shapes.

The pressure drop across a packed bed can be obtained from the *Ergun* equation [41]:

$$\frac{dP}{dx} = \frac{150\mu\mathscr{V}}{\mathscr{L}^2} + \frac{1.75\rho\mathscr{V}^2}{\mathscr{L}}; \quad 1 < \mathrm{Re} < 10^4 \tag{4.130}$$

The first term of Eq. (4.130) accounts for the viscous drag, and the second term accounts for form drag. The constants are based on experimental data for many shapes of particles, but the equation is most accurate for spherical particles. For flow of gases in a packed bed, an appropriate heat transfer correlation is [10]

$$\mathrm{Nu} = (0.5\mathrm{Re}^{1/2} + 0.2\mathrm{Re}^{2/3})\mathrm{Pr}^{1/3}; \quad 20 < \mathrm{Re} < 10^4 \tag{4.131}$$

Equation (4.131) can also be used for liquids of moderate Prandtl number with fair accuracy. The constants in Eq. (4.131) are based on experimental data for spheres and short cylinders, as well as for commercial packings used in mass-transfer operations, such as Raschig rings and Berl saddles. Variable-property effects can be accounted for using a viscosity ratio with $n = -0.14$. Entrance effects are negligible for packed beds; thus, Eq. (4.131) gives both the local and average Nusselt number.

Finally, we note that the transfer perimeter $\mathscr{P}$ [m] is often used to characterize the surface area for transfer in a packed bed:

$$\mathscr{P} = \frac{\text{Total surface area of particles}}{\text{Bed length}}$$

It follows from Eq. (4.122) that

$$\mathscr{P} = aA_c$$

where A_c is the cross-sectional area of the bed.

Perforated-Plate Packings

A novel form of packing consists of perforated plates separated by gaskets of low thermal conductivity, as shown in Fig. 4.46. Heat exchangers using such packing are popular for cryogenic refrigeration systems, where the characteristic low gas flow rates require effects of conduction in the flow direction to be reduced. Geometrical parameters that may affect pressure drop and heat transfer include the open-area ratio ε_p, plate spacing p, plate thickness t, and the hole diameter d. The Reynolds number is defined in terms of flow through the holes: $\text{Re} = Gd/\mu$, where G is the mass velocity through the holes. Recommended correlations for pressure drop and heat transfer are based on data for air of Shevyakova and Orlov [42]. The pressure drop across a single plate is correlated in terms of the *Euler number*, $\text{Eu} = \Delta P\rho/G^2$, and is given by

$$\text{Eu} = 8.17\text{Re}^{-0.55}(1.707 - \varepsilon_p)^2; \quad 20 < \text{Re} < 150 \tag{4.132a}$$

$$\text{Eu} = 0.5(1.707 - \varepsilon_p)^2; \quad 150 < \text{Re} < 3000 \tag{4.132b}$$

The heat transfer is correlated in terms of a Stanton number, $\text{St} = h_c/Gc_p$. The

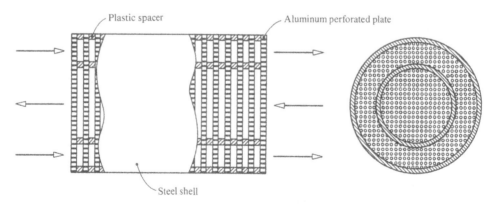

Figure 4.46 A perforated-plate heat exchanger.

heat transfer coefficient is based on the specific surface area a, which includes both the plate faces and hole interiors:

$$St = CRe^n Pr^{-2/3}; \quad 300 < Re < 3000; \quad Pr > 0.5 \tag{4.133}$$

where

$$C = 3.6 \times 10^{-4}[(1 - \varepsilon_p)\varepsilon_p - 0.2]^{-2.07}$$
$$n = -4.36 \times 10^{-2}\varepsilon_p^{-2.34}$$

These correlations are valid for plates 0.5 mm thick, with $0.3 < \varepsilon_p < 0.6, 0.4 < p < 1.6$ mm, and $0.625 < d < 1.65$ mm. The relative location of holes in adjacent plates is arbitrary.

EXAMPLE 4.13   A Pebble Bed Thermal Store

A pebble bed is used to store thermal energy in a solar energy utilization system. The bed is 2 m long, 1 m in diameter, and the pebbles can be approximated as 2 cm-diameter spheres packed with a void fraction of 0.44. During the storage phase of the cycle, 0.8 kg/s of air at 360 K and 1 atm comes from a solar collector. If the bed is initially at a uniform temperature of 300 K, determine the initial rate of heat storage in the bed.

Solution

Given: Hot air flowing through a packed bed of pebbles.

Required: Initial rate of heat storage.

Assumptions: Bed initially at a uniform temperature of 300 K.

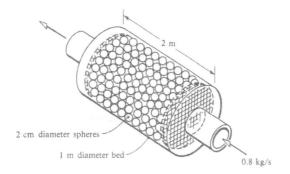

Guess an outlet air temperature of $\sim$ 300 K, and evaluate air properties at an average bulk air temperature of $(1/2)(360 + 300) = 330$ K. From Table A.7, $k = 0.0287$ W/m K, $\rho = 1.073$ kg/m^3, $\mu = 19.71 \times 10^{-6}$ kg/m s, $c_p = 1006$ J/kg K, and Pr $= 0.69$. Equation (4.131) applies.

The characteristic velocity and length are

$$\mathcal{V} = \frac{\dot{m}}{\rho A_c \varepsilon_v} = \frac{0.8}{(1.073)(\pi/4)(1^2)(0.44)} = 2.16 \text{ m/s}$$

$$\mathcal{L} = \frac{d_p \varepsilon_v}{1 - \varepsilon_v} = \frac{(0.02)(0.44)}{1 - 0.44} = 0.0157 \text{ m}$$

The Reynolds number is

$$\text{Re} = \frac{\mathcal{V}\mathcal{L}\rho}{\mu} = \frac{(2.16)(0.0157)(1.073)}{19.71 \times 10^{-6}} = 1848$$

Using Eq.(4.131),

$$\text{Nu} = (0.5\text{Re}^{1/2} + 0.2\text{Re}^{2/3})\text{Pr}^{1/3}$$

$$= [0.5(1848)^{1/2} + 0.2(1848)^{2/3}](0.69)^{1/3} = 45.6$$

Correcting for variable-property effects,

$$\mu_b = 19.71 \times 10^{-6} \text{ kg/m s}$$

$$\mu_s = 18.43 \times 10^{-6} \text{ kg/m s}$$

$$(\mu_s/\mu_b)^{-0.14} = (18.43/19.71)^{-0.14} = 1.009$$

$$\text{Nu} = (1.009)(45.6) = 46.0$$

$$h_c = (k/\mathcal{L})\text{Nu} = (0.0287/0.0157)(46.0) = 84.1 \text{ W/m}^2 \text{ K}$$

At time $t = 0$, the pebbles are at a uniform temperature; thus, Eq. (4.11) applies with $2\pi R$ replaced by the transfer perimeter $\mathcal{P}$.

For spherical particles, the specific surface area is

$$a = \frac{6(1 - \varepsilon_v)}{d_p} = \frac{6(1 - 0.44)}{0.02} = 168 \text{ m}^{-1}$$

and the transfer perimeter $\mathcal{P}$ is

$$\mathcal{P} = aA_c = (168)(\pi/4)(1)^2 = 131.9 \text{ m}$$

$$\frac{\bar{h}_c \mathcal{P} L}{\dot{m} c_p} = \frac{(84.1)(131.9)(2)}{(0.8)(1006)} = 27.6$$

which is very large. Our guess of $T_{\text{out}} \simeq 300$ K is certainly correct.

The rate of heat storage is equal to the rate at which the air gives up energy:

$$\dot{Q} = \dot{m} c_p (T_{\text{in}} - T_{\text{out}}) = (0.8)(1006)(360 - 300) = 48.3 \text{ kW}$$

Comments

1. As the bed heats up, the temperature along the bed will no longer be uniform, and calculation of the rate of heat storage requires an analysis of the transient response of the bed.

2. Use CONV to check h_c.

EXAMPLE 4.14   A Perforated-Plate Heat Exchanger

A perforated-plate two-stream heat exchanger for a helium refrigeration system has 0.5 mm-thick plates spaced 1.0 mm apart. The holes are 1.0 mm in diameter and are arranged in square arrays with an open-area ratio of 0.4. The flow rate and cross-sectional area for each stream are 0.02 kg/s and 0.004 m^2, respectively. Determine the pressure drop per plate and the heat transfer coefficient. Evaluate the helium properties at 200 K, 1 atm.

Solution

Given: Helium flowing through a perforated-plate heat exchanger.

Required: Pressure drop per plate and heat transfer coefficient.

Assumptions: Fluid properties may be evaluated at 200 K.

Using Table A.7 for helium at 200 K, $k = 0.116$ W/m K, $\rho = 0.244$ kg/m^3, $c_p = 5200$ J/kg K, $\mu = 15.6 \times 10^{-6}$ kg/m s, and Pr = 0.70.

The Reynolds number is calculated as follows:

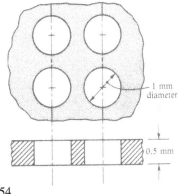

$$\text{Hole area} = \varepsilon_p A_c$$

$$= (0.4)(0.004) = 0.0016 \text{ m}^2$$

$$\text{Mass velocity } G = 0.02/0.0016$$

$$= 12.5 \text{ kg/m}^2 \text{ s}$$

$$\text{Re} = \frac{Gd}{\mu} = \frac{(12.5)(0.001)}{15.6 \times 10^{-6}} = 801$$

Equation (4.132b) gives the Euler number:

$$\text{Eu} = 0.5(1.707 - \varepsilon_p)^2 = 0.5(1.707 - 0.4)^2 = 0.854$$

$$\Delta P = \frac{\text{Eu}G^2}{\rho} = \frac{(0.854)(12.5)^2}{0.244} = 547 \text{ Pa}$$

The Stanton number is obtained from Eq. (4.133).

$$C = 3.6 \times 10^{-4}[(1 - \varepsilon_p)\varepsilon_p - 0.2]^{-2.07} = 3.6 \times 10^{-4}[(1 - 0.4)0.4 - 0.2]^{-2.07}$$

$$= 0.282$$

$$n = -4.36 \times 10^{-2}\varepsilon_p^{-2.34} = -(4.36 \times 10^{-2})(0.4)^{-2.34} = -0.372$$

$$\text{St} = C\text{Re}^n\text{Pr}^{-2/3} = 0.282(801)^{-0.372}(0.7)^{-2/3} = 2.97 \times 10^{-2}$$

$$h_c = \text{St}Gc_p = (2.97 \times 10^{-2})(12.5)(5200) = 1930 \text{ W/m}^2 \text{ K}$$

Comments

1. Usually in exchanger design we are interested in the $h_c \mathscr{P}$ product in order to calculate the number of transfer units, N_{ta}. The perimeter $\mathscr{P}$ is calculated as follows:

$$\text{Plate face area} = (\text{Flow area})(2 \text{ faces})(1 - \text{Open-area ratio})$$
$$= (0.004)(2)(1 - 0.4)$$
$$= 0.0048 \text{ m}^2$$

$$\text{Hole interior area} = (\text{Number of holes})(\pi dt)$$
$$= \left[\frac{A_c \varepsilon_p}{(\pi/4)d^2}\right]\pi dt = \frac{4A_c \varepsilon_p t}{d}$$
$$= \frac{(4)(0.004)(0.4)(0.0005)}{0.001} = 0.0032 \text{ m}^2$$

$$\text{Area/plate} = 0.0048 + 0.0032 = 0.008 \text{ m}^2$$

The perimeter is the heat transfer area per unit length; since the plate spacing is 1 mm, the pitch is 1.5 mm and $\mathscr{P} = 0.008/0.0015 = 5.33$ m.

2. Alternatively, the specific surface area a is often quoted for packed bed exchangers: $a = \mathscr{P}/A_c = (5.33)/(0.004) = 1333$ m^{-1}.

3. Use CONV to check ΔP and h_c.

4.6   ROTATING SURFACES

The design of cooling systems for rotating machinery, such as turbines and electric motors, as well as the design of rotating heat exchangers, high-speed gas bearings, and a variety of other equipment, requires data for convective heat transfer in rotating systems. Many complex configurations are encountered, and there is an extensive literature on the subject. In this section, correlations are presented only for some very simple configurations.

4.6.1 Rotating Disks, Spheres, and Cylinders

Figure 4.47 shows a disk, sphere, and cylinder rotating in an infinite quiescent fluid. For the disk, there is a transition from laminar to turbulent flow at a Reynolds number $\text{Re}_r = \Omega r_{\text{tr}}^2 / \nu \simeq 2.4 \times 10^5$, where r_{tr} is the radius at which transition occurs and, $\Omega[\text{s}^{-1}]$ is the angular velocity. In the laminar region, the local Nusselt number at radius r is given by Edwards et al. [5] as

$$\text{Nu}_r = \frac{0.585 \text{Re}_r^{1/2}}{0.6/\text{Pr} + 0.95/\text{Pr}^{1/3}} \tag{4.134}$$

and is valid for all values of Prandtl number. Notice that the resulting heat transfer coefficient is independent of r. For the turbulent region of the disk (if present), the local Nusselt number based on data of Cobb and Saunders [43] is

$$\text{Nu}_r = 0.021 \text{Re}_r^{0.8} \text{Pr}^{1/3}; \quad \text{Re}_r \gtrsim 2.4 \times 10^5; \quad \text{Pr} > 0.5 \tag{4.135}$$

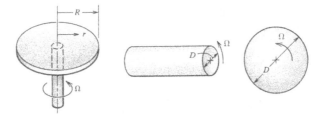

Figure 4.47   Schematic of a rotating disk, cylinder, and sphere.

The correlation given for air has been extrapolated to other fluids by including an ad hoc Prandtl number correction. The average Nusselt number can be found by integration in a similar way as for the flat plate in Section 4.3.2.

For spheres and cylinders, the behavior of the local Nusselt number is complicated; therefore, only correlations for the average value $\overline{\mathrm{Nu}}_D$ will be given. If the Reynolds number is defined as $\mathrm{Re}_D = \Omega D^2/\nu$, then for a sphere in fluids of $\mathrm{Pr} > 0.7$, Kreith et al. [44] recommend

$$\overline{\mathrm{Nu}}_D = 0.43\mathrm{Re}_D^{0.5}\mathrm{Pr}^{0.4}; \quad 10^2 < \mathrm{Re}_D < 5 \times 10^5 \qquad (4.136a)$$

$$\overline{\mathrm{Nu}}_D = 0.066\mathrm{Re}_D^{0.67}\mathrm{Pr}^{0.4}; \quad 5 \times 10^5 < \mathrm{Re}_D < 7 \times 10^6 \qquad (4.136b)$$

For a cylinder,

$$\overline{\mathrm{Nu}}_D = 0.133\mathrm{Re}_D^{2/3}\mathrm{Pr}^{1/3}; \quad \mathrm{Re}_D < 4.3 \times 10^5, \quad 0.7 < \mathrm{Pr} < 670 \qquad (4.137)$$

Mixed natural- and forced-convection effects may be expected to become significant for $\mathrm{Re}_D < 4.7(\mathrm{Gr}_D^3/\mathrm{Pr})^{0.137}$.

EXAMPLE 4.15   Heat Loss from a Centrifuge

The coverplate of a thermostatically controlled centrifuge is a horizontal disk 40 cm in diameter. The centrifuge rotates at 18,000 revolutions per minute. What is the convective heat loss from the coverplate when it is at 305 K and the ambient air is at 295 K?

Solution

Given: Heated horizontal disk rotating at 18,000 rpm.

Required: Convective heat transfer to ambient air.

Assumptions: Still ambient air.

Evaluate fluid properties at a mean film temperature of 300 K, $k = 0.0267$ W/m K, $\nu = 15.66 \times 10^{-6}$ m^2/s, $\mathrm{Pr} = 0.69$. The angular velocity is

$$\Omega = (2\pi)(18{,}000/60) = 1885 \text{ s}^{-1}$$

If the transition Reynolds number is taken as 2.4×10^5, the radius at which transition occurs is

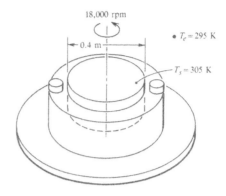

$$r_{tr} = \left(\frac{2.4 \times 10^5 \nu}{\Omega} \right)^{1/2}$$

$$= \left(\frac{(2.4 \times 10^5)(15.66 \times 10^{-6})}{1885} \right)^{1/2}$$

$$= 0.0447 \text{ m}$$

In the laminar region, the heat transfer coefficient is obtained from Eq. (4.134) and is independent of r.

$$\frac{h_c r}{k} = \frac{0.585(\Omega r^2/\nu)^{1/2}}{0.6/\text{Pr} + 0.95/\text{Pr}^{1/3}}$$

$$h_c = \frac{(0.585)k(\Omega/\nu)^{1/2}}{0.6/\text{Pr} + 0.95/\text{Pr}^{1/3}} = \frac{(0.585)(0.0267)(1885/15.66 \times 10^{-6})^{1/2}}{0.6/0.69 + 0.95/(0.69)^{1/3}} = 88.1 \text{ W/m}^2 \text{ K}$$

$$\dot{Q}_{\text{laminar}} = h_c A \Delta T = (88.1)(\pi)(0.0447)^2(10) = 5.53 \text{ W}$$

In the turbulent region, Eq. (4.135) applies:

$$\frac{h_c r}{k} = 0.021 \left(\frac{\Omega r^2}{\nu} \right)^{0.8} \text{Pr}^{1/3}$$

$$h_c = 0.021 k (\Omega/\nu)^{0.8} \text{Pr}^{1/3} r^{0.6}$$

$$\dot{Q}_{\text{turbulent}} = \int_{r_{tr}}^{R} h_c \Delta T 2\pi r \, dr$$

$$= 0.021 k (\Omega/\nu)^{0.8} \text{Pr}^{1/3} \Delta T (2\pi) \int_{r_{tr}}^{R} r^{1.6} \, dr$$

$$= 0.021 k (\Omega/\nu)^{0.8} \text{Pr}^{1/3} \Delta T (2\pi/2.6)(R^{2.6} - r_{tr}^{2.6})$$

$$= (0.021)(0.0267)(1885/15.66 \times 10^{-6})^{0.8}(0.69)^{1/3}(10)(2\pi/2.6)(0.2^{2.6} - 0.0447^{2.6})$$

$$= 520 \text{ W}$$

$$\dot{Q}_{\text{total}} = \dot{Q}_{\text{laminar}} + \dot{Q}_{\text{turbulent}} = 5 + 520 = 525 \text{ W}$$

Comments

1. An upper limit on the radiative loss is for a black surface,

$$\dot{Q}_{\text{rad}} = h_r A \Delta T \simeq (6)(\pi/4)(0.4)^2(10) = 7.5 \text{ W}$$

which is 1.5% of the convective loss.

2. Check using CONV and $\dot{Q} = \bar{h}_c A \Delta T$.

EXAMPLE 4.16   Heat Loss from a Shaft

A 3 cm-diameter shaft rotates at 15,000 rpm in air at 1 atm pressure and 300 K. If the shaft temperature is estimated to be 420 K, calculate the heat loss per unit length.

Solution

Given: A hot cylinder rotating in air.

Required: Heat loss per unit length.

Assumptions: 1. No end effects.
2. The shaft temperature is uniform.

Evaluate fluid properties at the mean film temperature of 360 K; $k = 0.0306$ W/m K, $\nu = 21.30 \times 10^{-6}$ m^2/s, Pr $= 0.69$. The angular velocity is

$$\Omega = (15{,}000/60)(2\pi) = 1571 \text{ s}^{-1}$$

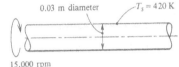

Hence, the Reynolds number is

$$\text{Re}_D = \frac{\Omega D^2}{\nu} = \frac{(1571)(0.03)^2}{21.3 \times 10^{-6}} = 6.64 \times 10^4$$

Equation (4.137) applies:

$$\overline{\text{Nu}}_D = 0.133\text{Re}_D^{2/3}\text{Pr}^{1/3} = (0.133)(6.64 \times 10^4)^{2/3}(0.69)^{1/3} = 193$$

$$\overline{h}_c = (k/D)\overline{\text{Nu}}_D = (0.0306/0.03)(193) = 197 \text{ W/m}^2 \text{ K}$$

$$\dot{Q} = \overline{h}_c A \Delta T = (197)(\pi)(0.03)(1)(420 - 300) = 2.23 \text{ kW}$$

Comments

1. Check to see if free convection is negligible:

$$\text{Gr}_D = \frac{\beta \Delta T g D^3}{\nu^2} = \frac{(120/360)(9.81)(0.03)^3}{(21.3 \times 10^{-6})^2} = 1.946 \times 10^5$$

$$4.7\left(\frac{\text{Gr}_D^3}{\text{Pr}}\right)^{0.137} = 4.7\frac{(1.946 \times 10^5)^3}{(0.69)^{0.137}} = 738 < \text{Re}_D$$

Free convection can be ignored.

2. Use CONV to check h_c.

4.7   ROUGH SURFACES

All the correlations presented so far are for smooth surfaces. In practice, surfaces are often rough, perhaps due to poor surface finish, corrosion, or deposits. Sometimes surfaces are roughened on purpose to increase the heat transfer in a turbulent flow. For example, transverse ribs may be provided on the outside of the fuel element cladding in a gas-cooled nuclear reactor. Figure 4.48 shows two applications of transverse ribs. Such surfaces are called **enhanced surfaces**.

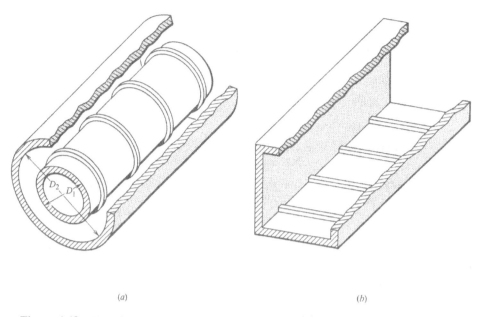

(a)   *(b)*

Figure 4.48   Use of transverse ribs to augment convective heat transfer.
(*a*) Flow in an annulus with the inner surface heated (e.g., in a gas cooled
nuclear reactor core), (*b*) Flow in a square channel with heating on one side
only (e.g., in a cooling panel of an actively cooled scram-jet engine inlet).

4.7.1 Effect of Surface Roughness

Wall roughness has no effect on skin friction or heat transfer in laminar flows, provided that the height of the roughness elements is small compared to an appropriate
length scale—for example, the pipe diameter or boundary layer thickness. However, in turbulent flows, the roughness elements need only protrude outside the thin
viscous sublayer to have a significant effect on skin friction and heat transfer. For
example, if water at 300 K flows through a 3 cm-diameter pipe at 5 m/s, the thickness of the viscous sublayer is only about 20 μm. The effect of roughness must
be accounted for if the mean roughness height significantly exceeds this value. The
effect of roughness is to increase skin friction and heat transfer, *but the effect on
skin friction is always greater*, except for high-Pr fluids at low Re. Reynolds-type
analogies between friction and heat transfer [e.g., Eq. (4.53)] are invalid for rough
walls. The increase in skin friction is due to form drag on the roughness elements:
vortex shedding causes a higher pressure on the front than on the rear of the elements.
But there is no heat transfer analog to form drag, and on a rough surface heat must
still be transferred by molecular conduction through small viscous sublayers on and
between the roughness elements.

When surfaces are intentionally roughened to increase heat transfer rates, the
design needs to be carefully optimized, since the pumping power required is also
increased.

Skin Friction Correlations

Figure 4.49 shows an adaptation of the widely used **Moody chart**, which gives the friction factor for flow in rough pipes. The characteristic velocity V is the *bulk velocity* u_b; for a constant density flow, $u_b = \dot{m}/\rho A_c = G/\rho$. In the *fully rough* regime, the friction factor is independent of Reynolds number $\mathrm{Re}_D = u_b D/\nu$, that is, independent of viscosity, since the viscous drag is negligible compared to form drag. Between the *hydrodynamically smooth* regime and the fully rough regime is the *transitionally rough* regime. The pioneering experimental work was done by J. Nikuradse in 1933 [46], who used a surface uniformly coated with closely packed sand grains of mean diameter k_s. As a consequence, other roughness patterns are often characterized in terms of an **equivalent sand grain roughness** k_s, which gives the same friction factor in the fully rough regime. Table 4.8 gives values of equivalent sand grain roughness for selected rough surfaces, and these values can be used to calculate the parameter k_s/D of the Moody chart. Whereas use of the equivalent sand grain roughness ensures that the correct friction factor is obtained in the fully rough regime, the values of f in the transitionally rough regime given by the Moody chart are applicable only to commercial surfaces, for which there is a wide range of protuberance sizes. Regular roughness patterns, such as Nikuradse's sand grain roughness, exhibit a different behavior in the transitionally rough regime.

Zigrang and Sylvester [49] recommend an explicit formula that can be used in place of the Moody chart:[5]

$$f = \left\{ -2.0 \log \left[\frac{(k_s/R)}{7.4} - \frac{5.02}{\mathrm{Re}_D} \log \left(\frac{k_s/R}{7.4} + \frac{13}{\mathrm{Re}_D} \right) \right] \right\}^{-2} \tag{4.138}$$

which has as its asymptote in the fully rough regime Nikuradse's famous formula [46],

$$f = [1.74 + 2.0 \log (R/k_s)]^{-2} \tag{4.139}$$

We can define a dimensionless sand grain size by the relations

$$k_s^+ = \frac{u_b k_s}{\nu} \left(\frac{f}{8} \right)^{1/2} \quad \text{for a pipe} \tag{4.140a}$$

$$= \frac{u_e k_s}{\nu} \left(\frac{C_{fx}}{2} \right)^{1/2} \quad \text{for a plate} \tag{4.140b}$$

The criteria for the flow regimes are then as follows:

$0 < k_s^+ \leq 5:$ hydrodynamically smooth

$5 < k_s^+ < 60:$ transitionally rough

$60 < k_s^+:$ fully rough

The rationale underlying Eqs. (4.140) will become evident following study of the analysis of turbulent flows.

---

[5] Note that the logarithms in Eqs. (4.138) and (4.139) are to base 10.

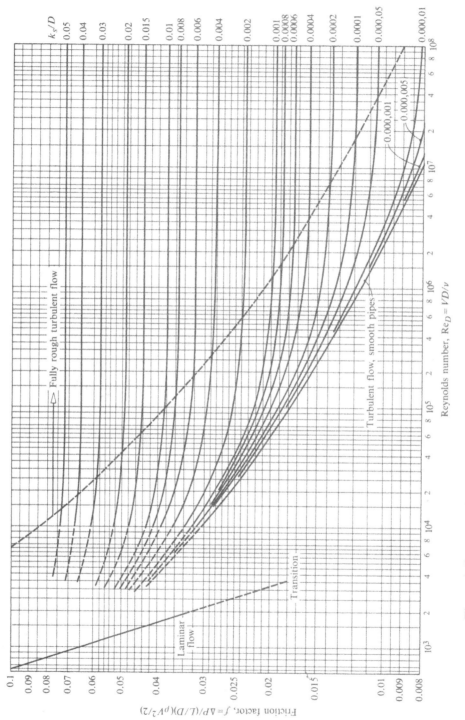

Figure 4.49 Friction factor for commercial pipes (Moody chart). The parameter k_s is the equivalent sand grain roughness: a commercial pipe with equivalent sand grain roughness k_s will have the same value of f in the fully rough regime as a pipe roughened with closely packed sand grains of average diameter k_s [45]. (Adapted with permission.)

Table 4.8   Equivalent sand grain roughness k_s for various surfaces [47,48].

Roughness Pattern	D mm	p mm	h mm	k_s mm
1. Spheres—staggered array	4.1	10	4.1	0.492
	4.1	20	4.1	1.68
	4.1	10	4.1	9.96
	4.1	6	4.1	10.6
	4.1	Densest	4.1	1.55
	2.1	10	2.1	0.903
	2.1	5	2.1	5.19
2. Spherical segments—staggered array	8.0	40	2.6	0.0468
	8.0	30	2.6	0.0884
	8.0	20	2.6	0.723
	8.0	Densest	2.6	2.48
3. Cones—staggered array	8.0	40	3.75	0.173
	8.0	30	3.75	0.458
	8.0	20	3.75	1.77

Roughness Pattern				k_s mm
4. Welded steel				
New				0.04–0.10
Uniformly rusted				0.13–0.41
Cleaned after long use				0.10–0.20
5. Commercial tubes				
Glass				0.00031
Drawn tubing				0.0015
Steel or wrought iron				0.05
Asphalted cast iron				0.12
Galvanized steel				0.1–0.3
Cast iron				0.26
Concrete				0.3–3
Riveted steel				0.9

6. Transverse square ribs

$2 < p/h < 6.3$:
$$k_s = h\exp[3.4 - 3.7(p/h)^{-0.73}]$$
$6.3 < p/h < 20$:
$$k_s = h\exp[3.4 - 0.42(p/h)^{0.46}]$$

For flow along a rough plate, Mills and Hang [50] recommend for the fully rough regime

$$C_{fx} = \left(3.476 + 0.707 \ln\frac{x}{k_s}\right)^{-2.46}, \quad 150 < \frac{x}{k_s} < 1.5 \times 10^7 \tag{4.141}$$

and

$$\overline{C}_f = \left(2.635 + 0.618 \ln\frac{L}{k_s}\right)^{-2.57}, \quad 150 < \frac{L}{k_s} < 1.5 \times 10^7 \tag{4.142}$$

when boundary layer transition occurs at the leading edge. Reliable correlations are unavailable for the transitionally rough regime.

Heat Transfer Correlations

The effect of roughness on heat transfer is very much dependent on roughness pattern. A characteristic height h is defined for the roughness patterns (e.g., the height of a rib for transverse square ribs, or the sand grain size for sand grain roughness). In the fully rough regime, heat transfer correlations adapted from the work of Dipprey and Sabersky [51,52] are

$$St = \frac{f/8}{0.9 + (f/8)^{1/2}[g(h^+, Pr) - 7.65]} \quad \text{for a pipe} \tag{4.143a}$$

$$St_x = \frac{C_{fx}/2}{0.9 + (C_{fx}/2)^{1/2}[g(h^+, Pr) - 7.65]} \quad \text{for a flat plate} \tag{4.143b}$$

where the function $g(h^+, Pr)$ is given in Table 4.9 for a variety of roughness patterns. The dimensionless characteristic height h^+ is defined similarly to the definition of

Table 4.9   The function $g(h^+, Pr)$ for use In Eqs. (4.143): heat transfer to fully rough walls [52,53].

	Roughness Pattern, h	$g(h^+, Pr)$	Prandtl Number Range
1.	Sand grain indentation: $h = k_s$, the sand grain diameter	$4.8(h^+)^{0.2}Pr^{0.44}$	$1 < Pr < 6$
2.	Transverse square ribs of height h, pitch p: $10 < p/h < 40$, $0.02 < h/R$, or $\delta < 0.08$ where R is the pipe radius and δ the boundary layer thickness $h = h$ the rib height	$4.3(h^+)^{0.28}Pr^{0.57}$	$0.7 < Pr < 40$
3.	General (hence, not too accurate): $h = $ Mean height of protrusions	$0.55(h^+)^{1/2}(Pr^{2/3} - 1) + 9.5$	$Pr > 0.5$

k_s^+ by Eqs. (4.140); alternatively, since k_s^+ is usually calculated first, one can simply use $h^+ = (k_s^+/k_s)h$. Simple Stanton number correlations are not available for the transitionally rough regime.

Variable-property effects for rough walls are not well established. The research literature contains a few correlations for specific applications.

EXAMPLE 4.17   Flow of Helium in a Roughened Tube

Helium at 5×10^5 Pa pressure flows in a 4 cm–I.D. tube at 60 m/s. Calculate the pressure gradient and heat transfer coefficient for fully developed conditions when (i) the tube wall is smooth, and (ii) the tube is artificially roughened with transverse ribs of height 0.8 mm and pitch 8 mm. Evaluate fluid properties at 500 K.

Solution

Given: Helium flow in a tube.

Required: Pressure gradient and heat transfer coefficient for (i) smooth wall and (ii) transverse rib roughness.

Assumptions: 1. Helium behaves as an ideal gas.
2. Fully developed conditions (no entrance effect).

For an ideal gas, k, c_p, and μ are independent of pressure, and the helium properties at 1 atm pressure in Table A.7 can be used to obtain properties at 5×10^5 Pa: $k = 0.205$ W/m K, $\rho = 0.481$ kg/m^3, $\nu = \mu/\rho = 58.7 \times 10^{-6}$ m^2/s, and Pr = 0.72. The Reynolds number is $\mathrm{Re}_D = u_b D/\nu = (60)(0.04)/58.7 \times 10^{-6} = 4.09 \times 10^4$.

(i) For a smooth wall, Eq. (4.42) gives the friction factor:

$$f = (0.790\ln\mathrm{Re}_D - 1.64)^{-2} = (0.790 \ln 4.09 \times 10^4 - 1.64)^{-2} = 0.0219$$

$$\frac{dP}{dx} = \left(\frac{f}{D}\right)\frac{1}{2}\rho u_b^2 = \left(\frac{0.0219}{0.04}\right)(0.5)(0.481)(60)^2 = 474 \text{ N/m}^2 \text{ per meter}$$

Equation (4.45) gives the Nusselt number:

$$\mathrm{Nu}_D = \frac{(f/8)(\mathrm{Re}_D - 1000)\mathrm{Pr}}{1 + 12.7(f/8)^{1/2}(\mathrm{Pr}^{2/3} - 1)} = \frac{(0.0219/8)(40{,}900 - 1000)(0.72)}{1 + 12.7(0.0219/8)^{1/2}(0.72^{2/3} - 1)} = 90.5$$

$$h_c = (k/D)\mathrm{Nu}_D = (0.205/0.04)(90.5) = 464 \text{ W/m}^2 \text{ K}$$

(ii) For the transverse rib roughness $h = 0.8$ mm, $p = 8$ mm, $p/h = 10$.

The equivalent sand grain size is obtained from Table 4.8, item 6:

$$k_s = h\exp[3.4 - 0.42(p/h)^{0.46}] = 0.8 \times 10^{-3} \exp[3.4 - 0.42(10)^{0.46}]$$

$$= 7.14 \times 10^{-3} \text{ m}$$

We will assume fully rough conditions and check later. Equation (4.139) gives the friction factor as

$$f = [1.74 + 2.0\log(R/k_s)]^{-2} = [1.74 + 2.0\log(0.02/0.00714)]^{-2} = 0.144$$

The dimensionless equivalent sand grain size is obtained from Eq. (4.140a):

$$k_s^+ = \left(\frac{u_b k_s}{\nu}\right)\left(\frac{f}{8}\right)^{1/2} = \frac{(60)(7.14 \times 10^{-3})}{58.7 \times 10^{-6}}\left(\frac{0.144}{8}\right)^{1/2} = 979 > 60$$

and the conditions are fully rough as assumed.

$$\frac{dP}{dx} = (f/D)(1/2)\rho u_b^2 = (0.144/0.04)(0.5)(0.481)(60)^2 = 3120 \text{ N/m}^2 \text{ per meter}$$

Equation (4.143a) gives the Stanton number, with $g(h^+, \text{Pr})$ from Table 4.9, item 2:

$$h^+ = (h/k_s)k_s^+ = (0.8/7.14)979 = 109.7$$

$$g(h^+, \text{Pr}) = 4.3(h^+)^{0.28}\text{Pr}^{0.57} = 4.3(109.7)^{0.28}(0.72)^{0.57} = 13.29$$

$$\text{St} = \frac{f/8}{0.9 + (f/8)^{1/2}[g(h^+, \text{Pr}) - 7.65]} = \frac{0.144/8}{0.9 + (0.144/8)^{1/2}[13.29 - 7.65]} = 0.0109$$

$$\text{Nu}_D = \text{StRe}_D\text{Pr} = (0.0109)(4.09 \times 10^4)(0.72) = 320$$

$$h_c = (k/D)\text{Nu}_D = (0.205/0.04)(320) = 1640 \text{ W/m}^2 \text{ K}$$

Comments

1. The effect of the ribs is to increase the pressure drop (3120/474) = 6.6 times and the heat transfer (1640/464) = 3.5 times.

2. Check using CONV.

EXAMPLE 4.18 Air Flow over a Sandblasted Flat Plate

Air at 1 atm pressure flows at 20 m/s over 2 m-long flat plate that has been sandblasted to give a surface of equivalent sand grain roughness size of 1 mm. Compare the drag force and average heat transfer coefficient with that for a smooth plate. Evaluate fluid properties at 300 K, and take $\text{Re}_{tr} = 100,000$ and 50,000 for the smooth and rough plates, respectively.

Solution

Given: Air flowing along a roughened flat plate.

Required: Comparison of drag force and $\bar{h}_c$ with smooth-plate values.

Assumptions: 1. $\text{Re}_{tr} = 100,000$ and 50,000 for the smooth and rough plates, respectively.
2. For the heat transfer calculation, the roughness pattern can be approximated as similar to a sand grain indentation pattern.

Air properties at 300 K and 1 atm are: $k = 0.0267$ W/m K, $\rho = 1.177$ kg/m^3, $\nu = 15.7 \times 10^{-6}$ m^2/s, $c_p = 1005$ J/kg K, and Pr = 0.69. The Reynolds number at the end of the plate is

$$\text{Re}_L = \frac{VL}{\nu} = \frac{(20)(2)}{15.7 \times 10^{-6}} = 2.55 \times 10^6$$

(i) *Smooth plate:*

Equation (4.62) gives the average skin friction coefficient:

$$\overline{C}_f = 1.328 \mathrm{Re}_{tr}^{-1/2}\left(\frac{\mathrm{Re}_{tr}}{\mathrm{Re}_L}\right) + \frac{0.523}{\ln^2 0.06\mathrm{Re}_L} - \left(\frac{\mathrm{Re}_{tr}}{\mathrm{Re}_L}\right)\frac{0.523}{\ln^2 0.06\mathrm{Re}_{tr}}$$

$$= 1.328(10^5)^{-1/2}\left(\frac{0.1}{2.55}\right) + \frac{0.523}{\ln^2 0.06(2.55 \times 10^6)} - \left(\frac{0.1}{2.55}\right)\frac{0.523}{\ln^3 0.06(10^5)}$$

$$= 0.00356$$

The drag force is

$$F = \overline{C}_f\left(\frac{1}{2}\rho V^2\right)WL = (0.00356)(0.5)(1.177)(20)^2(1)(2) = 1.68 \text{ N per meter width}$$

Equation (4.65) gives the average Nusselt number:

$$\overline{\mathrm{Nu}} = 0.664\mathrm{Re}_{tr}^{1/2}\mathrm{Pr}^{1/3} + 0.036\mathrm{Re}_L^{0.8}\mathrm{Pr}^{0.43}\left[1 - \left(\frac{\mathrm{Re}_{tr}}{\mathrm{Re}_L}\right)^{0.8}\right]$$

$$= (0.664)(10^5)^{1/2}(0.69)^{1/3} + 0.036(2.55 \times 10^6)^{0.8}(0.69)^{0.43}\left[1 - \left(\frac{0.1}{2.55}\right)^{0.8}\right]$$

$$= 3970$$

$$\overline{h}_c = (k/L)\overline{\mathrm{Nu}} = (0.0267/2)(3970) = 53.0 \text{ W/m}^2 \text{ K}$$

(ii) *Rough plate:*

Check the dimensionless equivalent sand grain roughness at the end of the plate. At $x = L = 2$ m, Eq. (4.141) gives

$$C_{fx} = (3.476 + 0.707 \ln L/k_s)^{-2.46} = (3.476 + 0.707 \ln 2/10^{-3})^{-2.46} = 0.00468$$

From Eq. (4.140b),

$$k_s^+ = \left(\frac{u_e k_s}{\nu}\right)\left(\frac{C_{fx}}{2}\right)^{1/2} = \frac{(20)(1 \times 10^{-3})}{15.7 \times 10^{-6}}\left(\frac{0.00468}{2}\right)^{1/2} = 61.6 > 60$$

that is, fully rough. Since C_{fx} decreases with x, the complete turbulent region will be fully rough if the end of the plate is fully rough. We can assume transition at the leading edge without incurring a significant error, because

$$\frac{\mathrm{Re}_{tr}}{\mathrm{Re}_L} = \frac{0.05 \times 10^6}{2.55 \times 10^6} = 0.0196 \simeq 2\%$$

Equation (4.142) gives the average skin friction coefficient:

$$\overline{C}_f = (2.635 + 0.618 \ln L/k_s)^{-2.57} = (2.635 + 0.618 \ln 2/10^{-3})^{-2.57} = 0.00597$$

and the drag force is

$$F = \overline{C}_f\left(\frac{1}{2}\rho V^2\right)WL = 0.00597(0.5)(1.177)(20)^2(1)(2) = 2.81 \text{ N per meter width}$$

Equation (4.143b) gives the local Stanton number:

$$St_x = \frac{C_{fx}/2}{0.9 + (C_{fx}/2)^{1/2}[g(h^+, Pr) - 7.65]}$$

Approximating the roughness pattern as item 1 in Table 4.9,

$$g(h^+, Pr) = 4.8(k_s^+)^{0.2}Pr^{0.44}, \quad \text{where } k_s^+ = \left(\frac{u_e k_s}{\nu}\right)\left(\frac{C_{fx}}{2}\right)^{1/2}$$

$$= (4.8)\left[\frac{(20)(0.001)}{15.7 \times 10^{-6}}\right]^{0.2}(0.69)^{0.44}\left(\frac{C_{fx}}{2}\right)^{0.1}$$

$$= 17.0\left(\frac{C_{fx}}{2}\right)^{0.1}$$

$$St_x = \frac{C_{fx}/2}{0.9 + 17.0(C_{fx}/2)^{0.6} - 7.65(C_{fx}/2)^{1/2}}$$

The average Stanton number is obtained by integration:

$$\overline{St} = \frac{1}{L}\int_0^L St_x dx = \frac{1}{2}\int_0^2 \frac{C_{fx}/2 \, dx}{0.9 + 17.0(C_{fx}/2)^{0.6} - 7.65(C_{fx}/2)^{1/2}}$$

Equation (4.141) gives $C_{fx}/2 = (0.5)[3.476 + 0.707\ln(x/0.001)]^{-2.46}$: substituting and integrating numerically for $L = 2$ m gives $\overline{St} = 0.00304$. The average heat transfer coefficient is

$$\overline{h}_c = \rho c_p u_e \overline{St} = (1.177)(1005)(20)(0.00304) = 71.9 \text{ W/m}^2 \text{ K}$$

Solution using CONV

The required input for the rough wall case is:

> Configuration number = 25 (fully rough flat plate)
> Fluid number = 21 (air)
> $T_s = 300$
> $T_e = 300$
> $P = 1.013 \times 10^5$
> $L = 2$
> $k_s = 0.001$
> Roughness pattern = 1 (sand grain indentation)
> $u_e = 20$

The output is:

> Air properties at 300 K, 1.013×10^5 Pa
> $Re_L = 2.56 \times 10^6$
> $k_s^+ = 61.8$
> $C_{fL} = 4.68 \times 10^{-3}$
> $\overline{C}_f = 5.97 \times 10^{-3}$

$$\tau_{sL} = 1.103 \text{ N/m}^2$$
$$\overline{\tau}_s = 1.41 \text{ N/m}^2$$
$$\text{St}_L = 2.39x10^{-3}$$
$$\overline{\text{St}} = 2.97 \times 10^{-3}$$
$$\text{Nu}_L = 4210$$
$$\overline{\text{Nu}} = 5230$$
$$h_{cL} = 56.3 \text{ W/m}^2 \text{ K}$$
$$\overline{h}_c = 69.8 \text{ W/m}^2 \text{ K}$$

Comments

1. The effect of the roughness is to increase the drag force $(2.81/1.68) = 1.67$ times and the heat transfer $(71.9/53.0) = 1.36$ times.

2. Since numerical integration is required, the use of a computer is indicated. CONV performs this integration.

3. Note that in CONV we put $T_e = T_s = 300$ K, the given temperature, for evaluation of fluid properties.

4.8   THE COMPUTER PROGRAM CONV

CONV calculates the heat transfer coefficient (and the pressure gradient or wall shear stress, where appropriate) for the 25 flow configurations listed in Table 4.10. CONV has a menu of 5 gases and 10 dielectric liquids, with thermophysical properties based on the data in Tables A.7, A.8, A.10, and A.13a of Appendix A. Liquid metals are not included because of the need to use special correlations for most configurations. Properties are evaluated according to the rules described in Section 4.2.4. Configurations 1 through 24 require straightforward evaluation of algebraic correlation formulas. Configuration 25 requires numerical integration of the local Stanton number to obtain the average value, as described in Example 4.18.

4.9   CLOSURE

The objective of this chapter was to develop expertise in the calculation of heat transfer coefficients and, where appropriate, in the calculation of pressure drop or skin friction as well. Flows of concern generally can be classified in three ways: (1) forced versus natural flow, (2) internal versus external flows, and (3) laminar versus turbulent flows. Relevant parameters, rules for evaluation of fluid properties, and even the definition of the heat transfer coefficient may differ from case to case. Dimensional analysis based on the Buckingham pi theorem was used to obtain the dimensionless groups pertinent to a given flow. The student should now be thoroughly familiar with dimensionless groups such as the Nusselt, Stanton, Prandtl, Reynolds, Grashof, and Rayleigh numbers, as well as those that were used less frequently. The pi theorem and method of indices have served their purpose, which was to efficiently introduce the dimensionless groups used to correlate experimental and numerical convection data. A wide range of flows were discussed and appropriate correlations presented. The computer code CONV should have proven to be a most useful tool for executing the required calculations.

The correlations presented in Chapter 4 are but a small fraction of those that can be found in the heat transfer literature. The flow configurations were chosen to include those most commonly encountered in thermal design, and also to illustrate the wide range of possibilities. For many of the cases considered, the literature contains many alternative correlations; those presented here often reflect a compromise between simplicity and accuracy. The question of accuracy is a particularly vexing one. Very few statements concerning the accuracy of a particular correlation have been made in Chapter 4, because it is almost impossible to do so unambiguously. This point requires further elaboration.

Sometimes a correlation is based on data obtained from a single test rig. The engineer will usually report an accuracy reflecting the scatter in the data, such as a standard deviation or equivalent measure. But this accuracy reflects only the random error in the experiment. *Systematic* error due to various causes, such as faulty instrumentation or measurement techniques, is not reflected in this reported accuracy. More importantly, the test rig will usually match the primary variables of the model flow only; there are invariably secondary or "nuisance" variables present as well. For example, consider the wind-tunnel test of heat transfer from a cylinder described in Section 4.2.3. Unless the tunnel working section is sufficiently high, there will be blockage effects due to the flow being constrained as it accelerates past the cylinder. Also, unless the tunnel is sufficiently wide, there will be significant effects of vortices generated by the interaction of the boundary layer on the side walls with the cylinder. Time (and money) is seldom invested to carefully explore the effects of such secondary variables. The point is that the effects of secondary variables might be quite different in the particular engineering problem under consideration, and these effects are often not accounted for in the engineer's statement of accuracy.

In contrast, some correlations are based on data obtained by many different engineers using a great variety of test rigs. The stated possible error for such correlations is usually relatively large because of the effects of secondary variables, and often it is too conservative. First, there is the question of whether the data have been thoroughly screened to eliminate data of inferior quality. Second, there may be a particular data set obtained under conditions that match the problem of concern very closely and therefore should be more appropriate than the general correlation.

The bottom line is that the correlations in Chapter 4 should be adequate for most thermal design purposes. If accuracy is of particular concern, for example, in a research project, then the original literature should be studied carefully to obtain a meaningful accuracy estimate.

Table 4.10 The correlations contained in CONV. Evaluate all properties at the mean film temperature unless otherwise specified. All the heat transfer correlations are for isothermal walls. However, (1) item 1 can also be used for a uniform wall heat flux, and (2) values of Nu for external flows can also be used for a uniform wall heat flux, provided Eq. (4.82) is used to define the average heat transfer coefficient.

Item No	Configuration	Correlations	Comments
1	Turbulent flow in smooth ducts with fully developed hydrodynamics and heat transfer	$f = (0.790 \ln Re_{D_h} - 1.64)^{-2}$ $10^4 < Re_{D_h} < 5 \times 10^6$ **(4.42)** $Nu_{D_h} = \dfrac{(f/8)(Re_{D_h} - 1000)Pr}{1 + 12.7(f/8)^{1/2}(Pr^{2/3} - 1)}$; $3000 < Re_{D_h} < 10^6$ **(4.45)** $0.5 < Pr$	$D_h = \dfrac{4A_c}{\mathcal{P}}$ (= D for a circular tube) Exponents for property and temperature ratio corrections for duct flows (subscripts s and b refer to wall and bulk values, respectively): (see sub-table below)
2	Laminar flow in a pipe with fully developed hydrodynamics	$f = \dfrac{64}{Re_D}$; $Re_D < 2300$ **(4.39)** $\overline{Nu}_D = 3.66 + \dfrac{0.065(D/L)Re_D Pr}{1 + 0.04[(D/L)Re_D Pr]^{2/3}}$; $Re_D < 2300$ **(4.50)**	
3	Laminar flow between parallel plates with fully developed hydrodynamics	$f = \dfrac{96}{Re_{D_h}}$; $Re_{D_h} < 2800$ **(Table 4.5)** $\overline{Nu}_{D_h} = 7.54 + \dfrac{0.03(D_h/L)Re_{D_h} Pr}{1 + 0.016[(D_h/L)Re_{D_h} Pr]^{2/3}}$; $Re_{D_h} < 2800$ **(4.51)**	$D_h = \dfrac{4A_c}{\mathcal{P}} = 2 \times$ Plate spacing
4	Laminar boundary layer on a flat plate	$\overline{C}_f = 1.328 Re_L^{-1/2}$; $10^3 < Re_L \lesssim 5 \times 10^5$ **(4.55)** $\overline{Nu} = 0.664 Re_L^{1/2} Pr^{1/3}$; $10^3 < Re_L \lesssim 5 \times 10^5$, $Pr > 0.5$ **(4.57)**	
5	Turbulent boundary layer on a smooth flat plate	$\overline{C}_f = 1.328 Re_{tr}^{-1/2}\left(\dfrac{Re_{tr}}{Re_L}\right) + \dfrac{0.523}{(\ln 0.06 Re_L)^2}\left(1 - \dfrac{Re_{tr}}{Re_L}\right)$ **(4.62)** $Re_{tr} < Re_L < 10^9$ $\overline{Nu} = 0.664 Re_{tr}^{1/2} Pr^{1/3} + 0.036 Re_L^{0.8} Pr^{0.43}[1 - (Re_{tr}/Re_L)^{0.8}]$ **(4.65)** $Re_{tr} < Re_L < 3 \times 10^7$, $0.7 < Pr < 400$	$Re_{tr} = 50,000 - 500,000$. Lower values are characteristic of practical situations where disturbing factors such as roughness and vibration are present

Sub-table for item 1 Comments:

Type of Flow	Fluid	Wall Condition	f m	Nu n
Laminar	Liquids (μ_s/μ_b)	Heating	0.58	−0.11
		Cooling	0.50	−0.11
	Gases (T_s/T_b)	Heating	1	0
		Cooling	1	0
Turbulent	Liquids (μ_s/μ_b)	Heating	0.25	−0.11
		Cooling	0.25	−0.25
	Gases (T_s/T_b)	Heating	−0.2	−0.55
		Cooling	−0.1	0.0

Table 4.10 (*Continued*)

6	Flow across a cylinder	$C_D = 1 + \dfrac{10}{Re_D^{2/3}};$ $\quad 1 < Re_D < 10^4$ (4.69)	$C_D =$ Drag force/$(1/2\rho V^2 A_f)$
		$Nu_D = 1.15 Re_D^{1/2} Pr^{1/3};$ $\quad Pr > 0.5$ (4.70)	Stagnation line local Nusselt number
		$\overline{Nu}_D = 0.3 + \dfrac{0.62 Re_D^{1/2} Pr^{1/3}}{[1+(0.4/Pr)^{2/3}]^{1/4}};$ $\quad Re_D < 10^4,$ $\quad Pr > 0.5$ (4.71a)	
		$= 0.3 + \dfrac{0.62 Re_D^{1/2} Pr^{1/3}}{[1+(0.4/Pr)^{2/3}]^{1/4}}\left[1+\left(\dfrac{Re_D}{282{,}000}\right)^{1/2}\right];$ $\quad 2\times10^4 < Re_D < 4\times10^5$ (4.71a)	
		$= 0.3 + \dfrac{0.62 Re_D^{1/2} Pr^{1/3}}{[1+(0.4/Pr)^{2/3}]^{1/4}}\left[1+\left(\dfrac{Re_D}{282{,}000}\right)^{5/8}\right]^{4/5};$ $\quad 4\times10^5 < Re_D < 5\times10^6$ (4.71c)	
		$\overline{Nu}_D = \dfrac{1}{0.8237 - \ln(Re_D Pr)^{1/2}};$ $\quad Re_D Pr < 0.2$ (4.72)	
7	Flow across a sphere	$C_D = \dfrac{24}{Re_D};$ $\quad Re_D < 0.5$ (4.73)	
		$C_D \simeq \dfrac{24}{Re_D}\left(1+\dfrac{Re_D^{2/3}}{6}\right);$ $\quad 2 < Re_D < 500$ (4.74)	
		$C_D \simeq 0.44;$ $\quad 500 < Re_D < 2\times10^5$ (**Fig. 4.28**)	
		$Nu_D = 1.32 Re_D^{1/2} Pr^{1/3};$ $\quad Pr > 0.5$ (4.75)	Stagnation point local Nusselt number
		$\overline{Nu}_D = 2 + (0.4 Re_D^{1/2} + 0.06 Re_D^{2/3}) Pr^{0.4};$ $\quad 3.5 < Re_D < 8\times10^4;$ $\quad 0.7 < Pr < 380$ (4.76)	Use mean film temperature; or better results can be obtained using a viscosity ratio correction with $n = -1/4$ applied to the convection contribution

8	Laminar natural-convection boundary layer on a vertical wall	$\overline{Nu}_L = 0.68 + 0.670(Ra_L\Psi)^{1/4}; \quad Ra_L < 10^9$ $$\Psi = \left[1 + \left(\frac{0.492}{Pr}\right)^{9/16}\right]^{-16/9}$$	(4.85)	
9	Turbulent natural-convection boundary layer on a vertical wall	$\overline{Nu}_L = 0.68 + 0.670(Ra_L\Psi)^{1/4}(1 + 1.6 \times 10^{-8}Ra_L\Psi)^{1/12}$ $10^9 < Ra_L < 10^{12}$	(4.86)	Ψ defined in item 8
10	Natural convection on a horizontal cylinder	$\overline{Nu}_D = 0.36 + \dfrac{0.518Ra_D^{1/4}}{[1 + (0.559/Pr)^{9/16}]^{4/9}}; \quad 10^{-4} < Ra_D \simeq 10^9$ $\overline{Nu}_D = \left\{0.60 + 0.387\left[\dfrac{Ra_D}{[1+(0.559/Pr)^{9/16}]^{16/9}}\right]^{1/6}\right\}^2; \quad Ra_D \gtrsim 10^9$	(4.87) (4.88)	
11	Natural convection on a sphere	$\overline{Nu}_D = 2 + \dfrac{0.589Ra_D^{1/4}}{[1 + (0.469/Pr)^{9/16}]^{4/9}}; \quad Ra_D \lesssim 10^{11}; \quad Pr > 0.5$	(4.89)	
12	Natural convection on a heated horizontal plate facing down, or a cooled plate facing up	$\overline{Nu}_L = 6.5\left[1 + 0.38\dfrac{L}{W}\right]\left[((1+X)^{0.39} - X^{0.39}]Ra_L^{0.13}\right.$ $X = 13.5Ra_L^{-0.16} + 2.2\left(\dfrac{L_a}{L}\right)^{0.7}$ $10^6 < Ra_L < 10^{10}; \quad 0.7 < Pr < 4800; \quad 0 < L_a/L < 0.2$	(4.90)	W is the length of the longer side, L is the length of the shorter side, L_a is the length of adiabatic extensions W is the length of the longer side, L is the length of the shorter side, L_a is the length of adiabatic extensions
13	Natural convection on a heated horizontal plate facing up, or a cooled plate facing down	$\overline{Nu}_L = 0.54Ra_L^{1/4}; \quad 10^5 < Ra_L < 2 \times 10^7$ $\overline{Nu}_L = 0.14Ra_L^{1/3}; \quad 2 \times 10^7 < Ra_L < 3 \times 10^{10}$	(4.91) (4.92)	L is the length of the shorter side

(Continued)

Table 4.10 *(Continued)*

14	Natural convection across thin enclosures ($H/L > 10$)	$0 \le \theta < 60°$		All properties to be evaluated at $T_r = (1/2)(T_H + T_c)$
		$\overline{Nu}_L = 1 + 1.44\left[1 - \dfrac{1708}{Ra_L\cos\theta}\right]\left\{1 - \dfrac{1708(\sin 1.8\theta)^{1.6}}{Ra_L\cos\theta}\right\}$		
		$\quad + \left[\left(\dfrac{Ra_L\cos\theta}{5830}\right)^{1/3} - 1\right]$; $\quad 0 < Ra_L < 10^5$	(4.98)	If a term in square brackets is negative, set equal to zero
		$\theta = 60°$		Strictly speaking, these correlations are for gases only; however, Eq. (4.98) can also be used for liquids of moderate Prandtl number
		$\overline{Nu}_{L60°} = \max\{Nu_1, Nu_2\};\quad 0 < Ra_L < 10^7$	(4.99)	
		$Nu_1 = \left\{1 + \left[\dfrac{0.0963 Ra_L^{0.314}}{1 + \frac{0.5}{[1+(Ra_L/3160)^{20.6}]^{0.1}}}\right]^7\right\}^{1/7}$		
		$Nu_2 = \left(0.104 + \dfrac{0.175}{H/L}\right) Ra_L^{0.283}$		
		$60° < \theta < 90°$		
		$\overline{Nu}_L = \left(\dfrac{90 - \theta}{30}\right)\overline{Nu}_{L60°} + \left(\dfrac{\theta - 60}{30}\right)\overline{Nu}_{L90°}$	(4.100)	
		$\theta = 90°$		
		$\overline{Nu}_{L90°} = \max\{Nu_1, Nu_2, Nu_3\};\quad 10^3 < Ra_L < 10^7$	(4.101)	For $Ra_L \le 10^3$, $\overline{Nu}_{L90°} = 1$
		$Nu_1 = 0.0605 Ra_L^{1/3}$		
		$Nu_2 = \left\{1 + \left[\dfrac{0.104 Ra_L^{0.293}}{1 + (6310/Ra_L)^{1.36}}\right]^3\right\}^{1/3}$		
		$Nu_3 = 0.242\left(\dfrac{Ra_L}{H/L}\right)^{0.272}$		

15	Natural convection across vertical cavities with insulated horizontal surfaces, $1 < H/L < 10$	$2 < H/L < 10:$ $$\overline{Nu}_L = 0.22 \left(\frac{Pr}{0.2+Pr} Ra_L\right)^{0.28} \left(\frac{H}{L}\right)^{-1/4}; \quad Ra_L < 10^{10}$$ (4.103a) $1 < H/L < 2:$ $$\overline{Nu}_L = 0.18 \left(\frac{Pr}{0.2+Pr} Ra_L\right)^{0.29}; \quad 10^3 < \frac{Pr}{0.2+Pr} Ra_L$$ (4.103b)	
16	Natural convection between concentric cylinders	$$\frac{k_{eff}}{k} = 0.386 \left(\frac{Pr}{0.861+Pr}\right)^{1/4} Ra_{cyl}^{1/4}; \quad 10^2 < Ra_{cyl} < 10^7$$ (4.104) $$Ra_{cyl} = \frac{[\ln(D_o/D_i)]^4}{L^3 \left(D_i^{-3/5} + D_o^{-3/5}\right)^5} Ra_L; \quad L = (D_o - D_i)/2$$	k_{eff} is the effective thermal conductivitfy for use in Eq. (2.14) Valid only for $k_{eff}/k > 1$
17	Natural convection between concentric spheres	$$\frac{k_{eff}}{k} = 0.74 \left(\frac{Pr}{0.861+Pr}\right)^{1/4} Ra_{sph}^{1/4}; \quad 10^2 < Ra_{sph} < 10^4$$ (4.105) $$Ra_{sph} = \frac{L}{(D_o D_i)^4} \frac{Ra_L}{\left(D_i^{-7/5} + D_o^{-7/5}\right)^5}; \quad L = (D_o - D_i)/2$$	k_{eff} is the effective thermal conductivity for use in Eq. (2.25a) Valid only for $k_{eff}/k > 1$
18	Flow through tube banks	$$\overline{Nu}_D^{10+} = \Phi \overline{Nu}_D^1; \quad N \geq 10$$ (4.106) $$\overline{Nu}_D = \frac{1 + (N-1)\Phi}{N} \overline{Nu}_D^1; \quad N < 10$$ (4.107) $$\Delta P = N \chi \left(\frac{\rho V_{max}^2}{2}\right) f$$ (4.108)	$\overline{Nu}_D^1 = \overline{Nu}_D$ from Item 6 with Re_D based on the average velocity in the space between two adjacent tubes (Eq. 4.113) Φ from Eqs. (4.116) and (4.117) V_{max} is the maximum velocity in the tube bank; Re_D for f based on V_{max}: f and χ from Fig.4.44a,b

(Continued)

Table 4.10 (*Continued*)

19	Flow through packed beds	$\dfrac{dP}{dx} = \dfrac{150\mu\mathcal{V}}{\mathcal{L}^2} + \dfrac{1.75\rho\mathcal{V}^2}{\mathcal{L}}$; $1 < Re < 10^4$	**(4.130)**	$d_p = 6\dfrac{V_p}{A_p}$; $\mathcal{L} = d_p\left(\dfrac{\varepsilon_v}{1-\varepsilon_v}\right)$
		$Nu = (0.5Re^{1/2} + 0.2Re^{2/3})Pr^{1/3}$; $20 < Re < 10^4$ $0.5 < Pr < 20$	**(4.131)**	$\mathcal{V} = \dfrac{\dot{m}}{\rho\varepsilon_v A_c}$ Viscosity ratio with $n = -0.14$ Specific surface area $a = \dfrac{A_p}{V_p}(1 - \varepsilon_v)$
20	Flow through perforated plates	$Eu = 8.17Re^{-0.55}(1.707 - \varepsilon_p)^2$; $20 < Re < 150$	**(4.132a)**	$Re = Gd/p$, where G is the mass velocity through the holes of diameter d
		$= 0.5(1.707 - \varepsilon_p)^2$; $150 < Re < 3000$	**(4.132b)**	$Eu = \Delta Pp/G^2$; ΔP for a single plate
		$St = CRe^n Pr^{-2/3}$; $300 < Re < 3000$; $Pr > 0.5$ $C = 3.6 \times 10^{-4}[(1 - \varepsilon_p)\varepsilon_p - 0.2]^{-2.07}$ $n = -4.36 \times 10^{-2}\varepsilon_p^{-2.34}$; $0.3 < \varepsilon_p < 0.6$	**(4.133)**	ε_p = Plate open-area ratio Eq. (4.133) is valid for gases and liquids of moderate Prandtl number
21	Rotating disk in a quiescent fluid	$Nu_r = \dfrac{0.585Re_r^{1/2}}{0.6/Pr + 0.95/Pr^{1/3}}$; $Re_r < 2.4 \times 10^5$	**(4.134)**	
		$Nu_r = 0.021Re_r^{0.8}Pr^{1/3}$; $Re_r > 2.4 \times 10^5$; $Pr > 0.5$	**(4.135)**	
22	Rotating sphere in a quiescent fluid	$\overline{Nu}_D = 0.43Re_D^{0.5}Pr^{0.4}$; $10^2 < Re_D < 5 \times 10^5$ $0.7 < Pr$	**(4.136a)**	
		$\overline{Nu}_D = 0.066Re_D^{0.67}Pr^{0.4}$; $5 \times 10^5 < Re_D < 7 \times 10^6$ $0.7 < Pr$	**(4.136b)**	
23	Horizontal rotating cylinder in a quiescent fluid	$\overline{Nu}_D = 0.133Re_D^{2/3}Pr^{1/3}$; $Re_D < 4.3 \times 10^5$ $0.7 < Pr < 670$	**(4.137)**	Lower limit on Re_D due to natural convection effects is $4.7(Gr_D^3/Pr)^{0.137}$

24	Turbulent flow in a rough pipe	$f = \left\{ -2.0 \log \left[\dfrac{(k_s/R)}{7.4} - \dfrac{5.02}{\mathrm{Re}_D} \log \left(\dfrac{k_s/R}{7.4} + \dfrac{13}{\mathrm{Re}_D} \right) \right] \right\}^{-2}$ (4.138) $$\mathrm{St} = \dfrac{f/8}{0.9 + (f/8)^{1/2}[g(h^+, \mathrm{Pr}) - 7.65]}; \quad k_s^+ > 60 \quad (4.143a)$$	Options for $g(h^+, \mathrm{Pr})$: **1.** Sand grain indentation: $4.8(h^+)^{0.2}\mathrm{Pr}^{0.44}; \quad 1 < \mathrm{Pr} < 6$ **2.** Rectangular transverse ribs, $10 < p/h < 40$: $4.3(h^+)^{0.28}\mathrm{Pr}^{0.57}; \quad 0.7 < \mathrm{Pr} < 40$ **3.** General: $0.55(h^+)^{1/2}(\mathrm{Pr}^{2/3} - 1) + 9.5; \mathrm{Pr} > 0.5$
25	Turbulent boundary layer on a fully rough flat plate	$$C_{fx} = \left(3.476 + 0.707 \ln \dfrac{x}{k_s} \right)^{-2.46} \quad (4.141)$$ $150 < \dfrac{x}{k_s} < 1.5 \times 10^7, \quad k_s^+ > 60$ $$\overline{C}_f = \left(2.635 + 0.618 \ln \dfrac{L}{k_s} \right)^{-2.57} \quad (4.142)$$ $150 < \dfrac{L}{k_s} < 1.5 \times 10^7, k_s^+ > 60$ $$\mathrm{St}_x = \dfrac{C_{fx}/2}{0.9 + (C_{fx}/2)^{1/2}[g(h^+, \mathrm{Pr}) - 7.65]}; \quad k_s^+ > 60 \quad (4.143b)$$	

■ REFERENCES

1. Petukhov, B. S., "Heat transfer and friction in turbulent pipe flow with variable physical properties," in *Advances in Heat Transfer,* vol. 6., eds. J. P. Hartnett and T. F. Irvine, Academic Press, New York (1970).

2. Gnielinski, V., "New equations for heat and mass transfer in turbulent pipe and channel flow," *Int. Chemical Engineering,* 16, 359–368 (1976).

3. Sparrow, E. M., and Ohadi, M. M., "Numerical and experimental studies of turbulent heat transfer in a tube," *Numerical Heat Transfer*, 11, 461–476 (1977).

4. Notter, R. H., and Sleicher, C. A., "A solution to the turbulent Graetz problem—III. Fully developed and entry region heat transfer rates," *Chem. Eng. Science,* 27, 2073–2093 (1972).

5. Edwards, D. K., Denny, V. E., and Mills, A. F., *Transfer Processes,* 2nd ed., Hemisphere, Washington, D.C. (1979).

6. Mills, A. F., "Experimental investigation of turbulent heat transfer in the entrance region of a circular conduit," *J. Mech. Eng. Sci.,* 4, 63–77 (1962).

7. Petukhov, B. S., and Roizen, L. I., "Generalized relationships for heat transfer in a turbulent flow of a gas in tubes of annular section," *High Temp.* (USSR), 2, 65–68 (1964).

8. White, F. M., *Viscous Fluid Flow,* 3rd. ed., McGraw-Hill, New York (2005).

9. White, F. M., *Heat Transfer,* Addison-Wesley, Reading, Mass. (1984).

10. Whitaker, S., "Forced convection heat transfer correlations for flow in pipes, past flat plates, single cylinders, single spheres, and for flow in packed beds and tube bundles," *AIChE Journal,* 18, 361–371 (1972).

11. Schlichting, H., and Gersten, K. *Boundary Layer Theory,* 8th revised ed., Springer-Verlag, Berlin (2000).

12. Van Meel, D. A., "A method for the determination of local convective heat transfer from a cylinder placed normal to an air stream," *Int. J. Heat Mass Transfer,* 5, 715–722 (1962).

13. Giedt, W. H., "Investigation of variation of point unit heat transfer coefficient around a cylinder normal to an air stream," *Trans. ASME,* 71, 375–381 (1949).

14. Achenbach, E., "Total and local heat transfer from a smooth circular cylinder in crossflow at high Reynolds number," *Int. J. Heat Mass Transfer,* 18, 1387–1396 (1975).

15. Kays, W. M., Crawford, M. E., and Wegand, B. *Convective Heat and Mass Transfer,* 4th ed., McGraw-Hill, New York (2004).

16. Churchill, S. W., and Bernstein, M., "A correlating equation for forced convection from gases and liquids to a circular cylinder in crossflow," *J. Heat Transfer,* 99, 300–306 (1977).

17. Nakai, S., and Okazaki, T., "Heat transfer from a horizontal circular wire at small Reynolds and Grashof numbers—I. Pure convection," *Int. J. Heat Mass Transfer,* 18, 387–396 (1975).

18. Bird, R. B., Stewart, W. E., and Lightfoot, E. N., *Transport Phenomena,* 2nd ed., John Wiley & Sons, New York (1962). [Eq. (2.6–14) rewritten in terms of drag coefficient.]

19. Reynolds, W. C, Kays, W. M., and Kline, S. J., "Heat transfer in the turbulent incompressible boundary layer, III. Arbitrary wall temperature and heat flux," NASA Memo 12-3-58 W (1958).

20. Mills, A. F, "Average Nusselt numbers for external flows," *J. Heat Transfer,* 101, 734–735 (1979).

21. Churchill, S. W, and Usagi, R., "A general expression for the correlation of rates of transfer and other phenomena," *AIChE Journal,* 18, 1121–1128 (1972).

22. Churchill, S. W., and Chu, H. H. S., "Correlating equations for laminar and turbulent free convection from a vertical plate," *Int. J. Heat Mass Transfer,* 18, 1323–1329 (1975).

23. Churchill, S. W., and Chu, H. H. S., "Correlating equations for laminar and turbulent free convection from a horizontal cylinder," *Int. J. Heat Mass Transfer,* 18, 1049–1053 (1975).

24. Yuge, T., "Experiments on heat transfer from spheres including combined natural and forced convection," *J. Heat Transfer,* 82, 214–220 (1960).

25. Churchill, S. W., "Free convection around immersed bodies," *Hemisphere Heat Exchanger Design Handbook,* ed. G. F. Hewitt, §2.5.7, Hemisphere, Washington, D.C. (1990).

26. Lienhard, J. H., "On the commonality of equations for natural convection from immersed bodies," *Int. J. Heat Mass Transfer,* 16, 2121–2123 (1973).

27. Hatfield, D. W, and Edwards, D. K., "Edge and aspect ratio effects on natural convection from the horizontal heated plate facing downwards," *Int. J. Heat Mass Transfer,* 24, 1019–1024 (1981). (Constants C_2, C_3, and C_4 have been corrected.)

28. Kadambi, V., and Drake, R. M., "Free convection heat transfer from horizontal surfaces for prescribed variations in surface temperature and mass flow through the surface," *Technical Report, Mech. Eng. HT-1,* Princeton University (1960).

29. Fugii, T., and Imura, H., "Natural convection heat transfer from a plate with arbitrary inclination," *Int. J. Heat Mass Transfer,* 15, 755–767 (1972).

30. McAdams, W. H., "Heat Transmission," 3rd ed., McGraw-Hill, New York (1954).

31. Barrow, H., and Sitharamaroa, T. L., "The effect of variable β on free convection," *Brit. Chem. Eng.,* 16, 704–705 (1971).

32. Hollands, K. G. T, Unny, T. E., Raithby, G. D., and Konicek, L., "Free convective heat transfer across inclined air layers," *J. Heat Transfer,* 98, 189–193 (1976).

33. ElSherbiny, S. M., Raithby, G. D., and Hollands, K. G. T, "Heat transfer by natural convection across vertical and inclined air layers," *J. Heat Transfer,* 104, 96–102 (1982).

34. Globe, S., and Dropkin, D., "Natural convection heat transfer in liquids confined between two horizontal plates," *J. Heat Transfer,* 81, 24–28 (1959).

35. Berkovsky, B. M., and Polevikov, V. K., "Numerical study of problems on high-intensive free convection," in *Heat Transfer and Turbulent Buoyant Convection,* eds. D. B. Spalding and N. Afgan, Hemisphere, Washington, D.C. (1977), pp. 443–455.

36. Raithby, G. D., and Hollands, K. G. T., "A general method of obtaining approximate solutions to laminar and turbulent free convection problems," in *Advances in Heat Transfer,* vol. 11, eds. J. R Hartnett and T. F. Irvine, Ir., Academic Press, New York (1975).

37. Metais, B., and Eckert, E. R. G., "Forced, mixed, and free convection regimes," *J. Heat Transfer,* 86, 295–296 (1964).

38. Churchill, S. W., "Combined free and forced convection around immersed bodies," *Hemisphere Heat Exchanger Design Handbook,* ed. G. F. Hewitt, §2.5.9., Hemisphere, Washington, D.C. (1990).

39. This text; developed from Gnielinski, V, Zukauskas, A., and Skrinska, A., "Banks of plain and finned tubes," *Hemisphere Heat Exchanger Design Handbook,* ed. G. F. Hewitt, §2.5.3., Hemisphere, Washington, D.C. (1990).

40. Zukauskas, A., and Ulinskas, R., "Efficiency parameters for heat transfer in tube banks," *Heat Transfer Engineering,* 6, No. 2, 19–25 (1985).

41. Ergun, S., "Fluid flow through packed columns," *Chem. Eng. Prog.,* 48, 89–94 (1952).

42. Shevyakova, S. A., and Orlov, V. K., "Study of hydraulic resistance and heat transfer in perforated-plate heat exchangers," *Inzhenerno-Eizicheskii Zhurnal,* 45, 32–36 (1983).

43. Cobb, E. C, and Saunders, O. A., "Heat transfer from a rotating disc," *Proc. , Roy. Soc. London,* ser. A, 236, 343–351 (1956).

44. Kreith, F, Roberts, L. G., Sullivan, J. A., and Sinha, S. N., "Convection heat

transfer and flow phenomena of rotating spheres," *Int. J. Heat Mass Transfer,* 6, 881–895 (1963).

45. Moody, L. F., "Friction factors for pipe flow," *Trans. ASME,* 66, 671–684 (1944).

46. Nikuradse, J., "Laws of flow in rough pipes," NACA TM 1292 (1950). (English translation of VDI-Forschungsheft 361, 1933.)

47. Coleman, H. W., Hodge, B. K., and Taylor, R. P., "A re-evaluation of Schlichting's surface roughness experiment," *J. Fluids Engineering,* 106, 60–65 (1984).

48. Dalle Donne, M., and Meyer, L., "Turbulent convective heat transfer from rough surfaces with two-dimensional rectangular ribs," *Int. J. Heat Mass Transfer,* 20, 583–620(1977).

49. Zigrang, D. J., and Sylvester, N. D., "Explicit approximations to the solution of Colebrook's friction factor equation," *AIChE Journal,* 28, 514–515 (1982).

50. Mills, A. F, and Xu Hang, "On the skin friction coefficient for a fully rough flat plate," *J. Fluids Engineering,* 105, 364–365 (1983).

51. Dipprey, D. F., and Sabersky, R. H., "Heat and momentum transfer in smooth and rough tubes at various Prandtl numbers," *Int. J. Heat Mass Transfer,* 6, 329–353 (1963).

52. Wassel, A. T., and Mills, A. F., "Calculation of variable property turbulent friction and heat transfer in rough pipes," *J. Heat Transfer,* 101, 469–474 (1979).

53. Yaglom, A.M., and Kader, B. A., "Heat and mass transfer between a rough wall and turbulent fluid flow at high Reynolds and Péclet numbers," *J. Fluid Mechanics,* 62, part 3, 601–623 (1974).

54. Buchberg, H., Catton, I., and Edwards, D. K., "Natural convection in enclosed spaces —A review of application to solar energy collection," *J. Heat Transfer,* 98, 182–188 (1976).

55. Sparrow, E. M., and Gregg, J. L., "Laminar free convection from a vertical plate with uniform surface heat flux," *Trans. Am. Soc. Mech. Engrs.,* 78, 435–440 (1956).

56. Jakob, M., *Heat Transfer,* vol. 1., John Wiley & Sons, New York, N.Y. (1949).

EXERCISES

4–1. A laminar flow in a 2 cm–I.D. tube has the following velocity and temperature profiles:

$$u = 0.1[1 - (r/0.01)^2]\,\text{m/s}$$

$$T = 400 - 3 \times 10^6(1.875 \times 10^{-5} - 0.25r^2 + 624r^4)\,\text{K}$$

for r in meters. Determine the bulk temperature.

4–2.   The following table gives velocity and temperature profiles for a turbulent flow of air between parallel plates 4 cm apart. The coordinate y is measured from the wall, and the profiles are symmetrical about the centerplane. Determine the bulk velocity and the bulk temperature.

y, cm	u, m/s	T, K	y, cm	u, m/s	T, K
0.0	0.0	373.0	1.2	21.4	302.6
0.2	16.3	322.4	1.4	21.8	301.0
0.4	18.2	315.2	1.6	22.1	299.8
0.6	19.4	310.7	1.8	22.2	299.1
0.8	20.2	307.3	2.0	22.3	298.8
1.0	20.9	304.7			

4–3.   For fully developed flow in a duct, show that the friction factor is four times the skin friction coefficient, $f = 4C_f$.

4–4.   A heated horizontal cylinder rotates rapidly about its axis in still air. Use dimensional analysis to obtain dimensionless groups pertinent to heat transfer.

4–5.   The central receiver of a solar power plant is in the form of a vertical cylinder, of height H and diameter D. Experimental data have been obtained for the effect of a crosswind on the convective heat loss from the receiver. Suggest appropriate dimensionless groups for correlating the data.

4–6.   Consider the packing of a perforated-plate heat exchanger for a cryogenic refrigeration system. Geometric parameters include the hole diameter d, hole pitch p (or, alternatively, the open-area ratio ε_p), the plate thickness t, and plate spacing s. Expected parameter values for a hydrogen flow include a mass velocity based on a hole cross-sectional area of 6×10^{-3} 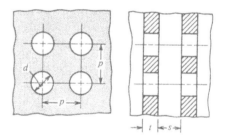 kg/m^2 s, a pressure of 100 Pa, temperatures of 10–350 K, and surface-to-gas temperature differences of 10 K. Hole diameters are in the range 1–3 mm. Suggest appropriate dimensionless groups for correlating experimental data for pressure drop and heat transfer.

4–7.   Consider a constant property, low-speed flow at velocity u_e along a heated flat plate. Use dimensional analysis to find the dimensionless groups governing the growth of the hydrodynamic boundary layer thickness δ along the plate, that , is, as a function of coordinate x measured from the leading edge. Obtain the corresponding result for the thermal boundary layer thickness Δ.

4–8.   Convective heat transfer is sometimes modeled as conduction across a thin stagnant layer of fluid adjacent to the surface. If this thickness is denoted δ_f, then Eq. (1.9) for conduction across a plane slab gives $q_s = (k/\delta_f)\Delta T$. Comparing with Eq. (4.2) or (4.9), $h_c = k/\delta_f$ or $\delta_f = k/h_c$. Calculate the equivalent stagnant film thickness for

(i) water at 300 K flowing at 5 m/s in a 2 cm–I.D. tube.

(ii) air in free convection on a vertical plate heated to 310 K in ambient air at 290 K and 1 atm, at a position 20 cm above the bottom of the plate.

Use the heat transfer coefficient formulas given in Section 1.3.3.

4–9. An experiment to study heat transfer from an elliptical cylinder was performed in a wind tunnel. Heat loss from a cylinder of major and minor axes 8 and 6 cm, respectively, to an air flow normal to the minor axis of the cylinder was measured. Heat loss from the cylinder consisted of both radiation and convection. The measured total heat loss, and a calculation of the radiation contribution for each test condition, are listed here. (The radiation calculation was based on an emittance of 0.8 for the cylinder surface.) Use the data to estimate the convective heat loss from a geometrically similar cylinder of 12 cm major axis at 30°C to a 4.5 m/s air flow at 23°C.

Test	V m/s	T_s K	T_e K	q_{rad} W/m^2	q_{total} W/m^2
1	1.37	327.5	304.3	133	585
2	2.83	309.4	301.5	42	261
3	3.54	304.7	291.6	65	553
4	6.10	306.1	301.4	23	261
5	8.63	302.8	297.9	24	403
6	8.50	311.5	305.9	32	410
7	8.81	309.2	304.4	26	402
8	8.32	299.5	292.7	35	552
9	11.5	308.2	304.5	18	466
10	11.4	307.0	301.8	26	548
11	14.8	306.3	302.4	19	546
12	17.5	302.2	298.8	16	577

4–10. A common problem in the design of electronic equipment is the need to control the temperatures of a number of heat-producing devices arranged in a symmetrical matrix. One effective method is to mount the devices on studs and to pump a liquid coolant over the studs. In a particular design there are 252 devices mounted on cylindrical studs, with generated heat fluxes as high as 5 W/cm^2 measured at the stud surface. Refrigerant-113 is a suitable coolant, having appropriate dielectric properties. To obtain design data, a model unit was made by attaching 1 cm–diameter, 4 cm–high copper studs, in a 2 cm–pitch square array, to a thin, electrically heated plate. Results of experiments are given. Suggest a correlation of the data suitable for engineering use. For R-113, use $k = 0.0747$ W/m K, $\nu = 0.426 \times 10^{-6}$ m^2/s, and Pr $= 8.52$.

Velocity V, m/s	1.0	1.6	2.4	3.1	3.6
Heat transfer coefficient h_c, W/m^2 K	1810	2430	3150	3730	4130

4–11.   Consider a 10 m length of 2 cm–I.D. tube. What is the average convective heat transfer coefficient and pressure gradient inside the tube when the tube wall is at 320 K and water enters at 300 K, 1 atm pressure, and flows at a velocity of 3 m/s?

4–12.   Consider a 10 m length of 2 cm–I.D. tube. What is the average heat transfer coefficient and pressure gradient inside the tube when the tube wall is at 320 K, and SAE 50 oil enters at 300 K, 1 atm pressure, and flows at a velocity of 3 m/s?

4–13.   Consider a 10 m length of 2 cm–I.D. tube. What is the average heat transfer coefficient and pressure gradient inside the tube when the tube wall is at 320 K, and mercury enters at 300 K, 1 atm pressure, and flows at a velocity of 3 m/s?

4–14.   Consider a 10 m length of 2 cm–I.D. tube. What is the average heat transfer coefficient and pressure gradient inside the tube when the tube wall is at 320 K, and air enters at 300 K, 1 atm pressure, and flows at a velocity of 3 m/s?

4–15.   Consider a 10 m length of 2 cm–I.D. tube. What is the average heat transfer coefficient and pressure gradient inside the tube when the tube wall is at 320 K, and helium enters at 300 K, 1 atm pressure, and flows at a velocity of 3 m/s?

4–16.   Therminol 60 is under consideration as a high-temperature coolant. Prepare tables of heat transfer coefficient and pressure gradient as a function of average temperature for flow at 3 m/s in a 5 cm–I.D. pipe. Ignore entrance effects. Take $T_s = T_b$.

4–17.   Liquid lithium flows at 5 m/s in a 1 cm–I.D. tube. Calculate the Nusselt number and heat transfer coefficient

   (i) if the wall temperature is uniform.
   (ii) if the wall heat flux is uniform.

Evaluate properties at 800 K.

4–18.   Air at 1×10^5 Pa and 300 K flows at 2 m/s through a 1 cm–I.D. pipe, 1 m long. An electric resistance heater surrounds the last 20 cm of the pipe and supplies a constant heat flux to raise the bulk temperature of the air to 330 K. What power input is required?

4–19.   SAE 50 oil flows in a 60 m-long, 25 mm-I.D. tube at 0.5 kg/s. The oil enters the tube at 300 K, and the walls of the tube are maintained at 370 K. Determine the exit bulk temperature of the oil.

4–20.   Calculate the Nusselt number and heat transfer coefficient for liquid sodium flowing at 3 m/s in a 5 cm–I.D. pipe if

   (i) the wall temperature is uniform.
   (ii) the wall heat flux is uniform.

Evaluate properties at 1000 K.

4–21. Aqueous solutions of ethylene glycol are used as "antifreeze" working fluids in many heat exchangers. For a 10 m-long, 3 cm–I.D. tube, prepare graphs of pressure drop and heat transfer coefficient for bulk velocities in the range 0.2–4.0 m/s. Use the properties in Table A.13a for 20, 30, 40, 50, and 60% glycol by mass. Assume $T_s = T_b = 270$ K.

4–22. Crude oil of specific gravity 0.862 flows at 0.2 kg/s through a 1.2 cm–I.D. horizontal tube. At a temperature of 370 K the pressure drop in a length of 3 m is 31.6 kPa. Calculate the dynamic and kinematic viscosities of the oil. Also, if the conductivity and specific heat of the oil are approximately the same as that for SAE 50 oil, estimate the convective heat transfer coefficient.

4–23. SAE 50 oil flows at 3 m/s between 0.25 m-long parallel plates spaced 3 mm apart. If the plates are at 400 K and the average of the inlet and outlet bulk temperatures is 360 K, determine the average heat transfer coefficient and pressure drop.

4–24. SAE 50 oil flows at 4 m/s through a 1 cm-square cross-section duct, 3 m long. If the oil enters at 290 K and the walls are at 330 K, determine the pressure drop, the average heat transfer coefficient, and the outlet bulk temperature.

4–25. Air flows at a rate of 0.05 kg/s through a 3 cm–I.D., 4 cm-O.D. annular duct, 9 m long. The air enters at 245 K, 10 atm pressure, and is heated by saturated steam condensing in the inner tube of the annulus at 1.1×10^5 Pa. Determine the average heat transfer coefficient, the air outlet temperature, and the pressure drop. The outer surface can be taken as insulated.

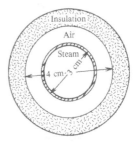

4–26. Air flows at 0.05 kg/s through a 3 cm–I.D., 4 cm-O.D. annular duct, 9 m long. The air enters at 240 K, 1 atm pressure, and is heated by saturated steam condensing in the inner tube of the annulus at 1.1×10^5 Pa. Determine the average heat transfer coefficient, the air outlet temperature, and the pressure drop. The outer surface can be taken to be insulated.

4–27. In the design of a shell-and-tube heat exchanger, the tube diameter must be chosen to satisfy a number of constraints. In particular, there is a trade-off between heat transfer and pressure drop considerations, since both increase as the tube diameter decreases. It is required to design an exchanger to heat 0.01 kg/s of helium from 100 to 300 K, and we wish to explore the effect of tube inside diameter in the range 3–30 mm for

  (i) laminar flow.
  (ii) turbulent flow.

In each case the flow cross-sectional area will be fixed, 0.007 m^2 for laminar (flow and 3×10^{-4} m^2 for turbulent flow. The nominal pressure is 10^5 Pa.

Prepare graphs of the heat transfer coefficient, the $h_c \mathscr{P}$ product, and pressure gradient as a function of tube size for each case. Assume fully developed conditions, and evaluate all properties at 200 K.

4–28. Saturated mercury vapor at 5.24 kPa pressure condenses on a 12 mm–O.D., 1.25 mm-wall-thickness copper tube coil in a small laboratory condenser. Coolant water is supplied at 0.05 kg/s from a supply at 290 K. The condenser shell is a stainless steel cylinder with stainless steel end plates, and the copper tubing enters and leaves the shell through Swagelok fittings in the end plates. What length of tube on the inlet side should be isolated to avoid skin burns on contact? For metallic surfaces, a temperature 45°C can be taken as the threshold for skin burn.

4–29. Liquid metals are attractive for high-temperature heat transfer applications owing to their characteristic high convective heat transfer coefficients. Liquid potassium flows at 4 m/s in a long, 2 cm–I.D. tube. Calculate the Nusselt number and heat transfer coefficient

  (i) for a uniform wall temperature.
  (ii) for a uniform wall heat flux.

Evaluate properties at 900 K. Repeat for liquid sodium.

4–30. Mercury flows at a rate of 60 kg/s through a 2 cm–I.D., 2.4 cm-O.D. annular duct, 2 m long. The mercury enters at 300 K, and the inner wall is maintained at 500 K by high-pressure steam condensing inside the inner tube. The heat loss from the outer surface is negligible. Determine the average heat transfer coefficient, the mercury outlet temperature, and the pressure drop.

4–31. In a heat exchanger for a hydrogen cryogenic refrigeration system, hydrogen gas flows at 2.0×10^{-3} kg/s in a 15 mm–I.D., 4 m long tube, wound in a coil. The hydrogen enters at 10^5 Pa and 320 K, and exits at 20 K. Determine the convective heat transfer coefficient at locations where the bulk hydrogen temperature is 320 K, 170 K, and 30 K. In each case, the wall temperature can be taken to be 5 K below the bulk temperature.

4–32. Aqueous ethylene glycol solution (30% glycol) flows at 3 m/s through a 2 cm-square-cross-section duct, 10 m long. If the glycol enters at 260 K and the walls are a uniform temperature of 290 K, determine the pressure drop, the average heat transfer coefficient, and the outlet bulk temperature.

4–33. Air enters a 3 cm–I.D. tube through a 90° elbow. The air flow rate is 0.4 kg/s, and it enters at 350 K, 1 atm. The tube walls are maintained at 360 K. Determine the local heat transfer coefficient in the entrance region for 3 cm $< x <$ 45 cm. Repeat for a 90° round-bend entrance.

4–34. In Chapter 8, it will be shown that the ratio St/f is an approximate indicator of the merit of a heat transfer surface for use in compact heat exchangers. Calculate St/f for air in the hydraulic Reynolds number range of $10^2 - 3 \times 10^4$ to compare the performance of ducts with the following cross sections. Assume a constant axial wall heat flux.

(i) Circular.
(ii) Square.
(iii) Equilateral triangular.
(iv) 1:3 rectangle.
(v) $1 : \infty$ rectangle.

4–35. Air at 300 K, 1 atm, enters a 5 cm–I.D., 2 m-long Inconel tube through a 90° elbow. The air flow rate is 0.14 kg/s. The tube walls are heated electrically to give a uniform wall heat flux of 9×10^4 W/m^2. Determine the variation of the local heat transfer coefficient in the entrance region. Account for property variations along the duct. *(Hint: First determine $T_b(x)$ and obtain $h_c/h_{c\infty}$ from Fig. 4.20. Then, using CONV, determine $h_{c\infty}(x)$ by iterating on $T_s(x)$.)*

4–36. In a double-tube heat exchanger (see Fig. 8.1c), 20% ethylene glycol aqueous solution flows at 0.4 kg/s in a 2 cm–I.D., 1 mm-wall-thickness copper tube, and water flows at the same rate in the outer annular duct. At an axial location where the bulk temperatures of the glycol and water are 280 K and 290 K, respectively, determine the outer diameter of the annular duct to give equal hot-and cold-side heat transfer coefficients.

4–37. Refrigerant-22 has been developed as an environment-friendly refrigerant to replace Freon 12. In refrigeration systems, boiling and condensation heat transfer is of concern to the design of the evaporators and condensers (see Chapter 7). However, liquid R-22 will also see use as a heat transfer fluid in conventional heat exchangers. For a 10 m-long, 2 cm–I.D. tube, prepare graphs of pressure drop and average heat transfer coefficient for bulk velocities in the range 0.1–1.0 m/s. Assume $T_s = T_b$ and consider $T_b = 260, 300, 320,$ K.

4–38. R-134a has been developed as an environment-friendly refrigerant. In refrigeration systems, boiling and condensation heat transfer is of concern to the design of the evaporator and condenser (see Chapter 7). However, liquid R-134a will also see use as a heat transfer fluid in conventional heat exchangers. For a 10 m-long, 2 cm–I.D. tube, prepare graphs of pressure drop and average heat transfer coefficient for bulk velocities in the range 0.01–1 m/s. Assume $T_s = T_b$ and consider $T_b = 260, 300, 320$ K.

4–39. Air at 400 K and 1250 kPa flows at 40 m/s along a 10 cm-long flat plate maintained at 300 K. Determine the drag force and heat transfer per unit width.

4–40. The roof of a building is flat and is 20 meters wide and long. When the wind speed over the roof is 10 m/s, determine

(i) the convective heat gain on a clear night when the roof temperature is 280 K and the air temperature is 290 K.
(ii) the convective heat loss on a hot, sunny day when the roof temperature is 320 K and the air temperature is 300 K.

4–41. An oil pan under an automobile is 30 cm wide, 70 cm long, and 15 cm deep. Estimate the total heat loss from the five exposed sides when the oil is at 120°C and the automobile is traveling at 80 km/h through ambient air at 20°C.

4–42. Insulating wet suits worn by scuba divers may be made of 3 mm–thick foam neoprene ($k = 0.05$ W/m K), which traps next to the skin a layer of water, say 1 mm thick. Estimate the rate of heat loss from a 1.8 m–tall diver swimming at 8 km/h in 286 K water if his skin temperature does not fall below 297 K.

4–43. A small submarine is to be as silent as possible. A proposed design requires the use of a thermoelectric generator rejecting heat directly to the hull, thus eliminating cooling water pumping machinery. An area 10 m wide and 10 m long at the bow is available for heat rejection. If the submarine cruises at 4 m/s and must reject 15 MW to 15°C water, estimate the average temperature of the heat rejection surface.

4–44. Consider flow along a flat plate. For free-stream velocities in the range 0.1–100 m/s determine the average heat transfer coefficient and drag force per unit width on a 1 m length of plate for

(i) air at 1 atm.
(ii) air at 0.01 atm.

Use a transition Reynolds number of 10^5, and evaluate properties at 295 K.

4–45. Consider flow along a flat plate. Determine the average heat transfer coefficient and drag force per unit width on a 1 m–long plate for free-stream velocities of 0.1, 1, and 10 m/s when the liquid is

(i) water.
(ii) SAE 50 oil.
(iii) mercury.
(iv) R-134a.

Evaluate properties at 300 K and use a transition Reynolds number of 10^5.

4–46. Helium at 1 atm pressure and 300 K flows at 30 m/s over a 1 m–long flat plate maintained at 300 K. Plot the local shear stress and the heat transfer coefficient as a function of distance from the leading edge. Use a transition Reynolds number of 1.5×10^5.

4–47. Air at 110 kPa and 360 K flows at 15 m/s over a 1 m-long flat plate maintained at 300 K. Plot the average shear stress and the heat transfer coefficient for transition Reynolds numbers in the range 5×10^4 to 5×10^5.

4–48. Air at -10°C blows at 10 m/s over the exterior of a 30 cm–O.D. pipeline spanning a river gorge. The line is heated to 40°C to reduce the viscosity of the liquid inside. What would be the heat loss per 100 m of pipe if the line was uninsulated?

4–49. What is the rise velocity of 1 mm–diameter oxygen bubbles 3 m below the surface of a water pool at 300 K?

4–50. A 6 mm-diameter hailstone falls at its terminal velocity through air at 287 K and 1 atm. Calculate the average heat transfer coefficient.

4–51. A 1.905 cm-diameter cylinder spans a small wind tunnel and is exposed to air at 97 kPa and 290 K flowing at 20 m/s. The cylinder contains an electrical heater over a length of 12 cm. Determine the stagnation line temperature and average cylinder temperature when the power input to the heater is 60 W. The cylinder wall is very thin and has a low conductivity.

4–52. A hot-film sensor to measure ocean speeds consists of a thin (0.1 μm) platinum film deposited on a 3 cm-long quartz cylinder with outside diameter of 9 mm. A known current is passed through the platinum film, and the resistance of the film is measured using an AC resistance bridge. From the known temperature dependence of resistivity for platinum, an average temperature of the sensor surface is determined. With the sensor perpendicular to the ocean flow, a power input of 1.101 W to the sensor is measured when the sensor is at 21.35°C and the ambient sea water temperature is 20.54°C.

   (i) Estimate the flow speed.
   (ii) If the individual temperature measurements are accurate only to ±0.01°C, what is the resulting expected error in flow speed?
   (iii) If bio-fouling results in deposits with thermal conductivity of approximately 1 W/m K, what thickness of deposit would give a 5% error in flow speed?

4–53. Water at 300 K flows at 2 m/s over a 3 cm-diameter sphere maintained at 340 K.

   (i) Determine the drag force on the sphere.
   (ii) Determine the ratio of the stagnation point heat transfer coefficient to the average value.
   (iii) Determine the heat loss from the sphere.

4–54. A horizontal aluminum rod, 1 cm in diameter and 10 cm long, has both ends maintained at 363 K. Air at 293 K and 1 bar is blown over the rod at 2.5 m/s. Calculate

   (i) the temperature at the midplane of the rod.
   (ii) the heat dissipated by the rod.

4–55. What is the cost per day, per meter length of pipe, of energy lost from an uninsulated 4 cm–O.D. hot water pipe with a surface temperature of 45°C exposed to air at 10°C blowing at 2 m/s normal to the pipe axis? Electric power in Los Angeles costs about 8 cents/kW h.

4–56. An alloy sphere 1 cm in diameter is suspended by a fine wire at the center of a laboratory furnace, the inside wall of which is maintained at 700 K. Helium at 350 K and 1.1 bar pressure is blown through the furnace at 5 m/s. The interior of the furnace may be assumed black, and the emittance of the alloy is 0.40. Calculate the steady-state temperature of the sphere.

4–57. An ice sensor is located on an airplane wing flying at 50 m/s through air at 61 kPa and $-26°C$. A heater is required to heat the sensor surface to $0°C$ to ascertain from the temperature-time response whether ice is present (if ice is present, the $T - t$ curve has a plateau corresponding to the supply of enthalpy of melting). If the sensor is in the form of a 2 cm-diameter disk, what is the minimum power required? The front of the wing can be modeled as a 30 cm-diameter cylinder with the sensor located on the stagnation line.

4–58. A 0.12 mm-diameter platinum wire 3 mm long is used for a constant-temperature (200°C) hot-wire anemometer to measure 20°C air velocities in the range 0.3 to 7 m/s. Tabulate the wire current versus air velocity. A value of $10 \times 10^{-8} \, \Omega$ m for the electrical resistivity of platinum should be used.

4–59. A thermistor is used to measure the air temperature in a forced-air heating system duct. The air velocity is 1.3 m/s, and the thermistor records a temperature of 46.8°C when the duct walls are at 41.2°C. Determine the true air temperature. The thermistor can be modeled as a 3 mm sphere with an emittance of 0.8.

4–60. The rate of heat loss from an animal is proportional to its surface area, whereas the rate of heat generation is approximately proportional to its volume. Thus, smaller warm-blooded animals must consume proportionately more food in order to maintain a constant body temperature. It is impossible for a warmblooded animal to be smaller than a mouse. The Arctic is home to bears, not mice, and the smallest dolphins are restricted to tropical oceans. Thermal analysis of aquatic mammals such as dolphins can provide insights into naturally occurring size distributions. The following data are pertinent to a particular species:

Length to diameter ratio $L/D = 12$
Blubber thickness to diameter ratio $t/D = 0.05$
Average swimming speed $V = 3$ m/s
Metabolic heat generation rate $\dot{Q}_v''' = 23.5$ kW/m^3
Internal temperature $T_i = 37°C$
Blubber thermal conductivity $= 0.2$ W/m K

Prepare a graph of minimum possible length as a function of ocean temperature for $0°C < T_e < 25°C$.

4–61. A 1 cm-diameter horizontal tube is located transverse to a flow of water in a wide duct. The outer wall of the tube is maintained at 320 K, and the water temperature is 300 K. Determine the convective heat loss per unit length for water speeds in the range 0–1 m/s.

4–62. Equation 4.71 gives the average Nusselt number for forced convection across a cylinder of circular cross section. For noncircular cross sections, Jakob [56] recommends a simple power law correlation

$$\overline{Nu}_L = B\,Re_L^n\,Pr^{1/3}$$

with B and n given in the following table:

Cross Section	Re_L	B	n
	$5 \times 10^3 - 10^5$	0.104	0.675
	$5 \times 10^3 - 10^5$	0.251	0.588
	$4 \times 10^3 - 1.5 \times 10^4$	0.232	0.731
	$5 \times 10^3 - 1.95 \times 10^4$ $1.95 \times 10^4 - 10^5$	0.163 0.0396	0.638 0.782
	$5 \times 10^3 - 10^5$	0.156	0.638

Prepare graphs of $\overline{Nu}_L$ versus Re_L for $5 \times 10^3 < Re_L < 10^5$ and $Pr = 0.7$ for each geometry as well as for a circular cross section. Comment on the behavior seen in the graphs.

4–63. An 8 in Schedule 40 steel pipe transports 70 kg/s of hot oil between two units of a process plant. In order to choose an insulation thickness, nominal ambient air conditions are specified, including a temperature of 290 K and a velocity transverse to the pipe of 5 m/s. The insulation to be used has a thermal conductivity of 0.13 W/m K and is sheathed with aluminum foil for protection and to minimize radiation heat loss. At a location where the bulk oil temperature is 420 K, determine the heat loss per meter if an insulation thickness of 12 cm is chosen.

4–64. A mercury-in-glass thermometer is to be used to measure the temperature of hot helium flowing in a duct. To protect the thermometer, a pocket is made from 7 mm-O.D., 0.5 mm-wall-thickness AISI 316 stainless steel tube, with one end sealed and the other welded to the duct wall. The small gap between the thermometer and the pocket wall is filled with oil to ensure good thermal contact, and the thermometer bulb is in contact with the sealed end. The helium stream is at 80°C and 1 kPa, and the duct wall is at 60°C. If the error in the thermometer reading is to be less than 1°C, determine the length of pocket required for gas velocities in the range 1 m/s to 30 m/s.

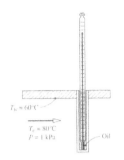

4–65. A thermistor is to be used to measure the temperature of a 1 atm–pressure superheated steam flow in a large pipe. Nominal conditions are a steam velocity of 5 m/s, a thermistor bead in the form of a 3 mm-diameter sphere, and a pipe wall at 500 K. If the thermistor reads 526.6 K, what is the true steam temperature? The emittance of the thermistor bead is 0.8, and conduction along the thermistor leads is negligible.

4–66. An urgent project required average convective heat transfer coefficients to be measured for a semicylinder in transverse flow. A thermocouple was installed in a 2 cm-diameter polished aluminum cylinder, and the cylinder was heated in an oven to a uniform temperature of 340 K and quickly inserted in a wind tunnel operating at 300 K. Using the thermocouple temperature-time response and the lumped thermal capacity model, the following heat transfer coefficients were determined:

$$V_1 = 15 \text{ m/s}, \qquad \overline{h}_{c1} = 88.1 \text{ W/m}^2 \text{ K}$$

$$V_2 = 30 \text{ m/s}, \qquad \overline{h}_{c2} = 134.2 \text{ W/m}^2 \text{ K}$$

Assuming that the Nusselt number has a simple power law dependence on the Reynolds number, determine the average heat transfer coefficient for a similar cross section stainless steel cylinder of 6 cm diameter for an air velocity of 7 m/s in the same air temperature range.

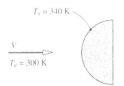

4–67. Very thin stainless steel shim stock is bonded to a 10 cm-square Teflon plate to serve as a heater element. Water at 290 K flows as a laminar boundary along the plate. The voltage drop and current flow in the heater are measured to obtain the power dissipated; also, from the known dependence of electrical resistance on temperature, the average plate temperature can be deduced. In a particular test, the power is 1025 W, and the average plate temperature is 311.5 K. Estimate the water velocity, and the temperature distribution along the plate.

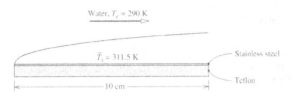

4–68. The concept of a critical radius for maximum heat loss was developed in Section 2.3.2 for a cylinder; a corresponding analysis for a sphere is required in Exercise 2-37. In both cases, analytical results are obtained for $h_c = $ constant and $h_c \propto r_o^{-1/2}$. These results can be misleading for very small values of the outer radius r_0: for a sphere, the limiting Nusselt number of 2 implies $h_c \propto r_o^{-1}$. More accurate evaluation of h_c is then required. To investigate these issues, perform the following tasks:

(i) Show that the heat loss from a sphere with insulation of inner radius r_i and outer radius r_o in a fluid at temperature T_e is

$$\frac{\dot{Q}}{4\pi(T_i - T_e)} = \frac{h_o r_o^2}{1 + \dfrac{h_o r_o}{k}\left(\dfrac{r_o}{r_i} - 1\right)}$$

(ii) Calculate the natural convection heat transfer coefficient on a sphere at 305 K in air at 295 K, 1 bar, in the range $0.1 < r_o < 100$ mm. Also calculate $h_o = \bar{h}_c + h_r$ for an assumed $h_r = 5$ W/m^2 K.

(iii) For $k = 0.03$ W/m K and $r_i = 0.1$ mm, 0.5 mm, calculate $\dot{Q}/4\pi(T_i - T_e)$ for $r_i < r_o < 10$ mm.

(iv) For $k = 0.1$ W/m K and $r_i = 1$ mm, 10 mm, calculate $\dot{Q}/4\pi(T_i - T_e)$ for $r_i < r_o < 100$ mm.

(v) Comment on the results, particularly as they relate to the concept of a critical radius of insulation.

4–69. A high-temperature argon plasma is used in a spray-coating process. In a research and development project, temperature profiles are to be measured. Owing to the high temperatures (up to 30,000 K), conventional temperature measurement schemes cannot be used. A proposed new method is to insert a tungsten-rhenium thermocouple for a very short period of time and obtain the temperature-time response at temperatures below the melting temperatures of the thermocouple pair.

From the data and an estimated heat transfer coefficient for the spherical bead of the thermocouple, the argon temperature can be deduced. For the following data estimate the argon temperature:

Time for thermocouple temperature to change from 500 K to 1500 K = 0.42 s
Plasma jet velocity = 200 m/s
Thermocouple bead diameter = 3 mm
Argon properties at an estimated mean film temperature $\nu = 7 \times 10^{-3}$ m^2/s, $k = 0.17$ W/m K, Pr = 0.67
Density-specific heat product for bead = 7.3×10^6 J/m^3 K

4–70. Consider a length of horizontal pipe of outside diameter 3 cm. What is the average convective heat transfer coefficient on the outside of the pipe when the pipe is at 320 K and

    (i) it is immersed in still air at 300 K, 2 atm pressure?
    (ii) it is immersed in still water at 300 K?

4–71. A 100 liter covered tank full of water at 290 K is to be heated to 340 K by means of steam condensing inside a 1 cm–O.D. copper tube coil having 10 turns of 0.5 m diameter. If the steam-side thermal resistance is negligible and the tank well insulated, estimate the time required for heating and the total amount of steam condensed. Assume a steam pressure of 1 atm.

1 cm–diameter copper tube

4–72. An uninsulated aluminum overhead electrical transmission line has a diameter of 2 cm and a resistance of $8.33 \times 10^{-5}\ \Omega/$m. If the current flowing is 600 A, determine the surface temperature of the wire on a still day with a low cloud cover at the air temperature of 290 K. The emittance of the wire can be taken as 0.7, and of the clouds 1.0.

4–73. Consider a sphere of 1 mm diameter in air at 300 K and 1 atm pressure.

    (i) At what air velocity will it have a heat transfer coefficient 50% greater than the conduction limit value ($\mathrm{Nu}_D = 2.0$)?
    (ii) If the air is stationary so that only natural convection occurs, at what temperature will it have a heat transfer coefficient 50% greater than the conduction limit value?

Repeat for water at 300 K.

4–74. A 60 W light bulb can be modeled as a 5 cm-diameter sphere. If the surface temperature is measured to be 135°C when the ambient air is at 20°C, estimate the fraction of the lamp power that is lost from the glass bulb by

    (i) natural convection.

    (ii) radiation emission.

Take $\varepsilon = 0.8$ for the glass.

4–75. W. H. McAdams [30] recommends a simple correlation for natural convection from horizontal cylinders:

$$\overline{\mathrm{Nu}}_D = 0.53\mathrm{Ra}_D^{1/4}$$

for $\mathrm{Pr} > 0.5, 10^3 < \mathrm{Gr}_D < 10^9$. Compare the values of Nusselt number given by this formula with those given by Eq. (4.87) for $\mathrm{Pr} = 0.7, 10,$ and 100.

4–76. A vertical plate 1 m high is immersed in a stagnant fluid. The plate is maintained at 350 K, and the fluid temperature ranges from 280 to 340 K. Determine the average heat transfer coefficient if the fluid is

    (i) air at 1 atm.

    (ii) air at 0.01 atm.

    (iii) water.

    (iv) SAE 50 oil.

4–77. Compare the convective heat loss from the top and bottom of a 10 cm–square plate immersed in water at 300 K when the plate temperature is

    (i) 310 K.

    (ii) 350 K.

4–78. Compare the convective heat loss from the top and the bottom of a heated 20 cm–square plate located in air at 1 atm pressure and 300 K when the plate temperature is

    (i) 320 K.

    (ii) 400 K.

4–79. A heated horizontal plate is 10 cm wide and 20 cm long and has 2 cm adiabatic extensions. What is the convective heat loss from the underside of the plate when it is at 320 K and is exposed to air at 300 K?

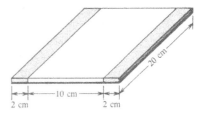

4–80. A steam supply pipe carries saturated steam at 1.2 MPa. The pipe has a 15 cm outer diameter and is lagged with 6 cm of 85% magnesia insulation. Determine the heat loss per meter length of pipe and the surface temperature of the insulation on a windless day when the air temperature is 20°C.

4–81. A 20 cm-square plate is inclined at 45° to the vertical in air at 1 atm and 300 K. If the plate temperature is 320 K, determine the convective heat loss from each side.

4–82. The sphere-cone shown in the sketch is maintained at 340 K in water at 300 K. Estimate the rate of heat loss if the top is well insulated.

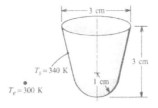

4–83. An 80 liter covered tank full of water at 300 K is to be heated to 360 K by steam at 1.1 atm pressure condensing inside a 1 cm–O.D. copper tube coil having 12 turns of 0.4 m diameter. If the steam-side thermal resistance is negligible, and the tank is well insulated, estimate the required time for heating, and the total amount of steam condensed.

4–84. An overhead aluminum electrical transmission line has wires of 2.2 cm diameter and resistance of $6.9 \times 10^{-5}\,\Omega/m$. The line is located in a canyon in Southern California where strong Santa Ana winds can cause adjacent lines to clash and arc. The resulting sparks (burning aluminum metal droplets) can ignite brush fires; thus, it is proposed to electrically insulate the lines. Since electrical insulation deteriorates at high temperatures, it is important that the insulation temperature not be excessive under normal conditions. If a 3 mm-thick layer of insulation of conductivity 0.2 W/m K is proposed, determine its maximum temperature for a current of 800 A on a still day when there is low cloud cover at the air temperature of 295 K. The emittance of the insulation can be taken as 0.85, and of the clouds 1.0.

4–85. A 2 cm-diameter polished copper sphere originally at 20°C is suddenly immersed in air at 81°C, 1 atm. Determine the time required for the sphere to reach 1 K of its equilibrium temperature. Assume that the fluid is not significantly disturbed in the immersion process.

4–86. Cooling systems sometimes involve heat removal by laminar natural convection from a uniformly heated vertical wall. Sparrow and Gregg [55] have shown that the average Nusselt number for an isothermal wall gives an excellent estimate of the temperature halfway up the wall, $T_s(x = L/2)$. Thus, Eq. (4.85) can be used to determine $T_s(x = L/2)$, though an iterative procedure is required because fluid properties must be evaluated at the unknown mean film temperature, $1/2(T_s + T_e)_{L/2}$. Sparrow and Gregg also showed that $(T_s - T_e) \propto x^{1/5}$ for a uniformly heated wall (see also Exercise 5–36). Of practical concern is the maximum wall temperature, which occurs at $x = L$, and it follows that $(T_s - T_e)_L = 2^{1/5}(T_s - T_e)_{L/2}$.

A vertical plate is 12 cm square and dissipates a total of 4 W distributed approximately uniformly over its surface. One side of the plate is exposed to quiescent ambient air at 300 K and 1 atm, and the heat loss from the other side is negligible. Determine the maximum temperature of the plate. The emittance of the exposed surface is small enough to ignore radiative heat transfer.

4–87. A 1 m-high vertical flat plate at 320 K is immersed in a fluid at 300 K. Estimate the characteristic natural convection velocity at the top of the plate for the following fluids:

(i) Air at 1 atm pressure.
(ii) Air at 10^3 Pa pressure.
(iii) Water.
(iv) SAE 50 oil.
(v) Therminol 60.
(vi) Mercury.

4–88. A vertical wall is maintained at 320 K in fluid at 300 K. If transition from a laminar to a turbulent boundary layer occurs at a Rayleigh number of 10^9, determine the location of transition for the following fluids:

(i) Air at 1 atm pressure.
(ii) Air at 10^3 Pa pressure.
(iii) Water.
(iv) SAE 50 oil.
(v) Therminol.
(vi) Mercury.

4–89. A bath of SAE 50 oil is to be heated from 20°C to 60°C by vertical cylindrical electric immersion heaters, each 14 cm high and 3 cm in diameter. Estimate the maximum allowable power per heater if the heater surface temperature should not exceed 150°C. The heaters contain electric resistance wire wound at a constant pitch (see Exercise 4–86).

4–90. Steel cylinders 1 m long and 10 cm in diameter at 200°C are to be cooled to 50°C in an oil bath maintained at 20°C.

(i) Should the cylinders be immersed vertically or horizontally to obtain the most rapid cooling?

(ii) Estimate the cooling time for the most rapid case.

Use properties of SAE 50 oil and AISI 4130 steel.

4–91. Exercise 2–30 requires determination of the critical radius of insulation on a 6 mm–O.D. tube using a simple power law relation for the average heat transfer coefficient. Rework that exercise using Eq. (4.87) to obtain $\bar{h}_c$ (The solution of Exercise 2–30 is $r_{cr} = 0.0046$ m, $T_s = 339.7$ K, $\dot{Q}_{max}/L = 12.1$ W/m.)

4–92. A 1 kW metal-sheathed electric immersion heater in a large tank of water is 30 cm long, has a diameter of 2 cm, and is located in a horizontal position.

(i) Calculate the surface temperature of the heater if the water is at 330 K and unstirred.

(ii) If the water level accidentally suddenly drops below the heater level, allowing ambient air at approximately 310 K to surround the heater, estimate the surface temperature the heater could attain (if it does not burn out). Take $\varepsilon = 0.8$ for the heater surface.

4–93. In a laboratory test rig, a test liquid to be cooled is passed through a copper coil immersed in a water tank: the water is maintained at 280 K by a refrigeration system. The 8 mm-O.D. copper tubing has a 0.75 mm wall thickness and is to be wound to give a mean coil diameter of 40 cm. The test liquid flow rate is 0.15 kg/s, and the liquid enters the coil at 360 K. Estimate the length of tube required to cool the liquid to 285 K, and the corresponding number of coil turns. The properties of the test liquid can be approximated by those for water, the water tank is unstirred, and the coil can be modeled as a horizontal tube. *(Hint:* With appropriate assumptions the system can be modeled as a single-stream heat exchanger.)

4–94. A 20 cm-square hot plate is maintained at 150°C in ambient air at 20°C, 1 atm. If its emittance is 0.8, determine the convective and radiative heat losses from its upper surface.

4–95. A long 2 m-high air cavity has insulated top and bottom surfaces. The side surfaces are at 320 K and 300 K. Calculate the convective heat flow across the cavity as a function of cavity thickness for 20 cm $< L <$ 80 cm. Comment on the values of heat transfer coefficient obtained.

4–96. A heated horizontal plate is 20 cm wide and 40 cm long, and has 4 cm-wide adiabatic extensions on the shorter side. The underside of the plate is at 300 K, and it is exposed to fluid at 280 K. Determine the convective heat loss from the plate underside if the fluid is

  (i) water.
  (ii) Therminol 60.
  (iii) SAE 50 oil.
  (iv) R-134a.

4–97. A 30 cm–square aluminum flat-plate heater is perfectly insulated on its lower surface and has a 1 cm–thick cover layer of an unknown material bonded to its upper surface. It is located in a room where the air is at 290 K, 1 atm; the walls of the room are also at 290 K. Thermocouples are located at the center of each side of the cover layer. With a power input of 100 W, the thermocouples give readings of 382.3 K and 369.7 K. Determine the surface emittance and thermal conductivity of the cover layer.

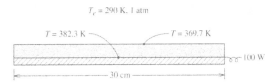

4–98. A vertical circuit board is 3 mm thick, 12 cm high, and 18 cm long. On one side are 100 closely spaced logic chips, each dissipating 0.04 W. The board is a composite containing copper and has an effective thermal conductivity of 15 W/m K. The heat generated by the chips is conducted across the board and dissipated from the backside to ambient air at 305 K, 1 atm, which flows along the board at 5 m/s. Determine the maximum temperature of the front side of the board. Use a transition Reynolds number of 1×10^5.

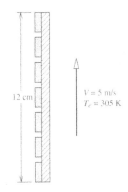

4–99. A 95 mm–high Styrofoam cup has 1.5 mm–thick walls, and its outside diameter varies from 77 mm at the top to 43 mm at the base. It is filled with 200 ml of coffee at 80°C, sealed with a 0.5 mm-thick plastic lid, and placed on a wooden table. If the ambient air is 24°C and 1000 mbar, estimate the time for coffee to cool to 60°C. Take $k = 0.033$ W/m K and 0.33 W/m K for the Styrofoam and plastic lid, respectively, and $\varepsilon = 0.85$ for both. Make reasonable assumptions but discuss their validity.

4–100. A 1 m-high double-glazed window has an air gap of 1 cm. In a test the facing glass surfaces were measured to be at 14.2 and -10.6°C. Calculate the convective heat transfer across the gap.

4–101. A flat-plate solar collector has an absorber plate at 350 K and a coverplate at 310 K. If the air gap is 5 cm, determine the convective heat flux across the gap when the collector is inclined at angles from the horizontal of

   (i) 0°.
   (ii) 30°.
   (iii) 60°.

4–102. An aluminum roof 30 m long and 10 m wide is inclined at and angle of 30°. A 2 cm-thick plasterboard ($k = 0.06$ W/m K) ceiling is located below the roof, giving a 5 cm air gap. On a clear winter night condensation on the outside of the roof maintains its temperature at 5°C, while heating within the room maintains the lower surface of the ceiling at 15°C. Neglecting the effects of structural members between the ceiling and the roof, estimate the heat loss through the ceiling.

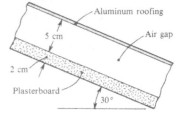

4–103. A long, 1 m–high, 30 cm–thick air cavity has insulated top and bottom surfaces. Calculate the convective heat transfer per meter when the side surfaces are at 450 and 350 K.

4–104. A horizontal cylindrical annulus has an outer diameter of 18 cm and an inner diameter of 10 cm. If the outer surface is maintained at 320 K and the inner surface at 280 K, determine the effective conductivity of enclosed air at 1 atm and the convective heat transfer per unit length.

4–105. Two concentric spheres of outer and inner diameters 18 cm and 10 cm, respectively, are separated by air at 1 atm pressure. If the outer surface and inner surface are maintained at 320 K and 280 K, respectively, determine the effective conductivity of the air and the convective heat transfer.

4–106. Buchberg et al. [54] proposed the following three-regime correlation for heat transfer across high-aspect-ratio inclined layers:

$$\overline{Nu}_L = 1 + 1.446 \left[1 - \frac{1708}{Ra_L \cos \theta} \right] \qquad 1708 < Ra_L \cos \theta < 5900$$

$$= 0.229 (Ra_L \cos \theta)^{0.252} \qquad 5900 < Ra_L \cos \theta < 9.23 \times 10^4$$

$$= 0.157 (Ra_L \cos \theta)^{0.285} \qquad 9.23 \times 10^4 < Ra_L \cos \theta < 10^6$$

where the term in square brackets is set equal to zero if negative. Compare the Nusselt numbers given by this correlation at $\theta = 30°$ and $50°$ for air at normal temperatures, with the values given by the correlations of Hollands et al., Eqs. (4.98)–(4.101).

4–107. Space heating in buildings accounts for about 20% of the total energy consumed in the United States. Energy conservation during cold weather requires reducing heat loss through large glass areas. Compare the heat loss through 3 mm–thick window glass for the two configurations shown in the sketch. Assume 600 m² of glass, an indoor temperature of 21°C, and an outdoor temperature of 3°C. The window drape has a thickness of 1 mm and a conductivity of 0.07 W/m K. To include thermal radiation, assume that the window glass and drape are opaque and that all surfaces have an emittance of 1.0.

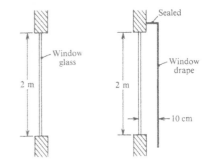

4–108. The sketch shows a test cell used to study natural convection in a circular crosssection cavity, 15.2 cm in diameter. The cold-plate assembly can be moved to vary the thickness of the liquid layer. The complete test cell can be rotated to incline the liquid layer at any angle θ. Referring to the thermal circuit, the rubber layers serve as heat flux meters. These meters are calibrated by rotating the cell to

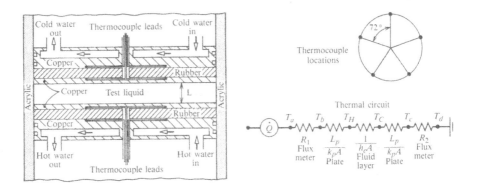

$\theta = 180°$, where the hot plate is above the cold plate, the liquid layer is stable, and hence heat transfer through the liquid is by conduction only. Its thermal resistance is then simply $1/h_cA = L/kA$, and is known. A series of tests measuring the temperatures T_a, T_b, T_c, and T_d allows the unknown meter resistances R_1 and R_2 to be determined as 0.346 and 0.327 K/W, respectively. Notice that the thermal resistances of the copper plates are negligible, so that $T_b = T_H, T_c = T_C$. It is advantageous to operate the system with the average fluid layer temperature approximately equal to the ambient temperature in order to minimize errors due to heat exchange with the surroundings. The plate spacing is varied to obtain a range of flow conditions. Typical results for water are shown in the table. Prepare graphs of Nusselt number versus Rayleigh number, and make comparisons with the correlations given in Section 4.4.2. (*Hint:* Use CONV to assist in the data processing.)

Angle, θ	L mm	T_a °C	T_b °C	T_c °C	T_d °C
0°	4	49.9	31.4	22.6	3.3
	7	52.6	33.5	23.5	3.7
	15	51.1	31.9	21.4	1.7
	30	53.3	33.8	22.3	2.3
90°	4	51.1	33.0	20.4	3.4
	15	50.6	32.3	19.9	2.3
	30	51.6	32.9	20.4	2.3
120°	4	51.6	33.3	20.3	3.4
	7	51.6	33.2	20.6	3.4
	15	50.6	32.3	19.2	2.2

4–109. A 3 cm–diameter horizontal cylinder is located transverse to the flow in a wind tunnel. The cylinder is maintained at 400 K in an air stream at 300 K and 1 atm. Determine the convective heat loss per meter length for air speeds in the range 0.05–0.30 m/s.

4–110. Water at 320 K flows at 0.15 m/s along a 10 cm–long, 1 m-wide flat plate. Determine the average heat transfer coefficient for plate temperatures in the range 280–360 K.

4–111. Consider a vertical fluid flow upward in a tube of diameter 2 cm and height 20 cm. The wall is maintained at 310 K, and the inlet temperature is 300 K. Below what velocity must mixed-convection effects be considered for the following fluids?

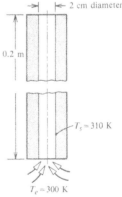

2 cm diameter

0.2 m

$T_s = 310$ K

$T_e = 300$ K

   (i) Water
   (ii) SAE 50 oil
   (iii) Therminol 60
   (iv) Air
   (v) Mercury

4–112. Water flows through a tube bank 10 rows deep, with 20 mm–O.D. tubes in an aligned arrangement of transverse pitch 30 mm and longitudinal pitch 25 mm. In a performance test the superficial velocity is 0.8 m/s, the water inlet and outlet temperatures are 324 and 336 K, respectively, and the tube outer wall temperature is approximately 453 K. Estimate the average heat transfer coefficient for the tube bank and the pressure drop across the bank.

4–113. An air heater consists of a staggered tube bank in which waste hot water flows inside the tubes, with air flow through the bank perpendicular to the tubes. There are 30 rows of 15 mm–O.D. tubes, with transverse and longitudinal pitches of 28 and 32 mm, respectively. The air is at 1 atm and flows at 5.36 kg/s in a duct of 1.0 m square cross section. Preliminary design calculations for this heat exchanger suggest average tube surface and bulk air temperatures of approximately 350 K and 310 K, respectively. Estimate the average heat transfer coefficient and pressure drop across the bank.

4–114. Air at 280 K and 1 atm pressure flows at 3.0 kg/s along a duct of cross section 1 m $\times$ 0.6 m. The duct contains a tube bank of 1 m-long, 10 mm–O.D. tubes in a staggered arrangement, with longitudinal pitch 24 mm and transverse pitch 20 mm. Steam at 1.29×10^5 Pa condenses inside the tubes. How many rows of tubes are required to heat the air to 360 K, and what is the pressure drop? Make reasonable assumptions.

4–115. A tube bank is 30 rows deep and has 20 mm-O.D. tubes in a staggered arrangement. Steam condenses in the tubes at 1.6×10^5 Pa, and air enters the bank at 4 m/s. Investigate the effect of transverse and longitudinal pitch on the average heat transfer coefficient and pressure drop. Take 25 mm $\leq S_T, S_L \leq$ 50 mm, and a bulk air temperature of 330 K. Prepare graphs with S_T as abscissa and S_L as a parameter.

4–116. Water enters a 20-row-deep tube bank at 1.5 m/s. The tubes have a 25 mm outside diameter and are in an aligned arrangement. Investigate the effect of transverse and longitudinal pitch on the average heat transfer coefficient for $30 \leq S_T, S_L \leq 60$ mm. Take a tube wall temperature of 350 K and a bulk water temperature of 300 K. Prepare graphs with S_T as abscissa and S_L as a parameter.

4–117. Helium at 250 K and 1 atm flows at 0.3 kg/s along a duct of cross section 1 m $\times$ 0.5 m. The duct contains a tube bank of 1 m–long, 15 mm–O.D. tubes in a staggered arrangement, with both transverse and longitudinal pitches of 30 mm. If steam is condensed in the tubes at 1.1×10^5 Pa, how many rows of tubes are required to heat the gas to 350 K? Also estimate the pressure drop.

4–118. A tube bank is 25 rows deep and has 16 mm–O.D., 1 mm–wall thickness brass tubes in a staggered arrangement, with a transverse pitch of 26 mm and a longitudinal pitch of 17 mm. Steam at 1.7×10^5 Pa condenses inside the tubes, and SAE 50 oil at 300 K enters the bank with a velocity of 1.23 m/s. If the steam-side heat transfer coefficient is estimated to be 8000 W/m^2 K, calculate the outlet oil temperature and the pressure drop.

4–119. Surplus 9 mm steel ball bearings are packed with a void fraction of 0.38 in a regenerative air heater. The bed cross-sectional area is 0.1 m^2 and the air flow is 0.05 kg/s. At a location along the bed where the balls are at 1500 K and the air is at 1000 K and 1 atm, determine the pressure gradient and heat transfer coefficient. Also calculate the specific area and transfer perimeter of the bed.

4–120. A perforated-plate heat exchanger for a hydrogen refrigeration system has flow cross-sectional areas of 0.002 m^2 for each 0.005 kg/s stream. The holes are 0.9 mm in diameter in square arrays with an open-area ratio of 0.35. The plates are 0.5 mm thick and are spaced 0.8 mm apart. At a location where the pressure is 0.8 atm and temperature is 200 K, determine the pressure drop per plate and heat transfer coefficient.

4–121. A regenerative air heater is to have a cross-sectional area of 0.2 m^2 and an air flow of 0.1 kg/s. Surplus steel ball bearings are to be used for the packing. A special feature of the design is that the bed must be relatively thin and a relatively large pressure drop can be accommodated. Thus, the use of small-diameter balls is indicated. Investigate the effect of ball diameter $(0.5 < d_p < 10$ mm) and volume void fraction $(0.3 < \varepsilon_v < 0.5)$ on the heat transfer coefficient, the heat transfer coefficient times specific area product, and pressure gradient. Evaluate properties at 1000 K and 110 kPa.

4–122. A regenerative heater has a cross-sectional area of 0.1 m^2 and is packed with pebbles, which can be approximated as 1 cm–diameter spheres. The void fraction is 0.41. Helium flows at 0.11 kg/s through the bed. At a location where the pebbles are at 1000 K and the helium is at 600 K and 1 atm, determine the heat transfer coefficient and the pressure gradient. Also calculate the specific area and transfer perimeter of the bed.

4–123. A two-stream helium-to-helium heat exchanger is an essential component of a cryogenic refrigeration system used to cool superconducting magnets on a magnetically levitated train. A perforated-plate packing is under consideration in the design process, and the effect of hole diameter d and open-area ratio ε_p are to be investigated. The plates are 0.5 mm thick and are spaced 1.0 mm apart. The holes are arranged in square arrays. The helium flow rate and cross-sectional area for each stream are 0.018 kg/s and 0.005 m^2, respectively. Prepare graphs of the heat transfer coefficient, the $h_c \mathscr{P}$ product, and pressure drop per plate, with hole diameter as abscissa and open-area ratio as a parameter. Evaluate the helium properties at 170 K, 1 atm.

4–124. A pebble bed is used to preheat 5 kg/s of air for a blast furnace. The bed is 2 m in diameter, is 0.5 m long, and is packed with pebbles that can be approximated as 2.2 cm–diameter spheres with a void fraction of 0.43. Calculate the heat transfer coefficient and pressure gradient at a location where the pebbles are at 1400 K and the air is at 1200 K and 1 atm.

4–125. An experimental high-temperature gas nuclear reactor is 8 cm in diameter and 24 cm long. It contains 8 mm–diameter spherical uranium oxide pellets, each with a 0.8 mm-thick graphite sheath, packed with a volume void fraction of 0.41. Helium enters the bed at 380 K, 1 atm with a velocity of 20 m/s. Calculate the maximum pellet surface temperature if the fission heat release is 3.6×10^7 W/m^3 within the uranium oxide fuel.

4–126. A rotary chemical reactor is in the form of a cylinder 5 m long and 1 m diameter, which rotates around a horizontal axis at 10 revolutions per minute. If the outer surface is measured to be at 380 K when the ambient air temperature is 300 K, estimate the convective heat loss from the cylindrical surface.

4–127. Estimate the heat loss from one side of a 1 m-diameter disk at 330 K rotating in still air at 300 K and 1 atm, if the rate of rotation is

   (i) 100 rpm.
   (ii) 10,000 rpm.

4–128. In a material processing experiment on a space station, a 1 cm-diameter alloy sphere spins at 3000 rpm in a nitrogen-filled enclosure at 800 K and 1 atm. The sphere is maintained at 1000 K by focusing a beam of infrared radiation on the sphere. At what rate must radiant energy be absorbed by the sphere at steady state? The emittance of the sphere is 0.15.

4–129. To estimate the effect of surface roughness on the skin drag of a high-speed underwater vehicle, model the vehicle as a flat plate 2.5 m long traveling at 70 km/h. What is the percentage increase in skin drag over the smooth-wall value if the equivalent sand grain roughness of the surface is 0.25 mm? Assume a sea temperature of 300 K.

4–130. An 8 cm–I.D., 2000 m–long pipeline with an equivalent sand grain roughness of 0.1 mm contains hot water at 350 K flowing from an elevation of 20 m to one of 60 m above sea level. The pump is 80% efficient and consumes 45 kW of power. What is the water flow rate and the pressure difference across the pump?

4–131. Air at 1 atm flows in a 4 cm–I.D. tube at 40 m/s. Investigate the effect of transverse square ribs on the heat transfer coefficient and pressure drop. Let the rib height vary from 0.2 mm to 2 mm for pitch/height ratios of 10, 15, and 20. Prepare graphs with rib height as abscissa and pitch/height as a parameter. Evaluate the air properties at 300 K.

4–132. Helium at 4×10^5 Pa pressure flows in a 3 cm–I.D. tube at 50 m/s. The tube is artificially roughened with transverse square ribs 0.6 mm high. Investigate the effect of pitch/height ratio on pressure gradient for $2 < p/h < 20$. Evaluate the helium properties at 400 K.

4–133. Water flows at 3 m/s in a badly fouled heat exchanger tube of 5 cm inside diameter. The deposits on the tube surface do not have a simple pattern, but the protrusions do have a mean height of 2.5 mm. Pressure drop measurements indicate an equivalent sand grain roughness of 3.2 mm. Estimate the pressure gradient and heat transfer coefficient at a location where the tube wall is at 360 K and the bulk water temperature is 400 K.

4–134. Air at 1 atm pressure flows at 40 m/s over a 1 m-long flat plate that has a sand grain indentation roughness pattern. The air is at 300 K, and the plate is maintained at 400 K. Investigate the effect of equivalent sand grain roughness size on the drag force and average heat transfer coefficient for 1 mm $< k_s <$ 5 mm.

4–135. Water flows under pressure at 6 m/s in a 4 cm–I.D. tube that is artificially roughened with transverse ribs of height 0.4 mm and pitch 4 mm. At a location where the wall and bulk fluid temperatures are 700 K and 300 K, respectively, calculate the equivalent sand grain roughness k_s, and k_s^+, f, τ_s, Nu, h_c, and q_s. Evaluate all properties at the mean film temperature.

4–136. Air at 295 K, 1 atm, flows at 20 m/s along a 1 m–long, 0.5 m–wide flat plate that dissipates 1600 W uniformly over its surface. The surface is textured and has a mean protrusion height of 2 mm. Hydraulic tests indicate an equivalent sand grain roughness of 1.5 mm. Estimate the maximum temperature of the plate. Assume that the boundary layer is turbulent from the leading edge.

CONVECTION ANALYSIS

CONTENTS

5.1   INTRODUCTION

The analysis of convection is a challenging task. The general equations governing convection are formidable, far more so than the general heat conduction equation. Exact analytical solutions can be obtained for the simplest flows only — for example, *Couette flow* and fully developed laminar flow in a tube — and such flows exhibit few of the essential features of convection. Approximate analytical methods, such as the *integral method* for boundary layers, work well for simple flows but become awkward and inaccurate when extended to more complicated flows. Series expansion methods have met with limited success. The governing equations for convection are a set of partial differential equations describing conservation of mass, momentum, and energy. The major mathematical difficulty is that the momentum equation is *non-linear*, and exact solution of such equations generally requires numerical methods. In some cases (e.g., *self-similar* boundary layers), the partial differential equations can be reduced to ordinary differential equations, which can be solved by rather simple numerical techniques, such as Runge-Kutta integration, or by iteration. In general, however, the partial differential equations must be solved using finite-difference or finite-element methods.

For laminar flows, the basic physics is simple, involving only conservation principles, Newton's law of viscosity, and Fourier's law of heat conduction. However, the flows of engineering concern are more often turbulent. Understanding the nature of turbulent flows remains one of the unsolved problems of physics. Only in the past few decades, with the aid of supercomputers, has there been significant progress at a basic level. Thus, out of necessity, engineers have relied on simple empirical models to describe turbulent transport of momentum and energy. Development of these models has relied heavily on experimental data, in particular, measured velocity and temperature profiles. An appreciation of the pertinent experimental data is essential. What engineers call turbulent flow theory is a partial theory only; but in science and engineering, it is possibly unique in its careful blending of theory and empiricism.

The difficulties just described dictate the choice of material for Chapter 5, which is only a brief introduction to the analysis of convection consistent with the general level of this book. In Sections 5.2 and 5.3, two simple and exact analytical solutions are presented. In both cases the governing equations are derived from first principles. The analysis of high-speed Couette flow in Section 5.2 leads to the *recovery factor* concept. This result allows generalization of the flat plate heat transfer correlations in Chapter 4 to include high-speed flow, for which viscous dissipation of mechanical energy (frictional or aerodynamic heating) is significant. The analysis of fully developed laminar flow in a tube in Section 5.3 should give the student insight into the various friction factor and Nusselt number formulas for laminar flow in tubes and ducts given in Section 4.3.1. In Section 5.4, the equations governing a forced-convection laminar boundary layer on a flat plate are derived, again from first principles. An exact analytical solution is obtained for the limit of zero Prandtl number that is useful for liquid metals. The *integral* method is used to obtain an appropriate analytical solution. The integral forms of the equations governing natural convection on a vertical wall are derived and solved. Turbulent flow is briefly dealt with in Section 5.5. The Prandtl mixing length model is developed, and the eddy viscosity and turbulent Prandtl number are introduced. The equations governing momentum and energy transport in a turbulent boundary layer on a flat plate are formulated and used to derive the *Reynolds analogy*. The complete solution of these equations is left to more advanced texts. The general conservation equations governing mass, momentum, and energy are presented in Section 5.6. Cartesian coordinates are used, but the results are also written in vector form for conciseness and generality.

5.2 HIGH-SPEED FLOWS

The work required to sustain the motion of a real fluid is converted irreversibly into heat by the action of viscosity, a process we call *viscous dissipation*. If the fluid is very viscous, significant heat can be produced at relatively low speeds, for example, in the extrusion of a plastic or in an oil-lubricated journal bearing. However, for a less viscous fluid such as air, significant heat is produced only at very high speeds. For flight through the atmosphere, *aerodynamic heating* becomes significant above a Mach number of about 3. Thus, an aluminum skin was adequate for a supersonic airliner such as the Concorde, whereas nickel or titanium alloys are used for hypersonic vehicles. In a compressible fluid such as air, heating by compression is also important in the hypersonic flow regime. The intense heating experienced by a blunt-nosed space vehicle upon reentry into the atmosphere is due primarily to the compression of the oncoming air by the bow shock wave.

5.2.1 A Couette Flow Model

Considerable insight into the phenomenon of aerodynamic heating can be obtained through analysis of a simple **Couette flow.** Figure 5.1 shows a fluid confined between two infinite parallel plates spaced L apart. The lower plate is stationary, and the upper plate moves with a uniform velocity u_e. There is no pressure gradient in the x direction. The upper plate may be imagined to be a conveyor belt, or it could be the outer of two concentric cylinders whose diameters are large compared to the gap between them. If the lower plate is maintained at a temperature T_s different from that of the upper plate, T_e, there will be heat conduction across the fluid layer. Also, if the upper plate velocity u_e is sufficiently large, there will be a significant heat generation in the fluid as a consequence of viscous dissipation. For simplicity, we will assume laminar flow and constant fluid properties: in this problem, the relevant properties are the viscosity μ and conductivity k.

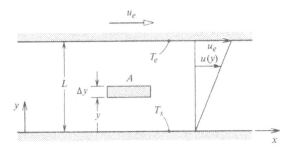

Figure 5.1 Schematic of simple Couette flow.

The procedure we follow is common to all analyses of forced convection in this chapter. The first step is to solve the fluid mechanics problem. The momentum conservation principle is used to derive the differential equation governing the velocity field. Solution of this equation subject to the given boundary conditions shows that the velocity profile $u(y)$ has a simple linear form. The next step is to solve the heat transfer problem. The energy conservation principle is used to derive the differential equation governing the temperature field. Solution of this equation subject to the given boundary conditions gives the temperature profile $T(y)$. Finally, Fourier's law is used to obtain the heat flux at the plate surface.

The Fluid Mechanics Problem

At steady state, the only forces acting in the fluid are shear forces due to the action of viscosity. According to Newton's law of viscosity, the shear stress on a plane at location y in a laminar flow is

$$\tau = -\mu \frac{du}{dy} \tag{5.1}$$

where μ is the *dynamic viscosity* with units [kg/m s], and our sign convention is such that the stress is applied on the fluid of greater y from below.[1] Since du/dy is positive, Eq. (5.1) states that τ is in the negative x direction, which is appropriate since the fluid below exerts a drag on the fluid of greater y. Of course, there is an equal and opposite shear stress exerted by the fluid above on the fluid of lesser y. The forces on an element of fluid $A\Delta y$ located between planes y and $y + \Delta y$ are shown in Fig. 5.2, in terms of both the shear stress τ and the velocity gradient du/dy.

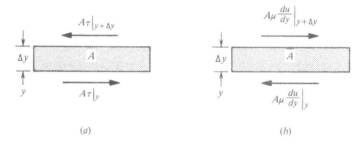

Figure 5.2   Force balance on a fluid element in a Couette flow, (a) In terms of the shear stress τ. (b) In terms of the velocity gradient du/dy.

---

[1] The opposite sign convention is used in many texts.

The momentum conservation principle reduces to a simple force balance:

$$A\mu \frac{du}{dy}\bigg|_{y+\Delta y} - A\mu \frac{du}{dy}\bigg|_y = \text{Rate of change of momentum} = 0$$

Hence,

$$\mu \frac{du}{dy}\bigg|_{y+\Delta y} = \mu \frac{du}{dy}\bigg|_y = \text{Constant} = C_1$$

or

$$\frac{du}{dy} = \frac{C_1}{\mu}$$

When the viscosity is assumed constant, integration yields

$$u = \frac{C_1}{\mu}y + C_2 \tag{5.2}$$

The constants are evaluated using the real fluid no-slip condition on each plate:

$$y = 0: \quad u = 0 \tag{5.3a}$$

$$y = L: \quad u = u_e \tag{5.3b}$$

to obtain

$$u = \frac{u_e}{L}y \tag{5.4}$$

The Couette flow velocity profile $u(y)$ is seen to have a simple linear form.

The Heat Transfer Problem

To obtain the differential equation governing the temperature profile in the fluid, we apply the first law of thermodynamics to an elemental control volume located between the planes y and $y + \Delta y$, as shown in Fig. 5.3. The other dimensions of the control volume will be taken as Δx and Δz, although, since there are no changes in temperature or velocity in the x and z directions, these dimensions could be taken as finite. The control volume is an open system, so that the steady-flow energy equation, Eq. (1.3), applies. Fluid flows in the left face of the volume at a rate $\dot{m} = \rho u \Delta y \Delta z$

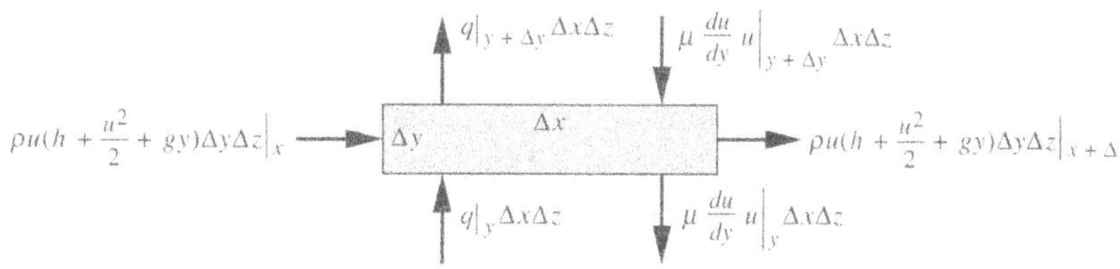

Figure 5.3 Application of the first law of thermodynamics to an elemental control volume in a Couette flow.

and out the right face at the same rate. Equation (1.3) becomes

$$\rho u \Delta y \Delta z \left[\left(h + \frac{u^2}{2} + gy \right)_{x+\Delta x} - \left(h + \frac{u^2}{2} + gy \right)_x \right] = 0 = \dot{Q} + \dot{W}$$

since T (and hence enthalpy h) and velocity u are functions of y only. The contributions to $\dot{Q}$ are the rates at which heat is transferred by conduction.

Into the system at y: $\dot{Q}|_y = q|_y \Delta x \Delta z$

Out of the system at $y + \Delta y$: $\dot{Q}|_{y+\Delta y} = q|_{y+\Delta y} \Delta x \Delta z$

The contributions to $\dot{W}$ are the rates at which work is done due to shear forces.

By the system at y: $\dot{W}|_y = \mu \frac{du}{dy} u \Big|_y \Delta x \Delta z$

On the system at $y + \Delta y$: $\dot{W}|_{y+\Delta y} = \mu \frac{du}{dy} u \Big|_{y+\Delta y} \Delta x \Delta z$

Substituting into $\dot{Q} + \dot{W} = 0$ and canceling the area $\Delta x \Delta z$ gives

$$(q|_y - q|_{y+\Delta y}) + \left(\mu \frac{du}{dy} u \Big|_{y+\Delta y} - \mu \frac{du}{dy} u \Big|_y \right) = 0$$

(recall that our steady-flow energy equation sign convention is that heat transfer *into* the system, and work done *on* the system, are positive).

Dividing by Δy and taking limits as $\Delta y \to 0$, we obtain

$$-\frac{dq}{dy} + \mu \left(\frac{du}{dy} \right)^2 = 0$$

since du/dy is a constant. Substituting Fourier's law $q = -kdT/dy$ then gives

$$k \frac{d^2 T}{dy^2} + \mu \left(\frac{du}{dy} \right)^2 = 0 \qquad\qquad (5.5)$$

if k is assumed constant. Notice that Eq. (5.5) is just the steady one-dimensional heat conduction equation with a heat source $\mu(du/dy)^2$, the viscous dissipation. The viscous dissipation is always positive, reflecting the fact that mechanical work is converted irreversibly into heat. There is no reverse process whereby heat is converted into work by the action of viscosity.

Equation (5.4) gives $du/dy = u_e/L$; substituting in Eq. (5.5),

$$k \frac{d^2 T}{dy^2} + \frac{\mu u_e^2}{L^2} = 0 \qquad\qquad (5.6)$$

Equation (5.6) is easily integrated:

$$\frac{d^2 T}{dy^2} = -\frac{\mu u_e^2}{kL^2}$$

$$T = -\frac{\mu u_e^2}{2k} \left(\frac{y}{L} \right)^2 + C_1 y + C_2$$

The integration constants are evaluated from the boundary conditions:

$$y = 0: \quad T = T_s \tag{5.7a}$$

$$y = L: \quad T = T_e \tag{5.7b}$$

to obtain the temperature profile as

$$\frac{T_s - T}{T_s - T_e} = \frac{y}{L}\left[1 - \frac{\mu u_e^2}{2k(T_s - T_e)}\left(1 - \frac{y}{L}\right)\right] \tag{5.8}$$

The heat flux at the lower plate can now be found using Fourier's law:

$$q_s = -k\frac{dT}{dy}\Big|_0 = \frac{k(T_s - T_e)}{L} - \frac{\mu u_e^2}{2L} \tag{5.9}$$

The first term is simply the conduction heat flux across a slab of thickness L; the second term gives the contribution of viscous dissipation to the heat flux.

Temperature Profiles

To examine the temperature profiles given by Eq. (5.8), consider the situation where the stationary plate temperature, T_s, is higher than that of the moving plate, T_e. Figure 5.4 shows how the temperature profiles depend on the dimensionless parameter $\mu u_e^2/k(T_s - T_e)$, which is the *Brinkman number*, introduced in Section 4.2.2. The effect of increasing the plate speed is to increase the Brinkman number and the magnitude of the viscous dissipation heat source, as evidenced by the change in shape of the temperature profile. For Br $\rightarrow$ 0, the viscous heating is negligible, and the temperature profile is linear, as for simple conduction across a slab. The heat transfer is away from the hot plate. For Br $= 2$, the heat transfer from the hot plate is zero, and for Br > 2, heat is transferred from the fluid to the hot plate.

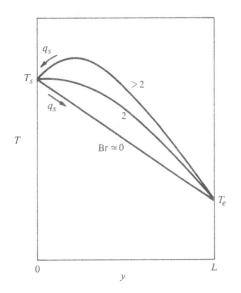

Figure 5.4 Temperature profiles for a Couette flow with $T_s > T_e$, showing the effect of Brinkman number.

5.2.2 The Recovery Factor Concept

Equation (5.9) for the wall heat flux can be rearranged as

$$q_s = \frac{k}{L}\left[T_s - \left(T_e + \mathrm{Pr}\frac{u_e^2}{2c_p}\right)\right] \tag{5.10}$$

With a little imagination, we can see how this result can be applied to a real boundary layer. Equation (5.10) shows that the driving potential for heat transfer when viscous heating is present is not $(T_s - T_e)$ but $T_s - [T_e + \mathrm{Pr}(u_e^2/2c_p)]$. If we define the *adiabatic wall temperature for a Couette flow* as $T_{aw} = T_e + \mathrm{Pr}(u_e^2/2c_p)$, then the effect of viscous dissipation on heat transfer can be accounted for by simply replacing T_e with T_{aw}. The adiabatic wall temperature is also the temperature the wall attains if $q_s = 0$, that is, is adiabatic. Thus, for real boundary layer flows with significant viscous heating, we simply replace Newton's law of cooling $q_s = h_c(T_s - T_e)$ with

$$q_s = h_c(T_s - T_{aw}) \tag{5.11}$$

and define a **recovery factor** r such that

$$T_{aw} = T_e + r\frac{u_e^2}{2c_p} \tag{5.12}$$

For the Couette flow, $r = \mathrm{Pr}$. More advanced analysis shows that the recovery factor for a laminar boundary layer on a flat plate is $r \simeq \mathrm{Pr}^{1/2}$, while experiment shows that for turbulent boundary layers $r \simeq \mathrm{Pr}^{1/3}$. If Eq. (5.12) is rewritten as

$$c_p T_{aw} = c_p T_e + r\frac{u_e^2}{2}$$

the recovery factor can be viewed as the fraction of kinetic energy of the free stream fluid, which is *recovered* as thermal energy in the fluid adjacent to an adiabatic wall. If the Prandtl number, and hence r, are less than unity, this concept seems reasonable. However, for most liquids, the Prandtl number, and hence r, are greater than unity, so the concept is too simplistic. Most high-speed flows of interest involve gases, particularly air, which explains the appeal of the recovery concept.

For engineering calculations of heat transfer in high-speed boundary layers, the formulas given in Section 4.3.2 for external flows are applicable if the free-stream temperature is replaced by the adiabatic wall temperature T_{aw}, with the appropriate recovery factor, namely, $\mathrm{Pr}^{1/2}$ for laminar flows and $\mathrm{Pr}^{1/3}$ for turbulent flows. Evaluation of fluid properties poses a special problem. For gases, fluid properties can be evaluated at the *Eckert reference temperature* [1]

$$T_r = T_e + 0.5(T_s - T_e) + 0.22(T_{aw} - T_e) \tag{5.13}$$

which is valid for both laminar and turbulent flows. Note also that most oils have viscosities that vary strongly with temperature; thus, more careful analysis allowing for a temperature-dependent viscosity may be needed for oils when temperature differences are large.

The student is cautioned that the recovery factors given here are, strictly speaking, accurate only for gas flows over flat, impermeable surfaces [2].

EXAMPLE 5.1   Temperature of a Helicopter Rotor

The main rotor of a helicopter has a radius of 5 m, has a blade width of 20 cm, and rotates at 380 rpm. Estimate the maximum temperature the rotor can attain if the forward speed is 90 m/s through air at 300 K.

Solution

Given: Helicopter flying at 90 m/s, rotor rotating at 380 rpm.

Required: An estimate of the maximum rotor temperature.

Assumptions: 1. Model the rotor as a flat plate.
                 2. The radiation heat loss and conduction within the rotor are negligible.

The rotor temperature varies with position and time in a complicated manner due to aerodynamic heating. The maximum temperature will occur at the rotor tip, where the surface energy balance is given by Eq. (1.34) as

$$q_{\text{cond}} - q_{\text{conv}} - q_{\text{rad}} = 0$$

An upper bound on the rotor surface temperature can be estimated by neglecting the radiation and conduction losses; then

$$q_{\text{conv}} = h_c(T_s - T_{\text{aw}}) = 0 \quad \text{or} \quad T_s = T_{\text{aw}}$$

From Eq. (5.12),

$$T_{\text{aw}} = T_e + r\frac{u_e^2}{2c_p}$$

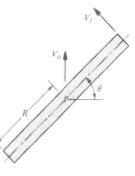

The tangential velocity component at the tip of the rotor is

$$V_t = R\omega = (5)(2\pi)(380/60) = 199.0 \text{ m/s}$$

The maximum speed is attained by the tip when $\theta = 0$ and is

$$u_e = V_0 + V_t = 90 + 199 = 289 \text{ m/s}$$

To evaluate the recovery factor r, we must check to see if the boundary layer is laminar or turbulent:

$$\text{Re} \simeq \frac{u_e L}{\nu} = \frac{(289)(0.20)}{(15.66 \times 10^{-6})} = 3.69 \times 10^6$$

That is, the boundary layer is turbulent over almost the whole blade in the tip region. Thus,

$$r \simeq \text{Pr}^{1/3} = (0.69)^{2/3} = 0.884$$

$$T_s = T_{\text{aw}} = 300 + \frac{(0.884)(289)^2}{(2)(1005)} = 300 + 36.7 = 336.7 \text{ K}$$

Comments

1. To check if radiation losses are negligible, we estimate the convective heat transfer co-efficient. Assuming that a turbulent boundary layer starts at the leading edge, Eq. (4.65) gives

$$\overline{\mathrm{Nu}} = 0.036\mathrm{Re}_L^{0.8}\mathrm{Pr}^{0.43} = 0.036(3.69 \times 10^6)^{0.8}(0.69)^{0.43} = 5500$$

$$\overline{h}_c = (k/L)\overline{\mathrm{Nu}} = (0.0267/0.20)(5500) = 734 \text{ W/m}^2 \text{ K}$$

where properties for air at 300 K have been used. The surface energy balance for convection-radiation equilibrium is

$$q_{\mathrm{conv}} + q_{\mathrm{rad}} = 0$$

Since $(T_s - T_e)$ is small, q_{rad} can be linearized as $4\sigma\varepsilon T_m^3(T_s - T_e) \simeq 7(T_s - T_e)$ for $T_m = 315$ K, $\varepsilon = 1$.

$$\overline{h}_c(T_s - T_{\mathrm{aw}}) + 7(T_s - T_e) = 0$$
$$734(T_s - 336.7) + 7(T_s - 300) = 0$$
$$T_s = 336.4 \text{ K}$$

which is called the *convection-radiation equilibrium temperature*. The effect of radiation is negligible in this situation.

2. The conduction losses are more difficult to estimate since the problem is both unsteady and two-dimensional.

5.3   LAMINAR FLOW IN A TUBE

Figure 5.5 shows a steady fluid flow in a tube sufficiently far downstream for entrance effects to be negligible. The flow is laminar, which usually requires a Reynolds number $u_b D/\nu < 2300$. Constant-property fully developed laminar flow in a tube is a simple flow, for which we can obtain the heat transfer coefficient by exact analysis. For this forced-convection problem, we follow the general procedure introduced in Section 5.2. The fluid mechanics problem is first solved to obtain the characteristic parabolic velocity profile $u(r)$. The student will no doubt already have encountered this flow in a physics or fluid mechanics course, where it was called *Poiseuille flow*.

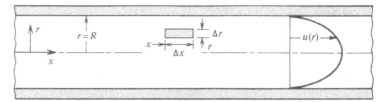

Figure 5.5   Schematic of fully developed laminar flow in a tube, showing the coordinate system and characteristic parabolic velocity profile.

The energy conservation principle is then used to derive the differential equation governing the temperature field. For uniform heating of the tube wall, this equation is solved for the temperature field $T(x, r)$ and the heat transfer coefficient and Nusselt number so obtained.

5.3.1 Momentum Transfer in Hydrodynamically Fully Developed Flow

At a long distance from the tube entrance, the flow "forgets" the precise nature of its initial velocity profile, and, irrespective of the entrance configuration (elbow, flare, etc.), there will be a characteristic axial velocity profile $u(r)$ unchanging along the tube (if the fluid properties remain constant). Correspondingly, there will be no velocity component in the radial direction. Also, the axial pressure gradient required to sustain the flow against the viscous forces will be constant along the tube. The flow is then said to be **hydrodynamically fully developed**.

The Governing Differential Equation

Since velocities are unchanging with time and unchanging along the tube, the momentum conservation principle reduces to a simple force balance. Accordingly, a force balance is made on an elemental fluid element, Δx long, located between radii r and $r + \Delta r$, as shown in Figs. 5.5 and 5.6a. The pressure and viscous forces acting

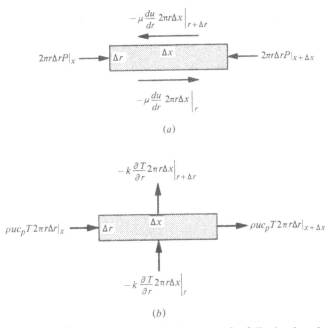

Figure 5.6 (a) Force balance on a fluid element for fully developed laminar flow in a tube. (b) The first law of thermodynamics applied to an elemental control volume for fully developed heat transfer in a tube.

on the element are in balance, and their sum equals zero:

$$[2\pi r \Delta r P]_x + \left[2\pi r \Delta x \left(-\mu \frac{du}{dr} \right) \right]_r - [2\pi r \Delta r P]_{x+\Delta x} - \left[2\pi r \Delta x \left(-\mu \frac{du}{dr} \right) \right]_{r+\Delta r} = 0$$

Dividing by $2\pi \Delta r \Delta x$ and rearranging,

$$\frac{r\mu(du/dr)|_{r+\Delta r} - r\mu(du/dr)|_r}{\Delta r} = \frac{rP|_{x+\Delta x} - rP|_x}{\Delta x}$$

Letting $\Delta r, \Delta x \to 0$ then gives

$$\frac{d}{dr}\left(r\mu \frac{du}{dr} \right) = r\frac{dP}{dx} \tag{5.14}$$

which is the desired differential equation. Since dP/dx is a constant for hydrodynamically fully developed flow, it is a second-order ordinary differential equation for $u(r)$, and the two boundary conditions required are

$$r = 0: \quad \frac{du}{dr} = 0, \quad \text{by symmetry} \tag{5.15a}$$

$$r = R: \quad u = 0, \quad \text{no slip at the wall} \tag{5.15b}$$

Solution for the Velocity Profile

Integrating Eq. (5.14) once with respect to r gives

$$r\mu \frac{du}{dr} = \frac{1}{2}r^2 \frac{dP}{dx} + C_1$$

or

$$\frac{du}{dr} = \frac{r}{2\mu}\frac{dP}{dx} + \frac{C_1}{r\mu}$$

Using the first boundary condition, Eq. (5.15a), $C_1 = 0$. Integrating again,

$$u = \frac{r^2}{4\mu}\frac{dP}{dx} + C_2$$

The second boundary condition, Eq. (5.15b), gives C_2.

$$0 = \frac{R^2}{4\mu}\frac{dP}{dx} + C_2$$

and substituting back gives the velocity profile $u(r)$:

$$u = \frac{R^2}{4\mu}\left(-\frac{dP}{dx} \right)\left[1 - \left(\frac{r}{R} \right)^2 \right] \tag{5.16}$$

The velocity profile is seen to be parabolic in shape. The maximum velocity is on the centerline, where $r = 0$, and is

$$u_{max} = \frac{R^2}{4\mu}\left(-\frac{dP}{dx}\right)$$ (5.17)

The bulk velocity u_b was introduced in Section 4.2.2 and is simply the area-weighted average velocity,

$$u_b = \frac{1}{\pi R^2}\int_0^R u2\pi r\,dr$$

Substituting from Eq. (5.16) for u and using Eq. (5.17) gives

$$u_b = \frac{1}{2}u_{max} = \frac{R^2}{8\mu}\left(-\frac{dP}{dx}\right)$$ (5.18)

The velocity profile $u(r)$ can then be conveniently expressed in terms of the bulk velocity as

$$u = 2u_b\left[1 - \left(\frac{r}{R}\right)^2\right]$$ (5.19)

In practice, the mass flow rate $\dot{m}$ [kg/s] is usually known; then the bulk velocity is obtained from the relation $\dot{m} = \rho u_b A_c$.

The Friction Factor and Skin Friction Coefficient

The Darcy friction factor for flow in a tube was defined by Eq. (4.15); writing $\Delta P/L$ as $-dP/dx$ and $D = 2R$ gives

$$f = \frac{(-dP/dx)(2R)}{(1/2)\rho u_b^2}$$

Substituting for dP/dx from Eq. (5.18),

$$f = \frac{64\mu}{(2R)\rho u_b} = \frac{64}{Re_D}$$ (5.20)

where the Reynolds number $Re_D = (2R)\rho u_b/\mu$; this result is the same as Eq. (4.39). The shear stress exerted by the fluid on the wall can be obtained from the velocity profile Eq. (5.19) by using Newton's law of viscosity:

$$\tau_s = -\mu\frac{du}{dr}\bigg|_{r=R} = -\mu(2u_b)\left(-\frac{2r}{R^2}\right)_{r=R} = \frac{4\mu u_b}{R}$$ (5.21)

The skin friction coefficient C_f is then

$$C_f = \frac{\tau_s}{(1/2)\rho u_b^2} = \frac{4\mu u_b/R}{(1/2)\rho u_b^2} = \frac{16\mu}{(2R)\rho u_b} = \frac{16}{Re_D} = \frac{f}{4}$$ (5.22)

Recall that the relation $f = 4C_f$ can also be obtained by a simple force balance (Exercise 4-3). (See also Exercise 5–19.)

5.3.2 Fully Developed Heat Transfer for a Uniform Wall Heat Flux

In Section 5.3.1, we saw that at distances sufficiently far from a tube entrance, the velocity profile does not change along the tube and has a characteristic parabolic form; the wall shear stress does not change either. If the tube wall is heated such that the wall heat flux is constant with axial position (e.g., by winding an electrical heating wire around a tube at a constant pitch), then the bulk fluid temperature increases steadily in the flow direction at a rate dependent on the power input to the heater. At a distance sufficiently far from where heating commences, the heat transfer coefficient for a constant-property fluid becomes constant, and the *shape* of the temperature profile also does not change along the tube: the flow is then said to be **thermally fully developed.**

The Governing Differential Equation

To derive the differential equation governing the temperature profile, we apply the first law of thermodynamics to an elemental control volume Δx long located between radii r and $r + \Delta r$, as shown in Figure 5.6b. We will assume constant fluid properties, low-speed flow so that work done by the viscous stresses can be ignored, and negligible change in potential energy.[2] Since the specific heat is assumed constant, we will write $h = c_p T$ for convenience, which implies an enthalpy datum state of $h = 0$ at $T = 0$. The steady-flow energy equation, Eq. (1.4), applies and requires that the net outflow of enthalpy equal the heat conducted into the volume. The required quantities are as follows:

Rate of enthalpy inflow at x: $\qquad \rho u c_p T 2\pi r \Delta r \big|_x$

Rate of enthalpy outflow at $x + \Delta x$: $\qquad \rho u c_p T 2\pi r \Delta r \big|_{x+\Delta x}$

Rate of heat conduction in at r: $\qquad -k\dfrac{\partial T}{\partial r} 2\pi r \Delta x \bigg|_r$

Rate of heat conduction out at $r + \Delta r$: $\qquad -k\dfrac{\partial T}{\partial r} 2\pi r \Delta x \bigg|_{r+\Delta r}$

Heat conduction in the x direction is neglected, as it is expected to be small compared to the enthalpy flow. We will return to this point after completing the analysis. Equation (1.4) becomes $\dot{m}\Delta h = \Delta \dot{Q}$ for an incremental change in enthalpy h

$$\rho u c_p T 2\pi r \Delta r \big|_{x+\Delta x} - \rho u c_p T 2\pi r \Delta r \big|_x = -k\frac{\partial T}{\partial r} 2\pi r \Delta x \bigg|_r + k\frac{\partial T}{\partial r} 2\pi r \Delta x \bigg|_{r+\Delta r}$$

Rearranging and dividing by $2\pi \Delta x \Delta r$ gives

$$\frac{\rho u c_p r (T\big|_{x+\Delta x} - T\big|_x)}{\Delta x} = \frac{k[r(\partial T/\partial r)\big|_{r+\Delta r} - r(\partial T/\partial r)\big|_r]}{\Delta r}$$

since $u = u(r)$ and is not a function of x for hydrodynamically fully developed flow; also, the fluid properties ρ, c_p, and k have been assumed constant. Let $\Delta x, \Delta r \to 0$;

---

[2] For constant density and a fully developed velocity profile, there is no change in kinetic energy along the tube.

then

$$\rho u c_p r \frac{\partial T}{\partial x} = k \frac{\partial}{\partial r} \left(r \frac{\partial T}{\partial r} \right) \qquad (5.23)$$

or

$$u \frac{\partial T}{\partial x} = \frac{\alpha}{r} \frac{\partial}{\partial r} \left(r \frac{\partial T}{\partial r} \right) \qquad (5.24)$$

where $\alpha = k/\rho c_p$ is the thermal diffusivity of the fluid. Substituting from Eq. (5.19) for the velocity $u(r)$ gives

$$2u_b \left[1 - \left(\frac{r}{R} \right)^2 \right] \frac{\partial T}{\partial x} = \frac{\alpha}{r} \frac{\partial}{\partial r} \left(r \frac{\partial T}{\partial r} \right) \qquad (5.25)$$

which is a partial differential equation since T is a function of both x and r. Fortunately, we can show that for fully developed heat transfer and uniform heating, $\partial T / \partial x$ is a constant, so that Eq. (5.25) reduces to an ordinary differential equation.

The Axial Temperature Gradient

To determine $\partial T / \partial x$, we apply the steady-flow energy equation to a slice of tube Δx long, as shown in Fig. 5.7.

$$\int_0^R \rho u c_p T 2\pi r dr \Big|_{x+\Delta x} - \int_0^R \rho u c_p T 2\pi r dr \Big|_x = q_s 2\pi R \Delta x \qquad (5.26)$$

But from the definition of bulk temperature, Eq. (4.7),

$$\int_0^R \rho u c_p T 2\pi r dr = \dot{m} c_p T_b = \rho u_b \pi R^2 c_p T_b \qquad (5.27)$$

hence,

$$\rho u_b c_p \pi R^2 T_b \Big|_{x+\Delta x} - \rho u_b c_p \pi R^2 T_b \Big|_x = q_s 2\pi R \Delta x$$

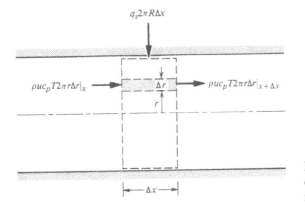

Figure 5.7 The steady-flow energy equation applied to flow in a length of tube Δx long.

Dividing by $\rho u_b c_p \pi R^2 \Delta x$ and letting $\Delta x \to 0$,

$$\frac{dT_b}{dx} = \frac{2q_s}{\rho u_b c_p R} = \text{Constant} \tag{5.28}$$

since q_s is a constant for a uniform wall heat flux. Figure 5.8 shows the expected temperature profiles for fully developed heat transfer. Far enough from the entrance, the *shape* of the profile does not change along the tube. Thus, radial temperature differences are constant; that is, $T_s - T_b, T_s - T(r)$, and $T(r) - T_b$ are independent of x. In particular,

$$T - T_b = \mathcal{R}(r) \tag{5.29}$$

is a function of r only. Hence,

$$\frac{\partial T}{\partial x} - \frac{dT_b}{dx} = 0$$

or, using Eq. (5.28),

$$\frac{\partial T}{\partial x} = \frac{dT_b}{dx} = \frac{2q_s}{\rho u_b c_p R} \tag{5.30}$$

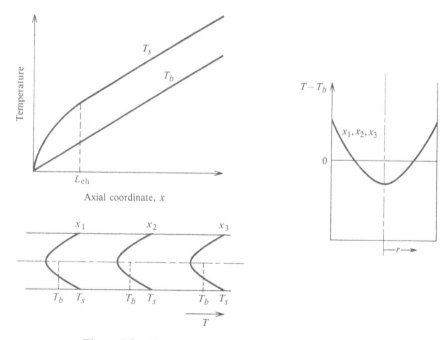

Figure 5.8   Temperature profiles for fully developed heat transfer in a tube with a uniform wall heat flux.

a constant. Also, differentiating Eq. (5.29) with respect to r,

$$\frac{\partial T}{\partial r} - 0 = \frac{d\mathscr{R}(r)}{dr}$$

that is, $\partial T/\partial r$ is not a function of x. By definition, the heat transfer coefficient is

$$h_c = \frac{q_s}{T_s - T_b} = \frac{-k(\partial T/\partial r)|_R}{T_s - T_b} \tag{5.31}$$

and since both the numerator and denominator are constants, so is h_c. Thus, we see that it is the unchanging shape of the temperature profile that gives the characteristic constant heat transfer coefficient for fully developed conditions.

Solution for the Radial Temperature Profile

From Eq. (5.29), we can write $T = T_b + \mathscr{R}(r)$ and substitute in Eq. (5.25) to obtain

$$2u_b \left[1 - \left(\frac{r}{R}\right)^2\right] \frac{dT_b}{dx} = \frac{\alpha}{r} \frac{d}{dr}\left(r\frac{d\mathscr{R}}{dr}\right)$$

where the partial derivatives have become total derivatives because $T_b = T_b(x)$ and $\mathscr{R} = \mathscr{R}(r)$. Then, using Eq. (5.28) and rearranging,

$$\left[1 - \left(\frac{r}{R}\right)^2\right] \frac{4q_s}{kR} = \frac{1}{r}\frac{d}{dr}\left(r\frac{d\mathscr{R}}{dr}\right) \tag{5.32}$$

which is an ordinary differential equation for $\mathscr{R}(r)$. Appropriate boundary conditions are

$$r = 0: \quad \frac{dT}{dr} = \frac{d\mathscr{R}}{dr} = 0 \quad \text{by symmetry} \tag{5.33a}$$

$$r = R: \quad T = T_s \quad \text{or} \quad \mathscr{R} = T_s - T_b \tag{5.33b}$$

Equation (5.33b) is used as the boundary condition at the wall since the specified constant heat flux condition has been used already in deriving the differential equation; the fact that T_s is unknown will present no difficulty. Integrating once,

$$r\frac{d\mathscr{R}}{dr} = \frac{4q_s}{kR}\left[\frac{r^2}{2} - \frac{r^4}{4R^2}\right] + C_1$$

$$\frac{d\mathscr{R}}{dr} = \frac{4q_s}{kR}\left[\frac{r}{2} - \frac{r^3}{4R^2}\right] + \frac{C_1}{r}$$

and the first boundary condition, Eq. (5.33a), requires that $C_1 = 0$. Integrating again,

$$\mathscr{R} = \frac{4q_s}{kR}\left[\frac{r^2}{4} - \frac{r^4}{16R^2}\right] + C_2$$

Evaluating C_2 from the second boundary condition, Eq. (5.33b), and noting that $T = T_b + \mathscr{R}$ from Eq. (5.29) gives

$$T = T_s - \frac{4q_s}{kR}\left[\frac{3R^2}{16} - \frac{r^2}{4} + \frac{r^4}{16R^2}\right] \tag{5.34}$$

a quartic form for the temperature profile $T(r)$. The bulk temperature can now be determined using its definition, Eq. (4.7), in the form

$$T_b = \frac{\int_0^R uT 2\pi r\,dr}{\pi R^2 u_b}$$

Substituting from Eqs. (5.19) and (5.34) and simplifying gives

$$T_b = \frac{4}{R^2}\int_0^R \left[1 - \left(\frac{r}{R}\right)^2\right]\left[T_s - \frac{4q_s}{kR}\left(\frac{3R^2}{16} - \frac{r^2}{4} + \frac{r^4}{16R^2}\right)\right]r\,dr$$

$$= T_s - \frac{11}{24}\frac{q_s R}{k}$$

or

$$T_s - T_b = \frac{11}{24}\frac{q_s R}{k} \tag{5.35}$$

Equation (5.35) shows that the difference between the wall and bulk temperatures is directly proportional to the wall heat flux and tube radius and inversely proportional to the fluid conductivity.

The Heat Transfer Coefficient and Nusselt Number

Since the temperature profile is in terms of the wall heat flux q_s, we do not need to use Fourier's law to find q_s and hence the heat transfer coefficient. The heat transfer coefficient is simply $h_c = q_s/(T_s - T_b)$; thus, using Eq. (5.35),

$$h_c = \frac{24k}{11R} = \frac{48k}{11D}, \quad \text{where } D = 2R$$

and the Nusselt number $\mathrm{Nu}_D = h_c D/k$ is

$$\mathrm{Nu}_D = \frac{48}{11} = 4.364 \tag{5.36}$$

which is the result given in Section 4.3.1 as Eq. (4.41). This result is perhaps surprising. The Nusselt number is not a function of the Reynolds and Prandtl numbers, as might have been expected from the dimensional analysis of Section 4.2.2. However, like the skin friction coefficient, Eq. (5.22), the Stanton number, $\mathrm{St} = \mathrm{Nu}/\mathrm{RePr}$, is inversely proportional to Reynolds number. The physical reason for why the heat transfer coefficient does not depend on velocity for this flow should become clear after boundary layer flows are analyzed in Section 5.4.

As a final comment, notice that Eq. (5.30) states that $\partial T/\partial x$ is a constant. Hence, the contribution of x-direction conduction to $\dot{Q}$ in the steady-flow energy equation is not just small, as was expected, but is in fact zero for this situation of fully developed heat transfer with a uniform wall heat flux. The rate of heat conduction into the elemental control volume at x exactly equals the rate of heat conduction out at $x + \Delta x$.

EXAMPLE 5.2   An Oil Heater

A special-purpose oil flows at 1.81×10^{-2} kg/s inside a 1 cm-diameter tube that is heated electrically at a rate of 76 W/m. At a particular location where the flow and heat transfer are fully developed, the wall temperature is 370 K. Determine (i) the oil bulk temperature, (ii) the centerline temperature, (iii) the axial gradient in bulk temperature, and (iv) the heat transfer coefficient. For properties, take $k = 0.139$ W/m K. $\rho = 854$ kg/m^3, $c_p = 2120$ J/kg K, and $\nu = 41 \times 10^{-6}$ m^2/s.

Solution

Given: Oil flowing inside a uniformly heated tube.

Required: (i) T_b, (ii) T_c, (iii) dT_b/dx, (iv) h_c.

Assumptions: Fully developed flow and heat transfer.

First check the Reynolds number:

$$u_b = \frac{\dot{m}}{\rho \pi R^2} = \frac{(1.81 \times 10^{-2})}{(854)(\pi)(0.005)^2} = 0.270 \text{ m/s}$$

$$\text{Re} = \frac{u_b D}{\nu} = \frac{(0.270)(0.01)}{(41 \times 10^{-6})} = 66$$

that is, laminar.

(i) From Eq. (5.35),

$$T_b = T_s - \frac{11}{24} \frac{q_s R}{k}$$

To obtain the wall heat flux q_s, the heat input per unit length must be divided by the tube perimeter πD:

$$q_s = \frac{76}{(\pi)(0.01)} = 2419 \text{ W/m}^2$$

Hence,

$$T_b = 370 - \frac{11}{24} \frac{(2419)(0.005)}{(0.139)} = 330.1 \text{ K}$$

(ii) Substituting $r = 0$ in Eq. (5.34),

$$T_c = T_s - \frac{4q_s}{kR}\frac{3R^2}{16} = T_s - \frac{3}{4}\frac{q_sR}{k}$$

$$T_c = 370 - \frac{3}{4}\frac{(2419)(0.005)}{(0.139)} = 304.7 \text{ K}$$

(iii) From Eq. (5.28),

$$\frac{dT_b}{dx} = \frac{2q_s}{\rho u_b c_p R} = \frac{(2)(2419)}{(854)(0.270)(2120)(0.005)} = 1.98 \text{ K/m}$$

(iv) From Eq. (5.36),

$$\text{Nu}_D = \frac{h_c D}{k} = 4.364$$

$$h_c = \frac{k\text{Nu}_D}{D} = \frac{(0.139)(4.364)}{(0.01)} = 60.7 \text{ W/m}^2 \text{ K}$$

5.4   LAMINAR BOUNDARY LAYERS

The analysis of laminar-flow boundary layers has played a central role in the development of convection theory. The results of such analysis have practical utility, but, perhaps more importantly, the laminar boundary layer is the simplest flow that exhibits the essential features of convection. There is a price to be paid: the analysis that follows is significantly more difficult than the preceding analyses of Couette and tube flow. Exact solution of the governing differential equations requires numerical rather than analytical methods. Fortunately, the approximate *integral method* can be used to obtain accurate results in many situations. Also, formulation and use of the integral method gives added insight into the physics of the problem. The boundary layer for forced flow along a flat plate will be analyzed first and given considerable attention. Subsequently, the boundary layer for natural convection on a vertical wall will be analyzed.

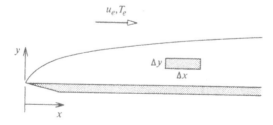

Figure 5.9 Schematic of a laminar boundary layer on a flat plate, showing the coordinate system.

5.4.1 The Governing Equations for Forced Flow along a Flat Plate

Figure 5.9 shows a schematic of the laminar boundary layer for forced flow along a flat plate, with a constant free-stream velocity u_e and constant free-stream temperature T_e. To perform an analysis of momentum and heat transfer, we need to first derive differential equations governing conservation of mass, momentum, and energy. For simplicity, we will assume steady flow, constant fluid properties, and low speeds such that there is negligible work done by the viscous stresses. Our elemental control volume is Δx by Δy by unity.

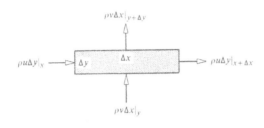

Figure 5.10 Conservation of mass applied to an elemental control volume in a laminar boundary layer on a flat plate.

Referring to Figs. 5.9 and 5.10, conservation of mass requires that the mass inflow minus the mass outflow equal zero:

$$\rho u \Delta y|_x + \rho v \Delta x|_y - \rho u \Delta y|_{x+\Delta x} - \rho v \Delta x|_{y+\Delta y} = 0$$

Dividing by $\rho \Delta x \Delta y$ and rearranging gives

$$\frac{u|_{x+\Delta x} - u|_x}{\Delta x} + \frac{v|_{y+\Delta y} - v|_y}{\Delta y} = 0$$

and letting $\Delta x, \Delta y \to 0$,

$$\frac{\partial u}{\partial x} + \frac{\partial v}{\partial y} = 0 \tag{5.37}$$

Equation (5.37) is the **mass conservation or continuity equation.**

Referring to Fig. 5.11, Newton's second law applied to the control volume requires that the net outflow of x-direction momentum equal the sum of the viscous forces

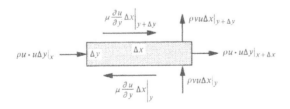

Figure 5.11 Newton's second law applied to an elemental control volume in a laminar boundary layer on a flat plate.

acting in the x-direction:[3]

$$\rho u^2 \Delta y\big|_{x+\Delta x} + \rho v u \Delta x\big|_{y+\Delta y} - \rho u^2 \Delta y\big|_x - \rho v u \Delta x\big|_y = \mu \frac{\partial u}{\partial y}\Delta x\bigg|_{y+\Delta y} - \mu \frac{\partial u}{\partial y}\Delta x\bigg|_y$$

Dividing by $\rho \Delta x \Delta y$ and rearranging gives

$$\frac{u^2\big|_{x+\Delta x} - u^2\big|_x}{\Delta x} + \frac{vu\big|_{y+\Delta y} - vu\big|_y}{\Delta y} = v \frac{(\partial u/\partial y)\big|_{y+\Delta y} - (\partial u/\partial y)\big|_y}{\Delta y}$$

Letting Δx and $\Delta y \to 0$,

$$\frac{\partial}{\partial x}(u^2) + \frac{\partial}{\partial y}(vu) = v\frac{\partial^2 u}{\partial y^2} \tag{5.38}$$

or

$$2u\frac{\partial u}{\partial x} + u\frac{\partial v}{\partial y} + v\frac{\partial u}{\partial y} = v\frac{\partial^2 u}{\partial y^2}$$

Multiplying the continuity equation by u and subtracting it from Eq. (5.38) gives

$$u\frac{\partial u}{\partial x} + v\frac{\partial u}{\partial y} = v\frac{\partial^2 u}{\partial y^2} \tag{5.39}$$

which is the **momentum conservation equation.** Notice that there are no pressure forces acting on the fluid since there is no x-direction pressure gradient in the flow outside the boundary layer for flow along a flat plate, and pressure changes across the thin boundary layer are negligible. Also, the viscous shear stress $-\mu \partial u/\partial x$ has been ignored since it is negligible in a thin boundary layer. Scaling arguments to justify these assumptions are given in advanced texts.

Referring to Fig. 5.12, the steady-flow energy equation, Eq. (1.4), requires that the net outflow of enthalpy equal the heat conducted into the volume. Writing $h = c_p T$ again,

$$\rho u c_p T \Delta y\big|_{x+\Delta x} - \rho u c_p T \Delta y\big|_x + \rho v c_p T \Delta x\big|_{y+\Delta y} - \rho v c_p T \Delta x\big|_y$$

$$= -k\frac{\partial T}{\partial y}\Delta x\bigg|_y + k\frac{\partial T}{\partial y}\Delta x\bigg|_{y+\Delta y}$$

Dividing by $\rho c_p \Delta x \Delta y$ and rearranging,

$$\frac{uT\big|_{x+\Delta x} - uT\big|_x}{\Delta x} + \frac{vT\big|_{y+\Delta y} - vT\big|_y}{\Delta y} = \alpha\left[\frac{(\partial T/\partial y)\big|_{y+\Delta y} - (\partial T/\partial y)\big|_y}{\Delta y}\right]$$

Letting $\Delta x, \Delta y \to 0$,

$$\frac{\partial}{\partial x}(uT) + \frac{\partial}{\partial y}(vT) = \alpha\frac{\partial^2 T}{\partial y^2} \tag{5.40}$$

---

[3] Alternatively, we could require that the net *inflow* of x-direction momentum equal the sum of the viscous forces acting in the x-direction to *oppose* the flow. Since the flow in the boundary layer is decelerating, the latter statement is more in line with our physical intuition.

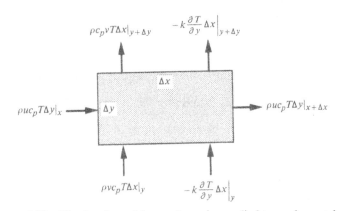

Figure 5.12 The first law of thermodynamics applied to an elemental control volume in a laminar boundary layer on a flat plate.

or

$$T\frac{\partial u}{\partial x} + u\frac{\partial T}{\partial x} + T\frac{\partial v}{\partial y} + v\frac{\partial T}{\partial y} = \alpha\frac{\partial^2 T}{\partial y^2} \tag{5.41}$$

Multiplying the continuity equation by T and subtracting from Eq. (5.41) gives the **energy conservation equation**

$$u\frac{\partial T}{\partial x} + v\frac{\partial T}{\partial y} = \alpha\frac{\partial^2 T}{\partial y^2} \tag{5.42}$$

Notice that conduction in the x direction, $-k\partial T/\partial x$, has been ignored since, like the shear stress $-\mu\partial u/\partial x$, it is negligible in a thin boundary layer. Scaling arguments to justify this assumption are given in advanced texts. In addition, since a low-speed flow has been assumed, there are no terms that account for production of thermal energy by viscous dissipation.

Equations (5.37), (5.39), and (5.42), subject to appropriate boundary conditions, are to be solved to obtain the velocity and temperature profiles and the associated rates of momentum and heat transfer.

5.4.2 The Plug Flow Model

We first consider a simple model in which the velocity, u, is assumed constant through the boundary layer at its free-stream value u_e, that is, as if the flow were inviscid. This model is often called a *plug* or *slug* flow model. Then,

$$\frac{\partial u}{\partial x} = \frac{du_e}{dx} = 0$$

and from Eq. (5.37),

$$\frac{\partial v}{\partial y} = 0, \quad v = \text{Constant} = 0 \quad \text{since } v\big|_{y=0} = 0$$

Thus, the energy equation Eq. (5.42) becomes

$$u_e \frac{\partial T}{\partial x} = \alpha \frac{\partial^2 T}{\partial y^2} \tag{5.43}$$

If the plate temperature is uniform, appropriate boundary conditions are

$$y = 0: \quad T = T_s \quad \text{and} \quad x = 0, y \to \infty: \quad T = T_e \tag{5.44}$$

where, like u_e, T_e is not a function of x. We now define a new independent variable $\zeta = x/u_e$ (which has the dimensions of time), and the mathematical problem becomes

$$\frac{\partial T}{\partial \zeta} = \alpha \frac{\partial^2 T}{\partial y^2} \tag{5.45}$$

$$y = 0: \quad T = T_s \quad \text{and} \quad \zeta = 0, y \to \infty: \quad T = T_e \tag{5.46}$$

This problem is identical to the problem of heat conduction in a semi-infinite solid, analyzed in Section 3.4.2. Replacing x by y, t by $\zeta = x/u_e$, and T_0 by T_e in Eq. (3.58) gives

$$\frac{T - T_e}{T_s - T_e} = \operatorname{erfc} \eta; \quad \eta = \frac{y}{(4\alpha x/u_e)^{1/2}} \tag{5.47}$$

and from Eq. (3.59),

$$q_s = \frac{k(T_s - T_e)}{(\pi \alpha x/u_e)^{1/2}} \tag{5.48}$$

The local heat transfer coefficient is then

$$h_{cx} = \frac{q_s}{T_s - T_e} = \frac{k}{\pi^{1/2}} \left(\frac{u_e}{\alpha x} \right)^{1/2} \tag{5.49}$$

and the local Nusselt number is

$$\mathrm{Nu}_x = \frac{h_{cx} x}{k} = \frac{1}{\pi^{1/2}} \left(\frac{u_e x}{\alpha} \right)^{1/2} = \frac{1}{\pi^{1/2}} \left(\frac{u_e x}{\nu} \right)^{1/2} \left(\frac{\nu}{\alpha} \right)^{1/2}$$

or

$$\mathrm{Nu}_x = 0.564 \mathrm{Re}_x^{1/2} \mathrm{Pr}^{1/2} (= 0.564 \mathrm{Pe}_x^{1/2}); \quad \mathrm{Pe}_x \gtrsim 10^3 \tag{5.50}$$

which is the basis for the correlation recommended for heat transfer to liquid metals in Section 4.3.2. Typical velocity and temperature profiles in a liquid-metal boundary layer are shown in Fig. 5.13, where it can be seen that the velocity is nearly constant throughout the region where the temperature changes from its wall to free-stream value. Or, to put it another way, the *hydrodynamic boundary layer* is appreciably thinner than the *thermal boundary layer*. It is the characteristic high thermal conductivity of liquid metals that causes this behavior, and this is evidenced in the characteristic low values of Prandtl number (0.001–0.05). In fact, Eq. (5.50) can be shown to be an exact solution in the limit of $\mathrm{Pr} \to 0$.

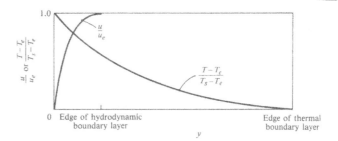

Figure 5.13 Velocity and temperature profiles for a liquid-metal laminar boundary layer.

5.4.3 Integral Solution Method

For fluids with Prandtl number of order unity or greater, such as gases, water, and oils, the simple plug flow model is inadequate. The hydrodynamic boundary layer is of the same or greater thickness than the thermal boundary layer. To avoid considering exact solutions that require numerical methods, we will consider the approximate **integral method**, which yields accurate results for the flat plate problem.

The Fluid Mechanics Problem

The *integral form* of the momentum conservation equation can be derived from first principles or by integrating the differential equation across the boundary layer; the former approach will be used here. Figure 5.14 shows an elemental control volume of unit depth located between x and $x + \Delta x$ and extending to a location $y = Y$, where Y is greater than the boundary layer thickness δ. Equating the net momentum outflow to the viscous force exerted by the wall on the fluid,

$$\int_0^Y \rho u^2 \, dy \Big|_{x+\Delta x} + \rho v u \Big|_Y \Delta x - \int_0^Y \rho u^2 \, dy \Big|_x = -\mu \frac{\partial u}{\partial y} \Big|_0 \Delta x$$

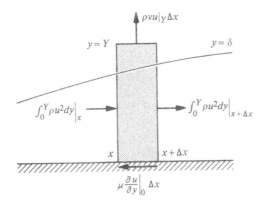

Figure 5.14 Elemental control volume for the integral method application of Newton's second law to a boundary layer on a flat plate.

Dividing by $\rho \Delta x$ and letting $\Delta x \to 0$ gives

$$\frac{d}{dx}\int_0^Y u^2 dy + vu|_Y = -v\frac{\partial u}{\partial y}\Big|_0$$

Now $u|_Y = u_e$, and from the continuity equation Eq. (5.37),

$$\frac{\partial v}{\partial y} = -\frac{\partial u}{\partial x}, \quad v(Y) = v|_0 - \int_0^Y \frac{\partial u}{\partial x}dy = -\int_0^Y \frac{\partial u}{\partial x}dy$$

Substituting back,

$$\frac{d}{dx}\int_0^Y u^2 dy - \int_0^Y u_e\frac{\partial u}{\partial x}dy = -v\frac{\partial u}{\partial y}\Big|_0$$

Interchanging the order of integration and differentiation in the second term, which allows the partial derivative to be replaced by a total derivative, and rearranging gives

$$\frac{d}{dx}u_e^2\int_0^Y \left(1 - \frac{u}{u_e}\right)\left(\frac{u}{u_e}\right)dy = v\frac{\partial u}{\partial y}\Big|_0$$

But the integrand is zero for $y > \delta$, so we can now let $Y \to \infty$:

$$\frac{d}{dx}u_e^2\int_0^\infty \left(1 - \frac{u}{u_e}\right)\left(\frac{u}{u_e}\right)dy = v\frac{\partial u}{\partial y}\Big|_0 \qquad (5.51)$$

which is the integral form of the momentum conservation equation.

It is convenient to define some characteristic thickness parameters of the boundary layer. Referring to Fig. 5.15, the deficit in mass flow due to retardation in the boundary layer is set equal to u_e times the *displacement thickness* δ_1:

$$\delta_1 u_e = \int_0^\infty (u_e - u)dy$$

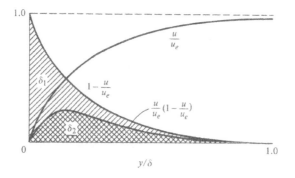

Figure 5.15   Displacement and momentum thickness integrals for a boundary layer.

or

$$\delta_1 = \int_0^\infty \left(1 - \frac{u}{u_e}\right) dy \tag{5.52}$$

That is, δ_1 is the distance the flow outside the boundary layer is displaced due to the presence of the wall. Similarly, the deficit in momentum is set equal to u_e^2 times the *momentum thickness* δ_2:

$$\delta_2 u_e^2 = \int_0^\infty u(u_e - u)dy$$

or

$$\delta_2 = \int_0^\infty \left(1 - \frac{u}{u_e}\right)\left(\frac{u}{u_e}\right) dy \tag{5.53}$$

Equation (5.51) can then be rewritten as

$$\frac{d\delta_2}{dx} = \frac{\mu(\partial u/\partial y)|_0}{\rho u_e^2}$$

or, introducing the skin-friction coefficient $C_{fx} = \tau_s/(1/2)\rho u_e^2$, where $\tau_s = \mu \left.\frac{\partial u}{\partial y}\right|_0$ is the shear stress exerted on the wall by the fluid from above,

$$\frac{d\delta_2}{dx} = \frac{C_{fx}}{2} \tag{5.54}$$

This simple equation states that the rate of increase of the momentum deficit in the boundary layer is equal to one-half the skin friction coefficient. Equation (5.54) is exact, but it cannot be used without knowledge of the velocity profile $u(x,y)$. An approximate solution to the equation can be obtained if the shape of the velocity profile is assumed. This approach, called the *profile method*, was pioneered by T. von Kármán in 1921 [3]. If a simple algebraic form is used for the profile. Eq. (5.54) can be solved analytically. For example, consider a cubic polynomial,

$$\frac{u}{u_e} = a + b\frac{y}{\delta} + c\left(\frac{y}{\delta}\right)^2 + d\left(\frac{y}{\delta}\right)^3 \tag{5.55}$$

where $\delta(x)$ is the boundary layer thickness, defined simply by the condition $u \simeq u_e$ to sufficient accuracy outside the boundary layer. The constants are determined from four appropriate boundary conditions. Two of these are obvious:

$$y = 0: \quad u = 0; \quad y = \delta: \quad u = u_e$$

A third comes from the requirement that the velocity profile be smooth at the boundary layer edge, that is,

$$y = \delta: \quad \frac{\partial u}{\partial y} = 0$$

The fourth comes from the differential form of the momentum equation, Eq. (5.39). At the wall, both u and v are zero; hence, $\partial^2 u/\partial y^2$ is zero also, that is,

$$y = 0: \quad \frac{\partial^2 u}{\partial y^2} = 0$$

Evaluating the constants gives the velocity profile as

$$\frac{u}{u_e} = \frac{3}{2}\frac{y}{\delta} - \frac{1}{2}\left(\frac{y}{\delta}\right)^3 \tag{5.56}$$

which is shown in Fig. 5.16. The momentum thickness can now be evaluated in terms of δ,

$$\delta_2 = \int_0^\delta \left[1 - \frac{3}{2}\frac{y}{\delta} + \frac{1}{2}\left(\frac{y}{\delta}\right)^3\right]\left[\frac{3}{2}\frac{y}{\delta} - \frac{1}{2}\left(\frac{y}{\delta}\right)^3\right]dy = \frac{39}{280}\delta$$

and the skin friction coefficient as

$$\rho u_e^2 \frac{C_{fx}}{2} = \mu\left.\frac{\partial u}{\partial y}\right|_0 = \mu u_e \frac{\partial}{\partial y}\left[\frac{3}{2}\frac{y}{\delta} - \frac{1}{2}\left(\frac{y}{\delta}\right)^3\right]\Bigg|_{y=0} = \frac{3}{2}\frac{\mu u_e}{\delta}$$

Substituting in Eq. (5.54) and rearranging gives an ordinary differential equation for δ:

$$\delta\frac{d\delta}{dx} = \frac{140}{13}\frac{\nu}{u_e} \tag{5.57}$$

which upon integration with $\delta = 0$ at $x = 0$ gives

$$\delta = 4.64(\nu x/u_e)^{1/2} \tag{5.58}$$

and, in particular, the skin friction coefficient becomes

$$C_{fx} = \frac{0.646}{Re_x^{1/2}} \tag{5.59}$$

Equation (5.59) is accurate within 3% of the exact numerical solution of the differential momentum conservation equation, which was given as Eq. (4.54). However, this accuracy is fortuitous. The higher the order of the polynomial used to represent the

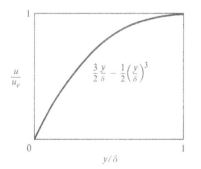

Figure 5.16 Cubic velocity profile for integral method analysis.

velocity profile, the greater the accuracy we might expect. Thus, we did not attempt to use a linear profile, and a quadratic is also not adequate. But if a quartic profile is used, the accuracy is also less than for a cubic! (See Exercise 5–28.)

The Heat Transfer Problem

We now derive the integral form of the energy conservation equation. Figure 5.17 shows an elemental control volume Δx long extending to $y = Y$, where Y is greater than the thermal boundary layer thickness Δ. In general, $\Delta \neq \delta$. The steady-flow energy equation, Eq. (1.4), requires that the net enthalpy outflow from the volume equal the heat transfer from the wall:

$$\int_0^Y \rho u c_p T \, dy \Big|_{x+\Delta x} + \rho v c_p T \big|_Y \Delta x - \int_0^Y \rho u c_p T \, dy \Big|_x = -k \frac{\partial T}{\partial y} \Big|_0 \Delta x$$

Again, we have neglected x-direction conduction, and y-direction conduction is zero outside the thermal boundary layer. Dividing by $\rho c_p \Delta x$ and letting $\Delta x \to 0$,

$$\frac{d}{dx} \int_0^Y u T \, dy + v T \big|_Y = -\alpha \frac{\partial T}{\partial y} \Big|_0$$

Now $T|_Y = T_e$, and from the continuity equation, $v(Y) = -\int_0^Y (\partial u / \partial x) dy$; thus,

$$\frac{d}{dx} \int_0^Y u T \, dy - \int_0^Y T_e \frac{\partial u}{\partial x} dy = -\alpha \frac{\partial T}{\partial y} \Big|_0$$

Interchanging the order of integration and differentiation in the second term and rearranging gives

$$\frac{d}{dx} \int_0^Y u(T - T_e) dy = -\alpha \frac{\partial T}{\partial y} \Big|_0$$

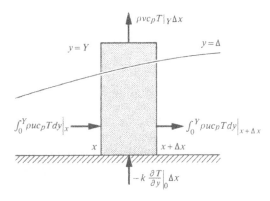

Figure 5.17 Elemental control volume for the integral method application of the first law of thermodynamics to a boundary layer on a flat plate.

The integrand is zero for $y > \Delta$, so we can let $y \to \infty$:

$$\frac{d}{dx} \int_0^\infty u(T - T_e)\,dy = -\alpha \frac{\partial T}{\partial y}\bigg|_0 \tag{5.60}$$

which is the integral form of the energy conservation equation. It can be rearranged as

$$\frac{d}{dx} \int_0^\infty \left(\frac{u}{u_e}\right)\left(\frac{T - T_e}{T_s - T_e}\right) dy = \frac{-k(\partial T/\partial y)|_0}{\rho u_e c_p (T_s - T_e)} \tag{5.61}$$

The *energy thickness* Δ_2 is defined as

$$\Delta_2 = \int_0^\infty \left(\frac{u}{u_e}\right)\left(\frac{T - T_e}{T_s - T_e}\right) dy \tag{5.62}$$

and the local Stanton number is

$$\mathrm{St}_x = \frac{h_{cx}}{\rho c_p u_e} = \frac{q_s}{\rho c_p u_e (T_s - T_e)}$$

Thus, Eq. (5.61) can be compactly written as

$$\frac{d\Delta_2}{dx} = \mathrm{St}_x \tag{5.63}$$

Use of the profile method to solve Eq. (5.63) requires consideration of whether the thermal boundary layer thickness Δ is greater than or less than the hydrodynamic boundary layer thickness δ. For high-conductivity fluids, such as liquid metals, we would expect $\Delta \gg \delta$, whereas for highly viscous fluids, such as oils, we would expect $\Delta \ll \delta$. Examination of the differential conservation equations shows that for $v = \alpha$, Eq. (5.39) is identical to Eq. (5.42), with T replacing u. Thus, for similar boundary conditions, the velocity and temperature profiles are identical, and $\Delta = \delta$, which is a statement of the Reynolds analogy for a laminar boundary layer. For $v = \alpha$ the Prandtl number is unity, and thus $\Delta \simeq \delta$ for gases. As is shown in advanced texts, careful scaling of the differential conservation equations gives $\Delta/\delta \simeq \mathrm{Pr}^{-1/3}$ for $\mathrm{Pr} \gg 1$. Also, the thermal boundary layer will be much thinner than the hydrodynamic boundary layer if the heating commences a long distance from the leading edge of the plate.

To show the power of analytical methods based on the integral conservation equations, we will now solve the *unheated starting length* problem, shown in Fig. 5.18. For $0 < x < \xi$, the plate is insulated and hence is at temperature T_e; heating commences at $x = \xi$ such that for $x > \xi$, the plate is isothermal at temperature T_s. We restrict our attention to fluids of Prandtl number unity or greater, so that always $\Delta < \delta$. As for the velocity profile, a cubic polynomial is chosen for the temperature profile and the constants chosen to satisfy analogous boundary conditions:

$$y = 0: \quad T = T_s, \quad \frac{\partial^2 T}{\partial y^2} = 0$$

$$y = \Delta: \quad T = T_e, \quad \frac{\partial T}{\partial y} = 0$$

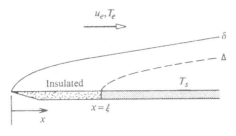

Figure 5.18 Coordinate system for the unheated starting length problem.

to give

$$\frac{T - T_e}{T_s - T_e} = 1 - \frac{3}{2}\frac{y}{\Delta} + \frac{1}{2}\left(\frac{y}{\Delta}\right)^3 \tag{5.64}$$

Then

$$\Delta_2 = \int_0^\Delta \left[\frac{3}{2}\frac{y}{\delta} - \frac{1}{2}\left(\frac{y}{\delta}\right)^3\right]\left[1 - \frac{3}{2}\frac{y}{\Delta} + \frac{1}{2}\left(\frac{y}{\Delta}\right)^3\right]dy = 3\delta\left(\frac{r^2}{20} - \frac{r^4}{280}\right)$$

where $r \equiv \Delta/\delta$. Differentiating with respect to x,

$$\frac{d\Delta_2}{dx} = 3\delta\left(\frac{r}{10} - \frac{r^3}{70}\right)\frac{dr}{dx} + 3\left(\frac{r^2}{20} - \frac{r^4}{280}\right)\frac{d\delta}{dx} \tag{5.65}$$

Also

$$\mathrm{St}_x = \frac{-k(\partial T/\partial y)|_{y=0}}{\rho u_e c_p(T_s - T_e)} = \frac{3\alpha}{2u_e\Delta} \tag{5.66}$$

Substituting into Eq. (5.63),

$$3\delta\left(\frac{r}{10} - \frac{r^3}{70}\right)\frac{dr}{dx} + 3\left(\frac{r^2}{20} - \frac{r^4}{280}\right)\frac{d\delta}{dx} = \frac{3\alpha}{2u_e\Delta}$$

To obtain an analytical solution, it is necessary to neglect the two highest-order terms in r. For $\Delta \ll \delta, r \ll 1$, and certainly such a simplification is justified. Also, when $\Delta \simeq \delta$ and $r \simeq 1$, either because the unheated starting length was short or because the location in question is far along the plate, $dr/dx \simeq 0$, and the error introduced cannot be large. Deleting these terms and rearranging,

$$2r^2\delta^2\frac{dr}{dx} + r^3\delta\frac{d\delta}{dx} = \frac{10\alpha}{u_e}$$

Substituting for δ^2 and $d\delta/dx$ from Eq. (5.57) and rearranging,

$$r^3 + 4r^2x\frac{dr}{dx} = \frac{13}{14\mathrm{Pr}}$$

which is an ordinary differential equation for $r(x)$. Let $\zeta = r^3$; then

$$r^2\frac{dr}{dx} = \frac{1}{3}\frac{d}{dx}(r^3) = \frac{1}{3}\frac{d\zeta}{dx}$$

and the differential equation becomes

$$\zeta + \frac{4}{3}x\frac{d\zeta}{dx} = \frac{13}{14Pr}$$

which is easily seen to have the solution

$$\zeta = r^3 = C_1 x^{-3/4} + \frac{13}{14Pr}$$

The constant C_1 is evaluated from the initial condition $\Delta = 0$ at $x = \xi$, that is, $r = 0$ at $x = \xi$, to obtain

$$r = 0.976Pr^{-1/3}[1 - (\xi/x)^{3/4}]^{1/3} \tag{5.67}$$

Finally, using Eqs. (5.58), (5.66), and (5.67), and the relation $Nu_x = St_x Re_x Pr$,

$$Nu_x = \frac{0.331Re_x^{1/2}Pr^{1/3}}{[1 - (\xi/x)^{3/4}]^{1/3}} \tag{5.68}$$

Notice that for $\xi = 0$, Eq. (5.68) is almost identical to Eq. (4.56), which is based on an exact numerical solution of the differential conservation equations for an isothermal plate. This agreement is reassuring, but it is also fortuitous, since our integral method is an approximate one. The various assumptions give compensating errors. Figure 5.19 shows typical variations in h_{cx} calculated from Eq. (5.68).

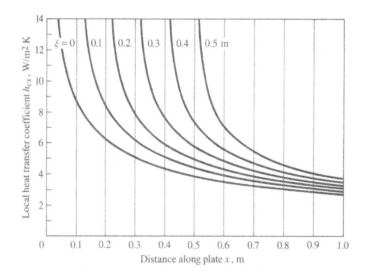

Figure 5.19   Effect of an unheated starting length on the local heat transfer coefficient for a laminar boundary layer on a flat plate. Air at 1 atm, 300 K; $u_e = 2$ m/s.

EXAMPLE 5.3   A Plate with an Unheated Starting Length

Air at 280 K and 1 atm pressure flows at 10 m/s along a flat plate. The plate is insulated for the first 10 cm from its leading edge and thereafter is maintained at 320 K. At a distance of 20 cm from the leading edge, determine $\delta, \delta_2, C_{fx}, \tau_s, r, \Delta, \Delta_2$, Nu_x, St_x, h_{cx}, and q_s. Take $Re_{tr} = 200{,}000$.

Solution

Given: Air flowing over a flat plate with an unheated starting length.

Required: At $x = 20$ cm: $\delta, \delta_2, C_{fx}, \tau_s, r, \Delta, \Delta_2$, Nu_x, St_x, h_{cx}, q_s.

Assumptions: A transition Reynolds number of 200,000.

Evaluate all properties at a mean film temperature of $(280 + 320)/2 = 300$ K. Then, from Table A.7, $k = 0.0267$ W/m K, $\rho = 1.177$ kg/m³, $c_p = 1005$ J/kg K, $v = 15.66 \times 10^{-6}$ m²/s, Pr = 0.69. Check the Reynolds number at $x = 0.2$ m,

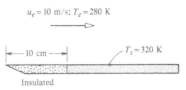

$$Re_x = \frac{(10)(0.2)}{(15.66 \times 10^{-6})} = 127{,}700 < 200{,}000$$

The boundary layer is laminar. Equation (5.58) gives the boundary layer thickness, δ:

$$\delta = 4.64 \left(\frac{vx}{u_e}\right)^{1/2} = 4.64 \left(\frac{15.66 \times 10^{-6} \times 0.2}{10}\right)^{1/2}$$

$$= 2.60 \times 10^{-3} \text{m} \ (2.60 \text{ mm})$$

The boundary layer momentum thickness is then

$$\delta_2 = \frac{39\delta}{280} = \frac{(39)(2.60 \times 10^{-3})}{280} = 3.62 \times 10^{-4} \text{ m}$$

Equation (5.59) gives the local skin friction coefficient as

$$C_{fx} = \frac{0.646}{Re_x^{1/2}} = \frac{0.646}{(127{,}700)^{1/2}} = 0.00181$$

Hence, the local shear stress is

$$\tau_s = \frac{1}{2}\rho u_e^2 C_{fx} = \left(\frac{1}{2}\right)(1.177)(10)^2(0.00181) = 0.106 \text{ N/m}^2$$

The ratio of the thermal-to-hydrodynamic boundary layer thickness $r = \Delta/\delta$ is obtained from Eq. (5.67) as

$$r = 0.975 Pr^{-1/3}[1 - (\xi/x)^{3/4}]^{1/3} = (0.975)(0.69)^{-1/3}[1 - (0.1/0.2)^{3/4}]^{1/3}$$

$$= 0.817$$

Hence, $\Delta = r\delta = (0.817)(2.60 \times 10^{-3}) = 2.12 \times 10^{-3}$m (2.12 mm).

The energy thickness Δ_2 is

$$\Delta_2 = 3\delta\left(\frac{r^2}{20} - \frac{r^4}{280}\right) = (3)(2.60 \times 10^{-3})\left(\frac{0.816^2}{20} - \frac{0.816^4}{280}\right)$$

$$= 2.47 \times 10^{-4} \text{ m}$$

Equation (5.68) gives the local Nusselt number:

$$Nu_x = \frac{0.331 Re_x^{1/2} Pr^{1/3}}{[1 - (\xi/x)^{3/4}]^{1/3}} = \frac{(0.331)(127,700)^{1/2}(0.69)^{1/3}}{[1 - (0.1/0.2)^{3/4}]^{1/3}} = 141$$

Hence, the local Stanton number is

$$St_x = \frac{Nu_x}{Re_x Pr} = \frac{141}{(127,700)(0.69)} = 0.00160$$

Finally, the local heat transfer coefficient and heat flux are

$$h_{cx} = (k/x)Nu_x = (0.0267/0.2)(141) = 18.8 \text{ W/m}^2 \text{ K}$$

$$q_s = h_{cx}(T_s - T_e) = 18.8(320 - 280) = 752 \text{ W/m}^2$$

Comments

If the plate had been isothermal at 320 K, Nu_x at $x = 20$ cm would be given by Eq. (5.68) for $\xi = 0$, with the value 105. The effect of the unheated starting length and thinner thermal boundary layer is to increase the Nusselt number by 35%.

5.4.4 Natural Convection on an Isothermal Vertical Wall

Figure 5.20 shows a schematic of a natural-convection boundary layer on a vertical, heated wall. If the Rayleigh number is less than about 10^9, the flow in the boundary layer will be laminar. For this natural-convection problem, the analysis strategy we used for forced-convection problems needs to be changed. First, we cannot assume

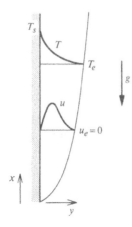

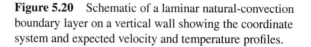

Figure 5.20   Schematic of a laminar natural-convection boundary layer on a vertical wall showing the coordinate system and expected velocity and temperature profiles.

all fluid properties to be constant. The flow is driven by a buoyancy force, which arises due to density gradients adjacent to the heated wall. Second, we cannot solve the fluid mechanics problem without reference to the heat transfer problem, since it is the temperature gradients adjacent to the wall that cause the density gradients. The differential equations governing the velocity and temperature profiles are coupled and must be solved simultaneously. The integral method of analysis gives a particularly good result for this problem.

Integral Conservation Equations

In deriving the integral form of the momentum conservation equation, the buoyancy force acting on the fluid must be included. According to Archimedes' principle, the force per unit volume acting on an element of warmer fluid in the boundary layer is $g(\rho_e - \rho)$ directed vertically upward, where g is the gravitational acceleration, ρ_e is the fluid density outside the boundary layer, which is constant, and ρ is the density of the fluid element that varies across the boundary layer. In Fig. 5.21, Newton's second law of motion is applied to the elemental control volume of unit depth and Δx long extending to $y = Y$, where Y is greater than the boundary layer thickness. The net momentum outflow from the volume is equal to the buoyancy force minus the viscous drag force exerted by the wall:

$$\int_0^Y \rho u^2 dy \Big|_{x+\Delta x} - \int_0^Y \rho u^2 dy \Big|_x = \int_0^Y g(\rho_e - \rho) dy \Delta x - \mu \frac{\partial u}{\partial y}\Big|_0 \Delta x$$

Notice that there is no momentum flow across the boundary at $y = Y$ since $u(Y) = 0$. Dividing by Δx and letting $\Delta x \to 0$,

$$\frac{d}{dx}\int_0^Y \rho u^2 dy = \int_0^Y g(\rho_e - \rho) dy - \mu \frac{\partial u}{\partial y}\Big|_0$$

We now introduce the **Boussinesq approximation** by taking the density to be constant except in the buoyancy term; dividing by ρ then gives

$$\frac{d}{dx}\int_0^Y u^2 dy = \int_0^Y g\left(\frac{\rho_e - \rho}{\rho}\right) dy - \nu \frac{\partial u}{\partial y}\Big|_0 \qquad (5.69)$$

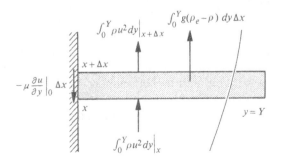

Figure 5.21 Elemental control volume for the integral method application of Newton's second law to a natural-convection boundary layer.

Since $\rho = \rho(P,T)$ we can write a Taylor expansion as $\rho_e - \rho = (\partial\rho/\partial T)_P(T_e - T) + (\partial\rho/\partial P)_T (P_e - P) +$ higher-order terms. The pressure variation across the boundary layer is negligible; thus, we can take $P = P_e$ to obtain

$$\frac{\rho_e - \rho}{\rho} = -\frac{1}{\rho}\left(\frac{\partial\rho}{\partial T}\right)_P (T - T_e) = \beta(T - T_e) \tag{5.70}$$

where β is the volumetric coefficient of thermal expansion. For an ideal gas, $\beta = 1/T$; selected data for liquids are given in Appendix A, Table A.10. Substituting Eq. (5.70) in Eq. (5.69) and letting $Y \to \infty$, since there is no contribution to the integrals for $y > Y$,

$$\frac{d}{dx}\int_0^\infty u^2\,dy = \int_0^\infty g\beta(T - T_e)\,dy - v\frac{\partial u}{\partial y}\bigg|_0 \tag{5.71}$$

The integral form of the energy conservation equation is identical to that for forced flow, Eq. (5.60):

$$\frac{d}{dx}\int_0^\infty u(T - T_e)\,dy = -\alpha\frac{\partial T}{\partial y}\bigg|_0 \tag{5.72}$$

Simultaneous Solution of the Equations

The two ordinary differential equations, Eqs. (5.71) and (5.72), are coupled since the variable T appears in both; hence, they must be solved simultaneously. Boundary conditions for the velocity and temperature profiles are

$$y = 0: \quad u = 0, \ T = T_s, \text{ a constant for an isothermal wall}$$

$$y \to \infty: \quad u = 0, \ T = T_e$$

We assume that the hydrodynamic and thermal boundary layers have the same thickness, $\Delta = \delta$, and assume the following forms for the velocity and temperature profiles:

$$\frac{u}{U} = \frac{y}{\delta}\left(1 - \frac{y}{\delta}\right)^2, \quad \frac{T - T_e}{T_s - T_e} = \left(1 - \frac{y}{\delta}\right)^2 \tag{5.73a,b}$$

where U is a scaling velocity, which is a function of x only and is yet to be determined. Figure 5.22 shows these profiles; the maximum velocity is $0.148U$ at $y = \delta/3$. In the forced-convection problem, the known free-stream velocity was used to scale the velocity profile; here the scaling velocity U, as well as the boundary layer thickness δ, are unknowns. Our profiles give $\partial u/\partial y = 0$ and $\partial T/\partial y = 0$ at $y = \delta$ (i.e., they are smooth at the edge of the boundary layer). However, the velocity profile does not have the correct limiting value of $\partial^2 u/\partial y^2$ at the wall; also, the assumption of $\Delta = \delta$ is not valid for high-Prandtl-number fluids. But our method is an approximate one, and use of these profiles gives surprisingly good results. Substituting the profiles in Eqs. (5.71) and (5.72) and performing the indicated integrations

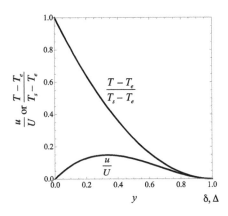

Figure 5.22 Velocity and temperature profiles for integral method analysis of laminar natural convection.

and differentiations gives

$$\frac{d}{dx}\left(\frac{U^2\delta}{105}\right) = -\frac{\nu U}{\delta} + \frac{g\beta(T_s - T_e)\delta}{3} \tag{5.74}$$

$$\frac{d}{dx}\left(\frac{U\delta}{30}\right) = \frac{2\alpha}{\delta} \tag{5.75}$$

which are a pair of ordinary differential equations for the unknowns U and δ. We will assume power law variations for δ and U:

$$\delta = Dx^m; \quad U = Xx^n$$

which give $\delta = 0$ and $u = 0$ at $x = 0$. Substituting in Eq. (5.74) gives

$$\frac{d}{dx}\left(\frac{X^2 D x^{2n+m}}{105}\right) = -\frac{\nu X x^{n-m}}{D} + \frac{g\beta(T_s - T_e)Dx^m}{3}$$

The x dependence cancels if $2n + m - 1 = n - m = m$, which requires $m = 1/4$, $n = 1/2$. Equations (5.74) and (5.75) then reduce to two algebraic equations for X and D:

$$\frac{5}{4}\frac{X^2 D}{105} = -\frac{\nu X}{D} + \frac{g\beta(T_s - T_e)D}{3} \tag{5.76}$$

$$\frac{3}{4}\frac{XD}{30} = \frac{2\alpha}{D} \tag{5.77}$$

Solving,

$$X = \frac{80\alpha}{D^2} \tag{5.78}$$

$$D = 3.94\left[\frac{(20/21)\alpha^2 + \nu\alpha}{g\beta(T_s - T_e)}\right]^{1/4} \tag{5.79}$$

The wall heat flux is obtained from the temperature profile, Eq. (5.73b), as

$$q_s = -k\frac{\partial T}{\partial y}\bigg|_0 = \frac{2k}{\delta}(T_s - T_e)$$

Thus,

$$\frac{h_{cx}}{k} = \frac{2}{\delta} = \frac{2}{Dx^{1/4}}$$

and

$$\text{Nu}_x = \frac{h_{cx}x}{k} = \frac{2x^{3/4}}{D}$$

Substituting for D and rearranging gives

$$\text{Nu}_x = 0.508 \left[\frac{\text{Pr}}{0.952 + \text{Pr}}\right]^{1/4} \text{Ra}_x^{1/4} \qquad \textbf{(5.80)}$$

where $\text{Ra}_x = \beta(T_s - T_e)gx^3/\nu\alpha$ is the Rayleigh number introduced in Section 4.2.2. This result agrees very well with exact numerical solutions to the differential conservation equations and was widely used before exact solutions became available.

EXAMPLE 5.4   A Natural-Convection Water Boundary Layer

A vertical plate at 320 K is immersed in water at 300 K. At a location 10 cm from the bottom of the plate, determine $\delta, U, \text{Nu}_x, h_{cx}$, and q_s. Also plot the velocity and temperature profiles.

Solution

Given: Natural-convection boundary layer on a vertical plate.

Required: At $x = 10$ cm: $\delta, U, \text{Nu}_x, h_{cx}, q_s, u(y), T(y)$.

Assumptions: Laminar flow.

Properties will be evaluated at a mean film temperature of 310 K; from Table A.8, $k = 0.628$ W/m K, $\rho = 993$ kg/m³, $\nu = 0.70 \times 10^{-6}$ m²/s, Pr = 4.6. Also, $\alpha = \nu/\text{Pr} = 1.52 \times 10^{-7}$ m²/s. From Table A.10b, $\beta = 3.62 \times 10^{-4}$ K⁻¹. The Rayleigh number is checked to see if the flow is laminar:

$$\text{Ra}_x = \frac{\beta(T_s - T_e)gx^3}{\nu\alpha} = \frac{(3.62 \times 10^{-4})(320 - 300)(9.81)(0.1)^3}{(0.70 \times 10^{-6})(1.52 \times 10^{-7})}$$

$$= 6.68 \times 10^8 < 10^9 \quad \text{(laminar)}$$

The boundary layer thickness is $\delta = Dx^{1/4}$, where D is given by Eq. (5.79):

$$D = 3.94 \left[\frac{(20/21)\alpha^2 + v\alpha}{g\beta(T_s - T_e)} \right]^{1/4}$$

$$= 3.94 \left[\frac{(20/21)(1.52 \times 10^{-7})^2 + (0.70 \times 10^{-6})(1.52 \times 10^{-7})}{(9.81)(3.62 \times 10^{-4})(320 - 300)} \right]^{1/4}$$

$$= 4.57 \times 10^{-3} \text{ m}^{3/4}$$

$$\delta = (4.57 \times 10^{-3})(0.1)^{1/4} = 2.57 \times 10^{-3} \text{ m} \ (2.57 \text{ mm})$$

The scaling velocity is $U = Xx^{1/2}$, where X is given by Eq. (5.78):

$$X = \frac{80\alpha}{D^2} = \frac{(80)(1.52 \times 10^{-7})}{(4.57 \times 10^{-3})^2} = 0.582 \text{ m}^{1/2}/\text{s}$$

$$U = (0.582)(0.1)^{1/2} = 0.184 \text{ m/s}$$

Equation (5.80) gives the local Nusselt number:

$$\text{Nu}_x = 0.508 \left[\frac{\text{Pr}}{0.952 + \text{Pr}} \right]^{1/4} \text{Ra}_x^{1/4}$$

$$= 0.508 \left[\frac{4.6}{0.952 + 4.6} \right]^{1/4} (6.68 \times 10^8)^{1/4} = 77.9$$

Hence, the local heat transfer coefficient and heat flux are

$$h_{cx} = (k/x)\text{Nu}_x = (0.628/0.1)(77.9) = 489 \text{ W/m}^2 \text{ K}$$

$$q_s = h_c(T_s - T_e) = 489(320 - 300) = 9790 \text{ W/m}^2$$

The velocity and temperature profiles are shown in the accompanying diagram.

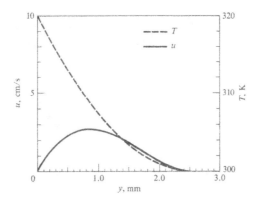

Comments

Notice that CONV does not give a value of the local heat transfer coefficient h_{cx}, since it uses Eq. (4.85) for $\overline{\text{Nu}}_L$ and is based on experimental data.

5.5 TURBULENT FLOWS

Reynolds, in 1883, introduced the concept of a critical Reynolds number, $\mathrm{Re} = u_b D/\nu$, for transition from laminar to turbulent flow in a pipe. A nominal value of 2300 is usually quoted, but in the absence of disturbances, the value can be much higher. The Reynolds number range from 2300 to about 10,000 is rather troublesome since the turbulence tends to have an intermittent character. Experimental data for skin friction and heat transfer in this regime are not easily reproduced; thus, this regime is usually avoided in engineering design, if at all possible. In contrast, in external flows, transition from laminar to turbulent flow often occurs on the surface of interest, and both the location of transition and the region immediately downstream of transition may play important roles in determining overall drag and heat transfer. Wind tunnel tests show that transition on a smooth, flat plate occurs at a critical Reynolds number $\mathrm{Re} = u_e x/\nu$ in the range of 300,000 to 500,000. In practice, factors such as free-stream turbulence and surface roughness tend to lower the critical Reynolds number, and values in the range 50,000 to 100,000 are often used for engineering calculations.

Turbulent flow is characterized by a complex eddying motion, which causes fluctuations of the velocity components, pressure, temperature, and in compressible flows, density as well. Consider a trace of the streamwise velocity component u taken with a hot-wire anemometer probe, as shown in Fig. 5.23. The instantaneous velocity u at time t_0 can be written as the sum of a mean component $\bar{u}$ and a fluctuating component u':

$$u = \bar{u} + u'; \quad \bar{u}(t_0) = \frac{1}{\tau} \int_{t_0 - \tau/2}^{t_0 + \tau/2} u \, dt \tag{5.81}$$

where the interval τ is longer than the period of any significant fluctuation but is much shorter than any mean flow time scale. The average value of the fluctuating component is zero, since

$$\bar{u} = \overline{\bar{u} + u'} = \bar{\bar{u}} + \overline{u'} = \bar{u} + \overline{u'}; \quad \overline{u'} = 0 \tag{5.82}$$

A mean flow may be steady, that is, at a given location $\bar{u}$ is not a function of time, but the instantaneous velocity $u = \bar{u} + u'$ is unsteady in a turbulent flow. The mean

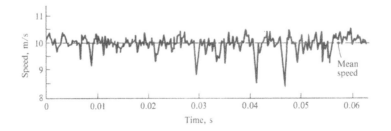

Figure 5.23 Hot-wire anemometer measurement of a turbulent velocity component.

component in Fig. 5.23 is steady. The fluctuating component in Fig. 5.23 can be counted to have a frequency of about 10^3 Hz, and its amplitude is about 5% to 10% of the mean value. Correlations of fluctuating components prove to be of considerable importance: consider, for example, the velocity components u and v, with mean values $\bar{u}$ and $\bar{v}$, respectively. The average of the product uv is

$$\overline{uv} = \overline{(\bar{u}+u')(\bar{v}+v')} = \bar{u}\bar{v} + \overline{\bar{u}v'} + \overline{u'\bar{v}} + \overline{u'v'}$$

But the average of a constant times a fluctuation whose average is zero, is also zero. That is, $\overline{\bar{u}v'} = 0$, $\overline{u'\bar{v}} = 0$. Hence,

$$\overline{uv} = \bar{u}\bar{v} + \overline{u'v'} \tag{5.83}$$

If u' and v' are uncorrelated, then $\overline{u'v'}$ is zero. However, if u' always tends to have the same or opposite sign as v', then $\overline{u'v'}$ is clearly nonzero.

It is quite impractical to attempt to predict a trace such as that shown in Fig. 5.23 by solving the equations governing mass and momentum conservation in a time-dependent flow. Thus, engineers have used experimental data to develop simple models of turbulent transport as a basis for the semi-empirical analysis of the behavior of mean quantities such as mean velocity and mean temperature, as well as the wall shear stress and heat transfer. The turbulent eddies and associated mixing motion are very effective in transporting momentum and heat if velocity or temperature gradients exist. The process is somewhat analogous to transport in a gas as described by the kinetic theory of gases, with eddies replacing molecules as momentum and thermal energy carriers, and a **mixing length** replacing the mean free path. One of the most popular and most successful approaches to modeling turbulent transport, the *Prandtl mixing length theory*, is based on such an analogy.

A large amount of experimental data is available for turbulent flows: the skin friction and heat transfer correlations given in Section 4.3 are examples of such data. Detailed measurements also have been made of mean velocity and mean temperature profiles, of the turbulent fluctuations, and of correlations of fluctuating components, such as $\overline{u'v'}$ and $\overline{T'v'}$. Such data form the essential basis of our knowledge about turbulent flows and have proven indispensable for the successful development and refinement of models for turbulent transport. Even today we cannot use theory alone to predict skin friction and heat transfer for the simplest of flows, for example, the flat plate boundary layer or fully developed pipe flow.

5.5.1 The Prandtl Mixing Length and the Eddy Diffusivity Model

Momentum Transport

In a turbulent flow, we can define an *effective viscosity* as the sum of the molecular viscosity μ and a turbulent viscosity μ_t. The turbulent viscosity accounts for momentum transport by eddies. A model of turbulent transport is required to determine μ_t, and a key step in the development of such models came in 1925 with Prandtl's mixing length hypothesis [4]. Its origin can be seen in the kinetic theory of gases, which gives the molecular viscosity as

$$\mu = \left(\frac{1}{3}\right)(\rho)(\text{Mean free path})(\text{Mean molecular speed}) \tag{5.84}$$

By analogy, the first part of the mixing length hypothesis is

$$\mu_t = \rho \ell v_t \tag{5.85}$$

where ℓ is the *mixing length*, and v_t is a characteristic turbulent speed. However, whereas the mean free path and mean molecular speed are a function of thermodynamic quantities such as temperature and pressure, ℓ and v_t can be expected to vary throughout the flow and have values dependent on the flow velocity and geometry.

Prandtl reduced the number of unknowns in Eq. (5.85) to one by introducing a second part of the hypothesis to estimate v_t. Figure 5.24 shows the mean velocity profile adjacent to a wall. An eddy moving upward has a positive v' and comes from near the wall, where $\bar{u}$ is less. Such an eddy will be surrounded by fluid of higher $\bar{u}$; thus, the difference between the eddy's x component of velocity u and $\bar{u}$ is a negative value of u'.[4] Since the eddy has moved a distance ℓ,

$$u' \simeq -\ell\frac{\partial \bar{u}}{\partial y} \quad \text{for } v' > 0 \tag{5.86}$$

and if the eddies are moving randomly,

$$|u'| \simeq |v'| \simeq |w'| \simeq \ell\left|\frac{\partial \bar{u}}{\partial y}\right| \tag{5.87}$$

Thus, if the absolute magnitude of the fluctuating components is taken as the characteristic turbulent speed v_t,

$$v_t = \ell\left|\frac{\partial \bar{u}}{\partial y}\right| \tag{5.88}$$

The absolute magnitude of the mean velocity gradient is used in order to generalize our result to coordinate systems in which the mean velocity gradient may be negative.

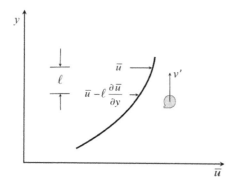

Figure 5.24   Prandtl's mixing length concept for a simple shear flow $\bar{u}(y)$.

---

[4] Notice that, in this shear flow, the correlation $\overline{u'v'}$ will be negative.

Substituting in Eq. (5.85),

$$\mu_t = \rho \ell^2 \left| \frac{\partial \overline{u}}{\partial y} \right| \tag{5.89}$$

The remaining task is to specify ℓ. If the mixing length model is to be useful, a fairly simple specification of ℓ must work for a variety of flow geometries. Prandtl's original suggestion for flow near any wall was a simple linear relation,

$$\ell = \kappa y \tag{5.90}$$

where $\kappa < 1$ since eddies cannot move through a solid wall; κ is often called the *von Kármán constant*, and it will be shown that experiments suggest a value $\kappa \simeq 0.4$. Specification of ℓ for external boundary layers will be discussed in Section 5.5.2.

Equation (5.89) gives the turbulent viscosity: in analogy to the molecular kinematic viscosity ν, we can divide μ_t by ρ to obtain the **eddy viscosity** or *eddy diffusivity of momentum*, ε_M :

$$\varepsilon_M = \ell^2 \left| \frac{\partial \overline{u}}{\partial y} \right| \tag{5.91}$$

Also, the total shear stress in the x direction on a plane of constant y is

$$\tau = -(\mu + \mu_t)\frac{\partial \overline{u}}{\partial y} = -\rho(\nu + \varepsilon_M)\frac{\partial \overline{u}}{\partial y} \tag{5.92}$$

where our sign convention is as discussed in Section 5.2.1.

Heat Transport

In a turbulent flow, the transport of heat is similar to the transfer of momentum. An eddy moving toward the wall transports momentum by virtue of its velocity surplus u'; likewise, if a hot fluid flows over a cold wall, the same eddy transports thermal energy by virtue of its temperature surplus T'. Hence, by analogy to Eq. (5.92), we write the total heat flux normal to the wall as the sum of molecular and turbulent contributions:

$$q_y = -(k + k_t)\frac{\partial \overline{T}}{\partial y} = -\rho c_p(\alpha + \varepsilon_H)\frac{\partial \overline{T}}{\partial y} \tag{5.93}$$

where k_t is the turbulent thermal conductivity and ε_H is the eddy diffusivity of heat. Also, we define a **turbulent Prandtl number**, Pr_t, in an analogy to the molecular Prandtl number, Pr:

$$Pr = \frac{\nu}{\alpha}; \quad Pr_t = \frac{\varepsilon_M}{\varepsilon_H} \tag{5.94}$$

Whereas ℓ (and hence ε_M) is found to vary throughout the flow, and in particular increases rapidly with distance away from a wall, Pr_t can often be taken as a constant of order unity. This result might have been expected from the physics involved, since eddies transport both momentum and heat.

5.5.2 Forced Flow along a Flat Plate

The analysis of even the simplest turbulent flow problem is made difficult by the need to use experimental data to develop an appropriate turbulent transport model and to indicate features that allow simplifications to be made. The general strategy for forced convection is similar to that used for constant-property laminar flows in Sections 5.2 through 5.4 and involves solving the fluid mechanics problem prior to solving the heat transfer problem.

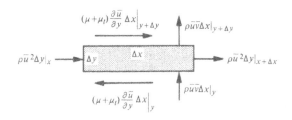

Figure 5.25 Elemental control volume for application of Newton's second law in a turbulent boundary layer on a flat plate.

The Momentum Conservation Equation

Figure 5.25 shows a fluid element Δx by Δy by unity in a turbulent boundary layer on a flat plate. Paralleling the derivation for the laminar case in Section 5.4.1, application of Newton's second law gives

$$\rho \bar{u}^2 \Delta y|_{x+\Delta x} + \rho \bar{u}\,\bar{v}\Delta x|_{y+\Delta y} - \rho \bar{u}^2 \Delta y|_x - \rho \bar{u}\,\bar{v}\Delta x|_y =$$

$$(\mu + \mu_t)\frac{\partial \bar{u}}{\partial y}\Delta x\Big|_{y+\Delta y} - (\mu + \mu_t)\frac{\partial \bar{u}}{\partial y}\Delta x\Big|_y$$

where $\bar{u}$ and $\bar{v}$ are mean velocity components. The effect of turbulence has been accounted for through use of the turbulent viscosity μ_t. *Use of the bar notation for a mean quantity is now discontinued since all the dependent variables are time averages.* Proceeding as in Section 5.4.1 and using Eq. (5.92) gives the differential momentum conservation equation as

$$u\frac{\partial u}{\partial x} + v\frac{\partial u}{\partial y} = \frac{\partial}{\partial y}\left[(v + \varepsilon_M)\frac{\partial u}{\partial y}\right] \tag{5.95}$$

where the essential difference from Eq. (5.39) for the laminar case is that $(v + \varepsilon_M)$ is left inside the $\partial/\partial y$ differential because ε_M is a function of distance from the wall. An empirical formula for the eddy viscosity $\varepsilon_M(y)$ is required in order to solve Eq. (5.95).

The Energy Conservation Equation

The steady-flow energy equation, Eq. (1.4), applied to the elemental control volume shown in Fig. 5.26 requires that the net outflow of enthalpy equal the heat conducted into the volume. Writing $\bar{h} = c_p \bar{T}$,

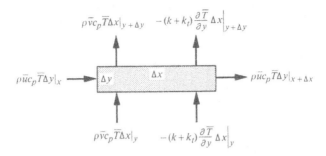

Figure 5.26 Elemental control volume for application of the first law of thermodynamics in a turbulent boundary layer.

$$\rho \bar{u} c_p \overline{T} \Delta y|_{x+\Delta x} - \rho \bar{u} c_p \overline{T} \Delta y|_x + \rho \bar{v} c_p \overline{T} \Delta x|_{y+\Delta y} - \rho \bar{v} c_p \overline{T} \Delta x|_y =$$
$$- (k + k_t) \frac{\partial \overline{T}}{\partial y} \Delta x|_y + (k + k_t) \frac{\partial \overline{T}}{\partial y} \Delta x|_{y+\Delta y}$$

where, as for laminar flow, constant-property low-speed flow is assumed, and the x-direction molecular conduction and turbulent transport are neglected. The bar notation for mean quantities can be discarded since turbulent transport is accounted for through use of the turbulent conductivity k_t. Proceeding as for the laminar case in Section 5.4.1 and using Eq. (5.93) gives the differential energy conservation equation as

$$u \frac{\partial T}{\partial x} + v \frac{\partial T}{\partial y} = \frac{\partial}{\partial y} \left[(\alpha + \varepsilon_H) \frac{\partial T}{\partial y} \right] \tag{5.96}$$

which is similar in form to Eq. (5.95) for momentum conservation. Since $Pr = v/\alpha$ and $Pr_t = \varepsilon_M / \varepsilon_H$, Eq. (5.96) can be rearranged as

$$u \frac{\partial T}{\partial x} + v \frac{\partial T}{\partial y} = \frac{\partial}{\partial y} \left[\left(\frac{v}{Pr} + \frac{\varepsilon_M}{Pr_t} \right) \frac{\partial T}{\partial y} \right] \tag{5.97}$$

Reynolds Analogy

Consider an isothermal plate: appropriate boundary conditions for Eq. (5.95) and (5.97) are

$$x = 0, \ y \rightarrow \infty; \quad u = u_e, \quad T = T_e \tag{5.98a}$$
$$y = 0; \quad u = 0, \quad T = T_s \tag{5.98b}$$

When the molecular Prandtl number Pr and the turbulent Prandtl number Pr_t are both equal to unity, Eqs. (5.95) and (5.97) are of the same form, and the normalized velocity and temperature profiles are identical. At a given x-location,

$$\frac{u(y)}{u_e} = \frac{T_s - T(y)}{T_s - T_e} \tag{5.99}$$

Turbulent fluctuations are damped out adjacent to a solid wall by the action of viscosity. Thus, the wall shear stress and heat flux are given by the laminar flow relations

$$\tau_s = \mu \frac{\partial \mu}{\partial y}\Big|_{y=0} \quad ; \quad q_s = -k \frac{\partial T}{\partial y}\Big|_{y=0}$$

Hence,

$$\frac{q_s}{\tau_s} = -\frac{k}{\mu} \frac{\partial T}{\partial u}\Big|_{y=0} \tag{5.100}$$

Evaluating $\partial T / \partial u$ from Eq. (5.99) and substituting in Eq. (5.100) gives

$$\frac{q_s}{\tau_s} = -\frac{k(T_s - T_e)}{\mu u_e} \tag{5.101}$$

Introducing the local Stanton number $St_x = q_s / \rho c_p u_e (T_s - T_e)$ and the friction coefficient $C_{fx}/2 = \tau_s / \rho u_e^2$, we obtain

$$St_x = \frac{C_{fx}}{2} \tag{5.102}$$

which is the famous **Reynolds analogy**, proposed by O. Reynolds in 1874. In his original derivation of Eq. (5.102), Reynolds ignored the *viscous sublayer* adjacent to the wall in which molecular viscosity damps out turbulence, and assumed $v \ll \varepsilon_M$ and $\alpha \ll \alpha_H$ throughout the boundary layer. Thus, he did not have to explicitly assume $\alpha = v(Pr = 1)$ to obtain his result. Our derivation is somewhat more satisfactory, since it clearly shows the need to assume $Pr = 1$ to obtain Eq. (5.102). It should be noted that our derivation is also valid for a laminar boundary layer.

The assumption of $Pr_t \simeq 1.0$ is quite good for all fluids, except high-conductivity liquid metals. Thus, Eq. (5.102) is reasonably accurate for fluids with $Pr \simeq 1$, for example, all gases. Based on experiment, a Prandtl number correction can be introduced in Eq. (5.102), giving Eq. (4.67),

$$St_x = \left(\frac{C_{fx}}{2}\right) Pr^{-0.57}, \quad 0.7 < Pr < 400 \tag{4.67}$$

Solution of the Momentum and Energy Equations

Solution of the momentum conservation equation, Eq. (5.95), requires a suitable expression for the eddy viscosity ε_M. Since ε_M is related to the mixing length ℓ by Eq. (5.89), we can equivalently specify ℓ. Prandtl's original hypothesis was given by Eq. (5.90) as $\ell = \kappa y$, and proves to be valid over a significant portion of the boundary layer. As the wall is approached, the effect of molecular viscosity is to reduce ℓ more rapidly than indicated by Eq. (5.90). In 1951, E. R. van Driest suggested a simple exponential damping factor for ℓ [5]

$$\ell = \kappa y \left[1 - \exp\left(-\frac{y}{y_t}\right)\right] \tag{5.103}$$

On the other hand, in the outer region of the boundary layer a constant value of ℓ is appropriate [6]

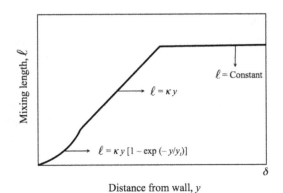

Figure 5.27 Schematic showing the variation of mixing length across a turbulent boundary layer.

$$\frac{\ell}{\delta} = \lambda \quad \text{for } \frac{y}{\delta} > \frac{\lambda}{\kappa} \tag{5.104}$$

where δ is the boundary layer thickness (defined here as where $u = 0.99u_e$). Figure 5.27 shows the variation of ℓ across the boundary layer. Appropriate values of the parameters κ, y_t, and λ have been deduced by comparing predicted and measured velocity profiles. Commonly used values include $\kappa = 0.41$, $y_t = 26\nu/(\tau_s/\rho)^{1/2}$, and $\lambda = 0.9$.

To obtain ε_M and ℓ using Eq. (5.89) and the simultaneous solution of Eq. (5.95) requires numerical procedures, of which the finite difference method has been proven most popular. Computer programs are readily available for this purpose, including SCSTREAM [7], TEXSTAN [8], COMSOL [9], PHOENICS [10], AcuSolve [11], ANSYS Fluent [12], and FLOW-3D [13]. However, it is important to understand that the solution of the momentum equation to obtain the local skin friction coefficient on a flat plate does not yield new information. The parameters κ, y_t, and λ were obtained using experimental data that include measured skin friction coefficients: thus, such a solution simply confirms the consistency of the turbulence model and the experimental data.

To solve the energy conservation equation, we must also specify the turbulent Prandtl number. Except for high-conductivity liquid metals, experimental data suggest that a constant value of 0.9 or 1.0 is quite adequate for engineering purposes. A value of 1.0 is, or course, in line with Reynolds analogy and the idea that eddies transport momentum and energy in an identical manner. As for the momentum equation, numerical methods must be used, and predictions of $St_x = f(Re_x, Pr)$ can be obtained that agree well with experiment at moderate Prandtl numbers. Predictions at very low Prandtl numbers are more difficult owing to uncertainty concerning the turbulent Prandtl number, and are also more difficult at very high Prandtl numbers owing to uncertainty concerning the variation of ℓ very close to the wall.

5.5.3 More Advanced Turbulence Models

The algebraic mixing length and eddy viscosity expressions described in Section 5.5.2 are adequate for the simple flows that were considered. However, for more complicated flows, such as those having recirculating regions, a more sophisticated approach is required. In recent years, considerable effort has been expended to develop more generally applicable turbulence models. Instead of using a simple algebraic specification of the mixing length, these models generally require the solution of additional partial differential equations, of similar form to the momentum and energy equation, in order to obtain the eddy viscosity. For example, one popular method solves equations governing the kinetic energy of the turbulence and its dissipation rate. Although considerable progress has been made by using such models, much remains to be done. Further development of such models will remain an important research area for many more years. The student should consult advanced texts and the journal literature for their current status.

5.6   THE GENERAL CONSERVATION EQUATIONS

Each time a new convection problem is to be studied, the experienced engineer need not start from scratch. The principles that apply to any problem are conservation of mass, momentum, and energy. It is advantageous to formulate general governing equations based on these principles; for a specific problem, the engineer then need only delete terms in the equations that are zero or negligible.

5.6.1 Conservation of Mass

For simplicity, we consider a Cartesian control volume element $\Delta x \Delta y \Delta z$ fixed in space, as shown in Fig. 5.28. The rate at which mass is stored in the volume equals the net rate of inflow of mass across the control volume boundary. Mass can be stored within the volume by a change in density; the rate of storage is

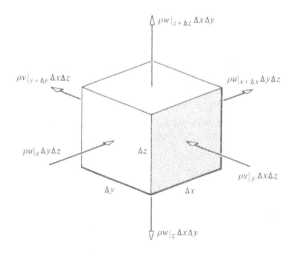

Figure 5.28   Conservation of mass applied to a Cartesian elemental control volume.

$$\frac{\partial \rho}{\partial t} \Delta x \Delta y \Delta z$$

Mass can cross the control volume boundaries by convection. The gross rate of inflow is

$$\rho u|_x \Delta y \Delta z + \rho v|_y \Delta x \Delta z + \rho w|_z \Delta x \Delta y$$

and the gross rate of outflow is

$$\rho u|_{x+\Delta x} \Delta y \Delta z + \rho v|_{y+\Delta y} \Delta x \Delta z + \rho w|_{z+\Delta z} \Delta x \Delta y$$

The net rate of inflow is found by subtracting the outflow from the inflow; expanding a quantity evaluated at $x+\Delta x, y+\Delta y$, or $z+\Delta z$ in a Taylor series; and canceling terms that subtract out. For example, the x-direction terms become

$$\rho u|_x \Delta y \Delta z - \rho u|_{x+\Delta x} \Delta y \Delta z$$

$$= \rho u|_x \Delta y \Delta z - \left[\rho u|_x + \frac{\partial}{\partial x}(\rho u)\Delta x + \frac{1}{2}\frac{\partial^2}{\partial x^2}(\rho u)\Delta x^2 + \cdots \right] \Delta y \Delta z$$

$$= -\frac{\partial}{\partial x}(\rho u)\Delta x \Delta y \Delta z - \frac{1}{2}\frac{\partial^2}{\partial x^2}(\rho u)\Delta x^2 \Delta y \Delta z + \cdots$$

with similar results in the other directions. Now substitute in our statement of the conservation principle, divide throughout by the volume $\Delta x \Delta y \Delta z$, and let Δx, Δy and Δz go to zero. Only first-order terms remain as the convective contribution; thus,

$$\frac{\partial \rho}{\partial t} = -\frac{\partial}{\partial x}(\rho u) - \frac{\partial}{\partial y}(\rho v) - \frac{\partial}{\partial z}(\rho w)$$

or

$$\frac{\partial \rho}{\partial t} + \frac{\partial}{\partial x}(\rho u) + \frac{\partial}{\partial y}(\rho v) + \frac{\partial}{\partial z}(\rho w) = 0 \tag{5.105}$$

Using the notation of vector calculus, Eq. (5.105) can be rewritten compactly as

$$\frac{\partial \rho}{\partial t} + \nabla \cdot (\rho \mathbf{v}) = 0 \tag{5.106}$$

where $\mathbf{v} = \mathbf{i}u + \mathbf{j}v + \mathbf{k}w$ and $\nabla = \mathbf{i}(\partial/\partial x) + \mathbf{j}(\partial/\partial y) + \mathbf{k}(\partial/\partial z)$ are the velocity vector and del operator for unit vectors, $\mathbf{i}, \mathbf{j}, \mathbf{k}$ in the x, y, and z directions, respectively. For coordinate systems other than Cartesian system, the divergence $\nabla\cdot$ can be expressed accordingly.

Equation (5.105) can be rearranged by performing the indicated differentiation to obtain

$$\frac{\partial \rho}{\partial t} + u\frac{\partial \rho}{\partial x} + v\frac{\partial \rho}{\partial y} + w\frac{\partial \rho}{\partial z} = -\rho \left(\frac{\partial u}{\partial x} + \frac{\partial v}{\partial y} + \frac{\partial w}{\partial z} \right) \tag{5.107}$$

or

$$\frac{D\rho}{Dt} = -\rho \nabla \cdot \mathbf{v} \tag{5.108}$$

The operator D/Dt is called the *substantial derivative* and is the time derivative for an observer moving with the fluid. Consider flow inside the cylinder of an automobile engine; in a time Δt, two things can happen. First, the compression of the gas by the piston causes the density at a given location to change by the amount

$$\frac{\partial \rho}{\partial t} \Delta t$$

Second, the swirling gas can flow to a colder region: in time Δt, it moves a distance $\Delta x = u\Delta t$, $\Delta y = v\Delta t$, $\Delta z = w\Delta t$, and the density change due to this flow is

$$\frac{\partial \rho}{\partial x}\Delta x + \frac{\partial \rho}{\partial y}\Delta y + \frac{\partial \rho}{\partial z}\Delta z = \frac{\partial \rho}{\partial x}u\Delta t + \frac{\partial \rho}{\partial y}v\Delta t + \frac{\partial \rho}{\partial z}w\Delta t$$

Thus, an observer moving with the gas observes a total density change

$$D\rho = \frac{\partial \rho}{\partial t}\Delta t + \frac{\partial \rho}{\partial x}u\Delta t + \frac{\partial \rho}{\partial y}v\Delta t + \frac{\partial \rho}{\partial z}w\Delta t$$

Divide by Δt and let Δt approach zero:

$$\frac{D\rho}{Dt} = \frac{\partial \rho}{\partial t} + u\frac{\partial \rho}{\partial x} + v\frac{\partial \rho}{\partial y} + w\frac{\partial \rho}{\partial z} \tag{5.109}$$

or

$$\frac{D\rho}{Dt} = \frac{\partial \rho}{\partial t} + \mathbf{v}\cdot\nabla\rho \tag{5.110}$$

For the special case of the fluid of constant density, Eq. (5.108) becomes

$$\nabla \cdot \mathbf{v} = 0 \tag{5.111}$$

Although no fluid is truly incompressible, Eq. (5.111) can often be used in engineering analysis with negligible error.

5.6.2 Conservation of Momentum

Choice of Control Volume

An *Eulerian* control volume is one that is fixed in space. Such a control volume was used for the preceding derivation of the mass conservation and for the derivation of the momentum equation for a boundary layer in Section 5.4.1. A *Lagrangian* control volume contains a fixed mass of fluid and moves with the fluid. We choose to use the Lagrangian approach here since the principle of conservation of momentum can be stated in its simplest form, that is, the time rate of change of momentum equals the sum of forces acting. Since the control volume contains a fixed mass and moves with the fluid, the time rate of change of momentum is the mass times acceleration. The acceleration is given by the substantial derivative introduced in Section 5.6.1; hence the rate of change of momentum is

$$\rho \Delta x \Delta y \Delta z \frac{D\mathbf{v}}{Dt}$$

where for the x component

$$\frac{Du}{Dt} = \frac{\partial u}{\partial t} + u\frac{\partial u}{\partial x} + v\frac{\partial u}{\partial y} + w\frac{\partial u}{\partial z}$$

Surface Forces

Figure 5.29 shows all the x-direction surface forces acting on a cube $\Delta x \Delta y \Delta z$, including forces due to both pressure and viscous stresses. To understand the sign and subscript notation for the viscous stresses, consider, for example, the shear stress τ_{yx}. This stress is applied to the fluid of greater y from below. The first subscript, y, denotes that the stress acts on a plane of constant y; the second subscript denotes that the stress is in the x direction. Similarly, the normal stress τ_{xx} is applied to the fluid of greater x from below: the first subscript denotes that the stress acts on a plane of constant x, and the second subscript denotes that the stress is in the x direction. Notice that τ_{xx} does not include the hydrostatic pressure P. The net x-direction surface force is

$$(\tau_{xx} + P)_x \Delta y \Delta z + (\tau_{yx})_y \Delta x \Delta z + (\tau_{zx})_z \Delta x \Delta y$$
$$- [(\tau_{xx} + P)_{x+\Delta x} \Delta y \Delta z + (\tau_{yx})_{y+\Delta y} \Delta x \Delta z + (\tau_{zx})_{z+\Delta z} \Delta x \Delta y]$$

Expanding the incremental quantities in Taylor series and canceling terms gives

$$\left(-\frac{\partial P}{\partial x} - \frac{\partial \tau_{xx}}{\partial x} - \frac{\partial \tau_{yx}}{\partial y} - \frac{\partial \tau_{zx}}{\partial z} \right) \Delta x \Delta y \Delta z$$

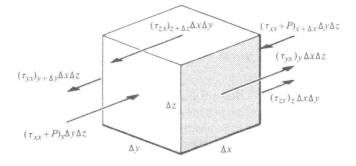

Figure 5.29 The x-direction forces acting on a Cartesian element $\Delta x \Delta y \Delta z$.

Volume Forces

Volume forces may also act on a fluid, for example, gravitational and electromagnetic forces; only the former will be considered here. If the gravitational force per unit mass is denoted by the vector $\mathbf{g}$, then the resulting force acting on the fluid within the control volume is $\mathbf{g}\rho \Delta x \Delta y \Delta z$, and its x component is $g_x \rho \Delta x \Delta y \Delta z$.

The Momentum Conservation Equation

We now equate the time rate of change of momentum to the sum of the forces acting. For the x direction,

$$\rho \Delta x \Delta y \Delta z \frac{Du}{Dt} = \left(-\frac{\partial P}{\partial x} - \frac{\partial \tau_{xx}}{\partial x} - \frac{\partial \tau_{yx}}{\partial y} - \frac{\partial \tau_{zx}}{\partial z} + \rho g_x \right) \Delta x \Delta y \Delta z$$

or

$$\rho \frac{Du}{Dt} = -\frac{\partial P}{\partial x} - \frac{\partial \tau_{xx}}{\partial x} - \frac{\partial \tau_{yx}}{\partial y} - \frac{\partial \tau_{zx}}{\partial z} + \rho g_x \tag{5.112a}$$

Similar equations can be obtained for the y and z directions:

$$\rho \frac{Dv}{Dt} = -\frac{\partial P}{\partial y} - \frac{\partial \tau_{xy}}{\partial x} - \frac{\partial \tau_{yy}}{\partial y} - \frac{\partial \tau_{zy}}{\partial z} + \rho g_y \tag{5.112b}$$

$$\rho \frac{Dw}{Dt} = -\frac{\partial P}{\partial z} - \frac{\partial \tau_{xz}}{\partial x} - \frac{\partial \tau_{yz}}{\partial y} - \frac{\partial \tau_{zz}}{\partial z} + \rho g_z \tag{5.112c}$$

The three equations can be summarized in matrix form:

$$\rho \frac{D}{Dt} \begin{bmatrix} u \\ v \\ w \end{bmatrix} = - \begin{bmatrix} P + \tau_{xx} & \tau_{yx} & \tau_{zx} \\ \tau_{xy} & P + \tau_{yy} & \tau_{zy} \\ \tau_{xz} & \tau_{yz} & P + \tau_{zz} \end{bmatrix} \begin{bmatrix} \partial/\partial x \\ \partial/\partial y \\ \partial/\partial z \end{bmatrix} + \rho \begin{bmatrix} g_x \\ g_y \\ g_z \end{bmatrix} \tag{5.113}$$

The convention for matrix multiplication of the operator is that row i of the desired one-column vector results from operating on the term in row i and column j of the matrix by the differential operator in row j of the operator vector, and summing the terms for $j = 1, 2, 3$. The matrix is called the *stress tensor*.

The Stokes Hypothesis

To use Eq. (5.113), it is necessary to express the viscous stresses in terms of the velocity gradients. Recall that a Newtonian fluid is one in which the shear stress is linearly proportional to the velocity gradient in a simple Couette-type shear flow. Stokes generalized the concept to a flow in which the velocity varies in more than one coordinate direction by assuming that the shear stress is linearly related to the rate of angular deformation. Most common fluids, such as water and air, are Newtonian, but many important fluids are not, for example, blood and polymer solutions.

Figure 5.30 shows how angular deformation occurs in simple shear and in the more general case. In time Δt, the upper left corner of what originally was a square of fluid will move a distance $(\partial u/\partial y)\Delta y \Delta t$ farther than the lower left corner. This

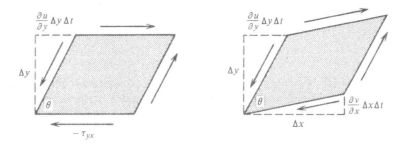

Figure 5.30 Rate of angular deformation in one and two dimensions.

distance divided by Δy gives the change in angle θ. Similarly, the movement of the lower right corner causes an angular change of $(\partial v/\partial x)\Delta t$. The combined rate of change of angle θ is then

$$-\frac{\Delta \theta}{\Delta t} = \frac{\partial u}{\partial y} + \frac{\partial v}{\partial x}$$

This angular rate of change was brought about by (or gives rise to) the negative shear stresses shown in Fig. 5.30. By the Stokes hypothesis,

$$\tau_{yx} = -\mu \left(\frac{\partial u}{\partial y} + \frac{\partial v}{\partial x} \right) = \tau_{xy} \tag{5.114a}$$

Similarly,

$$\tau_{zx} = -\mu \left(\frac{\partial u}{\partial z} + \frac{\partial w}{\partial x} \right) = \tau_{xz} \tag{5.114b}$$

$$\tau_{zy} = -\mu \left(\frac{\partial v}{\partial z} + \frac{\partial w}{\partial y} \right) = \tau_{yz} \tag{5.114c}$$

It can be shown, by some rather involved analysis involving coordinate transformations, that the compatible relations for the normal viscous stresses are

$$\tau_{xx} = -\mu \left(2\frac{\partial u}{\partial x} - \frac{2}{3}\nabla \cdot \mathbf{v} \right) \tag{5.115a}$$

$$\tau_{yy} = -\mu \left(2\frac{\partial v}{\partial y} - \frac{2}{3}\nabla \cdot \mathbf{v} \right) \tag{5.115b}$$

$$\tau_{zz} = -\mu \left(2\frac{\partial w}{\partial z} - \frac{2}{3}\nabla \cdot \mathbf{v} \right) \tag{5.115c}$$

In some fluids, such as liquids with small gas bubbles, a *second coefficient of viscosity* is needed. It adds a term proportional to $\nabla \cdot \mathbf{v}$ to each normal stress. But in most situations, we can simply approximate the normal viscous stresses as

$$\tau_{xx} = -2\mu \frac{\partial u}{\partial x} \tag{5.116a}$$

$$\tau_{yy} = -2\mu \frac{\partial v}{\partial y} \tag{5.116b}$$

$$\tau_{zz} = -2\mu \frac{\partial w}{\partial z} \tag{5.116c}$$

If Eqs. (5.114) and (5.116) are substituted in Eq. (5.113) and a constant density and viscosity assumed, the resulting three components of the momentum conservation equation are

$$\rho \left(\frac{\partial u}{\partial t} + u\frac{\partial u}{\partial x} + v\frac{\partial u}{\partial y} + w\frac{\partial u}{\partial z} \right) = -\frac{\partial P}{\partial x} + \mu \left(\frac{\partial^2 u}{\partial x^2} + \frac{\partial^2 u}{\partial y^2} + \frac{\partial^2 u}{\partial z^2} \right) + \rho g_x \tag{5.117a}$$

$$\rho \left(\frac{\partial v}{\partial t} + u\frac{\partial v}{\partial x} + v\frac{\partial v}{\partial y} + w\frac{\partial v}{\partial z} \right) = -\frac{\partial P}{\partial y} + \mu \left(\frac{\partial^2 v}{\partial x^2} + \frac{\partial^2 v}{\partial y^2} + \frac{\partial^2 v}{\partial z^2} \right) + \rho g_y \tag{5.117b}$$

$$\rho \left(\frac{\partial w}{\partial t} + u\frac{\partial w}{\partial x} + v\frac{\partial w}{\partial y} + w\frac{\partial w}{\partial z} \right) = -\frac{\partial P}{\partial z} + \mu \left(\frac{\partial^2 w}{\partial x^2} + \frac{\partial^2 w}{\partial y^2} + \frac{\partial^2 w}{\partial z^2} \right) + \rho g_z \tag{5.117c}$$

Equations (5.117) can be written in the compact form

$$\rho \frac{D\mathbf{v}}{Dt} = -\nabla P + \mu \nabla^2 \mathbf{v} + \rho \mathbf{g} \tag{5.118}$$

These equations are often called the **Navier-Stokes equations.**

5.6.3 Conservation of Energy

Control Volume

We return to the Eulerian viewpoint and consider a Cartesian elemental control volume $\Delta x \Delta y \Delta z$ located in a pure fluid. The energy conservation principle requires that the rate at which energy is stored within the volume equal the net rate of inflow of energy by convection, plus the rate of heat transfer across the boundary, plus the rate at which work is done on the fluid within the volume, plus the rate at which energy is produced within the volume.

Energy Storage

Energy can be stored within the volume as internal energy and as kinetic energy of the mass as a whole. The rate of energy storage per unit time is thus

$$\frac{\partial}{\partial t} \left(\rho u + \frac{1}{2}\rho v^2 \right) \Delta x \Delta y \Delta z$$

where u is the specific internal energy,[5] and $v^2 = \mathbf{v} \cdot \mathbf{v} = u^2 + v^2 + w^2$.

---

[5] Care must be taken not to confuse the specific internal energy u with the velocity component u in this derivation.

Energy Inflow

Energy can flow into the control volume by convection. As mass flows across the volume boundaries, it brings with it energy $[u+(1/2)v^2]$ per unit mass. For example, the convection across the y face per unit time is

$$\left[\rho v\left(u+\frac{1}{2}v^2\right)\right]_y \Delta x\Delta z$$

and the convection out across the $y+\Delta y$ face is

$$\left[\rho v\left(u+\frac{1}{2}v^2\right)\right]_{y+\Delta y} \Delta x\Delta z$$

The net rate of inflow for the two faces is

$$-\frac{\partial}{\partial y}\left[\rho v\left(u+\frac{1}{2}v^2\right)\right]\Delta x\Delta y\Delta z$$

Similar terms arise for the other two pairs of faces. The sum of all these terms can be written

$$-\nabla\cdot\left[\rho\mathbf{v}\left(u+\frac{1}{2}v^2\right)\right]\Delta x\Delta y\Delta z$$

Heat Transfer

Heat is transferred into the control volume by conduction. The rate of heat inflow across the x face by conduction is

$$\left(-k\frac{\partial T}{\partial x}\right)_x \Delta y\Delta z$$

and the rate of outflow across the $x+\Delta x$ face is

$$\left(-k\frac{\partial T}{\partial x}\right)_{x+\Delta x} \Delta y\Delta z$$

The net rate of inflow for the two faces is

$$\frac{\partial}{\partial x}\left(k\frac{\partial T}{\partial x}\right)\Delta x\Delta y\Delta z$$

Again, the other two pairs of faces give similar terms. The sum for all faces can be written

$$\nabla\cdot k\nabla T\Delta x\Delta y\Delta z$$

Work

The rate at which work is done by the volume force $\mathbf{g}$ is simply the vector dot product of force and velocity:

$$\rho\mathbf{g}\cdot\mathbf{v}\Delta x\Delta y\Delta z = (\rho g_x u+\rho g_y v+\rho g_z w)\Delta x\Delta y\Delta z$$

Work can also be done by the surface forces acting on a volume element. The vector force acting on the x face is the first column of the stress tensor times the area of the face. The rate at which work is done *on* the element at this face is the vector dot product of velocity and force. Denote the first, second, and third columns of the stress tensor as $\mathbf{F}_x, \mathbf{F}_y$, and $\mathbf{F}_z$, respectively. The subscripts denote the direction of the normal to the face on which the force acts. Then, for example,

$$\mathbf{F}_x = \mathbf{i}(P + \tau_{xx}) + \mathbf{j}(\tau_{xy}) + \mathbf{k}(\tau_{xz})$$

and the rate at which work is done on the element by this force is

$$\mathbf{v} \cdot \mathbf{F}_x \Delta y \Delta z = [(P + \tau_{xx})u + \tau_{xy}v + \tau_{xz}w]\Delta y \Delta z$$

At the $x + \Delta x$ face, work is done *by* the element, on the fluid of greater x, at the rate

$$(\mathbf{v} \cdot \mathbf{F}_x)_{x+\Delta x}\Delta y \Delta z$$

The net rate at which work is done *on* the element for these two faces is

$$-\frac{\partial}{\partial x}(\mathbf{v} \cdot \mathbf{F}_x)\Delta x \Delta y \Delta z$$

The net work done on the element for all six faces is then

$$-\left[\frac{\partial}{\partial x}(\mathbf{v} \cdot \mathbf{F}_x) + \frac{\partial}{\partial y}(\mathbf{v} \cdot \mathbf{F}_y) + \frac{\partial}{\partial z}(\mathbf{v} \cdot \mathbf{F}_z)\right]\Delta x \Delta y \Delta z$$

Heat Generation

Internal heat generation may arise due to resistive (I^2R) heating, neutron and fission fragment slowing, or photon absorption and emission. These sources may be represented by an equivalent volumetric source term $\dot{Q}_v'''$ [W/m^3], and the production rate within the elemental volume is

$$\dot{Q}_v''' \Delta x \Delta y \Delta z$$

Total Energy Equation

Substituting all the preceding terms in our statement of conservation of energy and dividing by the volume $\Delta x \Delta y \Delta z$ gives the **total energy equation,**

$$\frac{\partial}{\partial t}\rho\left(u + \frac{1}{2}v^2\right) = -\nabla \cdot \left[\rho\mathbf{v}\left(u + \frac{1}{2}v^2\right)\right] + \nabla \cdot k\nabla T + \rho\mathbf{g} \cdot \mathbf{v} - \frac{\partial}{\partial x}(\mathbf{v} \cdot \mathbf{F}_x)$$

$$-\frac{\partial}{\partial y}(\mathbf{v} \cdot \mathbf{F}_y) - \frac{\partial}{\partial z}(\mathbf{v} \cdot \mathbf{F}_z) + \dot{Q}_v''' \tag{5.119}$$

With the aid of the mass conservation equation, Eq. (5.106), Eq. (5.119) can be written in terms of the substantial derivative as

$$\rho\frac{D}{Dt}\left(u + \frac{1}{2}v^2\right) = \nabla \cdot k\nabla T + \rho\mathbf{g} \cdot \mathbf{v} - \frac{\partial}{\partial x}(\mathbf{v} \cdot \mathbf{F}_x)$$

$$-\frac{\partial}{\partial y}(\mathbf{v} \cdot \mathbf{F}_y) - \frac{\partial}{\partial z}(\mathbf{v} \cdot \mathbf{F}_z) + \dot{Q}_v''' \tag{5.120}$$

Thermal Energy Equations

Since Eq. (5.120) accounts for conservation of both thermal and mechanical energy, it is not the most convenient starting point for the solution of most heat transfer problems. We prefer to have a conservation equation that isolates thermal phenomena, which can be derived by subtracting the conservation equation for mechanical energy from the total energy equation, as follows.

The mechanical energy equation is obtained by taking the dot product of velocity $\mathbf{v}$ and the momentum equation to obtain the rate at which mechanical work is done. In Cartesian coordinates, we add u times the x-momentum equation, v times the y-momentum equation, and w times the z-momentum equation. The result is

$$\rho \frac{D}{Dt}\left(\frac{1}{2}v^2\right) = -\mathbf{v}\cdot\frac{\partial \mathbf{F}_x}{\partial x} - \mathbf{v}\cdot\frac{\partial \mathbf{F}_y}{\partial y} - \mathbf{v}\cdot\frac{\partial \mathbf{F}_z}{\partial z} + \rho\mathbf{v}\cdot\mathbf{g} \tag{5.121}$$

Using the product rule of differentiation in Eq. (5.120) and subtracting Eq. (5.121) gives

$$\rho \frac{Du}{Dt} = \nabla\cdot k\nabla T - \frac{\partial \mathbf{v}}{\partial x}\cdot\mathbf{F}_x - \frac{\partial \mathbf{v}}{\partial y}\cdot\mathbf{F}_y - \frac{\partial \mathbf{v}}{\partial z}\cdot\mathbf{F}_z + \dot{Q}_v''' \tag{5.122}$$

Substituting for $\mathbf{F}_x, \mathbf{F}_y$, and $\mathbf{F}_z$ using Eqs. (5.114) and (5.116) and rearranging gives

$$\rho \frac{Du}{Dt} = -P\nabla\cdot\mathbf{v} + \nabla\cdot k\nabla T + \mu\Phi + \dot{Q}_v''' \tag{5.123}$$

where the quantity $\mu\Phi$ represents the portion of the mechanical work that is irreversibly converted into heat by the viscous stresses. In terms of the velocity components,

$$\Phi = 2\left[\left(\frac{\partial u}{\partial x}\right)^2 + \left(\frac{\partial v}{\partial y}\right)^2 + \left(\frac{\partial w}{\partial z}\right)^2\right] + \left(\frac{\partial u}{\partial y} + \frac{\partial v}{\partial x}\right)^2 + \left(\frac{\partial u}{\partial z} + \frac{\partial w}{\partial x}\right)^2 + \left(\frac{\partial v}{\partial z} + \frac{\partial w}{\partial y}\right)^2 \tag{5.124}$$

The quantity $\mu\Phi$ is the viscous dissipation; Φ is often simply called the **dissipation function.** The term $-P\nabla\cdot\mathbf{v}$ represents the portion of the mechanical work that is reversibly converted into heat by compression.

We often prefer to work in terms of enthalpy h rather than internal energy u. Substituting the thermodynamic relation $h = u + P/\rho$ into Eq. (5.123) and rearranging,

$$\rho \frac{Dh}{Dt} = \frac{DP}{Dt} + \nabla\cdot k\nabla T + \mu\Phi + \dot{Q}_v''' \tag{5.125}$$

For the special case of an ideal gas, $dh = c_p dT$, and Eq. (5.125) becomes

$$\rho c_p \frac{DT}{Dt} = \frac{DP}{Dt} + \nabla\cdot k\nabla T + \mu\Phi + \dot{Q}_v''' \tag{5.126}$$

For an incompressible liquid with specific heat $c = c_p = c_v$, we go back to Eq. (5.123): since $\nabla\cdot\mathbf{v} = 0$,

$$\rho c \frac{DT}{Dt} = \nabla\cdot k\nabla T + \mu\Phi + \dot{Q}_v''' \tag{5.127}$$

Equations (5.123) through (5.127) are all forms of the **thermal energy equation** for a Newtonian fluid. One of these equations is the usual starting point of a convection analysis.

5.6.4 Use of the Conservation Equations

Solving the general conservation equations, even by direct numerical means, is difficult and usually impractical. Fortunately, however, many problems of engineering interest are adequately described by simplified forms of these equations, and these forms often can be solved more easily. Sections 5.2 through 5.4 contained examples of such problems involving laminar flows, and there are many more. Earlier in this chapter we set up the governing equations from first principles; now we are in a position to obtain the governing equations by simply deleting superfluous terms in the general equations. The following discussion applies directly to laminar flows. In the case of turbulent flows, the remarks apply only to the time-averaged equations. For example, on average, a turbulent flow might be steady and two-dimensional, but there are instantaneous fluctuations of all three velocity components.

The manner in which a particular problem is posed may immediately imply that some terms are identically zero. For example, if a timewise steady-state solution is sought, the time derivatives are set equal to zero. Some resulting classes of simplified flows are:

1. Constant density

2. Constant transport properties

3. Timewise steady flow (or quasi-steady flow)

4. Two-dimensional flow

5. One-dimensional flow

6. Fully developed flows

Terms may also be negligibly small. Often, intuition or experimental evidence will suffice to make a decision. But a more rigorous approach is to use order-of-magnitude estimates based on scaling arguments, as is done in more advanced texts. Some classes of flow that result are:

1. Creeping flows

2. Forced flows

3. Natural flows

4. Boundary layer flows

5. Low-speed flows

A complete mathematical statement of a convection problem requires specification of boundary conditions. Each boundary condition is based on a physical statement or principle. The student can examine the analyses of Sections 5.2 through 5.5 for examples.

5.7   CLOSURE

The early part of this chapter introduced the student to the analysis of convection by considering the simplest laminar flows only. The analyses should have provided insight into some of the essential features of convection, such as the effect of viscous dissipation in a high-speed flow, the nature of fully developed flow in a duct, and the important role played by the growth and thickness of boundary layers. The analytical results show why some of the correlations in Chapter 4 took their particular forms.

Modeling of turbulence was briefly introduced, but detailed analysis of turbulent flows is left to more advanced texts. The chapter closed with a presentation of the general conservation equations: familiarity with these equations is necessary to proceed further with the study of convection analysis.

The evolution of computer use for convection analysis makes an interesting story. In the late 1950s, the then-new digital computer was used to solve ordinary differential equations governing simpler flows. During the 1960s, efficient numerical methods were developed to solve the partial differential equations of boundary layer and duct flows; by the end of the decade such solutions were routine. The 1970s saw attention directed toward solving the general conservation equations governing flows with vortices, shock waves, and other complicated features. Many of these flows remain very challenging today, requiring sophisticated numerical methods and large, fast computers. The most recent development has been the increased availability of packaged computational fluid dynamics (CFD) software, for example PHOENICS [10], AcuSolve [11], ANSYS Fluent [12], and FLOW-3D [13]. Such programs vary in degree of generality, but all can be used to solve a wide variety of flows, even by users having little knowledge of (and sometimes no access to) the numerical methods incorporated in the program. Sophisticated interactive programming allows new geometric configurations and boundary conditions to be easily implemented, and standard procedures are pre- scribed to ensure a properly converged solution. Unquestionably, the availability of such programs implies that portions of the convection literature are now obsolete. On the other hand, the difficulties and subtleties of convection make it impossible to use such programs blindly. When new problems are attempted, the first solutions obtained are often far from accurate. A good understanding of the fundamentals is required, as well as a familiarity with benchmark experimental data and previous analytical and numerical solutions of established validity.

REFERENCES

1. Eckert, E. R. G., "Engineering relations for heat transfer and friction in high-velocity laminar and turbulent boundary layer flow over surfaces with constant pressure and temperature," *Trans. ASME*, 78, 1273–1284 (1956).

2. Wortman, A., Mills, A. F., and Soo Hoo, G., "The effect of mass transfer on recovery factors in laminar boundary layer flows," *Int. J. Heat Mass Transfer*, 15, 443–156 (1972).

3. von Kármán, T., "Über laminare und turbulente Reibung," ZAMM, 1, 233–252 (1921); see also NACA TM 1092 (1946).

4. Prandtl, L., "Bericht über Untersuchungen zur ausgebildeten Turbulenz," *ZAMM*, 5, 136 (1925).

5. van Driest, E. R., "On turbulent flow near a wall," *J. Aero. Sci.*, 23, 1007–1011 (1951).

6. Escudier, M. P., "The distribution of the mixing length in turbulent flows near walls," Imperial College, London, *Mechanical Engineering Department Report TWF/TN/1* (1965).

7. SCSTREAM, www.cradle.cfd.com.

8. Crawford, M. E., TEXSTAN, available from www.textstan.com.

9. COMSOL Multiphysics 5.0, www.comsol.com.

10. PHOENICS (Parabolic, Hyperbolic, or Elliptic Numerical Integration Code Series); available from CHAM Ltd., 40 High Street, Wimbledon, London, SW19 5AU, England, www.cham.co.uk.

11. AcuSolve, www.altairhyperworks.com.

12. ANSYS Fluent, www.ansys.com.

13. FLOW-3D, FlowScience, Santa Fe, N.M., www.flow3d.com.

14. Sparrow, E. M., and Patankar, S. V., "Relationships among boundary conditions and the Nusselt numbers for thermally developed duct flows," *J. Heat Transfer*, 99, 483–485 (1977).

15. Ede, A. J., "Advances in free convection" in *Advances in Heat Transfer*, vol. 4, eds. Hartnett, J. P., and Irvine, T. E, Jr., Academic Press, New York (1967).

16. Fugii, T, and Fugii, M. "The dependence of local Nusselt number on Prandtl number in the case of free convection along a vertical surface with uniform heat flux," *Int. J. Heat Mass Transfer*, 19, 121–122 (1976).

17. Lykoudis, P. S., "Non-dimensional numbers as ratios of characteristic times," *Int. J. Heat Mass Transfer*, 33, 1560–1570 (1990).

EXERCISES

5–1. Two 1 m-long concentric cylinders form an annular gap 1 mm wide. The outer cylinder is stationary. The inner cylinder has a radius of 10 cm and rotates at

1000 rpm. Determine the power dissipated by viscous dissipation in the gap if the fluid is

    (i) air,
   (ii) water,
  (iii) SAE 50 oil,

all at 300 K and 1 atm pressure.

5–2.   Prepare a table of recovery factors for the following fluids:

     (i) Air at 300 K, 1 atm.
    (ii) Saturated steam at 380 K.
   (iii) Water at 300 K.
   (iv) Water at 456 K.
    (v) SAE 50 oil at 350 K.
   (vi) Liquid mercury at 600 K.

In each case, calculate the values for a Couette flow, a laminar boundary layer on a flat plate, and a turbulent boundary layer.

5–3.   A fighter aircraft's mission necessitates a high-speed penetration behind enemy lines and beneath radar detection altitude. Prior to making this run, the aircraft cruises at high altitude, and most of the structure — in particular the wings — reaches a uniform temperature of 273 K. By modeling a wing as a 16 m × 2.5 m flat plate, estimate:

    (i) the rate of heat input to the wing during the initial stages of a high-speed run, at 1500 km/h.
   (ii) the fraction of the heat input that can be attributed to viscous dissipation.

Take the ambient air to be at 303 K and 1 bar.

5–4.   At an altitude of 30,000 m the atmospheric pressure is 1197 Pa, and the temperature is 227 K. For a turbulent boundary layer on a flat plate, prepare a graph of adiabatic wall temperature versus Mach number for $0 < M < 5$. Also calculate the convection-radiation equilibrium temperature at $x = 0.5$ m for a surface emittance of 0.9. Assume transition takes place at the leading edge.

5–5.   Recovery factors for turbulent flows are based on experimental measurements. To determine recovery factors and heat transfer coefficients on a curved surface geometry, a heated copper model is tested in a high-speed wind tunnel. Miniature heat flux sensors are mounted flush with the surface and also contain thermocouples for measurement of surface temperature. The air is at 1 atm, 300 K, and has a speed of 240 m/s. Tests are done at two power inputs to the model heater. In the first test, the surface is maintained at 340.0 K and the heat flux meter at the location under consideration records a heat flux of 7160 W/m^2. In the second test, the surface is at 370.0 K and the heat flux is 22,200 W/m^2. Use these data to obtain the recovery factor and heat transfer coefficient.

5–6. An 8 cm-long flat copper plate is part of a structure undergoing tests in a high-speed wind tunnel. The test conditions are an air velocity of 1025 m/s, a pressure of 50 kPa, and a temperature of 280 K. Determine the equilibrium temperature of the plate if its surface emittance is 0.26 and the tunnel walls are at 355 K. Take $\text{Re}_{tr} = 0$.

5–7. Consider low-speed, hydrodynamically fully developed flow in a parallel-plate duct. For $x < 0$, both plates are insulated, and the fluid is at temperature T_1; for $x > 0$, one plate is maintained at $T = T_2 > T_1$, and the other remains insulated. Draw sketches of

(i) $T_b(x)$
(ii) $T(y)$ at various values of x

to illustrate the development of the temperature field.

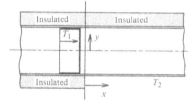

5–8. Consider low-speed, constant-property, fully developed laminar flow between two parallel plates at $y = \pm b$. The plates are electrically heated to give a uniform wall heat flux. Show that

(i) the velocity profile is $u = (3/2)u_b[1 - (y/b)^2]$.
(ii) the friction factor is $f = 96/\text{Re}_{D_h}$.
(iii) the Nusselt number is $\text{Nu} = 140/17 = 8.24$.

5–9. As in Exercise 5–8, consider constant-property, fully developed laminar flow between parallel plates, but now one plate is insulated, the other is isothermal, and the velocity is high enough for viscous dissipation to be significant. Determine the temperature profile.

5–10. Consider low-speed, constant-property, fully developed laminar flow between parallel plates, with one plate insulated and the other uniformly heated. Show that the Nusselt number is $\text{Nu} = 140/26 = 5.385$.

5–11. Consider laminar flow of high-viscosity oil in a circular tube with a uniform wall temperature. If viscous heating is significant, determine the temperature profile a long distance from the inlet. Assume constant properties.

5–12. A gas flows between parallel porous plates. The same gas is blown through one wall and exhausted through the other, such that the normal velocity component v_s is equal at both walls. At a location far from the entrance, determine

(i) the velocity profile
(ii) the temperature profile

if both walls are isothermal at different temperatures T_1 and T_2. Assume constant-property, low-speed flow.

5–13. Lithium enters a parallel-plate duct of spacing $2b$, with a uniform velocity U and a uniform temperature T_0. The walls are maintained at a uniform temperature T_s. If laminar plug flow is assumed, determine the temperature distribution downstream of the entrance. Also, if $Re_{D_h} = 1500$ and $Pr = 0.03$, determine the Nusselt number variation along the duct. (*Hint:* There is an analogous conduction problem in Chapter 3.)

5–14. Show that for fully developed laminar flow in an annular duct the velocity profile is given by

$$u = \frac{R_o^2}{4\mu}\left(-\frac{dP}{dx}\right)\left[1 - \left(\frac{r}{R_o}\right)^2 + \frac{1-\gamma^2}{\ln(1/\gamma)}\ln\frac{r}{R_o}\right]$$

where $\gamma = R_i/R_o$ and R_i, R_o are the inner and outer radii, respectively. Hence show that the bulk velocity is

$$u_b = \frac{R_o^2}{8\mu}\left(-\frac{dP}{dx}\right)\left[1+\gamma^2 - \frac{1-\gamma^2}{\ln(1/\gamma)}\right]$$

and the location where the maximum velocity occurs is given by

$$\frac{r}{R_o} = \left(\frac{1-\gamma^2}{2\ln(1/\gamma)}\right)^{1/2}$$

5–15. Calculate the pressure drop per meter length when air at 300 K, 1 atm pressure flows at 7×10^{-4} kg/s through an annulus of inner and outer radii 6 and 10 mm, respectively.

5–16. In the footnote to the steady-flow energy equation, Eq. (1.3), it is noted that an integration is required to give an appropriate value of the velocity V in the kinetic energy term $V^2/2$. The bulk velocity u_b is an area weighted average velocity (for a constant-density fluid), and the mass flow rate is $\rho u_b A_c$: u_b is not the appropriate value of V in the kinetic energy term. Determine appropriate expressions for V in the terms of u_b for the following three constant-density flows:

  (i) Laminar flow in a circular pipe.
  (ii) Laminar flow between parallel plates (see Exercise 5–8).
  (iii) The turbulent flow given in Exercise 4–2.

5–17. Thermally fully developed tube and duct flows are characterized by a constant Nusselt number. An analysis for laminar flow in a tube with a uniform wall heat flux (UHF) is given in Section 5.3.2. Results for both a uniform wall temperature (UWT) and UHF are given in Table 4.5 for various duct cross sections. A constant Nusselt number is also obtained for an exponential wall heat flux variation, $q_s \propto e^{\beta x}$: UHF and UWT are special cases. Show that for UWT the constant β is $-7.314/(D/2)Re_D Pr$.

5–18. Air at 290 K, 3 bar, and a Mach number of 0.15 flows inside a microtube that has a 20 μm inside diameter. If the tube is insulated, estimate the rate of change in bulk temperature due to viscous dissipation.

5–19. The Darcy friction factor for Poiseuille flow is $f = (-dP/dx)D/(1/2)\rho u_b^2 = 64/\mathrm{Re}_D$. In this definition of the friction factor, the pressure gradient is scaled by the turbulent shear stresses in the flow. For a laminar flow it is more appropriate to scale the pressure force by the viscous force exerted on the wall. Show that the friction factor so obtained is independent of the Reynolds number. Compare this result to Eq. (5.36) for the Nusselt number, and comment.

5–20. Referring to Exercise 5–17, consider convective heat transfer from the external surface of a tube to surroundings at temperature T_e with a constant external heat transfer coefficient $h_{c,o}$. Sparrow and Patankar [14] show that this boundary condition also yields a constant Nusselt number: they obtained the results in the table below, where k is the conductivity of the fluid in the tube.

$\mathrm{Bi} = h_{c,o}R/k$	$-\mathrm{Re}_D \mathrm{Pr} \beta R$	Nu	$\widehat{\mathrm{Nu}}$
0	0	4.364	0
0.1	0.3818	4.330	0.1909
0.25	0.8943	4.284	0.4471
0.5	1.615	4.221	0.8075
1	2.690	4.112	1.345
2	3.995	3.997	1.998
5	5.547	3.840	2.773
10	6.326	3.758	3.163
100	7.195	3.663	3.597
∞	7.314	3.657	3.657

The two Nusselt numbers are defined in terms of heat transfer coefficients $h_c = q_s/(T_s - T_b)$ and $\widehat{h}_c = q_s/(T_e - T_b)$, respectively.

Helium at 1 atm pressure flows at 10^{-3} kg/s in a 3 cm-I.D. copper tube. There is a cross-flow of air at 300 K across the tube, giving an outside heat transfer coefficient of 68.3 W/m² K. Calculate the fully developed overall heat transfer coefficient and the rate of heat loss at a location where the helium bulk temperature is 500 K. Evaluate helium properties at 500 K.

5–21. If in Exercise 5–9 both plates are insulated, determine the rate of increase in bulk temperature dT_b/dx and the temperature profile $T(y)$ for thermally fully developed conditions.

5–22. Derive the differential momentum conservation equation for a laminar boundary layer on a flat plate using a Lagrangian approach: that is, apply Newton's second law of motion to an element of fluid of fixed mass moving along a stream-line.

5–23. Derive the integral momentum equation for a laminar boundary layer on a flat plate, Eq. (5.51), by integrating the differential momentum equation, Eq. (5.39), across the boundary layer.

5–24. Determine the skin friction coefficient for a laminar boundary layer on a flat plate by assuming a quartic polynomial for the velocity profile when using the von Kármán profile method to solve the integral momentum equation.

5–25. Air at 300 K and 1 atm pressure flows along a flat plate at 2 m/s. For $x < 5$ cm, $T_s = 300$ K, whereas for 5 cm $< x < 20$ cm, $T_s = 330$ K. Calculate the heat loss from the plate and compare your result with the heat loss if the plate were isothermal at 330 K. Assume a laminar boundary layer.

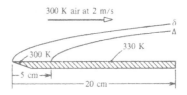

5–26. Equation (5.50) gives the local Nusselt number for plug flow along an isothermal flat plate as

$$\text{Nu}_x = 0.564 \text{ Pe}_x^{1/2}$$

(i) Using Eq. (3.60), show that the corresponding result for a uniformly heated plate is

$$\text{Nu}_x = 0.886 \text{ Pe}_x^{1/2}$$

(ii) Using Eqs. (1.26) and (4.82) where appropriate, obtain corresponding expressions for $\overline{\text{Nu}}_L$ and compare the results.

5–27. Water at 290 K flows at 1 m/s along a flat plate. For $x < 10$ cm, $T_s = 290$ K, whereas for $10 < x < 30$ cm, $T_s = 310$ K. Calculate the local heat flux at $x = 0.3$ m and compare your result with the heat flux if the plate were isothermal at 310 K.

5–28. Determine the skin friction for a laminar boundary layer on a flat plate by assuming a quadratic velocity profile $u/u_e = 2(y/\delta) - (y/\delta)^2$ when using the von Kármán profile method to solve the integral momentum equation. Also determine the Nusselt number for an isothermal surface using a quadratic temperature profile $(T - T_e)/(T_s - T_e) = 1 - 2(y/\Delta) + (y/\Delta)^2$.

5–29. Repeat Exercise 5–28 with sine function velocity and temperature profiles, namely, $u/u_e = \sin(\pi y/2\delta); (T - T_e)/(T_s - T_e) = 1 - \sin(\pi y/2\Delta)$.

5–30. Derive the integral momentum and energy equations for a laminar boundary layer on a flat plate in a pure fluid with blowing or suction at the wall.

5–31. Repeat the integral method analysis of Section 5.4.4, but allow the wall temperature to vary as $(T_s - T_e) = Cx^\gamma$. Show that the new expressions for X and D are

$$X = \frac{60\alpha}{D^2} \frac{4}{3 + 5\gamma}; \quad D^4 = \frac{720}{Cg\beta} \left[\frac{12\alpha^2}{21} \frac{5 + 3\gamma}{(3 + 5\gamma)^2} + \frac{\alpha v}{3 + 5\gamma} \right]$$

Hence, show that if the wall heat flux is uniform, the wall temperature increases proportional to $x^{1/5}$.

5–32. A vertical plate at 320 K is located in air at 300 K and 1 atm. At a location 10 cm from the bottom of the plate determine $\delta, U, \text{Nu}_x, h_{cx}$ and q_s.

5–33. Consider natural convection from the underside of a heated horizontal cylinder. Derive an integral form of the boundary layer momentum equation valid for laminar or turbulent flow.

5–34. The integral method solution for laminar natural convection on a vertical wall given in Section 5.4.4 shows that the local heat transfer coefficient is proportional to $x^{-1/4}$, where x is measured from the leading edge. Use this result to derive a formula for the local Nusselt number consistent with the correlation for the average Nusselt number, Eq. (4.85). Compare your result with Eq. (5.80).

5–35. Ede [15] suggests the following correlation of exact numerical solutions for laminar natural convection on a vertical isothermal wall:

$$\mathrm{Nu}_x = 0.5964 \left(\frac{\mathrm{Pr}}{1 + 2\mathrm{Pr}^{1/2} + 2\mathrm{Pr}} \right)^{1/4} \mathrm{Ra}_x^{1/4}$$

(i) Compare this result with Eq. (5.80), which was the result of an integral method analysis.
(ii) Obtain the corresponding expression for the average Nusselt number, and compare it with the Churchill and Chu correlation, Eq. (4.85).

5–36. Cooling systems for electronic equipment may be required to remove heat by laminar natural convection from a uniformly heated vertical wall. Since the heat flux is usually specified, it is the variation of temperature along the wall (and, in particular, its maximum value) that is of primary concern. Fugii and Fugii [16] correlated the results of exact numerical solutions as

$$\mathrm{Nu}_x = \left(\frac{\mathrm{Pr}}{4 + 9\mathrm{Pr}^{1/2} + 10\mathrm{Pr}} \right)^{1/5} \mathrm{Ra}_x^{*1/5}$$

where Ra_x^* is a modified Rayleigh number constructed by replacing the temperature difference in the usual definition of Ra_x by $q_s x/k$ to obtain $\mathrm{Ra}_x^* = g\beta q_s x^4/kv\alpha$.

A vertical panel is 12 cm square and contains devices that dissipate a total of 24 W distributed approximately uniformly over the panel. One side of the board is exposed to quiescent Therminol 60 at 300 K, and the heat loss from the other side is negligible. Determine the maximum temperature of the panel. Evaluate properties at the mean film temperature halfway up the panel. (See also Exercise 4–86.)

5–37. Referring to Exercise 5–36, obtain an expression for the appropriately averaged Nusselt number defined by Eq. (4.82) and comment on its utility.

5–38. A vertical panel is 12 cm square and 1 cm thick, and contains electronic components that dissipate a total of 4.67 W distributed approximately uniformly inside the panel. The panel is exposed to quiescent air at 300 K, 1 atm, on both sides, and the edges are insulated. Determine the maximum surface temperature if the surface emittance is small enough for radiation heat transfer to be negligible. Evaluate properties at the mean film temperature halfway up the panel (see Exercises 4–86 and 5–36).

5–39. The equations governing an unsteady boundary layer flow along a flat plate are

$$\frac{\partial \rho}{\partial t} + \frac{\partial}{\partial x}(\rho u) + \frac{\partial}{\partial y}(\rho v) = 0 \tag{5.128}$$

$$\rho \frac{\partial u}{\partial t} + \rho u \frac{\partial u}{\partial x} + \rho v \frac{\partial u}{\partial y} = \frac{\partial}{\partial y}(-\tau_{xy}) \tag{5.129}$$

$$\rho \frac{\partial h}{\partial t} + \rho u \frac{\partial h}{\partial x} + \rho v \frac{\partial h}{\partial y} = \frac{\partial}{\partial y}(-q_y) \tag{5.130}$$

In a turbulent flow we would expect Newton's and Fourier's laws to apply instantaneously, that is, $\tau_{xy} = -\mu \partial u/\partial y$; $q_y = -k \partial T/\partial y$. The momentum and energy equations, Eqs. (5.129) and (5.130), can be written as

$$\rho \frac{\partial \phi}{\partial t} + \rho u \frac{\partial \phi}{\partial x} + \rho v \frac{\partial \phi}{\partial y} = \frac{\partial}{\partial y} F \tag{5.131}$$

Multiplying the continuity equation, Eq. (5.128), by ϕ and adding to Eq. (5.131) gives

$$\frac{\partial}{\partial t}(\rho \phi) + \frac{\partial}{\partial x}(\rho u \phi) + \frac{\partial}{\partial y}(\rho v \phi) = \frac{\partial}{\partial y} F \tag{5.132}$$

Equation (5.132) is a suitable starting point for deriving time-averaged conservation equations for a turbulent boundary layer on a flat plate. All terms should be written as sums of mean and fluctuating components [see Eq. (5.81)]. Perform the necessary algebra for a constant-density, steady, mean flow. Identify the eddy diffusivities in the time-averaged equations.

5–40. Introduce appropriate dimensionless variables into the thermal energy equation, Eq. (5.127) to show that viscous dissipation for flow in a rectangular cross-section duct is negligible for Br $\ll$ 1, where Br is the Brinkman number.

5–41. Starting with the total energy equation, Eq. (5.120), derive the thermal energy equation, Eq. (5.123).

5–42. Starting with the thermal energy equation in the form of Eq. (5.123), derive the alternative forms, Eqs. (5.125), (5.126), and (5.127).

5–43. It is often useful to interpret dimensionless groups as ratios of characteristic times. Such times include the following:

Convective time: $t_c \sim L/V$
Free-fall time: $t_f \sim [L/g(\Delta\rho/\rho)]^{1/2}$
Momentum diffusion time: $t_M \sim L^2/\nu$
Heat diffusion time: $t_H \sim L^2/\alpha$

Express the Reynolds, Peclet, Prandtl, Grashof, Rayleigh, and Boussinesq numbers in terms of these characteristic times. (See also reference [17].)

5–44.   A constant-viscosity incompressible fluid flows steadily between parallel plates $2b$ apart, and is heated by a uniform wall heat flux.

   (i) Write down the differential thermal energy equation in terms of specific heat for fully developed hydrodynamic and thermal conditions, with viscous dissipation.
   (ii) Integrate this equation across the duct to obtain a one-dimensional thermal energy equation in terms of bulk quantities.
   (iii) Write down the two-dimensional mechanical energy equation, and integrate across the duct to obtain a one-dimensional equation in terms of bulk velocity.
   (iv) Add the equations obtained in parts (ii) and (iii) to obtain the steady-flow energy equation of thermodynamics.

THERMAL RADIATION

CONTENTS

6.1   INTRODUCTION

Radiation can be viewed either in terms of electromagnetic waves or in terms of transport of photons. The medium involved can be either *nonparticipating* or *participating*. Nonparticipating media include outer space and atmospheric air over short distances, in which photons can travel almost unimpeded from one surface to another. Radiation heat exchange between such surfaces depends only on surface temperatures, surface radiation properties, and the geometry of the configuration. Participating media include combustion gases containing H_2O and CO_2, as well as gases containing *aerosols*, such as dust, soot, and small liquid droplets. Radiation exchange between surfaces separated by a participating medium depends also on the radiation properties of the medium. In the case of a participating gas species, the emissivity and absorptivity depend strongly on temperature.

The first half of Chapter 6, Sections 6.2 through 6.4, is concerned with the engineering calculation of radiation exchange between surfaces separated by a nonparticipating medium. Section 6.2 reviews the physics of radiation and defines the *gray surface* model. Section 6.3 is the key section of the chapter. It deals with methods for calculating radiation heat exchange between surfaces, first for black surfaces and then for diffuse gray surfaces. Section 6.4 extends the diffuse gray surface model to problems involving solar and atmospheric radiation by allowing for a different value of absorptance to short-wavelength solar radiation.

The second half of Chapter 6, Sections 6.5 through 6.7, deals with more advanced topics. Section 6.5 examines *directional* characteristics of surface radiation. The *shape factor* concept, used in Section 6.3, is developed rigorously, and some simple problems involving *specularly* reflecting (mirror-like) surfaces are analyzed. *Spectral* characteristics of surface radiation are further examined in Section 6.6, with the limited objective of demonstrating how to calculate *total hemispherical* radiation properties from experimental data for spectral values. Finally, in Section 6.7, the subject of radiation transfer through gases is introduced. Data for total radiation properties are given, and analytical methods are developed for some simple radiation exchange problems in enclosures containing combustion gases.

Three computer programs accompany Chapter 6. RAD1 calculates shape factors for some simple configurations. RAD2 solves the problem of radiation heat exchange in an enclosure of diffuse gray surfaces. RAD3 calculates the radiation properties of combustion gases. The examples used to illustrate the theory in this chapter are diverse, including a solar collector, a spacecraft, and the combustion chamber of a hydrogen-fueled scramjet.

6.2   THE PHYSICS OF RADIATION

Electromagnetic radiation is characterized by its wavelength. *Thermal radiation* is associated with thermal agitation of molecules, that is, with atomic or molecular transitions, and has wavelengths in the range of about 0.1 to 1000 μm. In Section 6.2.2, the ideal *black surface* (blackbody) is defined. Also in this section, *Planck's law*, which describes the spectrum of radiant energy emitted by a black surface, is reviewed. It is subsequently related to the *Stefan-Boltzmann law*, which gives the total radiant energy emitted by a black surface. Section 6.2.3 deals with real opaque surfaces, and the important concept of a *gray surface* is introduced. In Section 6.3, the gray surface is used as a model of real surfaces for the engineering analysis of radiant heat exchange between surfaces.

6.2.1 The Electromagnetic Spectrum

All matter continuously emits electromagnetic radiation, which travels through a vacuum at the speed of light, $c_0 = 2.9979 \times 10^8$ m/s $\simeq 3 \times 10^8$ m/s. Radiation exhibits both a wave nature (e.g., interference effects) and a particle nature (e.g., the photoelectric effect). The wavelength of the radiation λ is related to its frequency ν_f and speed of propagation c as

$$c = \nu_f \lambda \tag{6.1}$$

The units used for λ are micrometers, or microns (1μm $= 10^{-6}$ m); meters; or nanometers (1 nm $= 10^{-9}$ m). The unit used for ν_f is the hertz (1 Hz $= 1$ s^{-1}). The wavenumber ν of the radiation is related to its wavelength and frequency as

$$\nu = \frac{1}{\lambda} = \frac{\nu_f}{c} \tag{6.2}$$

The units for ν are usually cm^{-1}; hence $\nu[\text{cm}^{-1}] = 10^4/\lambda[\mu\text{m}]$. Figure 6.1 shows the *electromagnetic spectrum*. Thermal effects are associated with radiation in the band of wavelengths from about 0.1 to 1000 μm, and visible radiation is in the very narrow band from about 0.4 to 0.7 μm.

According to quantum mechanics, radiation interacts with matter in discrete quanta called *photons*, with each photon having an energy E given by

$$E = \hbar \nu_f \tag{6.3}$$

where $\hbar = 6.6261 \times 10^{-34}$ J s is *Planck's constant*. Each photon also has a momentum p given by

$$p = \frac{\hbar \nu_f}{c} \tag{6.4}$$

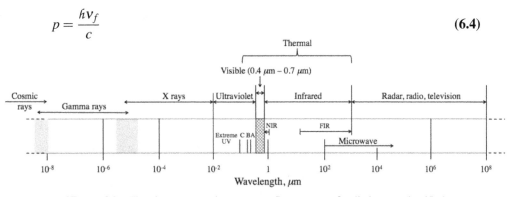

Figure 6.1 The electromagnetic spectrum. Some types of radiation are classified by source (or origin) and not by wavelength, thus the overlap regions in light gray.

In most solids and liquids, the radiation emitted by one molecule is strongly absorbed by surrounding molecules. Thus, the radiation emitted or absorbed by these liquids or solids involves only a layer of molecules close to the surface; for metals, this layer is a few molecules thick, whereas for nonmetals, it is about a few micrometers thick. For these materials, radiation emission and absorption can be regarded as *surface* phenomena. On the other hand, in gas mixtures containing species such as water vapor and carbon dioxide, or in a semi transparent solid, the absorption is weak, and radiation leaving the body can originate from anywhere in the body. Emission and absorption of radiation are then *volumetric* phenomena. Radiation transfer between surfaces through a nonparticipating medium is relatively simple to analyze and is the primary concern of Chapter 6.

6.2.2 The Black Surface

The black surface (or blackbody), introduced in Section 1.3.2, is an ideal surface. It is defined as a surface that absorbs all radiation falling upon it, irrespective of wavelength or angle of incidence: no radiation is reflected. An obvious consequence of this definition is that all the radiation leaving the surface is emitted by the surface. A subtler consequence is that no surface can emit more radiation than a black surface at a given temperature or wavelength, as will be derived in Section 6.2.3. A black surface also has no preferred direction for emission of radiation; that is, the emission is **diffuse**.

Many surfaces encountered in engineering can be approximated as being black; furthermore, the black surface is a useful standard against which to compare real surfaces, as was seen when the gray surface was defined in Section 1.3.2. A black surface can be closely approximated by a hole into a cavity, as shown in Fig. 6.2. Radiation entering the hole is trapped, since each reflection absorbs part of the remaining energy. A black surface appears black because it absorbs all visible radiation falling upon it and reflects none. However, there are many surfaces that absorb nearly all incident thermal radiation but do not appear black. These surfaces reflect sufficient radiation in the visible range for the eye to detect; examples are ceramics such as aluminum or magnesium oxides. Conversely, surfaces that appear black to the eye may reflect radiation outside the visible spectrum.

A black surface emits a spectrum of radiant energy. We define the **monochromatic emissive power** for a black surface, $E_{b\lambda}$, as the radiant power emitted per unit area of surface per unit wavelength; it commonly has the units W/m^2 μm. The

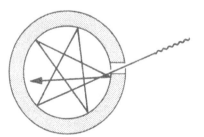

Figure 6.2   A cavity: the hole approximates a blackbody.

spectral density distribution of $E_{b\lambda}$, that is, $E_{b\lambda}$ as a function of wavelength λ for a given temperature T, is given by **Planck's law**,

$$E_{b\lambda} = \frac{C_1 \lambda^{-5}}{e^{C_2/\lambda T} - 1} \tag{6.5}$$

where for λ in μm, $C_1 = 3.7418 \times 10^8$ W μm^4/m^2, and $C_2 = 1.4389 \times 10^4 \mu$m K.[1] Figure 6.3a shows a plot of Planck's law for various temperatures: as the temperature increases, the emitted flux per unit of wavelength also increases, and the peak of the distribution shifts to shorter wavelengths. *Wien's displacement law* is an equation relating the peak wavelength to temperature

$$\lambda_{\max} T = 2898 \, \mu\text{m K} \tag{6.6}$$

that is often used incorrectly. Planck's law is a density distribution (describing power flux per unit of wavelength interval). Both the shape and the peak of Planck's distribution depend on the variable used to build the distribution. When Planck's density distribution is rewritten in terms of a variable that has a nonlinear relationship with λ, for example $\nu_f = c/\lambda$, equally spaced intervals of λ correspond to varying intervals of ν_f along the spectrum since $d\nu_f = -c/\lambda^2 d\lambda$. Also, by conservation of energy it follows that $E_{b\lambda} d\lambda = E_{b\nu_f} d\nu_f$. For the frequency-based version of Planck's law, $E_{b\nu_f} = 2\pi \hbar \nu_f^3 / c^2 (e^{(\hbar \nu_f / \kappa_B T)} - 1)$, the peak wavelength is given by $\lambda_{\max} T = 5100 \, \mu$m K, as shown in Fig. 6.3b. A derivation of Wien's displacement law and a discussion of its proper use is required in Exercises 6–1 and 6–2.

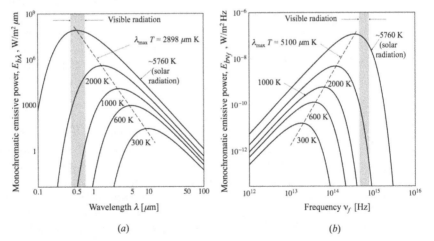

Figure 6.3 (a) Monochromatic emissive power of a black surface at various temperatures according to Planck's law, Eq. (6.5). (b) The same power flux distribution as in (a) expressed in terms of ν_f, $E_{b\nu_f}(T) = 2\pi \hbar \nu_f^3 / c^2 (e^{(\hbar \nu_f / \kappa_B T)} - 1)$.

---

[1] In SI-consistent units, $E_{b\lambda}$ has units [W/m^3], $C_1 = 3.7418 \times 10^{-16}$ W m^2, and $C_2 = 1.4389 \times 10^{-2}$ m K. In terms of more fundamental constants, $C_1 = 2\pi \hbar c^2$ and $C_2 = \hbar c / \kappa_B$.

Figure 6.3a shows that the peak wavelength for solar radiation ($T \sim 5760$K) coincides with the center of the visible radiation ($\lambda_{max} \sim 0.5 \mu$m, $\nu_{f\,max} \sim 6 \times 10^{14}$ Hz) for the wavelength-based distribution $E_{b\lambda}$, whereas the peak of the frequency-based distribution $E_{b\nu_f}$ in Fig. 6.3b is in the infrared region ($\nu_{f\,max} \sim 3.3 \times 10^{14}$ Hz, $\lambda_{max} \sim 0.9 \mu$m). A consistent method to determine relevant wavelengths (and associated frequencies) for different temperatures is presented in Section 6.6.1.

The radiant energy emitted by a surface at all wavelengths is its **total emissive power**, denoted E_b for a black surface. It has a fourth-power temperature dependence given by the **Stefan-Boltzmann law**,

$$E_b = \sigma T^4 \tag{6.7}$$

where $\sigma = C_1 \pi^4 / 15 C_2^4$ is 5.6697×10^{-8} W/m^2 K^4. Stefan discovered this law in 1879 based on experimental data and thermodynamic arguments, but it can also be derived from Planck's law by integrating over all wavelengths:

$$E_b = \int_0^\infty E_{b\lambda} d\lambda \tag{6.8}$$

This integration is required as Exercise 6–3.

6.2.3 Real Surfaces

Most surfaces of engineering importance are opaque (not transparent), although there are some important exceptions, such as glass and plastic films. If radiation falls on a real opaque surface, some will be absorbed, and the remainder will be reflected. The fraction of the incident radiation that is absorbed or reflected depends on the material and surface condition, the wavelength of the incident radiation, and the angle of incidence. There is also a small dependence on the surface temperature. However, for simple engineering calculations, it is often adequate to assume that appropriate average values can be used. If the absorptance (absorptivity) α is defined as the fraction of all incident radiation absorbed, and the reflectance (reflectivity) ρ is defined as the fraction of all incident radiation reflected, then for an opaque surface,

$$\alpha + \rho = 1 \tag{6.9}$$

Now consider an isothermal evacuated enclosure at temperature T, as shown in Fig. 6.4. If a black object of surface area A_b is placed in the enclosure and allowed to come to equilibrium, its temperature will also be T, and it will absorb as much radiation as it emits. Since, according to the Stefan-Boltzmann law, it emits $\sigma T^4 A_b$, it also absorbs $\sigma T^4 A_b$. By definition, a black surface absorbs all radiation falling upon it, so $\sigma T^4 A_b$ is also the radiation incident on the object. Since the shape, size, and location of the object is arbitrary, the irradiation G on any surface in the enclosure is

$$G = \sigma T^4 \tag{6.10}$$

If an object of surface area A and absorptance α is also placed in the enclosure and allowed to come to equilibrium, then it emits as much radiation as it absorbs, and an energy balance requires that

$$\text{Absorption} = AG\alpha = AE = \text{Emission} \tag{6.11}$$

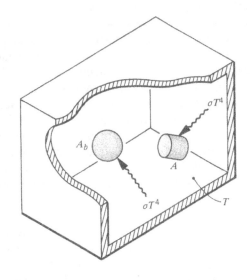

Figure 6.4 An isothermal enclosure at temperature T containing a black object of surface area A_b and a gray object of surface area A.

where E is the total emissive power of the surface. We now define the total emittance (emissivity) ε of a real surface as the ratio of its emissive power to that of a black surface at the same temperature, $\varepsilon = E/E_b$; then $E = \varepsilon E_b$, and substituting Eq. (6.10) in Eq. (6.11) gives

$$A\sigma T^4 \alpha = A\varepsilon E_b = A\varepsilon\sigma T^4$$

or

$$\alpha = \varepsilon \qquad\qquad\qquad\qquad\qquad\qquad (6.12)$$

As was mentioned earlier, the absorptance of a real surface depends on the wavelength of the incident radiation: hence, the absorptance depends on the temperature of the radiation source. Likewise, the emittance of a real surface depends on the wavelength of the emitted radiation, so it also is a function of temperature. Equation (6.12) states only that the absorptance equals the emittance when the surface is at the same temperature as the enclosure. However, α and ε vary rather weakly with temperature, so it is convenient to define a **gray surface** as one for which α and ε are constant and equal over the range of source and surface temperatures of concern. Much useful engineering analysis can be performed by modeling real surfaces as gray surfaces.[2] An important exception consists of problems involving solar radiation, which will be discussed in Section 6.4.

In summary, we will model real opaque surfaces as gray surfaces, the radiation properties of which can be characterized by a single average property value [since given α, ε, or ρ, the other two can be obtained from Eqs. (6.9) and (6.12)]. An appropriate table of property data is one that gives the *total hemispherical emittance*, since these values also have been properly averaged over all directions (e.g., Table 1.3 and Tables A.5a,b in Appendix A).[3]

---

[2] The gray surface will be defined more rigorously in Section 6.6.

[3] The significance of the terms total and hemispherical will be explained in Section 6.6.

6.3   RADIATION EXCHANGE BETWEEN SURFACES

An analysis of radiation heat exchange between black surfaces is simple because a black surface reflects none of the energy falling on it. In simple configurations, such as two large plates facing each other, all radiation leaving one surface is intercepted by the other. The shape (view) factor is then unity. In more complex configurations, only a portion of the radiation leaving one surface is intercepted by a second surface, and the shape factor is less than unity. Calculation of shape factors from first principles is a difficult problem involving solid geometry and surface integration. Thus, in Section 6.3.2, shape factors are simply compiled, and rules are given for their use. Derivation of shape factors is deferred to Section 6.5. Section 6.3.3 introduces an electrical network analogy to radiant energy exchange, which is shown to be a useful conceptual aid.

The analysis of radiation exchange between real surfaces is based on the **diffuse gray surface** model. When only two surfaces form an enclosure, algebraic formulas can be obtained for the radiation exchange, as shown in Section 6.3.4. If there are many surfaces involved, the problem becomes one of solving a system of simultaneous linear algebraic equations, for which standard methods and computer subroutines are available. The computer program RAD2, introduced in Section 6.3.5 for this purpose, is based on the Gauss-Jordan elimination method.

6.3.1   Radiation Exchange between Black Surfaces

Consider a convex black object (1) in a black isothermal enclosure (2) as shown in Fig. 6.5. When both are in equilibrium at temperature T_2, the radiation incident on and absorbed by surface 1 is $A_1 \sigma T_2^4$ in order to balance the emission of $A_1 \sigma T_2^4$. If the temperature of the object is raised to T_1, then the emission becomes $A_1 \sigma T_1^4$ while the absorption remains $A_1 \sigma T_2^4$. Thus, the rate at which energy is lost by radiation from the object is $A_1 (\sigma T_1^4 - \sigma T_2^4)$. Since this energy must be gained by the enclosure, we can say that the net rate of heat transfer from the object to the enclosure, or the net

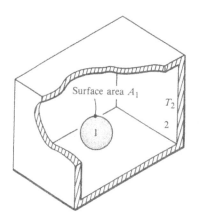

Figure 6.5   A convex black object, 1, in a black enclosure, 2.

radiant energy exchange, is

$$\dot{Q}_{12} = A_1(\sigma T_1^4 - \sigma T_2^4) \tag{6.13}$$

Object 1 could be the inner sphere of two spheres or the inner cylinder of two long coaxial cylinders. If surfaces 1 and 2 are large, parallel walls facing each other, with areas $A_1 = A_2 = A$, then the same argument is valid: Eq. (6.13) still applies, but is more appropriately written as

$$\frac{\dot{Q}_{12}}{A} = \sigma T_1^4 - \sigma T_2^4 \tag{6.14}$$

The preceding situations were particularly simple to analyze because all the radiation leaving surface 1 reached surface 2. Now consider radiation exchange between two finite black surfaces A_1 and A_2, as shown in Fig. 6.6. By inspection, only part of the radiation leaving surface 1 is intercepted by surface 2 and vice versa. We define the **shape factor** (or **view factor**) F_{12} as the fraction of energy leaving A_1 that is intercepted by A_2; likewise, F_{21} is the fraction of energy leaving A_2 that is intercepted by A_1. The shape factor is a geometrical concept and depends only on the size, shape, and orientation of the surfaces. Radiation leaves surface 1 at the rate of $E_{b1}A_1$ [W]; the portion that is intercepted by surface 2 is then $E_{b1}A_1F_{12}$. Likewise, the radiation leaving surface 2 that is intercepted by surface 1 is $E_{b2}A_2F_{21}$. Since both surfaces are black, all incident radiation is absorbed, and the net radiant energy exchange is

$$\dot{Q}_{12} = E_{b1}A_1F_{12} - E_{b2}A_2F_{21} \tag{6.15}$$

If both surfaces are at the same temperature, the second law of thermodynamics requires that there be no net energy exchange; that is, if $E_{b1} = E_{b2}$, $\dot{Q}_{12} = 0$ and

$$A_1F_{12} = A_2F_{21} \tag{6.16}$$

which is the **reciprocal rule** for shape factors. Note that since the shape factors

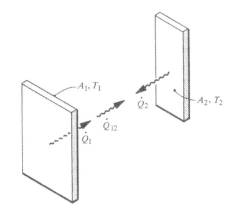

Figure 6.6   Radiation energy exchange between two finite surfaces.

depend only on geometry, this relationship is true even when the surfaces are at different temperatures. Substituting in Eq. (6.15),

$$\dot{Q}_{12} = A_1 F_{12}(E_{b1} - E_{b2}) = A_1 F_{12}(\sigma T_1^4 - \sigma T_2^4) \tag{6.17}$$

The reciprocal rule can also be derived using a purely geometrical argument, as will be shown in Section 6.5.2.

EXAMPLE 6.1   Heat Gain by an Ice Rink

A circular ice rink is 20 m in diameter and is to be temporarily enclosed in a hemispherical dome of the same diameter. The ice is maintained at 270 K, and on a particular day the inner surface of the dome is measured to be 290 K. Estimate the radiant heat transfer from the dome to the rink if both surfaces can be taken as black.

Solution

Given: Circular ice rink enclosed in a hemispherical dome.

Required: Radiant heat transfer from dome to ice.

Assumptions: Both ice and dome are black.

Denote the rink as surface 1 and the dome as surface 2. The area of the rink is $A_1 = (\pi/4)(20)^2 = 314.2$ m^2. All the radiation leaving the rink is intercepted by the dome; thus, Eq. (6.13) applies:

$$\dot{Q}_{12} = A_1(\sigma T_1^4 - \sigma T_2^4)$$

$$= (314.2)[(5.67)(2.70)^4 - (5.67)(2.90)^4]$$

$$= (314.2)[301.3 - 401.0]$$

$$= -31.3 \times 10^3 \text{ W}$$

But $\dot{Q}_{12} = -\dot{Q}_{21}$; hence $\dot{Q}_{21} = 31.3 \times 10^3$ W (31.3 kW).

Comments

Notice that we could have started with $\dot{Q}_{21} = A_2 F_{21}(\sigma T_2^4 - \sigma T_1^4)$ and used the reciprocal rule to write $A_2 F_{21} = A_1 F_{12} = A_1$, obtaining the same result.

6.3.2 Shape Factors and Shape Factor Algebra

Determining shape factors generally requires the evaluation of a double surface integral, which is not easy. However, shape factors have been determined for a great variety of configurations and are available in the form of formulas and graphs. Table 6.1 gives formulas for shape factors of some simple 2-D and 3-D configurations; some of these can be written down by inspection. Item 1 consists of a long cylinder

Table 6.1 Shape factors for two- and three-dimensional configurations.

Configuration	Shape Factors
Two-dimensional	

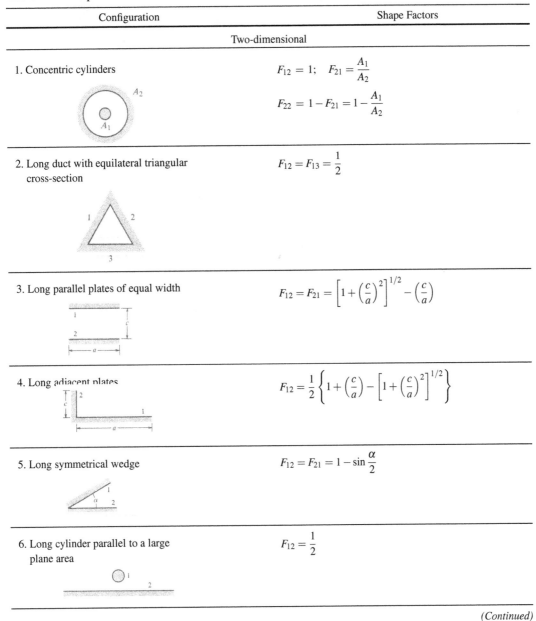

1. Concentric cylinders

$$F_{12} = 1; \quad F_{21} = \frac{A_1}{A_2}$$

$$F_{22} = 1 - F_{21} = 1 - \frac{A_1}{A_2}$$

2. Long duct with equilateral triangular cross-section

$$F_{12} = F_{13} = \frac{1}{2}$$

3. Long parallel plates of equal width

$$F_{12} = F_{21} = \left[1 + \left(\frac{c}{a}\right)^2\right]^{1/2} - \left(\frac{c}{a}\right)$$

4. Long adjacent plates

$$F_{12} = \frac{1}{2}\left\{1 + \left(\frac{c}{a}\right) - \left[1 + \left(\frac{c}{a}\right)^2\right]^{1/2}\right\}$$

5. Long symmetrical wedge

$$F_{12} = F_{21} = 1 - \sin\frac{\alpha}{2}$$

6. Long cylinder parallel to a large plane area

$$F_{12} = \frac{1}{2}$$

(Continued)

Table 6.1   *(Continued)*

Configuration	Shape Factors
7. Long cylinder parallel to a plate 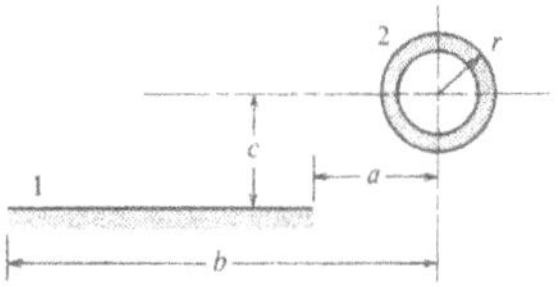	$$F_{12} = \frac{r}{b-a}\left(\tan^{-1}\frac{b}{c} - \tan^{-1}\frac{a}{c}\right)$$
8. Long adjacent parallel cylinders of equal diameters 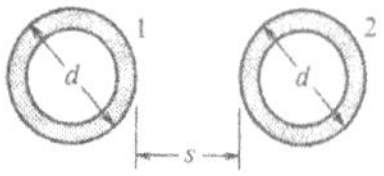	Let $X = 1 + \dfrac{s}{d}$, then $$F_{12} = F_{21} = \frac{1}{\pi}\left[(X^2-1)^{1/2} + \sin^{-1}\frac{1}{X} - X\right]$$
Three-dimensional	
9. Concentric spheres 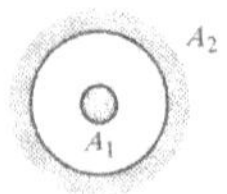	$$F_{12} = 1; \quad F_{21} = \frac{A_1}{A_2}$$ $$F_{22} = 1 - F_{21} = 1 - \frac{A_1}{A_2}$$
10. Regular tetrahedron 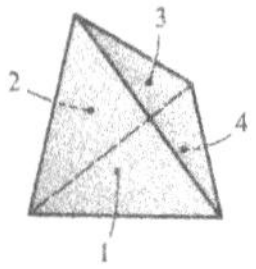	$$F_{12} = F_{13} = F_{14} = \frac{1}{3}$$
11. Sphere near a large plane area 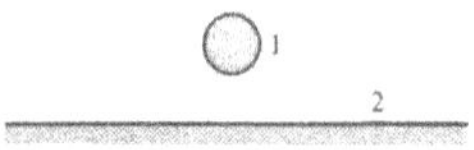	$$F_{12} = \frac{1}{2}$$
12. Small area perpendicular to the axis of a surface of revolution 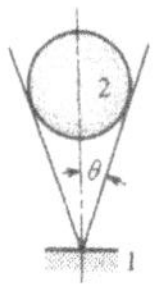	$$F_{12} = \sin^2\theta$$

(Continued)

Table 6.1 *(Continued)*

Configuration	Shape Factors

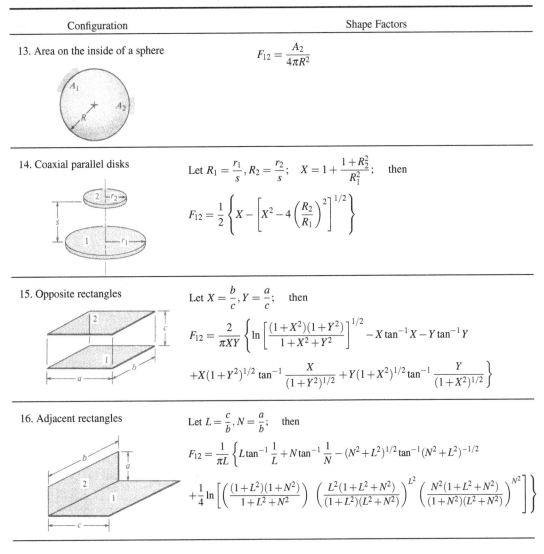

13. Area on the inside of a sphere

$$F_{12} = \frac{A_2}{4\pi R^2}$$

14. Coaxial parallel disks

Let $R_1 = \dfrac{r_1}{s}$, $R_2 = \dfrac{r_2}{s}$; $X = 1 + \dfrac{1+R_2^2}{R_1^2}$;   then

$$F_{12} = \frac{1}{2} \left\{ X - \left[X^2 - 4 \left(\frac{R_2}{R_1} \right)^2 \right]^{1/2} \right\}$$

15. Opposite rectangles

Let $X = \dfrac{b}{c}$, $Y = \dfrac{a}{c}$;   then

$$F_{12} = \frac{2}{\pi XY} \left\{ \ln \left[\frac{(1+X^2)(1+Y^2)}{1+X^2+Y^2} \right]^{1/2} - X\tan^{-1}X - Y\tan^{-1}Y \right.$$
$$\left. + X(1+Y^2)^{1/2}\tan^{-1}\frac{X}{(1+Y^2)^{1/2}} + Y(1+X^2)^{1/2}\tan^{-1}\frac{Y}{(1+X^2)^{1/2}} \right\}$$

16. Adjacent rectangles

Let $L = \dfrac{c}{b}$, $N = \dfrac{a}{b}$;   then

$$F_{12} = \frac{1}{\pi L} \left\{ L\tan^{-1}\frac{1}{L} + N\tan^{-1}\frac{1}{N} - (N^2+L^2)^{1/2}\tan^{-1}(N^2+L^2)^{-1/2} \right.$$
$$\left. + \frac{1}{4}\ln\left[\left(\frac{(1+L^2)(1+N^2)}{1+L^2+N^2} \right) \left(\frac{L^2(1+L^2+N^2)}{(1+L^2)(L^2+N^2)} \right)^{L^2} \left(\frac{N^2(1+L^2+N^2)}{(1+N^2)(L^2+N^2)} \right)^{N^2} \right] \right\}$$

of surface area A_1 surrounded by a cylinder of area A_2. Since all the radiation leaving the inner cylinder is incident on the outer cylinder, $F_{12} = 1$. Use of the reciprocal rule then gives $F_{21} = A_1/A_2$. The radiation leaving the outer cylinder that is not intercepted by the inner cylinder returns to itself; thus, $F_{22} = 1 - A_1/A_2$, that is, surface 2 "sees" itself because it is concave. Identical relations hold for the spheres shown as item 9. Item 2 is a long duct with an equilateral triangular cross section; by symmetry, $F_{12} = F_{13} = 1/2$. Similarly, for the regular tetrahedron shown as item 10, the shape factor from one side to another is 1/3. If a small area, 1, is adjacent and perpendicular to a large area, 2, $F_{12} \simeq 1/2$, and this is the limit form for the shape factors given for items 4 and 16. Item 6 shows a long cylinder (1) parallel to a large plane area (2), for which $F_{12} \simeq 1/2$, and the same result holds for the sphere shown as item 11. If a small area, 1, is perpendicular to the axis of any surface of revolution, 2, then $F_{12} = \sin^2 \theta$, where θ is the angle between the axis and the limiting ray tangent to or at the edge of the surface. Item 12 shows an example. Finally, the shape factor from an area A_1 on the inside of a sphere of radius R to another area A_2 also on the inside of the sphere is $F_{12} = A_2/4\pi R^2$, as shown for item 13.

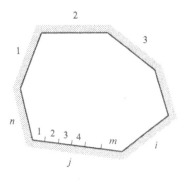

Figure 6.7   An enclosure of n surfaces, with surface j subdivided into m surfaces.

The utility of these shape factor formulas and graphs can be significantly extended by the use of shape factor algebra, including the reciprocal rule, Eq. (6.16), and the **summation rule**, which was just introduced to obtain F_{22} for concentric cylinders. More generally, if surface i is one surface in an enclosure of n surfaces, as shown in Fig. 6.7, then all energy leaving surface i must be intercepted by some surface of the enclosure (including i itself if it is concave); thus,

$$\sum_{j=1}^{n} F_{ij} = 1 \tag{6.18}$$

And if surface j is subdivided into m subareas, as is also shown in Fig. 6.7,

$$F_{ij} = \sum_{k=1}^{m} F_{ik} \tag{6.19}$$

Figure 6.8 shows a simple application of this relation.

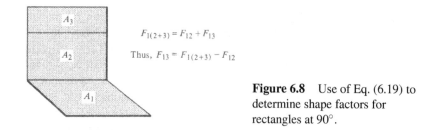

$$F_{1(2+3)} = F_{12} + F_{13}$$

Thus, $F_{13} = F_{1(2+3)} - F_{12}$

Figure 6.8   Use of Eq. (6.19) to determine shape factors for rectangles at 90°.

If Eq. (6.19) is multiplied by area A_i and the reciprocal rule is applied to each term, two more useful relations are obtained:

$$A_j F_{ji} = \sum_{k=1}^{m} A_k F_{ki} \tag{6.20}$$

$$F_{ji} = \frac{\displaystyle\sum_{k=1}^{m} A_k F_{ki}}{\displaystyle\sum_{k=1}^{m} A_k} \tag{6.21}$$

As another application, consider two opposing rectangles subdivided as shown in Fig. 6.9a. We introduce the notation $\mathcal{G}_{ij} = A_i F_{ij}$ for convenience. By symmetry, $\mathcal{G}_{14} = \mathcal{G}_{32}$ or, using the reciprocal rule,

$$\mathcal{G}_{14} = \mathcal{G}_{23} \tag{6.22}$$

The shape factor F_{14} can now be determined by considering the radiation leaving surfaces 1 and 2 that is intercepted by surfaces 3 and 4. Using Eq. (6.19),

$$\begin{aligned}
\mathcal{G}_{(1+2)(3+4)} &= \mathcal{G}_{1(3+4)} + \mathcal{G}_{2(3+4)} \\
&= \mathcal{G}_{13} + \mathcal{G}_{14} + \mathcal{G}_{23} + \mathcal{G}_{24} \\
&= \mathcal{G}_{13} + 2\mathcal{G}_{14} + \mathcal{G}_{24}
\end{aligned} \tag{6.23}$$

Hence,

$$\mathcal{G}_{14} = \frac{1}{2}[\mathcal{G}_{(1+2)(3+4)} - \mathcal{G}_{13} - \mathcal{G}_{24}] \tag{6.24}$$

which can be evaluated directly using item 15 of Table 6.1. Equations (6.22) and (6.24) also hold for the adjacent rectangles shown in Fig. 6.9b, but when the sides are not of equal size, the proof is more difficult.

$$A_1 F_{14} = \frac{1}{2}[A_{(1+2)} F_{(1+2)(3+4)} - A_1 F_{13} - A_2 F_{24}]$$

$$\mathcal{G}_{14} = \frac{1}{2}[\mathcal{G}_{(1+2)(3+4)} - \mathcal{G}_{13} - \mathcal{G}_{24}]$$

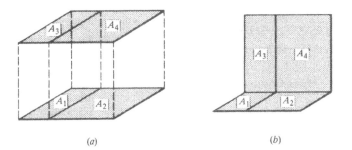

(a)   (b)

Figure 6.9 Applications of shape factor algebra to (a) opposing and (b) adjacent rectangles.

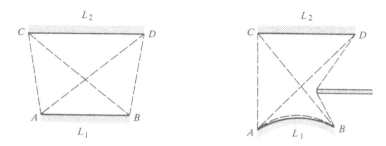

Figure 6.10   The "string rule" for shape factors of two-dimensional configurations.

Hottel's "string rule" [1] is a useful method for determining shape factors to two-dimensional configurations. Referring to Fig. 6.10, F_{12} is given by

$$F_{12} = \frac{1}{2L_1}[AD + BC - AC - BD] \tag{6.25}$$

where the diagonal distances AD, BC and the side distances AC, BD are to be evaluated as the lengths of strings stretched tightly between the respective vertices. The derivation of the string rule is given as Exercise 6–9.

Computer Program RAD1

RAD1 calculates shape factors using the formulas listed in Table 6.1.

Shape Factor Graphs

Figure C.3 in Appendix C gives shape factors in graphical form for items 14, 15, and 16 of Table 6.1, namely, coaxial parallel disks, opposed rectangles, and adjacent rectangles.

EXAMPLE 6.2   Shape Factor Determination

A 20 cm-diameter disk is coaxial and 20 cm distant from a 40 cm-I.D., 20 cm-long cylindrical tube. Determine the shape factor from the disk to the tube.

Solution

Given: Disk A_1 and tube A_2

Required: Shape factor F_{12}.

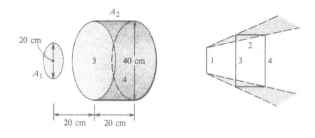

Locate hypothetical disk surfaces 3 and 4 as shown; then

$$F_{12} = F_{13} - F_{14}$$

since the radiation passing through surface 3, but not through surface 4, is intercepted by surface 2. Item 14 of Table 6.1 gives the required shape factors.

$$F_{13} : R_1 = \frac{10}{20} = 0.5; \quad R_3 = \frac{20}{20} = 1; \quad X = 1 + \frac{1+(1)^2}{(0.5)^2} = 9$$

$$F_{13} = \frac{1}{2} \left\{ X - \left[X^2 - 4 \left(\frac{R_3}{R_1} \right)^2 \right]^{1/2} \right\} = \frac{1}{2} \left\{ 9 - \left[81 - 4 \left(\frac{1}{0.5} \right)^2 \right]^{1/2} \right\} = 0.469$$

$$F_{14} : R_1 = \frac{10}{40} = 0.25; \quad R_4 = \frac{20}{40} = 0.5; \quad X = 1 + \frac{1+(0.5)^2}{(0.25)^2} = 21$$

$$F_{14} = \frac{1}{2} \left\{ 21 - \left[441 - 4 \left(\frac{0.5}{0.25} \right)^2 \right]^{1/2} \right\} = 0.192$$

Thus, $F_{12} = F_{13} - F_{14} = 0.469 - 0.192 = 0.277$.

Comments

RAD1 could be used to obtain F_{13} and F_{14}.

6.3.3 Electrical Network Analogy for Black Surfaces

Equation (6.17) gives the net radiation energy exchange between two finite black surfaces. If the blackbody emissive power $E_b = \sigma T^4$ is viewed as a potential (rather than the temperature, T), then the linear form of Eq. (6.17) suggests an electrical analogy [2]; for surfaces i and j,

$$\dot{Q}_{ij} = \frac{E_{bi} - E_{bj}}{1/A_i F_{ij}} \tag{6.26}$$

Thus, E_b is equivalent to electrical potential, $\dot{Q}$ is equivalent to current, and $1/A_i F_{ij}$

is equivalent to electrical resistance: $R_{ij} = 1/A_i F_{ij}$ will be termed a *space radiation resistance* between *nodes i* and *j*. As an illustration of a simple series circuit, consider two large, black parallel walls, 1 and 2, separated by a thin black plate, 3, as shown in Fig. 6.11. At steady state, there can be no energy stored in the plate, so $\dot{Q}_1 = \dot{Q}_{13} = \dot{Q}_{32} = -\dot{Q}_2$. From the equivalent circuit,

$$\dot{Q}_1 = \frac{E_{b1} - E_{b2}}{1/A_1 F_{13} + 1/A_3 F_{32}} \quad \text{with } A_1 = A_2 = A_3 = A; F_{13} = F_{32} = 1$$

Thus,

$$\dot{Q}_1 = \frac{1}{2} A(\sigma T_1^4 - \sigma T_2^4) \tag{6.27}$$

That is, the effect of the plate is to halve the radiation energy exchange, and the plate is thus called a **radiation shield**. For *m* black shields,

$$\frac{\dot{Q}_1}{A} = \frac{1}{m+1}(\sigma T_1^4 - \sigma T_2^4) \tag{6.28}$$

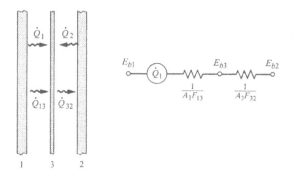

Figure 6.11   Radiation exchange between two parallel black surfaces separated by a black shield.

As another illustration, consider an enclosure of *n* black surfaces. The net radiant energy leaving surface $i, \dot{Q}_i$, can be obtained from Eq. (6.17) by summing over all surfaces forming the enclosure:

$$\dot{Q}_i = \sum_{j=1}^{n} A_i F_{ij}(E_{bi} - E_{bj}) = \sum_{j=1}^{n} \frac{E_{bi} - E_{bj}}{1/A_i F_{ij}}, \quad i = 1, 2, \ldots, n \tag{6.29}$$

$\dot{Q}_i$ can also be termed the *radiation heat transfer* from surface *i*. When many surfaces are involved, the equivalent circuit becomes a complicated network of resistances. Figure 6.12 shows a cylindrical enclosure of isothermal surfaces consisting of a base (1), side walls (2), and a roof (3). Notice that in the network, there is no need to include a resistance $1/A_2 F_{22}$: although $F_{22} \neq 0$, since the concave cylinder sees itself, there is no current flowing in the resistance because the cylinder has a uniform temperature and experiences no heat exchange with itself. Application of Kirchhoff's

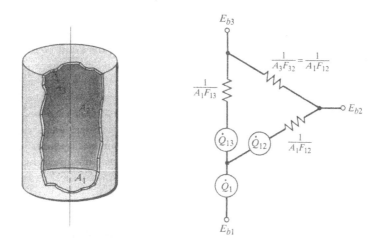

Figure 6.12   Equivalent electrical network for an enclosure of three black surfaces.

current law to node 1, for example, gives

$$\dot{Q}_1 = \dot{Q}_{12} + \dot{Q}_{13}$$

which is equivalent to Eq. (6.29).

EXAMPLE 6.3   Heat Loss from a Melt

A graphite block has a cylindrical cavity 10 cm in diameter and serves as a crucible for laboratory experiments. It is heated from below, and the side walls are well insulated. The cavity is filled with melt at 600 K to 5 cm below the opening. What will be the rate of heat loss from the melt by radiation if the surrounds are at 300 K and all surfaces are approximated as being black?

Solution

Given: Cylindrical crucible containing a melt at 600 K.

Required: Heat loss by radiation.

Assumptions: 1. Side walls are adiabatic.

2. All surfaces are isothermal and black, $\varepsilon = \alpha = 1$.

Let surface 1 be the top of the melt, surface 2 be the exposed side walls of the cavity, and surface 3 be the surrounds: surface 3 can be relocated at the opening of the cavity as shown without altering the radiation exchange problem. Equation (6.29) written for A_1 gives the heat flow through surface 1:

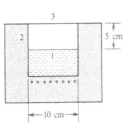

$$\dot{Q}_1 = A_1 F_{12}(E_{b1} - E_{b2}) + A_1 F_{13}(E_{b1} - E_{b3})$$

$$A_1 = (\pi/4)D^2 = (\pi/4)(0.1)^2 = 0.00785 \text{ m}^2$$

Using Table 6.1 or RAD1, item 14, $F_{13} = 0.38$, and the summation rule,

$$F_{12} = 1 - F_{13} = 0.62$$

The blackbody emissive powers are

$$E_{b1} = \sigma T_1^4 = (5.67)(6.00)^4 = 7348 \text{ W/m}^2$$

$$E_{b3} = \sigma T_3^4 = (5.67)(3.00)^4 = 459 \text{ W/m}^2$$

Since T_2 is unknown, E_{b2} cannot be calculated directly. Instead, we make use of the fact that $\dot{Q}_2 = 0$ since the sidewalls are adiabatic; Eq. (6.29) written for A_2 is

$$\dot{Q}_2 = A_2 F_{21} (E_{b2} - E_{b1}) + A_2 F_{23} (E_{b2} - E_{b3}) = 0$$

Now $F_{21} = F_{23}$ by symmetry; solving for E_{b2},

$$E_{b2} = \frac{1}{2}(E_{b1} + E_{b3}) = \frac{1}{2}(7348 + 459) = 3903 \text{ W/m}^2 \text{ (and } T_2 = \left(\frac{E_{b2}}{\sigma}\right)^{1/4} = 512.2 \text{ K)}$$

Thus,

$$\dot{Q}_1 = (0.00785)[0.62(7348 - 3903) + 0.38(7348 - 459)]$$

$$= (0.00785)[2136 + 2618]$$

$$= 37.3 \text{ W}$$

Comments

1. Note how the surroundings are replaced by the hypothetical surface 3 as a conceptual aid.

2. An approximate estimate of the surroundings temperature will suffice: the effect on $\dot{Q}_1$ is small.

6.3.4 Radiation Exchange between Two Diffuse Gray Surfaces

Radiation exchange between black surfaces is easy to analyze because all the radiation falling on a surface is absorbed. With nonblack surfaces, the radiation can be reflected back and forth many times. These multiple reflections must be properly accounted for, which complicates the analysis. We will limit our attention to isothermal opaque gray surfaces, so that the radiative properties of each surface can be characterized by a single value of emittance ε. We also assume that the surfaces are diffuse emitters and that the reflection is diffuse (*Specular reflection*, for which angles of incidence and reflection are equal, is treated in Section 6.5).

The Ray Tracing Method

To see how multiple reflections affect radiative heat transfer, consider two large, parallel gray surfaces facing each other, as shown in Fig. 6.13. We first trace how rays of radiant energy leaving surface 1 are absorbed and reflected. For $A_1 = A_2 = A$,

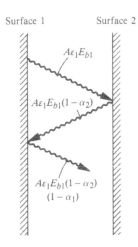

Figure 6.13 Radiation exchange between two parallel gray surfaces: ray tracing.

Surface 1 emits $A\varepsilon_1 E_{b1}$

Surface 2 absorbs $A\varepsilon_1 E_{b1}\alpha_2$

Surface 2 reflects $A\varepsilon_1 E_{b1}(1-\alpha_2)$

Surface 1 absorbs $A\varepsilon_1 E_{b1}(1-\alpha_2)\alpha_1$

Surface 1 reflects $A\varepsilon_1 E_{b1}(1-\alpha_2)(1-\alpha_1)$

Surface 2 absorbs $A\varepsilon_1 E_{b1}(1-\alpha_2)(1-\alpha_1)\alpha_2$

Surface 2 reflects $A\varepsilon_1 E_{b1}(1-\alpha_2)(1-\alpha_1)(1-\alpha_2)$

Surface 1 absorbs $A\varepsilon_1 E_{b1}(1-\alpha_2)(1-\alpha_1)(1-\alpha_2)\alpha_1$

and so on. Let $\gamma = (1-\alpha_2)(1-\alpha_1)$; then radiation emitted by surface 1 and absorbed by itself is

$$A\varepsilon_1 E_{b1}(1+\gamma+\gamma^2+\cdots)(1-\alpha_2)\alpha_1 = \frac{A\varepsilon_1 E_{b1}(1-\alpha_2)\alpha_1}{1-\gamma}$$

Similarly, the radiation emitted by surface 2 that is absorbed by surface 1 is found to be

$$A\varepsilon_2 E_{b2}(1+\gamma+\gamma^2+\cdots)\alpha_1 = \frac{A\varepsilon_2 E_{b2}\alpha_1}{1-\gamma}$$

Thus, the net radiant energy leaving surface 1 is

$$\dot{Q}_1 = A\varepsilon_1 E_{b1} - \frac{A\varepsilon_1 E_{b1}(1-\alpha_2)\alpha_1}{1-\gamma} - \frac{A\varepsilon_2 E_{b2}\alpha_1}{1-\gamma}$$

Rearranging with $\varepsilon = \alpha$ gives the net radiant energy exchange as

$$\dot{Q}_{12} = \dot{Q}_1 = \frac{A(E_{b1}-E_{b2})}{1/\varepsilon_1 + 1/\varepsilon_2 - 1} \tag{6.30}$$

The Energy Balance Method

Although the preceding method of analysis shows how multiple reflections affect radiative heat transfer, it is not a practical method for more complicated configurations. A more useful method is based on conservation of energy rather than on ray tracing. Our starting point is the definitions of irradiation and radiosity introduced in Section 1.3.2 for gray surfaces. Figure 6.14 shows a gray surface with an irradiation G [W/m^2], that is, G accounts for all the radiation incident on the surface. The radiosity of the surface, J [W/m^2], is all the radiation that leaves the surface, whether emitted or reflected; that is, $J = \varepsilon E_b + \rho G$. The heat flux through the surface can be written in two ways. First, referring to the imaginary s-surface located just above the real surface,

$$q = J - G \qquad\qquad (6.31)$$

which uses the sign convention that q is positive away from the surface. Second, referring to the imaginary u-surface located just below the real surface and indefinitely close to it, so that absorption and emission take place *below* the u surface,

$$q = \varepsilon E_b - \alpha G \qquad\qquad (6.32)$$

That is, the radiative heat transfer from a gray surface can be expressed either as radiosity minus irradiation, or as emission minus absorption. The radiative heat transfer can also be expressed in terms of radiosity and blackbody emissive power, as follows. Starting with the definition of J,

$$J = \varepsilon E_b + \rho G$$

substitute for G from Eq. (6.31):

$$J = \varepsilon E_b + \rho (J - q)$$

Solving for q,

$$\rho q = \varepsilon E_b - (1 - \rho)J$$

and substituting $1 - \rho = \alpha = \varepsilon$ gives

$$q = \frac{\varepsilon}{1 - \varepsilon}(E_b - J) \qquad\qquad (6.33)$$

Also, $\dot{Q} = qA$, so

$$\dot{Q} = \frac{\varepsilon A}{1 - \varepsilon}(E_b - J) \qquad\qquad (6.34)$$

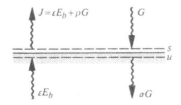

Figure 6.14   Radiant heat fluxes for a gray surface. The imaginary s- and u-surfaces are adjacent to the real interface, with all absorption and emission taking place below the u-surface.

Equation (6.34) gives the net radiant energy leaving a gray surface in terms of its blackbody emissive power, $E_b = \sigma T^4$, and its radiosity, J.

Now consider an enclosure formed by two long concentric cylinders or concentric spheres, as shown in Fig. 6.15. The inner surface has area A_1 and is at uniform temperature T_1; the outer surface has area A_2 and is at a uniform temperature T_2. The limit case is two large parallel surfaces facing each other. By symmetry, the irradiation and radiosity of the surfaces are also uniform. The fraction of energy leaving surface 1 that is intercepted by surface 2 is $J_1 A_1 F_{12}$, where the shape factor F_{12} is the same as that introduced for black surfaces in Section 6.3.1, because we have assumed that our gray surfaces are *diffuse* emitters and reflectors. Likewise, the radiant energy leaving surface 2 that is intercepted by surface 1 is $J_2 A_2 F_{21}$; thus, the net radiant energy exchange is

$$\dot{Q}_{12} = J_1 A_1 F_{12} - J_2 A_2 F_{21}$$

and using the reciprocal rule,

$$\dot{Q}_{12} = A_1 F_{12}(J_1 - J_2) \tag{6.35}$$

Also, from Eq. (6.34), the net radiant energy leaving each surface is

$$\dot{Q}_1 = \frac{\varepsilon_1 A_1}{1 - \varepsilon_1}(E_{b1} - J_1); \quad \dot{Q}_2 = \frac{\varepsilon_2 A_2}{1 - \varepsilon_2}(E_{b2} - J_2) \tag{6.36a,b}$$

and conservation of energy requires

$$\dot{Q}_1 = \dot{Q}_{12} = -\dot{Q}_2 \tag{6.37}$$

These equations can be solved for the radiant energy exchange as follows:

$$\dot{Q}_1 \left(\frac{1 - \varepsilon_1}{\varepsilon_1 A_1} \right) = E_{b1} - J_1$$

$$\dot{Q}_{12} \left(\frac{1}{A_1 F_{12}} \right) = J_1 - J_2$$

$$-\dot{Q}_2 \left(\frac{1 - \varepsilon_2}{\varepsilon_2 A_2} \right) = J_2 - E_{b2}$$

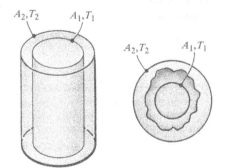

Figure 6.15 An enclosure formed by two long concentric cylinders or two concentric spheres.

Adding to eliminate the radiosities, and rearranging using Eq. (6.37) gives

$$\dot{Q}_{12} = \frac{E_{b1} - E_{b2}}{\dfrac{1-\varepsilon_1}{\varepsilon_1 A_1} + \dfrac{1}{A_1 F_{12}} + \dfrac{1-\varepsilon_2}{\varepsilon_2 A_2}} \qquad (6.38)$$

If the temperatures of both surfaces are specified, then $E_{b1} = \sigma T_1^4$, $E_{b2} = \sigma T_2^4$, and $\dot{Q}_{12}$ can be calculated.

Special Forms of Eq. (6.38)

Equation (6.38) is a very useful formula. It can be further simplified since, for the configurations considered, $F_{12} = 1$; thus,

$$\dot{Q}_{12} = \frac{\varepsilon_1 A_1}{1 + \dfrac{\varepsilon_1 A_1}{\varepsilon_2 A_2}(1-\varepsilon_2)}(E_{b1} - E_{b2}) \qquad (6.39)$$

When $(\varepsilon_1 A_1/\varepsilon_2 A_2)(1-\varepsilon_2)$ is small compared to unity, perhaps because $\varepsilon_1 A_1/\varepsilon_2 A_2$ is small or $(1-\varepsilon_2)$ is small, Eq. (6.39) becomes

$$\dot{Q}_{12} = \varepsilon_1 A_1 (\sigma T_1^4 - \sigma T_2^4) \qquad (6.40)$$

which was introduced as Eq. (1.18). Equation (6.40) is of great practical importance since it can be applied to the common situation of a small object in large, nearly black surrounds. Notice also that for two large parallel walls facing each other, $F_{12} = 1$ and $A_1 = A_2 = A$, and Eq. (6.38) becomes equivalent to Eq. (6.30),

$$\frac{\dot{Q}_{12}}{A} = \frac{\sigma T_1^4 - \sigma T_2^4}{1/\varepsilon_1 + 1/\varepsilon_2 - 1} \qquad (6.41)$$

Electrical Network Analogy

An electrical network analogy is also useful for gray surfaces. Equation (6.35) can be rewritten as

$$\dot{Q}_{12} = \frac{J_1 - J_2}{1/A_1 F_{12}} \qquad (6.42)$$

and the *space resistance* $R_{12} = 1/A_1 F_{12}$ is the same as for black surfaces, but the potential difference across it is the difference in radiosities rather than the difference in blackbody emissive powers. Equation (6.34) can be rewritten as

$$\dot{Q} = \frac{E_b - J}{(1-\varepsilon)/\varepsilon A} \qquad (6.43)$$

which defines a *surface resistance* $R = (1-\varepsilon)/\varepsilon A$, with a corresponding potential difference of blackbody emissive power minus radiosity. Considering again the enclosure of two gray surfaces, Eq. (6.37) interpreted in terms of a circuit simply states that the current flowing through the space and the two surface resistances is the same; that is, the resistances are in series, and the circuit is as shown in Fig. 6.16. Equation (6.38) is then immediately obtained by inspection.

Figure 6.16 Equivalent electrical circuit for radiation exchange in an enclosure of two gray surfaces.

We now return to the radiation shield problem shown in Fig. 6.11. For gray surfaces, the equivalent circuit is shown in Fig. 6.17, and the heat flow is

$$\frac{\dot{Q}_{12}}{A} = \frac{\sigma T_1^4 - \sigma T_2^4}{\dfrac{1-\varepsilon_1}{\varepsilon_1} + 1 + \dfrac{2(1-\varepsilon_3)}{\varepsilon_3} + 1 + \dfrac{1-\varepsilon_2}{\varepsilon_2}} \tag{6.44}$$

A shield with a low emittance is desirable to maximize the surface resistances $(1 - \varepsilon_3)/\varepsilon_3$.

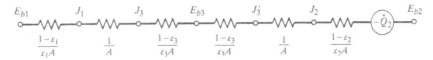

Figure 6.17 Equivalent electrical circuit for a gray radiation shield between parallel gray surfaces.

The major results obtained in this section are summarized as items 1–3 in Table 6.2 in the form of the radiation transfer factor $\mathscr{F}_{12}$ defined by Eq. (1.17), namely,

$$\dot{Q}_{12} = A_1 \mathscr{F}_{12}(\sigma T_1^4 - \sigma T_2^4) \tag{6.45}$$

Table 6.2   Transfer factors $\mathscr{F}_{12}$. (Item 4 is derived in Section 6.3.5.)

	Configuration	Transfer Factors $\mathscr{F}_{12}$
1.	Two large parallel walls facing each other	$\dfrac{1}{1/\varepsilon_1 + 1/\varepsilon_2 - 1}$
2.	Concentric cylinders or spheres: A_1 surrounded by A_2	$\dfrac{\varepsilon_1}{1 + (\varepsilon_1 A_1/\varepsilon_2 A_2)(1 - \varepsilon_2)}$
3.	Small object A_1 in large, nearly black surrounds	ε_1
4.	Surfaces A_1 and A_2 with the enclosure completed by refractory surface A_3	$\dfrac{1}{\dfrac{1-\varepsilon_1}{\varepsilon_1} + \dfrac{(1-\varepsilon_2)A_1}{\varepsilon_2 A_2} + \dfrac{1}{F_{12} + 1/(1/F_{13} + A_1/A_3 F_{32})}}$

The transfer factors for concentric cylinders and spheres are reasonably accurate even when the surfaces are not concentric, provided that the lack of symmetry does not give a markedly nonuniform radiosity distribution on either surface.

EXAMPLE 6.4   Boil-off from a Cryogenic Dewar Flask

Liquid oxygen is stored in a thin-walled spherical container, 96 cm in diameter, which in turn is enclosed in a concentric container 100 cm in diameter. The surfaces facing each other are plated and have an emittance of 0.05, and the space in between is evacuated. The inner surface is at 95 K, and the outer surface is at 280 K. (i) What is the oxygen boil-off rate? (ii) If a thin radiation shield also of emittance 0.05 is placed midway between the containers, what is the new boil-off rate?

Solution

Given: Spherical Dewar flask containing liquid oxygen.

Required: Effect of radiation shield on boil-off rate, $\dot{m}$ [kg/s].

Assumptions: 1. A perfect vacuum, and hence no conduction across the space between the shells.
2. Negligible conduction through filler neck.
3. Gray diffuse surfaces.

(i) We first calculate the heat leakage into the container. The equivalent circuit is as shown. The surface areas and blackbody emissive powers are as follows:

Inner sphere:

$$A_1 = \pi(0.96)^2 = 2.895 \text{ m}^2$$

$$E_{b1} = (5.67)(0.95)^4 = 4.6 \text{ W/m}^2$$

Outer sphere:

$$A_2 = \pi(1.0)^2 = 3.142 \text{ m}^2$$

$$E_{b2} = (5.67)(2.8)^4 = 348.5 \text{ W/m}^2$$

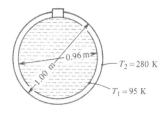

$T_2 = 280$ K

$T_1 = 95$ K

0.96 m

1.00 m

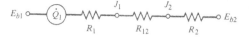

The resistances are

$$R_1 = \frac{1-\varepsilon_1}{\varepsilon_1 A_1} = \frac{1-0.05}{(0.05)(2.895)} = 6.563$$

$$R_{12} = \frac{1}{A_1 F_{12}} = \frac{1}{(2.895)(1)} = 0.345$$

$$R_2 = \frac{1-\varepsilon_2}{\varepsilon_2 A_2} = \frac{1-0.05}{(0.05)(3.142)} = 6.047$$

$$\sum R = 6.563 + 0.345 + 6.047 = 13.0$$

The heat transfer between the surfaces is then

$$\dot{Q}_{12} = \frac{E_{b1} - E_{b2}}{\sum R} = \frac{4.6 - 348.5}{13.0} = -26.5 \text{ W}$$

The boil-off rate $\dot{m}$ is $-\dot{Q}_{12}/h_{fg}$. From Table A.8, the enthalpy of vaporization of oxygen is 0.213×10^6 J/kg; thus,

$$\dot{m} = -\frac{\dot{Q}_{12}}{h_{fg}} = \frac{26.5}{0.213 \times 10^6} = 1.24 \times 10^{-4} \text{ kg/s}$$

that is, about 0.1 gram per second.

(ii) The new equivalent circuit is as shown. The shield area is $A_3 = (\pi)(0.98)^2 = 3.017 \text{ m}^2$.

The new resistances are

$$R_3 = \frac{1 - \varepsilon_3}{\varepsilon_3 A_3} = \frac{1 - 0.05}{(0.05)(3.017)} = 6.298$$

$$R_{13} = \frac{1}{A_1 F_{13}} = \frac{1}{(2.895)(1)} = 0.345 (= R_{12})$$

$$R_{32} = \frac{1}{A_3 F_{32}} = \frac{1}{(3.017)(1)} = 0.331$$

Thus,

$$\sum R = 6.563 + 0.345 + (2)(6.298) + 0.331 + 6.047 = 25.9$$

$$\dot{Q}_{12} = \frac{4.6 - 348.5}{25.9} = -13.3 \text{ W}; \quad \dot{m} = 6.24 \times 10^{-5} \text{ kg/s}$$

Comments

The effect of the radiation shield is to halve the boil-off rate.

6.3.5 Radiation Exchange between Many Diffuse Gray Surfaces

It is possible to obtain algebraic formulas for radiative heat transfer between diffuse gray surfaces only when the configuration is particularly simple, such as those considered in Section 6.3.4. The general problem of determining the radiation exchange in an enclosure formed by n gray diffuse surfaces requires the solution of n linear algebraic equations. Cramer's rule, matrix inversion, or successive substitution methods can be used. For this purpose, it is useful to formulate a standard calculation procedure, subject to the following restrictions:

1. Each surface is opaque and gray.

2. The emission from each surface is diffuse.

3. The reflection from each surface is diffuse.

4. The radiosity of each surface is uniform, which requires a uniformly irradiated and isothermal surface.

The first three of these restrictions have been stated already; the fourth is implicit in the analyses of Section 6.3.4 due to symmetry. Figure 6.18 shows two surfaces of uniform temperature forming a wedge. Since the shape factor $F_{\Delta A_1 A_2}$ is not uniform and increases towards the vertex of the wedge, it follows that the irradiation on surface 1 increases toward the vertex, as does the radiosity $J_1 = \varepsilon \sigma T_1^4 + (1 - \varepsilon) G_1$. However, restriction 4 is not a serious one: any surface can be subdivided to obtain surfaces of sufficiently uniform radiosity, depending on the required accuracy of the result and whether or not the work required to calculate the shape factors and organize the calculations is justified. The calculation method is independent of the number of surfaces.

Consider an enclosure of n surfaces. The radiation incident on the ith surface is

$$A_i G_i = J_1 A_1 F_{1i} + J_2 A_2 F_{2i} + J_3 A_3 F_{3i} + \cdots$$

$$= J_1 A_i F_{i1} + J_2 A_i F_{i2} + J_3 A_i F_{i3} + \cdots \quad \text{(using the reciprocal rule)}$$

$$= A_i \sum_{k=1}^{n} J_k F_{ik}$$

or

$$G_i = \sum_{k=1}^{n} J_k F_{ik}$$

the irradiation of surface i. The radiosity of surface i is $J_i = \varepsilon_i E_{bi} + (1 - \varepsilon_i) G_i$; substituting for G_i gives

$$J_i = \varepsilon_i E_{bi} + (1 - \varepsilon_i) \sum_{k=1}^{n} J_k F_{ik}; \quad i = 1, 2, \ldots, n \tag{6.46}$$

If the temperatures of all the surfaces are specified, then $E_{bi} = \sigma T_i^4$, and Eq. (6.46)

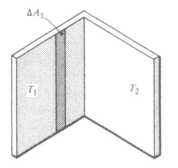

Figure 6.18 A wedge formed from two isothermal gray surfaces.

is a system of n linear equations in the n unknowns J_i. Upon solving and obtaining the J_i, the radiant heat transfer from each surface is obtained from Eq. (6.34) as

$$\dot{Q}_i = \frac{\varepsilon_i A_i}{1 - \varepsilon_i}(E_{bi} - J_i) \qquad \text{(6.47a)}$$

If a surface is black, Eq. (6.47a) is replaced by

$$\dot{Q}_i = A_i \left(E_{bi} - \sum_{k=1}^{n} J_k F_{ik} \right) \qquad \text{(6.47b)}$$

If the temperatures of some surfaces and the heat transfer from other surfaces are specified, Eq. (6.47a) must be used to eliminate the unknown values of E_{bi} in Eq. (6.46).

It is often useful to draw the equivalent network as a conceptual aid and, in the case of simple problems, to assist in their solution. For example, consider an enclosure of three surfaces with one surface being adiabatic, that is, well insulated. Note that adiabatic surfaces are often called **refractory** surfaces in thermal radiation problems, since a historically important topic has been the design of furnaces in which a surface lined with refractory brick can usually be taken to be adiabatic. Figure 6.19 shows a simple configuration and the equivalent network. Recognizing that the space resistances R_{13} and R_{32} are in series and that R_{12} and $(R_{13} + R_{32})$ are in parallel, the net exchange between surfaces 1 and 2 is

$$\dot{Q}_{12} = \frac{E_{b1} - E_{b2}}{R_1 + R_2 + \left(\dfrac{1}{R_{12}} + \dfrac{1}{R_{13} + R_{32}} \right)^{-1}} \qquad \text{(6.48)}$$

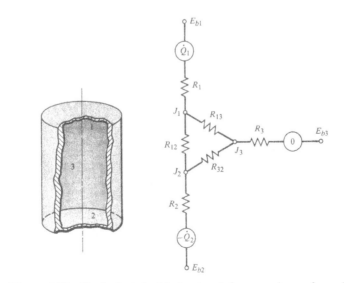

Figure 6.19 Equivalent electrical network for an enclosure formed by three gray surfaces; surface 3 is a refractory (adiabatic) surface.

When surfaces 1 and 2 are the only surfaces in an enclosure through which there is heat flow (other surfaces, if present, are adiabatic), it is convenient to determine the net exchange between surfaces 1 and 2, which we designate $\dot{Q}_{12}$, as was done in Section 6.3.4. When there are more than two nonadiabatic surfaces, it is more useful to determine the heat flow through surface i, as given by Eq. (6.47). Usually it is of little value to calculate $\dot{Q}_{ij} = (J_i - J_j)/(1/A_i F_{ij})$, which is the heat flow in the network between nodes J_i and J_j.

The Computer Program RAD2

RAD2 solves the problem of radiation exchange between many gray surfaces. For each surface, either the temperature or heat flux can be specified. The resulting system of n linear algebraic equations derived from Eqs. (6.46) and (6.47) are solved using Gauss-Jordan elimination. This method might give a large truncation error for some ill-conditioned matrices but will always give a solution if the matrix is nonsingular. The required matrix of shape factors should be calculated first using RAD1. RAD2 will check to ensure that input shape factors satisfy the reciprocal and summation rules.

EXAMPLE 6.5   Radiant Transfer in a Furnace

A long furnace used for stress relieving and annealing is 3 m × 3 m in cross section, with side walls and roof at 1700 K and 1400 K, respectively. What is the radiant heat transfer to the work floor when it is at 600 K? All the surfaces can be taken to be gray and diffuse and have an emittance of 0.5.

Solution

Given: Long furnace with a square cross section.

Required: Radiant heat transfer to work floor.

Assumptions: 1. Diffuse gray surfaces.

2. 2-D geometry so that Table 6.1, item 3 can be used for shape factors.

Equation (6.46) is used for each surface in turn, with the side walls treated as a single surface for convenience:

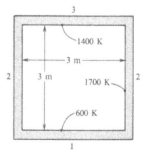

$$J_1 = \varepsilon_1 E_{b1} + (1 - \varepsilon_1)[J_1 F_{11} + J_2 F_{12} + J_3 F_{13}]$$

$$J_2 = \varepsilon_2 E_{b2} + (1 - \varepsilon_2)[J_1 F_{21} + J_2 F_{22} + J_3 F_{23}]$$

$$J_3 = \varepsilon_3 E_{b3} + (1 - \varepsilon_3)[J_1 F_{31} + J_2 F_{32} + J_3 F_{33}]$$

The emittances, blackbody emissive powers, and shape factors are as follows.

$$\varepsilon_1 = \varepsilon_2 = \varepsilon_3 = 0.5$$

$$E_{b1} = \sigma T_1^4 = (5.67 \times 10^{-8})(600)^4 = 7.348 \text{ kW/m}^2$$

$$E_{b2} = \sigma T_2^4 = (5.67 \times 10^{-8})(1700)^4 = 473.6 \text{ kW/m}^2$$

$$E_{b3} = \sigma T_3^4 = (5.67 \times 10^{-8})(1400)^4 = 217.8 \text{ kW/m}^2$$

From Table 6.1, item 3,

$$F_{13} = (1+1^2)^{1/2} - 1 = 0.414 = F_{31} = F_{22}$$

Using the summation rule

$$F_{12} = 1 - F_{11} - F_{13} = 1 - 0 - 0.414 = 0.586 = F_{32}$$

Using the reciprocal rule

$$F_{21} = (A_1/A_2)F_{12} = 0.5F_{12} = (0.5)(0.586) = 0.293 = F_{23}$$

Substituting in the radiosity equations gives

$$J_1 = (0.5)(7.348) + (1-0.5)[0 + 0.586J_2 + 0.414J_3]$$

$$J_2 = (0.5)(473.6) + (1-0.5)[0.293J_1 + 0.414J_2 + 0.293J_3]$$

$$J_3 = (0.5)(217.8) + (1-0.5)[0.414J_1 + 0.586J_2 + 0]$$

Rearranging,

$$J_1 - 0.293J_2 - 0.207J_3 = 3.674$$

$$0.147J_1 - 0.793J_2 + 0.147J_3 = -236.8$$

$$0.207J_1 + 0.293J_2 - J_3 = -108.9$$

Solving,

$$J_1 = 166.4 \text{ kW/m}^2$$

$$J_2 = 376.2 \text{ kW/m}^2$$

$$J_3 = 253.6 \text{ kW/m}^2$$

From Eq. (6.47) written for surface 1

$$\frac{\dot{Q}_1}{A_1} = \frac{\varepsilon_1}{1-\varepsilon_1}(E_{b1} - J_1)$$

$$= \frac{0.5}{1-0.5}(7.348 - 166.4)$$

$$= -159.0 \text{ kW/m}^2$$

which is the radiant heat flux across the work floor.

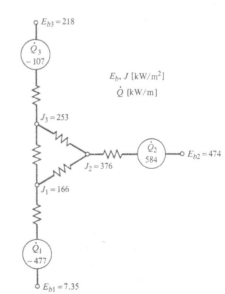

Solution using RAD2

The required input is

Surface	ε	A m^2/m	T K	q W/m^2
1	0.5	3	600	—
2	0.5	6	1700	—
3	0.5	3	1400	—

where the areas are per unit length of furnace.

$F_{13} = 0.414$ can be obtained from RAD1, and the remaining shape factors calculated as before. The shape factor matrix is

$$F_{11} = 0 \qquad F_{12} = 0.586 \qquad F_{13} = 0.414$$

$$F_{21} = 0.293 \qquad F_{22} = 0.414 \qquad F_{23} = 0.293$$

$$F_{31} = 0.414 \qquad F_{32} = 0.586 \qquad F_{33} = 0$$

RAD2 gives the following output:

Surface	Temperature K	Heat Flux W/m^2	Radiosity W/m^2
1	600	-1.5903×10^5	1.6638×10^5
2	1700	97,392	3.7617×10^5
3	1400	$-35,750$	2.5357×10^5

$$\sum \dot{Q}_i \simeq 10^{-10} \text{ W}$$

Comments

1. The heat loss through the roof is excessive: better insulation is indicated.

2. The equivalent network for the furnace is shown in the accompanying figure; the heat flows are per unit length of furnace.

EXAMPLE 6.6   A Radiant Heater Panel

A radiant heater panel used in a high-vacuum process consists of a row of cylindrical electrical heating elements 1 cm in diameter, 150 cm long, spaced at a 3 cm pitch, and backed by a well insulated wall to act as a reflector and reradiator. The panel has dimensions 30 × 150 cm and is located 30 cm above a workpiece, which is also 30 × 150 cm. The heater elements are rated at 5 kW each. Estimate the operating temperature of the elements when the workpiece and surroundings are at 300 K. Take the emittances of the elements and back wall to be 0.9 and 0.8, respectively; the workpiece and surroundings can be assumed black.

Solution

Given: Radiant heating panel.

Required: Operating temperature of the 5 kW heater elements.

Assumptions: 1. Diffuse gray surfaces.
2. The radiosities of the heating elements and insulated back wall are uniform.
3. Negligible end losses, to give an upper-bound (conservative) estimate of the operating temperature.

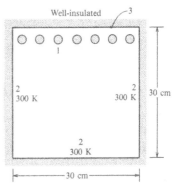

Since their emittance is high and pitch not too small, the radiosity of the heating elements is relatively uniform around their periphery and thus can be represented by a single node, 1. Likewise, since the back (refractory) wall also has a high emittance, if it is not too close to the elements and is a good conductor, it too will have a relatively uniform radiosity and can be represented by a single node, 3. The workpiece and surroundings are both at the same temperature and black, so they can be represented by a single node, 2. Thus, we have the problem shown in Fig. 6.19, for which Eq. (6.48) applies:

$$\dot{Q}_{12} = \frac{E_{b1} - E_{b2}}{R_1 + R_2 + \left(\dfrac{1}{R_{12}} + \dfrac{1}{R_{13} + R_{32}} \right)^{-1}}$$

The panel is relatively long, so the shape factor F_{11} between one cylinder and two adjacent cylinders is twice the shape factor given by item 8 of Table 6.1:

$$F_{11} = \frac{2}{\pi} \left[(X^2 - 1)^{1/2} + \sin^{-1} \frac{1}{X} - X \right]; \quad X = 1 + \frac{s}{d} = \frac{d+s}{d} = \frac{3}{1} = 3$$

$$= \frac{2}{\pi} \left[(3^2 - 1)^{1/2} + \sin^{-1} \frac{1}{3} - 3 \right] = 0.107$$

By symmetry, and from the summation rule,

$$F_{12} \simeq F_{13} = \frac{1}{2}(1 - F_{11}) = 0.446$$

Using the reciprocal rule,

$$F_{31} = \left(\frac{A_1}{A_3} \right) F_{13} = \left(\frac{\pi d}{d+s} \right) F_{13} = \left(\frac{\pi}{3} \right) F_{13} = 0.468$$

and from the summation rule,

$$F_{32} = 1 - F_{31} - F_{33} = 1 - 0.468 - 0 = 0.532$$

Equation (6.48) can be solved for the unknown heater element temperature T_1:

$$\sigma T_1^4 = \sigma T_2^4 + \dot{Q}_{12}\left[R_1 + R_2 + \left(\frac{1}{R_{12}} + \frac{1}{R_{13}+R_{32}}\right)^{-1}\right]$$

Number of elements $= 30/3 = 10$; $\dot{Q}_{12} = (10)(5000) = 50,000$ W

$$A_1 = (10)(\pi)(0.01)(1.5) = 0.471 \text{ m}^2$$

$$A_2 = (3)(0.3)(1.5) = 1.35 \text{ m}^2$$

$$A_3 = (0.3)(1.5) = 0.45 \text{ m}^2$$

$$R_1 = \frac{1-\varepsilon_1}{\varepsilon_1 A_1} = \frac{1-0.9}{(0.9)(0.471)} = 0.236; \quad R_2 = \frac{1-\varepsilon_2}{\varepsilon_2 A_2} = \frac{1-1}{(1)(1.35)} = 0$$

$$R_{12} = \frac{1}{A_1 F_{12}} = \frac{1}{(0.471)(0.446)} = 4.76 = R_{13}$$

$$R_{32} = \frac{1}{A_3 F_{32}} = \frac{1}{(0.45)(0.532)} = 4.18$$

$$\sigma T_1^4 = (5.67 \times 10^{-8})(300)^4 + (50,000)\left[0.236 + \left(\frac{1}{4.76} + \frac{1}{4.76+4.18}\right)^{-1}\right]$$

$$T_1 = 1311 \text{ K}$$

Solution using RAD2

The required input is

Surface	ε	A	T	q
		m^2	K	W/m^2
1	0.9	0.471	—	1.062×10^5
2	1.0	1.35	300	—
3	0.8	0.45	—	0

where $q_1 = \dot{Q}_{12}/A_1 = 50,000/0.471 = 1.062 \times 10^5$ W/m^2, and $q_3 = 0$ since surface 3 is a refractory surface. The additional shape factors required are

$$F_{21} = \frac{A_1}{A_2}F_{12} = \frac{0.471}{1.35}(0.446) = 0.156; \quad F_{23} = \frac{A_3}{A_2}F_{32} = \frac{0.45}{1.35}(0.532) = 0.177$$

$$F_{22} = 1 - F_{21} - F_{23} = 1 - 0.156 - 0.177 = 0.667$$

The shape factor matrix is then

$F_{11} = 0.107$	$F_{12} = 0.446$	$F_{13} = 0.446$
$F_{21} = 0.156$	$F_{22} = 0.667$	$F_{23} = 0.177$
$F_{31} = 0.468$	$F_{32} = 0.532$	$F_{33} = 0$

RAD2 gives the following output:

Surface	Temperature K	Heat Flux W/m^2	Radiosity W/m^2
1	1310.9	1.062×10^5	1.557×10^5
2	300.0	$-37,068$	459
3	1065.5	0	73,093

$$\sum \dot{Q}_i = -21 \simeq 0 \text{ W}$$

Comments

A number of assumptions were made to simplify this problem: how confident are you that the result is satisfactory?

6.3.6 Radiation Transfer through Passages

An interesting engineering problem is the calculation of radiation heat transfer through narrow passages. Applications include regenerative heat exchangers and cracks in a layer of insulation. In general, such problems involve all three modes of heat transfer: conduction, convection, and radiation. However, in some cases, only radiation is important. An example is a crack in a layer of high-temperature insulation for which both axial conduction and convection in the gas-filled crack and conduction in the poorly conducting insulation are negligible: then the walls of the crack can be assumed to be perfectly insulated, that is, they are refractory surfaces. Such problems can be formulated as enclosures of gray diffuse surfaces by subdividing the walls into a number of nodes and using the methods of Section 6.3.5. In practice, more sophisticated methods are favored. The discrete sums of Eq. (6.46) for the wall nodes can be converted into integrals to yield an integral equation, which then can be solved approximately or numerically. But modern practice is to use the *Monte Carlo* method, because of the ease with which it can be generalized to apply to nongray or specular surfaces. As applied to radiation exchange problems, this method is simply a computerized statistical sampling approach to ray tracing.

Results for various passage shapes can be presented in terms of a transfer factor $\mathscr{F}_{12}^b$, where surfaces 1 and 2 are the ends of the passage, each assumed black, such that $\dot{Q}_{12} = A_1 \mathscr{F}_{12}^b (\sigma T_1^4 - \sigma T_2^4)$. Figure 6.20 gives $\mathscr{F}_{12}^b$ for diffuse, refractory-walled passages as a function of L/D_h, where L is the passage length and D_h is its hydraulic diameter ($D_h = 4A_c/\mathscr{P}$, where A_c is the cross-sectional area and $\mathscr{P}$ is the perimeter: for a cylinder, D_h is simply the diameter; and for a slot, D_h is twice the wall spacing). When the end surfaces, 1 and 2, are not black, the transfer factor is given by

$$\mathscr{F}_{12} = \frac{1}{\dfrac{1-\varepsilon_1}{\varepsilon_1} + \dfrac{1}{\mathscr{F}_{12}^b} + \dfrac{1-\varepsilon_2}{\varepsilon_2}} \tag{6.49}$$

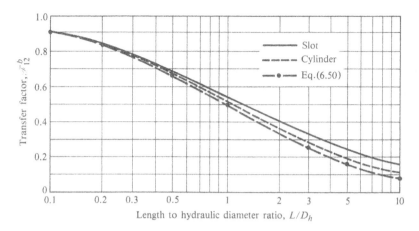

Figure 6.20 The transfer factor $\mathscr{F}_{12}^b$ for diffuse refractory-walled passages of length L and hydraulic diameter D_h [3].

If the passage wall is treated as a single-node refractory surface of area A_3, that is, of uniform radiosity, then Eq. (6.48) applies and gives

$$A_1\mathscr{F}_{12}^b = \frac{1}{R_{12}} + \frac{1}{R_{13}+R_{32}}$$

$$= A_1F_{12} + \frac{1}{1/A_1F_{13}+1/A_3F_{32}}$$

Following Edwards [3], we set $A_1 = A_2 = A_c$ and use symmetry and shape factor algebra to give

$$\mathscr{F}_{12}^b = F_{12} + \frac{1}{2}F_{13}$$

$$= (1-F_{13}) + \frac{1}{2}F_{13}$$

$$= 1 - \frac{1}{2}F_{13}$$

$$= 1 - \frac{1}{2}\frac{A_3}{A_c}F_{31}$$

Equation (6.48) will be accurate for a short passage, that is, $L/D_h \ll 1$; then F_{33} is small so that $F_{31} = F_{32} \simeq 1/2$, and

$$\mathscr{F}_{12}^b \simeq 1 - \frac{1}{4}\frac{A_3}{A_c} = 1 - \frac{1}{4}\frac{\mathscr{P}L}{A_c} = 1 - \frac{L}{D_h}$$

But for small L/D_h, $1-L/D_h=1/(1+L/D_h)$, where the latter expression is preferred, since then $\mathscr{F}_{12}^b$ will also have the correct asymptote of zero as $L/D_h \to \infty$. Thus,

$$\mathscr{F}_{12}^b \simeq \frac{1}{1+L/D_h} \qquad (6.50)$$

a result that is independent of passage shape. Equation (6.50) is also plotted in Fig. 6.20, where it is seen to be quite accurate even for values of L/D_h larger than unity.

EXAMPLE 6.7 Heat Loss through a Crack

Estimate the heat leak through a 2 mm crack in a 4 cm-thick layer of insulation sandwiched between steel plates at 800 K and 400 K. Take the emittance of the plates as 0.6.

Solution

Given: 2 mm-wide crack in a layer of insulation.

Required: Heat transfer through crack by radiation.

Assumptions: 1. The crack walls are refractory surfaces and reflect diffusely.

2. The steel plates are diffuse gray surfaces.

Equation (6.49) applies:

$$\mathscr{F}_{12} = \frac{1}{(1-\varepsilon_1)/\varepsilon_1 + 1/\mathscr{F}_{12}^b + (1-\varepsilon_2)/\varepsilon_2}$$

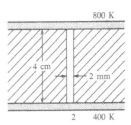

To obtain the transfer factor $\mathscr{F}_{12}^b$ from Fig. 6.20, we evaluate L/D_h, where L is the crack length and D_h its hydraulic diameter:

$$D_h = 2 \times \text{Wall spacing} = (2)(0.002) = 0.004 \text{ m}$$

$$L/D_h = 0.04/0.004 = 10$$

From Fig. 6.20, $\mathscr{F}_{12}^b \simeq 0.16$. Hence,

$$\mathscr{F}_{12} = \frac{1}{(1-0.6)/0.6 + 1/0.16 + (1-0.6)/0.6} = 0.132$$

The heat leak for a meter of crack is

$$\dot{Q}_{12} = A_1 \mathscr{F}_{12} \sigma (T_1^4 - T_2^4)$$

$$= (1)(0.002)(0.132)(5.67 \times 10^{-8})(800^4 - 400^4)$$

$$= 5.75 \text{ W/m}$$

Comments

For a square meter of insulation of $k \sim 0.1$ W/m K, the design heat flow is $\dot{Q} = (k/L)\Delta T = (0.1/0.04)(400) = 1000$ W. If there are 20 meters of cracks, the heat leak is $(20)(5.75)/(1000) = 11.5\%$ of the design heat flow.

6.4   SOLAR RADIATION

Methods for calculating absorption of solar radiation are required for the design of solar collectors, temperature control systems for spacecraft, and air-conditioning systems for transit cars and buildings. In the case of a spacecraft, the solar radiation flux at the edge of the atmosphere is required and is easily determined. However, for a solar collector, the incident solar radiation depends on many variables, including the time of the day, the season, the latitude, and the weather conditions. The practice then is to use standard data found in appropriate design handbooks. In Section 6.4, the diffuse gray surface model used in Section 6.3 is extended to problems involving solar radiation. A different value of absorptance to short-wavelength solar radiation is allowed, while retaining a constant value for all long-wavelength radiation characteristic of surfaces at lower temperatures. This simple model gives results that are adequate for most engineering purposes.

6.4.1 Solar Irradiation

The *solar constant* G_0 is the average flux of solar energy incident normally on the outer fringes of the Earth's atmosphere when the Earth is at its mean distance from the Sun of 1.495×10^{11} m (1 astronomical unit) and has a commonly used value of 1.353 kW/m². Owing to the elliptical nature of the Earth's orbit, the actual solar flux varies from 1.31 kW/m² in June (aphelion) to 1.40 kW/m² in January (perihelion). The uncertainty in these values for G_0 is about 1 or 2%, and they are continually revised as new remote sensing data become available.[4] The total emissive power of the Sun equals that of a blackbody at 5762 K, but its spectral distribution differs somewhat from blackbody behavior.[5] As the solar radiation passes through the Earth's atmosphere, some energy is scattered by gas molecules and aerosols, some is absorbed or reflected back to space by ice crystals and droplets in the clouds, and some is absorbed by gas molecules, particularly by H_2O and CO_2. The solar radiation incident on the Earth's surface thus consists of both a direct component and a diffuse component of scattered radiation. The sum of these is the ground level solar irradiation, G_s, and its value for a horizontal surface is termed the *insolation*. Figure 6.21 shows a schematic of the absorption and scattering process in the atmosphere, and Fig. 6.22 shows the resulting effect on the spectral distribution of radiant energy. The *solar altitude* is the angle of the Sun above the horizon. At lower solar altitudes, the path length of the radiation through the atmosphere is larger, and there is more attenuation, particularly at smaller wavelengths. Thus, insolation depends on the hour of the day, the time of the year, and latitude. Calculation of solar irradiation on surfaces oriented at an arbitrary angle to the Sun's rays—for example, the walls of a building—is a complex geometrical problem that is made more complicated by the presence of clouds. Careful design of equipment such as solar collectors and solar concentrators also requires that the solar irradiation be separated into its direct and diffuse components, which is a complicated task.

---

[4] The term *solar constant* is widely used in engineering. A more precise technical term is *total solar irradiance* (TSI). Here we adopt the commonly used value of 1.353 kW/m² for G_0 even though measurements taken during the 2008 solar minimum revised the baseline value to 1.361 kW/m². Variations in sunspot activity account for about 0.1% of fluctuations in G_0, or much less than variations due to the eccentricity of the Earth's orbit.

[5] The actual spectral distribution based on remote sensing measurements is readily available at the Solar Radiation and Climate Experiment (SORCE) website: lasp.colorado.edu/home/sorce.

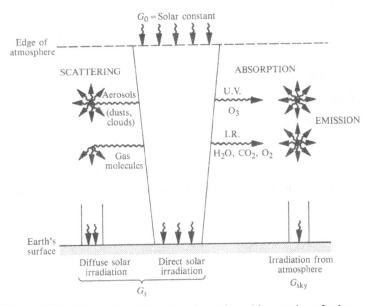

Figure 6.21 Schematic of clear sky absorption and scattering of solar radiation, and the components of irradiation incident on the Earth's surface.

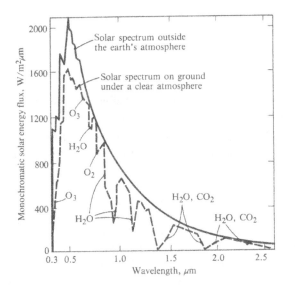

Figure 6.22 Solar spectra outside the Earth's atmosphere and on the ground.

EXAMPLE 6.8   Effective Temperature of the Sun

If the Sun has a diameter of 1.39×10^6 km and is assumed to radiate like a blackbody, what is its effective temperature?

Solution

Given: The Sun.

Required: Estimate of temperature.

Assumptions: 1. The Sun radiates like a blackbody.

   2. The solar constant $G_0 = 1353$ W/m^2.

If the Sun radiates like a blackbody at temperature $T_\odot$, the rate at which radiant energy leaves the Sun is

$$\dot{Q}_\odot = A_\odot \sigma T_\odot^4$$

$$= (\pi)(1.39 \times 10^9)^2 (5.67 \times 10^{-8}) T_\odot^4$$

$$= 3.44 \times 10^{11} T_\odot^4$$

The radiant energy flux at the Earth's mean distance from the Sun is $\dot{Q}_\odot$ divided by the area of a sphere with radius equal to one astronomical unit:

$$G_0 = 1353 = \frac{3.44 \times 10^{11} T_\odot^4}{(4)(\pi)(1.495 \times 10^{11})^2}$$

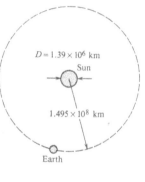

Solving, $T_\odot = 5765$ K.

Comments

1. There are other criteria that could be used to define an effective temperature of the Sun's photosphere, so a number of values are in common use.

2. Here we used the astronomical symbol $\odot$ to denote properties of the Sun. In Section 6.4.3 we will use the subscript s to refer to the shortwave solar radiation on the Earth.

6.4.2 Atmospheric Radiation

Molecules in the atmosphere, chiefly H_2O and CO_2, not only absorb solar radiation but also absorb radiation from the Earth's surface. The radiation emitted by surfaces on Earth is mainly from sources at 250 to 320 K, which produce radiation in the mid-infrared range with wavelengths ranging from about 4 to 40 μm. Atmospheric emission and absorption is mainly at wavelengths of 5 to 8 μm and above 13 μm, and this radiation is not distributed like blackbody radiation. Nevertheless, for engineering calculations, it is possible to approximate the emission from the atmosphere or sky as a fraction of blackbody radiation corresponding to the temperature of the air near the ground, T_e. The sky emittance ε_{sky} is defined such that the sky radiosity is

$$J_{sky} = \varepsilon_{sky} \sigma T_e^4 \tag{6.51}$$

Brunt [4] recommended an approximate correlation for ε_{sky} that accounts for atmospheric CO_2, H_2O, and dust. For a clear sky,

$$\varepsilon_{sky} \simeq 0.55 + 1.8(P_{H_2O}/P)^{1/2} \leq 1 \qquad (6.52)$$

where P_{H_2O} is the partial pressure of water vapor in the atmosphere, and P is the total atmospheric pressure. Alternatively, based on data obtained at a number of locations in the United States, Berdahl and Fromberg [5] recommend the following correlations for a clear sky:

Nighttime: $\qquad \varepsilon_{sky} \simeq 0.741 + 0.0062 T_{DP}(^\circ C) \qquad$ **(6.53a)**

Daytime: $\qquad \varepsilon_{sky} \simeq 0.727 + 0.0060 T_{DP}(^\circ C) \qquad$ **(6.53b)**

where T_{DP} is the dewpoint temperature. Cloud cover has a strong effect on atmospheric radiation, but the effect decreases in importance with cloud elevation because higher clouds are usually colder than low clouds.

An alternative approach that is occasionally used is to define an effective sky temperature, T_{sky}, assuming that it emits radiation like a blackbody. The effective sky temperature is always lower than the air temperature. When there is a low cloud cover we may assume $T_{sky} \simeq T_e$.

EXAMPLE 6.9   Calculation of Sky Emittance and Effective Sky Temperature

For a clear night sky with ambient air at 25°C, 10^5 Pa, and relative humidity of 50%, what is the sky emittance and effective sky temperature?

Solution

Given: Atmospheric conditions at night.

Required: Sky emittance, ε_{sky}, and effective sky temperature, T_{sky}.

Assumptions: No clouds.

Estimates can be made using either Eq. (6.52) or Eq. (6.53a). At 25°C = 298.15 K, steam tables (see Table A.12a) give $P_{sat} = 3168$ Pa; thus,

$$P_{H_2O} = (RH)(P_{sat}) = (0.5)(3168) = 1584 \text{ Pa}$$

Going back to the steam tables, $P_{sat} = 1584$ Pa corresponds to a temperature of 13.9°C, which is the dewpoint. From Eq. (6.52),

$$\varepsilon_{sky} = 0.55 + 1.8(1584/100,000)^{1/2} = 0.777$$

and from Eq. (6.53a),

$$\varepsilon_{sky} = 0.741 + (0.0062)(13.9) = 0.827$$

The effective sky temperature is obtained by equating blackbody emissive power to the actual sky radiosity:

$$\sigma T_{sky}^4 = J_{sky} = \varepsilon_{sky} \sigma T_e^4$$

or

$$T_{sky} = [\varepsilon_{sky} T_e^4]^{1/4}$$

Using the value of ε_{sky} obtained from Eq. (6.53a),

$$T_{\text{sky}} = [(0.827)(298.15)^4]^{1/4} = 284.3 \text{ K}$$

Comments

The different results given by Eqs. (6.52) and (6.53a) indicate that these equations are approximate in nature. Eq. (6.52) gives a lower bound for ε_{sky}. Eq. (6.53a) fits well modern experimental data [6].

6.4.3 Solar Absorptance and Transmittance

The absorptance of real surfaces depends on the wavelength of the incident radiation. In calculations of radiant exchange for gray surfaces, the values of absorptance used are suitable averages over the longer wavelengths characteristic of terrestrial applications, that is, in the range 300–2000 K, perhaps. For many surfaces, the average absorptance to short-wavelength solar radiation is very different from its value for longer wavelengths. For example, white epoxy paint has an absorptance to 270 K radiation of 0.85, whereas its absorptance to solar radiation is only 0.25. Thus, for engineering analysis involving solar radiation, it is necessary to modify the gray surface model by allowing for a different value of absorptance to solar radiation. Table A.5*a* in Appendix A gives values of the solar absorptance as for this purpose. A high value of α_s coupled with a low value of emittance at terrestrial temperatures is desirable for a surface that is required to collect solar heat. The reverse is true for a surface that is required to reject solar heat. Table 6.3 lists some surface finishes suitable for solar heat collection and rejection.

Table 6.3   Some surface finishes suitable for solar heat collection or rejection: total hemispherical emittance ε at ~ 300 K, and solar radiation absorptance α_s.

Surface Finish	ε	α_s
Black chrome electrodeposited on copper, aluminum, stainless steel, or mild steel	0.1–0.2	0.94–0.97
Black nickel electrodeposited on mild steel	0.15	0.90
Copper oxide, chemical conversion of copper sheet surface	0.10–0.15	0.87
Blue stainless steel	0.20	0.89
Nickel, Tabor solar absorber, electro-oxidized on copper, 110–30	0.05	0.85
4022 aluminum–201 nickel(TI)	0.21	0.97
Electroblack	0.05	0.86
PbS, vacuum deposit	0.20	0.98
Aluminum, vacuum-deposited on Mylar	0.03	0.10
Aluminum, hard-anodized	0.80	0.03
White epoxy paint	0.85	0.25
White potassium zirconium silicate spacecraft coating	0.87	0.13
Teflon	0.85	0.12

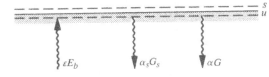

Figure 6.23 Radiant heat fluxes through a surface with solar radiation: absorption and emission takes place below the u-surface.

The heat flux through a surface receiving solar radiation is best formulated by referring to fluxes through a u-surface, as shown in Fig. 6.23. A surface energy balance requires that

$$q = \varepsilon E_b - \alpha_s G_s - \alpha G \tag{6.54}$$

where G_s is the solar irradiation, and the irradiation G includes radiation received from the sky (with radiosity J_{sky}) and from surrounding surfaces:

$$G_i = \sum_{k=1}^{n} J_k F_{ik}$$

The transmittance of real semitransparent materials is also dependent on wavelength; hence, the transmittance to short-wavelength visible radiation can be quite different from that for long-wavelength infrared radiation. The most important practical example is glass, which has a very high transmittance at wavelengths below 2 μm but is nearly opaque at wavelengths greater than 3.4 μm. Thus, solar radiation can readily enter a greenhouse and be absorbed by the plants, while the radiation emitted by the plants is reflected and absorbed by the glass. Table 6.4 gives the long-wavelength ($\lambda > 2.8\mu$m) and short-wavelength ($\lambda < 2.8\mu$m) transmission for some glazing materials.[6]

Table 6.4 Long-wavelength ($\lambda > 2.8\mu$m) and short-wavelength ($\lambda < 2.8\mu$m) transmission for selected glazing materials [7].

Cover Material	Long-Wavelength Transmission (%)	Solar Radiation Transmission (%) Versus Angle of Incidence (Degrees)					
		0	14	30	45	60	67
Glass, double-strength, 3 mm	3	89	89	88	85	82	74
Flat fiberglass, 25 mil	12	87	84	82	80	73	57
Corrugated fiberglass, 40 mil	8	83	83	82	80	62	43
Polycarbonate, 1.58 mm	6	86	85	84	83	80	72
Polyester, weatherable, 5 mil	32	88	88	88	86	81	72
Polyethylene, 4 mil, UV-resistant	80	92	91	88	86	82	67

---

[6] It is beyond the scope of this text to develop the theory of radiation transmission through a medium such as glass, for which the refractive index $n = c/c_0$ is not unity. The student can consult the advanced texts listed in the bibliography at the end of the book.

EXAMPLE 6.10 Temperature of an Airplane Roof

An airplane is left parked in the sun on a calm day. Calculate the roof temperature if the solar irradiation is 870 W/m², the air temperature is 24°C, the sky emittance is 0.77, and the top surface is (i) weathered aluminum alloy 75S-T6, and (ii) white epoxy paint. Take the convective heat transfer coefficient as $h_c = 1.24\Delta T^{1/3}$ W/m² K.

Solution

Given: Airplane parked outdoors.

Required: Roof temperature if it is (i) bare aluminum, and (ii) covered with white epoxy paint.

Assumptions: 1. The underside of the roof is well insulated.
2. The sky emittance is $\varepsilon_{sky} = 0.77$.
3. The roof-to-sky shape factor can be taken as unity.

Let surface 1 be the plane roof. The surface energy balance for a unit area is given by Eq. (1.34) as

$$q_{cond} - q_{conv} - q_{rad} = 0$$

where $q_{cond} = 0$ since the underside of the roof can be taken to be well insulated. The net radiation q_{rad} is obtained using Eq. (6.54):

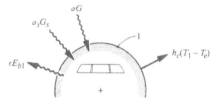

$$q_{rad} = \varepsilon E_b - \alpha_s G_s - \alpha G$$

where $E_b = \sigma T_1^4$, $G_s = 870$ W/m² (assumed normal to the roof to give a maximum value of T_1), and $G \simeq J_{sky} = \varepsilon_{sky}\sigma T_e^4$, since F_{1sky} is approximately unity; so

$$q_{rad} = \varepsilon\sigma T_1^4 - \alpha_s(870) - \alpha(0.77)\sigma(297)^4$$

The convective heat loss is

$$q_{conv} = h_c(T_1 - T_e) = 1.24(T_1 - T_e)^{1/3}(T_1 - T_e) = 1.24(T_1 - 297)^{4/3}$$

Substituting in the surface energy balance,

$$1.24(T_1 - 297)^{4/3} + \varepsilon\sigma T_1^4 - \alpha_s(870) - \alpha(0.77)\sigma(297)^4 = 0$$

Setting $\varepsilon = \alpha$ and rearranging gives

$$1.24(T_1 - 297)^{4/3} + \varepsilon(5.67 \times 10^{-8}T_1^4 - 340) - 870\alpha_s = 0$$

Radiation properties are obtained from Table A.5a. Substituting these values and solving numerically using Newton's method, or drawing a graph, gives the desired result:

	ε	α_s	T_1
(a) Weathered aluminum	0.20	0.54	364 K (91°C)
(b) White paint	0.85	0.25	312 K (39°C)

Comments

1. Use of white paint will reduce the load on the cabin air-conditioning system.
2. Assuming a well insulated roof gives an upper bound for T_1.

EXAMPLE 6.11 Temperature Control of a Spacecraft

A cubical spacecraft is one astronomical unit distant from the Sun, and one face is exactly perpendicular to the Sun's rays. It is coated with stripes of aluminum paint over a fraction F_1 of its area and with vacuum-deposited aluminum over a fraction $F_2 = 1 - F_1$. What value of F_1 will give an equilibrium temperature for the spacecraft of 300 K?

Solution

Given: Spacecraft coated with strips of aluminum paint over vacuum-deposited aluminum.

Required: Fraction of coated area to give a desired equilibrium temperature.

Assumptions: 1. Only one side of the spacecraft sees the Sun.
2. Each side is isothermal.
3. The solar constant is 1353 W/m^2.

Table A.5a gives the required values of the emittance, ε, and solar absorptance, α_s:

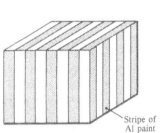

Stripe of
Al paint

A1 paint : $\qquad$ $\varepsilon_1 = 0.20; \alpha_{s1} = 0.27$

Vacuum-deposited Al: $\qquad$ $\varepsilon_2 = 0.03; \alpha_{s2} = 0.10$

If only one side of the spacecraft sees the Sun, then an energy balance on the spacecraft of side length L gives

$$\text{Absorption of solar radiation} = \text{Emission of long-wavelength radiation}$$

$$(F_1\alpha_{s1} + F_2\alpha_{s2})G_0L^2 = (F_1\varepsilon_1 + F_2\varepsilon_2)\sigma T^4(6L^2)$$

$$\frac{F_1\alpha_{s1} + F_2\alpha_{s2}}{F_1\varepsilon_1 + F_2\varepsilon_2} = \frac{6\sigma T^4}{G_0} = \frac{(6)(5.67 \times 10^{-8})(300)^4}{1353} = 2.04$$

$F_2 = 1 - F_1$; hence,

$$\frac{F_1\alpha_{s1} + F_2\alpha_{s2}}{F_1\varepsilon_1 + F_2\varepsilon_2} = \frac{F_1(\alpha_{s1} - \alpha_{s2}) + \alpha_{s2}}{F_1(\varepsilon_1 - \varepsilon_2) + \varepsilon_2} = \frac{0.17F_1 + 0.10}{0.17F_1 + 0.03} = 2.04$$

Solving, $F_1 = 0.219, F_2 = 0.781$.

EXAMPLE 6.12 A Flat-Plate Solar Collector

A flat-plate solar collector used to heat water is 1 m wide and 3 m long and has no coverplate. The absorbing surface is selective, with an emittance of 0.14 and a solar absorptance of 0.93. Water enters the collector at 25°C and a flow rate of 0.015 kg/s. Estimate the outlet water temperature and the collector efficiency at noon when the solar irradiation is 800 W/m^2 and the air temperature is 20°C. Use a sky emittance of 0.6 and an overall heat transfer coefficient for heat transfer from the absorbing surface to the bulk water as $U = 25.0$ W/m^{2}K. The relation $h_c = 1.1(T_s - T_e)^{1/3}$ W/m^{2}K can be used to estimate the convective heat transfer from the absorbing surface to the ambient air. For the water, take $c_p = 4180$ J/kg K.

Solution

Given: Flat-plate solar collector with no coverplate.

Required: Outlet water temperature and collector efficiency.

Assumptions: 1. Steady state.
2. A well-insulated collector base.
3. Shape factor from collector to sky of unity.

The first step is to derive a differential equation governing the change in water temperature. The steady-flow energy equation, Eq. (1.4), applied to an element of the absorber plate W wide and Δx long gives

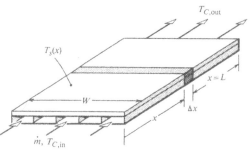

$$\dot{m}c_p\Delta T_C = U(T_s - T_C)W\Delta x$$

where $T_s(x)$ is the temperature of the absorbing surface and T_C is the water temperature (the subscript C is used since the water is the *coolant* stream for this *heat exchanger*). Dividing by Δx and letting $\Delta x \to 0$,

$$\dot{m}c_p\frac{dT_C}{dx} = U(T_s - T_C)W \qquad (1)$$

Equation (1) is a first-order ordinary differential equation and requires one boundary condition, namely,

$$T_C = T_{C,\text{in}} \quad \text{at } x = 0 \qquad (2)$$

The absorber surface temperature T_s is determined from a surface energy balance on the absorber. The energy fluxes are shown in the sketch, taken to be positive as indicated.

$$\begin{array}{ccc} \text{Net radiant energy} & \text{Convective heat} & \text{Heat transfer} \\ \text{flux across } u\text{-surface} & - \text{ loss to ambient air} & = \text{ to water} \end{array} \qquad (3)$$

$$q_{\text{rad}} - q_{\text{conv}} = U(T_s - T_C)$$

From Eq. (6.54),

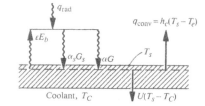

$$q_{\text{rad}} = \alpha_s G_s + \alpha G - \varepsilon E_b$$

$$= \alpha_s G_s + \alpha\varepsilon_{\text{sky}}\sigma T_e^4 - \varepsilon\sigma T_s^4$$

if the shape factor from the collector to the sky is assumed to be 1. Also,

$$q_{\text{conv}} = h_c(T_s - T_e)$$

$$= 1.1(T_s - T_e)^{1/3}(T_s - T_e)$$

$$= 1.1(T_s - T_e)^{4/3}$$

Substituting in Eq. (3) with $\alpha = \varepsilon$,

$$\alpha_s G_s + \varepsilon \varepsilon_{sky} \sigma T_e^4 - \varepsilon \sigma T_s^4 - 1.1(T_s - T_e)^{4/3} = U(T_s - T_C) \qquad (4)$$

The absorber surface temperature should be regarded as the unknown in Eq. (4). Numerical methods are required to solve the problem. For example, the Runge-Kutta method can be used to solve Eq. (1), with T_s, evaluated at each Δx step from Eq. (4) using Newton's method. Substituting numerical values, we have

$$\text{Eq.(1):} \quad (0.015)(4180)\frac{dT_C}{dx} = (25.0)(T_s - T_C)(1)$$

$$\frac{dT_C}{dx} = 0.399(T_s - T_C)$$

$$\text{Eq.(2):} \quad x = 0: \quad T_C = 298 \text{ K}$$

$$\text{Eq.(4):} \quad (0.93)(800) + (0.14)(0.6)(5.67 \times 10^{-8})(293)^4 - (0.14)(5.67 \times 10^{-8})T_s^4$$

$$- 1.1(T_s - 293)^{4/3} = 25.0(T_s - T_C)$$

$$779 - 7.94 \times 10^{-9} T_s^4 - 1.1(T_s - 293)^{4/3} = 25.0(T_s - T_C)$$

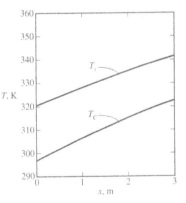

The graph shows the resulting variation of T_C and T_s along the collector. In particular, $T_{C,\text{out}} = 323.5$ K. The collector efficiency is defined as the heat transfer to the water divided by the radiation incident upon the collector:

$$\eta = \frac{\dot{m}c_p(T_{C,\text{out}} - T_{C,\text{in}})}{(G_s + \varepsilon_{sky}\sigma T_e^4)WL}$$

$$= \frac{(0.015)(4180)(323.5 - 298.0)}{[800 + (0.6)(5.67 \times 10^{-8})(293)^4](1)(3)}$$

$$= 50.8\%$$

Comments

A good introduction to the thermal design of solar collectors is given by Goswami, Kreith and Kreider [9].

6.5 DIRECTIONAL CHARACTERISTICS OF SURFACE RADIATION

In Section 6.3, consideration was restricted to gray surfaces that were assumed to emit and reflect diffusely, that is, in no preferred direction. The shape factor F_{12} was defined as the fraction of radiant energy leaving surface 1 that is intercepted by surface 2. Formulas for F_{12} were given, but how such formulas are obtained was not shown. We will now proceed to examine the directional characteristics of surface radiation more carefully. In Section 6.5.1, the concept of *radiation intensity* is introduced, which allows *Lambert's law* for diffuse surfaces to be stated. In Section 6.5.2, a general formula for the shape factor is derived, and examples of its use are subsequently given. In Section 6.5.3, the directional properties of real surfaces are briefly discussed; in particular, the radiation exchange problem is discussed for surfaces that reflect **specularly**, that is, with angle of reflection equal to angle of incidence, as for a mirror.

6.5.1 Radiation Intensity and Lambert's Law

Recall that a differential plane angle $d\alpha$ is the arc length ds on a circle divided by the radius of the circle, that is, $d\alpha = ds/R$, as shown in Fig. 6.24a. Similarly, a differential solid angle $d\omega$ is the area element dA_s on a sphere divided by the sphere radius squared:

$$d\omega = \frac{dA_s}{R^2} \tag{6.55}$$

as shown in Fig. 6.24b. A solid angle is dimensionless, but it is said to be measured in *steradians* (sr), just as a plane angle is said to be measured in radians. Notice that the area of a surface element on a sphere of unit radius has the same numerical value as the corresponding solid angle. A good grasp of the solid angle concept is essential to the understanding of radiation interchange between surfaces.

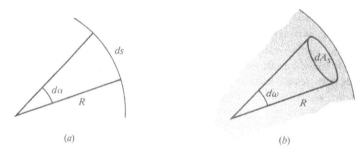

(a)   *(b)*

Figure 6.24   Plane and solid angles, (*a*) Differential plane angle, (*b*) Differential solid angle.

A circular surface-area element dA is shown in Fig. 6.25a. Radiant heat leaves dA in all directions: the **radiation intensity** I^+ is a quantity introduced to describe how this heat flow varies with respect to direction, that is, how it is distributed with respect to zenith angle θ and azimuthal angle ϕ in a spherical coordinate system. Consider how this heat flow might be measured. If a sensor of area dA_s is placed on the surface of a hemisphere surrounding the radiating area dA as shown, then the heat flow to the sensor will be proportional to the solid angle $d\omega$ it occupies. For a spherical surface, $dA_s = Rd\theta R \sin\theta d\phi$, and thus

$$d\omega = dA_s/R^2 = \sin\theta d\theta d\phi \tag{6.56}$$

It is perhaps less obvious that the heat flow to the sensor will also depend on angle θ. Figure 6.25b shows how the radiating area "seen" by the sensor decreases as θ increases from 0° to 90° as a simple cosine variation $dA\cos\theta$. When the sensor is at

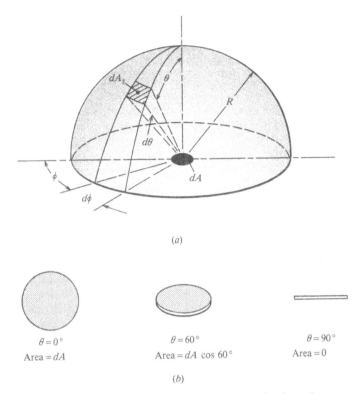

Figure 6.25   (*a*) Elemental areas used to define intensity.
(*b*) The radiating area seen by a sensor at angle θ.

0°, the maximum radiation is received, whereas at 90°, no radiation can be received. The intensity of radiation *leaving* the surface in a given direction $I^+(\theta, \phi)$ is defined as the heat flow from the surface per unit solid angle, per unit area projected normal to the direction. Thus, if our sensor records a heat flow $d\dot{Q}(\theta, \phi)$,

$$I^+(\theta, \phi) = \frac{d\dot{Q}}{dA \cos\theta\, d\omega} \text{ W/m}^2 \text{ sr} \tag{6.57}$$

If $I^+(\theta, \phi)$ is independent of direction, the radiation is said to obey **Lambert's law**; this is the condition for diffuse radiation. Also, Lambert's law implies that radiation in an isothermal black enclosure is *isotropic*. In illumination engineering, the intensity defined by Eq. (6.57) is analogous to the *brightness* of the surface. Imagine a planar light source on the ceiling of a room: as the viewer's eye moves from 0° to 90°, the brightness of the light does not vary if Lambert's law is obeyed, although the amount of light received decreases proportional to $\cos\theta$ until, at 90°, the light cannot be seen. (Note that in some older texts, intensity is defined in terms of actual, rather than projected, area. Lambert's law then states that the intensity is the value at $\theta = 0$ multiplied by $\cos\theta$.)

From Eq. (6.57), the heat flux $q^+(\theta, \phi)$ leaving the surface is

$$q^+(\theta, \phi) = \frac{d\dot{Q}}{dA} = I^+ \cos\theta \, d\omega$$

Substituting for $d\omega$ from Eq.(6.56) and integrating over the hemisphere gives the radiosity J:

$$J = \int_0^{2\pi} \int_0^{\pi/2} I^+(\theta, \phi) \cos\theta \sin\theta d\theta d\phi \qquad (6.58)$$

For a diffuse surface, I^+ is a constant; noting that $\int_0^{2\pi}\int_0^{\pi/2} \cos\theta \sin\theta d\theta d\phi = \pi$ gives

$$J = \pi I^+ \qquad (6.59)$$

A black surface radiates diffusely with a radiosity $J = E_b$; thus

$$E_b = \pi I_b^+ \qquad (6.60)$$

That is, the intensity of radiation from a black surface I_b^+ is simply $\sigma T^4/\pi$.

The intensity of radiation *incident* on a surface from a given direction is defined in a similar manner: $I^-(\theta, \phi)$ is the heat flow to the surface per unit solid angle, per unit area projected normal to the direction of incidence. Then the irradiation G is given by the integral

$$G = \int_0^{2\pi} \int_0^{\pi/2} I^-(\theta, \phi) \cos\theta \sin\theta d\theta d\phi \qquad (6.61)$$

and for diffuse incident radiation,

$$G = \pi I^- \qquad (6.62)$$

Intensity also can be defined on a spectral basis, as will be shown in Section 6.6.2.

EXAMPLE 6.13 Calculation of Irradiation from Intensity

Two surfaces, each of area 1 cm^2, are located 10 cm apart, as shown in the sketch. Surface 1 is black and is maintained at 1000 K. Calculate the rate at which surface 2 is irradiated by surface 1.

Solution

Given: Two surfaces, 1 and 2.

Required: $d\dot{Q}_{12}$, the irradiation of surface 2 by surface 1.

Assumptions: Since both surfaces are small compared to the distance between them, they can be approximated by elemental areas dA_1 and dA_2.

Surface 1 is black and is thus a diffuse emitter: the intensity of radiation leaving A_1 is given by Eq. (6.60) as

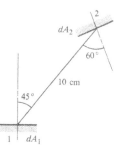

$$I_1^+ = \frac{E_{b1}}{\pi} = \frac{\sigma T_1^4}{\pi} = \frac{(5.67 \times 10^{-8})(1000)^4}{\pi} = 1.805 \times 10^4 \text{ W/m}^2 \text{ sr}$$

The heat flow in the solid angle $d\omega_{12}$ subtended by surface dA_2 at dA_1 is obtained from Eq. (6.57) as

$$d\dot{Q}_{12} = I_1^+ dA_1 \cos\theta_1 d\omega_{12}$$

But from Eq. (6.55),

$$d\omega_{12} = \frac{dA_s}{R^2} = \frac{dA_2 \cos\theta_2}{R^2}$$

since dA_s is the projection of dA_2 perpendicular to the direction of the incident radiation from dA_1. Thus,

$$d\dot{Q}_{12} = \frac{I_1^+ dA_1 dA_2 \cos\theta_1 \cos\theta_2}{R^2} = \frac{(1.805 \times 10^4)(10^{-4})(10^{-4})(0.707)(0.500)}{(0.1)^2}$$

$$= 6.38 \times 10^{-3} \text{ W}(6.38 \text{ mW})$$

6.5.2 Shape Factor Determination

The shape factor F_{12} introduced in Section 6.3.2 was defined as the fraction of radiation leaving a black surface A_1 that is intercepted by surface A_2. Since a black surface emits diffusely, the shape factor is a geometrical concept that depends only on the size, shape, and orientation of the surfaces. In Section 6.3.4, the same shape factor was used for interchange between diffuse gray surfaces. In general, the shape factor F_{12} applies to any surface for which the intensity I^+ obeys Lambert's law. To derive a formula for the shape factor, we first consider the radiation leaving an elemental area dA_1 that is intercepted by finite area A_2, as shown in Fig. 6.26. If the element dA2 subtends a solid angle $d\omega$ at element dA_1, Eq. (6.57) gives the radiation intercepted by dA_2 as

$$d\dot{Q} = I_1^+ dA_1 \cos\theta_1 d\omega$$

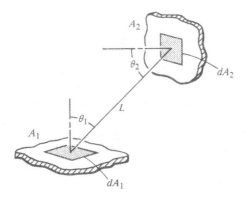

Figure 6.26 Geometry for derivation of the shape factor formula.

Substituting $I_1^+ = J_1/\pi$ from Eq. (6.59), and integrating over the solid angle subtended by area A_2 at dA_1, gives the radiation leaving dA_1 that is intercepted by surface A_2:

$$d\dot{Q}_{12} = \frac{J_1 dA_1}{\pi} \int_{\omega_{12}} \cos\theta_1 d\omega$$

The radiation leaving dA_1 in all directions is

$$d\dot{Q}_1 = \frac{J_1 dA_1}{\pi} \int_0^{2\pi} \cos\theta_1 d\omega = J_1 dA_1$$

The shape factor F_{12} is the fraction of the radiation leaving dA_1 that is intercepted by area A_2, $F_{12} = d\dot{Q}_{12}/d\dot{Q}_1$, or

$$F_{12} = \frac{1}{\pi} \int_{\omega_{12}} \cos\theta_1 d\omega \qquad (6.63)$$

Equation (6.63) is convenient for application to simple configurations, as illustrated in Example 6.14. Introduction of $d\omega = dA_2 \cos\theta_2/L^2$ into Eq. (6.63) gives an alternative form:

$$F_{12} = \frac{1}{\pi} \int_{A_2} \frac{\cos\theta_1 \cos\theta_2}{L^2} dA_2 \qquad (6.64)$$

To obtain the shape factor for a finite area A_1, F_{12} given by Eq. (6.64) is averaged over area A_1:

$$F_{12} = \frac{1}{A_1} \int_{A_1} \int_{A_2} \frac{\cos\theta_1 \cos\theta_2}{\pi L^2} dA_2 dA_1 \qquad (6.65)$$

From the symmetry of the problem, we can also write

$$F_{21} = \frac{1}{A_2} \int_{A_2} \int_{A_1} \frac{\cos\theta_2 \cos\theta_1}{\pi L^2} dA_1 dA_2 \qquad (6.66)$$

It follows from Eqs. (6.65) and (6.66) that

$$\int_{A_1} \int_{A_2} \frac{\cos\theta_1 \cos\theta_2}{\pi L^2} dA_2 dA_1 = A_1 F_{12} = A_2 F_{21} \qquad (6.67)$$

which is both a formula for the shape factor and a statement of the reciprocal rule for shape factors. The summation rule follows directly from Eq. (6.63). Notice that in deriving Eq. (6.63), the heat flows $d\dot{Q}_{12}$ and $d\dot{Q}_1$ are indicated as differential elements of heat flow because they are leaving the elemental area dA_1. However, perhaps we could have written $d^2\dot{Q}$ since this heat flow is also a flow in an elemental solid angle $d\omega$; such is the practice in some texts.

Unless the geometry is very simple, analytical evaluation of the double integral in Eq. (6.67) is a challenging mathematical problem. Computer codes for the evaluation of Eq. (6.67) are available, many of them based on NASA's TRASYS [8].

EXAMPLE 6.14 Shape Factor between an Elemental Area and a Disk

Determine the shape factor F_{12} for the surfaces shown in the following sketch. Surface 2 is a disk of diameter D, and surface 1 is small $(A_1 \ll A_2)$, parallel to surface 2, and a distance H away.

Solution

Given: Elemental area 1 parallel to disk 2.

Required: Shape factor F_{12}.

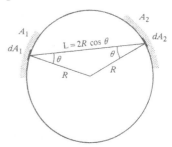

To use Eq. (6.63), we write $d\omega = \sin\theta_1 d\theta_1 d\phi$, to give

$$F_{12} = \frac{1}{\pi} \int_{\omega_{12}} \cos\theta_1 d\omega$$

$$= \frac{1}{\pi} \int_0^{2\pi} \int_0^{\Theta} \cos\theta_1 \sin\theta_1 d\theta_1 d\phi$$

$$= 2 \int_0^{\Theta} \cos\theta_1 \sin\theta_1 d\theta_1$$

$$= \sin^2 \Theta$$

$$\sin\Theta = \frac{D/2}{[(D/2)^2 + H^2]^{1/2}}; \quad \text{thus } F_{12} = \frac{D^2}{D^2 + 4H^2}$$

Alternatively, to use Eq. (6.64), we note that $\theta_1 = \theta_2 = \theta$, $\cos\theta = H/L$, $L^2 = r^2 + H^2$, $dA_2 = 2\pi r\, dr$.

$$F_{12} = \frac{1}{\pi} \int_{A_2} \frac{\cos\theta_1 \cos\theta_2}{L^2} dA_2 = 2H^2 \int_0^{D/2} \frac{r\, dr}{(r^2 + H^2)^2} = \frac{D^2}{D^2 + 4H^2}$$

EXAMPLE 6.15 Shape Factor between Areas on a Sphere

Determine the shape factor between two areas A_1 and A_2 on the inside of a sphere.

Solution

Given: Two areas on the inside of a sphere, A_1 and A_2.

Required: Shape factor F_{12}.

The figure shows that $\theta_1 = \theta_2 = \theta$, and $L = 2R\cos\theta$. Substituting into Eq. (6.65),

$$F_{12} = \frac{1}{A_1} \int_{A_1} \int_{A_2} \frac{\cos^2\theta}{4\pi R^2 \cos^2\theta} dA_2 dA_1 = \frac{1}{A_1} \int_{A_1} \int_{A_2} \frac{dA_2 dA_1}{4\pi R^2}$$

$$= \frac{A_2}{4\pi R^2} \left(= \frac{A_2}{\text{Total sphere area}} \right)$$

Comments

Notice that the shape factor is independent of the relative location of the areas A_1 and A_2: any location on the inner surface of a sphere "sees" all other locations in the same way.

6.5.3 Directional Properties of Real Surfaces

A diffuse gray surface obeys Lambert's law; that is, the intensity of radiation leaving the surface is independent of direction, which requires that the emittance and reflectance are independent of angle. The emittance of a real surface usually has negligible variation with azimuthal angle ϕ, but it does vary somewhat with zenith angle θ. Experimental data for the directional variation of emittance are sparse; Fig. 6.27

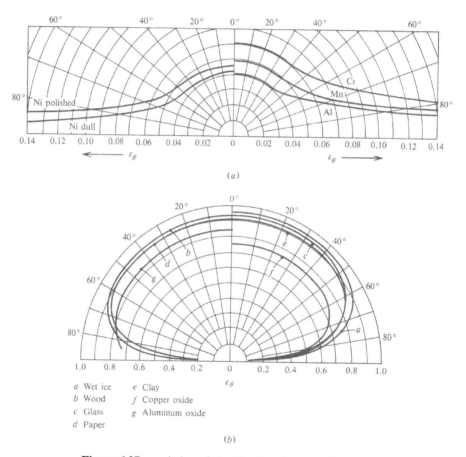

Figure 6.27   variation of directional emittance with zenith angle θ. (a) Metals, (b) Nonmetals [10].

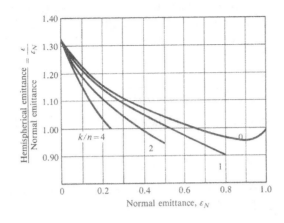
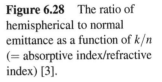

Figure 6.28 The ratio of hemispherical to normal emittance as a function of k/n (= absorptive index/refractive index) [3].

shows data for some selected surfaces. Electromagnetic theory gives the results shown in Fig. 6.28, where ε is the *hemispherical emittance*, that is, the emittance averaged over a hemisphere of solid angle 2π steradians above the surface, and ε_N is the emittance in the direction normal to the surface. The parameter k/n is the ratio of the absorptive index to the refractive index of the medium: for metals, a value of 1 can be used; 0 is more appropriate for nonmetals. A simple rule based on Fig. 6.28 is $\varepsilon/\varepsilon_N \simeq 1.2$ for low-emittance polished metals, and $\varepsilon/\varepsilon_N \simeq 0.96$ for high-emittance nonmetallic surfaces. Notice that it is the hemispherical emittance that is used in the diffuse gray surface model.

Specular Reflectors

Of some interest are surfaces for which reflection is not perfectly diffuse. Highly polished metal surfaces reflect thermal radiation in the same manner as a mirror reflects visible light, that is, with the angle of reflection equal to the angle of incidence. Such a surface is termed a *specular* reflector. Mixed specular and diffuse reflection can occur. For example, the surface of a gloss enamel paint tends to reflect specularly, while the underlying pigment particles reflect diffusely. The calculation of radiation exchange when some surfaces are specular reflectors is not as simple as for diffuse reflectors, unless the geometry is very simple. In the case of infinite parallel walls, all radiation reflected by one wall reaches the other directly, irrespective of whether the reflection is diffuse or specular: Eq. (6.41) and Table 6.2, item 1 are thus valid for specular surfaces as well. The same holds true for infinite concentric cylinders and concentric spheres when the inner surface is a specular reflector: Eq. (6.39) and Table 6.2, item 2 are valid for this situation. However, when the outer surface is specular and the inner surface is specular or diffuse, a new result is obtained, which can be derived as follows.

The inner surface is denoted A_1 and the other surface A_2. We will first use ray tracing to obtain the result. Using Eq. (6.32) for a gray surface,

$$\dot{Q}_{12} = \varepsilon_1 A_1 E_{b1} - \varepsilon_1 A_1 G_1 \qquad (6.68)$$

All radiation leaving surface 1 is intercepted by surface 2. In addition, all radiation *specularly* reflected by surface 2 returns to surface 1 (the angle of incidence equals the angle of reflection). Thus, the radiation incident on surface 1 due to radiation emitted by surface 1 is

$$\varepsilon_1 E_{b1} A_1 (\rho_2 + \rho_1 \rho_2^2 + \cdots) = \varepsilon_1 E_{b1} A_1 \frac{\rho_2}{1 - \rho_1 \rho_2} \tag{6.69}$$

The radiation incident on surface 1 due to radiation emitted by surface 2 is

$$\varepsilon_2 E_{b2} A_2 F_{21} (1 + \rho_1 \rho_2 + \cdots) = \varepsilon_2 E_{b2} A_1 \frac{1}{1 - \rho_1 \rho_2} \tag{6.70}$$

where the reciprocal rule $A_2 F_{21} = A_1 F_{12} = A_1$ has been used for the diffuse emission from surface 2. Substituting for $A_1 G_1$ in Eq. (6.68) gives

$$\dot{Q}_{12} = \varepsilon_1 A_1 E_{b1} - \varepsilon_1 \left(\frac{\varepsilon_1 A_1 E_{b1} \rho_2}{1 - \rho_1 \rho_2} + \frac{\varepsilon_2 E_{b2} A_1}{1 - \rho_1 \rho_2} \right)$$

Setting $\rho = 1 - \alpha = 1 - \varepsilon$ and rearranging,

$$\dot{Q}_{12} = \frac{A_1 (E_{b1} - E_{b2})}{1/\varepsilon_1 + 1/\varepsilon_2 - 1} \tag{6.71}$$

which is identical to Eq. (6.41), the result for infinite parallel walls. Alternatively, this result can be obtained by applying the energy conservation principle to obtain G_1. The radiation emitted (diffusely) by surface 2 and intercepted by surface 1 is $\varepsilon_2 E_{b2} A_2 F_{21}$ ($= \varepsilon_2 E_{b2} A_1$ from the reciprocal rule for diffuse radiation shape factors). The radiation emitted and reflected by surface 1 that returns to surface 1 is $A_1 (\varepsilon_1 E_{b1} + \rho_1 G_1) \rho_2$, since *all* radiation leaving surface 1 and reflected by surface 2 returns to surface 1. Hence, the radiation incident on surface 1, $A_1 G_1$, is

$$A_1 G_1 = \varepsilon_2 E_{b2} A_1 + A_1 (\varepsilon_1 E_{b1} + \rho_1 G_1) \rho_2$$

$$G_1 = \frac{\varepsilon_2 E_{b2} + \varepsilon_1 \rho_2 E_{b1}}{1 - \rho_1 \rho_2}$$

Substituting for G_1 in Eq. (6.68)

$$\dot{Q}_{12} = \varepsilon_1 A_1 \left(E_{b1} - \frac{\varepsilon_2 E_{b2} + \varepsilon_1 \rho_2 E_{b1}}{1 - \rho_1 \rho_2} \right) = \frac{A_1 (E_{b1} - E_{b2})}{1/\varepsilon_1 + 1/\varepsilon_2 - 1}$$

as before.

If an enclosure contains only a few specular plane surfaces, the mirror image concept can be used to modify the formulation developed in Section 6.3.5 for diffuse gray surface enclosures. The important idea is that a ray coming from a diffuse surface i and reflected by a specular surface m to a diffuse surface k, is equivalent to an uninterrupted straight-line ray from the mirror image of i to k, as shown in Fig. 6.29. With more than one specular surface, multiple reflections must be accounted for, and the method is feasible only if these can be easily added. However, in advanced engineering practice, the designer might wish to consider many complicating factors in addition to specular components of reflectance. For example, it might be desirable

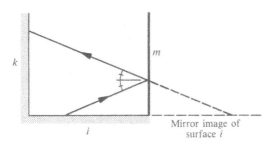

Figure 6.29 Use of the mirror image concept for a single specularly reflecting surface.

to consider directional and spectral variations of emittance. The *Monte Carlo* ray tracing method is then the most popular numerical method used.

Specular Reflecting Passages

A situation in which the effect of specular reflection is particularly important is the transmission of radiation along a passage with specular side walls. For a square or circular cross section, the problem is straightforward; but for a slot, **polarization** of the reflected radiation plays an important role. Radiation is polarized in the sense of having two wave components at right angles to each other and to the propagation direction; for black radiation, the two components are equal. Effects of polarization on radiant heat transfer are dealt with in more advanced texts. Figure 6.30 shows

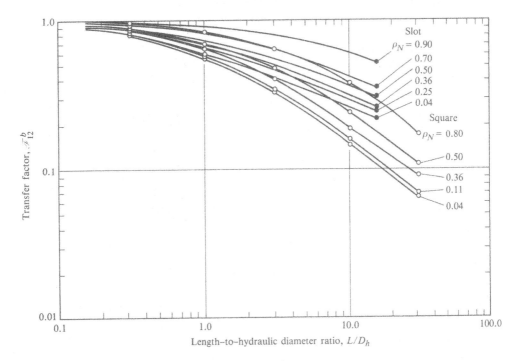

Figure 6.30 Transfer factor $\mathscr{F}_{12}^{b}$ for specular refractory walled passages of length L and hydraulic diameter D_h [3].

the transfer factor $\mathscr{F}_{12}^b$ for the slot and square passage with refractory side walls. A comparison with Fig. 6.20 for diffuse walls leads to the following observations:

1. The diffuse wall transfer factor does not depend on side wall reflectance (because there is no difference between diffusely reflected and diffusely reemitted radiation).

2. If the specular reflectance is unity, the transfer factor is 1.0, no matter how long the passage.

3. With specular reflection, the transfer factor is more sensitive to passage cross section.

EXAMPLE 6.16   Heat Gain by a Liquid Nitrogen Line

Liquid nitrogen flows through a 3 cm-O.D. tube, which is contained in an evacuated 5 cm-I.D. cylindrical shell. The nitrogen is at its normal boiling point of 77.4 K, and the shell inner surface is at 260 K. Estimate the heat gain per meter length if the emittances of the surfaces are 0.04 and (i) the surfaces are diffuse reflectors, (ii) the surfaces are specular reflectors.

Solution

Given: Dewar-type insulated liquid nitrogen line.

Required: Heat gain per meter length for diffuse and specular coatings.

Assumptions: 1. Emittance independent of temperature.

2. Completely diffuse or completely specular surfaces.

(i) For diffuse surfaces, item 2 of Table 6.2 gives the transfer factor as

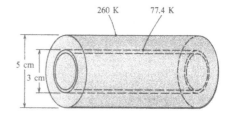

$$\mathscr{F}_{12} = \frac{\varepsilon_1}{1 + (\varepsilon_1 A_1 / \varepsilon_2 A_2)(1 - \varepsilon_2)}$$

where the tube is surface 1 and the shell is surface 2. Using Eq. (6.45),

$$\dot{Q}_{12} = A_1 \mathscr{F}_{12} (\sigma T_1^4 - \sigma T_2^4) = \frac{A_1 \varepsilon_1 \sigma (T_1^4 - T_2^4)}{1 + (\varepsilon_1 A_1 / \varepsilon_2 A_2)(1 - \varepsilon_2)}$$

$$= \frac{(\pi)(0.03)(1)(0.04)(5.67 \times 10^{-8})(77.4^4 - 260^4)}{1 + (0.04/0.04)(3/5)(1 - 0.04)}$$

$$= -0.615 \text{ W/m}$$

(ii) For specular surfaces, Eq. (6.71) gives

$$\dot{Q}_{12} = \frac{A_1(E_{b1} - E_{b2})}{1/\varepsilon_1 + 1/\varepsilon_2 - 1} = \frac{(\pi)(0.03)(1)(5.67 \times 10^{-8})(77.4^4 - 260^4)}{1/0.04 + 1/0.04 - 1}$$

$$= -0.494 \text{ W/m}$$

Comments

The heat gain for specular surfaces is 20% less than for diffuse surfaces.

EXAMPLE 6.17   Radiation Transmission through a Spacecraft Shutter

A louvered shutter for temperature control of a spacecraft is 60 cm wide and 31 cm high.
When fully open, the passages are 3 cm high and 6 cm long. The louvers are 1 mm thick.
The spacecraft interior can be taken to be a black enclosure at 300 K, and the exterior is outer
space at ~ 0 K. Determine the heat loss through the shutter (i) if the louvers are diffuse reflectors, and (ii) if the louvers have a coating that reflects specularly with a normal reflectance of
0.9.

Solution

Given: Louvered shutter for spacecraft temperature control.

Required: Effect of wall reflection properties on heat loss.

Assumptions: The louvers have a low thermal conductivity, so they can be taken to be adiabatic.

The heat loss through the shutter is given by

$$\dot{Q}_{12} = A_1 \mathscr{F}_{12}^b \sigma (T_1^4 - T_2^4)$$

where the transfer factors $\mathscr{F}_{12}^b$ are obtained from Fig.
6.20 for diffuse reflection and from Fig. 6.30 for specular reflection. The hydraulic diameter of the slot-
shaped passages between the louvers is

$$D_h = (2)(3) = 6 \, \text{cm}$$

$$L/D_h = 6/6 = 1.0$$

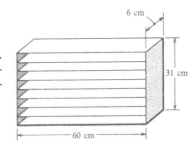

The number of passages is $31/(3+0.1) = 10$, so that the area for radiant energy transmission
is $A_1 = (10)(0.03)(0.6) = 0.18 \, \text{m}^2$.

(i) Diffuse reflection: From Fig. 6.20, $\mathscr{F}_{12}^b \simeq 0.54$,

$$\dot{Q}_{12} = (0.18)(0.54)(5.67 \times 10^{-8})(300^4 - 0^4) = 44.6 \, \text{W}$$

(ii) Specular reflection: From Fig. 6.30, $\mathscr{F}_{12}^b \simeq 0.91$, for $\rho_N = 0.90$.

$$\dot{Q}_{12} = (0.18)(0.91)(5.67 \times 10^{-8})(300^4 - 0^4) = 75.2 \, \text{W}$$

Comments

1. The effect of the specularly reflecting coating is to increase the radiation transmission by
   69%.

2. How the louver transmits as a function of angular rotation also depends significantly on
   whether the reflection is specular or diffuse.

6.6   SPECTRAL CHARACTERISTICS OF SURFACE RADIATION

In Section 6.3, the assumption of a gray surface allowed the calculation of radiant energy exchange without the need to consider the spectral characteristics of emitted or reflected radiation. Only the *total* energy emitted or reflected at all wavelengths was involved; thus, properties such as emittance were implicitly assumed to be appropriate average values. In order to solve simple problems involving solar radiation in Section 6.4, we crudely approximated the spectral distribution of radiant energy by using values of the absorptance that were different for solar and terrestrial radiation, otherwise treating the surface as if it were gray. We now proceed to examine the spectral characteristics of surface radiation more carefully. In Section 6.6.1, Planck's law is restated and the fractional functions defined. In Section 6.6.2, spectral-directional properties are defined and a fundamental statement of *Kirchhoff's law* given. Hemispherical and total hemispherical properties are obtained by appropriate averaging.

6.6.1 Planck's Law and Fractional Functions

Planck's law, Eq. (6.5), gives the spectral distribution of the monochromatic emissive power for a black surface, that is, $E_{b\lambda}$ as a function of λ:

$$E_{b\lambda} = \frac{C_1 \lambda^{-5}}{e^{C_2/\lambda T} - 1} \tag{6.5}$$

where for λ in μm, $C_1 = 3.7418 \times 10^8$ W μm^4/m^2, and $C_2 = 1.4389 \times 10^4$ μm K. The fraction of the total blackbody emissive power for wavelengths between 0 and λ is called the **external fractional function** f_e:

$$f_e(\lambda, T) = \frac{\int_0^\lambda E_{b\lambda}(\lambda, T) d\lambda}{\sigma T^4} \tag{6.72}$$

Substituting from Eq. (6.5) and then changing the integration variable from λ to $\zeta = C_2/\lambda T$,

$$f_e(\lambda T) = \frac{C_1}{\sigma C_2^4} \int_\zeta^\infty \frac{\zeta^3 d\zeta}{e^\zeta - 1} \tag{6.73}$$

Table 6.5 shows this function. To obtain the fraction of power between two wavelengths, we simply subtract the fraction below the lower value from the fraction below the higher value:

$$\Delta f = f(\lambda T)_1 - f(\lambda T)_2$$

The use of the external fraction is demonstrated in Example 6.18. At this point, it is also appropriate to introduce the **internal fractional function** f_i, which is defined as

$$f_i(\lambda, T) = \frac{\int_0^\lambda [\partial E_{b\lambda}(\lambda, T)/\partial T] d\lambda}{4\sigma T^3} \tag{6.74}$$

The function $f_i(\lambda T)$ is also shown in Table 6.5. The importance of the internal function will be demonstrated in Section 6.6.2.

Table 6.5 Values of λT at specified values of the external and internal fractions of blackbody radiation [3].

f	Δf	$\Sigma \Delta f$	$(\lambda T)_e$ μm K $\times 10^{-4}$	$(\lambda T)_i$ μm K $\times 10^{-4}$	f	Δf	$\Sigma \Delta f$	$(\lambda T)_e$ μm K $\times 10^{-4}$	$(\lambda T)_i$ μm K $\times 10^{-4}$
0.0025	0.005	0.005	0.1230	0.1073	0.51	0.02	0.52	0.416	0.326
0.0075	0.005	0.010	0.1395	0.1209	0.53	0.02	0.54	0.428	0.334
0.0125	0.005	0.015	0.1495	0.1291	0.55	0.02	0.56	0.440	0.342
0.0175	0.005	0.02	0.1573	0.1352	0.57	0.02	0.58	0.453	0.351
0.025	0.01	0.03	0.1662	0.1423	0.59	0.02	0.60	0.467	0.361
0.035	0.01	0.04	0.1762	0.1501	0.61	0.02	0.62	0.482	0.371
0.045	0.01	0.05	0.1848	0.1565	0.63	0.02	0.64	0.497	0.382
0.055	0.01	0.06	0.1922	0.1625	0.65	0.02	0.66	0.513	0.393
0.07	0.02	0.08	0.202	0.1701	0.67	0.02	0.68	0.531	0.405
0.09	0.02	0.10	0.214	0.1791	0.69	0.02	0.70	0.550	0.417
0.11	0.02	0.12	0.225	0.1874	0.71	0.02	0.72	0.570	0.430
0.13	0.02	0.14	0.235	0.1949	0.73	0.02	0.74	0.593	0.446
0.15	0.02	0.16	0.245	0.202	0.75	0.02	0.76	0.616	0.462
0.17	0.02	0.18	0.254	0.209	0.77	0.02	0.78	0.642	0.479
0.19	0.02	0.20	0.263	0.216	0.79	0.02	0.80	0.671	0.498
0.21	0.02	0.22	0.272	0.222	0.81	0.02	0.82	0.704	0.521
0.23	0.02	0.24	0.281	0.229	0.83	0.02	0.84	0.741	0.546
0.25	0.02	0.26	0.290	0.235	0.85	0.02	0.86	0.783	0.574
0.27	0.02	0.28	0.299	0.242	0.87	0.02	0.88	0.825	0.609
0.29	0.02	0.30	0.308	0.248	0.89	0.02	0.90	0.899	0.653
0.31	0.02	0.32	0.317	0.255	0.91	0.02	0.92	0.982	0.690
0.33	0.02	0.34	0.326	0.261	0.93	0.02	0.94	1.090	0.774
0.35	0.02	0.36	0.335	0.268	0.945	0.01	0.95	1.200	0.846
0.37	0.02	0.38	0.344	0.275	0.955	0.01	0.96	1.298	0.909
0.39	0.02	0.40	0.353	0.281	0.965	0.01	0.97	1.433	0.997
0.41	0.02	0.42	0.363	0.288	0.975	0.01	0.98	1.63	1.123
0.43	0.02	0.44	0.373	0.295	0.9825	0.005	0.985	1.87	1.28
0.45	0.02	0.46	0.384	0.303	0.9875	0.005	0.990	2.22	1.43
0.47	0.02	0.48	0.394	0.310	0.9925	0.005	0.995	2.56	1.70
0.49	0.02	0.50	0.405	0.318	0.9975	0.005	1.000	7.34	2.49

EXAMPLE 6.18   Radiation Wavelengths of Practical Significance

Determine the wavelength range encompassing the 10–90% fraction of radiant energy emitted by black surfaces at the following temperatures: (i) a room temperature of 300 K, (ii) a cherry red-hot temperature of 1000 K, (iii) the temperature of a tungsten lamp filament, 3000 K, and (iv) the Sun's temperature, 5700 K.

Solution

Given: Black surfaces at various temperatures.

Required: Wavelength range encompassing the 10–90% fraction of the emitted radiation.

From Table 6.5, the values of $(\lambda T)_e$ corresponding to $f = 0.1$ and $f = 0.9$ are

$$f = 0.1: \quad (\lambda T)_e = 0.220 \times 10^4 \ \mu\text{m K}$$

$$f = 0.9: \quad (\lambda T)_e = 0.940 \times 10^4 \ \mu\text{m K}$$

and $\Delta f_e = f_e(\lambda T)_2 - f_e(\lambda T)_1 = 0.9 - 0.1 = 0.8 \,(80\%)$.

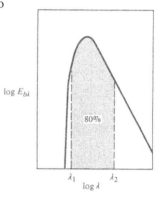

(i) $T = 300$ K:

$$\lambda_1 = (\lambda T)_1/T = 0.220 \times 10^4/300 = 7.33 \ \mu\text{m}$$

$$\lambda_2 = (\lambda T)_2/T = 0.940 \times 10^4/300 = 31.3 \ \mu\text{m}$$

(ii), (iii), (iv). Similarly, for $T = 1000$, 3000, and 5700 K, and the results are summarized below.

Source Temperature K	Wavelength Range μm
300	7.33 – 31.3
1000	2.20 – 9.40
3000	0.733 – 3.13
5700	0.386 – 1.65

Comments

1. As the temperature increases, the wavelength range of practical concern for radiant energy transfer shifts to shorter wavelengths and becomes narrower.

2. At room temperature, wavelengths of concern are on the order of 10 μm; at the Sun's temperature, wavelengths are on the order of 1 μm.

3. The external fractional function provides a variable-independent method to determine relevant wavelength ranges for different temperatures, as opposed to the ubiquitous but incorrect use of Eq. (6.6). Notice that $f = 0.25$ for $(\lambda T)_e = 2898 \ \mu\text{m K}$ and $f = 0.50$ (median value of the distribution) for $(\lambda T)_e = 4110 \ \mu\text{m K}$.

6.6.2 Spectral Properties

Surface radiation properties, such as the emittance, depend on the nature of the surface. The chemical composition, physical structure (as perhaps modified by heat treatment or cold work), and roughness all influence the emittance. Nuclear irradiation may also be used to modify the surface. Once the surface is defined, its thermodynamic state is essentially fixed by its temperature, T_s; thus, the emittance may depend on T_s as well. In addition, the emittance depends on the direction and wavelength of the emitted radiation. The absorptance, reflectance, and transmittance may also depend on the state of polarization of the incident radiation, but this aspect will not be considered here.

For a given surface, we define **spectral-directional** radiation properties at temperature T_s. For example, the spectral-directional emittance of a surface at temperature T_s is the ratio of the intensity of radiation emitted at wavelength λ in the direction (θ, ϕ) to the intensity of radiation of the same wavelength emitted by a black surface at the same temperature:

$$\varepsilon(\theta, \phi, \lambda, T_s) = \frac{I_\lambda^+(\theta, \phi, \lambda, T_s)}{I_{b\lambda}^+(\lambda, T_s)} \tag{6.75}$$

There are similar definitions for the spectral-directional absorptance, reflectance, and transmittance. The **principle of detailed balancing** of statistical thermodynamics requires that

$$\varepsilon(\theta, \phi, \lambda, T_s) = \alpha(\theta, \phi, \lambda, T_s) \tag{6.76}$$

which is usually taken as the most fundamental statement of **Kirchhoff's law**. The net heat flux through a surface is

$$q = \frac{\dot{Q}}{A} = \int_0^\infty \int_0^{2\pi} \int_0^{\pi/2} \{\varepsilon(\theta, \phi, \lambda, T_s)I_{b\lambda}^+(\lambda, T_s)$$
$$- \alpha(\theta, \phi, \lambda, T_s)I_\lambda^-(\theta, \phi, \lambda)\} \cos\theta \sin\theta \, d\theta \, d\phi \, d\lambda \tag{6.77}$$

If the Monte Carlo method is to be used to calculate radiation exchange between surfaces, it is relatively straightforward to allow for directional and spectral variations in surface properties. However, often the engineer must use the simpler diffuse gray surface approximation, either because time or funds are limited or because the necessary data for the spectral-directional properties are unavailable for the surfaces of interest. Then suitable averaged surface properties are required.

Total Hemispherical Emittance

A **hemispherical property** is a value directionally averaged over the 2π steradian hemisphere of solid angle above the surface. A **total property** is a value spectrally averaged over all wavelengths from zero to infinity. Thus, the *total hemispherical emittance* is

$$\varepsilon(T_s) = \frac{1}{\sigma T_s^4} \int_0^\infty \int_0^{2\pi} \int_0^{\pi/2} \varepsilon(\theta, \phi, \lambda, T_s)I_{b\lambda}(\lambda, T_s) \cos\theta \sin\theta \, d\theta \, d\phi \, d\lambda \tag{6.78}$$

It is convenient to separate the averaging into two steps. In Section 6.5.3, we saw that, except for mirror-like surfaces, directional variations of the properties are not large, and data for such variations are sparse. More data are available for the hemispherical properties. Thus, as a first step, we directionally average by integrating with respect to θ and ϕ to give the *spectral* (or *monochromatic*) *hemispherical emittance* as

$$\varepsilon_\lambda(\lambda, T_s) = \frac{1}{\pi} \int_0^{2\pi} \int_0^{\pi/2} \varepsilon(\theta, \phi, \lambda, T_s) \cos\theta \sin\theta \, d\theta \, d\phi \qquad (6.79)$$

and subsequently average with respect to λ to give

$$\varepsilon(T_s) = \frac{1}{\sigma T_s^4} \int_0^\infty \varepsilon_\lambda(\lambda, T_s) \pi I_{b\lambda}(\lambda, T_s) \, d\lambda \qquad (6.80)$$

But from Eq. (6.72), the differential external fractional function is

$$df_e(\lambda T_s) = \frac{\pi I_{b\lambda}(\lambda, T_s) \, d\lambda}{\sigma T_s^4}$$

Thus,

$$\varepsilon(T_s) = \int_0^1 \varepsilon_\lambda(\lambda, T_s) \, df_e(\lambda T_s) \qquad (6.81)$$

Notation

There is no standard notation for the various surface properties, namely, spectral-directional, hemispherical, spectral, and total properties. The notation used here is consistent with many texts but has been simplified to avoid triple subscripts. An unsubscripted symbol, (e.g., ε) is a total hemispherical property and may be written $\varepsilon(T_s)$ when it is necessary to indicate its dependence on surface temperature. But unsubscripted ε is also used to denote the spectral-directional emittance: in this case, it is always written $\varepsilon(\theta, \phi, \lambda, T_s)$ to indicate its dependence on these parameters. Symbols subscripted λ [e.g., ε_λ or $\varepsilon_\lambda(\lambda, T_s)$] are spectral values that have been hemispherically averaged. Note also that subscripts θ and N are used for directional total properties in Section 6.5.3.

Total Hemispherical Absorptance

Averaging the spectral-directional absorptance (or reflectance) presents a more difficult problem, since the incident radiation is not unique but depends on the nature of the enclosure. If the enclosure surrounding the surface is taken to be black and isothermal at temperature T_e, then

$$\alpha(T_s, T_e) = \frac{1}{\sigma T_e^4} \int_0^\infty \int_0^{2\pi} \int_0^{\pi/2} \alpha(\theta, \phi, \lambda, T_s) I_{b\lambda}(\lambda, T_e) \cos\theta \sin\theta \, d\theta \, d\phi \, d\lambda$$

$$(6.82)$$

Notice that the important temperature here is T_e, which characterizes the spectrum of the incident radiation; temperature T_s characterizes the surface thermodynamic state and usually has a less significant effect on $\alpha(T_s, T_e)$. Again, the averaging is conveniently done in two steps:

$$\alpha_\lambda(\lambda, T_s) = \frac{1}{\pi} \int_0^{2\pi} \int_0^{\pi/2} \alpha(\theta, \phi, \lambda, T_s) \cos\theta \sin\theta d\theta d\phi \tag{6.83}$$

$$\alpha(T_s, T_e) = \frac{1}{\sigma T_e^4} \int_0^\infty \alpha_\lambda(\lambda, T_s) \pi I_{b\lambda}(\lambda, T_e) d\lambda = \int_0^1 \alpha_\lambda(\lambda, T_s) df_e(\lambda T_e) \tag{6.84}$$

Kirchoff's law requires $\alpha_\lambda(\lambda, T_s) = \varepsilon_\lambda(\lambda, T_s)$, so that data for only one of these properties are required.[7]

The heat flux through a surface at temperature T_s surrounded by a black isothermal enclosure at temperature T_e is

$$q = \int_0^\infty \int_0^{2\pi} \int_0^{\pi/2} \{\varepsilon(\theta, \phi, \lambda, T_s) I_{b\lambda}^+(\lambda, T_s)$$

$$- \alpha(\theta, \phi, \lambda, T_s) I_{b\lambda}^-(\lambda, T_e)\} \cos\theta \sin\theta d\theta d\phi d\lambda$$

$$= \varepsilon(T_s)\sigma T_s^4 - \alpha(T_s, T_e)\sigma T_e^4 \tag{6.85}$$

and we see that the emittance ε used for diffuse gray surfaces in Section 6.3 is correctly interpreted as the total hemispherical value.

Internal Emittance

As T_e approaches T_s, $\alpha(T_s, T_e)$ does not approach $\varepsilon(T_s)$. Edwards [3] shows how to obtain the correct result. When T_e approaches T_s, Eq. (6.85) may be written

$$q = \int_0^\infty \alpha_\lambda(\lambda, T_s)\{\pi I_{b\lambda}(\lambda, T_s) - \pi I_{b\lambda}(\lambda, T_e)\}d\lambda$$

$$= \int_0^\infty \alpha_\lambda(\lambda, T_s)\frac{\partial E_{b\lambda}(T_s)}{\partial T_s}(T_s - T_e)d\lambda \tag{6.86}$$

The *internal total hemispherical emittance* is defined as

$$\varepsilon^i(T_s) = \frac{1}{4\sigma T_s^3} \int_0^\infty \alpha_\lambda(\lambda, T_s)\frac{\partial E_{b\lambda}(T_s)}{\partial T_s}d\lambda \tag{6.87}$$

Using Eq. (6.74) gives $\varepsilon^i(T_s)$ in terms of the internal fractional function:

$$\varepsilon^i(T_s) = \int_0^1 \alpha_\lambda(\lambda, T_s)df_i(\lambda T_s) \tag{6.88}$$

---

[7] Often the irradiation is from a source that cannot be reasonably approximated by a black or gray surface, for example, a mercury lamp. Exercises 6–110 and 6–111 show how to calculate total absorptance in such situations.

Substituting Eq. (6.87) in Eq. (6.86) gives

$$q = \varepsilon^i(T_s) 4\sigma T_s^3 (T_s - T_e) = \alpha^i(T_s) 4\sigma T_s^3 (T_s - T_e) \qquad \textbf{(6.89)}$$

and we see that the internal total hemispherical emittance is the appropriate limit for an isothermal enclosure. The situation where T_s is nearly equal to T_e occurs often inside enclosures such as rooms and space vehicles: hence the use of the adjective *internal*.

The Diffuse Gray Surface Approximation

We are now in a position to define a *gray* surface rigorously: it is a surface for which the spectral radiation properties are independent of wavelength. In the case of a *diffuse gray* surface, $\varepsilon_\lambda(\lambda, T_s)$ and $\alpha_\lambda(\lambda, T_s)$ are independent of wavelength. It follows that when these properties are spectrally averaged, the resulting properties are independent of temperature, that is, $\varepsilon(T_s)$ and $\alpha(T_s, T_e)$ are independent of temperature. Notice that, in general, the radiation incident on a surface is not directionally and spectrally distributed as blackbody radiation, and thus $\alpha(T_s, T_e)$ as given by Eq. (6.84) is not necessarily an appropriate value of the total hemispherical absorptance. However, in using the diffuse gray surface approximation we take $\alpha = \varepsilon$. The engineer must then simply choose a suitable common value for the conditions under consideration.

Spectral Property Data

Measuring surface radiation properties clearly requires considerable effort owing to the great variety of surfaces encountered in engineering practice. The choice of which property to measure depends on both the desired application and convenience. Data available in the heat transfer literature might be spectral or total values,

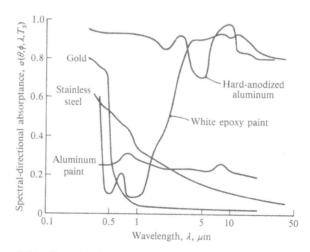

Figure 6.31   Spectral absorptance $\alpha_{\lambda,N}$ as a function of wavelength λ for selected surfaces. $T_s \approx 300\,\text{K}$, and $\theta = 25°$.

directional or hemispherical, and the total values may be internal or external averages. Usually, either the emittance or reflectance is measured, and the other properties are obtained by subtraction or from Kirchhoff's law. Table A.6a gives normal-incidence spectral and total absorptances for metals, from which approximate hemispherical values can be obtained using Fig. 6.28, with a value of $k/n = 1$ being appropriate for metals. Table A.6b and Fig. 6.31 give spectral absorptances of some selected surfaces. The data in Tables A.6a and A.6b are for a surface at room temperature, $T_s \simeq 300$ K. However, for many engineering purposes, spectral values of α_λ (or ε_λ) can be taken to be independent of temperature over quite a wide temperature range. Hence, if ε_λ increases with decreasing wavelength, ε will increase with temperature owing to the relatively greater emission at smaller wavelengths with increasing temperature.

EXAMPLE 6.19   Calculation of the Total Absorptance of White Epoxy Paint

Calculate the total absorptance of a surface at 300 K coated with white epoxy paint and exposed to a blackbody radiation source at 2000 K. Use the data in Table A.6b.

Solution

Given: White epoxy paint surface at 300 K.

Required: Total absorptance to blackbody radiation at 2000 K.

Assumptions: Incident radiation at 25° from the normal to the surface, so as to permit use of Table A.6b.

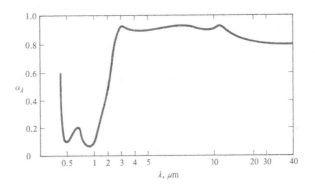

The data for α_λ of white epoxy paint from Table A.6b are plotted in the accompanying graph. We will assume that, for a nonconductor such as epoxy paint, the hemispherically averaged absorptance is approximately equal to the value at 25°. Then Eq. (6.84) gives

$$\alpha(T_s, T_e) = \frac{1}{\sigma T_e^4} \int_0^\infty \alpha_\lambda(\lambda, T_s)\pi I_{b\lambda}(\lambda, T_e)d\lambda = \int_0^1 \alpha_\lambda(\lambda, T_s)df_e(\lambda T_e)$$

Using Table 6.5 and the graph of α_λ, the following table is constructed to effect integration using the trapezoidal rule.

f	Δf	$\dfrac{(\lambda T)_e}{\mu m\,K}$	$\dfrac{\lambda}{\mu m}$	$\alpha_\lambda(\lambda, 300\,K)$
0.0025	0.005	0.1230×10^4	0.615	0.16
0.0075	0.005	0.1395×10^4	0.698	0.21
0.0125	0.005	0.1495×10^4	0.748	0.15
$\vdots$				
0.9925	0.005	2.56×10^4	12.8	0.92
0.9975	0.005	7.34×10^4	36.7	0.80

$$\alpha(300\,K, 2000\,K) = \sum \alpha_\lambda(\lambda, 300\ K)\Delta f_e = 0.53$$

Comments

Since epoxy paint is a nonconductor, the hemispherical total absorptance will be at most 4% lower than this value.

EXAMPLE 6.20   Radiant Heat Transfer inside a Spacecraft

An evacuated space inside a spacecraft is cylindrical in shape with a diameter of 20 cm and a height of 30 cm. The cylindrical walls and top surface can be taken to be black and are maintained at 280 K. The base is hard-anodized aluminum. Estimate the heat loss through the base when it is 300 K.

Solution

Given: Cylindrical enclosure, anodized aluminum base.

Required: Heat flow through base.

Assumptions: Walls and top surface are perfectly black.

Equation (6.89) gives the heat flux through the base surface as

$$q = \varepsilon^i(T_s)4\sigma T_s^3(T_s - T_e)$$

We will first assume that the data in Table A.6*b* for absorptance to radiation incident at 25° can be used to approximate the hemispherically averaged value, and we will make a correction later. From Eq. (6.88),

$$\varepsilon^i(T_s) = \int_0^1 \alpha_\lambda(\lambda, T_s)df_i(\lambda T_s)$$

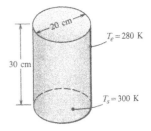

Using Tables 6.5 and A.6*b*, the following table is constructed to effect integration using the trapezoidal rule:

f	Δf	$\frac{(\lambda T)_i}{\mu m\,K}$	$\frac{\lambda}{\mu m}$	$\alpha_\lambda(\lambda, 300\,K)$
0.0025	0.005	0.1073×10^4	3.58	0.82
0.0075	0.005	0.1209×10^4	4.03	0.74
0.0125	0.005	0.1291×10^4	4.30	0.73
.				
.				
.				
0.9925	0.005	1.70×10^4	56.7	0.80
0.9975	0.005	2.49×10^4	83	0.80

$$\varepsilon^i(300\,K) = \sum \alpha_\lambda(\lambda, 300\,K)\Delta f_i$$
$$= 0.87$$

For a high-emittance surface, the hemispherical average will be about 0.03 lower than this value; hence, we have

$$\varepsilon^i(300\,K) = 0.87 - 0.03 = 0.84$$

$$\dot{Q} = qA = \varepsilon^i(T_s)4\sigma T_s^3(T_s - T_e)(\pi D^2/4)$$
$$= (0.84)(4)(5.67 \times 10^{-8})(300)^3(300 - 280)(\pi/4)(0.2)^2$$
$$= 3.2\,W$$

Comments

Note that Table A.5a gives $\varepsilon = 0.80$ for hard-anodized aluminum.

6.7 RADIATION TRANSFER THROUGH GASES

In the preceding sections of Chapter 6, we considered radiation exchange between surfaces separated by perfectly transparent or *nonparticipating* media. Such media include a vacuum, as well as air at normal temperatures and pressures. Symmetrical gas molecules, such as N_2 and O_2, emit and absorb negligible radiation unless temperatures are high enough for electronic excitation or ionization to occur–for example, behind the bow shock wave of a reentry vehicle. Gas species with nonsymmetrical molecules can emit and absorb radiation; the most important of such species are H_2O, CO_2, CH_4, CO, SO_2, and NH_3. (However, the small amounts of H_2O and other trace gases in atmospheric air means that their presence is usually ignored unless the path length is long, e.g., on the order of tens of meters; or if the wavelengths of the radiation under consideration coincide with strongly absorbing or emitting bands). Gas radiation is particularly significant to the design of furnaces and combustion chambers with carbon, hydrogen, or hydrocarbon fuels. A gas can also participate in the radiation exchange process by virtue of containing an *aerosol*. Such aerosols include liquid droplets, dust particles, and, of particular importance, soot particles in combustion products. Not only can radiation be absorbed and emitted by aerosols, but radiation can also be *scattered*. The participation of aerosols in radiation exchange will not be treated here.

Section 6.7.1 derives the *equation of transfer* for radiation propagating through a participating medium. Section 6.7.2 presents data and a methodology for evaluating total gas properties for CO_2, H_2O, and CO_2-H_2O mixtures. In Section 6.7.3, the effective and mean beam length concepts are formulated, and a simple prescription is given for evaluating the mean beam length. Section 6.7.4 analyzes radiation exchange between an isothermal nongray gas and a black enclosure, and Section 6.7.5 analyzes radiation exchange between an isothermal gray gas and a gray enclosure. Finally, in Section 6.7.6, an approximate result is given for radiation exchange between a nongray gas and a single-surface gray enclosure.

6.7.1 The Equation of Transfer

Consider a beam of radiation leaving a surface dA_s and propagating in direction x through a nonscattering medium, as shown in Fig. 6.32. The intensity of the beam is attenuated due to absorption by the medium and is augmented as a result of emission by the medium. A spectral energy balance on an elemental volume Δx thick requires that

$$I_\lambda \Delta\lambda|_{x+\Delta x} - I_\lambda \Delta\lambda|_x = \left[\begin{array}{cc}\text{Emission per} & \text{Absorption per} \\ \text{unit length} & \text{unit length}\end{array}\right]\Delta\lambda\,\Delta x \qquad \textbf{(6.90)}$$

The absorption per unit length is $\kappa_\lambda I_\lambda$, where κ_λ is the *spectral absorption coefficient* of the medium and has units m^{-1}. If the medium is in local thermodynamic equilibrium, the emission per unit length must be $\kappa_\lambda I_{b\lambda}$ so that if there is complete thermodynamic equilibrium, with $I_\lambda = I_{b\lambda}$, absorption will equal emission. Substituting in Eq. (6.90),

$$I_\lambda \Delta\lambda|_{x+\Delta x} - I_\lambda \Delta\lambda|_x = (\kappa_\lambda I_{b\lambda} - \kappa_\lambda I_\lambda)\Delta\lambda\,\Delta x$$

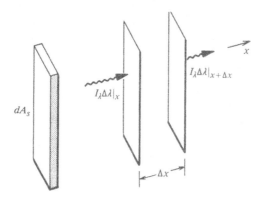

Figure 6.32   Attenuation of a beam of radiation in a nonscattering medium.

Dividing by $\Delta\lambda\Delta x$ and letting $\Delta x \to 0$ gives

$$\frac{dI_\lambda}{dx} = -\kappa_\lambda I_\lambda + \kappa_\lambda I_{b\lambda} \tag{6.91}$$

which is the *equation of transfer* for a nonscattering medium on a spectral basis.

For an isothermal medium, $I_{b\lambda}$ and κ_λ are constant so that Eq. (6.91) can be integrated to give

$$I_\lambda(x) = I_{\lambda,s}e^{-\kappa_\lambda x} + I_{b\lambda}\left(1 - e^{-\kappa_\lambda x}\right) \tag{6.92}$$

where $I_{\lambda,s}$ is the spectral intensity of the radiation leaving the surface at $x = 0$. In particular, at $x = L$,

$$I_\lambda(L) = I_{\lambda,s}e^{-\kappa_\lambda L} + I_{b\lambda}\left(1 - e^{-\kappa_\lambda L}\right) \tag{6.93}$$

The first term on the right-hand side is the radiation intensity at $x = L$ due to transmission of radiation leaving the surface at $x = 0$; the second term is the radiation intensity at $x = L$ due to emission along the path. Thus, we write

$$I_\lambda(L) = \tau_{g\lambda}I_{\lambda,s} + \varepsilon_{g\lambda}I_{b\lambda} \tag{6.94}$$

where $\tau_{g\lambda} = e^{-\kappa_\lambda L}$ and $\varepsilon_{g\lambda} = 1 - e^{-\kappa_\lambda L}$ are the *spectral* (or monochromatic) *transmissivity* and *emissivity*, respectively, of gas along the path $(0, L)$. Since there is no reflection, the *spectral absorptivity* $\alpha_{g\lambda} = 1 - \tau_{g\lambda}$, and Kirchhoff's law $\varepsilon_{g\lambda} = \alpha_{g\lambda}$ is obeyed, as required by the formulation of the equation of transfer.

In general, the absorption coefficient κ_λ is strongly dependent on wavelength, as will be discussed for gaseous mixtures in Section 6.7.2. Nevertheless, it is often possible to perform satisfactory engineering calculations using total properties averaged over all wavelengths. Equation (6.93) can be written on a total basis as

$$I(L) = I_s e^{-\kappa L} + I_b\left(1 - e^{-\kappa L}\right) \tag{6.95}$$

or

$$I(L) = \tau_g I_s + \varepsilon_g I_b \tag{6.96}$$

For continuum radiation, where κ_λ varies smoothly with wavelength, as is the case for a cloud containing a distribution of particle sizes, the calculation of a total property is straightforward. The total emissivity is simply

$$\varepsilon_g = \frac{\int_0^\infty I_\lambda d\lambda}{\int_0^\infty I_{b\lambda} d\lambda} = \int_0^1 \left[1 - e^{-\kappa_\lambda L}\right] df_e(\lambda T) \tag{6.97}$$

where f_e is the external fractional function. For the radiation emitted by a molecular gas, the averaging process to obtain ε_g is more complicated.

6.7.2 Gas Radiation Properties

Gaseous radiation is particularly relevant to heat transfer in furnaces and combustion chambers burning carbon, hydrogen, or hydrocarbon fuels. Absorption and emission by carbon dioxide and water vapor is of particular concern due to their relatively high concentrations in the products of combustion, and due to their strong radiation absorption. Figure 6.33 shows some spectral absorptivity data for CO_2, H_2O, and

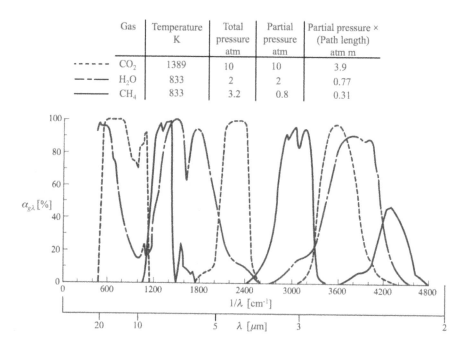

Gas	Temperature K	Total pressure atm	Partial pressure atm	Partial pressure × (Path length) atm m
------ CO_2	1389	10	10	3.9
--- H_2O	833	2	2	0.77
—— CH_4	833	3.2	0.8	0.31

Figure 6.33 Spectral absorptivity $\alpha_{g\lambda}$ [%] as a function of wavenumber $1/\lambda$ and wavelength λ for carbon dioxide, water vapor, and methane [3].

CH_4. The absorption (or emission) does not take place continuously over the entire spectrum but rather occurs in a number of moderately wide *bands* of relatively strong absorption. The radiation properties are obviously nongray. Accurate radiation exchange calculations generally require that the nongray behavior be properly accounted for, albeit approximately, using *band models*. As mentioned before, often satisfactory engineering calculations can be made on a total basis, and for this purpose total gas properties are required. The absorption bands of gases are actually arrays of lines at discrete wavelengths. The averaging process to obtain total gas properties is complicated, and only the results will be presented here. The *Hottel charts*, first presented by H. Hottel in 1927, have been widely used but are of limited accuracy [11–13]. Edwards and Matavosian [14,15] have developed calculation procedures that give more accurate results, and their charts and methods are presented here. A recent review on the treatment of participating gases, including a historical perspective, can be found in [16].

Total Emissivity Charts

The total emissivity of a gas, ε_g, is a function of its temperature T_g due to nongrayness and density variation. Also, since the absorption coefficient κ is defined on a length basis, emissivity depends on partial pressure of the emitting gaseous species, P_a, times the thickness of the gas layer, L. The total emissivity can also depend on total pressure, P, due to a phenomenon called *collision line broadening*. Figures 6.34a and 6.35a give ε_g for CO_2 and H_2O in N_2 at an **equivalent broadening**

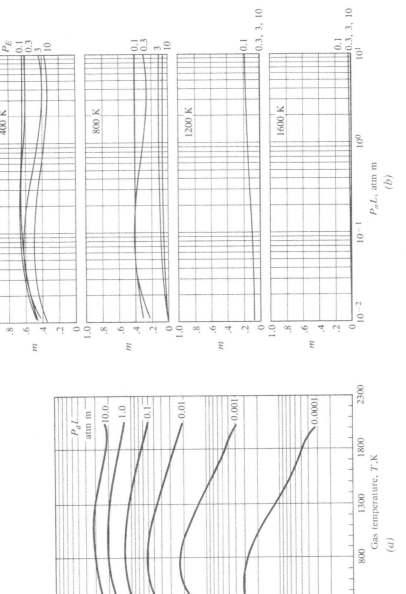

Figure 6.34 Gas radiation properties for CO_2 in N_2. (*a*) Gas emissivity ε_g at an equivalent broadening pressure ratio $P_E = 1$. (*b*) Total pressure scaling exponent m.

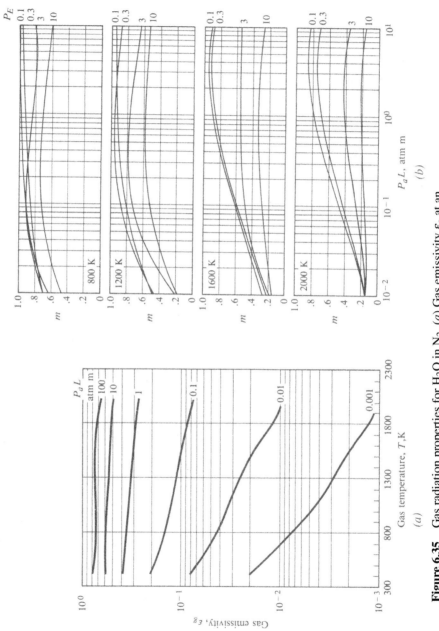

Figure 6.35 Gas radiation properties for H_2O in N_2. (a) Gas emissivity ε_g at an equivalent broadening pressure ratio $P_E = 1$. (b) Total pressure scaling exponent m.

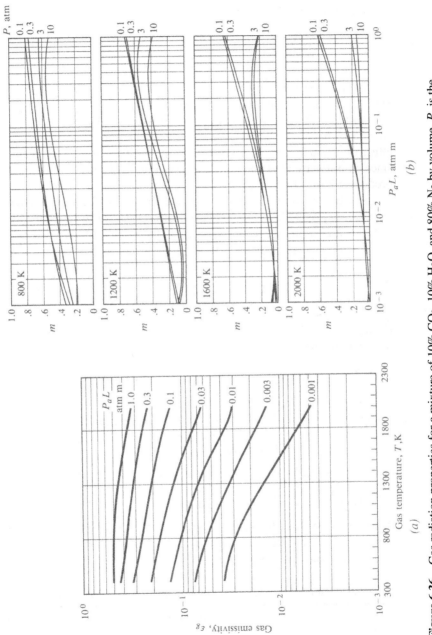

Figure 6.36 Gas radiation properties for a mixture of 10% CO_2, 10% H_2O, and 80% N_2 by volume. P_a is the partial pressure of H_2O only. (a) Gas emissivity at 1 atm total pressure, (b) Total pressure scaling exponent m.

pressure ratio P_E of unity; P_E is defined as

$$P_E = \left\{ \frac{P}{P_0} \left[1 + (b-1)\frac{P_a}{P} \right] \right\}^n \tag{6.98}$$

where P_0 is one atmosphere; b and n for CO_2 and H_2O are, respectively,

$$CO_2 : b = 1.3; \quad n = 0.8$$

$$H_2O : b = 8.6(T_0/T_g)^{1/2} + 0.5; \quad n = 1$$

where $T_0 = 100$ K. Figure 6.36a gives ε_g for a mixture of 10% CO_2, 10% H_2O and 80% N_2, by volume, at a total pressure of one atmosphere. For this *equimolar* CO_2-H_2O mixture, P_a is defined as the partial pressure of H_2O only. Changing the relative amount of N_2 or the presence of other nonradiating species, such as O_2, has little effect on ε_g; also, the precise relative amount of CO_2 is of secondary importance, since emission by H_2O dominates. Hence, Fig. 6.36a can be used for the combustion products of most hydrocarbon fuels.

The effect of total pressure can be accounted for by using a simple scaling rule. For CO_2 and H_2O, the desired value of ε_g for a given $P_a L$ is obtained from the values of ε_g for $P_a L'$ in Figs. 6.34 and 6.35 using the scaling rule (see Example 6.21):

$$\varepsilon_g(T_g, P_a L, P_E) = \varepsilon_g(T_g, P_a L', P_E = 1) \tag{6.99a}$$

where

$$P_a L' = P_a L P_E^m \tag{6.99b}$$

For the CO_2–H_2O mixture, the rule is of the same form, with P_E replaced by P [atm], the total pressure. The exponent m is given in Figs. 6.34b, 6.35b, and 6.36b.

Total Absorptivity

The total absorptivity of a gas at temperature T_g to radiation from a black source at temperature T_s can be estimated using a second scaling law. For CO_2 and H_2O,

$$\alpha_g(T_g, T_s, P_a L, P_E) = (T_g/T_s)^{1/2} \varepsilon_g(T_s, P_a L'_\alpha, P'_E) \tag{6.100a}$$

$$P_a L'_\alpha = P_a L(T_s/T_g)^r \tag{6.100b}$$

$$P'_E = P_E(T_g/T_s)^s \tag{6.100c}$$

Again, P_E is replaced by total pressure P for the CO_2-H_2O mixture. The exponents r and s are given in Table 6.6. The two-step scaling required to obtain the absorptivity is illustrated in Examples 6.21 and 6.22.

Table 6.6 Path and pressure exponents for gas total absorptivity scaling.

Gas	Path Exponent r	Pressure Exponent s
CO_2	1.0	2.4
H_2O	1.5	1.6
CO_2/H_2O	1.5	1.6

The Computer Program RAD3

RAD3 calculates total gas properties using the data and methodology recommended by Edwards and Matavosian. The original data were curve-fitted, and these curve fits were also used to prepare Figs. 6.34 through 6.36. Thus, hand calculations using these figures should agree with the output of RAD3. The output of RAD3 is ε_{g1}, ε_{g2}, α_{g1}, and α_{g2}, where subscripts 1 and 2 refer to emission or absorption for path lengths L and $2L$, respectively. Example 6.26 describes the use of these properties for the calculation of radiation exchange between an isothermal nongray gas and a single gray surface enclosure. Note that RAD3 sets $P_E = 10$ if $P_E > 10$, in order to calculate the total pressure scaling exponent m: extrapolation of the data for m is not recommended (see Example 6.26). RAD3 cannot be used to extrapolate ε_g outside of the temperature ranges of Figs. 6.34a, 6.35a and 6.36a.

EXAMPLE 6.21   Total Properties of Hydrogen Combustion Products

Exhaust gases from a combustor burning hydrogen are at 1200 K and 2 atm pressure, and they contain 10% (by volume) H_2O, the remainder being N_2. Calculate the total emissivity of the gas and its total absorptivity to black radiation from a surface at 800 K. Use a path length of 0.5 m.

Solution

Given: Mixture of H_2O and N_2.

Assumptions: The scaling laws of Section 6.7.2 are valid.

We first calculate the equivalent broadening pressure ratio from Eq. (6.98):

$$P_E = \left\{ \frac{P}{P_0} \left[1 + (b-1)\frac{P_a}{P} \right] \right\}^n$$

$$P_0 = 1\,\text{atm}$$

$$b = 8.6(T_0/T_g)^{1/2} + 0.5 = 8.6(100/1200)^{1/2} + 0.5 = 2.98$$

$$n = 1$$

$$P_E = \left\{ \frac{2}{1} \left[1 + (2.98 - 1)\frac{(0.1)(2.0)}{2.0} \right] \right\}^1 = 2.40$$

Next we calculate P_aL' from Eq. (6.99b):

$$P_aL' = P_aLP_E^m$$

$P_aL = (0.1)(2.0)(0.5) = 0.1$ atm m; from Fig. 6.35b, the exponent m is 0.74; thus

$$P_aL' = (0.1)(2.40)^{0.74} = 0.191 \text{ atm m}$$

Then, from Eq. (6.99a) and Fig. 6.35a,

$$\varepsilon_g = \varepsilon_g(T_g, P_aL') = \varepsilon_g(1200, 0.191) = 0.16$$

To determine the total absorptivity, we first calculate P'_E from Eq. (6.100c);

$$P'_E = P_E(T_g/T_s)^s, \quad s = 1.6 \text{ from Table 6.6}$$

$$P'_E = 2.40(1200/800)^{1.6} = 4.59$$

We then calculate $P_aL'_\alpha$ from Eq. (6.100b):

$$P_aL'_\alpha = P_aL(T_s/T_g)^r, \quad r = 1.5 \text{ from Table 6.6}$$

$$P_aL'_\alpha = 0.1(800/1200)^{1.5} = 0.0544 \text{ atm m}$$

Equation (6.100a) for α_g, requires that we next calculate from Eqs. (6.99a) and (6.99b)

$$\varepsilon_g(T_s, P_aL'_\alpha, P'_E) = \varepsilon_g(T_s, P_aL''_\alpha, P'_E = 1)$$

$$P_aL''_\alpha = P_aL'_\alpha(P'_E)^{m'}$$

where the notations L''_α and m' denote a second application of the scaling rule. For $T_s = 800$ K, $P_aL'_\alpha = 0.0544$ atm m and $P'_E = 4.59$, Fig. 6.35b gives $m' = 0.81$.

$$P_aL''_\alpha = (0.0544)(4.59)^{0.81} = 0.187 \text{ atm m}$$

Figure 6.35a gives $\varepsilon_g(800, 0.187) = 0.20$. Equation (6.100a) gives

$$\alpha_g = (T_g/T_s)^{1/2}(0.20) = (1200/800)^{1/2}(0.20) = 0.24$$

Comments

1. This result would remain valid if the exhaust gases contained oxygen as a result of using excess air in the combustion process.

2. Interpolation on the graphs can be rather difficult, and use of RAD3 is thus recommended: RAD3 gives $\varepsilon_g = 0.164$, $\alpha_g = 0.237$.

EXAMPLE 6.22   Total Properties of a Hydrocarbon Fuel Combustion Products

Exhaust gas from a combustor burning a hydrocarbon fuel is at 1600 K and 3 atm pressure. The composition can be approximated as 10% CO_2, 10% H_2O, and 80% N_2, by volume. Calculate the total emissivity for a path length of 0.34 m and the total absorptivity for radiation from a black wall at 800 K.

Solution

Given: Combustion products of a hydrocarbon fuel.

Required: Total gas emissivity and absorptivity.

Assumptions: Scaling laws of Section 6.7.2 are valid.

We first calculate P_aL' from Eq. (6.99b), where for the CO_2-H_2O mixture, P_E is replaced by total pressure, P.

$$P_aL' = P_aLP^m$$

$P_a = (0.1)(3) = 0.3 \text{ atm}; P_a L = (0.3)(0.34) = 0.102 \text{ atm m}$. From Fig. 6.36b, $m = 0.26$.

$$P_a L' = (0.102)(3)^{0.26} = 0.136 \text{ atm m}$$

Then, from Eq. (6.99a) and Fig. 6.36a,

$$\varepsilon_g = \varepsilon_g(T_g, P_a L') = \varepsilon_g(1600, 0.136) = 0.19$$

To determine the total absorptivity, we first calculate P' from Eq. (6.100c), with total pressure P replacing P_E:

$$P' = P(T_g/T_s)^s, \quad s = 1.6 \text{ from Table 6.6}$$

$$P' = 3(1600/800)^{1.6} = 9.1 \text{ atm}$$

We then calculate $P_a L'_\alpha$ from Eq. (6.100b):

$$P_a L'_\alpha = P_a L(T_s/T_g)^r, \quad r = 1.5 \text{ from Table 6.6}$$

$$P_a L'_\alpha = (0.102)(800/1600)^{1.5} = 0.036 \text{ atm m}$$

Equation (6.100a) for α_g requires that we next calculate

$$\varepsilon_g(T_s, P_a L'_\alpha, P') = \varepsilon_g(T_s, P_a L''_\alpha, P' = 1)$$

$$P_a L''_\alpha = P_a L'_\alpha (P')^{m'}$$

For $T_s = 800$ K, $P_a L'_\alpha = 0.036$ atm m, $P' = 9.1$ atm, Fig. 6.36b gives $m' = 0.41$.

$$P_a L''_\alpha = (0.036)(9.1)^{0.41} = 0.0890 \text{ atm m}$$

Figure 6.36a gives $\varepsilon_g(800, 0.0890) = 0.24$. Then Eq. (6.100a) gives

$$\alpha_g = (T_g/T_s)^{1/2}(0.240) = (1600/800)^{1/2}(0.24) = 0.34$$

Comments

1. This result would be unaffected if the relative amount of CO_2 were somewhat different. An equimolar mixture of CO_2 and H_2O in the products requires that the overall chemical formula of the hydrocarbon fuel be $C_n H_{2n}$, which is not strictly true for most hydrocarbon fuels.

2. RAD3 gives $\varepsilon_g = 0.194$, $\alpha_g = 0.353$.

6.7.3 Effective Beam Lengths for an Isothermal Gas

We will develop the concept of an effective beam length for radiation in a participating medium on a total basis; however, identical results can be obtained on a band or spectral basis. Consider two surface elements dA_1 and dA_2 of an enclosure containing an isothermal gas at temperature T_g as shown in Fig. 6.37. The contribution to irradiation on surface dA_1 due to radiation received in the cone of solid angle $d\omega$ subtended by dA_2 is

$$dG_1 = I_1^- \cos\theta_1 d\omega \tag{6.1}$$

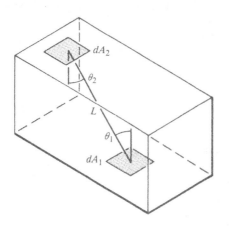

Figure 6.37   Elemental surfaces for radiation in an enclosure containing an isothermal gas.

Equation (6.95) gives

$$I_1^- = I_2^+ e^{-\kappa L} + I_{bg}(1 - e^{-\kappa L})$$

Thus,

$$dG_1 = [I_2^+ e^{-\kappa L} + I_{bg}(1 - e^{-\kappa L})]\cos\theta_1 d\omega \tag{6.2}$$

For a finite surface A_2, Eq. (6.2) is integrated over the solid angle containing surface A_2:

$$dG_1 = \int_{\omega_2} [I_2^+ e^{-\kappa L} + I_{bg}(1 - e^{-\kappa L})]\cos\theta_1 d\omega \tag{6.3}$$

Introducing $d\omega = dA_2 \cos\theta_2/L^2$, $I_2^+ = J_2/\pi$, $I_{bg} = E_{bg}/\pi$ gives

$$dG_1 = \int_{A_2} [J_2 e^{-\kappa L} + E_{bg}(1 - e^{-\kappa L})]\frac{\cos\theta_1 \cos\theta_2}{\pi L^2}dA_2 \tag{6.4}$$

For a finite area A_1, Eq. (6.4) is averaged over area A_1:

$$G_{12} = \frac{1}{A_1}\int_{A_1}\int_{A_2} [J_2 e^{-\kappa L} + E_{bg}(1 - e^{-\kappa L})]\frac{\cos\theta_1 \cos\theta_2}{\pi L^2}dA_2 dA_1 \tag{6.5}$$

where G_{12} is the contribution to irradiation of surface 1 by radiation along beams from surface 2. The beam length L varies over the surface: we define a *constant effective beam length* $\mathcal{L}_{12}$ such that

$$G_{12} = [J_2 e^{-\kappa\mathcal{L}_{12}} + E_{bg}(1 - e^{-\kappa\mathcal{L}_{12}})]\frac{1}{A_1}\int_{A_1}\int_{A_2}\frac{\cos\theta_1 \cos\theta_2}{\pi L^2}dA_2 dA_1$$

Using Eq. (6.65), which defines the shape factor F_{12},

$$G_{12} = [J_2 e^{-\kappa\mathcal{L}_{12}} + E_{bg}(1 - e^{-\kappa\mathcal{L}_{12}})]F_{12} \tag{6.6}$$

and comparing Eqs. (6.5) and (6.6) shows that the effective beam length $\mathcal{L}_{12}$ is given by

$$e^{-\kappa \mathscr{L}_{12}} F_{12} = \frac{1}{A_1} \int_{A_1} \int_{A_2} e^{-\kappa L} \frac{\cos \theta_1 \cos \theta_2}{\pi L^2} dA_2 dA_1 \tag{6.7}$$

Sometimes is it possible to model a furnace or combustion chamber as a single-surface enclosure, that is, with a uniform wall temperature and emittance: then, with $A_1 = A_s$, $F_{12} = 1$, Eq. (6.6) becomes

$$G_s = [J_s e^{-\kappa \mathscr{L}_m} + E_{bg}(1 - e^{-\kappa \mathscr{L}_m})] \tag{6.8}$$

where $\mathscr{L}_m$ is called the **mean beam length** of the enclosure and is given by

$$e^{-\kappa \mathscr{L}_m} = \frac{1}{A_s} \int_{A_s} \int_{A_s'} e^{-\kappa L} \frac{\cos \theta \cos \theta'}{\pi L^2} dA_s' dA_s \tag{6.9}$$

The mean beam length is simply the effective beam length for a complete enclosure.

The introduction of an effective or mean beam length will be useful only if simple rules for their evaluation can be formulated. In general, $\mathscr{L}$ depends on the *optical depth* of the gas, which is the absorption coefficient times a characteristic path length. In the limit $\kappa \to 0$, $\mathscr{L}$ is independent of κ, as can be seen by making the approximation $e^{-x} \simeq 1 - x$ in either Eq. (6.7) or Eq. (6.9), to give *geometric* effective or mean beam lengths $\mathscr{L}^0$,

$$\mathscr{L}_{12}^0 = \lim_{\kappa \to 0} \mathscr{L}_{12} = \frac{1}{A_1 F_{12}} \int_{A_1} \int_{A_2} \frac{\cos \theta_1 \cos \theta_2}{\pi L} dA_2 dA_1 \tag{6.110a}$$

$$\mathscr{L}_m^0 = \lim_{\kappa \to 0} \mathscr{L}_m = \frac{1}{A_s} \int_{A_s} \int_{A_s'} \frac{\cos \theta \cos \theta'}{\pi L} dA_s' dA_s \tag{6.110b}$$

Equation (6.110b) can be integrated to give the very simple result,

$$\mathscr{L}_m^0 = \frac{4V_g}{A_s} \tag{6.11}$$

where V_g is the volume of the gas in the enclosure. For a long duct of cross-sectional area A_c and perimeter $\mathscr{P}$, $\mathscr{L}_m^0 = 4A_c/\mathscr{P} = D_h$, the hydraulic diameter. The geometric mean beam length proves to be a good approximation for the actual mean beam length. For a sphere, the error is at worst 5.2% (high); for an infinite cylinder or slab, use of $\mathscr{L}_m = 0.9 \mathscr{L}_m^0$ for $\kappa \mathscr{L}_m^0 > 0.1$ gives an error of less than 7%. Table 6.7 presents formulas for the geometric effective beam length $\mathscr{L}_{12}^0$ for opposite and adjacent rectangles. As for the mean beam length, $\mathscr{L}_{12}$ is not much less than $\mathscr{L}_{12}^0$. The rules of shape factor algebra illustrated in Fig. 6.9 also apply to geometric effective beam lengths, as can be seen by rewriting Eq. (6.110a) as

$$\mathscr{L}_{ij}^0 A_i F_{ij} = \int_{A_i} \int_{A_j} \frac{\cos \theta_i \cos \theta_j}{\pi L} dA_j dA_i = \mathscr{L}_{ji}^0 A_j F_{ji}$$

Thus, to use Figure 6.9, set $\mathscr{G}_{ij}$ equal to $\mathscr{L}_{ij}^0 A_i F_{ij}$. The rules $\mathscr{G}_{ij} = \mathscr{G}_{ji}$ and $\mathscr{G}_{i(j+k)} = \mathscr{G}_{ij} + \mathscr{G}_{ik}$ follow.

Table 6.7 Geometric effective beam lengths $\mathscr{L}_{12}^0$ for opposite and adjacent rectangles [3, 17]. (See Table 6.1 for F_{12}).

1. Opposite rectangles

$X = a/c; \ Y = b/c$

$$\mathscr{L}_{12}^0 F_{12} = \frac{4c}{\pi XY}\left\{ XY\tan^{-1}\frac{XY}{(1+X^2+Y^2)^{1/2}} + X\ln\frac{X+(1+X^2+Y^2)^{1/2}}{[X+(1+X^2)^{1/2}](1+Y^2)^{1/2}} \right.$$

$$\left. + Y\ln\frac{Y+(1+X^2+Y^2)^{1/2}}{[Y+(1+Y^2)^{1/2}](1+X^2)^{1/2}} + (1+X^2)^{1/2} + (1+Y^2)^{1/2} - (1+X^2+Y^2)^{1/2} - 1 \right\}$$

2. Adjacent rectangles

$X = a/c; \ Y = b/c$

$$\mathscr{L}_{12}^0 F_{12} = \frac{c}{3\pi X}\left\{ 3X^2\ln\frac{[1+(1+X^2)^{1/2}][(X^2+Y^2)^{1/2}]}{X[1+(1+X^2+Y^2)^{1/2}]} + 3Y^2\ln\frac{[1+(1+Y^2)^{1/2}](X^2+Y^2)^{1/2}}{Y[1+(1+X^2+Y^2)^{1/2}]} \right.$$

$$+ 3X^2[(1+X^2+Y^2)^{1/2} - (X^2+Y^2)^{1/2} - (1+X^2)^{1/2}]$$

$$+ 3Y^2[(1+X^2+Y^2)^{1/2} - (X^2+Y^2)^{1/2} - (1+Y^2)^{1/2}]$$

$$\left. + (1+X^2)^{3/2} + (1+Y^2)^{3/2} + (X^2+Y^2)^{3/2} - (1+X^2+Y^2)^{3/2} + 2X^3 + 2Y^3 - 1 \right\}$$

EXAMPLE 6.23   Mean Beam Length in a Tube Bank

A tube bank has 4 cm–O.D. tubes with centers forming equilateral triangles of side length 7 cm. Determine the mean beam length for combustion products of coke at 1200 K and 6 atm total pressure.

Solution

Given: Tube bank with tubes in a triangular pattern.

Required: Mean beam length for coke combustion products.

Assumptions: Stoichiometric combustion.

The first step is to calculate the geometric mean beam length from Eq. (6.11):

$$\mathscr{L}_m^0 = \frac{4V_g}{A_s}$$

$$V_g = \left[\frac{1}{2}(7)(7^2 - 3.5^2)^{1/2} - \frac{1}{2}(\pi/4)(4)^2\right](1)$$

$$= 21.22 - 6.28 = 14.94 \text{ cm}^3$$

$$A_s = (3)(1/6)(\pi)(4)(1) = 6.28 \text{ cm}^2$$

$$\mathscr{L}_m^0 = \frac{4(14.94)}{6.28} = 9.51 \text{ cm}$$

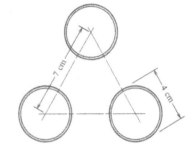

Next we must see if the product of absorption coefficient times path length is greater than or less than 0.1.

$$\varepsilon_g = 1 - e^{-\kappa L}, \quad \text{or} \quad \kappa L = \ln\left(\frac{1}{1 - \varepsilon_g}\right)$$

Where $L = \mathscr{L}_m^0$ for this purpose. The combustion reaction is

$$C + O_2 + (3.76N_2) \rightarrow CO_2 + (3.76N_2)$$

Assuming complete combustion, the partial pressure of CO_2 is

$$P_{co_2} = (1/4.76)(6) = 1.26 \text{ atm} = P_a$$

RAD3 can be used to calculate ε_g. The required input is:

$$2 : CO_2 \text{ in } N_2$$

$$P = 6$$

$$P_a = 1.62$$

$$T_g = 1200$$

$$T_s = \text{ any value}$$

$$\mathscr{L}_m^0 = 0.0951$$

The output is:

$$\varepsilon_g = 0.121$$

$$\kappa L = \ln\left(\frac{1}{1 - \varepsilon_g}\right) = 0.13 > 0.1$$

Hence, $\mathscr{L}_m \simeq 0.9\mathscr{L}_m^0 = (0.9)(0.0951) = 0.086$ m.

6.7.4 Radiation Exchange between an Isothermal Gas and a Black Enclosure

In a furnace or combustion chamber, turbulence ensures good mixing, and temperature variations are confined to thin boundary layers adjacent to the walls. Thus, as a first approximation, it is reasonable to assume that the gas as a whole is isothermal. Also, due to oxidation and soot deposits, the walls can be approximated as black surfaces when hydrocarbon fuels are used. Consider the enclosure of n black surfaces containing an isothermal gas at temperature T_g shown in Fig. 6.38. The radiant heat flux across the ith surface is

$$q_i = E_{bi} - \sum_{k=1}^{n} G_{ik} \tag{6.12}$$

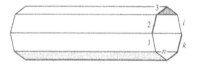

Figure 6.38 An enclosure of n black surfaces containing an isothermal gas.

Equation (6.6) can be written for the (i, k) black-surface pair as

$$G_{ik} = F_{ik}\left(\tau_g^{ik} E_{bk} + \varepsilon_g^{ik} E_{bg}\right)$$

where $\tau_g^{ik} = e^{-\kappa\mathscr{L}_{ik}}, \varepsilon_g^{ik} = 1 - e^{-\kappa\mathscr{L}_{ik}}$. Substituting in Eq. (6.12),

$$q_i = E_{bi} - \sum_{k=1}^{n} F_{ik}\left(\tau_g^{ik} E_{bk} + \varepsilon_g^{ik} E_{bg}\right) \tag{6.13}$$

which are n linear equations in the n unknown heat fluxes or wall temperatures. The total gas emissivity ε_g^{ik} is for gas at temperature T_g over an effective beam length of $\mathscr{L}_{ik}$; the total transmissivity τ_g^{ik} is for radiation from a black source at temperature T_k transmitted by a gas at temperature T_g over an effective beam length of $\mathscr{L}_{ik}$.

For an enclosure consisting of a single surface at temperature T_s, Eq. (6.13) reduces to

$$q = E_{bs} - \tau_g E_{bs} - \varepsilon_g E_{bg}$$

or

$$q = \alpha_g \sigma T_s^4 - \varepsilon_g \sigma T_g^4 \tag{6.14}$$

where ε_g is the total gas emissivity at temperature T_g over the mean beam length of the enclosure, and α_g is the total gas absorptivity for radiation from a black source at temperature T_s absorbed over the mean beam length by a gas at temperature T_g.

EXAMPLE 6.24   A Kerosene Combustor

Exhaust gas from a kerosene-fueled combustor is at 1600 K and 3 atm pressure, with a composition of 10% CO_2, 10% H_2O, and 80% N_2 by volume. If the mean beam length is 0.34 m, calculate the radiant heat flux to walls at 800 K. Assume that the walls are black.

Solution

Given: A combustor burning a hydrocarbon fuel.

Required: Radiant heat flux to wall.

Assumptions: The walls are black.

The total emissivity and absorptivity for this combustor were calculated in Example 6.22; RAD3 gave

$$\varepsilon_g = 0.194$$

$$\alpha_g = 0.353$$

Equation (6.14) gives the radiant heat flux:

$$q = \alpha_g \sigma T_s^4 - \varepsilon_g \sigma T_g^4$$

$$= (0.353)(5.67 \times 10^{-8})(800)^4 - (0.194)(5.67 \times 10^{-8})(1600)^4$$

$$= (8.2 - 72.1)10^3$$

$$= -63.9 \text{ kW/m}^2$$

Comments

Since the walls are relatively cold, radiation emitted by the walls and absorbed by the gas is of minor importance; hence, an accurate value of α_g is not required.

6.7.5 Radiation Exchange between an Isothermal Gray Gas and a Gray Enclosure

The calculation method of Section 6.7.4 applies to a nongray gas but is restricted to an enclosure of black surfaces to allow evaluation of the transmissivity τ_g^{ik}. When surface j is black, the spectral distribution of the emitted radiation is fixed by its temperature, and the transmissivity can be calculated using the rules given in Section 6.7.2. When surface j is gray, its radiosity has components of reflected radiation from other surfaces at different temperatures, and from the gas. Thus, the spectral distribution is complicated and unknown. In order to calculate the radiation exchange in an enclosure of gray surfaces containing an isothermal gas on a total basis, it is necessary to assume that the gas is gray as well, that is, $\varepsilon_g^{ik} = \alpha_g^{ik} = 1 - \tau_g^{ik}$. The heat flux across the ith surface is now

$$q_i = J_i - \sum_{k=1}^{n} G_{ik} \tag{6.15}$$

where

$$J_i = \varepsilon_i E_{bi} + (1 - \varepsilon_i)G_i \tag{6.16}$$

and

$$G_{ik} = F_{ik}(\tau_g^{ik} J_k + \varepsilon_g^{ik} E_{bg}) \tag{6.17}$$

Rearranging these equations into the form used for a gray enclosure in Section 6.3.5 gives

$$J_i = \varepsilon_i E_{bi} + (1 - \varepsilon_i) \sum_{k=1}^{n} F_{ik}[\tau_g^{ik} J_k + \varepsilon_g^{ik} E_{bg}] \tag{6.18}$$

$$\dot{Q}_i = \frac{\varepsilon_i A_i}{1 - \varepsilon_i}(E_{bi} - J_i) \tag{6.19}$$

Equations (6.18) and (6.19) can be solved for the unknown $\dot{Q}_i$ or T_i, if T_g and either $\dot{Q}_i$ or T_i are specified for each surface.

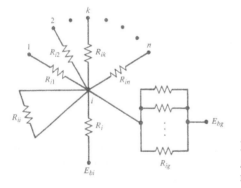

Figure 6.39   An isothermal gray gas in a gray enclosure: network connections for node i.

An equivalent electrical network can be constructed for the gray enclosure containing a gray gas. The surface resistances from Eq. (6.19) are as before,

$$R_i = \frac{1 - \varepsilon_i}{\varepsilon_i A_i} \tag{6.20}$$

Equations (6.15) and (6.17) can be arranged as

$$\dot{Q}_i = \sum_{k=1}^{n} A_i F_{ik} \tau_g^{ik}(J_i - J_k) + \sum_{k=1}^{n} A_i F_{ik} \varepsilon_g^{ik}(J_i - E_{bg}) \tag{6.21}$$

since for a gray gas $1 - \tau_g^{ik} = \alpha_g^{ik} = \varepsilon_g^{ik}$. The space resistances are thus

$$R_{ik} = \frac{1}{A_i F_{ik} \tau_g^{ik}} \tag{6.22}$$

and there is a parallel set of radiosity node to gas resistances:

$$R_{ig} = \frac{1}{\sum\limits_{k=1}^{n} A_i F_{ik} \varepsilon_g^{ik}} \tag{6.23}$$

It is essential that R_{ii} resistances be included when F_{ii} is not zero, that is, when a surface sees itself. Figure 6.39 shows the network connections for node i, while Fig. 6.40 shows a complete network for a two-surface enclosure.

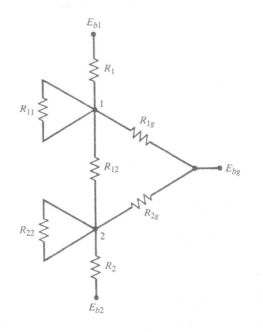

Figure 6.40 An isothermal gray gas in a two-gray-surface enclosure: the equivalent network.

EXAMPLE 6.25 Radiation Exchange in a Two-Surface Enclosure Containing a Gray Gas

An absorbing gas is contained between two large parallel plates, 1 and 2, with $T_1 = 1200$ K, $\varepsilon_1 = 0.8; T_2 = 800$ K, $\varepsilon_2 = 0.7$. The gas is assumed to be gray with $\varepsilon_g = 0.4$. Determine the effect of the gas on the radiation heat transfer between the two plates.

Solution

Given: Radiating gas contained between parallel plates.

Required: Effect of gas on radiation heat transfer.

Assumptions: 1. Gray gas, $\varepsilon_g^{12} = 1 - \tau_g^{12}$
2. Large plates, $F_{12} = 1$
3. Convection has a minor effect on the bulk gas temperature.

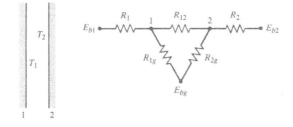

Referring to the equivalent network, the various resistances on a unit-area basis with $F_{12} = 1$ are

$$R_1 = \frac{1 - \varepsilon_1}{\varepsilon_1} = \frac{1 - 0.8}{0.8} = 0.25; \quad R_2 = \frac{1 - 0.7}{0.7} = 0.429$$

$$R_{12} = \frac{1}{1 - \varepsilon_g^{12}} = \frac{1}{1 - 0.4} = 1.667$$

$$R_{1g} = \frac{1}{\varepsilon_g^{12}} = \frac{1}{0.4} = 2.5; \quad R_{2g} = \frac{1}{\varepsilon_g^{12}} = \frac{1}{0.4} = 2.5$$

Also,

$$E_{b1} = \sigma T_1^4 = (5.67 \times 10^{-8})(1200)^4 = 117.6 \text{ kW/m}^2$$

$$E_{b2} = \sigma T_2^4 = (5.67 \times 10^{-8})(800)^4 = 23.2 \text{ kW/m}^2$$

Combining the space and gas resistances gives

$$R = \frac{1}{1/1.667 + 1/(2.5 + 2.5)} = 1.25$$

The heat transfer per unit area of plate is then

$$\frac{\dot{Q}_{12}}{A} = \frac{E_{b1} - E_{b2}}{\Sigma R} = \frac{117.6 - 23.2}{0.25 + 1.25 + 0.429} = 48.9 \text{ kW/m}^2$$

If there is no gas present, the heat transfer is given by Eq. (6.30):

$$\frac{\dot{Q}_{12}}{A} = \frac{E_{b1} - E_{b2}}{1/\varepsilon_1 + 1/\varepsilon_2 - 1} = \frac{117.6 - 23.2}{1/0.8 + 1/0.7 - 1} = 56.2 \text{ kW/m}^2$$

(or set $\varepsilon_g^{12} = 0$ above to obtain the same result).

Comments

1. The effect of the gas on the heat flux is not large, even though its emissivity is relatively large.

2. The gas acts somewhat like a radiation shield.

6.7.6 Radiation Exchange between an Isothermal Nongray Gas and a Single-Gray-Surface Enclosure

Equation (6.14) for radiation exchange between a *nongray* gas and a single-surface *black* enclosure is

$$\frac{\dot{Q}}{A} = \alpha_g \sigma T_s^4 - \varepsilon_g \sigma T_g^4 \qquad (6.24)$$

where α_g is the gas absorptivity for radiation from a black source at temperature T_s absorbed by gas at temperature T_g. For radiation exchange between a single-surface *gray* enclosure and a *gray* gas, Eqs. (6.18) and (6.19) become, with $F_{11} = 1$,

$$J_1 = \varepsilon_1 E_{b1} + (1 - \varepsilon_1)(\tau_g^{11} J_1 + \varepsilon_g^{11} E_{bg}) \qquad (6.25)$$

$$\frac{\dot{Q}_1}{A_1} = \frac{\varepsilon_1}{1 - \varepsilon_1}(E_{b1} - J_1) \qquad (6.26)$$

Solving for J_1 from Eq. (6.25) and substituting in Eq. (6.26) gives

$$\frac{\dot{Q}_1}{A_1} = \frac{\varepsilon_1 \varepsilon_g^{11}}{1 - (1 - \varepsilon_1)\tau_g^{11}}(E_{b1} - E_{bg}) \qquad (6.27)$$

Simplifying the notation to be consistent with Eq. (6.24) gives

$$\frac{\dot{Q}}{A} = \frac{\varepsilon_s \varepsilon_g \sigma T_s^4}{1 - (1 - \varepsilon_s)\tau_g} - \frac{\varepsilon_s \varepsilon_g \sigma T_g^4}{1 - (1 - \varepsilon_s)\tau_g} \qquad (6.28)$$

where for a gray gas $1 - \tau_g = \varepsilon_g$. The dilemma to be resolved is that Eq. (6.24) applies only to black walls, that is, it does not allow for reflection of radiation by the walls, whereas Eq. (6.28) assumes a gray gas. Equation (6.28) gives unsatisfactory results when the wall temperature is close to, or greater than, the gas temperature. Fortunately, it is possible to obtain an exact solution for the problem of exchange between a nongray gas and a single-gray-surface enclosure [14]. A convenient and adequate approximation to this exact solution is

$$\frac{\dot{Q}}{A} = \frac{\varepsilon_s \alpha_{g1} \sigma T_s^4}{1 - (1 - \varepsilon_s)[(\alpha_{g2} - \alpha_{g1})/\alpha_{g1}]} - \frac{\varepsilon_s \varepsilon_{g1} \sigma T_g^4}{1 - (1 - \varepsilon_s)[(\varepsilon_{g2} - \varepsilon_{g1})/\varepsilon_{g1}]} \qquad (6.29)$$

where α_{g1} is the gas absorptivity for the mean beam length of the enclosure, and α_{g2} is for two mean beam lengths (that is, including the effect of one reflection). The emissivities ε_{g1} and ε_{g2} are defined similarly. The absorptivities are evaluated for absorption of radiation from a black surface at temperature T_s. The exact solution accounts for an infinite number of reflections; the approximation of Eq. (6.29) retains the effect of the first reflection only. The result proves satisfactory for most engineering applications.

EXAMPLE 6.26   A Scramjet Combustor for a Hypersonic Aircraft

A Mach 10 hypersonic drone for testing a hydrogen-fueled scramjet engine has a combustor with a 0.40 m square cross section. The walls have an emittance of 0.2 and are cooled to a temperature of 800 K. At a location where the pressure is 8 atm and the combustion gases are at 2000 K, estimate the radiative heat transfer to the walls. Take the excess air ratio as 2.

Solution

Given: Hydrogen-fueled scramjet combustor.

Required: Radiative heat transfer to walls.

Assumptions: 1. The combustor is long enough to ignore end effects.
2. The one-reflection approximation, Eq. (6.29), is adequate.
3. The presence of O_2 in the combustion products has a negligible effect on gas radiation properties calculated from Fig. 6.35 for a H_2O–N_2 mixture.

The first task is to determine the mean beam length: For a long combustor, we assume a 2-D geometry.

$$\mathcal{L}_m^0 = \frac{4V_g}{A_s} = \frac{(4)(0.40)^2}{(4)(0.40)} = 0.40 \text{ m}$$

$$\mathcal{L}_m \simeq 0.9\mathcal{L}_m^0 = (0.9)(0.4) = 0.36 \text{ m}$$

Next the partial pressure of the water vapor must be calculated. The combustion reaction is

$$1 \text{ kmol } O_2 + 2 \text{ kmol } H_2 \rightarrow 2 \text{ kmol } H_2O$$

and, with 100% excess air, there are 1 kmol O_2 and 2×3.76 kmol N_2 in the products. Hence, for a total pressure of 8 atm, the partial pressure of H_2O is

$$P_{H_2O} = 8 \left(\frac{2}{2 + 1 + (2)(3.76)} \right) = 1.52 \text{ atm}$$

RAD 3 gives $\varepsilon_{g1} = 0.248, \alpha_{g1} = 0.683, \varepsilon_{g2} = 0.319, \alpha_{g2} = 0.796$. The wall heat flux is obtained from Eq. (6.29) as

$$\frac{\dot{Q}}{A} = \frac{\varepsilon_s \alpha_{g1} \sigma T_s^4}{1 - (1 - \varepsilon_s)[(\alpha_{g2} - \alpha_{g1})/\alpha_{g1}]} - \frac{\varepsilon_s \varepsilon_{g1} \sigma T_g^4}{1 - (1 - \varepsilon_s)[(\varepsilon_{g2} - \varepsilon_{g1})/\varepsilon_{g1}]}$$

$$= \frac{(0.2)(0.683)(5.67 \times 10^{-8})(800)^4}{1 - (1 - 0.2)[(0.796 - 0.683)/0.683]} - \frac{(0.2)(0.24)(5.67 \times 10^{-8})(2000)^4}{1 - (1 - 0.2)[(0.319 - 0.248)/0.248]}$$

$$= (3.66 - 58.4)10^3 = -54.7 \text{ kW/m}^2$$

Comments

1. The convective heat transfer for this scramjet combustor has been estimated to be 300 kW/m^2; thus, the radiation contribution is about 15%. The importance of having a low wall emittance (absorptance) to reduce the radiation flux is clearly evident.

2. $\kappa L = \ln[1/(l - \varepsilon_g)] \simeq \ln[1/(1 - 0.24)] = 0.27 > 0.1$; hence, the assumption $\mathscr{L}_m \simeq 0.9\mathscr{L}_m^0$ is justified.

3. The low wall temperature (800 K) and small emittance (0.2) cause the absorbed radiation to play a minor role. Hence, the error introduced by not extrapolating Fig. 6.36b past $P_E = 10$ is small.

4. Try a hand calculation to obtain the total properties: it is not easy! Set $P_E = 10$ if $P_E \geq 10$ to calculate exponent m.

6.8   CLOSURE

The physics of radiation is quite different from that of conduction and convection. Many new concepts were introduced in Chapter 6. Analysis of radiation transfer involved mainly algebra, rather than the solution of differential equations.

Sections 6.2 through 6.4 were concerned with the engineering calculation of radiation energy exchange between surfaces separated by a nonparticipating medium. Real engineering surfaces were modeled as diffuse gray surfaces. An exception was solar radiation, where allowance was made for a different value of absorptance to short-wavelength solar radiation. Engineering problem solving was aided by two computer programs. RAD1 calculates shape factors for often-used configurations. RAD2 solves for the radiation energy exchange in enclosures of diffuse gray surfaces; in practice, the number of surfaces used is limited by the work required to assemble the matrix of shape factors.

Sections 6.5 through 6.7 dealt with more advanced topics. Section 6.5 examined directional characteristics of surface radiation. The concept of radiation intensity, which was earlier avoided, was defined and used to rigorously derive shape factors. Effects of specular reflection were examined briefly, and, in particular, radiation transmission through cracks with either diffuse or specular reflecting walls was compared. Spectral characteristics of surface radiation were examined in Section 6.6, with the limited objective of demonstrating how total radiation properties may be obtained from spectral values. Total hemispherical radiation properties are required when the diffuse gray surface model is used, but for a particular surface, or surface coating, only spectral data may be available to the engineer. Chapter 6 closes with an introduction to radiation transfer through gases in Section 6.7. Both H_2O and CO_2 absorb and emit radiation strongly and are found in combustion products. A procedure and data for calculating total radiation properties for these species were given. Since the required calculations are complicated and lengthy, the computer program RAD3, which performs this task, proved to be a useful tool. Analytical methods were developed for some simple radiation exchange problems in enclosures containing combustion gases. In particular, the approximate formula given for a single-gray-surface enclosure containing a nongray gas was shown to be useful for making preliminary estimates of radiation exchange in furnaces and combustion chambers.

In keeping with the introductory nature of this text, the material in Chapter 6 has been selected to allow a wide range of engineering problems to be analyzed, without facing the difficult problem of nongray behavior head-on. In principle, it is not difficult to properly account for nongray behavior of both surfaces and participating gases. However, computer programs are required to make the calculations practical, and often spectral data are unavailable for the surfaces involved. Thus, the gray surface model should always be used to make an initial assessment of a problem. If such an assessment shows that an accurate estimate of radiation exchange is essential to develop a successful design, then advanced texts should be consulted. Another problem that was avoided is the participation of aerosols in radiation transfer. Of particular importance are the effects of soot in combustion chambers burning hydrocarbon fuels. This topic has received considerable attention in recent years, and much useful information is available in current journal papers.

REFERENCES

1. Hottel, H. C, and Sarofim, A. E, *Radiative Transfer*, McGraw-Hill, New York (1967).

2. Oppenheim, A. K., "Radiation analysis by the network method," *Transactions of the ASME*, 725–735 (1956).

3. Edwards, D. K., *Radiation Heat Transfer Notes*, Hemisphere, Washington, D.C. (1981).

4. Brunt, D., "Radiation in the atmosphere," *Quart. J. R. Meteorol. Soc,* 66 (Suppl.), 34–40 (1940).

5. Berdahl, P., and Fromberg, R., "The thermal radiance of clear skies," *Solar Energy*, 29, 299–314 (1982).

6. Bilbao J., and de Miguel, A. H., "Estimation of daylight downward longwave atmospheric irradiance under clear sky and all-sky conditions," *Journal of Applied Meteorology and Climatology*, 46, 878–889 (2007).

7. Garg, H. P., *Treatise on Solar Energy, Vol. 1: Fundamentals of Solar Energy,* John Willy & Sons, New York (1982).

8. TRASYS User's Manual, Version P22. Manual prepared for NASA Johnson Space Center by Lockheed Eng. Man. Services, April 1988.

9. Goswami, D. Y., Kreith, F., and Kreider, J.F., *Principles of Solar Engineering*, 2nd ed., Taylor and Francis, New York, N.Y. (2000).

10. Schmidt, E., and Eckert, E. R. G., "Über die Richtungsverteilung der Wärmestrahlung von Oberflachen," *Forsch. Gebiet Ingenieurw.*, 6, 175–183 (1935).

11. Hottel, H. C, "Heat transmission by radiation from non-luminous gases," *Trans. AIChE*, 19, 173–205 (1927).

12. Hottel, H. C, "Radiant Heat Transmission," Chapter 3 of *Heat Transmission*, by W. H. McAdams, 3rd ed., McGraw-Hill, New York (1954).

13. Hottel, H. C, and Egbert, R. B., "Radiant heat transmission from water vapor," *Trans. AIChE*, 38, 531–568 (1942).

14. Edwards, D. K., and Matavosian, R., "Scaling rules for total absorptivity and emissivity of gases," *J. Heat Transfer*, 106, 684–689 (1984).

15. Edwards, D. K., and Matavosian, R., "Emissivity data for gases," Section 5.5.5, *Hemisphere Handbook of Heat Exchanger Design*, ed. G. F. Hewitt, Hemisphere, New York (1990).

16. Modest, M. F., "The treatment of nongray properties in radiative heat transfer: from past to present,"*J. Heat Transfer*, 135, 061801-1 – 061801-12 (2013).

17. Oppenheim, A. K., and Bevans, J. T., "Geometric factors for radiant heat transfer through an absorbing medium in Cartesian coordinates," *J. Heat Transfer*, 82, 360–368 (1960).

18. Soffer, B. H., and Lynch, D. K., "Some paradoxes, errors, and resolutions concerning the spectral optimization of human vision," *American Journal of Physics*, 67, 946–953 (1999).

19. Heald, M. A., "Where is the Wien's peak?," *American Journal of Physics*, 71, 1322–1323 (2003).

EXERCISES

6-1. Calculate the blackbody emissive power E_b for surfaces at 300 K, 1000 K, 2000 K, and 5000 K. Also estimate the peak wavelength λ_{max} for each surface using:

(i) $\lambda_{max}T = 2898\,\mu$m K (from Eq. 6.6 and Fig. 6.3*a*)
(ii) $\lambda_{max}T = 5100\,\mu$m K (from Fig. 6.3*b*)

6-2. Derive Wien's displacement law, Eq. (6.6), from Planck's law, Eq. (6.5). In a similar manner, determine the maximum of $E_{bv_f}(T) = 2\pi h v_f^3/c^2(e^{(hv_f/\kappa_B T)} - 1)$ in terms of v_f/T. This maximum corresponds to $\lambda_{max}T = 5100\,\mu$m K as shown in Fig. 6.3*b*. Explain why a different peak wavelength is found when the density distribution is expressed in terms of v_f. [*Hint*: Equal intervals of λ in the wavelength density distribution do not correspond to equal intervals of frequency because $dv_f = -c/\lambda^2 d\lambda$ (see [18, 19] for more details)].

6-3. Show that the Stefan-Boltzmann law, Eq. (6.7), can be obtained from Planck's law, Eq. (6.5), by integrating over all wavelengths. [*Hint:* Let $x = 1/\lambda T$, and hence $d\lambda = -(1/x^2 T)dx$ to effect the integration.]

6-4. Plot graphs of monochromatic emissive power versus wavelength for emission from black surfaces at

(i) 300 K.
(ii) 5800 K.

Approximately what fraction of the energy is in the visible range in each case?

6-5. A gray opaque surface at 500 K has an emittance of 0.3 and is exposed to a high-temperature heat source such that the irradiation on the surface is 30,000 W/m². Calculate the following heat fluxes:

    (i) Absorbed flux
    (ii) Reflected flux
    (iii) Emitted flux
    (iv) The radiosity

6-6. Useful approximations to Planck's law are the Wien and Rayleigh-Jeans spectral distributions, which are valid for very small and very large values of λT, respectively. These distributions are

$$E_{b\lambda} \simeq \frac{C_1}{\lambda^5} e^{-C_2/\lambda T} \quad \text{(Wien's distribution)}$$

$$E_{b\lambda} \simeq \frac{C_1}{C_2} \frac{T}{\lambda^4} \quad \text{(Rayleigh-Jeans distribution)}$$

Derive these distributions from Planck's law, Eq. (6.5), by in turn letting $C_2/\lambda T \gg 1$ and $C_2/\lambda T \ll 1$. Show all three laws on a graph of $\lambda^5 E_{b\lambda}$ versus λT.

6-7. Referring to the sketch, calculate the shape factors F_{14}, F_{41}, F_{23}, and F_{32}.

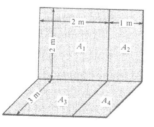

6-8. Referring to the sketch, calculate the shape factors F_{14}, F_{41}, F_{23}, and F_{32}.

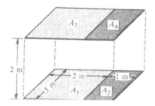

6-9. Prove the string rule for two-dimensional configurations, Eq. (6.25).

6-10. Derive the shape factors given as items 3, 4, and 5 of Table 6.1 using the string rule.

6-11. The string rule is convenient to use for 2-D configurations, for example, long furnaces. But how long must such a furnace be for the 2-D assumption to be accurate? Consider a furnace of width W, height H, and length L. For $W = H$ plot the percent error in the string rule shape factor between the roof and floor, as a function of L/W.

6-12.   A cavity is in the form of a vertical cylinder 8 cm in diameter and 16 cm in length and is open at the top to black surroundings at 300 K. The bottom end is heated electrically and is maintained at 1900 K. If the side walls are at a uniform temperature of 1500 K, calculate the power input to the heater and the heat loss to the surroundings through the open end. Take the inner surfaces to be black.

6-13.   Two 20 cm-diameter parallel disks are 10 cm apart. The facing sides of the two disks are maintained at 500 K and 400 K, while the back faces are well insulated. The disks are located in an enclosure with walls at 300 K. If all the surfaces can be taken to be black, determine

   (i)   the net radiation heat exchange between the disks.
   (ii) the rate of radiative heat loss from each disk.

6-14.   A long duct has the cross section shown in the figure.

   (i)   Calculate the 4 × 4 shape factor matrix.
   (ii) If the surfaces 1 and 4 are combined into a single surface 1′, calculate the 3 × 3 shape factor matrix.

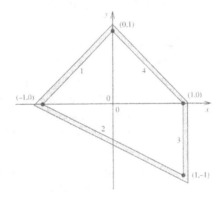

6-15.   A spherical nuclear fireball is 100 m in diameter. Calculate the shape factor for diffuse radiation from a one square meter segment of ground directly beneath the fireball when it is at altitudes of 100 m, 200 m, and 300 m.

6-16.   A cavity has a 8 cm × 8 cm cross section and is 16 cm deep. The bottom end surface is heated electrically and maintained at 1500 K, while the open end is exposed to black surroundings at 300 K. If the side walls are insulated and the heated surface is black, determine the power supply to the heater. Also determine the side wall temperature.

6-17.   Rework Exercise 6–16 with the side walls subdivided into two surfaces of equal areas, each assumed isothermal at different temperatures.

6-18.   The cross section of a long evacuated equilateral triangular duct with 1 m sides is shown. The inner surfaces are black. Surface 1 is maintained at 800 K by a cooling system and thermostat. Surface 2 is insulated. Surface 3 is provided with a heater. What must be the rate of heat transfer into the duct through surface 3 if it is desired to have surface 2 at 1200 K? Also, determine the temperature of surface 3.

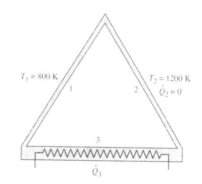

6-19. A circular cross-section heating duct of 50 cm diameter runs horizontally through the basement garage of a large apartment complex. The ambient air and walls of the garage are at 17°C. At a location where the surface temperature of the duct is 40°C, determine the radiation contribution to the total heat loss per meter length if

(i) the duct has an emittance of 0.7.
(ii) the duct is painted with aluminum paint with $\varepsilon = 0.2$.

6-20. Liquid oxygen flows through a tube of outside diameter 3 cm, the outer surface of which has an emittance of 0.03 and a temperature of 85 K. This tube is enclosed by a larger concentric tube of inside diameter 5 cm, the inner surface of which has an emittance of 0.05 and a temperature of 290 K. The space between the tubes is evacuated.

(i) Determine the heat gain by the oxygen per unit length of inner tube.
(ii) How much is the heat gain reduced if a thin-wall radiation shield with an emittance of 0.03 on each side is placed midway between the tubes?

6-21. Hot combustion gases flow in a duct whose walls are at 600 K. A thermocouple located in the center of the duct records a temperature of 873 K. If the emittance of the thermocouple is 0.8 and the heat transfer coefficient between the gas and the thermocouple is 110 W/m² K, determine the true gas temperature. How much gain in accuracy is obtained if the thermocouple is surrounded by a radiation shield in the form of a cylinder 1 cm in diameter and 4 cm long, made from a thin-wall stainless steel tube with its axis in the flow direction? Take the emittance of the shield as 0.6 and the heat transfer coefficient between the gas and the shield as 40 W/m² K.

6-22. A thermocouple is to be utilized to measure the temperature of exhaust gases from an automobile engine, and the effect of cylindrical radiation shields, installed as shown in the sketch, is to be investigated. The inner and outer shield diameters are 8 mm and 12 mm, respectively. The emittance of the thermocouple bead and the shields can be taken to be 0.4. If the exhaust gases are at 1200 K and the exhaust pipe walls are at 600 K, determine the temperature registered by the thermocouple for

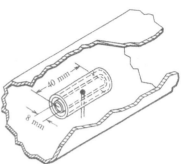

   (i) no shield.
   (ii) one shield.
   (iii) two shields.

Take the convective heat transfer coefficients as 300 and 100 W/m^2 K on the thermocouple and shields, respectively. Assume the pipe walls are black, and make reasonable engineering simplifications.

6-23. A 1 mm–diameter spherical thermocouple bead is to be used to measure the temperature of a low-density air flow. It is located in a 4 cm–diameter tube and is surrounded by a 4 cm–long, 1 cm-diameter thin-walled cylindrical radiation shield lined up with the flow. The emittance of the thermocouple bead is 0.2, of the shield 0.05, and of the tube walls 0.8. The air is at a pressure of 10 torr and has velocity of 100 m/s. When the tube walls are at 300 K and the air temperature is 325 K, estimate the expected error in the thermocouple reading.

6-24. A thermistor is used to measure the temperature of a hot-air stream in a forced-air heating system of a hospital. It is located in a 1 m-square duct through which air flows at 1.6 m/s. The thermistor can be modeled as a 3 mm-diameter sphere of emittance 0.8 and is located inside a radiation shield in the form of a 1 cm–diameter, 5 cm–long tube of emittance 0.7 (the axis of the tube is in the flow direction). If the thermistor records a temperature of 46.8°C when the duct walls are at 43.5°C, determine the true temperature of the air. Convective heat transfer coefficients on the thermistor and shield can be taken as 93 W/m^2 K and 22 W/m^2 K, respectively. (*Hint:* First estimate the shield temperature.)

6-25. A thermistor is used to measure the temperature of a low-density helium flow. It is located in a 10 cm–diameter tube through which the gas flows at 1.6 m/s. The thermistor can be modeled as a 3 mm–diameter sphere of emittance 0.8 and is located inside a radiation shield in the form of a 1 cm–diameter, 5 cm–long tube of emittance 0.08 (the axis of the tube is in the flow direction). If the thermistor records a temperature of 327.5°C when the tube walls are at 285°C, determine the true temperature of the helium. Convective heat transfer coefficients on the thermistor and shield can be taken as 19 W/m^2 K and 5 W/m^2 K, respectively. (*Hint:* First estimate the shield temperature.)

6-26. A radioisotope power source for a space vehicle is in the form of a 10 cm–diameter solid sphere, in which heat is generated uniformly. At steady state its surface temperature is 600 K. It is contained inside a solid spherical shell of inner and outer diameters 20 cm and 40 cm, respectively, and the space between is evacuated. The outer surface of the containment shell sees space, which can be taken to be a blackbody at 0 K. The thermal conductivities of the power source and containment shell are 2.0 and 3.2 W/m K, respectively. All the surfaces are gray with the same emittance ε. What value of ε must be specified for the surface finish in order to limit the heat loss from the assembly to 100 W?

6-27. Many layers of aluminized Mylar are used to form *superinsulation*. Suppose that a gas is present so that both molecular conduction and radiation transfer occur.

   (i) By linearizing Eq. (6.41) show that a "total" conductivity can be defined as

$$k_{total} = k_{gas} + \frac{4\delta\varepsilon\sigma T^3}{2-\varepsilon}$$

   and treat the heat transfer as an equivalent conduction process. The Mylar sheet spacing is δ, and the emittance of the aluminized surface is ε (both sides)

   (ii) Since k_{total} varies strongly with temperature, it is useful to account for a temperature-dependent total conductivity. For a cylindrical insulation, show that the heat transfer is

$$\dot{Q} = \frac{2\pi L(\Phi_1 - \Phi_2)}{\ln(r_2/r_1)}$$

   where $\Phi = \int_{T_0}^{T} k(T)dT$. Evaluate $(\Phi_1 - \Phi_2)$ for k_{gas} proportional to $T^{0.8}$.

   (iii) A 5 cm–O.D. pipeline is insulated with 20 sheets of aluminized Mylar contained in a 10 cm–I.D. vacuum jacket. Determine the radiation contribution to the total conductivity at $-50°C$ for an aluminum emittance constant at $\varepsilon = 0.03$.

   (iv) If the vacuum seal in part (iii) is broken and air at 1000 Pa fills the jacket, calculate the increase in total conductivity.

6-28. A long, 15 cm–O.D. horizontal steel exhaust pipe ($\varepsilon = 0.7$) is located in a
 workshop where the ambient air is at 24°C. Consider a location where the exhaust pipe is at 800 K.

   (i) Calculate the rate of heat loss per meter.

   (ii) If the pipe is contained in a thin, 25 cm–O.D. aluminum shield ($\varepsilon = 0.06$), how much is the heat loss reduced?

   In part (ii) the ends of the annular space are sealed. (*Hint:* Use CONV to prepare tables of relevant natural convection heat transfer coefficients before solving part (ii) by iteration.)

6-29. An outside wall of a cabin is 5 m long and 3 m high, and consists of two 1 cm–thick boards spaced 10 cm apart. The exterior surface temperatures are 20°C and $-10°C$.

   (i) Calculate the heat loss across the wall, neglecting any air leakage.

   (ii) If the air space is divided in half by a sheet of aluminum foil, calculate the reduction in the heat loss.

Take $\varepsilon = 0.8$ and $k = 0.17$ W/m K for the boards, and $\varepsilon = 0.05$ for the foil. (*Hint:* Use CONV to calculate required convective heat transfer coefficients.)

6-30. Modern ski wear is often made of several thin layers of fabric with air trapped between the layers. The resulting clothing is light and a good thermal insulator. Consider a jacket made from five layers of 0.2 mm-thick nylon fabric separated by 1 mm air gaps. Typical conditions on the ski slope give fabric outside and inside temperatures of $-13°C$ and $17°C$, respectively. Determine the thermal resistance of 1 m area of jacket. What thickness of wool jacket will give the same thermal protection? Take $\varepsilon = 0.85$ for the nylon.

6-31. A heating panel is to be placed on the ceiling of a hospital room so that the patient will not be uncomfortable in 290 K air. The panel is painted with an off-white matte paint of emittance 0.88. As a design criterion it is required that the patient's face—modeled as a dry, black adiabatic surface—attains a 305 K equilibrium temperature. The panel temperature is 355 K and is located 2 m above the patient. The convective heat transfer coefficient for the face can be taken as 4.0 W/m^2 K, and it can be assumed that the room walls are black at the air temperature. If the panel is to be a circular disk, determine its diameter.

6-32. A cylindrical cavity, with a diameter of 10 cm and 20 cm deep, is located in an enclosure with black walls at 300 K. Calculate the radiative heat loss from the cavity for the following conditions:

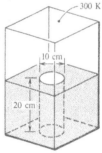

(i) All interior surfaces are black and are maintained at 1000 K.

(ii) All interior surfaces are diffuse-gray with emittance 0.5 and are maintained at 1000 K.

(iii) All interior surfaces are diffuse-gray with emittance 0.5; the base is maintained at 1000 K while the side walls are perfectly insulated.

(iv) As in case (iii), with the side walls having an emittance of 0.05.

6-33. A long furnace has a 3 m–square cross section. The roof is maintained at 2000 K by hot combustion gases, while the work floor is at 800 K. The side walls are well insulated with refractory brick. Calculate the radiant heat flux into the floor if the roof and side walls have an emittance of 0.7 and the floor has an emittance of 0.4.

6-34. A room is 3 m square and 2.5 m high. The walls can be taken as adiabatic and isothermal. The ceiling is at 35°C and has an emittance of 0.8, while the floor is at 20°C and has an emittance of 0.9. Denote the ceiling as surface 1, the floor 2, and the walls 3.

(i) Set up the radiosity equations. Determine and evaluate all the shape factors, and tabulate as a 3×3 array. Solve these equations to determine the heat flow into the floor, $\dot{Q}_2$.

(ii) Draw the radiation network. Use the network to obtain an expression for $\dot{Q}_2$, and solve for $\dot{Q}_2$ again.

6-35. Determine the heat transfer between the two surfaces shown in the sketch if $T_1 = 3000$ K, $\varepsilon_1 = 0.3; T_2 = 1500$ K, $\varepsilon_2 = 0.4$. Surface 1 is 1 m wide and 20 m long, and $A_2 = 2A_1$. Take the surroundings as black at 0 K.

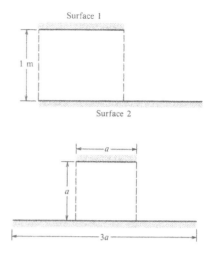

6-36. Two surfaces of length L are placed distance a apart as shown, with $L \gg a$. The smaller surface, 1, of width a, is at 2000 K and has an emittance of 0.2. The larger surface, 2, of width $3a$, is at 800 K and has an emittance of 0.6. Calculate the heat flux through surface 2 if

(i) the surroundings are black at 300 K.
(ii) the surfaces are joined by refractory surfaces.

6-37. A long furnace used for an enameling process has a 3 m × 3 m cross section with the roof maintained at 1900 K, and the side walls are well insulated. What is the radiant heat transfer to the floor when it is at 600 K? All surfaces may be taken to be gray and diffuse, with emittances of 0.5.

(i) Assume the side walls are isothermal.
(ii) Divide the side walls into two surfaces and allow these surfaces to be isothermal at different temperatures.

6-38. A long furnace is 40 cm wide and 25 cm high. Its roof is maintained at 1100°C, and its side walls and floor are lined with refractory brick. Stainless steel strip 36 cm wide and 0.5 mm thick passes through the furnace at a speed of 0.3 m/s.

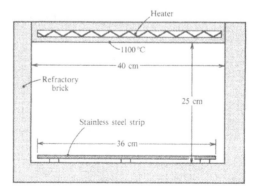

Estimate the length of furnace required to heat the strip from 100°C to 400°C. Take emittances for the roof and stainless steel as 0.8 and 0.25, respectively.

6-39. A long 90° wedge has 1 m sides, with surface 1 at 2000 K and surface 2 adiabatic. The surroundings can be taken as black at 0 K. If $\varepsilon_1 = 0.5$ and $\varepsilon_2 = 0.2$, determine the heat loss to the surroundings

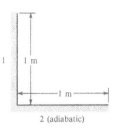

    (i) if the adiabatic wall is assumed to be isothermal.
    (ii) if the adiabatic wall is, in turn, subdivided into two, three, and four equal isothermal areas.

Plot your results.

6-40. Two 3 cm–diameter thin-walled tubes run parallel, with their centers 5 cm apart. Through one tube flows hot combustion gases at 800 K, while through the other flows water at 300 K. If both tubes have an emittance of 0.8, calculate the net radiant energy absorbed by the water per meter length of tube. The surroundings are also at 300 K. Assume that both inside convective heat transfer coefficients are large.

6-41. A small furnace for heat-treating metal alloy samples in space has an inside diameter of 6 cm and a length of 12 cm, as shown. A 9 cm length of the wall is to be maintained at 800°C by an electrical heater. The remaining sidewall and endwall

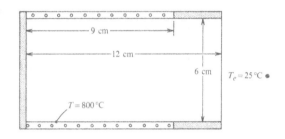

can be taken to be well insulated. The opening is exposed to black surroundings at 25°C. Owing to power supply limitations from a fuel cell source, a low operating power is desirable. Investigate the effect of heated wall emittance on the power required to balance radiative losses in steady operation.

6-42. Two large parallel plates are 4 cm apart and are at temperatures 2000 K and 1000 K. If both have an emittance of 0.5, what is the heat transfer between the plates? If a thin sheet also of emittance 0.5 is placed between the plates, what is the new heat transfer rate? Why is the exact location of the sheet unimportant?

6-43. Two 15 cm-diameter parallel disks, 10 cm apart, have facing sides at 500 K and 400 K, and insulated backs. For gray disk surfaces with $\varepsilon = 0.5$ and black surroundings at 0 K, determine the net radiation exchange between the disks and the rate of heat loss from each disk.

6-44. The base of the conical cavity shown in the sketch is maintained at a temperature of 400 K, while the side walls are perfectly insulated. For gray diffuse surfaces, of emittances shown, determine the radiant heat transfer to the base for a top-surface temperature ranging from 800 K to 2000 K. Tabulate your results.

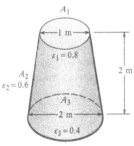

6-45. A cylindrical furnace has an inside diameter of 8 cm and is 16 cm high. The side walls are maintained at 1000 K, with power supplied by an electrical heater. The base and roof can be taken to be perfectly insulated. Determine the heat loss through the window in the roof when it is exposed to black surroundings at 300 K, for window diameters in the range 1 to 4 cm. The inside walls can be taken to be diffuse and gray with $\varepsilon = 0.6$; the transmittance of the window can be taken as unity.

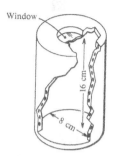

6-46. A square room of sides 8 m has a 3 m–diameter radiant heating panel centrally located on the ceiling, 3 m above the floor. The panel can be taken to be black and is maintained at 60°C. Calculate the irradiation on the floor due to the heater, at the center of the room, and in one corner.

6-47. Rework Exercise 6–18 for gray surfaces with emittances $\varepsilon_1 = 0.6$, $\varepsilon_2 = 0.4$, $\varepsilon_3 = 0.5$.

6-48. A long, square evacuated duct has 1 m sides. Two adjacent sides are perfectly insulated and have an emittance of 0.4. The other two sides have an emittance of 0.5. One of these sides is cooled and has a thermostat temperature control; the other contains an embedded electrical heater. If the thermostat maintains the cooled surface at 1000 K, at what rate must power be supplied to the electrical heater to maintain the two insulated sides at 2000 K? Also, what is the temperature of the heated side?

(i) Assume that the two insulated sides can be represented by a single isothermal surface.

(ii) Can the problem be solved if the two insulated sides are represented by separate isothermal surfaces?

(iii) Comment on the validity of the approach used in part (i).

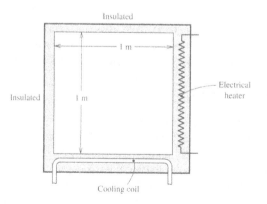

6-49. On an engine test rig, a 4 cm–O.D. exhaust gas pipe runs parallel to a 4 cm–O.D. water pipe at a centerline spacing of 8 cm. If the gas and water pipes are at 700 K and 295 K, respectively, calculate the rate of radiation heat transfer to the water pipe. Take the emittance of the exhaust pipe as 0.8 and of the water pipe as 0.6. Assume black surroundings at 300 K. First perform an "exact" calculation (or use RAD2). Then simplify the problem by recognizing that the exhaust pipe is much hotter than the water pipe and surroundings to obtain a quick approximate estimate.

6-50. Two 20 cm–diameter parallel disks, 10 cm apart, have facing sides at 1000 K and 700 K, and insulated backs. For gray disk surfaces with $\varepsilon = 0.4$, and black surroundings at 0 K, determine the net radiation exchange between the disks and the heat loss from each disk.

6-51. Two 5 mm–thick rectangular aluminum alloy plates are attached to the exterior surface of a spacecraft to form a channel 12 cm long, 2 cm wide, and 6 cm deep. If the plates and wall are at a uniform temperature of 310 K, calculate the radiation heat transfer from the inside of the channel to space (taken as black at 0 K). Take $\varepsilon = 0.23$ for the aluminum.

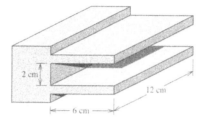

6-52. A long duct of length L has a triangular cross section with sides of width W. Each of the inside surfaces is isothermal, side 1 is gray, and the other two sides are black. Determine the radiant heat flow from each surface in terms of the surface temperatures $T_i, i = 1, 2, 3, \varepsilon$, and W. Check that the heat flows sum to zero.

6-53. Consider a long duct with a triangular cross section as shown. The angle between sides 1 and 2 is 90°, and between 1 and 3, 30°. The width of side 2 is 1 m. Sides 1 and 2 are isothermal with $T_1 = 2000$ K and $T_2 = 1000$ K, and surface 3 is perfectly insulated. Determine the radiative heat transfer from surface 1 to surface 2 (per unit length of duct). All surfaces have an emittance of 0.7.

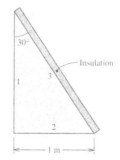

6-54. A small electric furnace is 60 cm long, 30 cm wide, and 30 cm high. Its top surface has an emittance of 0.8 and is maintained at 1500 K. The sides walls are refractory. The floor is covered with a material of emittance 0.5 that is to be processed at 500 K. Neglecting heat losses from the furnace exterior, what is the required electrical power input?

6-55. A sphere of diameter D and a cube of side length D are separated by a distance H, such that the shape factor from the sphere to the cube is $F_{12} = 1/5$. The initial temperature of the sphere is T_{10} and of the cube T_{20}. The objects are located in a vacuum chamber with black walls maintained at temperature T_3.

   (i) Assuming both objects are blackbodies, set up differential equations that give the temperature of each object as a function of time. Lumped thermal capacity models will be adequate, but specify an appropriate criterion.

  (ii) Generalize the result of (i) to the case where the cube is a blackbody, but the sphere has a diffuse gray surface of emittance 0.5.

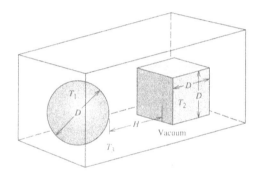

6-56. Write a computer program to solve case (ii) of Exercise 6–55, and obtain the temperature responses for the following parameter values:

$$D = 3 \text{ cm}$$
$$T_{10} = 600 \text{ K}; \rho_1 = 1800 \text{ kg/m}^3, c_1 = 1400 \text{ J/kg K}, \varepsilon_1 = 0.3$$
$$T_{20} = 500 \text{ K}; \rho_2 = 2770 \text{ kg/m}^3, c_2 = 900 \text{ J/kg K}$$
$$T_3 = 300 \text{ K}$$

6-57. A furnace is in the form of a cube with 3 m sides. The roof is heated to 1100°C, and the work floor is at 500°C. The walls are insulated. The emittance of the roof is 0.85. Calculate the radiant heat flux into the floor as a function of its emittance for $0.2 < \varepsilon < 1.0$.

6-58. A long furnace is 50 cm wide and 30 cm high. The roof is heated to 1200°C, and the side walls are insulated. A 50 cm–wide, 1.0 mm–thick, AISI 1042 steel strip is drawn along the floor of the furnace at 25 cm/s. How long must the furnace be to heat the strip from 50°C to 350°C? The emittances of the roof and stainless steel are 0.83 and 0.21, respectively. (*Hint:* Replace time by distance/speed in a lumped thermal capacity analysis.)

6-59. A 100 W, 10 cm–diameter disk heater is placed parallel to, and 5 cm from, a 10 cm–diameter disk receiver in an evacuated chamber. The backs of both disks are well insulated. If the disk surfaces have emittances of 0.8 and the walls of the chamber are black and maintained at 350 K, calculate the following quantities:

   (i) The heater temperature.
   (ii) The receiver temperature.
   (iii) The radiative heat transfer between the heater and receiver.
   (iv) The net radiative heat transfer to the chamber.

6-60. Consider a long duct with a triangular cross section as shown. The angle between sides 1 and 2 is 90°, and between sides 2 and 3 is 30°. The length of side 1 is L_1. If T_1, and T_2 are given with $T_1 > T_2$, and if surface 3 is perfectly insulated, obtain an explicit expression for the radiative heat transfer from surface 1 to surface 2. All sides have the same emittance ε.

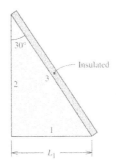

6-61. A kiln has an average inside surface temperature of 2330 K and has a small 15 × 15 cm square opening in its 20 cm–thick walls. If the sides of the opening can be assumed to be adiabatic, determine the rate of radiant energy loss through the opening.

6-62. A sliding door on a furnace does not close entirely but leaves a gap 1.6 cm wide and 1 m high. The door is 20 cm thick and the emittance of the refractory ceramic is 0.8. What is the heat loss through the gap when the furnace is at 1800 K and the surroundings are at 300 K?

6-63. A furnace has 15 cm–thick walls and a 1.5 cm–diameter peephole. If the inner walls of the furnace are at 1400 K, will there be a significant heat loss through the peephole to surroundings at 300 K?

6-64. The planet Mars has a diameter of 6772 km, and it orbits the Sun at a distance of 227.9×10^6 km. If the Sun is assumed to radiate like a blackbody at 5760 K, and Mars has an albedo of 0.15 (reflects 15% of incident radiation back to space), estimate the average temperature of the Martian surface. Ignore the effects of the thin Martian atmosphere.

6-65. Estimate the sky emittance and effective sky temperature for a clear daytime sky when the ambient air is at 1.01 bar and 16.8°C and has a 60% relative humidity.

6-66. Estimate the sky emittance and effective sky temperature for a clear night sky when the ambient air is at 1 bar and 13°C and has a 30% relative humidity.

6-67. On a hot sunny day the corrugated iron roof of a work shed is measured to be 50°C when the ambient air is at 30°C and 80% relative humidity. Calculate the heat flux into the shed if h_c for the roof is estimated to be 20 W/m² K. Take $\varepsilon = 0.6, \alpha_s = 0.8$ for the iron, a shape factor from the roof to the sky of unity, and a solar irradiation of 900 W/m².

6-68. Will dew form on the top of a transit car parked on a still, clear night when the air temperature is 10°C and the relative humidity is 90%? Take the surface as

   (i) weathered aluminum alloy 75S-T6.
   (ii) white epoxy paint.

The convective heat transfer coefficient is approximately $h_c = 1.24\Delta T^{1/3}$ W/m² K, for ΔT in kelvins.

6-69. Estimate the equilibrium temperature of a spherical satellite 1 m in diameter, exposed to the Sun, and in eclipse, at an altitude of 1000 km if it has an internal power dissipation of 300 W. Take the surface of the Earth to be black at 19°C. The emittance of the satellite skin is 0.15, and its solar absorptance is 0.10.

6-70. Calculate the equilibrium temperature of a small, flat plate with its top face exposed to an unobstructed view of the sky while air at 298 K, 1 atm and 20% relative humidity flows along both sides. The bottom face sees black surroundings at 298 K. The solar irradiation is 800 W/m², and the average convective heat transfer coefficient is 20 W/m² K. Obtain solutions for three different kinds of surfaces:

   (i) Representative of a very white paint, $\alpha_s = 0.2, \varepsilon = 0.9$
   (ii) A metallic paint (aluminum), $\alpha_s = 0.3, \varepsilon = 0.3$
   (iii) A black paint, $\alpha_s = 0.9, \varepsilon = 0.9$

6-71. A spacecraft is at 1 AU (astronomical unit) from the Sun and can be assumed spherical and isothermal. Internal power dissipation is negligible, and it is desired to maintain an equilibrium temperature of 300 K. It is planned to use a checkered surface with a fraction F_1 of vacuum-deposited aluminum and the remainder $(1 - F_1)$ coated with white epoxy paint. Determine F_1.

6-72. A spherical spacecraft is 1 AU from the Sun and can be assumed isothermal with negligible internal power dissipation. If the spacecraft is coated with Dow-Corning XP-310 aluminized silicone resin paint, determine its equilibrium temperature.

6-73. Calculate the irradiation of the area A_1 on a space vehicle as shown in the sketch. Surface A_2 is a solar cell array; the solar cells are at 50°C and can be taken to be black. (Area A_1 does not see the Sun.)

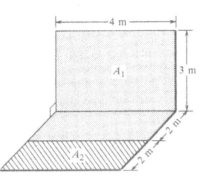

6-74. A flat-panel solar cell array on a three-axis-stabilized spacecraft in Earth orbit has one side directly facing the Sun and covered with solar cells; the other side is exposed to outer space at 0 K. The average solar absorptance on the front side is 0.8. The emittances of the front and back sides are 0.8 and 0.7, respectively. The solar cell operating efficiency is 15%, and the solar cell packing factor (the ratio of active solar cell area to total area) is 0.95. Determine the steady-state temperature of the panel.

6-75. A multilayer insulation (MLI) to be used on a spacecraft consists of 25 μm Kapton external foil sheets ($\alpha_s = 0.35, \varepsilon = 0.60$), aluminum-coated on the inside ($\varepsilon = 0.03$), sandwiching three aluminized Mylar sheets ($\varepsilon = 0.03$). The layers are separated by Dacron mesh to prevent contact of the sheets. If a MLI shield is faced to the Sun at 1 AU distance and the inner surface is maintained at 300 K, determine the heat flux through the MLI and the outer surface temperature.

6-76. Write a computer program to solve the flat-plate solar collector problem of Example 6.12. Allow $\alpha_s, \varepsilon, T_{C,in}, \dot{m}_C, G_s, T_e, \varepsilon_{sky}$, and U to be input parameters. Use the computer program to explore the effects of the various parameters on the collector efficiency.

6-77. Freezing of oranges in an orchard during winter can be catastrophic to the farmer. Estimate the rate at which an 8 cm–diameter orange hanging on a tree cools as a function of its temperature when the air has a temperature of 2°C and a relative humidity of 75%. Assume a convective heat transfer coefficient of 3 W/m^2 K, an emittance of orange peel of $\varepsilon_1 = 0.9$, and an emittance of surrounding leaves and grass of $\varepsilon_2 = 1.0$. Take the shape factor for the orange to surrounding leaves and grass as $F_{12} = 0.75$, and to the sky as $F_{13} = 0.25$. Estimate the temperature of the leaves and grass by requiring that they be adiabatic with $F_{23} = 0.5$. Will the convective heating when the orange is at 0°C exceed the radiative cooling? If not, will the use of fans to increase h_c from 3 to 12 W/m^2 K prevent freezing? For the orange, take $\rho = 900$ kg/m^3 and $c_p = 3600$ J/kg K. (*Hint:* First find T_2 from an energy balance on unit area $A_2 = 1$. Note that since $A_1 F_{12} = A_2 F_{21}, F_{21}$ is negligible. Then find $\dot{Q}_1$ and dT_1/dt.)

6-78. A Trombe wall is used for passive solar heating in winter. In the Northern Hemisphere the south wall of the house is made massive, say, 20 cm of concrete, and covered by 3mm-thick glass spaced 2 mm from the exterior surface. The exterior surface of the wall is painted black and the interior white. During the day the Sun warms the wall, the glazing reducing the heat losses from the exterior. The warm wall heats the interior of the house long after sunset. The glass is essentially opaque to long wavelength radiation, and has a transmittance τ_s and absorptance α_s, to solar radiation.

   (i) Set up a system of equations that can be used to determine the heat flux from the interior surface of the wall and its temperature. Carefully identify all the property data required.
   (ii) To get a feel for the magnitude of the heat flux, solve the following simplified problem: At time $t = 0$, the initial temperature of the wall is constant at 20°C, and for $t > 0$, the exterior surface of the concrete is at 100°C while the interior surface remains at 20°C. Determine the heat flux through the interior surface at $t = 4, 8$, and 12 hours.

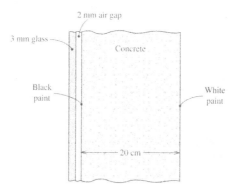

6-79. A solar cell array of an Earth satellite is a thin, 2 m–square flat panel. The solar absorptance α_s of the cells is 0.78 and the solar-to-electric conversion efficiency is 0.14. The long wavelength emittance ε of the active side of the panel is 0.78, and of the back side, 0.71. The panel can be assumed to be isothermal with a heat capacity per unit area of 8500 J/m² K. Solar radiation is incident on the panel at 90° before the satellite is eclipsed by the Earth. If the panel has attained a steady temperature, calculate its temperature 1200 s after being eclipsed.

6-80. A solar heater for a swimming pool consists of an unglazed copper plate 0.5 mm thick with a selective coating (solar absorptance = 0.92, total hemispherical emittance = 0.15). On the back side of the plate are attached 12 mm–O.D., 1 mm-wall-thickness copper tubes that are manifolded at the inlet and outlet to form a parallel tube matrix. The tubes have a centerline spacing of 20 cm and make perfect thermal contact with the plate. Water at 25°C is constantly recirculated from the pool at a flow rate per tube of 0.5 gallons/minute. On a given day the air temperature is 20°C, the air speed averages 4 m/s across a 3 m–wide array of solar collectors, the solar irradiation is 790 W/m², and the long wavelength radiation from the sky is about 340 W/m². Determine the tube length required to supply water back to the pool at 29°C. Assume that the heat loss due to conduction through the back insulation is 5% of the solar irradiation. Take $Re_{tr} = 10^5$. (*Hint:* Appropriate assumptions will allow an analytical solution to be obtained.)

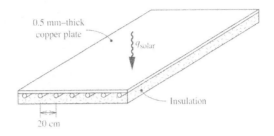

6-81. Space technologists often model spacecraft as isothermal spherical shells in order to compare the effect of different surface coatings for thermal control. Use such a model to evaluate the following four coatings:

	ε	α_s
White paint, new	0.9	0.2
White paint, aged	0.9	0.6
Black paint	0.9	0.9
Gold	0.03	0.08

Consider spacecraft at average distances from the Sun for the Earth (1 a.u.), Venus (0.723 a.u.), and Mars (1.523 a.u.), and determine their equilibrium temperatures (1 astronomical unit = 1.495×10^{11} m).

6-82. A flat circular plate is in a solar orbit one astronomical unit distant from the Sun, and is always oriented normal to the solar rays. Both sides have a Tabor finish that has spectral absorptance of 0.75 at wavelengths shorter than 3 μm, and 0.06 for wavelengths longer than 3 μm. Determine the equilibrium temperature of the plate if the Sun can be modeled as a blackbody at 5765 K. Assume that the Tabor finish is diffuse.

6-83. A spacecraft can be modeled as a thin cylindrical shell of 0.8 m diameter and 2.4 m length. It contains a payload of 4 m^2 surface area that is convex everywhere. The spacecraft is in a near-Earth orbit, and solar radiation is incident normal to the cylinder axis. The spacecraft spins about this axis fast enough so that the temperature of the shell is approximately uniform. The cylindrical portion of the shell is covered with solar cells with a packing factor of 95% (that accounts for gaps between cells). The cells have a solar absorptance of 0.80, and a long wavelength emittance of 0.35. The shell is made from Type 6061 aluminum alloy. The solar-electric conversion efficiency of the cells is 16%, and the electric power generated is dissipated in the payload. The inside of the shell and the outside of the payload are painted black with an emittance of 0.9, and conduction heat transfer along the payload supports is negligible. Calculate the temperatures of shell and the payload surface (assumed uniform).

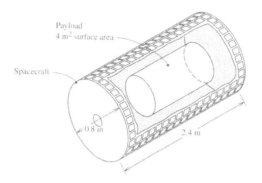

6-84. A rapid-transit car is left parked in the sun on a siding. The solar irradiation $G\cos\theta = 650$ W/m^2, and the ambient air is at 23°C and 50% RH. A wind blows across the 3 m–wide roof at 6 m/s. If the roof is well insulated, estimate the roof temperature for the following surfaces:

   (i) Black epoxy paint.
  (ii) White epoxy paint.
 (iii) Stainless steel ($\varepsilon = 0.4, \alpha_s = 0.8$).
 (iv) Weathered aluminum alloy 75S-T6.

Take $\mathrm{Re_{tr}} = 100{,}000$.

6-85. An alternative design for the solar collector of Example 6.12 is to have an inexpensive black absorbing surface, for which $\alpha_s = 0.9, \alpha_s/\varepsilon = 1$, and a 2 mm-thick cover glass spaced 1.8 cm from the absorbing surface. The glass is essentially opaque to longwave radiation; it transmits 85% and absorbs 7% of the solar irradiation, and for longwave radiation, $\varepsilon = \alpha = 0.85$. Estimate the outlet water temperature and collector efficiency for $G_s = 800$ W/m^2, $T_e = 20°C$, $\varepsilon_{\mathrm{sky}} = 0.6$, and compare your results to those obtained in Example 6.12. In both cases, the collectors are tilted at 50° to the horizontal. The relation $h_c = 0.84\Delta T^{0.38}$ W/m^2 K can be used to estimate the convective heat transfer between the absorber plate and cover plate for $5 < \Delta T < 35$ K.

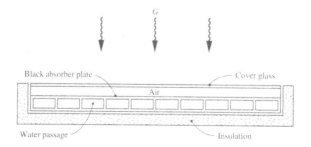

6-86. A 1 cm–thick redwood ($k = 0.1$ W/m K) patio cover is surfaced with a layer of black tar paper ($\varepsilon = 0.90, \alpha_s = 0.95$), and the underside is stained with redwood sealer-stain ($\varepsilon = 0.88$). Typical summer noon conditions include a solar irradiation of 950 W/m^2, a clear sky, and ambient air at 27°C and 50% RH. Wind blowing through the patio gives an estimated convective heat transfer coefficient on both sides of the cover of 8 W/m^2 K. An important factor determining the comfort of people sitting on the patio is the radiosity of the underside of the cover, since it is this radiosity that determines the radiant heating experienced by the people. Estimate the radiosity for the following situations:

(i) For the cover as designed.
(ii) If the underside is painted with aluminum paint ($\varepsilon = 0.22$).
(iii) If, instead, the top of the cover is painted with a white paint
　　　($\alpha_s = 0.30, \varepsilon = 0.85$) and kept clean.
(iv) The combination of (ii) and (iii).

6-87. A 10 cm–square horizontal plate coated with a black paint ($\varepsilon = 0.94$) is exposed to a clear night sky when the ambient air is still at 290 K, 1 atm, and 80% RH. If the backside of the plate is perfectly insulated, will dew form on the plate? Use the Brunt formula for sky emittance.

6-88. Two surfaces, each of area 1 m^2, are located 8 m apart, as shown in the sketch. Surface 1 is black and is maintained at 800 K. Calculate the irradiation on surface 2 due to surface 1.

6-89. A small spherical thermocouple is located in the center of an open-ended circular tube of length H and radius R. Determine the shape factor for radiation transfer between the thermocouple and the inside wall of the tube.

6-90. Consider an elemental area dA_1 and a rectangle A_2 in a plane parallel to the plane of dA_1. The normal through dA_1 passes through a corner of the rectangle, as shown in the sketch. Show that the shape factor F_{12} is

$$F_{12} = \frac{1}{2\pi} \left[\frac{X}{(X^2 + Z^2)^{1/2}} \sin^{-1} \frac{Y}{(X^2 + Y^2 + Z^2)^{1/2}} \right.$$
$$\left. + \frac{Y}{(Y^2 + Z^2)^{1/2}} \sin^{-1} \frac{X}{(X^2 + Y^2 + Z^2)^{1/2}} \right]$$

6-91. Liquid oxygen flows through a 2 cm–O.D. tube that is contained in an evacuated 4 cm–I.D. shell. The oxygen is at its normal boiling point of 90 K, and shell inner surface is at 250 K. Estimate the heat gain per meter length if the inner surface has an emittance of 0.03, the outer surface has an emittance of 0.5, and

   (i) the surfaces are both diffuse.
   (ii) the surfaces are both specular.
   (iii) the inner surface is diffuse, and the outer surface is specular.

6-92. Liquid oxygen is contained in a thin-walled spherical container 48 cm in diameter, which in turn is enclosed in a concentric container 50 cm in diameter. The space in between is evacuated. The inner surface is at 95 K, and the outer surface is at 250 K. Determine the oxygen boil-off rate if

   (i) both surfaces facing each other are gray with an emittance of 0.05 and reflect diffusely.
   (ii) both surfaces reflect specularly.
   (iii) the inner surface reflects diffusely, and the outer surface reflects specularly.

6-93. A louvered shutter on a spacecraft is 1 m wide and ∼50 cm high. When fully open, the passages are 4 cm wide and 7 cm long. The louvers are 1 mm thick. The spacecraft interior can be taken to be a black enclosure at 320 K, and the exterior is outer space at ∼0 K. Determine the heat loss through the shutter

   (i) if the louvers are diffuse reflectors.
   (ii) if the louvers have a coating that reflects specularly with a normal reflectance of 0.89.

6-94. Figure 6.2 shows a spherical shell with a small hole, called a *hohlraum* ("hollow space" in German). Since almost all energy entering the hole is trapped, the hole is said to approximate a blackbody, irrespective of the emittance of the interior surface. If the sphere is isothermal at T_1, the thermodynamic arguments of Section 6.2.3 require that the radiation leaving the hole is diffuse with intensity $I_b(T_1)$. To get some idea concerning the accuracy of this approximation, consider a 10 cm-diameter spherical aluminum shell ($\varepsilon = 0.06$) at 300 K with a 5 mm-diameter hole. Estimate the radiant energy flux leaving the hole, and use your result to evaluate the *hohlraum* blackbody approximation if the surroundings are black at

   (i) 600 K.
  (ii) 300 K.

6-95. Liquid sodium flows in a 4 cm–O.D. tube that is surrounded by a concentric tube of 5 cm I.D. The inner and outer tubes are at 800 K and 500 K, respectively, and the space between is evacuated. The outer surface of the inner tube has an emittance of 0.15, and the inner surface of the outer tube has an emittance of 0.25. Calculate the heat flow per unit length for the following situations:

   (i) The inner tube surface is a diffuse reflector and the outer tube surface is a specular reflector.
  (ii) The inner tube surface is a specular reflector and the outer tube surface is a diffuse reflector.

6-96. Determine the wavelength range encompassing the 20–80% fraction of radiant energy emitted by black surfaces at the following temperatures:

   (i) The normal boiling point of liquid oxygen, 90 K.
  (ii) A room temperature of 300 K.
 (iii) The normal boiling point of sodium, 1156 K.
 (iv) The temperature of a tungsten lamp filament, 3000 K.
  (v) A Sun temperature of 5700 K.

6-97. Use the data for stainless steel in Table A.6*b* to determine the variation of total absorptance of a surface at room temperature to black radiation from a source at temperatures $300 < T_e < 1600$ K.

6-98. Use the data in Table A.6*a* and Fig. 6.28 to estimate the total hemispherical absorptance at $T_s = 300$ K to black radiation from a source at $T_e = 2000$ K for

   (i) aluminum foil.
  (ii) Inconel X rolled plate.
 (iii) lapped 303 stainless steel.

6-99. Estimate the total hemispherical absorptance of a chromium-plated surface at 300 K when exposed to black radiation source at

   (i) 300 K.
  (ii) 1500 K.

Use the data in Table A.6*b*. Compare your results with values calculated from the formula in Table A.6*a*.

6-100. Estimate the total hemispherical absorptance for a gold-plated surface at 300 K exposed to radiation at 2000 K. Use the data in Table A6.*b*. Compare the result with the value calculated from the formula in Table A.6*a*.

6-101. Calculate the spectral absorptance to normally incident radiation for copper using the formula given in Table A.6*a*. Compare the result to the tabulated values for radiation incident at 25° given in Table A.6*b*.

6-102. Firebrick is often made of alumina and operates typically at 1600 K. Estimate the total emittance of firebrick at this temperature using the data for flame-sprayed alumina in Table A6.*b*.

6-103. Aluminum ingots are melted in a furnace for the purpose of alloying and recasting. The aluminum charge is covered with a layer of oxide (dross), which increases its absorptance and inhibits further oxidation. In practice the thickness of the dross must be carefully controlled so that it does not have too large a thermal resistance. Calculate the radiant heat transfer to 50 m^2 of dross surface at 1630 K from the furnace wall at 1600 K. Both the dross and the walls can be assumed to have the spectral characteristics at flame-sprayed alumina, as given in Table A.6*b*. Neglect furnace gas radiation.

6-104. Using the data in Table A.6*b*, calculate the absorptance of white epoxy paint for extraterrestrial solar radiation. Compare your result to value given in Table A.5*a*.

6-105. The effect of the Earth's atmosphere on the solar spectrum can be roughly approximated by assuming (*a*) the Sun radiates as a blackbody at 5760 K, (*b*) all radiation with wavelengths shorter than 0.4 μm and longer than 1.8 μm are absorbed, and (*c*) on a clear day 25% of the remaining radiation is absorbed.

  (i) Estimate the solar irradiation on the Earth's surface.
  (ii) Prepare a table of λT versus fractional function for terrestrial solar radiation based on these assumptions.
  (iii) Estimate the absorptance of white epoxy paint to terrestrial solar radiation.

6-106. Estimate the total hemispherical absorptance of a chromium-plated surface at 400 K to black radiation from a black source at the Sun temperature of 5700 K.

6-107. An isothermal furnace with a small aperture approximating a blackbody is a useful device for calibrating radiation thermometers and heat flux meters. A thermostat controlling the power input to the furnace is to be selected to maintain the furnace at a nominal temperature of 1800 K, such that the variation in spectral intensity at 0.7 μm is less than 1%. What is the allowable variation in the furnace temperature?

6-108. A stainless steel surface at room temperature is exposed to radiation from a black source at 2000 K. Estimate its total hemispherical absorptance using the following information:

  (i) Equation (6.84), Table A.6*b*, and Fig. 6.28
  (ii) The formula in Table A.6*a*, and Fig. 6.28

6-109. Estimate the total hemispherical emittance of a gold-plated surface at 600 K. Use the data in Table A.6b and assume that the spectral emittance is independent of surface temperature in this temperature range.

6-110. In a metals processing experiment on the space shuttle, it is required to melt a spherical 0.5 cm–diameter gold sample acoustically levitated in a furnace maintained at 800°C. The additional heating required is supplied by a focused beam of radiation from a xenon arc lamp. The following table gives the spectral distribution of intensity for the lamp. Available data show that the absorptance of gold near its melting point is not much different from room temperature values in the wavelength range of significance.

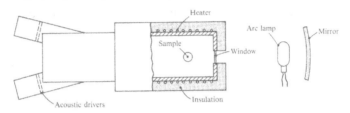

(i) Estimate an appropriate value of total hemispherical absorptance to be used in melting rate calculations.

(ii) In a proposed design the arc delivers 25 W to the sample, and the convective heat transfer coefficient on the sphere associated with a flow induced by the acoustical field is estimated to be 140 W/m^2 K. Can the sample be melted? If so, how long will it take to melt the sample once it reaches the melting temperature? The furnace walls can be taken to be black. The enthalpy of melting of gold is 6.44×10^4 J/kg.

$\lambda\,(\mu m)$	$\Delta I_\lambda/I\,(\%)$	$\lambda\,(\mu m)$	$\Delta I_\lambda/I\,(\%)$
0–0.25	2.49	0.60–0.70	12.12
0.25–0.35	3.50	0.70–0.80	9.36
0.35–0.40	4.58	0.80–1.00	24.86
0.40–0.45	5.25	1.00–1.50	13.23
0.45–0.50	5.63	1.50–2.00	4.24
0.50–0.60	10.94	2.00–2.50	1.04

6-111. Repeat part (i) of Exercise 6-110 for a mercury-xenon lamp with a spectral distribution of intensity as given in the following table.

$\lambda\,(\mu m)$	$\Delta I_\lambda/I\,(\%)$	$\lambda\,(\mu m)$	$\Delta I_\lambda/I\,(\%)$
0–0.26	0	0.60–0.70	3.47
0.26–0.30	11.99	0.70–0.80	2.04
0.30–0.40	23.96	0.80–1.00	6.13
0.40–0.45	9.19	1.00–1.50	16.35
0.45–0.50	1.31	1.50–2.00	10.67
0.50–0.60	14.88		

6-112. A long tunnel furnace has a 2 m square-cross section and is heated by combustion gases in a muffle enclosing the side walls. Machine parts to be heat-treated are conveyed the length of the furnace on a slowly moving conveyor belt: the closely packed parts form the floor of the furnace. A test of furnace performance is made by mounting thermocouples on the upper surfaces of some of the parts, and halfway along the furnace these thermocouples record a temperature of about 980 K. In addition, an infrared radiometer is used at the same location: sightings on the roof, sides, and floor yield 10 mV, 14 mV, and 4.5 mV, respectively. The radiometer was calibrated by sighting a heated black plate outside the furnace with a known emittance of 0.91, also at 980 K, and gave a reading of 3.4 mV (a linear relation between voltage output and T^4 can be assumed). Estimate the radiant heat flux into the parts, the emittance of the parts, and the roof temperature.

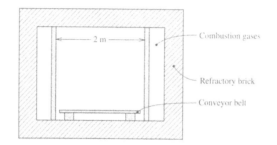

6-113. A solar sail is to be evaluated as a propulsion device in outer space at a distance of 1.11 astronomical units from the Sun, with the Sun modeled as a blackbody at 5765 K. The side of the sail facing the Sun is aluminized to give $\varepsilon_\lambda = 0.90$ for $\lambda < 0.3\mu$m, and $\varepsilon_\lambda = 0.02$ for $\lambda > 0.3\mu$m. The back of the sail is gray, with $\varepsilon = 0.5$. Determine the temperature of the sail.

6-114. Plain glass has a spectral transmittance that is nearly unity for 0.2 μm $< \lambda <$ 3.0 μm and nearly zero for other wavelengths. A tinted glass under consideration has a spectral transmittance of nearly unity for 0.5 μm $< \lambda <$ 1.0 μm and zero for other wavelengths. If the Sun can be modeled as a blackbody at 5765 K, find the ratio of solar energy transmitted through the tinted glass to that transmitted through the plain glass.

6-115. A window glass has a spectral transmittance of 0.95 for 0.33 μm $< \lambda <$ 2.1 μm and zero at other wavelengths. Determine the transmittance to black radiation from the following sources:

   (i) The interior of a room at 300 K.
   (ii) A solar absorber plate at 360 K, when used as a cover glass.
   (iii) The Sun into a room at 300 K.

6-116. Determine the rate at which energy is emitted by the Sun in the ultraviolet, visible, and infrared portions of the spectrum. Take the visible range of wavelengths as 0.4 μm $< \lambda <$ 0.7 μm. Model the Sun as a blackbody at 5700 K.

6-117. The spectral transmittance of the glass shell of a 100 W electric light bulb is approximately $\tau_\lambda = 0.9$ for wavelengths from 0.4 μm to 2.5 μm and zero outside this range.

    (i) Calculate the fraction of the energy leaving the lamp directly as radiation for a filament temperature of 3000 K.

    (ii) If the bulb can be approximated as a 7 cm–diameter sphere, estimate the glass temperature in 20°C air.

6-118. A gas mixture at 1500 K and 1 atm contains 20% H_2O and 80% N_2 by volume. Calculate the total emissivity for path lengths in the range 5 cm to 1 m. Plot your result.

6-119. Exhaust gas from a furnace burning a hydrocarbon fuel is at 2 atm pressure. The composition can be approximated as 10% CO_2, 10% H_2O, and 80% N_2 by volume. Calculate the total emissivity as a function of temperature in the range 1100 to 1800 K. Also calculate the absorptivity to radiation from a black wall at 800 K. Take a path length of 0.5 m.

6-120. A spherical enclosure of 1 m diameter contains 90% N_2 and 10% H_2O by volume, at 3 atm and 1200 K. The walls are black.

    (i) Calculate the gas emissivity.

    (ii) Calculate the gas absorptivity for wall temperatures in the range 400–2000 K.

Plot the results.

6-121. Radiation from a black source at 600 K passes through air at 400 K, 1 atm, and 15% relative humidity. Determine the fraction that is absorbed over path lengths of 0.3 m, 3 m, and 30 m.

6-122. Exhaust gas from a methane combustor is at 1400 K and 3 atm pressure for stoichiometric combustion with air. Estimate the radiant heat transfer to the walls of a 0.5 m-square duct maintained at 700 K. Assume that the walls are black. Repeat the calculation for an excess air ratio of 2.

6-123. A gray isothermal gas is contained in a long enclosure of square cross section, with 1 m sides. The walls are gray with emittance 0.60; one wall is at 350 K, and the others are refractory. The gas temperature is 3000 K, and its total absorption coefficient $\kappa = 0.3\,\text{m}^{-1}$. Calculate the rate of radiant heat transfer through the cold surface.

6-124. A cubical enclosure with 1 m sides contains 80% N_2, 10% H_2O, and 10% CO_2 by volume at 2.5 atm total pressure. The walls are black and are maintained at 600 K. If turbulence in the gas maintains an approximately uniform temperature, estimate the rate of temperature drop of the gas when it is at

    (i) 1600 K.

    (ii) 1000 K.

6-125. A long furnace with a 2 m–square cross section has three refractory walls, and one wall cooled to 400 K. Combustion gases at 1200 K flow through the furnace. The cooled wall is gray, with an emittance of 0.7, and the gases can be modeled as isothermal and gray, with a total absorption coefficient $\kappa = 0.15$ m^{-1}. Determine the radiant heat transfer to the cooled wall.

6-126. A kerosene combustor can be approximated as a long cylinder of 20 cm diameter. The exhaust gas contains 10% CO_2, 10% H_2O, and 80% N_2 by volume and is at 1400 K and 2 atm pressure. Calculate the radiant heat transfer to black walls at 700 K. Compare your result to an estimate of the convective heat transfer if the bulk velocity of the gas is 50 m/s.

6-127. Flue gas at 1100 K and 1 atm pressure contains 8% H_2O, 8% CO_2, and 84% N_2 by volume. The gas flows at 3 m/s along a 1 m–square flue with brick-lined walls. If the brick surface is at 1000 K, estimate the radiative and convective heat transfer to walls per meter length of flue. Take the brick surface to be black.

6-128. A cubical vessel has sides 2 m long with internal surfaces of emittance 0.8. The vessel contains a gas at 1000 K that can be assumed gray with a total absorption coefficient of 0.26. The side walls are lined with refractory brick and can be assumed adiabatic. The upper surface is maintained at 1200 K while the base is at 800 K. Calculate the radiant heat transfer to the base, and compare your answer to the result if no gas were present.

6-129. Steam at 4 atm pressure flows between black parallel plates spaced 5 cm apart. At a location where the bulk steam temperature is 1200 K and the plates are at 800 K, calculate the radiative and convective heat transfer to the plates.

6-130. Two large, black parallel plates are 40 cm apart and are maintained at temperatures of 1200 and 800 K. Between the plates is a gas mixture at 1000 K and 1 atm pressure, containing 15% H_2O, 10% CO_2, and 75% N_2 by volume. Calculate the radiant heat transfer to the cold plate, and compare your answer to the result if no gas were present.

6-131. A gray isothermal gas at 1600 K is contained in a 3 m–diameter spherical shell maintained at 700 K. The total absorption coefficient of the gas is 0.3 m^{-1}. Determine the radiant heat transfer to the shell if the inner surface

  (i) is black.
  (ii) has an emittance of 0.5.

6-132. A direct-fired furnace is cylindrical in shape, with a diameter of 5 m and a height of 3 m. The side walls and roof are refractory, and the gas is at 1600 K, with an effective gray absorption coefficient of 0.5 m^{-1}. The load covers the furnace floor and has an emittance of 0.5. Determine the radiant heat transfer to the load when it is at 1000 K.

Gas at 1600 K
5 m
3 m

6-133.   A direct-fired furnace is in the form of a cube with 3 m sides, with refractory side walls and roof. The furnace gases are at 1700 K and have an effective gray absorption coefficient of 0.3 m^{-1}. If the load covers the furnace floor and has an emittance of 0.7, determine the radiant heat transfer to the load when it is at 1100 K.

6-134.   A hydrogen-fueled combustor can be approximated as a long cylinder of diameter 0.2 m. The combustion gases are at 2000 K and 3 atm pressure for stoichiometric combustion with air. If the walls are to be cooled to 900 K, determine the radiant heat transfer to the walls as a function of their surface emittance in the range $0.2 < \varepsilon < 1.0$. Plot your result.

6-135.   A tube bank has 5 cm–diameter O.D. tubes in a square array at 10 cm pitch. Combustion products containing 11% H_2O, 11% CO_2, and 78% N_2, by volume, at 2 atm pressure flow across the tube bank. At a location in the bank where the gas temperature is 1200 K and the tube wall temperature is 500 K, calculate the radiant heat transfer to the tubes. Take $\varepsilon = 0.9$ for the tube surface. Also estimate the convective heat transfer if the gas velocity upstream of the bank is 8 m/s.

6-136.   A hydrogen-fueled combustor has a 0.3 m–diameter circular cross section. The combustion products contain 14% H_2O and 86% N_2 by volume and are at 1500 K and 5 atm pressure. The combustor walls are maintained at 800 K. Calculate the radiative heat flux to the walls for wall emittances in the range 0.1 to 0.7.

6-137.   A hydrogen-fueled combustor has a 0.5 m-square cross section. The walls have an emittance of 0.4 and are cooled by hydrogen from the fuel tank flowing in channels underneath the skin. The combustion products contain 15% H_2O and 85% N_2 by volume, and, at the location under consideration, are at 1600 K and 6 atm pressure. The total heat flux into the coolant cannot exceed 240 kW/m^2. What is the lowest allowable wall temperature if the convective heat transfer coefficient is estimated to be 200 W/m^2 K?

6-138.   A tube bank is 30 rows deep and has 15 mm–O.D. tubes in a staggered arrangement with both transverse and longitudinal pitches of 30 mm. Combustion products of coke with 20% excess air enter the bank at 5 m/s, and water boils in the tubes at 540 K. Halfway along the bank the gases are at 1100 K and 5 atm pressure.

   (i) Calculate the convective heat flux to the tubes.
   (ii) Estimate the radiative heat flux to the tubes, and hence determine whether gas radiation effects should be included in the design of this heat exchanger.

   Use a tube surface emittance of 0.85.

6-139.   The radiant section of a furnace can be modeled as a 4 m–diameter cylinder completely lined with water tubes at an average temperature of 600 K. The combustion gases are at 1 atm and an average temperature of 1400 K, and contain 10% CO_2 and 10% H_2O by volume. If the tube walls have an emittance of 0.8, estimate the radiant heat flux to the tubes.

CONDENSATION, EVAPORATION, AND BOILING

CONTENTS

7.1   INTRODUCTION

Condensation of vapors and the evaporation and boiling of liquids are commonplace in power and process engineering. In a fossil- or nuclear-fueled power plant, water is boiled in a boiler, and steam is condensed in a condenser. In an oil refinery, oil is evaporated in a distillation column to give a variety of vapor products, which are eventually condensed into liquid fuels such as gasoline and kerosene. In a multistage flash desalination plant, water vapor is produced by evaporation from brine and is subsequently condensed as pure water. Condensation processes require that the enthalpy of phase change be removed by a coolant, and evaporation and boiling processes require that this enthalpy be supplied from an energy source. Since the enthalpy of phase change is relatively large, particularly for water ($\sim 2.5 \times 10^6$ J/kg), the associated heat transfer rates are also usually large. In most industrial processes, the vapor and liquid phases both flow through the heat exchanger. Thus, the heat transfer to the phase interface is essentially a convective process, but it is often complicated by an irregular interface, for example, bubbles or drops.

In Section 7.2, *film condensation* is analyzed. Vapor condenses on a cooled surface, and the liquid drains off the surface under the action of gravity. An analysis of laminar film condensation on a vertical wall is presented first, followed by analyses of *wavy laminar* and *turbulent* film condensation on a vertical wall and laminar film condensation on a horizontal tube. The effects of vapor velocity and vapor superheat are analyzed using a Couette flow model of the vapor boundary layer. In Section 7.3 the reverse process is treated, in which a film of liquid flows down a heated wall while evaporation takes place from the film surface. Boiling on heater surfaces immersed in a pool of saturated liquid is dealt with in Section 7.4. After the regimes of *pool boiling* and the *boiling curve* are introduced, boiling inception is discussed, and correlations are presented for *nucleate boiling*, the *peak heat flux*, and *film boiling*.

Heatpipes are evaporator-condenser systems in which liquid is returned to the evaporator by capillary action. If the working fluid has a high enthalpy of vaporization, a small vapor flow can transport a large amount of thermal energy. Heatpipes are used extensively for space vehicle thermal control. In Section 7.5, a brief introduction to heatpipes is provided, and the performance of fixed-conductance and of variable-conductance heatpipes is analyzed. Two computer programs accompany Chapter 7, namely, PHASE for film condensation and evaporation, and BOIL for pool boiling. Both programs have a menu of six fluids: water, ammonia, nitrogen, mercury, R-22, and R-134a.

7.2   FILM CONDENSATION

When a heat exchanger wall exposed to a vapor is cooled below the saturation temperature of the vapor, condensation will occur on the surface. If the vapor is pure, the saturation temperature corresponds to the total pressure; however, if there is a vapor-noncondensable gas mixture, the saturation temperature corresponds to the partial pressure of the vapor. Two distinct modes of condensation are possible. If the condensate wets a vertical wall, a complete film of liquid will cover the wall, and the film thickness will grow as it flows down the wall under the action of gravity: this mode is called **film condensation**. Since the enthalpy of condensation given up at the surface of the film must be transferred across the film to the wall, the film presents a thermal resistance to heat transfer. Such condensate films are very thin, and the corresponding heat transfer coefficients are relatively large. For example, when steam at a saturation temperature of 305 K condenses on a 2 cm-O.D. tube with a wall temperature of 300 K, the average film thickness is $\sim 50 \mu$m, and the average heat transfer coefficient is 11,700 W/m^{2}K. If the condensate flow rate is small because either the temperature difference is small or the wall is short, the surface of the film will be smooth and the flow laminar. At higher flow rates, waves will form on the surface to give what is called *wavy laminar* flow, and at yet higher flow rates, the flow becomes *turbulent*.

If the condensate does not wet the wall, because either it is dirty or it has been treated with a nonwetting agent, droplets of condensate nucleate at small pits and other imperfections on the surface, and they grow rapidly by direct vapor condensation upon them and by coalescence. When the droplets become sufficiently large, they flow down the surface under the action of gravity and expose bare metal in their tracks, where further droplet nucleation is initiated. This mode is called **dropwise condensation** and is shown in Fig. 7.1. Most of the heat transfer is through drops of less than 100 μm diameter. The thermal resistance of such drops is small; hence, heat transfer coefficients for dropwise condensation can be very high: values of up to 300,000 W/m^{2}K have been measured. These high heat transfer coefficients are attractive to the design engineer; thus, considerable effort has been spent developing nonwetting heat exchanger surfaces, but no entirely satisfactory method has been found. If the surface is treated with a nonwetting agent such as stearic acid to promote dropwise condensation, the effect lasts but a few days until the promoter is washed off or oxidized. Continuous adding of the promoter to the vapor is expensive and contaminates the condensate. Bonding a polymer such as Teflon to the surface to promote dropwise condensation is expensive and adds an additional thermal resistance. Gold plating is also expensive and often does not have a lasting effect.

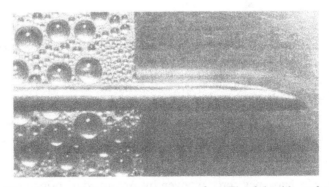

Figure 7.1 Steam condensing on a copper surface. The right-side surface is clean, giving filmwise condensation. The left side has a thin coating of cupric oleate, which causes dropwise condensation. The cylindrical temperature probe has a diameter of 1.7 mm. (Photograph courtesy of Professor J. W. Westwater, University of Illinois, Urbana.)

When an industrial or laboratory condenser is put into service for the first time, the initial mode of condensation is *dropwise* unless particular care is taken to clean the system. But in a matter of hours or days, a mixed mode of dropwise and film condensation will develop, and finally the condensation will be entirely filmwise. Thus, current design practice is to always assume film condensation for a conservative estimate of heat transfer surface area.

7.2.1 Laminar Film Condensation on a Vertical Wall

Figure 7.2 depicts a vertical wall exposed to a saturated vapor at pressure P and saturation temperature $T_{\text{sat}} = T_{\text{sat}}(P)$. The wall could be flat or could be the outside surface of a vertical tube. If the surface is maintained at a temperature $T_w < T_{\text{sat}}$, vapor will continuously condense on the wall and, if the liquid phase wets the surface well, will flow down the wall in a thin film. Provided the condensation rate is not too large, there will be no discernible waves on the film surface, and the flow in the film will be laminar. The first step in the analysis of this *laminar film condensation* process is to examine the dynamics of the flow of a thin liquid film.

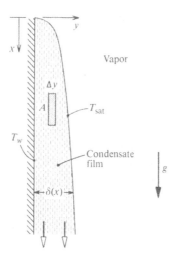

Figure 7.2   Coordinate system for film condensation on a vertical wall.

The Fluid Dynamics Problem

Since the velocity of a thin liquid film is relatively low, the forces required to accelerate the film are usually negligible, and the motion of the film is determined by a simple balance of buoyancy and viscous forces. Figure 7.3a depicts an element of fluid of area A, located between planes y and $y + \Delta y$, and the forces acting upon it. The buoyancy force is given by Archimedes' principle as $(\rho_l - \rho_v)gA\Delta y$ where ρ_l is the liquid density and ρ_v is the vapor density. The viscous forces are given by Newton's law of viscosity, with $\mu_l \left(\partial u/\partial y\right)|_y A$ acting upward and $\mu_l \left(\partial u/\partial y\right)|_{y+\Delta y} A$

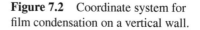

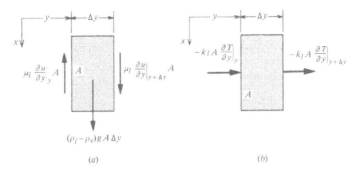

Figure 7.3 (*a*) Force balance on a fluid element $A\Delta y$ in a laminar falling film. (*b*) Energy balance on an elemental control volume $A\Delta y$ in a laminar falling film.

acting downward. Summing these forces gives

$$\mu_l \left.\frac{\partial u}{\partial y}\right|_{y+\Delta y} A - \mu_l \left.\frac{\partial u}{\partial y}\right|_y A + (\rho_l - \rho_v)gA\Delta y = 0$$

Dividing by $A\Delta y$ and assuming that the liquid viscosity μ is constant gives

$$\mu_l \frac{(\partial u/\partial y)|_{y+\Delta y} - (\partial u/\partial y)|_y}{\Delta y} + (\rho_l - \rho_v)g = 0$$

Then letting $\Delta y \to 0$,

$$\mu_l \frac{\partial^2 u}{\partial y^2} + (\rho_l - \rho_v)g = 0 \tag{7.1}$$

which is a differential equation for the velocity profile across the film, $u(y)$, at any particular value of x. Two boundary conditions are required. At the wall, there is the no-slip condition for a real fluid,

$$y = 0: \quad u = 0 \tag{7.2a}$$

and at the surface of the film, we assume that there is negligible drag of the vapor. If the film thickness is $\delta(x)$, the required boundary condition is

$$y = \delta: \quad \frac{\partial u}{\partial y} = 0 \tag{7.2b}$$

where $\delta(x)$ is a function still to be determined. The condition of negligible drag is applicable to many situations for which the vapor velocities are not too high, as will be discussed further in Section 7.2.4. Integrating Eq. (7.1) gives

$$\frac{\partial u}{\partial y} = -\frac{(\rho_l - \rho_v)g}{\mu_l}y + C_1$$

Applying the boundary condition Eq. (7.2b),

$$0 = -\frac{(\rho_l - \rho_v)g}{\mu_l}\delta + C_1$$

Solving for C_1 and substituting back,

$$\frac{\partial u}{\partial y} = \frac{(\rho_l - \rho_v)g}{\mu_l}(\delta - y)$$

Integrating again,

$$u = \frac{(\rho_l - \rho_v)g}{\mu_l}\left(\delta y - \frac{y^2}{2}\right) + C_2$$

Applying the boundary condition Eq. (7.2a) gives $C_2 = 0$. Rearranging,

$$u = \frac{(\rho_l - \rho_v)g\delta^2}{\mu_l}\left[\frac{y}{\delta} - \frac{1}{2}\left(\frac{y}{\delta}\right)^2\right] \tag{7.3}$$

Equation (7.3) gives the velocity profile $u(y)$, which is seen to be parabolic. The velocity is a maximum at the film surface and is obtained by setting $y = \delta$:

$$u_\delta = \frac{(\rho_l - \rho_v)g\delta^2}{2\mu_l} = \frac{g\delta^2}{2\nu_l}\frac{(\rho_l - \rho_v)}{\rho_l} \tag{7.4}$$

The *mass flow rate per unit width* of film is denoted Γ and has units kg/m s. It is obtained by integration of Eq. (7.3):

$$\Gamma = \int_0^\delta \rho_l u \, dy = \frac{g\delta^3(\rho_l - \rho_v)}{3\nu_l} \tag{7.5}$$

The Heat Transfer Problem

With the fluid dynamics of the film understood, we now turn to the heat transfer problem. Two special features of the problem are: (1) the film velocity is relatively low, and (2) temperature gradients in the x direction are negligible since both the wall and film surface are isothermal. Thus, in applying the steady-flow energy equation, Eq. (1.4), to the elemental control volume shown in Fig. 7.3b, the enthalpy convection term $\dot{m}\Delta h$ can be neglected, as can be the x-direction conduction contribution to $\dot{Q}$. Then the heat conducted out at y equals the heat conducted in at $y + \Delta y$:

$$-k_l A \frac{\partial T}{\partial y}\bigg|_y = -k_l A \frac{\partial T}{\partial y}\bigg|_{y+\Delta y}$$

Dividing through by $k_l A \Delta y$, for k_l assumed constant, and rearranging,

$$\frac{(\partial T/\partial y)|_{y+\Delta y} - (\partial T/\partial y)|_y}{\Delta y} = 0$$

Then letting $\Delta y \to 0$,

$$\frac{\partial^2 T}{\partial y^2} = 0 \tag{7.6}$$

Two boundary conditions are required. At the surface of the film, continuity of temperature requires that $T = T_{\text{sat}}$, where T_{sat} is the saturation temperature corresponding to the pressure P of the vapor:

$$y = \delta : \quad T = T_{\text{sat}} \tag{7.7a}$$

For the second boundary condition, we will assume that the wall is isothermal at temperature T_w:

$$y = 0 : \quad T = T_w \tag{7.7b}$$

Integrating Eq. (7.6) twice,

$$T = C_1 y + C_2$$

and fitting the boundary conditions,

$$T - T_w = \frac{y}{\delta}(T_{\text{sat}} - T_w) \tag{7.8}$$

which is a linear temperature profile since this problem is identical to that of conduction across a plane slab, which was treated in Section 1.3.1. The heat flux into the wall is simply the heat flux across the film, which from Eq. (1.9) is

$$k_l \left.\frac{\partial T}{\partial y}\right|_w = \frac{\dot{Q}}{A} = \frac{k_l(T_{\text{sat}} - T_w)}{\delta} \quad \text{(ignoring the } - \text{ sign)}$$

The local heat transfer coefficient h is defined as this heat flux divided by the temperature difference across the film $(T_{\text{sat}} - T_w)$:

$$h = \frac{k_l \left.(\partial T/\partial y)\right|_w}{T_{\text{sat}} - T_w} = \frac{k_l}{\delta} \tag{7.9}$$

Determination of the Film Thickness

To find how δ and hence h vary down the wall, we need to make mass and energy balances for an element of film Δx long, as shown in Fig. 7.4. If the condensation rate is denoted $\dot{m}''$ [kg/m^2 s] and taken to be a positive quantity, then mass conservation for unit width of film requires

$$\dot{m}'' \Delta x + \int_0^{\delta(x)} \rho_l u \, dy \Big|_x = \int_0^{\delta(x)} \rho_l u \, dy \Big|_{x+\Delta x}$$

Dividing by Δx and letting $\Delta x \to 0$,

$$\dot{m}'' = \frac{d}{dx} \int_0^{\delta(x)} \rho_l u \, dy = \frac{d\Gamma}{dx} \tag{7.10}$$

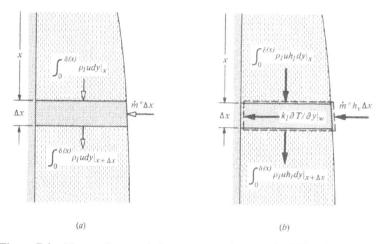

Figure 7.4   Mass and energy balances on an element of a falling film Δx long.

The steady-flow energy equation applied to the control volume shown in Fig. 7.4b requires that the enthalpy flowing into the volume equal the enthalpy flowing out plus the heat conducted to the wall:

$$\dot{m}'' h_v \Delta x + \int_0^{\delta(x)} \rho_l u h_l \, dy\Big|_x = \int_0^{\delta(x)} \rho_l u h_l \, dy\Big|_{x+\Delta x} + k_l \frac{\partial T}{\partial y}\Big|_w \Delta x$$

where h_v is the enthalpy of the vapor and h_l is the liquid enthalpy. Rearranging, dividing by Δx, and letting $\Delta x \to 0$,

$$k_l \frac{\partial T}{\partial y}\Big|_w = \dot{m}'' h_v - \frac{d}{dx} \int_0^{\delta(x)} \rho_l u h_l \, dy \tag{7.11}$$

Substituting Eq. (7.10) in Eq. (7.11) gives

$$k_l \frac{\partial T}{\partial y}\Big|_w = \frac{d}{dx} \int_0^{\delta(x)} \rho_l u (h_v - h_l) dy$$

which, for a constant liquid specific heat c_{pl}, becomes

$$k_l \frac{\partial T}{\partial y}\Big|_w = \frac{d}{dx} \int_0^{\delta(x)} \rho_l u [h_{fg} + c_{pl}(T_{sat} - T)] dy \tag{7.12}$$

The *subcooling* term $c_{pl}(T_{sat} - T)$ accounts for the sensible heat given up by the condensate as it cools below the saturation temperature T_{sat}. This term is usually quite small: for example, when steam condenses at 1 atm pressure and there is a 10 K temperature drop across the film, the subcooling term is at most 2% of the enthalpy of phase change h_{fg}. Thus, for simplicity, the subcooling term will be ignored for the present, and Eq. (7.12) can be rewritten using Eq. (7.5) as

$$k_l \left.\frac{\partial T}{\partial y}\right|_w = h_{fg} \frac{d}{dx} \int_0^{\delta(x)} \rho_l u \, dy = h_{fg} \frac{d\Gamma}{dx} \tag{7.13}$$

Now, from Eq. (7.9),

$$k_l \left.\frac{\partial T}{\partial y}\right|_w = \frac{k_l(T_{sat} - T_w)}{\delta}$$

and from Eq. (7.5),

$$\frac{d\Gamma}{dx} = \frac{g(\rho_l - \rho_v)}{\nu_l} \delta^2 \frac{d\delta}{dx}$$

Substituting in Eq. (7.13),

$$\frac{k_l(T_{sat} - T_w)}{\delta} = \frac{h_{fg} g(\rho_l - \rho_v)}{\nu_l} \delta^2 \frac{d\delta}{dx}$$

or

$$dx = \frac{h_{fg} g(\rho_l - \rho_v)}{k_l(T_{sat} - T_w)\nu_l} \delta^3 d\delta$$

If condensation begins at the top of the wall, $\delta = 0$ at $x = 0$, and integration is straightforward.

$$\int_0^x dx = \frac{h_{fg} g(\rho_l - \rho_v)}{k_l(T_{sat} - T_w)\nu_l} \int_0^\delta \delta^3 d\delta$$

$$x = \frac{h_{fg} g(\rho_l - \rho_v)\delta^4}{4k_l(T_{sat} - T_w)\nu_l}$$

$$\delta = \left[\frac{4xk_l(T_{sat} - T_w)\nu_l}{h_{fg} g(\rho_l - \rho_v)} \right]^{1/4} \tag{7.14}$$

The film thickness is seen to grow like $x^{1/4}$. Substituting for δ in Eq. (7.9) gives the heat transfer coefficient h,

$$h = \left[\frac{h_{fg} g(\rho_l - \rho_v)k_l^3}{4x(T_{sat} - T_w)\nu_l} \right]^{1/4} \tag{7.15}$$

which is seen to be proportional to $x^{-1/4}$ and $(T_{sat} - T_w)^{-1/4}$. The average heat transfer coefficient for a wall of height L is obtained by integration:

$$\bar{h} = \frac{1}{L} \int_0^L h \, dx = \frac{4}{3} h_{x=L}$$

$$\bar{h} = 0.943 \left[\frac{h_{fg} g(\rho_l - \rho_v)k_l^3}{L(T_{sat} - T_w)\nu_l} \right]^{1/4} \tag{7.16}$$

Equation (7.16) was first derived by W. Nusselt in 1916 [1]. The assumptions of negligible effects of fluid acceleration and heat convection are often called the *Nusselt assumptions* and have been used for many different falling film problems.

Effect of Liquid Subcooling

For liquids with relatively low enthalpies of phase change, such as many refrigerants, we should make a correction for the subcooling term, which was deleted in obtaining Eq. (7.13) from Eq. (7.12). If the subcooling term is retained in the analysis, the result is identical to Eq. (7.16) with h_{fg} replaced by a *modified latent heat* (enthalpy of phase change) h_{fg}' given by:

$$h_{fg}' = h_{fg} + \frac{3}{8}c_{pl}(T_{sat} - T_w) \tag{7.17a}$$

The derivation of Eq. (7.17a) is given as Exercise 7–4. More exact analysis, in which Nusselt assumptions are not invoked, gives [2]

$$h_{fg}' = h_{fg} + \left(0.683 - \frac{0.228}{Pr_l}\right)c_{pl}(T_{sat} - T_w); \ Pr > 0.6 \tag{7.17b}$$

Effect of Variable Fluid Properties

A reference temperature scheme to account for variable-property effects has been established by numerically solving the exact governing equations allowing for variable fluid properties [3]. The scheme requires that $h_{fg}' = h_{fg} + 0.35c_{pl}(T_{sat} - T_w)$ be used to account for liquid subcooling, with h_{fg} evaluated at the saturation temperature; all other liquid phase properties are evaluated at $T_r = T_w + \alpha(T_{sat} - T_w)$, where α is given in Table 7.1 for various fluids. Use of this scheme also accounts for the minor errors introduced through use of the Nusselt assumptions. For liquids not listed in Table 7.1, Eq. (7.17b) should be used for h_{fg}', with h_{fg} evaluated at T_{sat} and all other properties evaluated at the mean film temperature.

Table 7.1   Values of α in the reference temperature $T_r = T_w + \alpha(T_{sat} - T_w)$ for laminar film condensation on a vertical wall [3].

Fluid	T_{sat} K	ΔT K	α –
Carbon tetrachloride	290–320	5–20	0.07
Ethyl alcohol	300–350	5–20	0.12
n-propyl alcohol	330	10–20	0.15
n-butyl alcohol	290–330	20	0.25
t-butyl alcohol	330	10–20	0.29
Ethylene glycol	330	20	0.29
Glycerol	330	20	0.32
Water	280–380	1–30	0.33
Ammonia	270–320	20	0.61
Propane	350–370	20	1.00

EXAMPLE 7.1   Laminar Film Condensation of Steam

Saturated steam condenses on the outside of a 5 cm-diameter vertical tube, 50 cm high. If the saturation temperature of the steam is 302 K, and cooling water maintains the wall temperature at 299 K, calculate: (i) the average heat transfer coefficient, (ii) the total condensation rate, and (iii) the film thickness at the bottom of the tube.

Solution

Given: Film condensation of saturated steam on a vertical surface.

Required: (i) Average heat transfer coefficient $\overline{h}$, (ii) total condensation rate $\dot{m}$, and (iii) film thickness δ.

Assumptions: 1. Effect of tube curvature negligible.
2. Effect of liquid subcooling negligible.
3. Laminar flow.

(i) The average heat transfer coefficient is given by Eq. (7.16) with h_{fg} replaced by h'_{fg}:

$$\overline{h} = 0.943 \left[\frac{h'_{fg} g (\rho_l - \rho_v) k_l^3}{L(T_{sat} - T_w) v_l} \right]^{1/4}$$

However, for water, we let $h'_{fg} \simeq h_{fg}$ and evaluate h_{fg} at the saturation temperature of 302 K; from Table A.12a, $h_{fg} = 2.432 \times 10^6$ J/kg, and $\rho_v = 0.03$ kg/m^3. Table 7.1 gives $\alpha = 0.33$ for water, so the liquid phase properties are evaluated at $T_r = 299 + 0.33(302 - 299) = 300$ K: from Table A.8, $k_l = 0.611$ W/m K, $\rho_l = 996$ kg/m^3, $v_l = 0.87 \times 10^{-6}$ m^2/s.

$$\overline{h} = 0.943 \left[\frac{h_{fg} g (\rho_l - \rho_v) k_l^3}{L(T_{sat} - T_w) v_l} \right]^{1/4} = 0.943 \left[\frac{(2.432 \times 10^6)(9.81)(996 - 0.03)(0.611)^3}{(0.5)(3)(0.87 \times 10^{-6})} \right]^{1/4}$$

$$= 7570 \text{ W/m}^2 \text{ K}$$

(ii) The total condensation rate $\dot{m}$ is

$$\dot{m} = \frac{\dot{Q}}{h_{fg}} = \frac{\overline{h} \Delta T A}{h_{fg}} = \frac{(7570)(3)(\pi)(0.05)(0.5)}{(2.432 \times 10^6)} = 7.33 \times 10^{-4} \text{ kg/s}$$

(iii) The film thickness is obtained from Eq. (7.5): for $\rho_v \ll \rho_l$,

$$\delta = \left(\frac{3 v_l \Gamma}{\rho_l g} \right)^{1/3}$$

The mass flow rate per unit width of film Γ is

$$\Gamma = \frac{\dot{m}}{\pi D} = \frac{(7.33 \times 10^{-4})}{(\pi)(0.05)} = 4.67 \times 10^{-3} \text{ kg/m s}$$

Hence,

$$\delta = \left[\frac{3(0.87 \times 10^{-6})(4.67 \times 10^{-3})}{(996)(9.81)} \right]^{1/3} = 1.08 \times 10^{-4} \text{ m (0.108 mm)}$$

Comments

1. After reading Section 7.2.2, check the film flow to see if it is laminar as assumed.

2. In practice, the wall temperature will seldom be constant along the tube (see Exercise 7–6).

7.2.2 Wavy Laminar and Turbulent Film Condensation on a Vertical Wall

The Reynolds number of a falling film is defined in terms of the bulk velocity u_b and hydraulic diameter D_h of the film:

$$u_b = \frac{\Gamma}{\rho_l \times 1 \times \delta} = \frac{\Gamma}{\rho_l \delta}; \quad D_h = \frac{4A_c}{\mathscr{P}} = \frac{4 \times 1 \times \delta}{1}$$

for a unit width of film and recognizing that the only wetted surface is the wall. Thus,

$$\mathrm{Re} = \frac{\rho_l u_b D_h}{\mu_l} = \frac{\rho_l (\Gamma/\rho_l \delta) 4\delta}{\mu_l} = \frac{4\Gamma}{\mu_l} \tag{7.18}$$

The film Reynolds number is usually used to characterize the film flow.[1] Three regimes of film flow may be distinguished: *laminar, wavy laminar,* and *turbulent*. At low Reynolds numbers, the flow is laminar, and the surface of the film appears to be smooth. As the Reynolds number increases, ripples appear on the surface of the film, which at still higher Reynolds numbers develops into a complex three-dimensional wave pattern, as shown in Fig. 7.5. The waves cause some mixing of

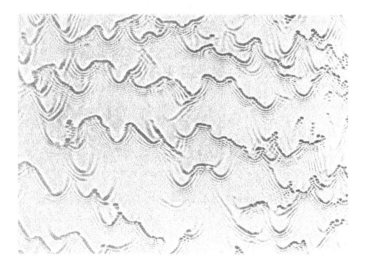

Figure 7.5 Surface waves on a laminar falling film, Reynolds number $\mathrm{Re} = 4\Gamma/\mu_l = 280$. (Photograph courtesy of Mr. B. Dooher, University of California, Los Angeles.)

---

[1] Note that in USSR literature, the factor of 4 is omitted in the film Reynolds number definition.

the liquid, but the base flow remains laminar until, at a sufficiently high flow rate, shear-induced instabilities result in a transition to turbulent flow throughout the film. For water at ~300 K, the onset of wavy laminar flow is at Re $\simeq$ 30; transition to turbulent flow in the outer region of the film occurs at Re $\simeq$ 1000, and transition is complete in the inner region at Re $\simeq$ 1800. As was demonstrated in Section 7.2.1, analysis of flow in the laminar regime is straightforward, and excellent agreement between theory and experiment is obtained. On the other hand, rigorous analysis of flow in the wavy laminar and turbulent regimes is very difficult, owing to the complexity of the wave motion on the surface and the nature of turbulent flow. There are large, long-wavelength waves caused by gravity, and superimposed small, short-wavelength waves caused by surface tension, and both play a role in heat transport near the liquid surface. Thus, for engineering purposes, it is necessary to use experimental data for the wavy laminar and turbulent regimes. If the problem is formulated correctly, the use of such data is straightforward.

The first step is to reorganize the analysis of laminar film condensation given in Section 7.2.1. For algebraic simplicity, we assume $\rho_v \ll \rho_l$, which is valid except near the critical point, so that Eqs. (7.5) and (7.18) can be rearranged to give the film thickness δ in terms of the film Reynolds number as

$$\frac{\delta}{(v_l^2/g)^{1/3}} = \left(\frac{3}{4}\mathrm{Re}\right)^{1/3} \tag{7.19}$$

where the grouping $(v_l^2/g)^{1/3}$ is seen to have the dimensions of length. The same grouping is used as a length scale to define a Nusselt number for condensation,

$$\mathrm{Nu} = \frac{h(v_l^2/g)^{1/3}}{k_l} \tag{7.20}$$

since other possible length scales are less suitable. The film thickness δ is appropriate on physical grounds but is an unknown and varies down the wall. Use of the distance down the wall x is inappropriate since the local heat transfer coefficient depends only on the local film thickness and not on the prior history of the film. Then, substituting for δ from Eq. (7.9) and using Eq. (7.19) gives

$$\mathrm{Nu} = \left(\frac{3}{4}\mathrm{Re}\right)^{-1/3}; \quad \mathrm{Re} < 30, \text{ laminar} \tag{7.21}$$

Equation (7.21) is valid for laminar films irrespective of wall temperature variation, whether or not the film has a zero or finite initial thickness, or, indeed, whether condensation or evaporation is taking place. It applies *locally* at a particular value of x and states that the local Nusselt number depends only on the local film Reynolds number or, equivalently, on the local value of Γ or δ.

The next step is to obtain relationships equivalent to Eq. (7.21) from experiment for wavy laminar and turbulent flows. Correlations of experimental data for water given by Chun and Seban [4] are

$$\mathrm{Nu} = 0.822\mathrm{Re}^{-0.22}; \quad 30 < \mathrm{Re} < \mathrm{Re_{tr}}, \text{ wavy laminar} \tag{7.22}$$

$$\mathrm{Nu} = 3.8 \times 10^{-3}\mathrm{Re}^{0.4}\mathrm{Pr}_l^{0.65}; \quad \mathrm{Re_{tr}} < \mathrm{Re}, \text{ turbulent} \tag{7.23}$$

where, for water, transition to a turbulent film takes place at

$$\mathrm{Re_{tr}} = 5800\mathrm{Pr}_l^{-1.06} \tag{7.24}$$

Since liquid Prandtl numbers are strongly temperature-dependent, so is $\mathrm{Re_{tr}}$. Equations (7.21), (7.22), and (7.23) are shown plotted in Fig. 7.6.

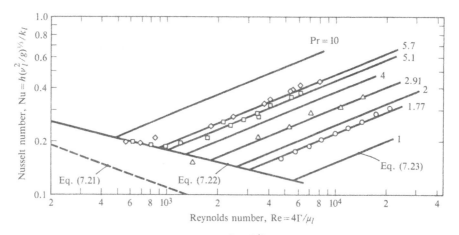

Figure 7.6   Local Nusselt number $\mathrm{Nu} = h(v_l^2/g)^{1/3}/k_l$ as a function of Reynolds number $\mathrm{Re} = 4\Gamma/\mu_l$, for the wavy laminar and turbulent regimes on a vertical surface: Eqs. (7.21), (7.22), and (7.23) and the experimental data for water of Chun and Seban [4].

To obtain a relation between average Nusselt number and film Reynolds number, we now rewrite the energy balance Eq. (7.13) using Eqs. (7.9), (7.18), and (7.20) as

$$\frac{dx}{(v_l^2/g)^{1/3}} = \frac{\mu_l h_{\mathrm{fg}}}{4k_l(T_{\mathrm{sat}} - T_w)\mathrm{Nu}} d\mathrm{Re}$$

or, introducing the Prandtl number $\mathrm{Pr}_l = c_{pl}\mu_l/k_l$ and the Jakob number $\mathrm{Ja}_l = c_{pl}(T_{\mathrm{sat}} - T_w)/h_{\mathrm{fg}}$,

$$\frac{dx}{(v_l^2/g)^{1/3}} = \frac{\mathrm{Pr}_l}{4\mathrm{Ja}_l} \frac{d\mathrm{Re}}{\mathrm{Nu}} \tag{7.25}$$

For an isothermal wall and condensation commencing at $x = 0, \Gamma$ and $\mathrm{Re} = 0$, and Eq. (7.25) can be integrated down a wall of length L as

$$\frac{L}{(v_l^2/g)^{1/3}} = \frac{\mathrm{Pr}_l}{4\mathrm{Ja}_l} \int_0^{\mathrm{Re_L}} \frac{d\mathrm{Re}}{\mathrm{Nu}} \tag{7.26}$$

Finally, an overall energy balance on unit width of film from $x = 0$ to $x = L$ requires that the heat transfer to the wall equal the enthalpy of phase change given up by the condensate:

$$\bar{h}L(T_{\mathrm{sat}} - T_w) = \Gamma_L h_{\mathrm{fg}} = \frac{\mu_l h_{\mathrm{fg}}}{4}\mathrm{Re}_L$$

then

$$\overline{Nu} = \frac{\overline{h}(v_l^2/g)^{1/3}}{k_l} = \frac{\mu_l h_{fg}(v_l^2/g)^{1/3}}{4k_l L(T_{sat} - T_w)} Re_L$$

or

$$\overline{Nu} = \frac{Pr_l}{4Ja_l} \frac{(v_l^2/g)^{1/3}}{L} Re_L \qquad (7.27)$$

Re_L can be obtained by integrating Eq. (7.26) using appropriate Nu(Re) formulas and, when substituted in Eq. (7.27), gives the desired result.

Laminar Film Condensation

Consider first the situation where $Re_L < 30$, so that the film is always laminar. Then substitution of Eq. (7.21) into Eq. (7.26) gives

$$\frac{L}{(v_l^2/g)^{1/3}} = \frac{Pr_l}{4Ja_l} \left(\frac{3}{4}\right)^{1/3} \int_0^{Re_L} Re^{1/3} dRe$$

$$= \frac{Pr_l}{4Ja_l} \left(\frac{3}{4}\right)^{4/3} Re_L^{4/3}$$

or

$$Re_L = \frac{4}{3} \left[\frac{4Ja_l}{Pr_l} \frac{L}{(v_l^2/g)^{1/3}} \right]^{3/4}$$

Substituting in Eq. (7.27) gives the average Nusselt number as

$$\overline{Nu} = \frac{4}{3} \left[\frac{Pr_l}{4Ja_l} \frac{(v_l^2/g)^{1/3}}{L} \right]^{1/4} \qquad (7.28)$$

Of course, Eq. (7.28) is simply Eq. (7.16) rearranged, with $\rho_v \ll \rho_l$.

Wavy Laminar Film Condensation

Next, consider the situation where $30 < Re_L < Re_{tr}$, with Re_{tr} for water given by Eq. (7.24). The integration in Eq. (7.26) is now performed for $0 < Re < 30$ using Eq. (7.21) for Nu, and for $30 < Re < Re_L$ using Eq. (7.22):

$$\frac{L}{(v_l^2/g)^{1/3}} = \frac{Pr_l}{4Ja_l} \left[\left(\frac{3}{4}\right)^{1/3} \int_0^{30} Re^{1/3} dRe + \frac{1}{0.822} \int_{30}^{Re_L} Re^{0.22} dRe \right]$$

$$= \frac{Pr_l}{4Ja_l} \left[\left(\frac{3}{4}\right)^{4/3} (30)^{4/3} + \frac{1}{(0.822)(1.22)} \left(Re_L^{1.22} - 30^{1.22}\right) \right]$$

$$= 1.00 \frac{Pr_l}{4Ja_l} Re_L^{1.22} \qquad (7.29)$$

The value of the coefficient 1.00 is fortuitous; however, the two constant terms cancel out because both Eqs. (7.21) and (7.22) give the same value of Nu at Re $= 30$. Solving,

$$\mathrm{Re}_L = \left[\frac{4\mathrm{Ja}_l}{\mathrm{Pr}_l} \frac{L}{(v_l^2/g)^{1/3}} \right]^{0.82}$$

Substituting in Eq. (7.27) gives the average Nusselt number as

$$\overline{\mathrm{Nu}} = \left[\frac{\mathrm{Pr}_l}{4\mathrm{Ja}_l} \frac{(v_l^2/g)^{1/3}}{L} \right]^{0.18} \tag{7.30}$$

Turbulent Film Condensation

Finally, consider the situation where $\mathrm{Re}_L > \mathrm{Re}_{\mathrm{tr}}$, that is, transition to a turbulent film takes place on the wall. First, from Eq. (7.29), the location of transition x_{tr} is given by

$$\frac{x_{\mathrm{tr}}}{(v_l^2/g)^{1/3}} = 1.00 \frac{\mathrm{Pr}_l}{4\mathrm{Ja}_l} \mathrm{Re}_{\mathrm{tr}}^{1.22} \tag{7.31}$$

Equation (7.26) is now integrated using Eq. (7.23) for $\mathrm{Re}_{\mathrm{tr}} < \mathrm{Re} < \mathrm{Re}_L$:

$$\frac{L}{(v_l^2/g)^{1/3}} = \frac{\mathrm{Pr}_l}{4\mathrm{Ja}_l} \left\{ \mathrm{Re}_{\mathrm{tr}}^{1.22} + \frac{1}{(3.8 \times 10^{-3})\mathrm{Pr}_l^{0.65}} \int_{\mathrm{Re}_{\mathrm{tr}}}^{\mathrm{Re}_L} \mathrm{Re}^{-0.4} d\mathrm{Re} \right\}$$

$$= \frac{x_{\mathrm{tr}}}{(v_l^2/g)^{1/3}} + \frac{\mathrm{Pr}_l^{0.35}}{4(3.8 \times 10^{-3})(0.6)\mathrm{Ja}_l} \left[\mathrm{Re}_L^{0.6} - \mathrm{Re}_{\mathrm{tr}}^{0.6} \right]$$

Solving for Re_L and substituting in Eq. (7.27),

$$\overline{\mathrm{Nu}} = \frac{\mathrm{Pr}_l}{4\mathrm{Ja}_l} \frac{(v_l^2/g)^{1/3}}{L} \left\{ \frac{9.12 \times 10^{-3}\mathrm{Ja}_l(L - x_{\mathrm{tr}})}{(v_l^2/g)^{1/3}\mathrm{Pr}_l^{0.35}} + \mathrm{Re}_{\mathrm{tr}}^{0.6} \right\}^{10/6} \tag{7.32}$$

which is the desired result.

Effect of Variable Properties

A reference temperature scheme to account for variable-property effects has not been established for wavy laminar and turbulent film condensation. Thus, the schemes developed for laminar film condensation described in Section 7.2.1 are tentatively recommended for use here.

Computer Program PHASE

Item 1 in the computer program PHASE calculates the average heat transfer coefficient and condensate flow rate per unit width for film condensation on an isothermal vertical surface according to the analysis of Section 7.2.2. Local heat transfer coefficients and transition Reynolds numbers for wavy laminar and turbulent flow can be calculated from the formulas given in Section 7.2.2, which are, strictly speaking, valid only for water. User-supplied constants and exponents for these formulas are also allowed. Variable-property and liquid subcooling effects are calculated following the recommendation given in Section 7.2.1. (The effect of vapor superheat is accounted for according to the recommendation in Section 7.2.4.) SI units, with temperature in kelvins, must be used.

There is a menu of six fluids: (1) water, (2) ammonia, (3) nitrogen, (4) mercury, (5) R-22, and (6) R-134a.

EXAMPLE 7.2   Condensation of Steam on a Long Vertical Tube

In a multiple-effect evaporator, saturated steam at 27,150 Pa condenses on the outside of a 5 cm-diameter, 8 m-long vertical tube. If the tube outer wall can be taken to be at 320 K, calculate the average heat transfer coefficient and the total condensation rate.

Solution

Given: Film condensation of steam on a vertical surface.

Required: Average heat transfer coefficient $\overline{h}$ and total condensation rate $\dot{m}$.

Assumptions: 1. Effect of wall curvature negligible.
2. Effect of liquid subcooling negligible.
3. Isothermal surface.

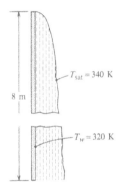

We first check to see if transition to turbulent flow takes place on the wall. For this purpose, and for later use, we evaluate water properties at the reference temperature, $T_r = T_w + \alpha(T_{\text{sat}} - T_w)$, where, from Table 7.1, $\alpha = 0.33$ for water. From Table A.12a, $T_{\text{sat}} = 340.0$ K; thus,

$$T_r = 320 + 0.33(340 - 320) = 327 \text{ K}$$

From Table A.8, $k_l = 0.649$ W/m K, $\rho_l = 986$ kg/m^3, $c_{pl} = 4177$ J/kg K, $\nu_l = 0.53 \times 10^{-6}$ m^2/s, $Pr_l = 3.36$. Also, at the saturation temperature of 340 K, Table A.12a gives $h_{\text{fg}} = 2.341 \times 10^6$ J/kg. Then from Eq. (7.24)

$$\text{Re}_{\text{tr}} = 5800Pr_l^{-1.06} = 5800(3.36)^{-1.06} = 1605$$

From Eq. (7.31), $x_{\text{tr}} = (\nu_l^2/g)^{1/3}(Pr_l/4Ja_l)\text{Re}_{\text{tr}}^{1.22}$, where

$$(\nu_l^2/g)^{1/3} = [(0.53 \times 10^{-6})^2/9.81]^{1/3} = 3.06 \times 10^{-5} \text{ m}$$

$$\text{Ja}_l = c_{pl}(T_{\text{sat}} - T_w)/h_{\text{fg}} = (4177)(20)/(2.341 \times 10^6) = 0.0357$$

Hence, $x_{\text{tr}} = (3.06 \times 10^{-5})[3.36/(4)(0.0357)](1605)^{1.22} = 5.86$ m.

Transition is seen to take place about three-quarters of the way down the wall. Equation (7.32) gives the average Nusselt number:

$$\overline{Nu} = \frac{Pr_l}{4Ja_l}\frac{(v_l^2/g)^{1/3}}{L}\left\{\frac{9.12\times10^{-3}Ja_l(L-x_{tr})}{(v^2/g)^{1/3}Pr_l^{0.35}}+Re_{tr}^{0.6}\right\}^{10/6}$$

$$=\frac{(3.36)(3.06\times10^{-5})}{(4)(0.0357)(8)}\left\{\frac{(9.12\times10^{-3})(0.0357)(8-5.86)}{(3.06\times10^{-5})(3.36)^{0.35}}+1605^{0.6}\right\}^{10/6}$$

$$=0.1897$$

$$\overline{h}=\frac{\overline{Nu}\,k_l}{(v_l^2/g)^{1/3}}=\frac{(0.1897)(0.649)}{3.06\times10^{-5}}=4020\text{ W/m}^2\text{ K}$$

$$\dot{m}=\frac{\dot{Q}}{h_{fg}}=\frac{\overline{h}\Delta TA}{h_{fg}}=\frac{(4020)(20)(\pi)(0.05)(8)}{2.341\times10^6}=4.32\times10^{-2}\text{ kg/s}$$

Solution using PHASE

The required input in SI units is:

Item 1
Y (correlations for water)
Fluid = 1 (H$_2$O)
Pressure = 27,150
Vapor temperature = 0 (gives saturation value)
T_w = 320
L = 8

PHASE gives the output:

Relevant property data
Re_L = 2080
Γ_L = 0.272 kg/m s
$\overline{h}$ = 4034 W/m^2 K

Comments

1. Using PHASE to obtain the condensation rate gives $\dot{m}=\Gamma_L\pi D=(0.272)(\pi)(0.05)=4.27\times10^{-2}$ kg/m s.

2. Notice that Nu values are $\sim 10^{-1}$, which is much smaller than the values for convection calculated in Chapter 4. Why?

7.2.3 Laminar Film Condensation on Horizontal Tubes

Shell-and-tube condensers, in which condensation takes place on a bank of horizontal tubes, are widely used in power plants and the process industries. Figure 7.7 depicts laminar film condensation on a single horizontal tube and a force balance on a liquid element. The buoyancy force per unit volume of liquid is now $(\rho_l-\rho_v)g\sin\phi$,

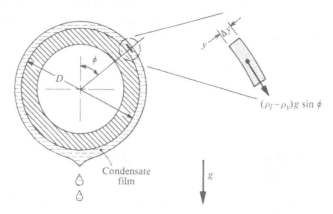

Figure 7.7   Coordinate system and gravitational force
for film condensation on a horizontal condenser tube.

with angle ϕ measured from the top of the tube. Thus, Eq. (7.1) becomes

$$\mu_l \frac{\partial^2 u}{\partial y^2} + (\rho_l - \rho_v)g\sin\phi = 0 \tag{7.33}$$

whereas Eq. (7.6) remains unchanged:

$$\frac{\partial^2 T}{\partial y^2} = 0 \tag{7.34}$$

Integrating Eq. (7.33) as before gives the mass flow rate per unit width as

$$\Gamma = \frac{(\rho_l - \rho_v)g\sin\phi\ \delta^3}{3\nu_l} \tag{7.35}$$

and the film Reynolds number $4\Gamma/\mu_l$ as

$$\mathrm{Re} = \frac{4}{3}\frac{(\rho_l - \rho_v)g\sin\phi\ \delta^3}{\rho_l \nu_l^2} \tag{7.36}$$

Integrating Eq. (7.34) as before gives the heat transfer coefficient

$$h = \frac{q}{T_{\mathrm{sat}} - T_w} = \frac{k_l\,(\partial T/\partial y)|_w}{T_{\mathrm{sat}} - T_w} = \frac{k_l}{\delta} \tag{7.37}$$

where q has been taken to be positive for condensation. Hence, the energy balance
Eq. (7.13) may be written as

$$q = k_l \left.\frac{\partial T}{\partial y}\right|_w = \frac{k_l(T_{\mathrm{sat}} - T_w)}{\delta} = h_{\mathrm{fg}}\frac{d\Gamma}{dx} \tag{7.38}$$

where coordinate x is measured from $\phi = 0$.

To proceed, the wall boundary condition must be specified: here we will obtain the solution for an isothermal wall T_w; the solution for a constant heat flux is given as Exercise 7–22. Equation (7.36) is solved for δ,

$$\delta = \left(\frac{3}{4} \frac{\rho_l v_l^2}{(\rho_l - \rho_v)g} \frac{\mathrm{Re}}{\sin\phi} \right)^{1/3}$$

and, together with the relations $\Gamma = \mu_l \, \mathrm{Re}/4$ and $x = (D/2)\phi$ is substituted in Eq. (7.38), to obtain

$$\frac{2k_l(T_{\mathrm{sat}} - T_w)D}{\mu_l h_{\mathrm{fg}}} \left(\frac{4}{3} \frac{(\rho_l - \rho_v)g}{\rho_l v_l^2} \right)^{1/3} \sin^{1/3}\phi \, d\phi = \mathrm{Re}^{1/3} d\mathrm{Re}$$

Integrating with $\mathrm{Re} = 0$ at $\phi = 0$, and $\mathrm{Re} = \mathrm{Re}_\pi$ at $\phi = \pi$ gives

$$\mathrm{Re}_\pi = \left[\left(\frac{4}{3} \right) \frac{2k_l(T_{\mathrm{sat}} - T_w)D}{\mu_l h_{\mathrm{fg}}} \left(\frac{4}{3} \frac{(\rho_l - \rho_v)g}{\rho_l v_l^2} \right)^{1/3} \int_0^\pi \sin^{1/3}\phi \, d\phi \right]^{3/4} \quad \textbf{(7.39)}$$

Notice that we choose to use Re rather than δ as the dependent variable. The Reynolds number Re is zero at $\phi = 0$ because Γ, the flow rate, is zero at the top of the tube by symmetry, whereas the film thickness δ is finite and unknown at $\phi = 0$.

An overall energy balance on half the tube gives the average heat transfer coefficient:

$$\bar{h}(\pi D/2)(T_{\mathrm{sat}} - T_w) = \Gamma_\pi h_{\mathrm{fg}} = \mu_l h_{\mathrm{fg}} \mathrm{Re}_\pi / 4 \quad \textbf{(7.40)}$$

Substituting for Re_π from Eq. (7.39) and rearranging gives

$$\bar{h} = \left(\frac{4}{3\pi} \right) \left(\frac{1}{2} \right)^{1/4} \left(\int_0^\pi \sin^{1/3}\phi \, d\phi \right)^{3/4} \left[\frac{(\rho_l - \rho_v)g h_{\mathrm{fg}} k_l^3}{v_l D(T_{\mathrm{sat}} - T_w)} \right]^{1/4}$$

From mathematical tables, we can find that

$$\int_0^{\pi/2} \sin^n x \, dx = \frac{\pi^{1/2}}{2} \frac{\Gamma(n/2 + 1/2)}{\Gamma(1 + n/2)}, \quad \text{for } n > -1$$

$$(\Gamma \text{ is the gamma function})$$

$$= 1.2936 \quad \text{for } n = 1/3$$

Hence, $\int_0^\pi \sin^{1/3}\phi \, d\phi = 2\int_0^{\pi/2} \sin^{1/3}\phi \, d\phi = 2.5872$, and finally

$$\bar{h} = 0.728 \left[\frac{(\rho_l - \rho_v)g h_{\mathrm{fg}} k_l^3}{v_l D(T_{\mathrm{sat}} - T_w)} \right]^{1/4} \quad \textbf{(7.41)}$$

Most commercial condensers condense steam on a large bundle of horizontal tubes, and condensate drips from one tube to the next, as shown in Fig. 7.8. If the analysis leading up to Eq. (7.41) is repeated for N tubes arranged vertically above one another, such that the condensate drips from one tube to the next, the result is

$$\bar{h} = 0.728 \left[\frac{(\rho_l - \rho_v)g h_{\mathrm{fg}} k_l^3}{N v_l D(T_{\mathrm{sat}} - T_w)} \right]^{1/4} \quad \textbf{(7.42)}$$

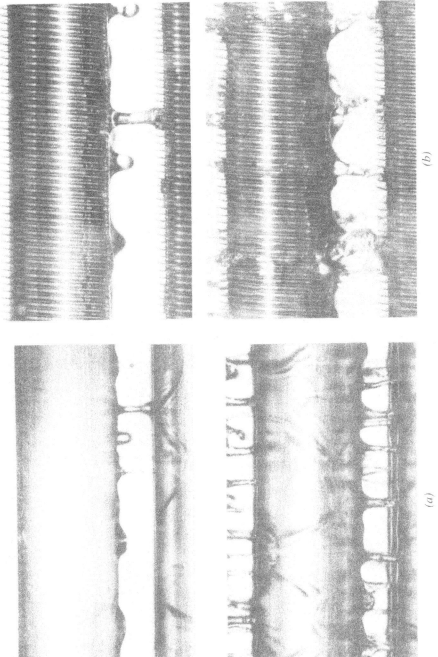

(a)

(b)

Figure 7.8 Condensation of downward-flowing R-113 vapor on in-line bundles of horizontal tubes, (*a*) Smooth tubes, 15.9 mm diameter. $P = 0.12$ MPa, $\Delta T = 10$ K. 1st row, $Re_D = 19$; 13th row, $Re_D = 150$. (*b*) Low-fin tubes (see Exercise 7–34) 15.6 mm diameter, $P = 0.11$ MPa, $\Delta T = 3.5$ K. 1st row, $Re_D = 50$; 13th row, $Re_D = 550$. (Photographs courtesy of Professor H. Honda, Kyushu University, Kasuga.)

Effect of Variable Fluid Properties

The reference property scheme described for laminar film condensation on a vertical wall is also applicable to condensation on a horizontal tube.

Computer Program PHASE

Item 2 of PHASE calculates the average heat transfer coefficient and condensate flow rate per unit length, for laminar film condensation on isothermal horizontal tubes, based on the analysis of Section 7.2.3. Variable-property and liquid subcooling effects are calculated following the recommendations given in Section 7.2.1; vapor superheat effects are accounted for following the recommendations in Section 7.2.4.

EXAMPLE 7.3   Condensation of Refrigerant-134a on a Single Horizontal Tube

Saturated R-134a vapor at 320 K condenses on the outside of a 2 cm-O.D., 1 mm-wall-thickness horizontal brass tube, through which flows coolant water. At an axial location where the bulk water temperature is 295 K and the inside heat transfer coefficient is 4500 W/m^{2}K, determine the average heat flux and condensation rate per unit length.

Solution

Given: Condensation of R-134a on a horizontal tube.

Required: Heat flux and condensation rate per unit length.

Assumptions: 1. The outer tube surface is isothermal.
2. Laminar film condensation.

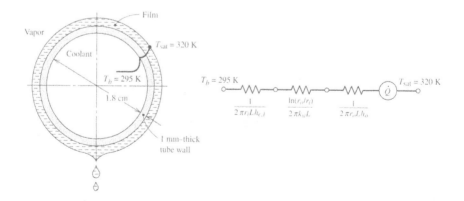

Referring to Section 2.3.1, the thermal circuit is as shown. Using Eq. (2.17), the overall heat transfer coefficient-area product is

$$\frac{1}{UA} = \frac{1}{2\pi r_i L h_{c,i}} + \frac{\ln(r_o/r_i)}{2\pi k_{\text{brass}} L} + \frac{1}{2\pi r_o L h_o}$$

If U is based on the outside area of the tube, $A = 2\pi r_o L$, and

$$\frac{1}{U} = \frac{1}{(r_i/r_o)h_{c,i}} + \frac{r_o \ln(r_o/r_i)}{k_{\text{brass}}} + \frac{1}{h_o}$$

$$= \frac{1}{(0.009/0.01)(4500)} + \frac{(0.01)\ln(0.01/0.009)}{111} + \frac{1}{h_o}$$

$$= 246.9 \times 10^{-6} + 9.49 \times 10^{-6} + \frac{1}{h_o}$$

where $k_{\text{brass}} = 111$ W/m K was obtained from Table A.1a. hence,

$$\dot{Q}/A = U(T_{\text{sat}} - T_b) = \frac{320 - 295}{2.56 \times 10^{-4} + 1/h_o}$$

To evaluate the condensation heat transfer coefficient h_o, we first guess a wall temperature T_w of 300 K and evaluate liquid R-134a properties at the mean film temperature of 310 K. From Table A.8, $k_l = 0.0763$ W/m K, $\rho_l = 1160$ kg/m³, $\nu_l = 0.160 \times 10^{-6}$ m²/s, $c_{pl} = 1484$ J/kg K, $\text{Pr}_l = 3.62$. From Table A.12f, at 320 K, $h_{fg} = 1.554 \times 10^5$ J/kg, and $\rho_v = 1/0.01466 = 68.2$ kg/m³. Equation (7.17b) gives h'_{fg}:

$$h'_{fg} = h_{fg} + \left(0.683 - \frac{0.228}{\text{Pr}_l}\right) c_{pl}(T_{\text{sat}} - T_w)$$

$$= 1.554 \times 10^5 + \left(0.683 - \frac{0.228}{3.62}\right)(1484)(320 - 300) = 1.738 \times 10^5 \text{ J/kg}$$

Substituting in Eq. (7.41), with h'_{fg} replacing h_{fg} to account for liquid subcooling,

$$h_o = 0.728 \left(\frac{(\rho_l - \rho_v)gh'_{fg}k_l^3}{\nu_l D(T_{\text{sat}} - T_w)}\right)^{1/4}$$

$$= 0.728 \left(\frac{(1160 - 68)(9.81)(1.738 \times 10^5)(0.0763)^3}{(0.160 \times 10^{-6})(0.02)(20)}\right)^{1/4}$$

$$= 1380 \text{ W/m}^2\text{K}$$

$$\frac{\dot{Q}}{A} = \frac{25}{(2.56 + 7.24)10^{-4}} = 2.550 \times 10^4 \text{ W/m}^2 \quad \text{(based on tube outside area)}$$

We now calculate T_w and compare it to our guessed value of 300 K.

$$T_{\text{sat}} - T_w = \frac{\dot{Q}/A}{h_o} = \frac{2.550 \times 10^4}{1380} = 18.5 \text{ K}; \quad T_w = 320 - 18.5 = 301.5 \text{ K}$$

For a second iteration, use of new values of h'_{fg} and properties is not justified; thus, $h_o = (18.5/20)^{-1/4}(1380) = 1407$ W/m² K; $\dot{Q}/A = 2.586 \times 10^4$ W/m². The condensation rate per unit length of tube is

$$\dot{m} = \frac{(\dot{Q}/A)\pi D}{h'_{fg}} = \frac{(2.586 \times 10^4)(\pi)(0.02)}{1.738 \times 10^5} = 9.35 \times 10^{-3} \text{ kg/s per meter}$$

Comments

1. Use PHASE to check h_o.

2. The film Reynolds number at the bottom of the tube is $(1/2)(4)(9.35 \times 10^{-3})/(1160)(0.160 \times 10^{-6}) = 101$. There is the possibility of wavy laminar flow, but precise criteria for ripple formation on R-134a films have yet to be developed.

3. Note the use of h'_{fg} to obtain $\dot{m}$ from $\dot{Q}/A$ [see Eq. (7.12)].

7.2.4 Effects of Vapor Velocity and Vapor Superheat

To obtain some insight into the effects of vapor velocity and vapor superheat, we return to the problem of laminar film condensation on a vertical wall, analyzed in Section 7.2.1. Figure 7.9 shows expected velocity and temperature profiles if the vapor flows down past the surface at a velocity U_e and has temperature T_e.

Effect of Vapor Velocity

To account for the effect of vapor drag on the liquid film, the boundary condition Eq. (7.2b) must be replaced by

$$y = \delta : \quad \mu_l \frac{\partial u}{\partial y} = \tau_s \tag{7.43}$$

where τ_s is obtained from an appropriate skin friction coefficient:

$$\tau_s = C_f \frac{1}{2} \rho_v U_e^2$$

It might appear that an appropriate expression for C_f can be obtained from Chapter 4, for example, Eq. (4.54) if the vapor boundary layer were laminar, or Eq. (4.60) if

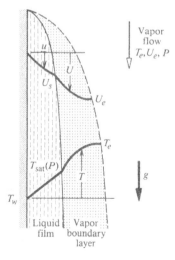

Figure 7.9   Velocity and temperature profiles for film condensation on a vertical wall from a flowing superheated vapor.

it were turbulent. However, these relations are valid only for an impermeable wall, whereas the surface of the liquid film has a velocity component normal to it due to condensation. In fluid mechanics texts, it would be said that there is *suction* at the surface: for typical condensation problems, the suction velocity is relatively large and causes the boundary layer thickness to become almost constant quite close to the leading edge, and the vapor velocity U becomes essentially a function of y only. Figure 7.10 shows a model laminar Couette flow that will be used to determine shear stress for a boundary layer under strong suction. The free-stream velocity is U_e, and the surface velocity is U_s, which will be assumed to be independent of x. In fact, $U_s = u_\delta$ increases with x in the actual film condensation problem, but since we are concerned with situations where $u_\delta \ll U_e$, the error introduced by this assumption is small. An elemental control volume Δy thick, of area A located inside the vapor boundary layer, is shown in Fig. 7.10. Mass conservation requires that

$$\rho_v V|_y A = \rho_v V|_{y+\Delta y} A$$

$$\rho_v V = \text{ Constant } = \rho_v V_s = -\dot{m}'' \tag{7.44}$$

where the condensation rate $\dot{m}''$ [kg/m^2 s] is again taken to be positive. Newton's second law requires that the sum of the forces acting on the volume equal the rate of change of momentum of the fluid flowing through the volume. In the x direction,

$$\mu_v \frac{dU}{dy}\bigg|_{y+\Delta y} A - \mu_v \frac{dU}{dy}\bigg|_y A = \rho_v VU|_{y+\Delta y} A - \rho_v VU|_y A$$

Substituting $\rho_v V = -\dot{m}''$ from Eq. (7.44), dividing by $A\Delta y$, and letting $\Delta y \to 0$,

$$\mu_v \frac{d^2U}{dy^2} = -\dot{m}''\frac{dU}{dy} \tag{7.45}$$

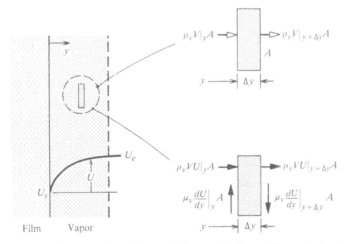

Figure 7.10   Model problem and elemental volumes for analysis of the effect of vapor velocity on condensation heat transfer.

which is the differential equation governing the vapor velocity $U(y)$ for constant vapor viscosity.[2] Appropriate boundary conditions are

$$y = 0: \quad U = U_s \tag{7.46a}$$

$$y \to \infty: \quad U \to U_e \tag{7.46b}$$

Rearranging Eq. (7.45) and integrating,

$$\frac{d^2U/dy^2}{dU/dy} = -\frac{\dot{m}''}{\mu_v}$$

$$\ln \frac{dU}{dy} = -\frac{\dot{m}''}{\mu_v}y + C_0$$

$$\frac{dU}{dy} = C_1 e^{-(\dot{m}''/\mu_v)y}$$

$$U = -C_1 \frac{\mu_v}{\dot{m}''} e^{-(\dot{m}''/\mu_v)y} + C_2$$

Evaluating the integration constants from the boundary conditions gives the velocity profile as

$$\frac{U - U_s}{U_e - U_s} = 1 - e^{-(\dot{m}''/\mu_v)y} \tag{7.47}$$

and the shear stress at the film surface is then found to be

$$\tau_s = \mu_v \frac{dU}{dy}\bigg|_{y=0} = \dot{m}''(U_e - U_s) \tag{7.48}$$

Thus, in this strong suction limit, *the shear stress exerted by the vapor on the liquid film is simply the momentum given up by the condensing vapor* as it decelerates from the free-stream velocity U_e to the film surface velocity U_s. It can be *many times greater* than the shear stress in the absence of suction.

To see when vapor drag might be important, we should compare τ_s with the shear stress at the wall when there is no vapor drag. Then a force balance on the film gives simply $\tau_w = (\rho_l - \rho_v)g\delta$; substituting for δ from Eq. (7.14) and rearranging gives

$$\frac{\tau_s}{\tau_w} = \left[\frac{\mathrm{Ja}_l}{4\mathrm{Pr}_l} \frac{\rho_l}{(\rho_l - \rho_v)} \frac{(U_e - U_s)^2}{gx} \right]^{1/2} \tag{7.49}$$

As expected, vapor drag becomes more important as U_e increases and as the quotient $\mathrm{Ja}_l/\mathrm{Pr}_l$ increases (since $\mathrm{Ja}_l/\mathrm{Pr}_l = k_l(T_{\mathrm{sat}} - T_w)/h_{\mathrm{fg}}\mu_l$ scales the rate of condensation). Vapor drag becomes less important as x increases down the plate, and there is no direct effect of vapor density ρ_v, for $\rho_v \ll \rho_l$.

---

[2] Equation (7.45) can also be obtained from Eq. (5.39) by setting $u = U(y), \rho v = -\dot{m}''$.

Substituting Eq. (7.48) in Eq. (7.43) gives the boundary condition required to account for vapor drag in the analysis of Section 7.2.1 when the condensation rate is sufficiently high; since the velocity profile is continuous, $U_s = u_\delta$, and

$$\mu_l \left.\frac{\partial u}{\partial y}\right|_{y=\delta} = \dot{m}''(U_e - u_\delta) \tag{7.50}$$

The algebra is now more complicated (see Exercise 7–29). For $\rho_v \ll \rho_l$ and $u_\delta \ll U_e$, the local heat transfer coefficient becomes [5]

$$h = \left[\frac{k_l^2 U_e}{8 v_l x} \left\{ 1 + \left(1 + \frac{16 \mathrm{Pr}_l}{\mathrm{Ja}_l} \frac{gx}{U_e^2} \right)^{1/2} \right\} \right]^{1/2} \tag{7.51}$$

Effect of Vapor Superheat

We turn now to the effect of vapor superheat. To account for the effect of vapor superheat in the analysis of Section 7.2.1, the energy balance on control volume must also include a term for heat transfer from the vapor to the surface of the film; Eq. (7.11) then becomes

$$k_l \left.\frac{dT}{dy}\right|_w = h_c(T_e - T_{\mathrm{sat}}) + \dot{m}'' \left. h_v \right|_s - \frac{d}{dx} \int_0^{\delta(x)} \rho_l u h_l \, dy \tag{7.52}$$

where h_c is the convective heat transfer coefficient for the vapor flowing over the film surface. As was the case for the skin friction coefficient, the Nusselt number correlations in Chapter 4, Eqs. (4.56) and (4.64), cannot be used since the boundary layer is under strong suction. Again we will analyze a model laminar flow problem, as shown in Fig. 7.11, to obtain an estimate for h_c. Convection and conduction in the x direction are neglected, and the steady-flow energy equation applied to an

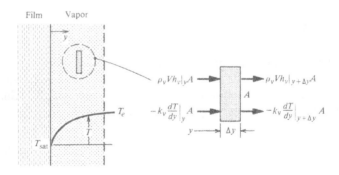

Figure 7.11 Model problem and elemental volume for analysis of the effect of vapor superheat on condensation heat transfer.

elemental control volume Δy thick of cross-sectional area A requires that

$$\rho_v V A \left[h_v|_{y+\Delta y} - h_v|_y \right] = -k_v A \left.\frac{dT}{dy}\right|_y + k_v A \left.\frac{dT}{dy}\right|_{y+\Delta y}$$

Dividing by $A\Delta y$, rearranging, and letting $\Delta y \to 0$ gives

$$k_v \frac{d^2 T}{dy^2} = \rho_v V \frac{dh_v}{dy}$$

From Eq. (7.44), $\rho_v V = -\dot{m}''$, and noting that

$$\frac{dh_v}{dy} = \frac{dh_v}{dT}\frac{dT}{dy} = c_{pv}\frac{dT}{dy}$$

we obtain

$$\frac{d^2 T/dy^2}{dT/dy} = -\frac{\dot{m}'' c_{pv}}{k_v} \tag{7.53}$$

which must be solved subject to the boundary conditions

$$y = 0: \quad T = T_{\text{sat}} \tag{7.54a}$$

$$y \to \infty: \quad T \to T_e \tag{7.54b}$$

The mathematical problem is seen to be identical to the vapor drag problem just solved, so the solution may be written down immediately after inspection of Eqs. (7.47) and (7.48). The temperature profile is

$$\frac{T - T_{\text{sat}}}{T_e - T_{\text{sat}}} = 1 - e^{(-\dot{m}'' c_{pv}/k_v)y} \tag{7.55}$$

and the heat transfer coefficient is

$$h_c = \frac{k_v \left(dT/dy \right)|_{y=0}}{T_e - T_{\text{sat}}} = \dot{m}'' c_{pv} \tag{7.56}$$

That is, in the limit of strong suction, *the heat transfer from the vapor to the film surface is simply the enthalpy given up by the condensing vapor* as it cools from the free-stream temperature T_e to the surface temperature T_{sat}. Substituting in Eq. (7.52) and rearranging using Eq. (7.10) gives

$$k_l \left.\frac{\partial T}{\partial y}\right|_w = \frac{d}{dx} \int_0^{\delta(x)} \rho_l u \left[\left\{ h_{\text{fg}} + c_{pv}(T_e - T_{\text{sat}}) \right\} + c_{pl}(T_{\text{sat}} - T) \right] dy \tag{7.57}$$

which replaces Eq. (7.12). Thus, it is seen that the effect of vapor superheat can be accounted for by simply adding $c_{pv}(T_e - T_{\text{sat}})$ to the enthalpy of phase change h_{fg}.

Effect of Variable Liquid Properties

The reference temperature scheme described in Section 7.2.1 was established for vapor velocities up to 60 m/s. Also, vapor superheat would not be expected to have any significant effect on the reference property scheme for the liquid film.

Effect of Turbulence in the Vapor Boundary Layer

Vapor boundary layers associated with film condensation are often laminar because condenser pressures are relatively low, and the strong suction has a marked effect in delaying transition from a laminar to a turbulent boundary layer. However, Eqs. (7.48) and (7.56) can be shown to be also valid for a turbulent boundary layer, so that the results obtained in this section are applicable whether the vapor boundary layer is laminar or turbulent.

EXAMPLE 7.4   Effect of Vapor Drag on Condensation of Refrigerant-134a

Saturated R-134a vapor at 320 K flows at 14 m/s down along a vertical tube of 5 cm O.D. Coolant flowing inside the tube maintains the outside wall at 300 K. Estimate the effect of vapor drag on the local heat transfer coefficient at a location 10 cm from the top of the tube.

Solution

Given: Forced-flow condensation of R-134a.

Required: Effect of vapor drag on h at $x = 10$ cm.

Assumptions: 1. Negligible effect of ripples.
2. Negligible effect of wall curvature.

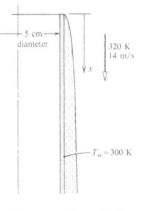

The film flow will prove to be in the wavy laminar regime, but we assume an absence of ripples in order to obtain an approximate estimate of the increase in h due to vapor drag. Since Table 7.1 does not have an α value for R-134a, liquid properties are evaluated at the mean film temperature of 310 K. From Table A.8: $k_l = 0.0763$ W/m K, $\rho_l = 1160$ kg/m^3, $v_l = 0.160 \times 10^{-6}$ m^2/s, $c_{pl} = 1484$ J/kg K, $\text{Pr}_l = 3.62$. From Table A.12e, $h_{fg} = 1.554 \times 10^5$ J/kg. Equation (7.51) gives the local heat transfer coefficient as

$$h = \left[\frac{k_l^2 U_e}{8 v_l x} \left\{ 1 + \left(1 + \frac{16 \text{Pr}_l}{\text{Ja}_l} \frac{gx}{U_e^2} \right)^{1/2} \right\} \right]^{1/2}$$

We first evaluate (gx/U_e^2) and Pr_l/Ja_l:

$$\frac{gx}{U_e^2} = \frac{(9.81)(0.1)}{(14.0)^2} = 5.01 \times 10^{-3}$$

From Eq. (7.17b),

$$h_{fg}' = h_{fg} + \left(0.683 - \frac{0.228}{\text{Pr}_l} \right) c_{pl} (T_{sat} - T_w)$$

$$h_{fg}' = 1.554 \times 10^5 + \left(0.683 - \frac{0.228}{3.62} \right) (1484)(320 - 300) = 1.738 \times 10^5 \text{ J/kg}$$

$$\frac{\text{Pr}_l}{\text{Ja}_l} = \frac{\rho_l v_l h_{fg}'}{k_l (T_{sat} - T_w)} = \frac{(1160)(0.160 \times 10^{-6})(1.738 \times 10^5)}{(0.0763)(20)} = 21.1$$

$$h = \left[\frac{(0.0763)^2(14)}{(8)(0.160 \times 10^{-6})(0.1)}\{1 + [1 + (16)(21.1)(5.01 \times 10^{-3})]^{1/2}\}\right]^{1/2}$$

$$= 1297 \ W/m^2 \ K$$

In the absence of vapor drag, h is given by Eq. (7.15); for $\rho_v \ll \rho_l$ and h'_{fg} replacing h_{fg},

$$h = \left[\frac{h'_{fg} g \rho_l k_l^3}{4x(T_{sat} - T_w)\nu_l}\right]^{1/4} = \left[\frac{(1.738 \times 10^5)(9.81)(1160)(0.0763)^3}{(4)(0.1)(20)(0.160 \times 10^{-6})}\right]^{1/4} = 910 \ W/m^2 \ K$$

Comments

The effect of vapor drag is to increase the heat transfer coefficient by 43%.

EXAMPLE 7.5   Effect of Vapor Superheat on Ammonia Condensation

Ammonia at 100°C and 1034 kPa is to be condensed on a 2 cm-O.D. horizontal tube maintained at 3°C. Estimate the condensation rate, and compare it to the condensation rate of saturated vapor at the same pressure.

Solution

Given: Ammonia condensing on a horizontal tube.

Required: Effect of vapor superheat on condensation rate.

Assumptions: The strong-suction limit for h_c, Eq. (7.56), is valid for condensation on a horizontal tube.

Equation (7.41) gives the heat transfer coefficient, provided we account for vapor superheat and liquid subcooling by replacing h_{fg} by h'_{fg}:

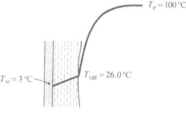

$$\bar{h} = 0.728 \left[\frac{(\rho_l - \rho_v)g h'_{fg} k_l^3}{\nu_l D(T_{sat} - T_w)}\right]^{1/4}$$

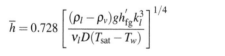

where $h'_{fg} = h_{fg} + c_{pv}(T_e - T_{sat}) + 0.35 c_{pl}(T_{sat} - T_w)$.

From Table A.12b for $P_{sat} = 1034$ kPa, $T_{sat} = 299.2$ K (26.0°C), $h_{fg} = 1.163 \times 10^6$ J/kg. From Table 7.1, $\alpha = 0.61$ for ammonia; hence, $T_r = 276.1 + 0.61(299.2 - 276.1) \simeq 290$ K. From Table A.8, properties of liquid ammonia at 290 K are: $k_l = 0.522$ W/m K, $\rho_l = 616$ kg/m³, $c_{pl} = 4800$ J/kg K, $\nu_l = 0.234 \times 10^{-6}$ m²/s. From Table A.7, for ammonia vapor at a mean temperature of 340 K, $c_{pv} = 2230$ J/kg K. Since $\rho_v \ll \rho_l$, it need not be calculated accurately: for $P = 1034$ kPa and $T = 350$ K, the ideal gas law gives $\rho_v = (1034 \times 10^3)(17)/(8314)(340) = 6.2$ kg/m³.

$$h'_{fg} = 1.163 \times 10^6 + 2230(100 - 26) + (0.35)(4800)(26 - 3)$$

$$= 1.163 \times 10^6 + 0.165 \times 10^6 + 0.039 \times 10^6 = 1.367 \times 10^6 \ J/kg$$

$$\bar{h} = 0.728 \left(\frac{(616 - 6)(9.81)(1.367 \times 10^6)(0.522)^3}{(0.234 \times 10^{-6})(0.02)(26 - 3)}\right)^{1/4} = 7420 \ W/m^2 \ K$$

$$\dot{Q} = (\pi D)\bar{h}(T_{sat} - T_w) = (\pi)(0.02)(7420)(26-3) = 10,720 \text{ W per meter length}$$

$$\dot{m} = \frac{\dot{Q}}{h'_{fg}} = \frac{10,720}{1.367 \times 10^6} = 7.84 \times 10^{-3} \text{ kg/s per meter length}$$

Repeating the above calculations for saturated vapor gives

$$h'_{fg} = h_{fg} + 0.35c_{pl}(T_{sat} - T_w) = (1.163 + 0.039)10^6 = 1.202 \times 10^6 \text{ J/kg}$$

$$\bar{h} = 7190 \text{ W/m}^2 \text{ K}; \quad \dot{Q} = 10,390 \text{ W per meter}; \quad \dot{m} = 8.64 \times 10^{-3} \text{ kg/s per meter}$$

Comments

1. The effect of vapor superheat is to increase the heat transfer coefficient $\bar{h}$ and the heat transfer $\dot{Q}$, but the condensation rate $\dot{m}$ is decreased. Why?

2. Use PHASE to check $\bar{h}$ and $\dot{m}$.

7.3 FILM EVAPORATION

Film or dropwise condensation takes place on a surface when it is cooled to a temperature below the saturation temperature, T_{sat}, corresponding to the pressure of the vapor. On the other hand, if a film of liquid is on a surface that is heated to a temperature above T_{sat}, evaporation from the surface of the film will occur. In industrial evaporators, evaporation may take place from a film falling down the insides of long vertical tubes or channels, as is shown in Fig. 7.12.

7.3.1 Falling Film Evaporation on a Vertical Wall

Heat transfer across an evaporating falling film is essentially identical to that for falling film condensation, except that it is in the opposite direction. There are, however, differences related to the circumstances in which these processes occur in industrial equipment. A major difference is that the film always starts with a finite thickness, and the film Reynolds number decreases down the wall as evaporation takes place; however, in many applications, the evaporation rate is small compared to the film flow rate, and the change in the Reynolds number might be small. Figure 7.13 shows the temperature profile across an evaporating turbulent falling film. Since $T_w > T_{sat}$, the liquid closer to the wall is superheated, and if the superheat is sufficient, *boiling* will occur on the wall (that is, bubbles will nucleate and grow on the wall), or bubbles entering with the feed liquid may *cavitate* (that is, grow explosively). In both cases, the film will be disrupted, and the heat transfer process will become much more complex. In Section 7.4.2 we will see how relatively small values of $(T_w - T_{sat})$ cause boiling. Often, a liquid evaporates into a vapor-gas mixture rather than into pure vapor; for example, water evaporates into an air stream in a great variety of industrial equipment. Then, because the total pressure is usually much higher than $P_{sat}(T_w)$, there is little possibility of boiling or cavitation. Since the evaporating vapor must diffuse away from the surface through the air, such equipment also involves a

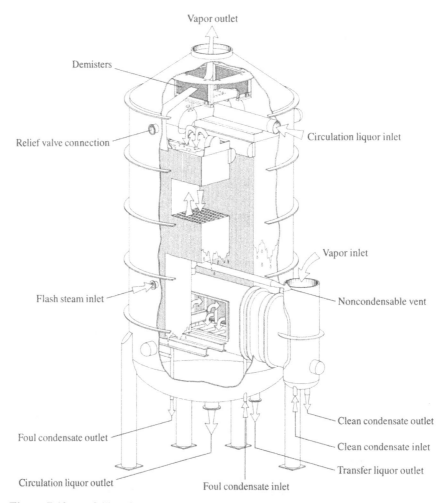

Figure 7.12 A falling-film evaporator used to concentrate black liquor in a pulp mill. (Courtesy Dr. J. W. Rauscher, Ahlstrom Recovery Inc., Roswell, Georgia.)

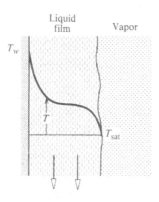

Figure 7.13 The temperature profile across an evaporating turbulent falling film.

mass transfer process. In contrast to film condensation, where the vapor boundary layer is under strong suction, for film evaporation there is a *blowing* effect on the vapor boundary layer since the vapor velocity normal to the film is directed away from the surface. Whereas suction increases vapor drag and heat transfer from the vapor, blowing reduces vapor drag and heat transfer to the vapor. If the evaporation rate is high enough, both can be negligible.

Since evaporating films are most often initially wavy laminar or turbulent, our analysis follows Section 7.2.2. If vapor drag is negligible, then Eqs. (7.21) through (7.24) again apply to water, namely,

$$\text{Nu} = \left(\frac{3}{4}\,\text{Re}\right)^{-1/3} \qquad\qquad 0 < \text{Re} < 30 \qquad \text{(laminar)}$$

$$\text{Nu} = 0.822\text{Re}^{-0.22} \qquad\qquad 30 < \text{Re} < \text{Re}_{\text{tr}} \ \text{(wavy laminar)}$$

$$\text{Nu} = 3.8 \times 10^{-3}\text{Re}^{0.4}\text{Pr}_l^{0.65} \qquad \text{Re}_{\text{tr}} < \text{Re} \qquad \text{(turbulent)}$$

where $\text{Re}_{\text{tr}} = 5800\text{Pr}_l^{-1.06}$. In fact, the turbulent correlation was developed using data for evaporating falling films. The energy balance Eq. (7.26) becomes

$$\frac{L}{(v_l^2/g)^{1/3}} = -\frac{\text{Pr}_l}{4\text{Ja}_l}\int_{\text{Re}_0}^{\text{Re}_L}\frac{d\text{Re}}{\text{Nu}} \qquad\qquad (7.58)$$

where Re_0 is the initial film Reynolds number at $x = 0$, and the expression for the average Nusselt number becomes

$$\overline{\text{Nu}} = \frac{\text{Pr}_l}{4\text{Ja}_l}\frac{(v_l^2/g)^{1/3}}{L}(\text{Re}_0 - \text{Re}_L) \qquad\qquad (7.59)$$

Effect of Variable Fluid Properties

The reference property schemes for laminar film condensation in Section 7.2.1 are tentatively recommended for falling film evaporation.

Entrance Effects

Equations (7.22) and (7.23) are based on data measured some distance downstream from the liquid distributor, where entrance effects were considered to be negligible. Thus, the foregoing analysis should not be applied to a very short evaporator. For turbulent films, such entrance effects are negligible for distances greater than about 20 times the film thickness. For wavy laminar films, entrance effects persist somewhat further. In fact, there is a question as to whether a truly fully developed condition is ever attained, since the long-wavelength gravity waves on the surface of the film may require a very large distance to become fully established. However, this entrance effect is of secondary importance since it leads to relatively small changes in the Nusselt number.

Computer Program PHASE

Item 3 of PHASE calculates the average heat transfer coefficient and evaporation rate for a falling film according to the analysis of Section 7.3.1. Local heat transfer coefficients and transition Reynolds numbers for wavy laminar and turbulent flow can be calculated from the formulas given in Section 7.3.1, which are, strictly speaking, valid only for water. User-supplied constants and exponents for these formulas are also allowed. Variable-property and liquid superheat effects are calculated according to the recommendations given in Section 7.2.1.

EXAMPLE 7.6   Evaporation from a Falling Water Film

In an experimental rig, water is fed at 0.01 kg/s to the top of a vertical 5 cm-O.D. tube, 5 m high. The outside of the tube wall is maintained at 311 K by steam condensing inside the tube, and the saturation temperature corresponding to the system pressure is 308 K. Calculate the vapor production rate.

Solution

Given: A water film flow outside a heated tube.

Required: Vapor production rate.

Assumptions: The reference temperature scheme for laminar films can be used to evaluate properties.

We first obtain the necessary property values. At 308 K, Table A.12a gives $h_{fg} = 2.418 \times 10^6$ J/kg. From Table 7.1, $\alpha = 0.33$; hence, $T_r = 311 - 0.33(311 - 308) = 310$ K. From Table A.8: $k_l = 0.628$ W/m K, $\rho_l = 993$ kg/m^3, $c_{pl} = 4174$ J/kg K, $\nu_l = 0.70 \times 10^{-6}$ m^2/s, $Pr_l = 4.6$. The initial film Reynolds number is

$$Re_0 = \frac{4\Gamma_0}{\mu_l} = \frac{4\dot{m}_0}{\pi D \rho_l \nu_l} = \frac{(4)(0.01)}{(\pi)(0.05)(993)(0.70 \times 10^{-6})} = 366$$

and $Re_{tr} = 5800(4.6)^{-1.06} = 1151 > 366$. We will assume that the film remains in the wavy laminar regime as it flows down the wall and check later. Then Eq. (7.58) becomes

$$\frac{L}{(\nu_l^2/g)^{1/3}} = -\frac{Pr_l}{4Ja_l}\int_{366}^{Re_L}\frac{dRe}{0.822Re^{-0.22}} = -\frac{Pr_l}{4Ja_l(0.822)(1.22)}\left[Re_L^{1.22} - 366^{1.22}\right]$$

Hence,

$$Re_L^{1.22} = 1341 - \frac{4.01Ja_l L}{(\nu_l^2/g)^{1/3}Pr_l}$$

$$Ja_l = \frac{c_{pl}(T_w - T_{sat})}{h_{fg}} = \frac{(4174)(3)}{(2.418 \times 10^6)} = 5.18 \times 10^{-3}$$

$$\left(\frac{\nu_l^2}{g}\right)^{1/3} = \left[\frac{(0.70 \times 10^{-6})^2}{9.81}\right]^{1/3} = 3.68 \times 10^{-5}\,m$$

Substituting,

$$\text{Re}_L^{1.22} = 1341 - \frac{(4.01)(5.18 \times 10^{-3})(5)}{(3.68 \times 10^{-5})(4.6)} = 727$$

and $\text{Re}_L = 222$. The film remains in the wavy laminar regime.

The water flow rate at the bottom of the tube is

$$\dot{m}_L = \frac{\pi D \rho_l v_l \text{Re}_L}{4} = \frac{(\pi)(0.05)(993)(0.70 \times 10^{-6})(222)}{4} = 6.05 \times 10^{-3} \text{ kg/s}$$

The vapor production rate is obtained from a mass balance:

$$\dot{m}_v = \dot{m}_0 - \dot{m}_L = 0.01 - 0.00605 = 3.95 \times 10^{-3} \text{ kg/s}$$

Comments

Use PHASE to check $\dot{m}_v$ and Re_L.

7.4   POOL BOILING

In industrial boilers, boiling may take place in a stationary pool of liquid, or the liquid may boil as it flows through a tube. We will examine **pool boiling** in this section and leave consideration of *forced-convection boiling* to more advanced texts. In Section 7.4.1, the regimes of pool boiling are described, and the *boiling curve* is introduced. In succeeding sections, the various regimes are described in detail, and correlations are given for the boiling heat flux or boiling heat transfer coefficient.

7.4.1   Regimes of Pool Boiling

We have all watched a pot of water come to a boil on the kitchen stove. When heating begins, the bulk liquid and the pot wall are at a temperature lower than the saturation temperature, T_{sat}, corresponding to the pressure P, and vapor cannot coexist with the liquid phase. The liquid and wall are then *subcooled*. Initially, no vapor bubbles are seen; however, gradients in the index of refraction caused by temperature gradients near the heated wall reveal fluid motion associated with natural-convection heat transfer. Some time later, stagnant bubbles attached to the wall become visible: these are not bubbles of pure water vapor but are mostly air that has come out of solution. The solubility of air in water decreases with increasing temperature; thus, the water becomes supersaturated with air, and mass transfer occurs to minute air spaces trapped in scratches, pits, and crevices in the wall. These air bubbles act as nuclei for growth of vapor bubbles.

    True boiling, which is the formation of pure vapor from superheated liquid, begins when the wall temperature T_w exceeds T_{sat} by a few degrees. The pot "sings" as vapor bubbles grow rapidly from nuclei on the superheated wall. The bubbles stop growing and collapse rapidly as they project through the thin natural-convection thermal boundary layer into the bulk subcooled liquid. At first, these bubbles are invisible to the eye, but as the bulk liquid temperature rises with continued heating, the growing and collapsing bubbles become large enough to be visible. This boiling is called *subcooled nucleate boiling*. With continued heating, the bulk liquid itself becomes mildly superheated: the bubbles grow from the wall nuclei until they are large enough for their buoyancy to tear them from the wall, and they rise to the pool surface. The heat transfer process is then one of transporting enthalpy of phase change by the vapor inside the bubbles; also, the bubbles sweep hot liquid away from the wall, causing a rapid increase in the heat transfer coefficient. **Saturated nucleate boiling** with isolated bubbles results, as shown in Fig. 7.14a.

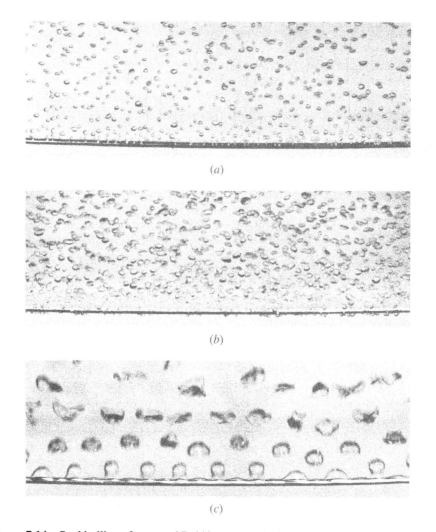

(*a*)

(*b*)

(*c*)

Figure 7.14   Pool boiling of saturated R-113 at atmospheric pressure on a horizontal platinum wire of 1 mm diameter, (*a*) Isolated bubbles, $q = 1 \times 10^4$ W/m^2, $\Delta T = 7$ K. (*b*) Slugs and columns, $q = 4.5 \times 10^4$ W/m^2, $\Delta T = 10$ K. (*c*) Film boiling, $q = 3.3 \times 10^4$ W/m^2, $\Delta T = 150$ K. (Photographs courtesy of Professor S. Nishio, University of Tokyo.)

Further heating produces a rapid increase in the number of active nucleation sites, and so many bubbles form that they merge to produce columns of vapor, which feed large slugs of vapor overhead, as shown in Fig. 7.14b. Finally, the vapor streams upward so fast that the liquid downflow to the surface is unable to sustain a higher evaporation rate. This condition is called the departure from nucleate boiling, or *boiling crisis*, and gives rise to a **peak heat flux** q_{max}. It is also often called the *burnout point*, which will be explained later. If the wall temperature is increased beyond the value corresponding to q_{max}, there is a transition regime where it becomes increasingly difficult for liquid to reach the wall, until a vapor film completely blankets the wall, preventing liquid from contacting the wall at all, as shown in Figure 7.14c. The corresponding heat fluxes are much lower than q_{max} since heat is transferred rather poorly across the low-conductivity vapor film, and there is a local minimum value q_{min}, characterizing this **film boiling**. Yet a further increase in wall temperature causes a modest increase in q, and if the temperature level is high enough, radiation heat transfer across the vapor film can become significant.

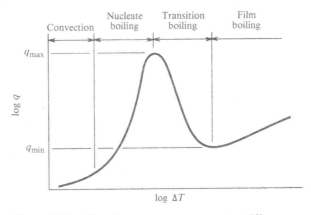

Figure 7.15   Heat flux, q, versus temperature difference, $\Delta T = T_w - T_{sat}$, for pool boiling: the *boiling curve*.

Figure 7.15 shows a typical *boiling curve*, in which the wall heat flux q is plotted versus $\Delta T = (T_w - T_{sat})$. We choose to plot q rather than the heat transfer coefficient because q_{max} and q_{min} can be relatively easily specified whereas the corresponding heat transfer coefficients cannot. The various regimes of boiling discussed above are indicated in Fig. 7.15. In practice, it is more often the case that the heat flux through the surface is controlled rather than the wall temperature; for example, when an electrical heater is used. If we imagine increasing the power input until q_{max} is reached, Fig. 7.16 shows that the situation becomes unstable, and a further small increase in q requires the wall temperature to jump from temperature T_A to a very high value in the film boiling regime, T_B, in order to effect the heat transfer. If T_B approaches or exceeds the melting point of the wall, the result can be catastrophic, and *burnout* is said to occur. Conversely, if during film boiling the heat flux is reduced, T_w decreases

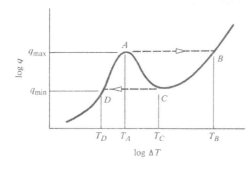

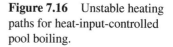

Figure 7.16 Unstable heating paths for heat-input-controlled pool boiling.

until q_{min} is reached. Figure 7.16 shows that the situation is again unstable, and a further decrease in q causes the wall temperature to jump from T_C to a very low value in the nucleate boiling regime, T_D.

7.4.2 Boiling Inception

Typical machined, drawn, or cast surfaces have a variety of pits, grooves, and crevices. When a liquid contacts the surface, surface tension forces prevent the liquid from entering the smaller cavities in which air or other gases are trapped. These cavities are the sites at which bubble nucleation occurs. To obtain an estimate of the superheat required to nucleate bubbles, we consider a force balance on a stationary spherical bubble, as shown in Fig. 7.17. The ambient pressure is denoted P_∞, P_v is the partial pressure of vapor inside the bubble, and P_g is the sum of the partial pressures of all the permanent gases inside the bubble. For static equilibrium, the surface tension force balances the net pressure force:

$$2\pi R_c \sigma = (P_v + P_g - P_\infty)\pi R_c^2$$

$$R_c = \frac{2\sigma}{P_v + P_g - P_\infty} \tag{7.60}$$

where σ is the surface tension and R_c is a **critical radius** of the bubble. It is reasonable to assume that the vapor in the bubble is in thermodynamic equilibrium with

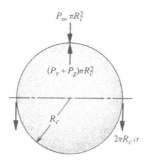

Figure 7.17 Force balance on a spherical bubble.

the liquid; thus, the vapor partial pressure P_v can be taken as $P_{sat}(T_s)$, where T_s is the temperature of the liquid surrounding the bubble. The partial pressure of the gas, P_g, is more difficult to estimate: the gas is not likely to be in equilibrium with dissolved gas in the liquid phase, since mass diffusion rates in liquids are very low. This equilibrium bubble is unstable to small perturbations. A small decrease in size causes an increase in $2\sigma/R$. Since P_∞ may be regarded as fixed, there is a resulting increase in P_v to give a supersaturated vapor that starts to condense, allowing a further decrease in bubble size and eventual collapse of the bubble. Conversely, a small increase in bubble size causes a decrease in $2\sigma/R$ and a decrease in P_v to give superheated vapor. Liquid then starts to evaporate, and the bubble continues to grow.

Figure 7.18a shows a bubble that has emerged from a surface cavity. For simplicity, we assume a contact angle of 90°, so that the bubble radius is equal to the cavity radius. If the bubble is able to continue to grow and detach from the surface, we say that the bubble has *nucleated* and that the cavity is a *nucleation site*. Experience shows that for typical boiler wall surface finishes, nucleation sites with radii in the range 2.5-7.5 μm are present when the wall is first wetted, but prolonged boiling in deaerated water will usually deactivate the larger sites. Application of Eq. (7.60) to bubble nucleation by a cavity is made difficult by the fact that the liquid surrounding the bubble may not be at uniform temperature. Figure 7.18b shows the temperature profile expected for laminar natural convection. Also, the profile might be disturbed by the bubble as it emerges from the cavity. For a very simple model, we can assume that the liquid surrounding the bubble is at the wall temperature T_w, that is, $T_s = T_w$. If the gas pressure is ignored in Eq. (7.60), it can be rearranged as

$$P_{sat}(T_s) = P_\infty + \frac{2\sigma}{R_c} \tag{7.61}$$

Equation (7.61) can be used to estimate the wall temperature $T_w = T_s$ at which bubbles will nucleate from a cavity of radius R_c. The corresponding wall superheat is $T_w - T_{sat}(P_\infty)$. If gas is present in the bubbles, Eq. (7.60) shows that nucleation will occur at lower wall superheat.

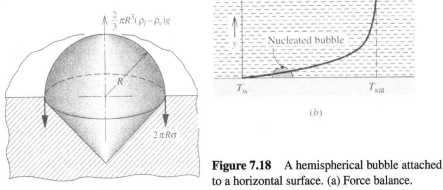

Figure 7.18 A hemispherical bubble attached to a horizontal surface. (a) Force balance. (b) Temperature profile in surrounding fluid.

The Computer Program BOIL

Item 1 in the program BOIL calculates the superheat required for boiling inception by a surface cavity of radius R_c, using Eq. (7.61). The fluid menu is identical to that contained in the computer program PHASE. SI units must be used, with temperature in kelvins.

EXAMPLE 7.7 Boiling Inception Superheat for Various Liquids

Estimate the superheat required for inception of boiling for a heater just below the surface of a pool of saturated liquid if the liquid is (i) water at 1 atm pressure, (ii) water at 300 K, and (iii) nitrogen at 1 atm pressure. Take the largest available nucleation sites to have a radius of 7.5 μm.

Solution

Given: Pool boiling.

Required: Boiling inception ΔT for (i) water at 1 atm, (ii) water at 300 K, and (iii) nitrogen at 1 atm.

Assumptions: 1. The largest available nucleation sites have a radius of 7.5 μm.
2. There is no dissolved gas.

Equation (7.61) gives the value of $P_{sat}(T_s)$ required to nucleate a bubble; the corresponding value of T_s is then found from vapor tables.

(i) Water at 1 atm:

$$P_{sat}(T_s) = P_\infty + \frac{2\sigma}{R_c}$$

$P_\infty = 1$ atm $= 1.0133 \times 10^5$ Pa; $T_{sat}(P_\infty) = 373.15$ K, from Table A.12a, and from Table A.11, $\sigma(T = 373.15$ K) is 58.9×10^{-3} N/m.

$$P_{sat}(T_s) = 1.0133 \times 10^5 + \frac{(2)(58.9 \times 10^{-3})}{(7.5 \times 10^{-6})} = 1.0133 \times 10^5 + 0.1571 \times 10^5$$

$$= 1.170 \times 10^5 \text{ Pa}$$

Table A.12a gives the corresponding value of T_s as 377.2 K. Thus, the required superheat is

$$\Delta T = T_s - T_{sat}(P_\infty) = 377.2 - 373.15 = 4.1 \text{ K}$$

(ii) Water at 300 K:

$T_{sat}(P_\infty) = 300$ K, $P_\infty = 3533$ Pa from Table A.12a; $\sigma = 71.7 \times 10^{-3}$ N/m from Table A.11.

$$P_{sat}(T_s) = 3533 + \frac{2(71.7 \times 10^{-3})}{(7.5 \times 10^{-6})} = 3533 + 19,120 = 22,650 \text{ Pa}$$

$$T_s = 336 \text{ K}, \quad \Delta T = 336 - 300 = 36 \text{ K}$$

(iii) Nitrogen at 1 atm:

$P_\infty = 1.0133 \times 10^5$ Pa, $T_{sat} = 77.4$ K from Table A.12c, $\sigma = 8.85 \times 10^{-3}$ N/m. The superheat is small in this case, and an accurate result cannot be obtained if Table A.12c is used to obtain T_s from $P_{sat}(T_s)$. Instead, we use the Clausius-Clapeyron relationship, which relates ΔP to ΔT along the saturation curve, and the ideal gas law:

$$\frac{\Delta P}{P} = \frac{h_{fg}M}{\mathscr{R}}\frac{\Delta T}{T^2}$$

$$\Delta T = \frac{T^2}{P}\frac{(\mathscr{R}/M)}{h_{fg}}\left(\frac{2\sigma}{R_c}\right)$$

At the normal boiling point of nitrogen, 77.4 K, $h_{fg} = 2.0 \times 10^5$ J/kg from Table A.12c. Hence,

$$\Delta T = \frac{(77.4)^2}{1.0133 \times 10^5}\frac{(8314/28)}{2.0 \times 10^5}\frac{(2)(8.85 \times 10^{-3})}{7.5 \times 10^{-6}} = 0.21 \text{ K}$$

Comments

1. In case (ii), a second iteration evaluating σ at $T_s = 336$ K is indicated.

2. Use BOIL to check these results.

7.4.3 Nucleate Boiling

As a matter of convenience, we define the heat transfer coefficient for boiling in terms of $T_w - T_{sat}(P_\infty)$, rather than $T_w - T_\infty$:

$$q = h(T_w - T_{sat}) \tag{7.62}$$

An appropriate characteristic length to form a Nusselt number for nucleate boiling is the size of a bubble as it breaks away from a wetted wall. Referring to Fig. 7.18a, the surface tension and buoyancy force are equated to give

$$2\pi R_b \sigma = \frac{2}{3}\pi R_b^3(\rho_l - \rho_v)g$$

where R_b is the bubble radius and subscripts l and v refer to the liquid vapor phase, respectively. Hence,

$$R_b = \left[\frac{3\sigma}{(\rho_l - \rho_v)g}\right]^{1/2}$$

and the characteristic length L_c can be taken simply as

$$L_c = \left[\frac{\sigma}{(\rho_l - \rho_v)g}\right]^{1/2}; \quad \text{Nu} = \frac{hL_c}{k_l} \tag{7.63}$$

The rate of bubble growth depends on convective heat transfer through the liquid to the liquid-vapor interface required to supply the enthalpy of vaporization. Thus, we would expect the Prandtl number of the liquid Pr_l to be a relevant dimensionless group. Also, we saw in Section 7.2.2 how the Jakob number is relevant to phase change problems. For nucleate boiling, it is appropriate to define the Jakob number as

$$Ja_l = \frac{c_{pl}(T_w - T_{sat})}{h_{fg}} \qquad (7.64)$$

since the superheated liquid has $c_{pl}(T_w - T_{sat})$ sensible heat, which can be given up to supply enthalpy of vaporization. Experimental data for nucleate boiling on a horizontal plate facing upward in a pool of liquid was correlated by W. M. Rohsenow [6] as

$$Nu = \frac{Ja^2}{C_{nb}^3 Pr_l^m} \qquad (7.65)$$

where values of the constant C_{nb} and exponent m are given in Table 7.2 for a variety of combinations of liquid and heater wall materials. All properties are to be evaluated at T_{sat}. Notice that a factor-of-2 difference in C_{nb} results in a factor-of-8 difference in Nu. Equation (7.65) implies that the heat flux q is proportional to ΔT^3; this strong dependence on temperature difference is due to the rapid increase in active nucleation sites with increase in superheat. Equation (7.65) is not very accurate, however, and errors of 100% in q and 25% in ΔT are typical. Fortunately, the design engineer is more interested in not exceeding q_{max}, so as to avoid burnout, and usually does not require a precise value of the heat transfer coefficient in the nucleate boiling regime.

The computer program BOIL, item 2, calculates the heat flux for nucleate boiling using Rohsenow's correlation, Eq. (7.65). The menu of surfaces is based on Table 7.2.

Table 7.2 The constant C_{nb} and exponent m for the nucleate boiling correlation, Eq. (7.65): $Nu = Ja^2/C_{nb}^3 Pr_l^m$.

Liquid	Surface	C_{nb}	m
Water	Copper, scored	0.0068	2.0
Water	Copper, polished	0.013	2.0
Water	Stainless steel, chemically etched	0.013	2.0
Water	Stainless steel, mechanically polished	0.013	2.0
Water	Stainless steel, ground and polished	0.008	2.0
Water	Brass	0.006	2.0
Water	Nickel	0.006	2.0
Water	Platinum	0.013	2.0
Carbon tetrachloride	Copper	0.013	4.1
Benzene	Chromium	0.010	4.1
n-pentane	Chromium	0.015	4.1
Ethanol	Chromium	0.0027	4.1
Isopropyl alcohol	Copper	0.0023	4.1
n-butyl alcohol	Copper	0.0030	4.1
35% K_2CO_3	Copper	0.0027	4.1

EXAMPLE 7.8   Nucleate Boiling of Water on Polished Copper

Determine the heat flux when water boils at 1 atm pressure on a polished copper surface at 390 K.

Solution

Given: Water boiling on a polished copper surface.

Required: Heat flux q.

Assumptions: Nucleate boiling regime.

Equation (7.65) applies: $\mathrm{Nu} = \mathrm{Ja}^2/C_{nb}^3 \, \mathrm{Pr}_l^m$.

     Evaluate all properties at $T_{\mathrm{sat}} = 373.15$ K: $\sigma = 58.9 \times 10^{-3}$ N/m, $h_{\mathrm{fg}} = 2.257 \times 10^6$ J/kg, $k_l = 0.681$ W/m K, $\rho_l = 958$ kg/m³, $c_{pl} = 4212$ J/kg K, $\mathrm{Pr}_l = 1.76$.

$$\mathrm{Ja} = \frac{c_{pl}(T_w - T_{\mathrm{sat}})}{h_{\mathrm{fg}}} = \frac{(4212)(390 - 373.15)}{(2.257 \times 10^6)} = 3.145 \times 10^{-2}$$

From Table 7.2, $C_{\mathrm{nb}} = 0.013, m = 2.0$. Substitute in Eq. (7.65):

$$\mathrm{Nu} = \frac{(3.145 \times 10^{-2})^2}{(0.013)^3 (1.76)^2} = 145$$

$$L_c = \left[\frac{\sigma}{(\rho_l - \rho_v)g} \right]^{1/2} = \left[\frac{58.9 \times 10^{-3}}{(958)(9.81)} \right]^{1/2} = 2.50 \times 10^{-3} \text{ m}$$

$$h = \left(\frac{k_l}{L_c} \right) \mathrm{Nu} = \left(\frac{0.681}{2.50 \times 10^{-3}} \right)(145) = 3.95 \times 10^4 \text{ W/m}^2 \text{ K}$$

$$q = h(T_w - T_{\mathrm{sat}}) = (3.95 \times 10^4)(390 - 373.15) = 6.66 \times 10^5 \text{ W/m}^2$$

Comments

1. Use BOIL to check q.

2. In general, ΔT for boiling inception and q_{max} should be calculated to bracket the nucleate boiling regime. For water at 1 atm, Example 7.7 shows that ΔT for inception is 4.1 K; Example 7.9 will show that $q < q_{\mathrm{max}}$.

7.4.4 The Peak Heat Flux

The peak heat flux is primarily determined by hydrodynamic considerations related to the maximum rate at which vapor can leave the wall. (An inability of the liquid to wet the wall may also play a role.) If we define a maximum vapor velocity V_{max}, then the peak heat flux q_{max} is given by

$$q_{\mathrm{max}} \sim \rho_v V_{\mathrm{max}} h_{\mathrm{fg}} \tag{7.66}$$

$$\frac{q_{\mathrm{max}}}{\rho_v V_{\mathrm{max}} h_{\mathrm{fg}}} = C_{\mathrm{max}} \tag{7.67}$$

where C_{max} may depend on such factors as geometry. One way to estimate V_{max} is to equate the kinetic energy of the vapor to work done by buoyancy forces over the characteristic length L_c, defined by Eq. (7.63):

$$\frac{1}{2}\rho_v V_{max}^2 = g(\rho_l - \rho_v)L_c$$

Substituting for L_c from Eq. (7.63) gives

$$V_{max} \sim \left(\frac{\sigma(\rho_l - \rho_v)g}{\rho_v^2}\right)^{1/4} \tag{7.68}$$

An alternative viewpoint that is often used is to consider the stability of a column of vapor, as shown in Fig. 7.19. If the column is subject to disturbances of wavelength L_c, the column is unstable when its velocity is

$$V_H = \left(\frac{2\pi\sigma}{\rho_v L_c}\right)^{1/2} \tag{7.69}$$

If this velocity is used as an estimate of V_{max}, and we write $V_{max} \sim (\sigma/\rho_v L_c)^{1/2}$, Eq. (7.68) is once again obtained. This phenomenon is known as *Helmholtz instability*, and the derivation of Eq. (7.69) can be found in advanced fluid mechanics texts. It is Helmholtz instability that also causes a flag to flap in the wind. Substituting Eq. (7.68) in Eq. (7.67) gives

$$q_{max} = C_{max} h_{fg} \left[\sigma\rho_v^2(\rho_l - \rho_v)g\right]^{1/4} \tag{7.70}$$

Equation (7.70) was first derived by S. Kutateladze in the USSR [7] and N. Zuber in the United States [8]. For large, flat heaters, Lienhard and Dhir [9] report experimental data showing that C_{max} is approximately 0.15, and lower values are more appropriate when the size of the heater is less than about $2L_c$. We define a dimensionless parameter L^* as the ratio of a characteristic length of heater L to L_c:

$$L^* = \frac{L}{L_c} = \frac{L}{[\sigma/(\rho_l - \rho_v)g]^{1/2}} \tag{7.71}$$

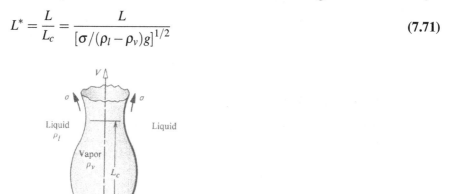

Figure 7.19   Schematic of Helmholtz instability of a vapor column.

Table 7.3   Values of C_{max} for Eq. (7.70) giving the peak heat flux [9]; $L^* = L/L_c = L/[\sigma/(\rho_l - \rho_v)g]^{1/2}$.

Geometry	C_{max}	Characteristic Heater Dimension	L^* Range
1. Infinite flat heater	0.15	Width or diameter	$L^* > 27$
2. Small flat heater	$0.15\frac{12\pi L_c^2}{\text{Area}}$	Width or diameter	$9 < L^* < 20$
3. Large horizontal cylinder	0.12	Cylinder radius	$L^* > 1.2$
4. Small horizontal cylinder	$0.12L^{*-1/4}$	Cylinder radius	$0.15 < L^* < 1.2$
5. Large sphere	0.11	Sphere radius	$4.26 < L^*$
6. Small sphere	$0.227L^{*-1/2}$	Sphere radius	$0.15 < L^* < 4.26$
7. Any large finite body	~ 0.12	—	

Table 7.3 gives values of C_{max} in terms of L^* for various heater shapes. All properties, including the vapor density, ρ_v, in Eq. (7.70) should be evaluated at T_{sat}. Notice that Eq. (7.70) does not allow the heat transfer coefficient corresponding to q_{max} to be calculated; hence, the superheat ($T_w - T_{sat}$) at q_{max} cannot be determined either. The computer program BOIL, item 3, calculates the peak heat flux for pool boiling using Eq. (7.70). A menu of seven configurations based on Table 7.3 is provided.

EXAMPLE 7.9   Peak Heat Flux for Pool Boiling of Water

Estimate the peak heat flux for boiling of water on a large, flat heater at saturation temperatures of (i) 100°C, (ii) 300 K.

Solution

Given: Pool boiling of water on a large, flat heater.

Required: Peak heat flux, q_{max}.

(i) At $T_{sat} = 100°$ C $= 373.15$ K, $h_{fg} = 2.257 \times 10^6$ J/kg, $\rho_v = 0.598$ kg/m^3, $\sigma = 58.9 \times 10^{-3}$ N/m, $\rho_l = 958$ kg/m^3. Substituting in Eq. (7.70) with $C_{max} = 0.15$ from Table 7.3,

$$q_{max} = C_{max}h_{fg}[\sigma\rho_v^2(\rho_l - \rho_v)g]^{1/4}$$
$$= (0.15)(2.257 \times 10^6)[(58.9 \times 10^{-3})(0.598)^2(958 - 0.598)(9.81)]^{1/4}$$
$$= 1.27 \times 10^6 \text{ W/m}^2$$

(ii) At $T_{sat} = 300$ K, $h_{fg} = 2.437 \times 10^6$ J/kg, $\rho_v = 0.0255$ kg/m^3, $\sigma = 71.7 \times 10^{-3}$ N/m, $\rho_l = 996$ kg/m^3.

$$q_{max} = (0.15)(2.437 \times 10^6)[(71.7 \times 10^{-3})(0.0255)^2(996)(9.81)]^{1/4}$$
$$= 3.00 \times 10^5 \text{ W/m}^2$$

Comments

1. The decrease in q_{max} at lower values of T_{sat} is due to the marked decrease in vapor density, which limits the mass flow in the vapor columns.

2. Use BOIL to check q_{max}.

7.4.5 Film Boiling

The phenomenon of film boiling on immersed cylinders, spheres, and plates is very similar in nature to film condensation. This fact was first exploited by L. Bromley in 1950 [10]. Figure 7.20 shows film boiling on a sphere. Instead of a liquid film flowing down over the sphere, as in film condensation, a vapor film flows upward over the sphere. There are some differences, however: (1) Since usually $\rho_v \ll \rho_l$ vapor films tend to be much thicker than liquid films; (2) although we were able to ignore vapor drag on a liquid film in the analysis of film condensation, liquid drag on the vapor film is not negligible; and (3) on larger objects, the vapor-liquid interface can become Helmholtz-unstable—in particular, on a vertical wall the interface becomes unstable quite close to the leading edge. Notwithstanding, correlations of the Nusselt number for laminar film condensation can be used for laminar film boiling on small objects, with liquid properties replaced by vapor properties and slight adjustments made in multiplying constants. Therefore, for laminar film boiling, we write

$$\overline{h} = C_{fb} \left[\frac{(\rho_l - \rho_v) g h'_{fg} k_v^3}{v_v L (T_w - T_{sat})} \right]^{1/4} \tag{7.72}$$

The characteristic length L is related to the length of vapor film: for a horizontal cylinder or sphere, L can be taken as the diameter D. The constant C_{fb} can be taken as 0.62 for a horizontal cylinder [10], 0.67 for a sphere [11], and approximately 0.71 for a plane vertical surface. These constants are somewhat lower than their

Figure 7.20   Film boiling of n-pentane on a 9.5 mm-diameter sphere. (Photograph courtesy of Professor T. K. Frederking, University of California, Los Angeles.)

counterparts for film condensation [cf. Eq. (7.41)] due to the influence of drag on the vapor flow. The modified latent heat is taken as $h'_{fg} = h_{fg} + 0.35c_{pv}(T_w - T_{sat})$ to account for vapor superheating.

Equation (7.72) is valid for a smooth vapor-liquid interface. At high vapor flow rates, waves form on the liquid-vapor interface, and the vapor flow can become turbulent, similar to the behavior for film condensation discussed in Section 7.2.2. Such conditions are usually encountered unless L is small and, in particular, for film boiling of cryogenic liquids at atmospheric pressure. Based on experimental data for liquid nitrogen, Frederking and Clark [13] recommend

$$\bar{h} = 0.15 \left[\frac{(\rho_l - \rho_v)gh'_{fg}k_v^2}{v_v(T_w - T_{sat})} \right]^{1/3} ; \quad \frac{L^3(\rho_l - \rho_v)gh'_{fg}}{k_v v_v (T_w - T_{sat})} > 5 \times 10^7 \qquad (7.73)$$

where $h'_{fg} = h_{fg} + 0.50c_{pv}(T_w - T_{sat})$. Notice that Eq. (7.73) gives a heat transfer coefficient that is independent of the size of the surface and applies to spheres, cylinders, and vertical surfaces.

For film boiling on a horizontal plate, the length of vapor film is related to the spacing between detaching bubbles, which is determined by the instability of the liquid-vapor interface. This mechanism is commonly called *Taylor instability*. As shown in Fig. 7.21, the interface is unstable to a wavelength λ_T on the order of the characteristic length given by Eq. (7.63), $L_c = [\sigma/(\rho_l - \rho_v)g]^{1/2}$. For one-dimensional waves, $\lambda_T = 2\pi\sqrt{3}L_c$; for two-dimensional waves, $\lambda_T = 2\pi\sqrt{6}L_c$. We will simply take $L = L_c$ and absorb the constant in the final correlation for $\bar{h}$; then for a large horizontal plate, $C_{fb} = 0.425$ in Eq. (7.72).

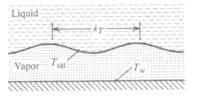

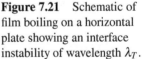

Figure 7.21 Schematic of film boiling on a horizontal plate showing an interface instability of wavelength λ_T.

Effect of Variable Properties

In Eqs. (7.72) and (7.73), h_{fg}, ρ_l, and σ are to be evaluated at T_{sat}; all other properties should be evaluated at the mean film temperature.

Minimum Heat Flux

If a surface operates in the film-boiling regime and q or ΔT is reduced, at a certain point the film breaks down, and the heat flux has a local minimum value q_{min}. If it is assumed that film breakdown occurs when the vapor generation rate becomes too low to sustain Taylor instability wave action on the interface, analysis yields [14]

$$q_{min} = C_{min}\rho_v h_{fg} \left[\frac{\sigma g(\rho_l - \rho_v)}{(\rho_l + \rho_v)^2} \right]^{1/4} \qquad (7.74)$$

where $C_{\min} = 0.09$ for large horizontal surfaces [12], and

$$C_{\min} = 0.0464 \left[\frac{18}{L^{*2}(2L^{*2}+1)} \right]^{1/4}$$

for horizontal wires [15], where $L^* = R/L_c$, and R is the wire radius. In Eq. (7.74), the vapor density ρ_v is evaluated at T_{sat}.

Effect of Radiation

Above about 600 K, radiation heat transfer across the vapor film becomes significant to film boiling. It is inaccurate to simply add the radiation transfer to the convective transfer, since the vapor film is thickened by the additional vapor produced. If radiation absorption in the vapor film is negligible, and if the absorptance of the liquid is unity, then from Eq. (1.19), the radiation transfer can be written in terms of a radiation heat transfer coefficient as

$$q_{\text{rad}} = h_r(T_w - T_{\text{sat}}); \quad h_r = 4\sigma\varepsilon \left(\frac{T_w + T_{\text{sat}}}{2} \right)^3 \tag{7.75}$$

where ε is the wall emittance. If the average boiling heat transfer coefficient is denoted h_b, then the total average heat transfer coefficient $\overline{h}$ is simply

$$\overline{h} = h_b + h_r \tag{7.76}$$

The similarity of film boiling to film condensation can be exploited to obtain a simple result for the effect of radiation on h_b. To the level of approximation desired, the result is independent of geometry, so for simplicity, laminar film boiling on a vertical wall will be considered. For film boiling, Eqs. (7.5) and (7.9) become

$$\Gamma = \frac{g\delta^3(\rho_l - \rho_v)}{3\nu_v} \tag{7.77}$$

$$h_b = \frac{k_v \, (\partial T/\partial y)|_w}{T_w - T_{\text{sat}}} = \frac{k_v}{\delta} \tag{7.78}$$

The energy balance Eq. (7.13), modified to include radiation, becomes

$$k_v \frac{\partial T}{\partial y}\bigg|_w + q_{\text{rad}} = h_{\text{fg}} \frac{d\Gamma}{dx} \tag{7.79}$$

Substituting Eqs. (7.75), (7.77), and (7.78) in Eq. (7.79) and rearranging,

$$\frac{k_v \nu_v (T_w - T_{\text{sat}})}{h_{\text{fg}} g(\rho_l - \rho_v)} dx = \frac{\delta^3 d\delta}{1 + h_r/(k_v/\delta)}$$

Integrating with $\delta = 0$ at $x = 0$ gives

$$\frac{4k_v \nu_v (T_w - T_{\text{sat}})x}{h_{\text{fg}} g(\rho_l - \rho_v)} = \delta^4 \left[1 - \frac{4}{5} \frac{h_r}{k_v/\delta} + \frac{4}{6} \left(\frac{h_r}{k_v/\delta} \right)^2 - \frac{4}{7} \left(\frac{h_r}{k_v/\delta} \right)^3 + \cdots \right]$$

If the film thickness for no radiation is denoted δ_0, and the corresponding heat transfer coefficient is $h_{b0}(=k_v/\delta_0)$, then

$$\frac{1}{h_{b0}^4} = \frac{1}{h_b^4}\left[1 - \frac{4}{5}\frac{h_r}{h_b} + \frac{4}{6}\left(\frac{h_r}{h_b}\right)^2 - \frac{4}{7}\left(\frac{h_r}{h_b}\right)^3 + \cdots\right] \tag{7.80}$$

which is an implicit equation for h_b that can be solved by iteration. As a first guess, the approximate result $h_b = h_{b0} - (1/5)h_r$ can be used. Equation (7.80) was first given by E. Sparrow [16].

The computer program BOIL, item 4, calculates the average heat transfer coefficient for film boiling according to the recommendations made in this section. The effect of radiation on film boiling is not included.

EXAMPLE 7.10   Film Boiling of a Cryogenic Liquid

A pure copper sphere of 1.5 cm diameter is suddenly immersed in a Dewar flask of saturated liquid nitrogen at 1 atm pressure. Since the temperature difference is large, boiling commences in the film-boiling regime. When $\Delta T = T_w - T_{sat} = 185$ K, determine (i) the average heat transfer coefficient and (ii) the rate of change of the sphere temperature.

Solution

Given: Film boiling of liquid N_2 on a copper sphere.

Required: (i) Average heat transfer coefficient when $\Delta T = 185$ K, and (ii) the corresponding rate of change of the sphere temperature.

Assumptions: The lumped thermal capacity model is adequate for part (ii).

(i) Equation (7.73) gives the average heat transfer coefficient for film boiling in liquid nitrogen:

$$\overline{h} = 0.15\left[\frac{(\rho_l - \rho_v)gh'_{fg}k_v^2}{v_v(T_w - T_{sat})}\right]^{1/3}$$

At 1 atm pressure, Table A.12c gives $T_{sat} = 77.4$ K, the normal boiling point, and $h_{fg} = 0.20 \times 10^6$ J/kg. From Table A.8, $\rho_l = 809$ kg/m³. Vapor properties are evaluated at the mean film temperature, $T_r = 77.4 + (185/2) = 170$ K. From Table A.7, $k_v = 0.0173$ W/m K, $c_{pv} = 1048$ J/kg K, $\rho_v = 2.0$ kg/m³, $v_v = 5.7 \times 10^{-6}$ m²/s.

$$h'_{fg} = h_{fg} + 0.5c_{pv}(T_w - T_{sat}) = 0.20 \times 10^6 + 0.5(1048)(185) = 0.297 \times 10^6 \text{ J/kg}$$

$$\overline{h} = 0.15\left[\frac{(809 - 2.0)(9.81)(0.297 \times 10^6)(0.0173)^2}{(5.7 \times 10^{-6})(185)}\right]^{1/3} = 131 \text{ W/m}^2 \text{ K}$$

(ii) The Biot number for the sphere is

$$\text{Bi}_{LTC} = \frac{\overline{h}(R/3)}{k_{Cu}} = \frac{(131)(0.0075/3)}{(400)} = 0.8 \times 10^{-3} \ll 0.1$$

Thus, a lumped thermal capacity model of the sphere temperature response is adequate. Using Eq. (1.36),

$$\frac{dT}{dt} = -\frac{\bar{h}A}{c\rho V}(T - T_{sat})$$

At $T = 77.4 + 185 = 262.4$ K, pure copper properties from Table A.1 are $c = 374$ J/kg K, $\rho = 8933$ kg/m^3. Also, $A/V = \pi D^2/(\pi D^3/6) = 6/D$.

$$\frac{dT}{dt} = -\frac{(131)(6/0.015)(185)}{(374)(8933)} = -2.90 \text{ K/s}$$

Solution using BOIL

The required input is:

> Item 4 (film boiling)
> Geometry 4 (sphere)
> Fluid = 3 (N$_2$)
> $P = 1.013 \times 10^5$
> $T_w = 262.4$
> $L = 0.015$

BOIL gives the output:

> Relevant property values
> $\bar{h} = 131$ W/m^2 K
> $q = 24,200$ W/m^2

Comments

1. Since Eq. (7.73) is based on experimental data for boiling N$_2$, the calculated heat transfer coefficient should be reliable.

2. BOIL does not use L_c for this case.

EXAMPLE 7.11   Film Boiling of Water on a Horizontal Plate

What heat flux can be rejected from a large horizontal plate by film boiling of water at 1 atm pressure, if the plate surface temperature is 827 K? The plate emittance is 0.8.

Solution

Given: Water boiling on a horizontal plate.

Required: Heat transfer rate q.

Assumptions: 1. Radiation absorption in the vapor film is negligible (check using the procedure given in Section 6.7).
2. The effect of radiation can be approximately estimated using the vertical wall result, Eq. (7.80).

We will first neglect radiation heat transfer; then Eq. (7.72) gives the average heat transfer coefficient as

$$\bar{h} = C_{\text{fb}} \left[\frac{(\rho_l - \rho_v)gh'_{\text{fg}}k_v^3}{v_v L(T_w - T_{\text{sat}})} \right]^{1/4}$$

The reference temperature for vapor properties is the mean film temperature, $T_r = (1/2)(373 + 827) = 600$ K: $k_v = 0.046$ W/m K, $c_{pv} = 2003$ J/kg K, $v_v = 58.5 \times 10^{-6}$ m²/s, $\rho_v = 0.366$ kg/m³. Also, at $T_{\text{sat}} = 373.15$ K, $h_{\text{fg}} = 2.257 \times 10^6$ J/kg, $\sigma = 58.9 \times 10^{-3}$ N/m, $\rho_l = 958$ kg/m³. The characteristic length to be used in Eq. (7.72) is L_c:

$$L = L_c = \left[\frac{\sigma}{(\rho_l - \rho_v)g} \right]^{1/2} = \left[\frac{(58.9 \times 10^{-3})}{(958)(9.81)} \right]^{1/2} = 2.50 \times 10^{-3} \text{ m (2.5 mm)}$$

The correction to the enthalpy of vaporization to account for vapor superheat is

$$h'_{\text{fg}} = h_{\text{fg}} + 0.35 c_{pv}(T_w - T_{\text{sat}}) = 2.257 \times 10^6 + (0.35)(2003)(454) = 2.575 \times 10^6 \text{ J/kg}$$

Substituting in Eq. (7.72), with $C_{\text{fb}} = 0.425$ for a large horizontal surface,

$$\bar{h} = 0.425 \left[\frac{(958)(9.81)(2.575 \times 10^6)(0.046)^3}{(58.5 \times 10^{-6})(2.5 \times 10^{-3})(827 - 373)} \right]^{1/4} = 184 \text{ W/m}^2 \text{ K}$$

$$\bar{q} = \bar{h}(T_w - T_{\text{sat}}) = (184)(827 - 373) = 8.35 \times 10^4 \text{ W/m}^2$$

We should verify that this value of $\bar{q}$ is greater than $q_{\min}$, which is given by Eq. (7.74) with $C_{\min} = 0.09$. In Eq. (7.74), ρ_v, is evaluated at T_{sat}: at 373.15 K, $\rho_v = 0.598$ kg/m³.

$$q_{\min} = C_{\min}\rho_v h_{\text{fg}} \left[\frac{\sigma g(\rho_l - \rho_v)}{(\rho_l + \rho_v)^2} \right]^{1/4}$$

$$= (0.09)(0.598)(2.257 \times 10^6) \left[\frac{(58.9 \times 10^{-3})(9.81)(958)}{(958)^2} \right]^{1/4} = 1.90 \times 10^4 \text{ W/m}^2$$

Thus, the film-boiling heat flux calculated above is about four times the minimum heat flux.

To include the effect of radiation transfer, we first calculate the radiation heat transfer coefficient h_r:

$$h_r = 4\sigma\varepsilon \left(\frac{T_w + T_{\text{sat}}}{2} \right)^3 = (4)(5.67 \times 10^{-8})(0.8) \left(\frac{827 + 373}{2} \right)^3 = 39.2 \text{ W/m}^2 \text{ K}$$

Retaining two terms in the series of Eq. (7.80) and rearranging,

$$1 - \left(\frac{h_b}{h_{b0}} \right) = 1 - \left[1 - \frac{4}{5}\frac{h_r}{h_b} + \frac{4}{6}\left(\frac{h_r}{h_b} \right)^2 \right]^{1/4} \quad ; \quad h_{b0} = \bar{h} = 184 \text{ W/m}^2 \text{ K}$$

Iteration is required to solve this equation. As a first guess, use $h_b = h_{b0} - (1/5)h_r = 184 - (0.2)(39.2) = 176.2$, and construct the following table.

h_b	l.h.s.	r.h.s.
176.2	0.0424	0.0384
177.0	0.0380	0.0383
176.8	0.0391	0.0383

Interpolating in the table,

$$h_b = 176.9 \text{ W/m}^2 \text{ K}; \quad \bar{h} = h_b + h_r = 176.9 + 39.2 = 216.1 \text{ W/m}^2 \text{ K}$$
$$\bar{q} = \bar{h}(T_w - T_{sat}) = (216.1)(827 - 373) = 9.81 \times 10^4 \text{ W/m}^2$$

Comments

1. The effect of radiation is to increase q by only 17%; hence, our approximate method for calculating the effect of radiation is adequate.

2. These film-boiling heat fluxes are only about 7% of the peak value calculated in Example 7.9.

7.5   HEATPIPES

A **heatpipe** is an evaporator-condenser system in which the liquid is returned to the evaporator by capillary action. In its simplest form, it is a hollow tube with a few layers of wire screen along the wall to serve as a wick, as shown in Fig. 7.22. The screen is filled with a wetting liquid such as sodium or lithium for high-temperature applications, or with water, ammonia, or methanol for moderate-temperature applications. If one end of the heatpipe is heated and the other end is cooled, the liquid evaporates at the hot end and condenses at the cold end. As the liquid is depleted in the evaporator section, cavities form in the surface due to the liquid clinging to the wires of the screen. In the condenser section, meanwhile, the screen becomes flooded. The surface tension acting on the concave liquid-vapor interface in the evaporator causes the pressure to be higher in the vapor than in the liquid. This pressure is transmitted by the vapor to the flooded condenser section, where the vapor and liquid pressures are nearly equal, so that liquid is driven from the condenser to the evaporator through the wick. In a gravity field, the evaporator may be placed below the condenser to assist the liquid flow. If a wick is unnecessary, the device is more properly called a *Perkins tube*. Heatpipes are particularly attractive for space

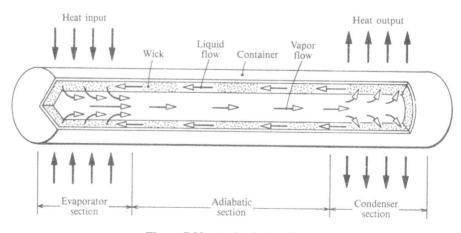

Figure 7.22   A simple heatpipe.

vehicle applications, where gravity fields are very weak, and an inherently reliable passive device is preferred. Of course, a pump could be used to return the liquid to the evaporator, but this would be an undesirable feature for many applications—in particular, space vehicle thermal control, where vibration and pump reliability would be major concerns.

The working fluids chosen for heatpipes have high enthalpies of vaporization. Thus, a small vapor flow along the tube can transport a large amount of thermal energy. The total temperature difference necessary is the sum of the differences required to conduct the heat through the evaporator wall and wick and through the condenser wall and wick, and the amount necessary to provide the vapor pressure difference that drives the vapor from the evaporator to the condenser. This temperature difference is quite small compared to the amount that would be necessary to transfer an equal amount of heat along the length of the pipe by conduction, even if the pipe were solid copper. In addition to a high enthalpy of vaporization, other desirable characteristics of working fluids include a high surface tension and a low viscosity, to improve capillary pumping in the wick, and a high liquid thermal conductivity, to reduce the temperature drops in the evaporator and condenser.

The wick must provide (1) surface pores to generate capillary pumping pressure, (2) internal flow passages for return of the liquid to the evaporator, and (3) a suitable heat flow path from the pipe inner wall to the liquid-vapor interface. Wick structures include wire screen, sintered metal, metal foam, metal felt, woven wire mesh, and axial grooves. Figure 7.23 shows some examples, and Table 7.4 lists the properties of some commonly used wicking materials. Wrapped screen wicks are simple to install and are widely used; however, the temperature drop across the wick tends to be large. Sintered powder metal wicks or axial grooves are preferred if the temperature drop across the wick is an important design constraint. A compromise is the screen-covered groove wick. The fine mesh screen gives a high capillary pumping pressure while the grooves give a low liquid flow resistance and a low resistance to heat flow across the wick. Graded-porosity wicks have a finer mesh in the evaporator to increase capillary pressure and a coarser mesh in the condenser to facilitate liquid flow.

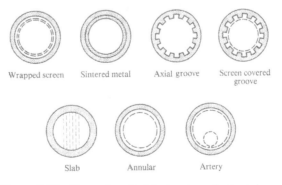

Figure 7.23   Examples of wick structures for heatpipes.

Table 7.4   Data for commonly used wick structures. (Courtesy of Mr. D. Antoniuk, TRW Systems Inc., Redondo Beach, California.)

Wick Type	Effective Pore Radius mm	Permeability $m^2 \times 10^{10}$	Porosity %
30 mesh screen	0.43	25	
100 mesh screen	0.12	1.8	58.5
200 mesh screen	0.063	0.55	67.6
Nickel felt	0.17	5.2	89.1
Nickel foam 210-5	0.23	36	94.4
33 μm-diameter stainless steel fibers	>0.034	2.0	80.8
75 μm-diameter stainless steel fibers	>0.041	11.6	82.8
Sintered fibers/powders	0.01–0.1	0.1–10	
Axial grooves	0.25–1.5	35-1250	
Open annulus	0.25–1.5	50–2000	
Open artery	0.50–1.5	300–3000	

Heatpipes that contain a single vapor are *fixed-conductance heatpipes*, since their thermal resistance is relatively insensitive to heat load. Also widely used are gas-loaded heatpipes, in which a small amount of noncondensable gas is added to the vapor to give a *variable-conductance heatpipe*. Figure 7.24 shows a schematic of a gas-loaded heatpipe with a wicked gas reservoir at the end of the condenser. When the heatpipe is off, that is, when there is no heat load, the evaporator is cold and the vapor pressure very low. The gas then expands to fill the heatpipe. As a heat load is applied, the evaporator temperature and vapor pressure increase, and the gas is pushed out of the evaporator toward the condenser. When the heatpipe is fully on, gas occupies only the very end of the condenser and the reservoir. Thus, the length of condenser in operation varies with heat load, and the heatpipe temperature can be maintained within a narrow temperature range over a wide variation of heat load and sink conditions. This feature is particularly attractive for some spacecraft applications, where it is important to keep electronic components within a prescribed temperature range. If a fixed-conductance heatpipe is used, as the heat load is reduced, the condenser temperature must drop closer to the sink temperature, and the evaporator temperature must fall accordingly. It is also easier to prevent

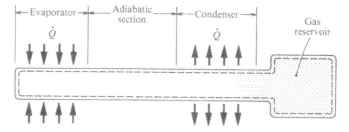

Figure 7.24   A gas-loaded heatpipe with a wicked gas reservoir.

gas-loaded heatpipes on satellites from freezing during extended "cold soak" periods. By supplying only a small amount of residual heat to the evaporator, the evaporator temperature can be maintained above the freezing point of the liquid.

7.5.1 Capillary Pumping

Figure 7.25 shows the pressures acting inside an annular wicked heatpipe. During steady operation, the balance is

$$\begin{array}{cc} \text{Capillary} \\ \text{pressure} \end{array} - \begin{array}{cc} \text{Gravitational} \\ \text{force/area} \end{array} = \begin{array}{cc} \text{Liquid-phase} \\ \text{pressure drop} \end{array} + \begin{array}{cc} \text{Vapor-phase} \\ \text{pressure drop} \end{array} \qquad (7.81)$$

$$\Delta P_C - \Delta P_G = \Delta P_l + \Delta P_v$$

To simplify the analysis, the balance is made for an effective heatpipe length L_{eff}, from the midpoint of the evaporator to the midpoint of the condenser. The gravitational force/unit area ΔP_G is zero for a horizontal heatpipe and is negative if the condenser is located above the evaporator (i.e., has a *favorable tilt*). In general, the pressure drops ΔP_l and ΔP_v increase with heat load due to the increase in flow rate; hence, the required capillary pressure ΔP_C also increases. However, there is a maximum capillary pressure $(\Delta P_C)_{max}$ that can be developed by a given liquid-wick combination. For a heatpipe to operate continuously, the required capillary pressure must not exceed this maximum value at any point along the pipe. If it does, the wick can dry out, a condition known as evaporator *burnout*. When burnout occurs, the load must be reduced to allow the wick to reprime.

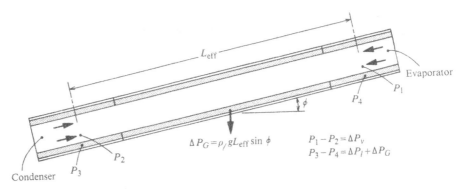

Figure 7.25   Pressures acting inside an annular wicked heatpipe with an adverse tilt.

Capillary Pressure

An elementary experiment in physics is the demonstration of a wetting liquid rising in a capillary tube, as shown in Fig. 7.26. A force balance gives the well-known result,

$$(\rho_l - \rho_v)gh = \frac{2\sigma \cos \theta}{r_p} \qquad (7.82)$$

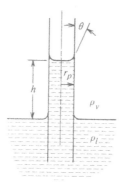

Figure 7.26 Rise of a wetting liquid in a capillary tube of radius r_p. Angle θ is the contact angle.

where θ is the *contact angle*. For wetting liquids, $0 < \theta < 90°$, and for nonwetting liquids, $\theta > 90°$. In heatpipes, it is essential to choose a liquid-wick pair that wets well. Figure 7.27 shows liquid menisci in the wick at the evaporator and condenser. Using the subscripts e to denote evaporator and c to denote condenser, the resultant capillary pressure available to pump the liquid along the wick is[3]

$$\Delta P_C = 2\sigma \left(\frac{\cos \theta_e}{r_p} - \frac{\cos \theta_c}{r_p} \right)$$

ΔP_C has a maximum value when $\cos \theta_e = 1$ and $\cos \theta_c = 0$:

$$(\Delta P_C)_{\text{max}} = \frac{2\sigma}{r_p} \tag{7.83}$$

Table 7.4 gives the pore radius r_p for various homogeneous wick materials.

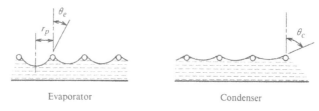

Evaporator   Condenser

Figure 7.27 Liquid menisci in a heatpipe wick, showing a smaller radius of curvature in the evaporator.

Gravitational Force

The gravitational force per unit area is the hydrostatic pressure differential between the evaporator and condenser and may be positive, zero, or negative depending on their relative elevations. Referring to Fig. 7.25, the gravitational head for $\rho_v \ll \rho_l$ is

$$\Delta P_G = \rho_l g L_{\text{eff}} \sin \phi \tag{7.84}$$

---

[3] These contact angles θ_e and θ_c depend on the rates of evaporation and condensation (rather than on the intrinsic wetting behavior).

Liquid-Phase Pressure Drop

The pressure drop for liquid flowing through a homogeneous wick can be calculated from Darcy's law for flow through porous media:

$$\Delta P_l = \frac{\mu_l L_{eff} \dot{m}}{\kappa \rho_l A_w} \tag{7.85}$$

where $\dot{m}$ [kg/s] is the liquid flow rate, A_w is the cross-sectional area of the wick, and κ [m^2] is the *permeability* of the wick material. Values of κ are given in Table 7.4. The pressure drop for liquid flowing along open grooves can be roughly estimated from Eq. (4.39) for laminar flow in a pipe using the hydraulic diameter:

$$\Delta P_l = \frac{64}{Re_{D_{hl}}} \frac{L_{eff}}{D_{hl}} \left(\frac{1}{2}\rho_l V_l^2\right); \quad D_{hl} = 4\frac{\text{Liquid flow area}}{\text{Wetted perimeter}} \tag{7.86}$$

At high vapor velocities, vapor drag may impede liquid flow in open grooves. A remedy is to cover the grooves with a screen to form a *composite* wick.

Vapor-Phase Pressure Drop

The vapor-phase pressure drop is generally much smaller than that for the liquid phase. Often it may be neglected, or a very approximate calculation may suffice. Provided the Mach number is less than about 0.3, incompressible flow may be assumed. The flow is usually laminar, and Eq. (4.39) for fully developed flow can be used:

$$\Delta P_v = \frac{64}{Re_{D_{hv}}} \frac{L_{eff}}{D_{hv}} \left(\frac{1}{2}\rho_v V_v^2\right); \quad D_{hv} = 4\frac{\text{Vapor flow area}}{\text{Wetted perimeter}} \tag{7.87}$$

Note that the flow is not fully developed at each end of the heatpipe, and there is a normal velocity component associated with phase change that decreases the wall shear stress in the evaporator and increases it in the condenser ("blowing" in the evaporator and "suction" in the condenser are discussed in Sections 7.2 and 7.3). In addition, there is the pressure drop required to accelerate the vapor in the evaporator, and a pressure recovery due to deceleration in the condenser. However, these effects are self-compensating and are thus generally ignored in heatpipe design.

Wicking Limitation

Equation (7.81) written for the maximum capillary pressure available, $(\Delta P_C)_{max}$, gives the *wicking limitation* of the heatpipe. For a homogeneous wick,

$$\frac{2\sigma}{r_p} - \rho_l g L_{eff} \sin\phi = \frac{\mu_l L_{eff} \dot{m}_{max}}{\kappa \rho_l A_w} + \frac{64}{Re_{D_{hv}}} \frac{L_{eff}}{D_{hv}} \left(\frac{1}{2}\rho_v V_v^2\right) \tag{7.88}$$

and

$$\dot{Q}_{max} = \dot{m}_{max} h_{fg} \tag{7.89}$$

If the vapor-phase pressure drop is neglected, Eqs. (7.88) and (7.89) can be rearranged as

$$\dot{Q}_{max} = \left(\frac{\rho_l \sigma h_{fg}}{\mu_l} \right) \left(\frac{A_w \kappa}{L_{eff}} \right) \left(\frac{2}{r_p} - \frac{\rho_l g L_{eff} \sin \phi}{\sigma} \right) \tag{7.90}$$

The fluid properties combination in the first set of parentheses is the **figure of merit** $\mathcal{M}$ of the fluid, usually expressed in kW/cm^2:

$$\mathcal{M} = \frac{\rho_l \sigma h_{fg}}{\mu_l} \, [kW/cm^2] \tag{7.91}$$

$\mathcal{M}$ is a function of temperature and is plotted in Fig. 7.28 for a selection of heat-pipe fluids. The clear superiority of water in the intermediate temperature range is due to its high latent heat and surface tension, but there are also other criteria for fluid selection, such as materials compatibility. Liquid metals are indicated for high-temperature heatpipes, with lithium being the first choice above 1400 K.

Heatpipes are often characterized by a **heat transfer factor**, defined as the product of the maximum heat flow and effective pipe length for a horizontal pipe:

$$(\dot{Q}_{max} L_{eff})_{\phi=0} = \left(\frac{\rho_l \sigma h_{fg}}{\mu_l} \right) (A_w \kappa) \left(\frac{2}{r_p} \right) = \frac{2 \mathcal{M} A_w \kappa}{r_p} \, [\text{W m}] \tag{7.92}$$

In practice, the engineer attempts to improve this factor by choosing a wick material with large permeability κ and small pore radius r_p.

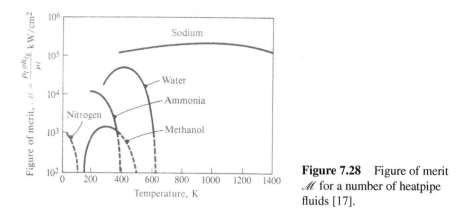

Figure 7.28   Figure of merit $\mathcal{M}$ for a number of heatpipe fluids [17].

EXAMPLE 7.12   An Ammonia Heatpipe

An ammonia heatpipe is constructed from an 0.5 inch-O.D. stainless steel tube and has an effective length of 1.40 m. The aluminum fibrous slab wick has a cross-sectional area of 4.7×10^{-5} m^2. To investigate the wick performance, an experiment was carried out in which the burnout heat load was determined as a function of heatpipe inclination. In these experiments, the adiabatic section was maintained at 22°C ±2°C.

Heatpipe angle, degrees	0	0.3	0.5	0.7	0.9
$\dot{Q}_{max}$, W		94	74	55	38	23

Estimate the effective pore radius and Darcy permeability of the aluminum wick.

Solution

Given: Burnout heat load as a function of inclination for an ammonia heatpipe.

Required: Estimate of wick pore radius, r_p, and permeability, κ.

Assumptions: Negligible vapor flow pressure drop, so that Eq. (7.90) applies.

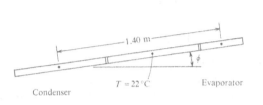

Equation (7.90) gives the burnout heat load:

$$\dot{Q}_{max} = \left(\frac{\rho_l \sigma h_{fg}}{\mu_l}\right)\left(\frac{A_w \kappa}{L_{eff}}\right)\left(\frac{2}{r_p} - \frac{\rho_l g L_{eff}\sin\phi}{\sigma}\right)$$

All ammonia properties will be evaluated at 22°C = 295 K. From Tables A.8, A.11, and A.12b, $\rho_l = 609$ kg/m³, $\mu_l = 1.38 \times 10^{-4}$ kg/m s, $\sigma = 21 \times 10^{-3}$ N/m, $h_{fg} = 1.179 \times 10^6$ J/kg. Substituting above,

$$\dot{Q}_{max} = \left[\frac{(609)(21 \times 10^{-3})(1.179 \times 10^6)}{1.38 \times 10^{-4}}\right]\left[\frac{(4.70 \times 10^{-5})\kappa}{1.40}\right]\left[\frac{2}{r_p} - \frac{(609)(9.81)(1.40)\sin\phi}{21 \times 10^{-3}}\right]$$

$$= 3.67 \times 10^6 \kappa \left(\frac{2}{r_p} - 3.98 \times 10^5 \sin\phi\right)$$

$\dot{Q}_{max}$ is seen to be a linear function of $\sin\phi$, which suggests a plot of $\dot{Q}_{max}$ versus ϕ as shown ($\phi = \sin\phi$ for small ϕ). The intercept for $\dot{Q}_{max} = 0$ gives r_p, and κ can be determined from the slope.

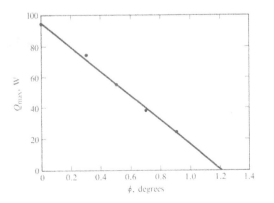

$$\dot{Q}_{max} = 0 \text{ at } \phi = 1.20 \text{ degrees } = 0.0209 \text{ radian}$$

$$r_p = 2/(3.98 \times 10^5)\sin\phi = 2/(3.98 \times 10^5)(0.0209) = 2.40 \times 10^{-4} \text{ m (0.24 mm)}$$

$$\frac{\dot{Q}_{max}}{\phi} = -\frac{95}{1.20} = -79.2 \text{ W/degree } = -4540 \text{ W/rad}$$

(from the intercept on the graph)

$$4540 = (3.67 \times 10^6)(3.98 \times 10^5)\kappa, \quad \kappa = 31 \times 10^{-10} \text{ m}^2$$

Comments

The heat transfer factor $(\dot{Q}_{max}L_{eff})_{\phi=0}$ for this heatpipe is $(95)(1.40) = 133$ W m.

7.5.2 Sonic, Entrainment, and Boiling Limitations

Apart from the fundamental wicking or capillary limitation on heat transport by a heatpipe, there are other factors that may, under some circumstances, limit heat transport. The most important of these factors are choking of the vapor flow, entrainment of liquid by the vapor flow, and boiling of the liquid in the evaporator wick. Choking, or the *sonic limitation*, is influenced only by the vapor core size. The entrainment limitation is increased by using wicks with a smaller pore size at the vapor-liquid interface. The boiling limitation can be increased by using wicks with a high effective thermal conductivity. Figure 7.29 shows how these limitations affect the performance of a typical heatpipe.

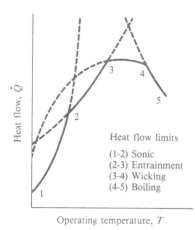

Heat flow limits

(1-2) Sonic
(2-3) Entrainment
(3-4) Wicking
(4-5) Boiling

Operating temperature, T

Figure 7.29   Schematic of limits on heatpipe performance.

Sonic Limitation

In a converging nozzle, a compressible gas accelerates to the sonic velocity due to area change; in a heatpipe, the vapor can accelerate to the sonic velocity by mass addition in the evaporator. If the vapor flow in a heatpipe is choked, a further decrease in the condenser temperature has no effect on the evaporator temperature; thus, a

Table 7.5 Sonic limitation for metal vapor heat pipes [18].

Evaporator exit temperature, °C	Sonic heat flux limit, kW/cm^2			
	Cesium	Potassium	Sodium	Lithium
400	1.0	0.5	—	—
500	4.6	2.9	0.6	—
600	14.9	12.1	3.5	—
700	37.3	36.6	13.2	—
800			38.9	1.0
900			94.2	3.9
1000				12.0
1100				31.1
1200				71.0
1300				143.8

sonic limitation to the axial heat transport per unit cross-sectional area of the core exists. The sonic limit is particularly relevant to the performance of liquid-metal heatpipes at the low end of their operating temperature range because of the low vapor density. Table 7.5 gives the sonic limitation for a selection of liquid metals. The sonic limitation may be encountered during start-up when the evaporator is cold, even though it may not be a factor at the design operating point.

Entrainment Limitation

If the momentum flux of the vapor flow is sufficiently high, the stress exerted by the vapor on the liquid surface can be sufficient to generate waves and entrain liquid droplets from the wave crests. Entrainment causes a substantial increase in the fluid circulation rate, and if the capillary pumping head is insufficient, a sudden dry out of the wick in the evaporator can occur. When there is significant entrainment, the droplets impinging at the condenser end may make an audible sound. The ratio of the vapor momentum flux to the surface tension forces restraining the liquid is proportional to the *Weber number*,

$$\text{We} = \frac{\rho_v V_v^2 L}{\sigma} \tag{7.93}$$

where the characteristic length L is taken to be the hydraulic diameter of the wick surface pores. This diameter is equal to wire spacing for screen wicks, to twice the groove width for groove wicks, and to 0.82 times the sphere radius for packed spheres. A value of We $\sim$ 1 is taken to indicate the entrainment limit.

Boiling Limitation

The formation of bubbles in the evaporator wick is undesirable because hot spots can be formed that obstruct the liquid flow. Whereas the wicking, sonic, and entrainment limitations are limitations on the axial heat flux, the boiling limitation is a limitation on the radial heat flux in the evaporator. From a practical viewpoint, the engineer increases the boiling limitation by choosing a wick of high effective thermal conductivity and by providing an adequate heat transfer area in the evaporator. Bubble nucleation can be predicted by the theory given for pool boiling in Section 7.4.2. However, once again, there are problems related to reliable estimation of nucleation site size, the role played by dissolved gases, and the effect of temperature gradients in the liquid phase. The boiling limitation proves to be unimportant for liquid-metal heatpipes but can be a major problem for water heatpipes, since the water does not easily fill nucleation sites.

7.5.3 Gas-Loaded Heatpipes

Gas-loaded variable-conductance heatpipes were briefly described earlier. Figure 7.30 is a schematic of a heatpipe that has a gas reservoir of volume V_r connected to the end of the condenser. Notice that the wick extends into the reservoir. The important operating characteristics of such a heatpipe can be obtained from a simple model based on the following assumptions.

1. There is a flat front between the vapor and the gas that divides the condenser of length L_c into an active length L_a and an inactive length $(L_c - L_a)$.

2. The total pressure in the condenser P_c is constant and equals the reservoir pressure.

3. Axial conduction along the wall and wick is negligible, so that there is a step change in temperature of the heatpipe and its contents at the vapor-gas front.

4. Since it is assumed that there is pure vapor in the active-length, the active-length temperature T_a is the saturation temperature corresponding to pressure $P_c : T_a = T_{\text{sat}}(P_c)$.

5. Under steady operating conditions, the inactive condenser and reservoir are in thermal equilibrium with the heat rejection sink at temperature T_o.

The active length of the condenser is obtained from an inventory of the gas. The vapor pressure in both the inactive condenser and the reservoir is $P_{\text{sat}}(T_o)$. Using Dalton's law of partial pressures and the ideal gas law, the mass of gas is

$$w_g = \frac{P_c - P_{\text{sat}}(T_o)}{R_g T_o} \left[A_v (L_c - L_a) + V_r \right] \tag{7.94}$$

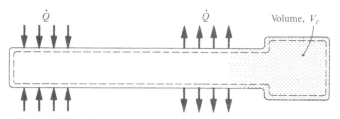

Figure 7.30   A gas-loaded heatpipe with a wicked reservoir.

where A_v is the cross-sectional area of the vapor core. Solving for L_a gives

$$L_a = L_c + \frac{V_r}{A_v} - \frac{w_g R_g T_o}{[P_c - P_{sat}(T_o)]A_v} \tag{7.95}$$

The heat rejected by the condenser can be written as

$$\dot{Q} = (U\mathscr{P})_c L_a (T_a - T_o) \tag{7.96}$$

where $U\mathscr{P}$ is the overall heat transfer coefficient-perimeter product for radial heat flow out of the condenser, and accounts for the thermal resistances of liquid-filled wick, wall, and convection or radiation to the sink at temperature T_o. Substituting Eq. (7.95) in Eq. (7.96) gives

$$\dot{Q} = (U\mathscr{P})_c (T_a - T_o) \left[L_c + \frac{V_r}{A_v} - \frac{w_g R_g T_o}{[P_c - P_{sat}(T_o)]A_v} \right] \tag{7.97}$$

The effect of the gas can be easily seen if we consider a situation where $V_r = 0$ (that is, no gas reservoir), as shown in Fig. 7.31. Recall that an objective of gas loading is to maintain a nearly constant evaporator temperature when the heat load drops below the design value. When there is no gas in the heatpipe and T_o is held constant, Eq. (7.97) shows that $(T_a - T_o)$ is nearly directly proportional to the heat load $\dot{Q}$, so that T_a must decrease as $\dot{Q}$ is reduced. But if there is an appropriate amount of gas present, a decrease in T_a decreases P_c, since $P_c = P_{sat}(T_a)$, and the last term in the square brackets of Eq. (7.97) increases as the gas expands. Thus, the effect of the gas is to limit the decrease in T_a required to match the decrease in heat load. The mass of gas added to the heatpipe depends on the size of reservoir, the sink temperature, and the desired design value for T_a. When the heatpipe is fully open, all the gas is in the reservoir, $\dot{Q} = \dot{Q}_{max}$, and the corresponding active length temperature is $T_{a,max}$. The total pressure in the reservoir is $P_c \simeq P_{sat}(T_{a,max})$. Using the ideal gas law,

$$w_g = \frac{[P_{sat}(T_{a,max}) - P_{sat}(T_o)]V_r}{R_g T_o} \tag{7.98}$$

which allows the required mass of gas to be calculated for a chosen reservoir volume V_r. To see how the reservoir volume affects the performance of the heatpipe, it is convenient to divide the actual heat load $\dot{Q}$ by the maximum value within its control range, $\dot{Q}_{max}$.

$$\dot{Q}_{max} = (U\mathscr{P})_c (T_{a,max} - T_o)L_c \tag{7.99}$$

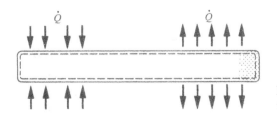

Figure 7.31 A gas-loaded heatpipe with no reservoir.

Dividing Eq. (7.97) by Eq. (7.99),

$$\frac{\dot{Q}}{\dot{Q}_{max}} = \frac{T_a - T_o}{T_{a,max} - T_o} \left[1 + \frac{V_r}{L_c A_v} - \frac{w_g R_g T_o}{[P_c - P_{sat}(T_o)]L_c A_v} \right] \qquad (7.100)$$

With $V_c = L_c A_v$ denoting the vapor core volume, and using Eq. (7.98), Eq. (7.100) can be rearranged as

$$\frac{T_a - T_o}{T_{a,max} - T_o} = \frac{\dot{Q}}{\dot{Q}_{max}} \left[\frac{V_c/V_r}{1 + \frac{V_c}{V_r} - \frac{P_{sat}(T_{a,max}) - P_{sat}(T_o)}{P_{sat}(T_a) - P_{sat}(T_o)}} \right] \qquad (7.101)$$

The reservoir-to-core volume ratio, V_r/V_c, determines the sensitivity of the pipe. As V_r/V_c increases, V_c/V_r decreases, and the decrease in T_a for a given reduction in heat load $\dot{Q}$ decreases. This is best illustrated with a numerical example (see Example 7.13).

The model used in the preceding analysis is useful because, in most heatpipes, there is indeed a relatively sharp transition between the active and inactive regions of the condenser. Figure 7.32 shows axial variations of vapor pressure and temperature along a gas-loaded heatpipe. Axial conduction in the wall and wick tends to smear the temperature step, and there is some interpenetration of gas and vapor by diffusion. More exact analysis that accounts for mass diffusion is left to advanced texts.

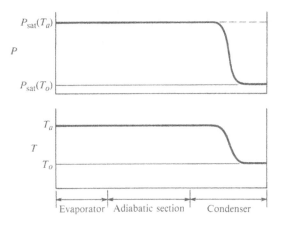

Figure 7.32   Axial variations of temperature and pressure along a gas-loaded heatpipe.

EXAMPLE 7.13   A Gas-Loaded Heatpipe for a Communications Satellite

A nitrogen-loaded methanol heatpipe is constructed from a 0.5 inch-O.D. stainless steel tube and is L-shaped, as shown in the sketch. A fibrous stainless steel slab wick extends into the gas reservoir. The cross-sectional area of the vapor space in the pipe is 50.3 mm², and the reservoir volume is 5.28×10^4 mm³. The condenser length is 35 cm. When the sink temperature is 264 K, the heatpipe is fully open for an adiabatic section temperature of 283 K and a heat load of 50 W. Determine the mass of gas in the heatpipe, and prepare a graph of heat load versus adiabatic section temperature. In addition, vary the reservoir volume to demonstrate its effect on performance. At a reference temperature $T_r = 5°C$, methanol has a vapor pressure of 5330 Pa and enthalpy of vaporization of 1.18×10^6 J/kg.

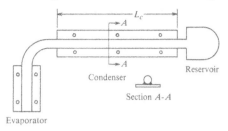

Solution

Given: Nitrogen-loaded methanol heatpipe, with a wicked gas reservoir.

Required: 1. Mass of gas in pipe, w_g.

2. $\dot{Q}$ versus T_a for three different reservoir volumes.

Assumptions: 1. A flat front between the vapor and gas.

2. Negligible pressure drop between adiabatic section and condenser.

When the heatpipe is fully open, the gas is contained in the reservoir, which is at $T_o = 264$ K. Using the ideal gas law,

$$w_g = \frac{[P_c - P_{\text{sat}}(T_o)]V_r}{R_g T_o}$$

where the total pressure has been taken to be equal to the pressure in the active condenser, $P_c = P_{\text{sat}}(T_a)$. The Clausius-Clapeyron relation will be used to obtain $P_{\text{sat}}(T)$ from $P_{\text{sat}}(T_r)$:

$$\frac{P}{P_r} \simeq \exp\left[-\frac{h_{\text{fg}}}{R_v}\left(\frac{1}{T} - \frac{1}{T_r}\right)\right]$$

$$P_r = 5330 \text{ Pa}, \quad T_r = 278.15 \text{ K}$$

$$h_{\text{fg}} = 1.18 \times 10^6 \text{ J/kg}, \quad R_v = 8314/32.04 = 259.5 \text{ J/kg K}$$

$$P = 5330 \exp\left[-4548\left(\frac{1}{T} - 3.595 \times 10^{-3}\right)\right]$$

$$T = 283 \text{ K}, \quad P = 7048 \text{ Pa}$$

$$T = 264 \text{ K}, \quad P = 2217 \text{ Pa}$$

$$w_g = \frac{(7048 - 2217)(5.28 \times 10^{-5})}{(8314/28)(264)} = 3.25 \times 10^{-6} \text{ kg}$$

which is the required mass of gas. Rearranging Eq. (7.100) gives the part load performance as:

$$\frac{T_a - T_o}{T_{a,\text{max}} - T_o} = \frac{\dot{Q}}{\dot{Q}_{\text{max}}}\left[\frac{V_c/V_r}{1 + V_c/V_r - (w_g/V_r)R_g T_o/[P_c - P_{\text{sat}}(T_o)]}\right]$$

$$V_c = L_c A_v = (0.35)(50.3 \times 10^{-6}) = 1.76 \times 10^{-5} \text{ m}^3$$

$$\frac{V_r}{V_c} = \frac{5.28 \times 10^{-5}}{1.76 \times 10^{-5}} = 3.00; \quad \frac{V_c}{V_r} = 0.333$$

$$T_{a,max} = 283K, \quad T_o = 264 \text{ K}$$

$$P_{sat}(T_o) = 2217 \text{ Pa}, \quad \frac{w_g R_g T_o}{V_r} = P_c(T_{a,max}) - P_{sat}(T_o) = 7048 - 2217 = 4831 \text{ Pa}$$

$$\frac{T_a - 264}{19} = \frac{\dot{Q}}{50} \left[\frac{0.333}{1 + 0.333 - 4831/(P_c - 2217)} \right]$$

For specified values of T_a, $P_c = P_{sat}(T_a)$ and $\dot{Q}$ can be calculated. The results are given in the table and graph that follow. Also given are results for $V_r/V_c = 2.0$ and 4.0.

T_a	P_c	$\dot{Q}[W]$		
K	Pa	$V_r/V_c = 2.0$	3.0	4.0
283	7048	50	50	50
282	6658	39.0	34.9	30.7
281	6286	28.0	19.6	11.2
280	5933	16.8	4.2	—
279	5598	5.6	—	—

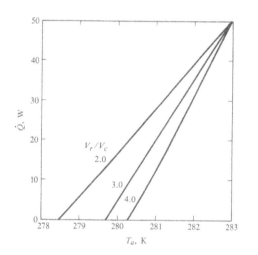

Comments

Notice that as V_r/V_c increases, the adiabatic section temperature becomes less sensitive to heat load.

7.6   CLOSURE

Heat transfer during phase change is essentially a convection process that is often made complicated by an irregular phase interface. Since enthalpies of phase change are large, the resulting heat transfer coefficients also tend to be large. Laminar film condensation and evaporation were satisfactorily analyzed using simplifying assumptions introduced by Nusselt in 1916. For situations where the film is wavy laminar or turbulent, we used empirical correlations for the local heat transfer coefficient and determined the change in film flow rate by satisfying mass and energy balances. The computer program PHASE allowed the rather complicated formulas to be evaluated efficiently and reliably.

We recognize two modes of boiling, namely, pool boiling and forced-convection boiling. The "boiling curve," that is, a graph of q versus $(T_w - T_{sat})$, defined the various regimes of pool boiling. Nucleate boiling is strongly influenced by surface condition but is relatively insensitive to surface geometry. On the other hand, the peak heat flux and film boiling depend primarily on hydrodynamic factors and hence are influenced by surface geometry. We considered only saturated pool boiling; the effects of liquid subcooling and forced convection are left to advanced texts. Boiling heat transfer formulas for engineering use are based on crude physical models and dimensional analysis, with adjustments made by curve-fitting experimental data. The computer program BOIL allowed these formulas to be evaluated efficiently.

The chapter concluded with a discussion of heatpipes. Both fixed-conductance pure-vapor heatpipes and variable-conductance gas-loaded heatpipes were analyzed. The key equation governing the performance of fixed-conductance heatpipes was derived by balancing the capillary pressure generated by the wick against the gravitational force and the liquid- and vapor-phase pressure drops. Heatpipe working fluids were characterized by a *figure of merit* $\mathcal{M} = \rho_l \sigma h_{fg} / \mu_l$, and heatpipe performance was characterized by a *heat transfer factor* $(\dot{Q}_{max} L_{eff})_{\phi=0}$. The characteristics of gas-loaded heatpipes were derived using a simple model that assumes a flat front between the vapor and the gas in the condenser. A worked example illustrated the use of this model for a heatpipe used to control the temperature of a communications satellite.

◼◼◼◼ REFERENCES

1. Nusselt, W., "Die Oberflachenkondensation des Wasserdampfes," *Z. Ver. D-Ing.*, 60, 541–546 (1916).

2. Sadasivan, P., and Lienhard, J. H., "Sensible heat correction in laminar film boiling and condensation," *J. Heat Transfer*, 109, 545–547 (1987).

3. Denny, V. E., and Mills, A. F., "Nonsimiliar solutions for laminar film condensation on a vertical surface," *Int. J. Heat Mass Transfer*, 12, 965–979 (1969).

4. Chun, K. R., and Seban, R. A., "Heat transfer to evaporating liquid films," *J. Heat Transfer*, 93, 391–396 (1971).

5. Shekriladze, I. G., and Gomelauri, V. I., "Theoretical study of laminar film condensation of flowing vapour," *Int. J. Heat Mass Transfer*, 9, 581–591 (1966).

6. Rohsenow, W. M., "A method of correlating heat transfer data for surface boiling of liquids," *Trans. ASME*, 74, 969–976 (1952).

7. Kutateladze, S. S., "On the transition to film boiling under natural convection," *Kotloturbostroenie*, 3, 10 (1948).

8. Zuber, N., "Hydrodynamic aspects of nucleate boiling," Ph.D. dissertation, Department of Engineering, University of California, Los Angeles (1959). [Also AEC Report AECU-4439 (1959).]

9. Lienhard, J. H., and Dhir, V. K., "Hydrodynamic prediction of peak pool-boiling heat fluxes from finite bodies," *J. Heat Transfer*, 95, 152–158 (1973).

10. Bromley, L. A., "Heat transfer in stable film boiling," *Chem. Eng. Prog.*, 46, 221–227 (1950).

11. Frederking, T. H. K., and Daniels, D. L, "The relation between bubble diameter and frequency of removal from a sphere during film boiling," *J. Heat Transfer*, 88, 87–93 (1966).

12. Berenson, P., "Film-boiling heat transfer from a horizontal surface," *J. Heat Transfer*, 83, 351–358 (1961).

13. Frederking, T. H. K., and Clark, J. A., "Natural convection film boiling on a sphere," *Advances in Cryogenic Engineering*, 8, 501–506 (1962).

14. Zuber, N., "On the stability of boiling heat transfer," *Trans. ASME*, 80, 711–720 (1958).

15. Lienhard, J. H., and Wong, P. T. Y., "The dominant unstable wavelength and minimum heat flux during film boiling on a horizontal cylinder," *J. Heat Transfer*, 86, 220–226 (1964).

16. Sparrow, E. M., "The effect of radiation on film-boiling heat transfer," *Int. J. Heat Mass Transfer*, 7, 229–238 (1964).

17. Reay, D. A., McGlen. R., and Kew, P. *Heat Pipes*, 6th ed., Elsevier, Waltham, M.A. (2014).

18. Faghri, A. *Heat Pipe Science and Technology*, Taylor and Francis, New York, N.Y., (1995).

19. Labuntsov, D. A., "Heat transfer in film condensation of pure steam on vertical and horizontal surfaces and tubes," *Teploenergetika*, 72–89 (July 1957).

20. Shmerler, J. A., and Mudawwar, L, "Local heat transfer coefficient in wavy free-falling turbulent liquid films undergoing uniform sensible heating," *Int. J. Heat Mass Transfer*, 31, 67–77 (1988).

21. Mills, A. F., Hubbard, G. L., James, R. K., and Tan, C, "Experimental study of film condensation on horizontal grooved tubes," *Desalination*, 16, 121–133 (1975).

22. Kim, S., and Mills, A. F., "Condensation on coherent turbulent liquid jets," *J. Heat Transfer*, 111, 1068–1074 (1989).

23. Hsu, Y. Y., "On the size range of active nucleation cavities on a heating surface," *J. Heat Transfer*, 84, 207–216 (1962).

EXERCISES

7-1. Saturated ammonia vapor at 310 K condenses on a vertical surface maintained at 305 K. If the surface is 2 cm high, determine the average heat transfer coefficient and the rate of condensation per unit width.

7-2. Consider laminar film condensation from a saturated vapor at its normal boiling point on a vertical surface maintained at a temperature 2 K below the boiling point. If the onset of wavy laminar flow is taken to be Re = 30, determine the location down the surface at which ripples might be expected to be seen for the following fluids: water, ammonia, nitrogen, mercury, R-22, and R-134a.

7-3. Saturated R-134a vapor at 1.25 MPa condenses on the outside of a 2 cm-O.D., 5 cm-high vertical tube maintained at 320 K. Determine the average heat transfer coefficient and rate of condensation.

7-4. Show that if the subcooling term is retained in Eq. (7.12), Eq. (7.16) remains valid if h_{fg} is replaced by a modified latent heat $h'_{fg} = h_{fg} + (3/8)c_{pl}(T_{sat} - T_w)$.

7-5. In practice, film condensation usually takes place on one side of a wall, and the enthalpy of condensation is taken up by a coolant flowing on the other side of the wall. Thus, there are three resistances in series in the thermal network: the condensate film resistance, the wall resistance, and the coolant convective resistance. If the sum of the latter two is much larger than the condensate film resistance, the heat flux along the wall will be nearly constant, rather than the wall temperature being constant, as assumed by Nusselt's analysis. Such a situation occurs in, for example, an air-cooled condenser. For laminar film condensation on a vertical wall with a constant heat flux q_w, show

(i) the film thickness increases proportional to $x^{1/3}$.

(ii) $\overline{Nu} = (4/3)^{4/3} Re_L^{-1/3}$, where $\overline{Nu} = q_w(v_l^2/g)^{1/3}/(\overline{T_{sat} - T_{wx}})k_l$, and show that this result is identical to that for the isothermal wall case.

7-6. Laminar film condensation occurs on a vertical wall with a temperature variation that can be approximated by a power law, $T_w - T_{sat} = ax^n$. Show that the average heat transfer coefficient for a wall of height L is

$$\overline{h} = \frac{4}{(3-n)}\left(\frac{n+1}{4}\right)^{1/4}\left[\frac{h_{fg}g(\rho_l - \rho_v)k_l^3}{L[T_{sat} - T_w(L)]v_l}\right]^{1/4}$$

where $T_w(L)$ is the wall temperature at $x = L$.

7-7. Consider laminar film condensation on the inside of a small-diameter vertical tube for which the film thickness cannot be assumed to be small compared to the tube radius. Extend the Nusselt-type analysis of Section 7.2.1 to this situation.

7-8. During a hypothetical accident in a liquid-metal fast-breeder reactor, saturated UO_2 vapor contacts a cooled vertical wall and condenses as a laminar film. Fission heating at a rate $\dot{Q}'''_v$ occurs in the condensate as it flows down the wall. If the wall is isothermal at temperature T_w, show how the film thickness and wall heat flux are related to the location x down the wall. At what location does the condensation cease?

7-9. Consider laminar film condensation of a saturated vapor on a vertical wall when there is volumetric heat generation at a rate $\dot{Q}_v'''$ in the liquid condensate as it flows down the wall. If heat is removed uniformly from the wall at a rate q_w, show how the film thickness and temperature difference $(T_{sat} - T_w)$ are related to location x down the wall. At what location does the condensation cease?

7-10. A thin mercury film runs down an isothermal vertical wall with a surface velocity u_o. A saturated alcohol vapor condenses on the surface of the mercury as a laminar film. Invoking the Nusselt assumptions and neglecting subcooling in the liquid film, determine the local condensate thickness and heat transfer coefficient.

7-11. Saturated R-22 vapor at 1.61 MPa condenses on the outside of a 4 cm-diameter, 3 m-high vertical tube. Determine the average heat transfer coefficient and total condensation rate for wall temperatures in the range 295–315 K.

7-12. Saturated steam at 1 atm condenses on the outside of a 3 cm-diameter, 2 m-high vertical tube. Determine the average heat transfer coefficient and total condensation rate for wall temperatures in the range 340–370 K.

7-13. D.A. Labuntsov [19] recommends the following correlation of the local Nusselt number for turbulent falling films:

$$\text{Nu} = 0.023\,\text{Re}^{0.25}\,\text{Pr}_l^{0.5}$$

Using a transition Reynolds number of 1600, show that the average Nusselt number for condensation commencing at the top of a vertical surface of length L, with $\text{Re}_L > 1600$, is

$$\overline{\text{Nu}} = \frac{\text{Re}_L}{8090 + 58\text{Pr}_l^{-0.5}(\text{Re}_L^{0.75} - 253)}$$

Recalculate $\overline{\text{Nu}}$ for Example 7.2 using this formula.

7-14. New experimental data for heat transfer across a turbulent falling film has been correlated by Shmerler and Mudawwar [20] as

$$\text{Nu} = 0.0106\,\text{Re}^{0.3}\,\text{Pr}_l^{0.63} \quad 2.55 < \text{Pr}_l < 6.87$$

Using this correlation as an alternative to Eq. (7.23), rework Example 7.2.

7-15. Saturated steam at 97 kPa condenses on the outside of a 6 cm-diameter vertical tube. Cooling water maintains the wall temperature at 369.0 K. Calculate the average heat transfer coefficient and condensate flow rate at the bottom of the tube for tube lengths of

(i) 10 cm.

(ii) 1 m.

Perform hand calculations and check against PHASE.

7-16. Saturated steam at 330 K condenses on the outside of a horizontal, 15 mm-O.D., 1 mm-wall-thickness copper tube, through which flows coolant water. At an axial location where the bulk water temperature is 310 K and the inside heat transfer coefficient is 7000 W/m^2 K, determine the rate of condensation per unit length of tube.

7-17. Saturated ammonia vapor at 310 K condenses on the outside of a horizontal, 2 cm-O.D., 1 mm-wall-thickness brass tube through which flows coolant water. At an axial location where the bulk coolant temperature is 290 K and the inside heat transfer coefficient is 6000 W/m^2 K, determine the heat transfer and condensation rates per unit length of tube.

7-18. Saturated mercury vapor at 0.010 MPa pressure condenses on a horizontal 2 cm-O.D., 1 mm-wall-thickness AISI 316 stainless steel tube through which flows pressurized water. At a location where the water bulk temperature is 490 K and the inside heat transfer coefficient is 12,000 W/m^2 K, determine the condensation heat transfer coefficient, the overall heat transfer coefficient, and the rate of condensation per unit length of tube.

7-19. Saturated R-22 vapor at 1.42 MPa pressure condenses on a horizontal 1.5 cm-O.D., 1 mm-wall-thickness brass tube through which coolant water flows at 2 m/s. At a location where the bulk water temperature is 290 K, determine the condensation heat transfer coefficient and the rate of condensation per unit length of the tube.

7-20. To assist in the design of a steam condenser, parametric data for the condensation
☐ heat transfer coefficient is required. Saturated vapor at 320 K is to be condensed on a bank of 15 mm-O.D. horizontal tubes in a square array. Prepare a graph of the average heat transfer coefficient versus $(T_{sat} - T_w)$, with the number of tubes in a column, N, as a parameter. Let $(T_{sat} - T_w)$ vary from 1 to 5 K, and N vary from 1 to 5.

7-21. Consider condensation of saturated R-134a vapor on a horizontal tube. For
☐ $(T_{sat} - T_w) = 5$ K, determine the average heat transfer coefficient for tube diameters in the range 1 to 5 cm, and pressures from 0.2 to 1 MPa. Prepare a graph and discuss the results.

7-22. Show that the average heat transfer coefficient for laminar film condensation on a horizontal tube with a constant heat flux q around its periphery is

$$\bar{h} = \frac{q}{T_w - T_{sat}} = 0.615 \left[\frac{(\rho_l - \rho_v)g h_{fg} k_l^3}{\nu_l D q} \right]^{1/3}$$

(*Hint:* The integral $\int_0^\pi (\phi / \sin \phi)^{1/3} d\phi$ must be evaluated carefully due to the singularity at $\phi = \pi$.)

7-23. For laminar film condensation on a horizontal tube with an isothermal wall, derive the differential equation governing the film thickness as a function of angle ϕ measured from the top of the tube. Obtain the limiting form of this equation as $\phi \to 0$, and hence solve for the film thickness at $\phi = 0$.

7-24. Show that the heat transfer coefficient for laminar film condensation on a horizontal tube can be expressed in terms of the Nusselt and film Reynolds numbers defined in Section 7.2.3 and Exercise 7-5 as

$$\overline{\mathrm{Nu}} = 1.209\,\mathrm{Re}_\pi^{-1/3} \quad \text{uniform wall temperature}$$

$$\overline{\mathrm{Nu}} = 1.135\,\mathrm{Re}_\pi^{-1/3} \quad \text{uniform wall heat flux}$$

7-25. Saturated ammonia vapor at 775 kPa condenses on the outside of a horizontal 3 cm-OD, 2 mm-wall-thickness brass tube, through which flows a glycol-water coolant. At an axial location where the bulk coolant temperature is 270 K and the inside heat transfer coefficient is 5000 W/m^2 K, determine the heat transfer and condensation rates per unit length of tube.

7-26. Laminar film condensation occurs on a vertical, isothermal cone of height H and angle α, as shown in the sketch. Show that the mass flow $\dot{m}(=\Gamma\mathscr{P})$ is given by

$$\dot{m} = \frac{2\pi x \sin\alpha(\rho_l - \rho_v)g\cos\alpha\delta^3}{3\nu_l}$$

and that the energy balance for the film is

$$\frac{k_l}{\delta}(T_{\mathrm{sat}} - T_w) = \frac{h_{\mathrm{fg}}}{2\pi x \sin\alpha}\frac{d\dot{m}}{dx}$$

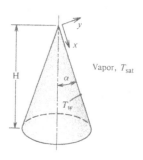

Hence show that the average heat transfer coefficient is given by

$$\overline{h} = 1.165\left[\frac{(\rho_l - \rho_v)g\cos^2\alpha h_{\mathrm{fg}}k_l^3}{\nu_l H(T_{\mathrm{sat}} - T_w)}\right]^{1/4}$$

7-27. Proceeding as in Exercise 7-26, show that the average heat transfer coefficient for laminar film condensation on an isothermal sphere is

$$\overline{h} = 0.828\left[\frac{(\rho_l - \rho_v)g h_{\mathrm{fg}}k_l^3}{\nu_l D(T_{\mathrm{sat}} - T_w)}\right]^{1/4}$$

(*Caution:* Incorrect values of the coefficient have appeared in the literature; the value of 0.828 given here is correct.)

7-28. A 15 mm-diameter copper sphere initially at 60°C is suddenly exposed to a flow of saturated steam at 1 atm pressure. Prepare a graph of the temperature-time response of the center of the sphere. Make reasonable simplifying assumptions.

7-29. Derive Eq. (7.51) for the effect of vapor drag on laminar film condensation on a vertical wall.

7-30. Saturated R-22 vapor at 0.65 MPa flows at 20 m/s down the outside of a 1 cm-O.D. vertical tube. Coolant inside the tube maintains the outside surface of the tube at 280 K. Estimate the effect of vapor drag on the local heat transfer coefficient at a location 5 cm from the top of the tube.

7-31. Steam at 10^4 Pa condenses on a 3 cm-O.D. horizontal tube maintained at 315 K. Determine the effect of vapor superheat on the average heat transfer coefficient and condensation rate, for superheats in the range 0–200 K. Graph your results.

7-32. Refrigerant-22 at 0.317 MPa condenses on a 2 cm-O.D. horizontal tube maintained at 250 K. Determine the effect of vapor superheat on the average heat transfer coefficient and condensation rate for superheats in the range 0–120 K.

7-33. Consider laminar film condensation on a vertical wall with heat removed uniformly along the surface, and vapor flow at a velocity U_e down parallel to the wall. If the shear stress exerted by the vapor on the liquid surface can be approximated as $\dot{m}''U_e$, where $\dot{m}''$ is the local condensation rate, show that the film thickness δ as a function of distance z from the top of the plate is given by a cubic equation. If saturated steam is condensing at 1 atm pressure and the vapor flow velocity is 60 m/s, calculate the wall temperature at distance 0.1 m down the plate if the wall heat flux is 40,000 W/m^2.

7-34. In experiments to investigate the effect of surface enhancement on film condensation, saturated steam at 6000 Pa was condensed on a 19.1 mm-diameter horizontal grooved brass tube, with 14.2 grooves/cm (36 tpi Standard American Screw Thread). The construction of the tube is shown in the drawing.

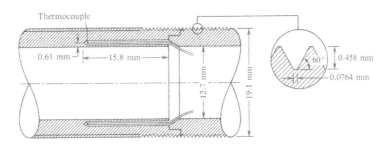

The average heat transfer coefficient was correlated as $\bar{h} = 89,700\Delta T^{-0.59}$ W/m^2 K, for a range of $\Delta T = T_{\text{sat}} - T_w$ from 1.1 K to 9 K. Prepare a graph comparing this result to the heat transfer coefficient for a smooth tube. What was the largest increase in $\bar{h}$ obtained? Details of the experiments may be found in reference [21].

7-35. Condenser tubes are often fluted to enhance the heat transfer. The flutes are contoured so that the radius of curvature $R(s)$ varies continuously with distance

s over a length $0 < s < S$. The effect of surface tension and the varying curvature induce a pressure gradient that causes the condensate to flow from the ridges into the troughs in such a way as to form a uniform thin film of thickness δ. Perform a Nusselt-type analysis to show that the equations governing the flow are

$$\frac{d\Gamma}{ds} = \frac{h\Delta T}{h_{fg}}; \quad h = \frac{k}{\delta}$$

$$\frac{dP}{ds} = -\frac{3\mu\Gamma(s)}{\rho\delta^3}; \quad P(s) = \frac{\sigma}{R(s)}$$

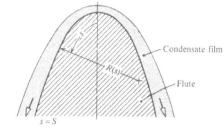

Condensate film

Flute

Tests are made using water to measure h versus ΔT. Use dimensional arguments to propose how these data should be correlated to allow application to ammonia condensation.

7-36. *Direct-contact* condensation involves condensation directly onto the cold liquid phase. James Watt's original steam condenser was of the direct-contact type, with steam condensing on a cold water spray. In an experiment, steam condenses on a vertical water jet. The nozzle diameter is 4 mm, and the jet collector is located 4 cm below the nozzle exit. In a particular test the water is supplied at 0.05 kg/s and 300 K, and the chamber pressure is maintained at 4200 Pa.

   (i) Assuming that the jet flow can be modeled as laminar plug flow, estimate the rate of condensation of the steam on the jet. (*Hint:* There is an analogous heat conduction problem in Chapter 3.)
   (ii) The jet is more likely to be turbulent. Kim and Mills [22] give the following correlation for the average Stanton number:

$$\overline{St} = 3.2 \text{Re}^{-0.20}\, \text{Pr}^{-0.7} (L/D)^{-0.57} (\sigma D/\rho v^2)^{-0.19}$$

   Using this correlation, make a new estimate of the condensation rate. (*Hint:* The turbulent jet can be modeled as a single-stream heat exchanger. See Section 8.4.1.)

7-37. An experimental vertical-tube evaporator consists of a single 4 cm-O.D. tube. The wall of the tube is maintained at 322 K, and the saturation temperature corresponding to the system pressure is 316 K. If water is fed at 0.028 kg/s to the top of the tube, over what length of tube is the film turbulent?

7-38. In a vertical-tube evaporator, liquid ammonia is distributed on the inside surfaces of 5 cm-I.D. tubes, at a rate of 0.1 kg/s per tube. Steam condenses on the outside of the tubes and maintains the tube wall temperature at 290 K. If the ammonia-side pressure is controlled to give a saturation temperature of 280 K, how long can the tubes be without having dryout of the film?

7-39. An array of 6 mm-square silicon chips on a ceramic baseplate is immersed in a pool of saturated R-134a at 1 MPa pressure to provide cooling. If operation at

60% of q_{max} provides a sufficient safety factor, what is the allowable power level per chip, and what will be the chip operating temperature? Take $C_{nb} = 0.004$ and $m = 4.1$ in Rohsenow's nucleate boiling correlation.

7-40. Prepare as complete as possible a "boiling curve" in the form of a graph of log q versus $\log(T_w - T_{sat})$ for saturated water boiling on a large, mechanically polished stainless steel surface at 10 atm pressure. Use a cavity size of 7.5 μm.

7-41. A 1000 W stainless steel sheathed electric heater is to be designed for a water kettle. How large should the heat transfer surface be if the heater is required to operate at 50% of the burnout heat flux? What will the surface temperature of the heater be at that heat flux? Model as a large horizontal surface.

7-42. Y. Y. Hsu [23] proposed a criterion for an active nucleation site that requires the temperature of the tip of a nucleating bubble to at least equal the saturation temperature corresponding to the pressure in the bubble (see Fig. 7.18b). Use this criterion to determine the size of a cavity that will nucleate first for pool boiling on a 10 cm-diameter horizontal surface. Consider the following situations:

   (i) water at 1 atm pressure.
   (ii) water at a saturation temperature of 300 K.
   (iii) R-134a at 0.69 MPa pressure.

Assume a contact angle of 90° and a linear temperature profile near the wall with a slope equal to the wall value obtained from an appropriate natural-convection heat transfer correlation. Also use the Clausius-Clapeyron relation to relate ΔT to ΔP (see Example 7.7(iii)). Discuss the relevance of this criterion to engineering systems.

7-43. A small laboratory water boiler is to be heated by a 3 kW stainless steel sheathed electric heater. Determine the surface area required if the surface heat flux should not exceed 50% of the burnout value. Model the heater as a horizontal cylinder of 2 cm radius. Also estimate the surface temperature.

7-44. Saturated ammonia is boiled at 300 kPa on a large horizontal surface.

   (i) Determine the peak heat flux for nucleate boiling.
   (ii) What wall temperature is required to have film boiling at the minimum heat flux?

7-45. Water boils at 5 atm pressure on a 0.8 mm-diameter platinum rod heater. Determine

   (i) the superheat required for boiling inception on 7.5 μm-radius nucleation sites.
   (ii) the peak heat flux.
   (iii) the rod temperature when the heat transfer is 50% of q_{max}.

7-46. A boiler is to be designed for a low-pressure condensation test rig and is required to supply 0.5×10^{-3} kg/s of saturated steam at 5000 Pa. How large must the surface area of the heater be if the heat flux is not to exceed 30% of the burnout heat flux? What will the surface temperature of the heater be at this heat flux if it is nickel-plated? Take $C_{nb} = 0.006$, and assume a large, flat surface.

7-47. A copper sphere is quenched in liquid nitrogen in an open Dewar flask. Calculate the average heat transfer coefficient and heat flux when $T_w - T_{sat} = 200$ K for spheres 1.5 and 2.5 cm in diameter.

7-48. Prepare a graph of q_{max} versus T_{sat} for water boiling on the outside of a 3 cm-diameter horizontal tube. Let $280 < T_{sat} < 550$ K. Explain the behavior of q_{max}.

7-49. Prepare a graph of q_{max} versus T_{sat} for R-22 boiling on the outside of a 3 cm-diameter horizontal tube. Let $0.2 < P < 1.8$ MPa. Explain the behavior of q_{max}.

7-50. In a test rig, film boiling is established on the outside of a 1 cm-diameter horizontal tube immersed in water. If the tube wall temperature is 1000 K and the system pressure is 0.5 MPa, determine the heat transfer per unit length of tube. Take an emittance of 0.5 for the tube surface.

7-51. Prepare a graph of q_{min} versus T_{sat} for water boiling on

  (i) a large horizontal surface.
  (ii) a 1 mm-diameter horizontal wire.

Let $280 < T_{sat} < 500$ K. Explain any significant trends in the results.

7-52. Prepare graphs of q_{max} and q_{min} versus T_{sat} for mercury boiling on a large horizontal surface. Let $400 < T_{sat} < 800$ K. Explain any significant trends in the results.

7-53. Hydrofluorocarbons (HFCs) are being used as replacements for chlorofluorocarbons (CFCs) as refrigerants (and as blowing agents, propellants, and cleaning agents). R-134a (BP = $-26.3°$C) is a replacement for R-12 (BP = $-30.2°$C). Compare the boiling peak heat flux values for these two fluids at 1 atm pressure. Property values for saturated R-12 at $-30.2°$C include $\rho_l = 1493$ kg/m^3, $\rho_v = 6.20$ kg/m^3, $h_{fg} = 0.165 \times 10^6$ J/kg, $\sigma = 16.1 \times 10^{-3}$ N/m.

7-54. A copper sphere is quenched in liquid nitrogen in an open Dewar flask. Calculate the average heat transfer coefficient and rate of heat loss when $\Delta T = T_w - T_{sat} = 180$ K for spheres 5 mm and 20 mm in diameter.

7-55. A 2 cm-diameter, 15 cm-long steel rod is heated to $460°$C in a furnace and then quenched in saturated water at 1 atm in a horizontal position. Estimate the initial cooling rate of the rod. Take $\varepsilon = 0.8$ for the steel.

7-56. A 20 cm-long, 1 mm-diameter platinum wire is stretched horizontally in a pool of saturated water at 1 atm. An electrical current is passed through the wire, and its surface temperature is maintained at 700 K. Determine the power input to the wire. Take $\varepsilon = 0.10$ for the platinum.

7-57. A 5 cm-long, 5 mm-diameter wire is placed in a horizontal position in an open Dewar flask containing liquid nitrogen. Calculate the average heat transfer coefficient and rate of heat loss when $\Delta T = T_w - T_{sat} = 180$ K.

7-58. Water is boiled on a large horizontal polished copper surface at 1 atm pressure. Referring to the boiling curve of Fig. 7.15, there are three different surface temperatures that will give a heat flux of 5×10^4 W/m^2: determine two of these temperatures.

7-59. An ammonia heatpipe is constructed from a 2 cm-I.D. stainless steel tube and has an effective length of 1.0 m. The nickel foam (210-5) wick has a cross-sectional area of 6.0×10^{-5} m^2. Estimate the burnout heat load for an adiabatic section maintained at 300 K and an adverse tilt of

(i) 0°.
(ii) 0.5°.

Also calculate the heat transfer factor for the heatpipe.

7-60. A water heatpipe is constructed from a 3 cm-I.D. stainless steel tube and has an effective length of 0.70 m. The wick is made from 30 mesh screen and has a cross-sectional area of 10.0×10^{-5} m^2. Estimate the burnout heat load for an adiabatic section maintained at 500 K and an adverse tilt of

(i) 0°.
(ii) 2°.

7-61. Prepare a plot of the figure of merit $\mathcal{M}$ for heatpipe application over an appropriate temperature range for

(i) mercury.
(ii) R-22.
(iii) R-134a.

7-62. A nitrogen heatpipe is constructed from a 2 cm-I.D. stainless steel tube and has an effective length of 0.50 m. The wick is made from nickel felt and has a cross-sectional area of 7×10^{-5} m^2. For an adiabatic section temperature in the vicinity of 85 K, prepare a plot of burnout heat load as a function of temperature. The heatpipe operates in a horizontal position.

7-63. A nitrogen-loaded methanol heatpipe is constructed from a 2 cm-I.D. stainless steel tube. A fibrous stainless wick is in the form of a 20 mm × 5 mm cross-section slab and extends into the gas reservoir. The condenser length is 40 cm, and the reservoir volume is 7×10^4 mm^3. The heatpipe is designed to be fully open for an adiabatic section temperature of 280 K, when the sink temperature is 260 K.

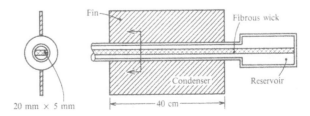

Determine the mass of gas in the heatpipe, and plot a graph of load versus adiabatic section temperature over an appropriate range of operation. At 5°C methanol has a vapor pressure of 5330 Pa and an enthalpy of vaporization of 1.18×10^6 J/kg.

7-64. A water heatpipe is 10 cm long and is made from a 2 cm-I.D. stainless steel tube. The wick consists of four layers of 30 mesh screen covered with one layer of 200 mesh screen, as shown in the sketch. Determine the maximum heat transport capability when the heat pipe is inclined at 10°, with the evaporator above the condenser, and the adiabatic section is at 100°C. The wire diameters of 30 mesh and 200 mesh screen are 0.86 mm and 0.126 mm, respectively.

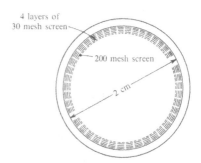

7-65. The condenser section of an ammonia heatpipe has three external radial fins per cm length. The fins are 5 cm in diameter, 0.3 mm thick, and made of aluminum. The pipe itself is aluminum with a 20 mm outside diameter, a 16 mm inside diameter, and a 0.5 mm-thick wick on the inner wall. The ammonia-filled wick has an effective thermal conductivity of 0.8 W/m K. If the outside convective heat transfer coefficient is 25 W/m² K, estimate the overall heat transfer coefficient from the condensing vapor to ambient cooling air. Make appropriate simplifying assumptions.

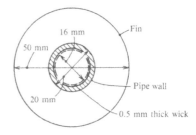

7-66. Thermosyphons (Perkins tubes) are to be used to transfer heat from a hot water stream flowing below a cold water stream. The tubes are 3 cm-O.D., 2 mm-wall-thickness stainless steel and are 2 m long. The lower and upper 50 cm lengths are immersed in the hot and cold streams, respectively, and the intervening 1 m length is insulated. At the location of concern the hot water stream is at 360 K, and the cold water stream is at 350 K. Both streams flow at 1 m/s. The working fluid in the tubes is water. At steady state, vapor condenses as a film in the condenser section of the tubes and evaporates as a film in the evaporator sections. The ends of the tubes

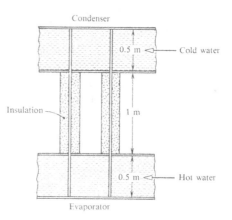

can be taken to be adiabatic. Determine the rate of heat transfer per tube. Startup of this device could be a problem: discuss.

7-67. A 19 mm-I.D., 25 mm-O.D. stainless steel Perkins tube is used to transfer heat from a hot stream of water at 360 K to colder water at 325 K. The hot stream flows at 0.3 m/s over the bottom 30 cm of the vertical tube, and the colder water exists under conditions of natural convection over the top 1 m of the 1.3 m-long tube.

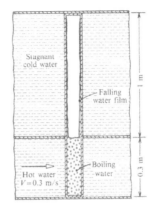

Within the tube is pure water and its vapor. The vapor condenses on the inside of the 1.0 m-long condenser section and falls as a film to maintain the level of water in the 0.3 m-long boiling section. What heat flow will the tube provide?

HEAT EXCHANGERS

CONTENTS

8.1 INTRODUCTION

A heat exchanger is a device that facilitates transfer of heat from one fluid stream to another. Power production, refrigeration, heating and air conditioning, food processing, chemical processing, oil refining, and the operation of almost all vehicles depend on heat exchangers of various types. The analysis and design of heat exchangers is the subject of Chapter 8. Design considerations were involved peripherally in many of the examples and exercises in Chapters 1 through 7. In Chapter 8, however, we will develop principles of heat exchanger design in a systematic manner, and formally introduce the student to the design process.

A great variety of heat exchangers are used. In Section 8.2, heat exchangers are classified with respect to flow configuration and heat transfer surface. In Section 8.3, the energy conservation principle is applied to an exchanger as a system to obtain *exchanger energy balances*. Also, the overall heat transfer coefficient concept is reviewed and extended. Section 8.4 presents an analysis of an evaporator as an example of a single-stream steady-flow exchanger. The key section in this chapter is Section 8.5, in which two-stream steady-flow exchangers are analyzed, using both the *logarithmic mean temperature difference* and the *effectiveness-number of transfer units* formulations. The analyses involve application of the steady-flow energy equation, and the results are presented in the form of formulas and charts that find extensive use in engineering practice.

Sections 8.3 through 8.5 deal with the **thermal analysis** of heat exchangers, that is, methods for calculating the heat transfer in the exchanger and the outlet fluid temperatures. Section 8.6 introduces the student to the broader subject of *heat exchanger design*. Two aspects of exchanger design are emphasized. First, the role played by pressure drop of the flow streams is examined, which leads to the concept of a **thermal-hydraulic design**: a heat exchanger is sized to give a required heat transfer performance, subject to a pressure drop constraint on one or both streams. Particular attention is given to **compact heat exchangers,** which have a large heat transfer surface area per unit volume. Second, some economic aspects of exchanger utilization are examined. In particular, operating costs are weighed against capital costs for waste heat recovery application.

Two computer programs accompany Chapter 8. HEX1 calculates thermal performance for nine flow configurations of steady two-stream exchangers. HEX2 introduces the student to computer-aided heat exchanger design. HEX2 performs the thermal-hydraulic design of a particular class of exchangers and also gives an economic evaluation of the resulting design.

8.2 TYPES OF HEAT EXCHANGERS

A great variety of heat exchangers are used in engineering practice. The geometry of the flow configuration, the type of heat transfer surface, and the materials of construction all vary according to the design requirements.

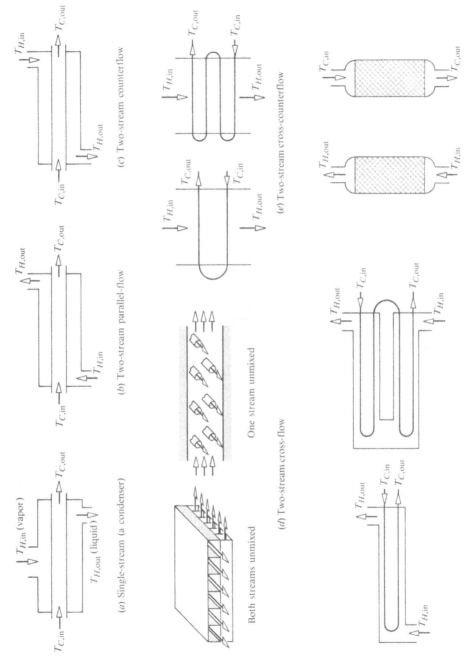

Figure 8.1 Schematics of common heat exchanger geometric flow configurations.

8.2.1 Geometric Flow Configurations

There are a number of possible geometric flow configurations for exchangers, the most important of which are the following.

Single-stream. In this configuration the temperature of only one stream changes in the exchanger; the direction in which the fluid flows is then immaterial. Examples include condensers and evaporators for power plants and refrigeration systems. A simple condenser is illustrated in Fig. 8.1a.

Parallel-flow two-stream. The two fluids flow parallel to each other in the same direction. In its simplest form, this type of exchanger consists of two coaxial tubes, as shown in Fig. 8.1b. In practice, a large number of tubes are located in a shell to form what is known as a *shell-and-tube* exchanger, as shown in Fig. 8.2. The shell-and-tube type is used most often for liquids and for high pressures. The *plate* type, shown in Fig. 8.3, consists of multiple plates separated by gaskets and is more suitable for gases at low pressures. This configuration is often alternatively termed a *cocurrent* exchanger.

Counterflow two-stream. The fluids flow parallel to each other in opposite directions. A simple coaxial tube type is shown in Fig. 8.1c, but, as with the cocurrent case, shell-and-tube or plate exchangers are most commonly used. We shall see that, for a given number of transfer units, the effectiveness of a counterflow exchanger is higher than that of a parallel-flow exchanger. Hence, counterflow exchangers are preferred in practice. Examples include feed water preheaters for boilers, and oil coolers for aircraft. This configuration is often alternatively termed a *counter-current* exchanger.

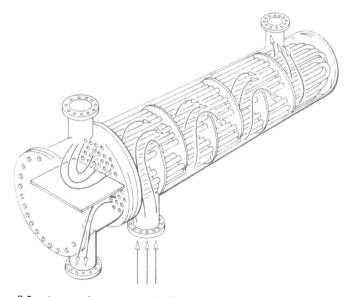

Figure 8.2   A two-tube-pass, one-shell-pass, shell-and-tube heat exchanger. The first tube pass gives parallel flow, and the second gives counterflow.

Figure 8.3 A plate-type heat exchanger during assembly. (Photograph courtesy of the Paul Mueller Company, Springfield, Missouri.)

Cross-flow two-stream. The two streams flow at right angles to each other, as shown in Fig. 8.1*d*. The hot stream may flow inside tubes arranged in a *bank* or *bundle*, and the cold stream may flow through the bank in a direction generally at right angles to the tubes. Either one or both of the streams may be *unmixed* as shown. This configuration is intermediate in effectiveness between parallel-flow and counterflow exchangers, but it is often simpler to construct owing to the relative simplicity of the inlet and outlet flow ducts. A common example is the automobile radiator shown in Fig. 8.4.

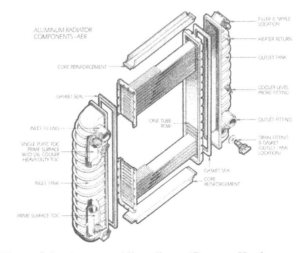
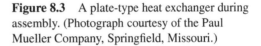

Figure 8.4 An automobile radiator. (Courtesy Harrison Radiator Division, General Motors, Lockport, New York.)

Cross-counterflow two-stream. In practice, exchanger flow configurations often approximate the idealizations shown in Fig. 8.1*e*; *two-* and *four-pass* types are shown, although more passes can be used. (In a two-pass exchanger, the tubes pass through the shell twice.) As the number of passes increases, the effectiveness approaches that of an ideal countercurrent exchanger.

Multipass two-stream. When the tubes of a shell-and-tube exchanger double back one or more times inside the shell, as shown in Fig. 8.1*f*, some passes are parallel flow while others are counterflow. The two-pass exchanger of this type is popular since only one end of the exchanger requires perforation to lead the tubes in and out, as shown schematically in Fig. 8.2.

Regenerators. The configurations discussed so far all involve steady flows and temperatures: such exchangers are traditionally called *recuperators*. On the other hand, in a *regenerator*, the two streams flow alternately through a *matrix* of substantial heat storage capacity. Heat transferred from the hot fluid is stored in the matrix, making its temperature rise, and is subsequently given up to the cold fluid when it, in turn, flows through the matrix, causing its temperature to fall. Regenerators can also have parallel-, counter-, or cross-flow configurations. The matrix may be in the form of a bed, as shown schematically in Fig. 8.1*g*, or in the form

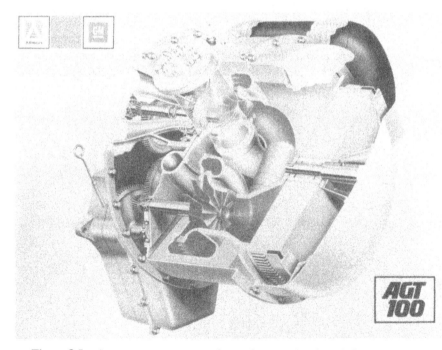

Figure 8.5   A rotary regenerator used to preheat combustion air for a gas turbine: the AGT 100 automotive engine. (Photograph courtesy of Allison Gas Turbine, Division of General Motors. Development sponsored by the U.S. Department of Energy and NASA-Lewis Research Center.)

of a rotating wheel, as shown in Fig. 8.5. The bed type has been used extensively in high-temperature applications, such as for the preheating of combustion air by combustion products using a matrix of refractory bricks.

The configurations just described are often only approximated in practice but are useful idealizations, or *models*, that can be analyzed to obtain an understanding of the essential features of heat exchanger thermal behavior. However, often a configuration is encountered for which a simple idealization is not feasible; in such cases, purely empirical data must be used for its thermal design. Figure 8.6 shows a commonly encountered example consisting of a steam coil used to heat a tank of oil. If there is no stirrer in the tank, natural convection will tend to give a cross-counterflow of oil up the steam coil, but the magnitude of this flow is difficult to estimate.

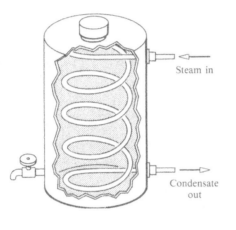

Steam in

Condensate out

Figure 8.6   An immersed coil heater.

8.2.2 Fluid Temperature Behavior

Associated with the geometric flow patterns are characteristic temperature variations within the exchanger, as is shown for the simpler cases in Fig. 8.7. The subscript H is used to denote the hot stream, and C is used to denote the cold stream. In Fig. 8.7a, which depicts a simple condenser, T_H is constant while T_C increases along the exchanger. In the parallel-flow two-stream exchanger shown in Fig. 8.7b, the temperature difference for heat transfer $(T_H - T_C)$ decreases along the exchanger in the flow direction. Also, it is clear that the outlet temperature of the cold stream cannot exceed that of the hot stream, that is, $T_{C,\text{out}} < T_{H,\text{out}}$. In the counterflow two-stream exchanger, shown in Fig. 8.7c, the temperature difference $(T_H - T_C)$ may decrease or increase along the exchanger, or, as a special case, it may be constant. Notice that the outlet temperature of the cold stream can exceed the outlet temperature of the hot stream. The temperature patterns in cross-flow exchangers are more complex since the temperature varies in two directions, as shown for the cold stream in Fig. 8.7d. In a regenerator, the outlet temperatures vary with time, as shown in Fig. 8.7e. Often there is a combination of these temperature variation patterns in a single exchanger, as is the case for the parallel-flow steam generator shown in Fig. 8.7f.

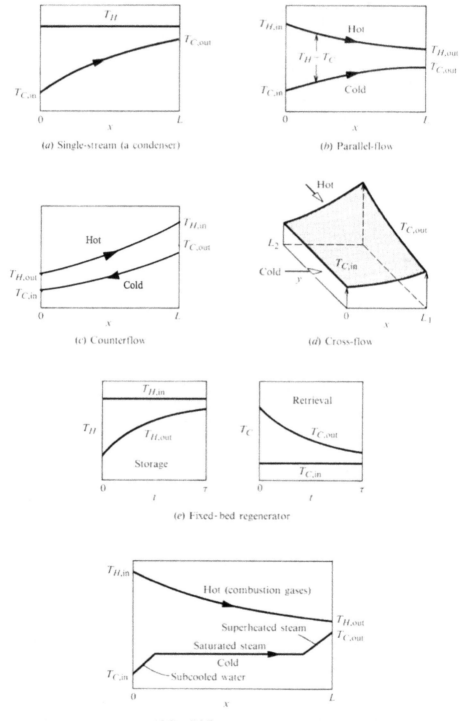

Figure 8.7 Characteristic fluid temperature variations for various exchanger configurations.

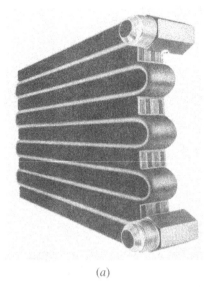

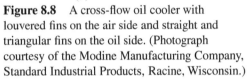

(b)

(a)

Figure 8.8 A cross-flow oil cooler with louvered fins on the air side and straight and triangular fins on the oil side. (Photograph courtesy of the Modine Manufacturing Company, Standard Industrial Products, Racine, Wisconsin.)

8.2.3 Heat Transfer Surfaces

The examples of heat exchangers shown in Figs. 8.2 through 8.6 show that the heat transfer surface can take many forms. The most common form is a plain tube, which is usually circular and straight but can be bent or coiled. When the heat transfer resistance on one side of the tube is much larger than that on the other, as is the case for a gas-to-liquid heat exchanger, the gas-side surface may be finned to increase the effective heat transfer area. Usually the fins are on the outside of the tube, but sometimes it is the internal surface that is provided with fins; an example is shown in Fig. 8.8. The manner in which fins are provided in plate- and compact-type heat exchangers can be quite complicated. Design engineers have shown considerable ingenuity in providing fins that increase the heat transfer area without incurring excessive pressure drop penalties, and that are inexpensive to fabricate. The automobile radiator in Fig. 8.4 is a good example.

8.2.4 Direct-Contact Exchangers

To complete this discussion of types of heat exchangers, it is appropriate to briefly mention an additional type of exchanger, in which there is *direct* contact between the two fluids. Direct-contact exchangers are more commonly used for mass transfer and simultaneous heat and mass transfer. A simple form of direct-contact heat exchanger is a water heater in which steam is bubbled through the water: the steam condenses and adds to the water inventory. Two-stream direct-contact heat exchangers may involve transfer between two immiscible liquids such as oil and water, as shown in Fig. 8.9. The expense of a heat transfer surface and surface-fouling problems are avoided, but the separation of the two liquids at each end of the exchanger is often difficult.

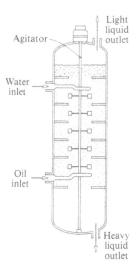

Figure 8.9   An oil-to-water direct-contact heat exchanger.

8.3   ENERGY BALANCES AND OVERALL HEAT TRANSFER COEFFICIENTS

Before commencing the analysis of heat exchanger performance, we need to develop some supporting concepts. First, we will apply the steady-flow energy equation to the complete exchanger, in order to relate outlet stream temperatures to inlet stream temperatures. Next, the overall heat transfer coefficient concept introduced in Chapters 1 and 2 will be extended to include the effects of deposits fouling a heat exchanger tube, and to include the effect of augmenting the heat transfer surface with fins.

8.3.1 Exchanger Energy Balances

Figure 8.10 shows a coaxial-tube parallel-flow heat exchanger. The steady-flow energy equation is applied to a control volume enclosing the complete exchanger, as shown in the figure. It reduces to an enthalpy balance since no external work is done, there is no heat transfer into the system if the exchanger is well insulated, and changes in kinetic and potential energy are usually negligible. Denoting the hot and cold streams with subscripts H and C, respectively,

$$(\dot{m}_H h_H + \dot{m}_C h_C)_{x=0} = (\dot{m}_H h_H + \dot{m}_C h_C)_{x=L}$$

or

$$\dot{m}_H(h_{H,0} - h_{H,L}) = \dot{m}_C(h_{C,L} - h_{C,0}) = \dot{Q} \qquad (8.1)$$

where $\dot{Q}$ is the heat transferred from the hot stream to the cold stream. If the specific heats can be assumed constant, then

$$(\dot{m}c_p)_H(T_{H,0} - T_{H,L}) = (\dot{m}c_p)_C(T_{C,L} - T_{C,0}) = \dot{Q} \qquad (8.2)$$

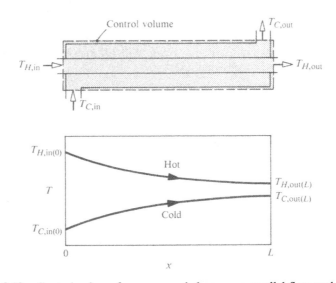

Figure 8.10   Control volume for an energy balance on a parallel-flow exchanger and the expected fluid temperature variations along the exchanger.

If, instead, the inlet and outlet states of each stream are denoted "in" and "out," respectively, then Eqs. (8.1) and (8.2) become

$$\dot{m}_H(h_{H,in} - h_{H,out}) = \dot{m}_C(h_{C,out} - h_{C,in}) = \dot{Q} \tag{8.3}$$

$$(\dot{m}c_p)_H(T_{H,in} - T_{H,out}) = (\dot{m}c_p)_C(T_{C,out} - T_{C,in}) = \dot{Q} \tag{8.4}$$

For a counterflow exchanger, it is convenient to adopt the convention that the flow rates $\dot{m}_H$ and $\dot{m}_C$ are positive, irrespective of direction. Then the **exchanger energy balance** in the form of Eqs. (8.3) and (8.4) applies to the counterflow configuration as well. Note also that, for multitube heat exchangers, it is usual to treat the exchanger as a whole rather than treating each tube separately; hence, $\dot{m}$ [kg/s] is the total flow rate through the exchanger of a given stream.

For a simple condenser that condenses saturated vapor, Eq. (8.3) becomes

$$\dot{m}_H h_{fgH} = (\dot{m}c_p)_C(T_{C,out} - T_{C,in}) \tag{8.5}$$

because $(h_{H,in} - h_{H,out}) = h_{fgH}$. For a simple evaporator generating saturated vapor,

$$(\dot{m}c_p)_H(T_{H,in} - T_{H,out}) = \dot{m}_C h_{fgC} \tag{8.6}$$

These exchanger energy balances, Eqs. (8.1)–(8.6), are also often called *overall* energy balances or, in the chemical engineering literature, *macroscopic* energy balances. Their use is straightforward: for example, in using Eq. (8.4), the flow rates, inlet temperatures, and one outlet temperature may be specified, and the unknown outlet temperature can then be found. Often they are used to find the maximum possible heat transfer in a given situation and to determine whether a specified goal is attainable. An example of this use follows.

EXAMPLE 8.1   Cooling Water Supply for a Steam Turbine Condenser

To obtain the required power output, a steam turbine must consume 11.5 kg/s of steam and exhaust to a back pressure of 4714 Pa. Will 300 kg/s of cooling water from a river at 290 K be sufficient to condense the steam?

Solution

Given: Cooling water supply.

Required: Evaluate adequacy for a steam turbine condenser.

From Table A.12*a*, at 4714 Pa, the saturation temperature T_H is 305.0 K and the enthalpy of vaporization h_{fgH} is 2.425×10^6 J/kg. The maximum possible outlet temperature of the cooling water is then 305.0 K and can be achieved in an infinitely long exchanger. Thus, the maximum energy that can be absorbed by the cooling water is

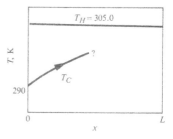

$$\dot{Q} = (\dot{m}c_p)_C[T_{C,\text{out}}(\text{max}) - T_{C,\text{in}}]$$
$$= (300)(4186)(305 - 290)$$
$$= 1.88 \times 10^7 \text{ W}$$

On the other hand, the enthalpy of vaporization to be given up by the condensing steam is

$$\dot{Q} = \dot{m}_H h_{fgH} = (11.5)(2.425 \times 10^6) = 2.79 \times 10^7 \text{ W}$$

which is greater than the amount that can be absorbed by the cooling water.

Comments

A larger flow of cooling water is required.

8.3.2 Overall Heat Transfer Coefficients

The overall heat transfer coefficient was introduced in Sections 1.4.1 and 2.3.1 for heat transfer across plane walls and cylindrical tube walls, respectively. Most of the heat exchangers discussed in Section 8.2 involve transfer of heat from one fluid to another across a plate or tube wall, with tube walls predominating. Figure 8.11 shows the cross section of a typical heat exchanger tube and the corresponding temperature profile and thermal circuit. For a clean tube, the product of overall heat transfer coefficient times perimeter is obtained from Eq. (2.17) by setting $A = \mathscr{P}L$ and canceling L throughout:

$$\frac{1}{U\mathscr{P}} = \frac{1}{h_{c,i}2\pi r_i} + \frac{\ln(r_o/r_i)}{2\pi k} + \frac{1}{h_{c,o}2\pi r_o} \tag{8.7}$$

where subscripts i and o denote the tube inside and outside walls, respectively. For an element of exchanger Δx long, the heat transfer rate is

$$\Delta\dot{Q} = U\mathscr{P}\Delta x(T_H - T_C) \tag{8.8}$$

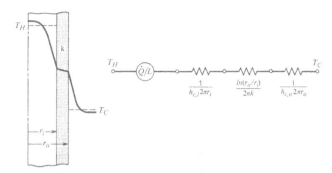

Figure 8.11   A local temperature profile and thermal circuit for heat flow through an exchanger tube. It is understood that the convective heat transfer coefficients may be average values: the bar notation is omitted consistent with our practice in Chapters 1–3.

The perimeter of the $U\mathscr{P}$ product in Eq. (8.7) can be chosen to be either $2\pi r_i$ or $2\pi r_o$. The corresponding value of U changes accordingly, since Eq. (8.7) specifies the $U\mathscr{P}$ product and not U itself. Often $h_{c,i}$ and $h_{c,o}$ are not accurately known, or the tube wall is relatively thin, so that a precise specification of $\mathscr{P}$ is unnecessary.

After a heat exchanger has been in service for some time, deposits may form on the heat transfer surfaces. Examples include soot deposits in furnace tubes, calcium or magnesium sulfate deposits in fresh water heaters, and rust. The heat transfer surfaces are then said to be *fouled:* Fig. 8.12 shows an extreme example. If the average thickness δ_f and thermal conductivity k_f of the deposit were known, one could estimate the additional thermal resistance per unit area as $R_f = \delta_f/k_f$. Generally,

Figure 8.12   An economizer tube fouled in a high-sulfur no. 6 fuel-oil exhaust. The deposits have a relatively high sulfur content, and the corroded fins are typical of sulfuric acid deposition. [Photograph courtesy of Dr. W. Marner, Jet Propulsion Laboratory, California Institute of Technology, Pasadena. See also W. J. Marner, "Gas-side fouling," Mechanical Engineering, 108, 3, 70–77 (March 1986).]

however, the extent of fouling is known only through a measured reduction in heat transfer performance of an exchanger, which is then attributed to a reduced value of the overall heat transfer coefficient. If U is the overall heat transfer coefficient for an unfouled heat exchanger, we can write

$$\frac{1}{U_f \mathscr{P}} = \frac{1}{U \mathscr{P}} + \frac{R_{fH}}{\mathscr{P}_H} + \frac{R_{fC}}{\mathscr{P}_C} \tag{8.9}$$

where U_f is the overall heat transfer coefficient for the fouled exchanger, and R_{fH} and R_{fC} are the hot stream and cold stream **fouling resistance,** respectively. Table 8.1 gives some representative values for fouling resistance per unit area. Clearly, the time-dependent nature of the fouling problem is such that it is very difficult to reliably estimate U values if fouling resistances are dominant.

Often one side of a tube wall is finned—for example, the air side of a water-air heat exchanger. The effect of the fins is conveniently included in the overall heat transfer coefficient through use of the fin efficiency. Thus, the overall heat transfer

Table 8.1   Recommended values of fouling resistances for heat exchanger design.

Fluid	Fouling Resistance, R_f $[\text{W/m}^2 \text{ K}]^{-1} \times 10^3$
Fuel oil	0.9
Transformer oil	0.2
Vegetable oils	0.5
Light gas oil	0.35
Heavy gas oil	0.5
Asphalt	0.9
Gasoline	0.2
Kerosene	0.2
Caustic solutions	0.35
Refrigerant liquids	0.2
Hydraulic fluid	0.2
Molten salts	0.1
Engine exhaust gas	1.8
Steam (non-oil-bearing)	0.1
Steam (oil-bearing)	0.2
Refrigerant vapors (oil-bearing)	0.35
Compressed air	0.35
Acid gas	0.2
Solvent vapors	0.2
Seawater	0.1–0.2
Brackish water	0.2–0.5
Cooling tower water (treated)	0.2–0.35
Cooling tower water (untreated)	0.5–0.9
River water	0.2–0.7
Distilled or closed-cycle condensate water	0.1
Treated boiler feedwater	0.1–0.2

Table 8.2 Approximate overall heat transfer coefficients.

Heat Exchanger Duty	U, W/m^2 K
Gas to gas	10–30
Water to gas (e.g., gas cooler, gas boiler)	10–50
Condensing vapor-air (e.g., steam radiator, air heater)	5–50
Steam to heavy fuel oil	50–180
Water to water	800–2500
Water to other liquids	200–1000
Water to lubricating oil	100–350
Light organics to light organics	200–450
Heavy organics to heavy organics	50–200
Air-cooled condensers	50–200
Water-cooled steam condensers	1000–4000
Water-cooled ammonia condensers	800–1400
Water-cooled organic vapor condensers	300–1000
Steam boilers	10–40 + radiation
Refrigerator evaporators	300–1000
Steam-water evaporators	1500–6000
Steam-jacketed agitated vessels	150–1000
Heating coil in vessel, water to water	
Unstirred	50–250
Stirred	500–2000

coefficient for a clean tube is given by

$$\frac{1}{U\mathscr{P}} = \frac{1}{h_{c,i}2\pi r_i} + \frac{\ln(r_o/r_i)}{2\pi k} + \frac{1}{h_{c,o}(A_f/L)\eta_f + h_{c,o}A_p/L} \tag{8.10}$$

where (A_f/L) is the surface area of fins per unit length of tube, η_f is the fin efficiency, and A_p/L is the *prime* (unfinned) surface area per unit length of tube. Alternatively, the total surface efficiency η_t, defined by Eq. (2.46), can be used. Some typical overall heat transfer coefficient values are listed in Table 8.2.

EXAMPLE 8.2   Overall Heat Transfer Coefficient for a Condenser

A brass condenser tube has a 30 mm outer diameter and 2 mm wall thickness. Sea water enters the tube at 290 K, and saturated low-pressure steam condenses on the outside of the tube. The inside and outside heat transfer coefficients are estimated to be 4000 and 8000 W/m^2 K, respectively, and a fouling resistance of 10^{-4} (W/m^2 K)$^{-1}$ on the water side is expected. Estimate the overall heat transfer coefficient based on inside area.

Solution

Given: Brass condenser tube with water-side fouling.

Required: Overall heat transfer coefficient U based on inside area.

Assumptions: The tube wall temperature is $\sim$300 K.

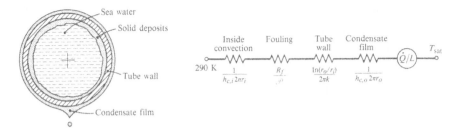

From Table A.1a, take k for brass as 111 W/m K. Equation (8.7) gives the $U\mathscr{P}$ product for a clean tube:

$$\frac{1}{U\mathscr{P}} = \frac{1}{h_{c,i}2\pi r_i} + \frac{\ln(r_o/r_i)}{2\pi k} + \frac{1}{h_{c,o}2\pi r_o}$$

$$= \frac{1}{(4000)(2\pi)(0.013)} + \frac{\ln(0.015/0.013)}{2\pi(111)} + \frac{1}{(8000)(2\pi)(0.015)}$$

$$= 10^{-3}(3.06 + 0.21 + 1.33) = 4.60 \times 10^{-3} \, (\text{W/m K})^{-1}$$

The inside perimeter is

$$\mathscr{P}_i = 2\pi r_i = (2\pi)(0.0130) = 0.0817 \text{ m}$$

Hence,

$$1/U = (0.0817)(4.60 \times 10^{-3}) = 3.76 \times 10^{-4} \, (\text{W/m}^2 \text{ K})^{-1}; \quad U = 2660 \text{ W/m}^2 \text{ K}$$

Then, from Eq. (8.9) for the fouled tube,

$$\frac{1}{U_f} = \frac{1}{U} + R_{fC} = 10^{-4}(3.76 + 1); \quad U_f = 2100 \text{ W/m}^2 \text{ K}$$

Comments

1. The fouling reduces the overall heat transfer coefficient by 21%.

2. Due to fouling, the use of $2\pi r_i$, in the inside convective resistance may be inappropriate.

EXAMPLE 8.3 Overall Heat Transfer Coefficient of a Finned Tube

An aluminum tube has a 3 cm outside diameter and a wall thickness of 2 mm. It has 100 pin fins per cm length, of 1.5 mm diameter and 4 cm long. The inside and outside heat transfer coefficients are 5000 and 7 W/m² K, respectively. Calculate the overall heat transfer coefficient-perimeter product. Take $k = 204$ W/m K for the aluminum.

Solution

Given: Aluminum tube with pin fins.

Required: The $U\mathscr{P}$ product.

Assumptions: The outside heat transfer coefficient is constant over the fins and bare surface.

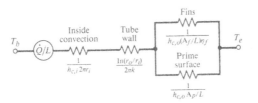

Equation (8.10) gives the $U\mathscr{P}$ product:

$$\frac{1}{U\mathscr{P}} = \frac{1}{h_{c,i}2\pi r_i} + \frac{\ln(r_o/r_i)}{2\pi k} + \frac{1}{h_{c,o}(A_f/L)\eta_f + h_{c,o}A_p/L}$$

If the pin fins have diameter d and length l,

$$(A_f/L) = (10{,}000)(\pi dl) = (10{,}000)(\pi)(0.0015)(0.04) = 1.885 \text{ m}$$

$$A_p/L = (2\pi r_o)(1) - (10{,}000)(\pi d^2/4) = (2)(\pi)(0.015) - (10{,}000)(\pi/4)(0.0015)^2$$

$$= 0.07658 \text{ m}$$

To use Eq. (2.41) we first calculate β:

$$\beta = \left(\frac{h_c\mathscr{P}}{kA_c}\right)^{1/2}; \quad \text{for one fin,} \quad \frac{\mathscr{P}}{A_c} = \frac{\pi d}{\pi d^2/4} = \frac{4}{d} = \frac{4}{(0.0015)} = 2667 \text{ m}^{-1}$$

$$\beta = \left(\frac{(7)(2667)}{204}\right)^{1/2} = 9.57 \text{ m}^{-1}; \quad \beta L = (9.57)(0.04) = 0.383$$

$$\eta_f = (1/\beta L)\tanh \beta L = (1/0.383)(0.366) = 0.956$$

$$\frac{1}{U\mathscr{P}} = \frac{1}{(5000)(2\pi)(0.013)} + \frac{\ln(0.015/0.013)}{(2\pi)(204)} + \frac{1}{(7)(1.885)(0.956) + (7)(0.07658)}$$

$$= 10^{-3}(2.45 + 0.11 + 76.04)$$

$$U\mathscr{P} = 12.7 \text{ W/m K}$$

Comments

1. Note that $\mathscr{P}$ is used for both the tube perimeter and the fin perimeter.

2. Despite the presence of the pin fins, inside and wall resistances are negligible.

8.4   SINGLE-STREAM STEADY-FLOW HEAT EXCHANGERS

Single-stream exchangers are defined as exchangers along which the temperature of only one stream varies. Examples include steam condensers of power plants, vapor condensers of vapor-compression refrigeration systems, power plant boilers, and evaporators for flash-desalination plants. We will perform an analysis for an evaporator, and obtain the result in a form that also applies to condensers.

8.4.1 Analysis of an Evaporator

Figure 8.13a shows a simple single-tube evaporator, or *boiler*. Saturated liquid enters the shell from the bottom, and vapor leaves from the top. Hot combustion gases flow through the tubes. The enthalpy of vaporization is transferred by convection from the hot gases to the inner tube wall, by conduction across the tube wall, and by a complex boiling process into the liquid. Figure 8.13a also shows the temperature variation of the combustion gases along the tube, and Fig. 8.13b shows the temperature variation across the tube wall. In Chapter 7, we defined boiling heat transfer coefficients with respect to the saturation temperature of the liquid, T_{sat}; T_{sat} is obtained from steam tables as the saturation temperature corresponding to the pressure maintained in the evaporator shell. The vapor flow rate is denoted $\dot{m}_C$ and the combustion gas flow rate $\dot{m}_H$ (the cold and hot streams, respectively). The exchanger energy balance was derived in Section 8.3.1 as Eq. (8.6):

$$(\dot{m}c_p)_H(T_{H,\text{in}} - T_{H,\text{out}}) = \dot{m}_C\, h_{\text{fg}\,C}$$

When the gas flow rate $\dot{m}_H$ and inlet temperature $T_{H,\text{in}}$ are known, Eq. (8.6) relates the outlet gas temperature, $T_{H,\text{out}}$, to the amount of vapor produced, $\dot{m}_C$.

To obtain the variation of the gas temperature along the exchanger, we make an energy balance on a differential element of the exchanger Δx long. When the steady-flow energy equation, Eq. (1.4), is applied to the control volume Δx long (shown

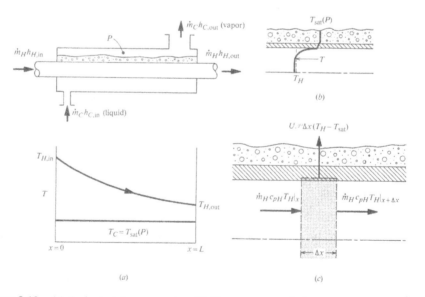

Figure 8.13   (a) A single-tube evaporator, (b) Temperature variation across an evaporator tube wall, (c) Elemental control volume for application of the steady-flow energy equation.

in Fig. 8.13c as a dotted line), the contribution due to x-direction conduction in the gas is small and can be neglected. The heat transfer across the tube wall must equal the gas flow rate times its enthalpy decrease: for the gas specific heat assumed to be constant,

$$U \mathscr{P} \Delta x (T_H - T_{\text{sat}}) = \dot{m}_H c_{pH} (T_H|_x - T_H|_{x+\Delta x}) \tag{8.11}$$

where U is the overall heat transfer coefficient for heat transfer from the gas to the steam, and $\mathscr{P} = \pi D$ is the perimeter of the tube wall (U and $\mathscr{P}$ must be based on the same diameter, either the I.D. or the O.D.). Dividing Eq. (8.11) by Δx, letting $\Delta x \to 0$, and rearranging,

$$\frac{dT_H}{dx} + \frac{U \mathscr{P}}{\dot{m}_H c_{pH}} (T_H - T_{\text{sat}}) = 0 \tag{8.12}$$

Equation (8.12) requires one boundary condition, which is

$$x = 0: \quad T_H = T_{H,\text{in}} \tag{8.13}$$

Integrating Eq. (8.12) and using Eq. (8.13) to evaluate the constant of integration gives

$$T_H - T_{\text{sat}} = (T_{H,\text{in}} - T_{\text{sat}}) e^{-(U \mathscr{P} / \dot{m}_H c_{pH}) x} \tag{8.14}$$

which shows an exponential decrease for $T_H(x)$. The outlet gas temperature is obtained by letting $x = L$ in Eq. (8.14):

$$T_{H,\text{out}} - T_{\text{sat}} = (T_{H,\text{in}} - T_{\text{sat}}) e^{-U \mathscr{P} L / \dot{m}_H c_{pH}} \tag{8.15}$$

Equation (8.15) can be rearranged as

$$\frac{T_{H,\text{in}} - T_{H,\text{out}}}{T_{H,\text{in}} - T_{\text{sat}}} = 1 - e^{-U \mathscr{P} L / \dot{m}_H c_{pH}} \tag{8.16}$$

or

$$\varepsilon = 1 - e^{-N_{\text{tu}}} \tag{8.17}$$

where ε is the exchanger *effectiveness*, and N_{tu} is the *number of transfer units* (NTU). The effectiveness is the actual temperature decrease of the gas $(T_{H,\text{in}} - T_{H,\text{out}})$ divided by the maximum possible decrease that could be obtained in an infinitely long exchanger $(T_{H,\text{in}} - T_{\text{sat}})$. From the exchanger energy balance, the heat transfer is $\dot{Q} = \dot{m}_H c_{pH} (T_{H,\text{in}} - T_{H,\text{out}})$. Thus, the effectiveness can be viewed as the actual heat transfer divided by the maximum heat transfer obtainable in an infinitely long exchanger. The dimensionless group NTU can be viewed as a measure of the heat transfer capability of the exchanger. Equation (8.17) indicates that the larger the number of transfer units, the higher the exchanger effectiveness. However, there are design tradeoffs to be considered, and in practice, values of ε between 0.6 and 0.9 are typical. Equation (8.17) also holds for condensers with the effectiveness evaluated as the actual temperature increase of the coolant water divided by its maximum possible increase (see Exercise 8–12).

EXAMPLE 8.4   An Open-Cycle Ocean Thermal Energy Conversion Pilot Plant

In a pilot open-cycle ocean thermal energy conversion plant, 1 kg/s of warm sea water at 300 K enters a direct contact evaporator maintained at 2619 Pa. The water is injected through an array of nozzles to give an estimated transfer area and liquid-side heat transfer coefficient of 0.80 m² and 17,000 W/m² K, respectively. At what rate is vapor produced?

Solution

Given: Evaporator to flash-evaporate sea water at 300 K.

Required: Rate of vapor production.

Assumptions: The water-vapor interface is at the saturation temperature.

Equation (8.17) applies:

$$\varepsilon = 1 - e^{-N_{tu}}; \quad N_{tu} = h_c A / \dot{m}_H c_{pH}$$

where h_c is the heat transfer coefficient for transfer of heat from the bulk water to the water-vapor interface, and $A = \mathscr{P}L$ Therefore,

$$N_{tu} = \frac{h_c A}{\dot{m}_H c_{pH}} = \frac{(17,000)(0.80)}{(1)(4178)} = 3.26$$

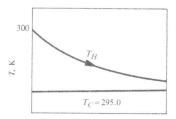

where c_{pH} was approximated by the value for pure water at 300 K using Table A.8.

$$\varepsilon = 1 - e^{-N_{tu}} = 1 - e^{-3.26} = 0.961$$

From the definition of effectiveness,

$$T_{H,\text{out}} = T_{H,\text{in}} - \varepsilon(T_{H,\text{in}} - T_{\text{sat}})$$

The saturation temperature corresponding to the evaporator pressure of 2619 Pa is 295.0 K from Table A.12*a*. Hence,

$$T_{H,\text{out}} = 300 - 0.961(300 - 295) = 295.2 \text{ K}$$

The rate of vapor production is obtained from the exchanger energy balance, Eq. (8.6):

$$(\dot{m}c_p)_H (T_{H,\text{in}} - T_{H,\text{out}}) = \dot{m}_C h_{fgC}$$

From Table A.12*a*, the enthalpy of vaporization at $T_{\text{sat}} = 295$ K is 2.449×10^6 J/kg. Thus,

$$(1)(4178)(300 - 295.2) = \dot{m}_C (2.449 \times 10^6)$$

$$\dot{m}_C = 8.20 \times 10^{-3} \text{ kg/s}$$

Comments

1. Examination of Table A.13*b* shows that the specific heat of sea water will be somewhat less than the pure water value.

2. Estimation of transfer area and heat transfer coefficient is not easy. In practice, we may often simply assume that a direct-contact evaporator of this type has an effectiveness in excess of 95%.

8.5 TWO-STREAM STEADY-FLOW HEAT EXCHANGERS

The most common type of heat exchanger is the two-stream steady-flow exchanger, with parallel, counter- or cross-flow of the two streams. Two methods will be used to analyze these exchangers: the logarithmic mean temperature method, in Section 8.5.1, and the effectiveness-number of transfer units method, in Section 8.5.2. Balanced-flow exchangers, for which the $\dot{m}c_p$ products for the streams are equal, are often required for air-conditioning systems and are examined in Section 8.5.3. For both methods, the model used assumes a constant overall heat transfer coefficient through the exchangers, and negligible heat conduction in the flow direction. If mass flow rates are small, heat conduction along the tube walls adversely affects heat exchanger performance. The effect is most significant when a high effectiveness is desired, as in the case for the heat exchangers used in many cryogenic refrigeration systems (see Exercise 8–78).

8.5.1 The Logarithmic Mean Temperature Difference

Figure 8.14*a* shows how the hot and cold stream temperatures, T_H and T_C, respectively, vary along heat exchangers for both parallel flow and counterflow. The temperature difference for heat transfer from the hot to the cold fluid, $T_H - T_C$, is seen to vary along the exchanger. In the case of parallel flow, for example, $T_H - T_C$ decreases continuously along the exchanger from the inlet to the outlet end. The total heat transfer in the exchanger will be written as

$$\dot{Q} = U\mathscr{P}L\Delta T_{\text{lm}} \tag{8.18}$$

where $U\mathscr{P}$ is the overall heat transfer coefficient $\times$ transfer perimeter product for the exchanger, L is the exchanger length, and ΔT_{lm} is an appropriate mean temperature difference between the hot and cold streams. In general, ΔT_{lm} must be determined by analysis. For parallel-flow and counterflow exchangers, the analysis is straight-forward, provided some simplifying assumptions are made.

To determine ΔT_{lm}, we first consider a parallel-flow exchanger and make an energy balance on an exchanger element Δx long, as shown in Fig. 8.14*b*. Application of the steady-flow energy equation, Eq. (1.4), to a control volume containing either stream requires that

$$\Delta \dot{Q} = \dot{m}\Delta h = \dot{m}c_p\Delta T; \quad \Delta T = T|_{x+\Delta x} - T|_x \tag{8.19}$$

since no external work is done and changes of kinetic energy and potential energy are negligible. Thus, for the cold stream,

$$\Delta \dot{Q} = U\mathscr{P}\Delta x(T_H - T_C) = (\dot{m}c_p)_C\Delta T_C \tag{8.20a}$$

and for the hot stream

$$\Delta \dot{Q} = U\mathscr{P}\Delta x(T_H - T_C) = -(\dot{m}c_p)_H\Delta T_H \tag{8.20b}$$

For convenience, we let $C = \dot{m}c_p$ [J/K s], the **flow thermal capacity** of the stream; Eq. (8.19) shows that C is the amount of heat a stream gains or loses for a temperature

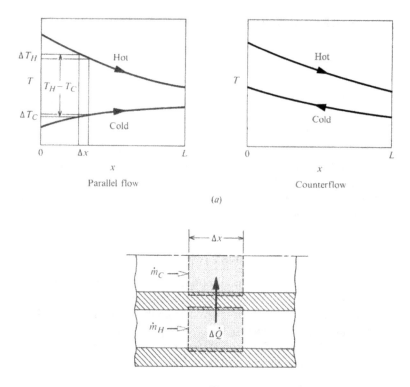

Figure 8.14 (*a*) Fluid temperature variations along parallel-flow and counterflow exchangers: notation for LMTD approach, (*b*) Elemental control volumes for analysis of a coaxial-tube parallel-flow exchanger.

change of 1 K. Dividing Eqs. (8.20) by Δx, rearranging, and letting $\Delta x \to 0$ gives

$$C_C \frac{dT_C}{dx} = U\mathscr{P}(T_H - T_C) \tag{8.21a}$$

$$C_H \frac{dT_H}{dx} = -U\mathscr{P}(T_H - T_C) \tag{8.21b}$$

Subtracting Eq. (8.21*a*) from Eq. (8.21*b*) and further rearranging,

$$\frac{d(T_H - T_C)}{T_H - T_C} = -U\mathscr{P}\left(\frac{1}{C_H} + \frac{1}{C_C}\right)dx$$

Integrating from $x = 0$ to $x = L$,

$$\ln\frac{T_{H,L} - T_{C,L}}{T_{H,0} - T_{C,0}} = -\int_0^L U\mathscr{P}\left(\frac{1}{C_H} + \frac{1}{C_C}\right)dx \tag{8.22}$$

If $U\mathscr{P}$ and the fluid specific heats are now assumed constant, then

$$\ln\frac{T_{H,L}-T_{C,L}}{T_{H,0}-T_{C,0}} = -U\mathscr{P}L\left(\frac{1}{C_H}+\frac{1}{C_C}\right) \tag{8.23}$$

But the exchanger energy balance, Eq. (8.2), is

$$(\dot{m}c_p)_H(T_{H,0}-T_{H,L}) = \dot{Q} = (\dot{m}c_p)_C(T_{C,L}-T_{C,0})$$

Hence,

$$C_H = \frac{\dot{Q}}{T_{H,0}-T_{H,L}}, \quad C_C = \frac{\dot{Q}}{T_{C,L}-T_{C,0}}$$

Substituting in Eq. (8.23) and rearranging gives

$$\dot{Q} = U\mathscr{P}L\frac{(T_{H,L}-T_{C,L})-(T_{H,0}-T_{C,0})}{\ln[(T_{H,L}-T_{C,L})/(T_{H,0}-T_{C,0})]} \tag{8.24}$$

Comparing Eqs. (8.24) and (8.18) gives the required formula for ΔT_{lm}:

$$\Delta T_{lm} = \frac{(T_H-T_C)_L-(T_H-T_C)_0}{\ln[(T_H-T_C)_L/(T_H-T_C)_0]} \tag{8.25}$$

ΔT_{lm} is called the **log mean temperature difference,** which is abbreviated as LMTD. If this analysis is repeated for a counterflow exchanger, exactly the same result is obtained.[1]

Often it is not appropriate to assume that the overall heat transfer coefficient is constant along the exchanger, perhaps due to entrance effects or due to fluid property variations. If concern is only with the effect of the entrance region on U, then U in Eq. (8.18) can be replaced by an average value, $\overline{U}$:

$$\dot{Q} = \overline{U}\mathscr{P}L\Delta T_{lm}; \quad \overline{U} = \frac{1}{L}\int_0^L U\,dx \tag{8.26}$$

If fluid property variations are also important, then numerical integration of Eq. (8.22) is required, because U, C_H, and C_C all vary along the exchanger.

F Factor Charts

The concept of an appropriate mean temperature difference between the two streams has an appealing simplicity, and, as a result, the LMTD is widely used in engineering practice. For two-stream configurations other than the ideal coaxial type considered here, a correction factor F is applied to the LMTD for a *counterflow configuration* with the same inlet and outlet temperatures:

$$\dot{Q} = UAF\Delta T_{lm} \tag{8.27}$$

Compilations of F factor charts may be found in References [1,2,3]; samples are given in Appendix C. The F factors obey the rule $F(P,R) = F(PR,1/R)$ where P,R

---

[1] In the limit $(T_H-T_C)_L \to (T_H-T_C)_0$ for counterflow exchangers, the common value equals the LMTD. This limit corresponds to *balanced* flow, $C_H = C_C$ (see Section 8.5.3).

are defined on the charts. For R large, the curves are nearly vertical and the curve for $1/R$ should be used. Improved performance charts for two-stream exchangers have been presented recently by Turton et al. [4]. Also, the performance of two-stream exchangers can be expressed in terms of effectiveness and number of transfer units, as will be shown in Section 8.5.2. The $\varepsilon - N_{tu}$ formulation is rapidly gaining popularity for various reasons, including its suitability for computer-aided design.

EXAMPLE 8.5 Counterflow Benzene Cooler

A coaxial-tube counterflow heat exchanger is to cool 0.03 kg/s of benzene from 360 K to 310 K with a counterflow of 0.02 kg/s of water at 290 K. If the inner tube outside diameter is 2 cm and the overall heat transfer coefficient based on outside area is 650 W/m² K, determine the required length of the exchanger. Take the specific heats of benzene and water as 1880 and 4175 J/kg K, respectively.

Solution

Given: Coaxial-tube counterflow exchanger.

Required: Exchanger length for specified performance.

Assumptions: U is constant along the exchanger.

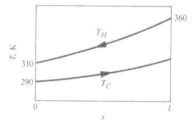

An exchanger energy balance is used to determine the required heat transfer and water outlet temperature. For a counterflow exchanger as shown,

$$(\dot{m}c_p)_H(T_{H,L} - T_{H,0}) = (\dot{m}c_p)_C(T_{C,L} - T_{C,0}) = \dot{Q}$$

$$(0.03)(1880)(360 - 310) = (0.02)(4175)(T_{C,L} - 290) = \dot{Q}$$

Solving, $\dot{Q} = 2820$ W, $T_{C,L} = 323.8$ K. Equation (8.25) gives the log mean temperature difference:

$$\Delta T_{lm} = \frac{(T_H - T_C)_L - (T_H - T_C)_0}{\ln[(T_H - T_C)_L/(T_H - T_C)_0]} = \frac{36.2 - 20}{\ln(36.2/20)} = 27.3 \text{ K}$$

The tube outside perimeter is $\mathscr{P} = \pi D = (\pi)(0.02) = 0.0628$ m. Solving Eq. (8.18) for the exchanger length L gives

$$L = \frac{\dot{Q}}{U \mathscr{P} \Delta T_{lm}} = \frac{2820}{(650)(0.0628)(27.3)} = 2.53 \text{ m}$$

Comments

A quick calculation of the arithmetic mean temperature difference provides a useful check of the LMTD, since it will not be very different. In this example, the arithmetic mean temperature difference is $(1/2)(36.2 + 20) = 28.1$ K.

EXAMPLE 8.6   Counterflow Oil Cooler

After a long time in service, a counterflow oil cooler is checked to ascertain if its performance has deteriorated due to fouling. In the test, SAE 50 oil flowing at 2.0 kg/s is cooled from 420 K to 380 K by a water supply of 1.0 kg/s at 300 K. If the heat transfer surface is 3.33 m^2 and the design value of the overall heat transfer coefficient is 930 W/m^2 K, how much has it been reduced by fouling?

Solution

Given: Performance data for a counterflow exchanger.

Required: Effect of fouling on overall heat transfer coefficient.

Assumptions: U is constant along the exchanger.

We first use an exchanger energy balance to find the heat transferred and the water outlet temperature. Evaluating c_{pH} at 400 K using Table A.8,

$$\dot{Q} = (\dot{m}c_p)_H(T_{H,L} - T_{H,0})$$
$$= (2.0)(2330)(420 - 380)$$
$$= 186,400 \text{ W}$$

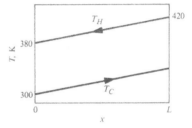

To find $T_{C,L}$, we guess an average water temperature of 320 K to evaluate c_{pC}

$$186,400 = (\dot{m}c_p)_C(T_{C,L} - T_{C,0})$$
$$= (1)(4174)(T_{C,L} - 300)$$

Hence,

$$T_{C,L} = 344.7 \text{ K}$$

and our guessed average temperature is close enough. From Eq. (8.25), the LMTD is

$$\Delta T_{\text{lm}} = \frac{(420 - 344.7) - (380 - 300)}{\ln[(420 - 344.7)/(380 - 300)]}$$
$$= \frac{75.3 - 80}{\ln(75.3/80)} = 77.6 \text{ K}$$

Then, from Eq. (8.18) and recognizing that $A = \mathscr{P}L$,

$$\dot{Q} = UA\Delta T_{\text{lm}} \quad \text{or} \quad U = \frac{\dot{Q}}{A\Delta T_{\text{lm}}} = \frac{186,400}{(3.33)(77.6)} = 721 \text{ W/m}^2 \text{ K}$$

Comments

1. The reduction in U due to fouling is $(930 - 721)/(930) = 22.5\%$.

2. Can you think of any other reasons why the performance might have deteriorated?

8.5.2 Effectiveness and Number of Transfer Units

The LMTD formula for heat exchanger performance is useful only when inlet and outlet temperatures are known, either because they have been measured in a test or because they have been specified in a design. If it is desired to calculate inlet or outlet temperatures for a given exchanger and flow rates, use of the LMTD method requires an iterative solution procedure or specially constructed charts. Iteration can be avoided if exchanger performance is expressed in terms of effectiveness and number of transfer units, as was done for the single-stream exchanger.

Temperature variations along parallel-flow and counterflow exchangers are shown in Fig. 8.15. Inlet and outlet temperature are designated by subscripts "in" and "out," respectively, rather than by the locations $x = 0$ and $x = L$, as was done in the LMTD analysis. The exchanger effectiveness is now defined as the ratio of the actual heat transferred to the maximum possible amount of heat that could be transferred in an infinitely long *counterflow* exchanger. This definition is more general than that used for single-stream exchangers. From the exchanger energy balance, Eq. (8.4), the heat transferred is

$$Q = C_H(T_{H,\text{in}} - T_{H,\text{out}}) = C_C(T_{C,\text{out}} - T_{C,\text{in}}) \tag{8.28}$$

In an infinitely long counterflow exchanger with $C_C < C_H$, Fig. 8.15b shows that $T_{C,\text{out}} \to T_{H,\text{in}}$; then $\dot{Q}_{\max} = C_C(T_{H,\text{in}} - T_{C,\text{in}})$. On the other hand, in an infinitely long parallel-flow exchanger with $C_C < C_H$, Fig. 8.15a shows $T_{C,\text{out}} \to T_{H,\text{out}} < T_{H,\text{in}}$, and the heat transfer $\dot{Q}$ will be less. Similarly, if $C_H < C_C$, the heat transfer will again be greater in an infinitely long counterflow exchanger, $\dot{Q}_{\max} = C_H(T_{H,\text{in}} - T_{C,\text{in}})$. If we write

$$C_{\min} = \min\ (C_H, C_C) \tag{8.29}$$

then the maximum heat transfer in an exchanger of any configuration is

$$\dot{Q}_{\max} = C_{\min}(T_{H,\text{in}} - T_{C,\text{in}}) \tag{8.30}$$

Note that it is the stream with the smaller flow thermal capacity that limits the amount of heat that can be transferred. The actual heat transfer is expressed in terms

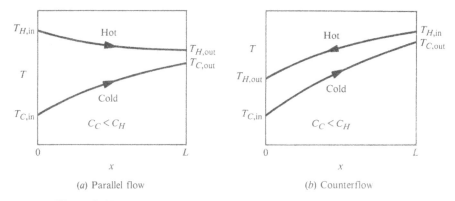

(a) Parallel flow   (b) Counterflow

Figure 8.15   Fluid temperature variations along parallel-flow and counterflow exchangers: notation for the $\varepsilon - N_{\text{tu}}$ approach.

of the enthalpy change of either stream, as given by Eq. (8.28); thus, the effectiveness is

$$\varepsilon = \frac{\dot{Q}}{\dot{Q}_{max}} = \frac{C_H(T_{H,in} - T_{H,out})}{C_{min}(T_{H,in} - T_{C,in})} = \frac{C_C(T_{C,out} - T_{C,in})}{C_{min}(T_{H,in} - T_{C,in})} \tag{8.31}$$

Which of these two expressions is more convenient depends on the problem specification.

We first determine the effectiveness of a parallel-flow exchanger. Changing subscripts in Eq. (8.23) from "$0, L$" to "in, out" gives

$$\ln\frac{T_{H,out} - T_{C,out}}{T_{H,in} - T_{C,in}} = -U\mathscr{P}L\left(\frac{1}{C_H} + \frac{1}{C_C}\right) = \frac{-U\mathscr{P}L}{C_C}\left(1 + \frac{C_C}{C_H}\right) \tag{8.32}$$

The number of transfer units, N_{tu}, is defined as

$$N_{tu} = \frac{U\mathscr{P}L}{C_{min}} \tag{8.33}$$

and the **capacity ratio** R_C as

$$R_C = \frac{C_{min}}{C_{max}} \quad (\leq 1) \tag{8.34}$$

Thus, if $C_C = C_{min}$, Eq. (8.32) becomes

$$\frac{T_{H,out} - T_{C,out}}{T_{H,in} - T_{C,in}} = e^{-N_{tu}(1+R_C)} \tag{8.35}$$

But from Eq. (8.28),

$$T_{H,out} = T_{H,in} - R_C(T_{C,out} - T_{C,in})$$

and substituting in Eq. (8.35) and rearranging gives

$$\frac{(T_{H,in} - T_{C,in}) - R_C(T_{C,out} - T_{C,in}) - (T_{C,out} - T_{C,in})}{T_{H,in} - T_{C,in}} = e^{-N_{tu}(1+R_C)} \tag{8.36}$$

Also for $C_C = C_{min}$, Eq. (8.31) gives

$$\varepsilon = \frac{T_{C,out} - T_{C,in}}{T_{H,in} - T_{C,in}} \tag{8.37}$$

so that Eq. (8.36) can be rewritten as

$$1 - (R_C + 1)\varepsilon = e^{-N_{tu}(1+R_C)} \tag{8.38}$$

Solving for ε gives

$$\varepsilon = \frac{1 - e^{-N_{tu}(1+R_C)}}{1+R_C} \tag{8.39}$$

If the hot stream is chosen to have the minimum capacity, that is, $C_H = C_{min}$, exactly the same result is obtained. For a given heat exchanger and flow rates, N_{tu} and R_C are known, so that the effectiveness ε can be calculated from Eq. (8.39), and the unknown outlet temperature can be determined. If, on the other hand, we are required to design a heat exchanger to have a specified outlet temperature, that is, a specified effectiveness, then the unknown is N_{tu}. Solving Eq. (8.39) for N_{tu} gives

$$N_{tu} = \frac{1}{1+R_C} \ln \frac{1}{1-(1+R_C)\varepsilon} \tag{8.40}$$

Using Eq. (8.40) is equivalent to using the LMTD approach, Eqs. (8.18) and (8.25).
A similar analysis can be performed for a counterflow exchanger to give

$$\varepsilon = \frac{1 - e^{-N_{tu}(1-R_C)}}{1 - R_C e^{-N_{tu}(1-R_C)}} \tag{8.41}$$

and

$$N_{tu} = \frac{1}{1-R_C} \ln \frac{1-\varepsilon R_C}{1-\varepsilon} \tag{8.42}$$

For a given N_{tu} and R_C, a counterflow exchanger is always more effective than a parallel-flow one, because a larger value of $(T_H - T_C)$ is sustained along a counterflow exchanger.

Figure 8.16 shows exchangers for which $C_C \ll C_H$; as a result, the temperature of the hot stream does not vary significantly along the exchanger. For $R_C \rightarrow 0$, both Eqs. (8.39) and (8.41) reduce to

$$\varepsilon = 1 - e^{-N_{tu}} \tag{8.43}$$

which is the same as the effectiveness of a single-stream exchanger. For $R_C = 0$, the temperature of one stream does not change through the exchanger, which satisfies the definition of a single-stream exchanger.

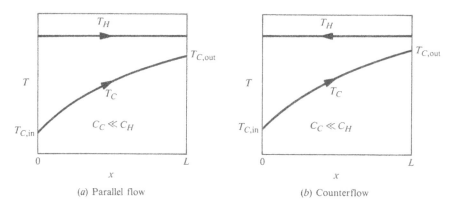

(a) Parallel flow   (b) Counterflow

Figure 8.16 Fluid temperature variations along parallel-flow and counterflow exchangers for $C_C \ll C_H$.

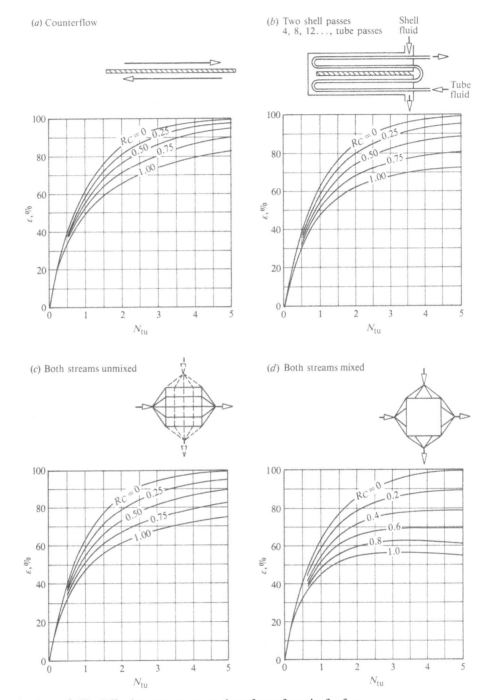

Figure 8.17 Effectiveness versus number of transfer units for four exchanger configurations, (*a*) Counterflow. (*b*) Multipass, (*c*) Cross-flow, both streams unmixed, (*d*) Cross-flow, both streams mixed.

Table 8.3a Effectiveness formulas for selected exchanger configurations.

Configuration	Effectiveness
1. Single-stream, and all exchangers when $R_C = 0$	$\varepsilon = 1 - \exp(-N_{tu})$
2. Parallel-flow	$\varepsilon = \dfrac{1 - \exp[-N_{tu}(1 + R_C)]}{1 + R_C}$
3. Counterflow	$\varepsilon = \dfrac{1 - \exp[-N_{tu}(1 - R_C)]}{1 - R_C \exp[-N_{tu}(1 - R_C)]}$
Single-pass cross-flow:	
4. Both fluids unmixed	$\varepsilon = 1 - \exp\left\{\dfrac{N_{tu}^{0.22}}{R_C}\left[\exp(-R_C N_{tu}^{0.78}) - 1\right]\right\}$
5. Both fluids mixed	$\varepsilon = \left[\dfrac{1}{1 - \exp(-N_{tu})} + \dfrac{R_C}{1 - \exp(-R_C N_{tu})} - \dfrac{1}{N_{tu}}\right]^{-1}$
6. C_{max} mixed, C_{min} unmixed	$\varepsilon = \dfrac{1}{R_C}\left\{1 - \exp\left[R_C\left(e^{-N_{tu}} - 1\right)\right]\right\}$
7. C_{max} unmixed, C_{min} mixed	$\varepsilon = 1 - \exp\left\{-\dfrac{1}{R_C}\left[1 - e^{-R_C N_{tu}}\right]\right\}$
Shell-and-tube:	
8. One shell pass; $2, 4, 6, \ldots$ tube passes[a]	$\varepsilon = \varepsilon_1 = 2\left\{1 + R_C + (1 + R_C^2)^{1/2}\dfrac{1 + \exp[-N_{tu}(1 + R_C^2)^{1/2}]}{1 - \exp[-N_{tu}(1 + R_C^2)^{1/2}]}\right\}^{-1}$
9. n shell passes; $2n, 4n, \ldots$ tube passes[a]	$\varepsilon = \left[\left(\dfrac{1 - \varepsilon_1 R_C}{1 - \varepsilon_1}\right)^n - 1\right]\left[\left(\dfrac{1 - \varepsilon_1 R_C}{1 - \varepsilon_1}\right)^n - R_C\right]^{-1}$

[a] In calculating ε_1, the N_{tu} *per shell pass* is used (i.e., N_{tu}/n).

Figure 8.17 gives $\varepsilon - N_{tu}$ graphs for four two-stream exchanger configurations; Table 8.3a and b gives $\varepsilon - N_{tu}$ and $N_{tu} - \varepsilon$ relations for a larger variety of configurations. Extensive compilations of $\varepsilon - N_{tu}$ graphs can be found in References [2,5]. It should be clearly understood that the effectiveness and LMTD formulations for two-stream exchangers are mathematically equivalent, and either is sufficient for the solution of a problem. Current practice tends to favor the effectiveness approach because both effectiveness and number of transfer units have a unique physical significance for a given exchanger and given flow thermal capacities. On the other hand, the log mean temperature difference depends also on the inlet temperatures of the streams and thus will vary according to the particular application. The design engineer using the effectiveness approach soon develops a "feel" for the number of transfer units and effectiveness to be desired or expected in a given situation.

Table 8.3b Formulas for number of transfer units for selected exchanger configurations.

Configuration	Number of Transfer Units
1. Single-stream, and all exchangers when $R_C = 0$	$N_{tu} = \ln\dfrac{1}{1-\varepsilon}$
2. Parallel-flow	$N_{tu} = \dfrac{1}{1+R_C}\ln\dfrac{1}{1-(1+R_C)\varepsilon}$
3. Counterflow	$N_{tu} = \dfrac{1}{1-R_C}\ln\dfrac{1-\varepsilon R_C}{1-\varepsilon}$
Single-pass cross-flow:	
4. Both fluids unmixed	
5. Both fluids mixed	
6. C_{max} mixed, C_{min} unmixed	$N_{tu} = -\ln[1+(1/R_C)\ln(1-\varepsilon R_C)]$
7. C_{max} unmixed, C_{min} mixed	$N_{tu} = -(1/R_C)\ln[R_C \ln(1-\varepsilon)+1]$
Shell-and-tube:	$N_{tu} = -(1+R_C^2)^{-1/2}\ln\left[\dfrac{E-1}{E+1}\right]$
8. One shell pass; 2, 4, 6, ... tube passes[a]	$E = \dfrac{2/\varepsilon - (1+R_C)}{(1+R_C^2)^{1/2}}$
9. n shell passes; $2n, 4n, ...$ tube passes[a]	$E = \dfrac{[2(F-R_C)/(F-1)]-(1+R_C)}{(1+R_C^2)^{1/2}}; F = \left(\dfrac{\varepsilon R_C - 1}{\varepsilon - 1}\right)^{1/n}$

[a] Substituting E gives the N_{tu} *per shell pass* (i.e., the N_{tu} for the exchanger is $n\times$ this value).

The Computer Program HEX1

HEX1 calculates the thermal performance of heat exchangers for the nine flow configurations listed in Table 8.3. There are two options:

1. *The rating problem.* Known are the mass flow rates $\dot{m}$, overall heat transfer coefficient U, and heat transfer area A. Required is the heat exchanger effectiveness ε, which is calculated from the formulas in Table 8.3a. Also, for input inlet temperatures, HEX1 calculates outlet temperatures.

2. *The design or sizing problem.* Known are the mass flow rates $\dot{m}$, inlet temperatures, and one outlet temperature. Required is the number of transfer units N_{tu}, which is calculated from the formulas in Table 8.3b. In the case of items 4 and 5, for which there are no explicit formulas for $N_{tu}(\varepsilon, R_C)$, the $\varepsilon(N_{tu}, R_C)$ formulas from items 4 and 5 of Table 8.3a are solved for the lowest N_{tu} value using Newton's method. HEX1 also calculates the unknown outlet temperature.

EXAMPLE 8.7   Cooling of a Distillation Column Product Stream

A 4 kg/s product stream from a distillation column is to be cooled by a 3 kg/s water stream in a counterflow exchanger. The hot and cold stream inlet temperatures are 400 and 300 K, respectively, and the heat transfer area of the exchanger is 30 m². If the overall heat transfer coefficient is estimated to be 820 W/m² K, determine the product stream outlet temperature. The specific heat of the product stream can be taken to be 2500 J/kg K.

Solution

Given: Counterflow heat exchanger to cool distillation product stream.

Required: Performance, in particular $T_{H,\text{out}}$.

Assumptions: U is constant along the exchanger.

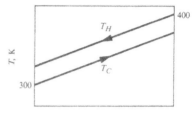

This is a *rating* problem since the $\dot{m}$ values, U, and A are known. We first calculate the effectiveness from Eq. (8.41):

$$\varepsilon = \frac{1 - e^{-N_{\text{tu}}(1-R_C)}}{1 - R_C e^{-N_{\text{tu}}(1-R_C)}}; \quad R_C = \frac{C_{\min}}{C_{\max}}; \quad N_{\text{tu}} = \frac{U\mathscr{P}L}{C_{\min}}$$

$$C_H = (\dot{m}c_p)_H = (4)(2500) = 10,000 \text{ W/K}$$

$$C_C = (\dot{m}c_p)_C = (3)(4180) = 12,540 \text{ W/K for } c_p \text{ evaluated at} \sim 330 \text{ K in Table A.8}$$

$$C_{\min} = \min(10,000, 12,540) = 10,000 \text{ W/K}$$

$$R_C = 10,000/12,540 = 0.797$$

$$N_{\text{tu}} = \frac{U\mathscr{P}L}{C_{\min}} = \frac{UA}{C_{\min}} = \frac{(820)(30)}{10,000} = 2.46$$

$$\varepsilon = \frac{1 - e^{-2.46(1-0.797)}}{1 - 0.797e^{-2.46(1-0.797)}} = 0.761$$

Equation (8.31) gives the effectiveness in terms of temperatures as

$$\varepsilon = \frac{C_H(T_{H,\text{in}} - T_{H,\text{out}})}{C_{\min}(T_{H,\text{in}} - T_{C,\text{in}})}$$

$$0.761 = \frac{10,000(400 - T_{H,\text{out}})}{10,000(400 - 300)}$$

Hence, $T_{H,\text{out}} = 323.9$ K.
   The exchanger energy balance is

$$C_H(T_{H,\text{in}} - T_{H,\text{out}}) = C_C(T_{C,\text{out}} - T_{C,\text{in}})$$

$$10,000(400 - 323.9) = 12,540(T_{C,\text{out}} - 300)$$

Hence, $T_{C,\text{out}} = 360.7$ K, and our guessed average temperature of 330 K is satisfactory.

Solution using HEX1

The required input in SI units is:

c (counterflow)
$\dot{m}_C = 3; c_{pC} = 4180$
$\dot{m}_H = 4; c_{pH} = 2500$
$T_{C,\text{in}} = 300, T_{H,\text{in}} = 400$
U and A known
$U = 820; A = 30$

The output is:

$N_{tu} = 2.46$
$\varepsilon = 0.761$
$T_{C,\text{out}} = 360.7$
$T_{H,\text{out}} = 323.9$

Comments

Use of the $\varepsilon - N_{tu}$ method is seen to be quite straightforward. All the student needs to do is to recognize from the given data whether it is a rating or design problem.

EXAMPLE 8.8   Cross-Flow Plate Exchanger Using the $\varepsilon - N_{tu}$ Approach

A cross-flow plate heat exchanger is to be designed for waste heat recovery from the exhaust streams of a metallurgical process. A flow of 5 kg/s of exhaust gases enters the exchanger at 240°C and must be cooled to 120°C by 5 kg/s of air supplied at 20°C. If the overall heat transfer coefficient is estimated to be 40 W/m² K, determine the required heat transfer area if both streams are unmixed. The specific heat of the exhaust gases can be taken as 1200 J/kg K.

Solution

Given: Cross-flow plate heat exchanger for waste heat recovery.

Required: Heat transfer area A for specified performance.

Assumptions: U is constant over the plates.

This is a *design* or *sizing* problem since the $\dot{m}$ values and three temperatures are known. Equation (8.31) defines the effectiveness as

$$\varepsilon = \frac{C_H(T_{H,\text{in}} - T_{H,\text{out}})}{C_{\min}(T_{H,\text{in}} - T_{C,\text{in}})}$$

For an estimated 400 K average temperature of the air stream, $c_{pC} = 1010$ J/kg K from Table A.7, and the flow thermal capacities are

$$C_C = (\dot{m}c_p)_C = (5)(1010) = 5050 \text{ W/K}$$

$$C_H = (\dot{m}c_p)_H = (5)(1200) = 6000 \text{ W/K}$$

$$C_{\min} = 5050 \text{ W/K}; \quad R_C = \frac{C_{\min}}{C_{\max}} = \frac{5050}{6000} = 0.842$$

Hence,

$$\varepsilon = \frac{6000(240 - 120)}{5050(240 - 20)} = 0.648$$

From Fig. 8.17c, the required number of transfer units is 2.0.

$$N_{tu} = \frac{UA}{C_{min}}; \quad A = \frac{C_{min}N_{tu}}{U} = \frac{(5050)(2.0)}{40} = 253 \text{ m}^2$$

Solution using HEX1

The required inputs in SI units with temperatures in °C are:

d (crossflow, both streams unmixed)
$\dot{m}_C = 5; c_{pC} = 1010$
$\dot{m}_H = 5; c_{pH} = 1200$
$T_{C,in} = 20, T_{H,in} = 240$
T_{out} known
$T_{H,out} = 120$

The output is:

$\varepsilon = 0.648$
$N_{tu} = 1.956$
$T_{C,out} = 162.6$
$UA = 9877$; hence, $A = 9877/40 = 247 \text{ m}^2$

Comments

Notice that HEX1 accepts temperatures in either kelvins or degrees Celsius.

8.5.3 Balanced-Flow Exchangers

Often exchangers have hot and cold streams of approximately equal flow thermal capacity, that is, $C_H = C_C$, or $R_C = 1$. Such exchangers are said to have *balanced* flow. Figure 8.18 shows a heatpipe air preheater used to recover heat from the exhaust of a gas turbine. The exhaust gas stream does have a larger $\dot{m}$ and c_p than the incoming air stream, but $R_C = 1$ is not a bad assumption. In many air-conditioning systems, warm, fresh ambient air is first cooled by colder, stale air leaving the system, and the exchanger can be exactly balanced. If $R_C = 1$ is substituted in Eq. (8.39) for the effectiveness of a parallel-flow exchanger, the result is

$$\varepsilon = \frac{1}{2}\left(1 - e^{-2N_{tu}}\right) \tag{8.44}$$

If, however, $R_C = 1$ is substituted in Eq. (8.41) for a counterflow exchanger, the result is indeterminate; application of L'Hôpital's rule then gives

$$\varepsilon = \frac{N_{tu}}{1 + N_{tu}} \tag{8.45}$$

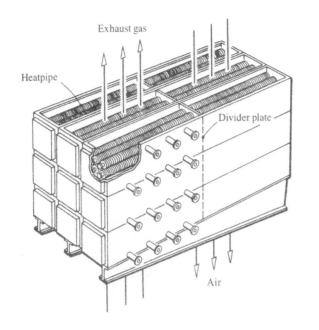

Figure 8.18 A heatpipe gas turbine recuperator.

The counterflow case is more interesting and of more practical importance than the parallel-flow case. Application of the steady-flow energy equation to a differential element of a balanced counterflow exchanger gives the differential equations

$$C_C \frac{dT_C}{dx} = U \mathscr{P} (T_H - T_C) \tag{8.46a}$$

$$C_H \frac{dT_H}{dx} = U \mathscr{P} (T_H - T_C) \tag{8.46b}$$

which are similar to Eqs. (8.21a,b) for the parallel-flow exchanger except for a sign change. Subtracting Eq. (8.46a) from (8.46b) with $C_C = C_H = C$ gives

$$\frac{d}{dx}(T_H - T_C) = 0 \tag{8.47}$$

Integrating,

$$T_H - T_C = \text{Constant} = T_{H,\text{in}} - T_{C,\text{out}} = T_{H,\text{out}} - T_{C,\text{in}}$$

as shown in Fig. 8.19. Substituting back in Eq. (8.46a),

$$\frac{dT_C}{dx} = \frac{U \mathscr{P}}{C} (T_{H,\text{in}} - T_{C,\text{out}}) \tag{8.48}$$

which states that T_C varies linearly with x if $U \mathscr{P}/C$ is constant along the exchanger. Similarly, T_H varies linearly with x.

Since $(T_H - T_C)$ is constant along a balanced-counterflow exchanger, the LMTD is equal to this constant difference. Notice that Eq. (8.25) is then indeterminate, and the logarithmic mean is simply equal to the constant value of $(T_H - T_C)$.

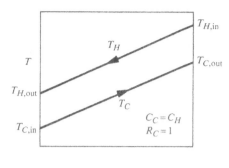

Figure 8.19   Fluid temperature variations along a balanced counterflow exchanger.

EXAMPLE 8.9   Recuperator for an Air-Conditioning System

In a solar-assisted air-conditioning system, 0.5 kg/s of ambient air at 270 K is to be preheated by the same amount of air leaving the system at 295 K. If a counterflow exchanger has an area of 30 m^2, and the overall heat transfer coefficient is estimated to be 25 W/m^2 K, determine the outlet temperature of the preheated air.

Solution

Given: Counterflow exchanger for balanced air flows.

Required: Performance, in particular $T_{C,\text{out}}$.

Assumptions: 1. Balanced flow, $C_{pC} = C_{pH}$.
2. U is constant along the exchanger.

This is a *rating* problem since the $\dot{m}$ values, U, and A are known. If we take $c_p = 1000$ J/kg K for air, the number of transfer units is

$$N_{\text{tu}} = \frac{UA}{\dot{m}c_p} = \frac{(25)(30)}{(0.5)(1000)} = 1.50$$

For balanced counterflow, Eq. (8.45) gives the effectiveness as

$$\varepsilon = \frac{N_{\text{tu}}}{1 + N_{\text{tu}}} = \frac{1.5}{1 + 1.5} = 0.6$$

$$T_{C,\text{out}} = T_{C,\text{in}} + \varepsilon(T_{H,\text{in}} - T_{C,\text{in}})$$

$$= 270 + 0.6(295 - 270)$$

$$= 285 \text{ K}$$

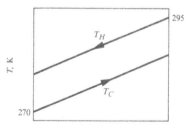

Comments

Use HEX1 to check N_{tu}, ε, and $T_{C,\text{out}}$.

EXAMPLE 8.10   Recuperator for a Gas Turbine

A flow of 0.1 kg/s of exhaust gases at 700 K from a gas turbine is used to preheat the incoming air, which is at the ambient temperature of 300 K. It is desired to cool the exhaust to 400 K, and it is estimated that an overall heat transfer coefficient of 30 W/m² K can be achieved in an appropriate exchanger. Determine the area required for a counterflow exchanger.

Solution

Given: Counterflow exchanger for a gas turbine.

Required: Area for specified performance.

Assumptions: 1. U is constant along the exchanger.

2. Balanced flow.

3. Specific heat of exhaust gases the same as for air, $\sim$1000 J/kg K.

This is a *sizing* problem since the $\dot{m}$ values and three temperatures are known. Equation (8.31) gives the effectiveness as

$$\varepsilon = \frac{T_{H,\text{in}} - T_{H,\text{out}}}{T_{H,\text{in}} - T_{C,\text{in}}}$$

$$= \frac{700 - 400}{700 - 300}$$

$$= 0.750$$

Equation (8.45) can be solved to give the number of transfer units:

$$\varepsilon = \frac{N_{\text{tu}}}{1 + N_{\text{tu}}}; \quad N_{\text{tu}} = \frac{\varepsilon}{1 - \varepsilon} = \frac{0.750}{1 - 0.750} = 3.0$$

$$N_{\text{tu}} = \frac{UA}{C_{\text{min}}} \quad \text{or} \quad A = \frac{CN_{\text{tu}}}{U} \simeq \frac{(0.1)(1000)(3.0)}{(30)} = 10 \text{ m}^2$$

Comments

1. Use HEX1 to check ε, N_{tu}, and the UA product.

2. This result is a rough estimate: the exchanger is not exactly balanced because both $\dot{m}$ and c_p are higher for the exhaust gases.

3. It would be impossible to cool the gases to 400 K in a parallel-flow exchanger: Eq. (8.44) shows that the maximum effectiveness of a balanced parallel-flow exchanger is 0.5.

8.6   ELEMENTS OF HEAT EXCHANGER DESIGN

The examples of heat exchanger calculations given in Section 8.5 show that heat transfer considerations alone do not determine the dimensions of a heat exchanger. For example, in Example 8.8, the required heat transfer area was calculated, but further constraints must be specified in order to fix the overall dimensions and number of plates. The usual heat exchanger design problem requires the engineer to specify a unit that will have a given heat transfer performance—that is, effectiveness—subject to a number of constraints, which might include: (1) a low capital cost; (2) a low operating cost; (3) limitations on size, shape, or weight; and (4) ease of maintenance. The most important operating cost may be the power required to pump the fluids. For a liquid, this power is usually rather low and does not have a major impact on the design. For gases, the power required per unit mass of working fluid is great, so it is often a critical design constraint. The pumping power is simply volume flow rate times pressure drop, divided by the blower efficiency. Thus, the need to have a low operating cost translates into a pressure drop constraint on the design.

In general, the engineer has the freedom to choose the exchanger configuration (counterflow, cross-flow, multiple-pass, etc.), the type of heat transfer surface (coaxial tube, plate-fin, tube bank, etc.), and characteristic dimensions of the surface (tube diameter, spacing in a tube bank, etc.). A low pressure drop requires a large cross-sectional area for flow, although appropriate selection of configuration and heat transfer surface also plays a role. A possible design strategy is as follows:

1. Specify the required heat transfer effectiveness.

2. Specify the allowable pressure drop for one or both streams.

3. Choose a configuration.

4. Choose a type of heat transfer surface.

5. Choose the dimensions of the surface.

6. Calculate the resulting dimensions of the unit.

7. Evaluate the design with respect to constraints such as capital cost, size, weight, and maintenance.

The dimensions obtained for step 6 may not be unique. Also, for some configurations, such as the coaxial-tube exchanger, it will not be possible to choose the surface dimensions since these will be fixed by the heat transfer and pressure drop requirements.

The complete exchanger design problem is one of optimization, for which advanced mathematical and computational methods are available. These methods should be used in any serious endeavor. But before such sophisticated tools are used, the engineer must have a clear understanding of principles underlying the design process. Of particular importance are the impact of a pressure drop constraint and economic considerations, and these topics are the primary focus of this section. The calculation of pressure drop in exchangers is discussed in Section 8.6.1, and the impact of a pressure drop constraint on sizing a heat exchanger is discussed in Section 8.6.2. Section 8.6.3 deals with the criteria for the selection of heat transfer surfaces for **compact heat exchangers**, which are defined somewhat arbitrarily as exchangers with a heat transfer surface area per unit volume greater than 700 m^2/m^3. The exchanger itself may not be small. Compact heat exchangers are used for gas-to-gas and gas-to-liquid transfer, because the large surface area gives a high h_cA product on the gas side, even though the heat transfer coefficient h_c is typically low for gases. In Section 8.6.4, a brief introduction to the economic analysis of exchanger utilization is given. The chapter concludes with an example of computer-aided heat exchanger design, in Section 8.6.5.

8.6.1 Exchanger Pressure Drop

For single-phase flow inside tubes, across tube banks, and through packed beds, formulas and data for pressure drop are given in Chapter 4. Compact heat exchangers, however, need special consideration.

Flow inside Tubes

For flow inside the tubes of a shell-and-tube or a coaxial-tube exchanger, pressure drop calculations may be made using the formulas and data given in Section 4.3. Care must be taken to ascertain whether the flow is laminar or turbulent. In addition, the low-Reynolds-number turbulent flow regime (Re < 10,000) is best avoided, if at all possible, owing to the uncertainty of the friction factor (and heat transfer coefficient) in this regime. Additional data may be found in the handbooks listed in the bibliography at the end of this text.

Flow across Tube Banks

Pressure drop across tube banks can be calculated from the formulas and graphs of Section 4.5. For a simple cross-flow exchanger, these data suffice. However, a complete calculation for the shell side of a shell-and-tube exchanger requires consideration of additional pressure drops across baffle windows and in the inlet and exit zones of the shell. Empirical formulas are used for this purpose. Examples may be found in handbooks, but often such data are company-proprietary, pertaining only to the manufacturer's own shell and baffle configurations.

Flow through Packed Beds

Packed beds are used in regenerators and also in some recuperators, such as the perforated-plate exchanger described in Example 4.14. Section 4.5 gives some appropriate formulas for pressure drop. The chemical engineering literature is a good source of additional data, owing to the extensive use of packed beds for mass exchangers.

Compact Heat Exchangers

Pressure drop is usually an important constraint on compact heat exchanger design, and for gas flows, these exchangers tend to have a large frontal area and short flow length. Figure 8.20 shows a schematic of a compact heat exchanger core. Following Kays and London [5], the total pressure drop between sections 1 and 2, ΔP, is written as the contraction pressure drop ΔP_{in} plus the core pressure drop ΔP_{core} minus the expansion pressure recovery ΔP_{out}:

$$\Delta P = \Delta P_{in} + \Delta P_{core} - \Delta P_{out} \tag{8.49}$$

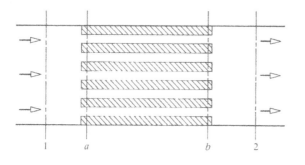

Figure 8.20   Schematic of a compact heat exchanger core.

The *inlet pressure drop* can be written as the sum of the pressure drop due to flow area change for an inviscid fluid, plus the irreversible pressure loss due to viscous effects. Assuming a constant density, since the pressure change is usually small compared to the total pressure,

$$\Delta P_{\text{in}} = \frac{1}{2}\rho_a V_a^2 \left[1 - \left(\frac{A_c}{A_{\text{fr}}} \right)^2 \right] + \frac{1}{2}\rho_a V_a^2 K_c \tag{8.50}$$

where A_c is the cross-sectional area for flow in the core, A_{fr} is the frontal area, and K_c is the *contraction coefficient*. Similarly, the outlet pressure rise is the sum of the pressure rise due to flow area change for an inviscid fluid, minus the pressure loss due to viscous effects. If A_c is constant and A_{fr} is the same at the inlet and outlet,

$$\Delta P_{\text{out}} = \frac{1}{2}\rho_b V_b^2 \left[1 - \left(\frac{A_c}{A_{\text{fr}}} \right)^2 \right] - \frac{1}{2}\rho_b V_b^2 K_e \tag{8.51}$$

where K_e is the *expansion coefficient*. There are two factors that contribute to the *core pressure drop*. First, there is the form and viscous drag of the heat transfer surface, and second, there is the pressure drop required to accelerate the fluid:

$$\Delta P_{\text{core}} = f\frac{1}{2}\rho_m V_m^2 (L/D_h) + (\rho_b V_b^2 - \rho_a V_a^2) \tag{8.52}$$

where ρ_m and V_m are suitable average values through the core, and L/D_h can be written as $A/4A_c$, where $A = \mathscr{P}L$ is the transfer surface area.

Introducing the mass velocity in the core, G [kg/m^2 s],

$$G = \rho_a V_a = \rho_m V_m = \rho_b V_b \tag{8.53}$$

and for a core-to-frontal area ratio $\sigma = A_c/A_{\text{fr}} < 1$,

$$V_a = \frac{V_1}{\sigma}; \quad V_b = \frac{V_2}{\sigma} \tag{8.54}$$

Then, assuming $\rho_1 = \rho_a$ and $\rho_2 = \rho_b$,

$$G = \rho_1 \frac{V_1}{\sigma} = \rho_2 \frac{V_2}{\sigma} \tag{8.55}$$

Substituting Eqs. (8.50) through (8.52) in Eq. (8.49) and rearranging gives the *exchanger pressure drop equation*:

$$\frac{\Delta P}{P_1} = \frac{G^2}{2\rho_1 P_1} \left[(1 - \sigma^2 + K_c) + \frac{f}{4} \frac{\rho_1}{\rho_m} \frac{A}{A_c} + 2 \left(\frac{\rho_1}{\rho_2} - 1 \right) - \frac{\rho_1}{\rho_2}(1 - \sigma^2 - K_e) \right] \tag{8.56}$$

The contraction and expansion coefficients, K_c and K_e, are functions of geometry and, to a lesser extent, functions of the Reynolds number in the core. Figure 8.21 shows some sample data. The contraction coefficient K_c also accounts for momentum changes associated with the developing velocity profile in the hydrodynamic entrance length: thus, the friction factor f in Eq. (8.56) is evaluated for fully developed hydrodynamic conditions.

For flow across tube banks, it is a common practice to account for inlet and outlet viscous pressure losses in the friction factor f. Then, setting $K_c = K_e = 0$ in Eq. (8.56) and rearranging gives

$$\frac{\Delta P}{P_1} = \frac{G^2}{2\rho_1 P_1} \left[\frac{f}{4} \frac{A}{A_c} \frac{\rho_1}{\rho_m} + (1 + \sigma^2) \left(\frac{\rho_1}{\rho_2} - 1 \right) \right] \tag{8.57}$$

For matrix-type surfaces, Eq. (8.56) also applies, but with σ replaced by the void fraction ε_v. Notice that in these equations, the frictional pressure drop $\Delta P/P_1$ is expressed in terms of a multiplier of the ratio of dynamic head to static head:

$$\frac{G^2}{2\rho_1 P_1} = \frac{V_a^2/2g}{P_1/\rho_1 g} = \frac{\text{Dynamic head}}{\text{Static head}}$$

The appropriate value of the mean density ρ_m to be used in Eq. (8.56) or (8.57) depends on the variation of temperature through the exchanger. When the capacity ratio $R_c \simeq 1$, a simple average of ρ_1 and ρ_2 can be used for all configurations except parallel flow.

In some applications, such as an automobile radiator mounted at the front of the vehicle, the total pressure drop is as given by Eq. (8.56). However, in most applications, the working fluids are ducted to the exchanger and must be distributed through a *manifold* or *header* to the core. Not only must the header provide a uniform distribution of fluid to the core, but it must also provide a transition from the duct cross-sectional area to the exchanger frontal area with a minimum pressure loss. When there is a large mismatch of duct and exchanger cross-sectional areas, the pressure losses in the headers may exceed those of the core. In such situations, an optimal design of the exchanger requires that header pressure drop be included as a constraint.

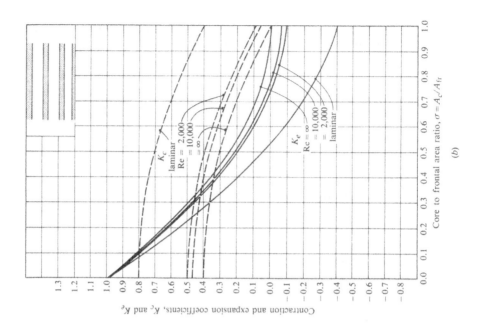

(a)

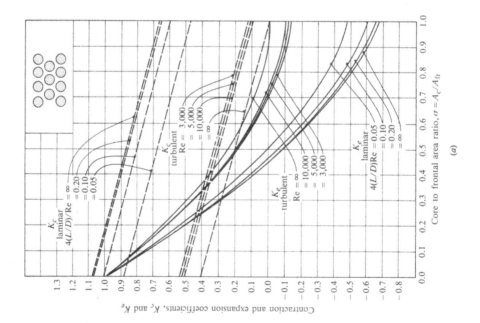

(b)

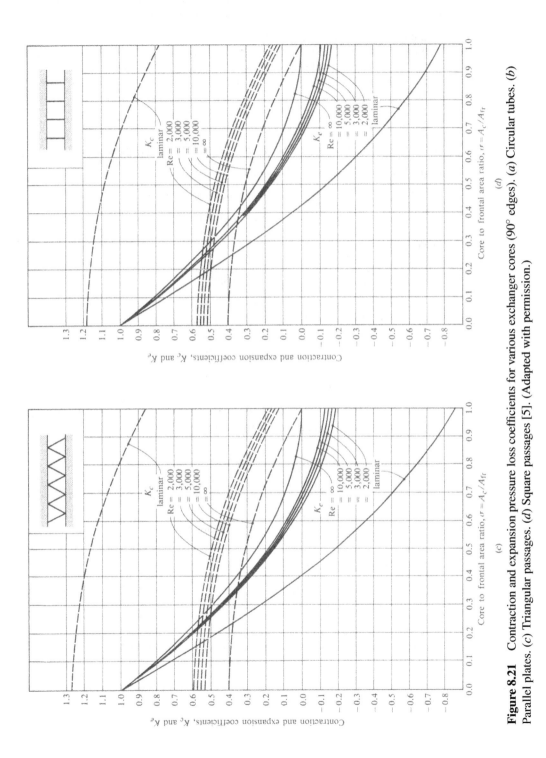

Figure 8.21 Contraction and expansion pressure loss coefficients for various exchanger cores (90° edges). (*a*) Circular tubes. (*b*) Parallel plates. (*c*) Triangular passages. (*d*) Square passages [5]. (Adapted with permission.)

EXAMPLE 8.11   Pressure Drop in a Multiple-Tube Exchanger

A multiple-circular-tube exchanger is constructed from 100 1 m-long, 6 mm–O.D., 0.5 mm-wall-thickness tubes in a square array, with a pitch of 8 mm. The frontal area of the exchanger is 0.0064 m². The tube walls are maintained at 350 K by condensing steam, while 5.66×10^{-3} kg/s of helium enters the tubes at 50 K and 15 kPa pressure. Determine the pressure drop of the helium flow.

Solution

Given: Multiple-circular-tube exchanger to heat helium by condensing steam.

Required: Pressure drop of helium flow.

Equation (8.56) gives the pressure drop. The cross-sectional area for helium flow is

$$A_c = (100)(\pi/4)(0.006 - 0.001)^2 = 0.00196 \text{ m}^2$$

The inlet helium density is

$$\rho_1 = \frac{PM}{\mathscr{R}T} = \frac{(15{,}000)(4)}{(8314)(50)} = 0.1443 \text{ kg/m}^3$$

To calculate the outlet helium density, we must estimate the outlet temperature. The number of transfer units and effectiveness of this single-stream exchanger are thus required. The Reynolds number is

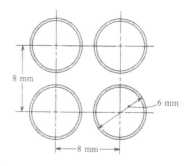

$$\text{Re} = \frac{\rho V D}{\mu} = \frac{\dot{m}D}{\mu A_c}$$

We will assume that the average bulk temperature of the helium is 200 K; then, from Table A.7, $\mu = 15.6 \times 10^{-6}$ kg/m s, $k = 0.116$ W/m K, $c_p = 5200$ J/kg K.

$$\text{Re} = \frac{(5.66 \times 10^{-3})(0.005)}{(15.6 \times 10^{-6})(0.00196)} = 926$$

The flow is laminar and the wall temperature is uniform; thus, from Table 4.5, Nu = 3.657.

$$h_c = \frac{k \, \text{Nu}}{D} = \frac{(0.116)(3.657)}{(0.005)} = 84.8 \text{ W/m}^2 \text{ K}$$

The heat transfer area is $A = (100)(\pi)(0.005)(1) = 1.57 \text{ m}^2$.

$$N_{tu} = \frac{h_c A}{\dot{m}c_p} = \frac{(84.8)(1.57)}{(5.66 \times 10^{-3})(5200)} = 4.52$$

Thus, the effectiveness is almost unity, and $T_{\text{out}} \simeq 350$ K, $T_{\text{avg}} \simeq (350 + 50)/2 = 200$ K, as assumed. Also, ignoring the entrance effect to calculate h_c is justified.

To use Eq. (8.56), we proceed as follows.

$$G = \frac{\dot{m}}{A_c} = \frac{5.66 \times 10^{-3}}{0.00196} = 2.89 \text{ kg/m}^2 \text{ s}$$

$$\frac{G^2}{2P_1\rho_1} = \frac{(2.89)^2}{2(15,000)(0.1443)} = 1.926 \times 10^{-3}$$

$$f = \frac{64}{Re_D}\left(\frac{T_s}{T_b}\right)^{1.0} = \left(\frac{64}{926}\right)\left(\frac{350}{200}\right) = 0.121 \quad \text{(using Tables 4.5 and 4.6)}$$

$$\frac{A}{A_c} = \frac{1.57}{0.00196} = 801$$

An iterative procedure is now required. After a few trials, $P_2 = 13.4$ kPa is seen to satisfy Eq. (8.56):

$$\rho_2 = \frac{P_2 M}{\mathscr{R}T_2} = \frac{(13,400)(4)}{(8314)(350)} = 0.0184 \text{ kg/m}^3$$

$$\rho_m = \frac{1}{2}(\rho_1 + \rho_2) = \frac{1}{2}(0.1443 + 0.0184) = 0.0814 \text{ kg/m}^3$$

$$\frac{f}{4}\frac{\rho_1}{\rho_m}\frac{A}{A_c} = \frac{0.121}{4}\left(\frac{0.1443}{0.0814}\right)(801) = 42.9$$

$$\sigma = \frac{A_c}{A_{fr}} = \frac{0.00196}{0.0064} = 0.306$$

$$2\left(\frac{\rho_1}{\rho_2} - 1\right) = 2\left(\frac{0.1443}{0.0184} - 1\right) = 13.7$$

$$\frac{4(L/D)}{Re} = \frac{4(1/0.005)}{926} = 0.864$$

From Fig. 8.21a, $K_c = 0.95, K_e = 0.27$.

$$(1 - \sigma^2 + K_c) = (1 - 0.306^2 + 0.95) = 1.9$$

$$\frac{\rho_1}{\rho_2}(1 - \sigma^2 - K_e) = \frac{0.1443}{0.0184}(1 - 0.306^2 - 0.27) = 5.0$$

Substituting in Eq. (8.56),

$$\frac{\Delta P}{P_1} = 1.926 \times 10^{-3}(1.9 + 42.9 + 13.7 - 5.0) = 0.103$$

$$\Delta P = (15.0)(0.103) = 1.55 \text{ kPa}$$

$$P_2 = P_1 - \Delta P = 15.0 - 1.55 \simeq 13.4 \text{ kPa}$$

Comments

1. The acceleration pressure drop is about one-third of the core viscous pressure drop in this situation.

2. Notice that the friction factor f varies appreciably along the exchanger, so that its evaluation at T_{avg} may be inadequate. A numerical integration of a differential momentum balance would give a more reliable result.

8.6.2 Thermal-Hydraulic Exchanger Design

Thermal-hydraulic design refers to the task of sizing a heat exchanger to give a required heat transfer effectiveness subject to a pressure drop constraint on one or both streams. The nature of the task depends on the geometric flow configuration of the exchanger, and it can also depend on whether the flow is laminar or turbulent; this is best illustrated by specific examples. In the case of the twin-tube exchangers to be covered in Examples 8.12 and 8.13, we will find that it is not possible to make an independent choice of the characteristic dimension of the heat transfer surface. On the other hand, for the plate-fin exchanger in Example 8.14, we will be free to choose this dimension, giving a family of exchangers that meet the heat transfer and pressure drop requirements. Criteria for choice of the characteristic dimension of the heat transfer surface, and for the choice of a particular type of surface, are discussed in Section 8.6.3.

EXAMPLE 8.12   A Laminar-Flow Twin-Tube Exchanger

A counterflow heat exchanger for a hydrogen cryogenic refrigeration system is to be fabricated by welding two steel tubes together. Each stream has a flow rate of 6.0×10^{-6} kg/s. The cold stream enters at 11 K and must be heated to 300 K. The hot stream enters at 310 K. The average pressure on the cold side is 667 Pa; on the hot side, it is 10,130 Pa. The exchanger is to be reversible; that is, cold and hot streams are passed alternately on each side. If the allowable pressure drop on the cold side is 80 Pa, determine suitable dimensions for the exchanger.

Solution

Given: Counterflow twin-tube exchanger to heat 6.0×10^{-6} kg/s of hydrogen from 11 K to 300 K, with an allowable pressure drop of 80 Pa.

Required: Tube inside diameter and exchanger length.

Assumptions: 1. Negligible axial conduction along the tube walls.

2. Fully developed laminar flow.

3. Balanced flow, $(\dot{m}c_p)_H = (\dot{m}c_p)_C$.

For an assumed balanced-flow exchanger, the required effectiveness is

$$\varepsilon = \frac{T_{C,\text{out}} - T_{C,\text{in}}}{T_{H,\text{in}} - T_{C,\text{in}}} = \frac{300 - 11}{310 - 11} = 0.9666$$

and from Eq. (8.45), the required number of transfer units is

$$N_{\text{tu}} = \frac{\varepsilon}{1 - \varepsilon} = \frac{0.9666}{1 - 0.9666} = 28.9$$

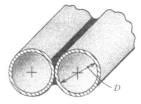

By definition,

$$N_{tu} = \frac{U\mathscr{P}L}{C_{min}}; \quad C_{min} = C_C = C_H = \dot{m}c_p$$

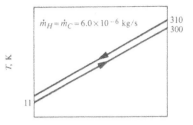

Solving for L,

$$L = \frac{\dot{m}c_p N_{tu}}{U\mathscr{P}}$$

Properties for both streams will be evaluated at an approximate average temperature of 160 K; from Table A.7, $c_p = 16,120$ J/kg K, $k = 0.131$ W/m K, $\mu = 5.84 \times 10^{-6}$ kg/m s.

$$L = \frac{(6.0 \times 10^{-6})(16,120)(28.9)}{U\mathscr{P}} = \frac{2.80}{U\mathscr{P}}$$

In order to proceed, we need to estimate the wall fin effectiveness: a value of 70% might be appropriate. Then, neglecting the resistance of the tubes where they are joined,

$$\frac{1}{U\mathscr{P}} = \frac{1}{0.7h_{c,H}\pi D} + \frac{1}{0.7h_{c,C}\pi D}$$

$$U\mathscr{P} = 0.35h_c\pi D$$

For laminar flow, the heat transfer coefficient can be obtained from Table 4.5. Since $\Delta T = T_H - T_C$ is almost constant along the exchanger, the wall heat flux can be taken to be constant, giving Nu = 4.364.

$$h_c = \text{Nu}(k/D) = 4.364(k/D)$$

$$U\mathscr{P} = (0.35)(4.364)(0.131)(\pi) = 0.629 \text{ W/m K}$$

$$L = \frac{2.80}{U\mathscr{P}} = \frac{2.80}{0.629} = 4.45 \text{ m}$$

The pressure drop is calculated using the friction factor:

$$\Delta P = f\left(\frac{L}{D}\right)\frac{1}{2}\rho V^2$$

From Table 4.5,

$$f = \frac{64}{\text{Re}_D} = \frac{64\mu}{\rho V D} \quad \text{and} \quad V = \frac{\dot{m}}{\rho A_c} = \frac{4\dot{m}}{\rho\pi D^2}$$

Hence if the acceleration pressure drop is neglected,

$$\Delta P = \frac{128L\dot{m}\mu}{\pi D^4\rho}; \quad \text{solving,} \quad D = \left(\frac{128L\dot{m}\mu}{\pi\Delta P\rho}\right)^{1/4}$$

On the cold side, the average pressure is 667 Pa,

$$\rho = \frac{PM}{\mathscr{R}T} = \frac{(667)(2.016)}{(8314)(160)} = 1.01 \times 10^{-3} \text{ kg/m}^3$$

$$D = \left[\frac{128(4.45)(6.0 \times 10^{-6})(5.84 \times 10^{-6})}{\pi(80)(1.01 \times 10^{-3})} \right]^{1/4} = 16.7 \times 10^{-3} \text{ m (16.7 mm)}$$

Thus, our preliminary estimate of the exchanger dimensions is a length of 4.45 m and a tube inner diameter of 16.7 mm. However, a number of the assumptions made in the calculation need to be validated.

1. *Is the flow laminar as assumed?* The Reynolds number is

$$\text{Re}_D = \frac{\rho V D}{\mu} = \frac{(\dot{m}/A_c)D}{\mu} = \frac{4\dot{m}}{\pi \mu D} = \frac{4(6.0 \times 10^{-6})}{\pi(5.84 \times 10^{-6})(0.0167)} = 77 \quad \text{(laminar)}$$

2. *Are entrance effects negligible?* The L/D ratio is

$$L/D = 4.45/0.0167 = 266$$

Equation (4.49) gives the thermal entrance length as

$$\frac{L_{\text{eh}}}{D} \simeq 0.033 \text{Re}_D \text{Pr} = (0.033)(77)(0.69) \simeq 2$$

so that the assumption of fully developed laminar flow is excellent.

3. *What is the tube wall fin effectiveness?* A 1 mm wall thickness will be used to check the assumed value of the tube wall fin effectiveness. The effective length is $\pi D/2 \simeq (\pi)(16.7 \times 10^{-3})/2 = 0.0262$ m. The heat transfer coefficient is $h_c = 4.364(k/D) = 4.364(0.131/0.0167) = 34.2$ W/m^2 K, and $k = 17$ W/m K will be used for the steel. Then $\beta = (h_c \mathscr{P}/kA_c)^{1/2} = [(34.2)/(17)(0.001)]^{1/2} = 44.9$ m^{-1}, $\beta L = (44.9)(0.0262) = 1.18, \eta_f = (1/\beta L) \tanh \beta L = (1/1.18) \tanh 1.18 = 0.70$, as assumed. (Fortunately! Had we not found $\eta_f = 0.7$, a second iteration would be required.)

4. *Is the acceleration pressure drop negligible?* The pressure drop required to accelerate the cold stream is equal to the change in momentum flux between inlet and outlet:

$$\Delta P(\text{Acceleration}) = (\rho V^2)_{\text{out}} - (\rho V^2)_{\text{in}} = G^2 \left(\frac{1}{\rho_{\text{out}}} - \frac{1}{\rho_{\text{in}}} \right)$$

$$G = \frac{\dot{m}}{A_c} = \frac{6.0 \times 10^{-6}}{(\pi/4)(16.7 \times 10^{-3})^2} = 0.0274 \text{ kg/m}^2 \text{ s}$$

If the acceleration pressure drop is small, $P_{\text{in}} \simeq 667 + 40 = 707$ Pa, $P_{\text{out}} \simeq 667 - 40 = 627$ Pa.

$$\rho_{\text{in}} = \frac{(707)(2.016)}{(8314)(11)} = 0.0156 \text{ kg/m}^3; \quad \rho_{\text{out}} = \frac{(627)(2.016)}{(8314)(300)} = 0.000507 \text{ kg/m}^3$$

$$\Delta P(\text{Acceleration}) = (0.0274)^2 \left(\frac{1}{0.000507} - \frac{1}{0.0156} \right) = 1.4 \text{ Pa}$$

which is indeed small compared to the viscous pressure drop of 80 Pa.

Comments

1. Notice that the diameter D cancels in the $U\mathscr{P}$ product, so that the required number of transfer units fixes the exchanger length, independent of tube diameter. The pressure drop constraint then fixes a minimum inside tube diameter of 16.7 mm.

2. The student is left to explore how to package and insulate the 4.45 m-long double tube.

3. The effect of tube-wall axial conduction is dealt with in more advanced texts.

4. For a design problem, it is usually necessary to make a number of initial assumptions (sometimes bold!) to be able to calculate a tentative design. Subsequently, the validity of the initial assumptions can be checked and changes made, if necessary. Such an approach was demonstrated in this example. The student should learn not to agonize over the initial assumptions: only when numerical values for a tentative design are established can the validity of the assumptions be established.

EXAMPLE 8.13 A Turbulent-Flow Twin-Tube Exchanger

A counterflow twin-tube heat exchanger is to be used for two flows of air at 2×10^{-3} kg/s. The cold stream enters at 280 K and must be heated to 330 K; the hot stream enters at 340 K. If the average pressure in each stream is 1 atm and the allowable pressure drop for the cold stream is 9000 Pa, determine suitable dimensions. Copper tubes should be used to give a high tube wall fin effectiveness.

Solution

Given: Counterflow twin-tube exchanger to heat 2×10^{-3} kg/s air from 280 K to 330 K, with an allowable pressure drop of 9000 Pa.

Required: Tube inside diameter and exchanger length.

Assumptions: 1. Fully developed turbulent flow.
2. A tube wall fin effectiveness of 100%.
3. Average pressure of 1 atm for both streams
4. Balanced flow, $(\dot{m}c_p)_H = (\dot{m}c_p)_C$

Proceeding as in Example 8.12,

$$\varepsilon = \frac{T_{C,\text{out}} - T_{C,\text{in}}}{T_{H,\text{in}} - T_{C,\text{in}}} = \frac{330 - 280}{340 - 280} = 0.833$$

$$N_{\text{tu}} = \frac{\varepsilon}{1 - \varepsilon} = \frac{0.833}{1 - 0.833} = 5.00$$

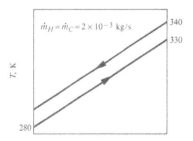

Evaluate properties for both air streams at 310 K:
$c_p = 1005$ J/kg K, $k = 0.0274$ W/m K, $\rho = 1.141$ kg/m^3, $\mu = 18.87 \times 10^{-6}$ kg/m s, Pr = 0.69.

$$L = \frac{\dot{m}c_p N_{\text{tu}}}{U\mathscr{P}} = \frac{(2.0 \times 10^{-3})(1005)(5.00)}{U\mathscr{P}} = \frac{10.05}{U\mathscr{P}}$$

$U\mathscr{P} \simeq (1/2)h_c \pi D$ for a tube wall fin effectiveness of unity.

The high flow rate indicates that the flow will be turbulent. For simplicity, use Eq. (4.44):

$$h_c = \text{Nu}_D(k/D) = 0.023\text{Re}_D^{0.8}\text{Pr}^{0.4}(k/D)$$

$$= 0.023\left(\frac{4\dot{m}}{\pi\mu D}\right)^{0.8}\text{Pr}^{0.4}\left(\frac{k}{D}\right)$$

$$= 0.023\left(\frac{(4)(2\times10^{-3})}{\pi(18.87\times10^{-6})}\right)^{0.8}(0.69)^{0.4}(0.0274)D^{-1.8}$$

$$= 2.75\times10^{-2}D^{-1.8}$$

$$U\mathscr{P} = \frac{1}{2}(2.75\times10^{-2}D^{-1.8})(\pi D) = 4.32\times10^{-2}D^{-0.8}$$

$$L = \frac{10.05}{4.32\times10^{-2}D^{-0.8}} = 233D^{0.8} \tag{1}$$

The pressure drop will give a second equation relating L and D:

$$\Delta P = f\left(\frac{L}{D}\right)\frac{1}{2}\rho V^2$$

Equation (4.43) gives $f = 0.184\text{Re}_D^{-0.2}$ for turbulent flow.

$$\Delta P = 0.184\left(\frac{4\dot{m}}{\pi\mu D}\right)^{-0.2}\left(\frac{L}{D}\right)\left(\frac{1}{2}\right)\rho\left(\frac{4\dot{m}}{\pi D^2\rho}\right)^2$$

$$9000 = 0.184\left[\frac{4(2.0\times10^{-3})}{\pi(18.87\times10^{-6})}\right]^{-0.2}\left(\frac{1}{2}\right)(1.141)\left[\frac{4(2.0\times10^{-3})}{\pi(1.141)}\right]^2\frac{L}{D^{4.8}}$$

$$L = 4.59\times10^{10}D^{4.8} \tag{2}$$

Solving Eqs. (1) and (2) gives the required dimensions of the exchanger:

$$233D^{0.8} = 4.59\times10^{10}D^{4.8}; \quad D = 8.44\times10^{-3} \text{ m} \simeq 8.5 \text{ mm}$$

$$L = 5.11 \text{ m}$$

We can now calculate the Reynolds number to check if the flow is turbulent, as assumed.

$$\text{Re}_D = \frac{4\dot{m}}{\pi\mu D} = \frac{4(2.0\times10^{-3})}{\pi(18.87\times10^{-6})(8.44\times10^{-3})} = 16,000 \quad \text{(turbulent)}$$

Comments

1. Equation (4.43) for f is not very accurate at this low Reynolds number; the problem could be reworked using Eq. (4.42).

2. The length-to-diameter ratio is $L/D = 5.11/0.00844 = 605$, which is much greater than the value of 30 required for fully developed heat transfer in a turbulent air flow.

3. In contrast to the laminar-flow exchanger of the previous example, heat transfer considerations alone do not fix the length of the exchanger. The heat transfer and pressure drop requirements together determine unique values of L and D.

EXAMPLE 8.14 A Cross-Flow Plate-Fin Exchanger

A cross-flow plate-fin heat exchanger with square cross-section passages is to be used for balanced flow of air, providing an effectiveness of 70%. When the flow rate is 0.330 kg/s, the allowable pressure drop is 330 Pa. Determine suitable dimensions for the exchanger. Evaluate fluid properties at 320 K and 1 atm, and neglect the thermal resistance of the plates and fins.

Solution

Given: Cross-flow plate fin exchanger with square passages for balanced flow of air.

Required: Cross section and length to give $\varepsilon = 0.7$ at $\Delta P = 330$ Pa.

Assumptions: 1. Evaluate properties at 320 K and 1 atm.
2. Negligible thermal resistance of the plates and fins.
3. Negligible acceleration, entrance and exit pressure drops.

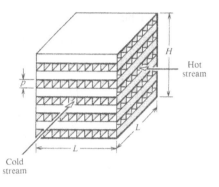

From Table A.7, air properties at 320 K and 1 atm are: $k = 0.0281$ W/m K, $\rho = 1.106$ kg/m^3, $c_p = 1006$ J/kg K, $\mu = 19.29 \times 10^{-6}$ kg/m s, Pr $= 0.69$.

This configuration corresponds to item 4 in Table 8.3a (cross-flow, both streams unmixed); solving by trial and error for $\varepsilon = 0.7$ gives $N_{tu} = 3.345$ (or use HEX1).

$$L = \frac{\dot{m}c_p N_{tu}}{U\mathscr{P}} = \frac{(0.33)(1006)(3.345)}{U\mathscr{P}} = \frac{1110}{U\mathscr{P}}$$

Let the pitch of the plates be p, and assume that the wall thickness $t \ll p$. Also, we will assume that the fin effectiveness is unity. Then the perimeter for one passage is $4p$, and the number of passages for each stream is $(1/2)(H/p)(L/p)$. Thus, $\mathscr{P} = 2HL/p$

$$L = \frac{1110}{(2HL/p)U}; \quad L^2 H = \frac{555p}{U} \tag{1}$$

The pressure drop is $\Delta P = f(L/D_h)(1/2)\rho V^2$; $D_h = p$, the side of a square passage; $V = \dot{m}/\rho A_c = 2\dot{m}/\rho LH$.

$$\Delta P = f\left(\frac{L}{p}\right)\frac{1}{2}\frac{4\dot{m}^2}{\rho L^2 H^2}; \quad LH^2 = \frac{2f\dot{m}^2}{\rho p \Delta P} \tag{2}$$

Since the perimeter is the same for both streams,

$$\frac{1}{U} \simeq \frac{1}{h_{c,H}} + \frac{1}{h_{c,C}}; \quad U = \frac{1}{2}h_c$$

Assume laminar flow, fully developed thermal conditions, and a uniform wall heat flux; then

Table 4.5 gives

$$h_c = \text{Nu}(k/D_h) = 3.6(k/p)$$

$$U = \left(\frac{1}{2}\right)(3.6)(k/p) = 1.8(k/p)$$

Substituting in Eq. (1) gives

$$L^2 H = \frac{555p}{1.8(k/p)} = \frac{555p^2}{(1.8)(0.0281)} = 1.097 \times 10^4 p^2 \,[\text{m}^3] \tag{3}$$

Table 4.5 also gives $f = 57/\text{Re}_{D_h}$.

$$\text{Re}_{D_h} = \frac{Vp\rho}{\mu} = \frac{(2\dot{m}/\rho LH)p\rho}{\mu} = \frac{2\dot{m}p}{LH\mu}$$

$$f = \frac{57LH\mu}{2\dot{m}p}$$

Substituting in Eq. (2) gives

$$H = \frac{57\mu\dot{m}}{\rho p^2 \Delta P} = \frac{(57)(19.3 \times 10^{-6})(0.33)}{(1.106)(330)p^2} = \frac{0.995 \times 10^{-6}}{p^2} \,[\text{m}] \tag{4}$$

Choosing various values of pitch p and substituting into Eqs. (3) and (4) gives the following results. Values of Re_{D_h} are shown to confirm that the flow is laminar.

| p | H | L | Re_{D_h} |
mm	m	m	–
1.0	0.995	0.105	327
1.2	0.691	0.152	391
1.4	0.508	0.206	456
1.6	0.389	0.268	524
1.8	0.307	0.340	589
2.0	0.249	0.420	655

Comments

1. Notice that Eqs. (3) and (4) are two equations in three unknowns: p, H, and L. Thus, we were free to vary p and then determine resulting values of H and L. All the exchangers in the table satisfy the heat transfer and pressure drop specification. In fact, all have the same frontal area, $LH = 0.1045$ m^2, as can be deduced by eliminating p^2 between Eqs. (3) and (4). Often a nearly cubical shape will be preferred to simplify the manifold construction, which indicates a pitch somewhat less than 1.8 mm.

2. An essential difference between the twin-tube exchangers of the previous examples and the plate-fin exchanger is clear. In the case of the twin-tube exchangers, the characteristic dimensions of the flow cross-sectional area and the heat transfer surface were the same, that is, the tube diameter D. The result was that heat transfer and pressure drop requirements completely determined the exchanger. In contrast, the characteristic lengths for the plate-fin exchanger were L and H for the flow area and p for the heat transfer surface, which are independent and gave an extra degree of freedom with which to meet the heat transfer and pressure drop requirements.

8.6.3 Surface Selection for Compact Heat Exchangers

A large range of heat transfer surfaces is used in compact heat exchangers, including smooth tubes, externally finned tubes, internally enhanced tubes, and plate-fin surfaces. Examples are shown in Fig. 8.22. Useful insight into a strategy for selecting a surface appropriate to a given application can be obtained by writing down an approximate equation for the core velocity. Following Kays and London [5], we neglect the acceleration and inlet and outlet pressure drops, and we assume a fin effectiveness of unity. Then, from Eq. (8.56),

$$\frac{V_a^2/2}{P_1/\rho_1} \simeq \left(\frac{\Delta P/P_1}{N_{\text{tu}}^1} \right) \frac{\rho_m}{\rho_1} \frac{4St}{f} \tag{8.58}$$

where N_{tu}^1 is the number of transfer units for the one side of the exchanger under consideration. For a heat transfer area $A = \mathscr{P}L$ and flow cross-sectional area A_c, it is defined as

$$N_{\text{tu}}^1 = \frac{h_c A}{\dot{m} c_p} = St \frac{A}{A_c} \tag{8.59}$$

since the Stanton number is $St = h_c/c_p \rho V = h_c/c_p(\dot{m}/A_c)$. The number of transfer units of the exchanger is related to the one-side N_{tu} as follows:

$$N_{\text{tu}} = \frac{UA}{C_{\min}}$$

$$\frac{1}{UA} = \frac{1}{(h_c A)_H} + \frac{1}{(h_c A)_C} \quad \text{for a negligible wall and fin resistance}$$

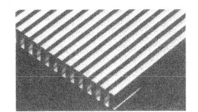

(a)

(b)

(c)

Figure 8.22 Some plate-fin surfaces used in compact heat exchangers. (a) Plain, (b) Serrated, (c) Perforated. (Photographs courtesy ALTEC, Inc., La Crosse, W.I.)

Substituting Eq. (8.59) and rearranging,

$$\frac{1}{N_{\text{tu}}} = \frac{1}{(C_H/C_{\min})N_{\text{tu}}^{1,H}} + \frac{1}{(C_C/C_{\min})N_{\text{tu}}^{1,C}} \tag{8.60}$$

For a given effectiveness ε and capacity flow rates, N_{tu} can be calculated and then N_{tu}^1 obtained from Eq. (8.60) by estimating or specifying a desirable ratio of hot side-to-cold side thermal resistance.

The utility of Eq. (8.58) is based on the fact that, for a given fluid, St/f is a weak function of Reynolds number and, also, does not vary more than about fivefold for a wide range of surface types. Figure 8.23 shows representative data that allow St/f to be readily estimated for a preliminary calculation of the frontal area. As a first approximation, the Reynolds number dependence can be ignored and the result used to estimate Re for a second approximation. Since the velocity is proportional to the square root of St/f, only a $2\frac{1}{2}$-fold variation of velocity can be achieved by surface type selection. Also, if a small frontal area is a design requirement, Fig. 8.23 allows an appropriate surface type to be selected: the higher the value of St/f, the smaller the flow area A_c required. Since $A_{\text{fr}} = A_c/\sigma$, the frontal area will also be smaller, provided σ is not anomalously large. The ratio St/f can be viewed as an approximate indicator of the merit of the heat transfer surface. However, the fin effectiveness was assumed to be unity in this analysis, so that for values of η_f significantly lower than unity, the merit of a heat transfer surface must be evaluated more carefully (see, for example, Soland et al. [6]). Table 8.4c in Section 8.6.5 gives examples of correlations for both f and St for use in computer calculations.

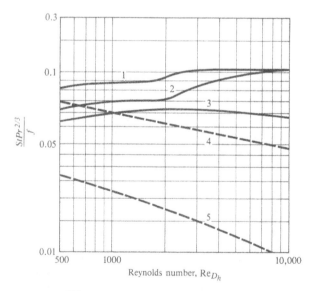

Figure 8.23 The ratio $\text{StPr}^{2/3}/f$ for five heat transfer surfaces. (1) Perforated plate-fin (13.95P). (2) Plain plate-fin (11.94T). (3) Louvered plate-fin (3/8–11.1). (4) Tube bank (S150.125). (5) Packed particle bed. [Numbers in parentheses are designations in *Compact Heat Exchangers*, W. M. Kays and A. L. London, McGraw-Hill, New York (any edition)].

EXAMPLE 8.15   Design of a Cross-Flow Plate-Fin Exchanger

Rework Example 8.14 using the equations developed in Section 8.6.3.

Solution

Given: Cross-flow plate-fin exchanger with square passages for balanced flow of air.

Required: Dimensions to give $\varepsilon = 0.7$ for $\Delta P = 330$ Pa.

Assumptions: 1. Properties evaluated at 320 K and 1 atm.

   2. Negligible thermal resistance of plates and fins.

   3. Negligible acceleration, entrance and exit pressure drops.

From Example 8.14, $\rho_m = \rho_1 = 1.106$ kg/m^3, Pr $= 0.69$, $P_1 = 1.013 \times 10^5$ Pa, and $N_{tu} = 3.345$. For balanced flow with $C_{min} = C_H = C_C$, Eq. (8.60) becomes

$$\frac{1}{N_{tu}} = \frac{1}{N_{tu}^{1,H}} + \frac{1}{N_{tu}^{1,C}} = \frac{2}{N_{tu}^1}$$

Hence, $N_{tu}^1 = 2N_{tu} = (2)(3.345) = 6.69$. Using Table 4.5,

$$\frac{St}{f} = \left(\frac{Nu}{Re_{D_h}Pr} \right)\left(\frac{1}{f} \right) = \left(\frac{3.6}{Re_{D_h}Pr} \right)\left(\frac{Re_{D_h}}{57} \right) = \frac{0.063}{Pr}$$

$$\frac{4\,St}{f} = \frac{(4)(0.063)}{0.69} = 0.365$$

Rearranging Eq. (8.58),

$$V_a^2 = \frac{2\Delta P}{\rho_1 N_{tu}^1}\frac{4\,St}{f} = \frac{(2)(330)(0.365)}{(1.106)(6.69)} = 32.6; \quad V_a = 5.71 \text{ m/s}$$

Using $\sigma = 0.5$ for thin fins and plates, the frontal velocity and area are

$$V_1 = (A_c/A_{fr})V_a = \sigma V_a = (0.5)(5.71) = 2.85 \text{ m/s}$$

$$A_{fr} = \frac{\dot{m}}{\rho_1 V_1} = \frac{0.330}{(1.106)(2.85)} = 0.1045 \text{ m}^2$$

At this point, we can see that specification of the *type* of heat transfer surface fixes the frontal area of the exchanger. To finalize the design, the pitch of the plates must be chosen and Eq. (3) of Example 8.14 can be used to obtain L and H.

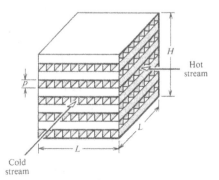

$$A_{fr} = LH = 0.1045 \text{ m}^2$$

$$L^2H = 1.097 \times 10^4 p^2 \text{ m}^3$$

Hence, $L = 1.050 \times 10^5 p^2$. For example, for $p = 1.8$ mm $= 1.8 \times 10^{-3}$ m, $L = 0.340$ m, and $H = 0.307$ m, as before.

Comments

1. In this calculation procedure, the effect of heat transfer surface *type* on exchanger frontal area can be easily seen. The ratio St/f depends on the type of surface, and A_{fr} changes accordingly. Note also that St/f is independent of Reynolds number, so no iteration is required.

2. The volume of the exchanger is $L^2 H$ and hence increases proportionally to pitch squared. But, as mentioned in Example 8.14, a nearly cubical shape is often desired to simplify header construction.

3. The plate and fin areas are $L^2 H / p$, which is proportional to pitch; hence, the weight of the exchanger core increases proportionally to pitch (for fixed plate and fin thickness).

8.6.4 Economic Analysis

The designer is often able to quantify both the costs and the benefits of a heat exchanger. Then optimization of the design requires maximization of the amount by which benefits exceed costs. Costs may be divided into three categories: *initial, annual,* and *operating* costs. The initial cost is the purchase price of the equipment. Annual costs include taxes, insurance, and maintenance. The major operating cost is usually the pump or fan power cost.

The initial cost, C_i, can be converted to an equivalent annual cost by assuming that the capital is borrowed from a bank at an annual interest rate i and must be repaid, or *amortized,* with n yearly payments.[2] The annual charge, or *annuity,* to repay a loan of $1 over n years is

$$r = \frac{i}{1 - [1/(1+i)]^n} \tag{8.61}$$

If the exchanger is expected to have a salvage price C_s at the end of its working lifetime, the *present value* of the salvage price should be subtracted from the capital cost. The present value of $1 received m years in the future is $[1/(1+i)]^m$; thus, the annual cost of the exchanger attributed to the initial cost is

$$\mathscr{C}_i = r \left[C_i - \left(\frac{1}{1+i} \right)^m C_s \right] \tag{8.62}$$

For example, the annual cost of a heat exchanger with initial cost $10,000 and a salvage value after 15 years of $1000, with a loan at 9% amortized over 15 years, is

$$\mathscr{C}_i = \frac{0.09}{1 - (1/1.09)^{15}} \left[10,000 - \left(\frac{1}{1.09} \right)^{15} 1000 \right]$$

$$= 0.124(10,000 - 275)$$

$$= \$1206$$

Notice that the salvage value has little impact on the annual cost.

---

[2] In this economic analysis, the following notation scheme is used: C is a cost [$], c is a unit cost, e.g., per unit heat transfer area [$/m^2], $\mathscr{C}$ is an annual cost [$/yr], and e is a unit annual cost, e.g., per unit heat transfer area [$/m^2 yr].

For preliminary design purposes, it is often sufficient to assume that the initial cost is linearly proportional to the heat transfer surface area $\mathscr{P}L$. Referring to Fig. 8.24,

$$C_i = C_{if} + c_i \mathscr{P}L \tag{8.63}$$

where C_{if} is a fixed cost accounting for installation and other setup expenses, and c_i is the cost per unit transfer area. Substituting Eq. (8.63) in Eq. (8.62) gives

$$\mathscr{C}_i = r \left[C_{if} + c_i \mathscr{P}L - \left(\frac{1}{1+i} \right)^m C_s \right] \tag{8.64}$$

Since taxes and insurance are usually related to purchase price, these annual costs can also be taken to be linearly proportional to the transfer area:

$$\mathscr{C}_t = \mathscr{C}_{tf} + c_t \mathscr{P}L \tag{8.65}$$

Adding Eqs. (8.64) and (8.65) gives the total annual cost, $\mathscr{C}_a$, as

$$\mathscr{C}_a = \left[rC_{if} - r \left(\frac{1}{1+i} \right)^m C_s + \mathscr{C}_{tf} \right] + (rc_i + c_t)\mathscr{P}L \tag{8.66}$$

The operating cost associated with the pump or fan power for each stream is

$$\mathscr{C}_p = \frac{e_p \tau}{\eta_p} \cdot \frac{\dot{m}\Delta P}{\rho} \tag{8.67}$$

where e_p [\$/W h] is the unit pumping power cost, τ is the operating time per year [h/yr], and η_p is the pump or fan efficiency (including the efficiencies of the motor and drive box). The pressure drop can be expressed as a linear function of exchanger length:

$$\Delta P = \Delta P_0 + f \left(\frac{L}{D_h} \right) \frac{1}{2}\rho V^2$$

where ΔP_0 accounts for inlet and outlet losses and acceleration of a gas. Substituting in Eq. (8.67) gives the annual power cost as

$$\mathscr{C}_p = \frac{e_p \tau}{\eta_p} \cdot \frac{\dot{m}\Delta P_0}{\rho} + \frac{e_p \tau}{\eta_p} f \left(\frac{L}{D_h} \right) \frac{\dot{m}V^2}{2}$$

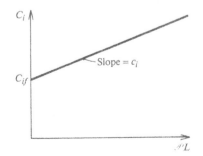

Figure 8.24 Initial cost C_i of a heat exchanger as a function of heat transfer surface area $\mathscr{P}L$.

or

$$\mathscr{C}_p = \mathscr{C}_{pf} + c_{pL}L \tag{8.68}$$

where c_{pL} is the annual cost of pumping power per unit length of exchanger.

In common process heating or heat recovery applications, the annual value of the benefit is directly proportional to the value of the heat transferred,

$$\mathscr{V} = \nu\dot{Q}\tau \tag{8.69}$$

where ν [\$/W h] is the value of the heat per watt hour. On the other hand, if the exchanger is used in a power cycle for generation of mechanical or electrical power, the available energy transfer (the integral of heat transfer times Carnot efficiency) is a more appropriate measure of value. In such applications, a careful analysis of the system using the second law of thermodynamics is advisable.

Balanced-Counter flow Exchangers

Optimization of an exchanger design is generally an involved task requiring the use of appropriate computational tools. However, the important issues involved can be illustrated by a simple example problem. Consider a balanced-counterflow exchanger for which initial and operating costs are linear with length. We will specify a heat transfer surface and the cross-sectional area for each stream. The effectiveness and heat transfer increase with exchanger length, but so do the initial and operating costs; the problem is to determine the exchanger length that maximizes the benefit-cost differential.

Substituting for $\dot{Q}$ in Eq. (8.69) using the definition of effectiveness gives

$$\mathscr{V} = \nu\dot{m}c_p(T_{H,\text{in}} - T_{C,\text{in}})\varepsilon\tau \tag{8.70}$$

For algebraic convenience, we will write the total annual costs $(\mathscr{C}_a + \mathscr{C}_p)$ as

$$\mathscr{C} = \mathscr{C}_f + cL \tag{8.71}$$

where $\mathscr{C}_f$ is the total of the fixed costs, and

$$c = (rc_i + c_t + c_{pL}/\mathscr{P})\mathscr{P} \tag{8.72}$$

is the total cost per unit length of exchanger. Figure 8.25 shows $\mathscr{V}$ and $\mathscr{C}$ plotted

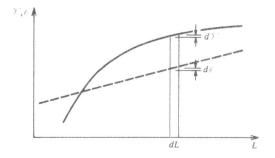

Figure 8.25   Annual benefit $\mathscr{V}$ and total annual costs $\mathscr{C}$ as a function of exchanger length.

versus L. The designer should maximize $(\mathcal{V} - \mathcal{C})$, that is, increase the exchanger length until the incremental annual cost $d\mathcal{C}$ equals the incremental annual value of the benefit $d\mathcal{V}$,

$$d\mathcal{V} = \nu \dot{m} c_p (T_{H,\mathrm{in}} - T_{C,\mathrm{in}}) \tau d\varepsilon \qquad (8.73)$$

But

$$d\varepsilon = \frac{d\varepsilon}{dN_{\mathrm{tu}}} \frac{dN_{\mathrm{tu}}}{dL} dL = \frac{d\varepsilon}{dN_{\mathrm{tu}}} \frac{U\mathcal{P}}{\dot{m}c_p} dL$$

Hence,

$$d\mathcal{V} = \nu \tau (T_{H,\mathrm{in}} - T_{C,\mathrm{in}}) U\mathcal{P} \frac{d\varepsilon}{dN_{\mathrm{tu}}} dL \qquad (8.74)$$

Also

$$d\mathcal{C} = c\,dL = (rc_i + c_t + c_{pL}/\mathcal{P})\mathcal{P}\,dL \qquad (8.75)$$

Equating Eqs. (8.74) and (8.75) gives

$$\frac{d\varepsilon}{dN_{\mathrm{tu}}} = \frac{(rc_i + c_t + c_{pL}/\mathcal{P})}{\nu \tau (T_{H,\mathrm{in}} - T_{C,\mathrm{in}})U} = \mathcal{E} \qquad (8.76)$$

where $\mathcal{E}$ is a dimensionless *thermoeconomic* parameter. Examination of Fig. 8.25 shows that an exchanger should not be used at all if the length is such that $\mathcal{V}$ is less than $\mathcal{C}$. Equation (8.45) gives the effectiveness of a balanced-counterflow exchanger as

$$\varepsilon = \frac{N_{\mathrm{tu}}}{1 + N_{\mathrm{tu}}}$$

Hence

$$\frac{d\varepsilon}{dN_{\mathrm{tu}}} = \frac{1}{(1 + N_{\mathrm{tu}})^2} \qquad (8.77)$$

Equating Eqs. (8.76) and (8.77) and solving for N_{tu} gives

$$N_{\mathrm{tu,opt}} = \frac{1 - \mathcal{E}^{1/2}}{\mathcal{E}^{1/2}}. \qquad (8.78)$$

The optimal length is then obtained from the definition $N_{\mathrm{tu}} = U\mathcal{P}L/\dot{m}c_p$.

Large integrated systems for heat and power supply with heat recovery exchangers are more difficult to optimize; see, for example, Linnhoff et al. [7,8].

EXAMPLE 8.16 Energy Conservation in a Brewery

In a bottle washing operation, 75°C wash water is dumped in a drain, and 80°C clean water is required. Currently, 20°C water is heated in an electric water heater, and the water requirement is 10,000 kg/h, 8 h/day, 250 days/yr. The cost of electricity is 8 cents/kW h. In order to conserve energy, it has been proposed that a counterflow heat exchanger be installed to preheat the water feed to the electric water heater. The projected cost of the unit is $30,000 plus $1,800 per square meter of exchanger surface. The interest rate to amortize the investment over 15 years is 9% per annum. Taxes and insurance are expected to have a fixed cost of $1100 per annum plus $70/yr per square meter of exchanger surface. The overall heat transfer coefficient can be taken as 1200 W/m² K. What is the optimal heat transfer area of the exchanger, and what are the corresponding net annual savings?

Solution

Given: Counterflow exchanger for preheating feed water.

Required: Optimal heat transfer area and net annual savings.

Assumptions: 1. Balanced flow.
2. The exchanger has a negligible salvage value.
3. The additional pumping power requirements are negligible.
4. Exchanger maintenance costs are negligible.

First we calculate the thermoeconomic parameter, $\mathscr{E}$. From Eq. (8.76),

$$\mathscr{E} = \frac{r c_i + c_t + c_{pL}/\mathscr{P}}{\nu\tau(T_{H,\text{in}} - T_{C,\text{in}})U}$$

$$c_i = \$1800/\text{m}^2 \quad \nu = 8\times 10^{-5} \ \$/\text{Wh}$$

$$c_t = \$70/\text{m}^2 \ \text{yr} \quad \tau = (250)(8) = 2000 \ \text{h/yr}$$

$$c_{pL} = 0 \quad U = 1200 \ \text{W/m}^2 \ \text{K}$$

$$T_{H,\text{in}} - T_{C,\text{in}} = 75 - 20 = 55 \ \text{K}$$

$$r = \frac{1}{1 - [1/(1+i)]^n} = \frac{0.09}{1 - (1/1.09)^{15}} = 0.124 \ \text{yr}^{-1}$$

$$\mathscr{E} = \frac{(0.124)(1800) + 70}{(8\times 10^{-5})(2000)(55)(1200)} = 0.0278$$

Substituting in Eq. (8.78),

$$N_{\text{tu,opt}} = \frac{1 - \mathscr{E}^{1/2}}{\mathscr{E}^{1/2}} = \frac{1 - (0.0278)^{1/2}}{0.0278^{1/2}} = 5.00$$

The heat exchanger area $\mathscr{P}L$ is found from the definition of N_{tu}:

$$\mathscr{P}L = \frac{\dot{m}c_p N_{\text{tu}}}{U} = \frac{(10,000/3600)(4180)(5.00)}{1200} = 48.4 \ \text{m}^2$$

Also, from Eq. (8.45), the effectiveness for a balanced-flow exchanger is

$$\varepsilon = \frac{N_{\text{tu}}}{1 + N_{\text{tu}}} = \frac{5.00}{1 + 5.00} = 0.833$$

The annual value of the energy saved is given by Eq. (8.70):

$$\mathscr{V} = e\dot{m}c_p(T_{H,\text{in}} - T_{C,\text{in}})\varepsilon\tau$$
$$= (8 \times 10^{-5})(10,000/3600)(4180)(55)(0.833)(2000)$$
$$= \$85,100/\text{yr}$$

If the salvage value of the exchanger is ignored, the total annual cost is given by Eq. (8.66) as

$$\mathscr{C}_a = [rC_{if} + \mathscr{C}_{tf}] + (rc_i + e_t)\mathscr{P}L$$
$$= [(0.124)(30,000) + 1100] + [(0.124)(1800) + 70](48.4)$$
$$= \$19,010/\text{yr}$$

The extra pumping power requirements for the waste stream will add negligibly to the operating cost and have been ignored. Thus, the net annual savings are

$$\mathscr{V} - \mathscr{C}_a = 85,100 - 19,010 = \$66,100/\text{yr}$$

Comments

1. The annual savings suggest that the exchanger is well worth installing.

2. Although the ratio of average annual savings to annual cost is $85,100/19,010 \simeq 4.5$, the last dollar of the $19,010 annual cost gave exactly $1 of benefit.

8.6.5 Computer-Aided Heat Exchanger Design: HEX2

An essential element of design is investigation of the effects of changes in the design's key parameters. Optimization is a special case of this process, in which a particular function of the parameters is maximized or minimized—for example, the benefit-cost differential was maximized in the economic design analysis of Section 8.6.4. The computer is an ideal tool for this purpose, and there is an increasing availability of computer software to aid the design engineer. To demonstrate the utility and potential of the computer for design calculations, a simple example is presented here to complete our introduction to heat exchanger design.

We will restrict our attention to gas-to-gas heat exchange in a single-pass cross-flow configuration with a plate-fin surface, similar to the exchanger analyzed in Examples 8.14 and 8.15. The flows will be balanced and unmixed. A typical application of such an exchanger is for heat recovery in an air-conditioning system. With these restrictions, it is possible to focus on the effects of such parameters as the type of plate-fin surface, the dimensions of the surface, and the wall and fin material. Table 8.4c gives dimensions, and correlations for f and St, for six plate-fin heat transfer surfaces. The design problem is to size the exchanger for a required heat transfer performance, subject to a pressure drop constraint. Subsequently, through economic analysis, the cost-benefit implications of the pressure drop constraint can be explored. The computer program HEX2 accomplishes these objectives.

Table 8.4a Six plate-fin heat transfer surfaces (all dimensions in mm).

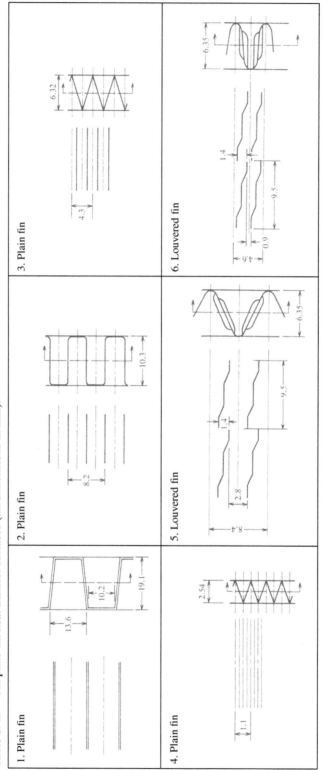

| 1. Plain fin | 2. Plain fin | 3. Plain fin |
| 4. Plain fin | 5. Louvered fin | 6. Louvered fin |

Table 8.4b Geometrical data for six plate-fin heat transfer surfaces.

Item	Surface Type	Plate Spacing, b mm	Fins per Meter m^{-1}	Fin Height, L_f mm	Fin Thickness, t_f mm	Fin Area Total Heat Transfer Area, A_f/A	Hydraulic Diameter, D_h mm	Heat Transfer Area Volume between Plates, β' m^{-1}
1	Plain fin (2.0)	19.1	78.7	9.14	0.813	0.606	14.5	250
2	Plain fin (6.2)	10.3	244	5.02	0.254	0.728	5.54	665
3	Plain fin (11.94 T)	6.32	470	3.00	0.152	0.769	2.87	1290
4	Plain fin (46.45 T)	2.54	1829	1.24	0.0508	0.837	0.805	4371
5	Louvered fin (3/8–6.06)	6.35	239	2.39	0.152	0.640	4.45	840
6	Louvered fin (3/8–11.1)	6.35	437	2.96	0.152	0.756	3.08	1204

Notes: Numbers in parentheses are designations in *Compact Heat Exchangers* (W. M. Kays and A. L. London, McGraw-Hill, New York, any edition). The core-to-frontal-area ratio $\sigma(=A_c/A_{fr})$ can be calculated from $\sigma = b\beta'D_h/[8(b+t_w)]$, where t_w is the plate thickness.

Table 8.4c Friction factor and Stanton number data for six heat transfer surfaces (correlations of data given by Kays and London [5]).

Surface	Reynolds Number Range	Friction Factor, $f = a_0 + a_1 Re^{-1} + a_2 Re^{-0.5} + a_3 Re^{-0.2}$				Stanton Number, $StPr^{2/3} = b_0 + b_1 Re^{-1} + b_2 Re^{-0.5} + b_3 Re^{-0.2}$			
		a_0	a_1	a_2	a_3	b_0	b_1	b_2	b_3
1	$4{,}000 < Re < 60{,}000$	-1.483×10^{-2}	172.60	-5.2079	0.4974	1.589×10^{-3}	-9.2526	0.3478	-0.005146
2	$800 < Re < 12{,}000$	-7.782×10^{-2}	155.28	-9.7484	1.1947	-1.234×10^{-2}	17.426	-1.3971	0.1748
3	$300 < Re < 10{,}000$	5.187×10^{-2}	64.348	0.1556	-0.1886	1.453×10^{-2}	0.3589	0.5123	-0.1037
4	$500 < Re < 2{,}000$	4.523×10^{-1}	14.611	13.601	-3.278	2.590×10^{-2}	-2.5448	1.0758	-0.2092
5	$500 < Re < 10{,}000$	-2.732×10^{-4}	141.37	-7.392	1.216	-1.770×10^{-2}	11.491	-1.1057	0.2086
6	$500 < Re < 10{,}000$	-2.340×10^{-1}	245.09	-18.574	3.090	-1.066×10^{-2}	7.8027	-0.5942	0.1343

Note: The Reynolds number is based on passage hydraulic diameter.

A schematic of the exchanger is shown in Fig. 8.26. HEX2 first asks for the following inputs:

Mass flow rate: $\dot{m}_H = \dot{m}_C = \dot{m}$ [kg/s]

Hot-stream inlet temperature: $T_{H,\text{in}}$ [K]

Cold-stream inlet temperature: $T_{C,\text{in}}$ [K]

Which outlet temperature is known, $T_{H,\text{out}}$ or $T_{C,\text{out}}$, and its value [K]

HEX2 then calculates the effectiveness ε from Eq. (8.31) and the required N_{tu} by solving the $\varepsilon - N_{\text{tu}}$ relation given by item 4 of Table 8.3a using Newton iteration. Also calculated is a reference temperature for evaluation of c_p, μ, and Pr, which is taken as the average of the average hot and cold stream temperatures,

$$T_r = \frac{1}{2}\left\{\frac{1}{2}\left(T_{H,\text{in}} + T_{H,\text{out}}\right) + \frac{1}{2}\left(T_{C,\text{in}} + T_{C,\text{out}}\right)\right\} \tag{8.79}$$

HEX2 then asks for the following gas properties:

Molecular weight: M [kg/kmol]

Specific heat: c_p [J/kg K]

Viscosity: μ [kg/m s]

Prandtl number: Pr

Next the heat transfer surface is selected. Referring to Table 8.4b for typical data, HEX2 asks for the following inputs:

Plate spacing: b [mm]

Passage hydraulic diameter: D_h [mm]

Plate thickness: t_w [mm]

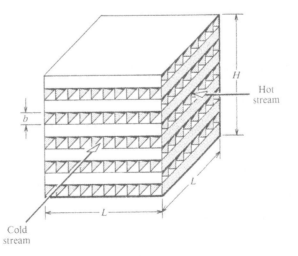

Figure 8.26   Schematic of a single-pass, cross-flow exchanger with a plate-fin surface.

Fin thickness: t_f [mm]

Fin height: L_f [mm]

Fin area/total area: A_f/A

Heat transfer area/volume between plates $= \beta'$ [m^{-1}]

Scale factor (e.g., a value of 2 will double all surface dimensions)

Plate and fin thermal conductivity: k [W/m K]

The performance of the heat transfer surface is supplied in the form of the correlations given in Table 8.4c (notice that these are independent of the scale factor):

Friction factor constants: a_0, a_1, a_2, a_3

StPr$^{2/3}$ constants: b_0, b_1, b_2, b_3

Finally, the pressure drop data are supplied:

Choice of stream for pressure drop constraint: hot or cold

Inlet pressure: P_1 [Pa]

Allowable pressure drop: ΔP [Pa]

Choice of geometry (c) or (d) of Fig. 8.21 for the contraction and expansion coefficients, K_c and K_e

An iterative calculation procedure is required. To obtain a reasonable first guess, HEX2 uses the approximate Eq. (8.58) rearranged as

$$G^2 \simeq \frac{8\rho_m}{\text{Pr}^{2/3}} \left(\frac{\Delta P}{N_{\text{tu}}^1}\right) \left(\frac{\text{StPr}^{2/3}}{f}\right) \tag{8.80}$$

The one-side N_{tu}^1 is estimated assuming a negligible wall and fin resistance; from Eq. (8.60) for balanced flow,

$$N_{\text{tu}}^1 = 2N_{\text{tu}} \tag{8.81}$$

Based on Fig. 8.23, (StPr$^{2/3}/f$) is set equal to 0.1. All other quantities on the right-hand side of Eq. (8.80) are known from the input data, and, hence, G can be calculated. This value is the initial guess for G. The calculation steps are now as follows:

1. Reynolds number:

$$\text{Re} = \frac{GD_h}{\mu}$$

2. Core and frontal area:

$$A_c = \frac{\dot{m}}{G}; \quad A_{\text{fr}} = \frac{A_c}{\sigma}, \quad \text{where } \sigma = \frac{b\beta'D_h}{8(b+t_w)}$$

3. Friction factor: f

4. Stanton number: $St = (StPr^{2/3})/Pr^{2/3}$

5. Heat transfer coefficient: $h_c = Gc_pSt$

6. Fin effectiveness, Eq. (2.42): $\eta_f = (1/\beta L_f)\tanh \beta L_f$

7. Overall heat transfer coefficient:

$$\frac{1}{U} = \frac{2}{h_c[\eta_f(A_f/A)+(1-A_f/A)]} + \frac{t_w}{k(1-A_f/A)}$$

8. Required heat transfer area:

$$A = \frac{\dot{m}c_pN_{tu}}{U}$$

9. Flow length:

$$L = \frac{A}{\mathscr{P}} = \frac{AD_h}{4A_c}$$

10. Contraction and expansion coefficients: K_c and K_e

11. Pressure drop: ΔP, calculated from Eq. (8.56).

This value of ΔP is then compared with the required value, and a new value of G is estimated using $\Delta P \propto G^2$. The next iteration commences with step 1 to obtain a new Reynolds number, *et seq*. The iteration process is repeated until the calculated ΔP agrees with the required value, to within a specified error tolerance.

HEX2 now asks whether an economic analysis is required. If the answer is yes, the following additional input is required:

Interest rate: i [%]

Loan period: n [yr]

Fixed cost for exchanger: C_{if} [$]

Salvage value of exchanger: C_s [$]

Working lifetime: m [yr]

Cost of heat transfer surface per unit area: c_i [$/m^2]

Fixed cost for taxes and insurance: $\mathscr{C}_{tf}$ [$/yr]

Cost of taxes, maintenance, and insurance per unit area: c_t [$/m^2 yr]

Value of heat: ν [$/W h]

Operating time per year: τ [h/yr]

Unit power cost: c_p [$/W h]

Fan efficiency: η_f

HEX2 then calculates the benefit-cost differential $(\mathscr{V} - \mathscr{C})$ [$/yr].

EXAMPLE 8.17   A Balanced-Cross-Flow Plate-Fin Recuperator for an Air-Heating System

A recuperator is required for an air-heating system that processes 0.5 kg/s air. The hot air enters at 295 K, and the cold air enters at 270 K and is to be heated to 287 K. The inlet pressure for the hot air is 986 mbar. Determine the dimensions of a cross-flow exchanger, with the louvered-fin surface given as item 6 in Table 8.4a, that will give a pressure drop of 330 Pa on the hot side. The core is to be fabricated from AISI 1010 steel with a 0.30 mm plate thickness.

Solution

Given: Balanced-cross-flow exchanger with a louvered-fin heat transfer surface.

Required: Dimensions for a specified heat transfer performance and pressure drop.

Assumptions: 1. The density will be taken to be constant and equal to its average value ρ_m in the pressure drop equation; hence, the acceleration pressure drop is assumed to be negligible.
2. The exchanger will be sized for the hot stream.

The purpose of this example is to illustrate the calculation procedure of HEX2. Following the steps given in Section 8.6.5, we have

$$\dot{m} = 0.5 \text{ kg/s}$$
$$T_{H,\text{in}} = 295 \text{ K}$$
$$T_{C,\text{in}} = 270 \text{ K}$$
$$T_{C,\text{out}} = 287 \text{ K}$$

Using HEX1, $\varepsilon = 0.680, N_{\text{tu}} = 2.92, T_{H,\text{out}} = 278$ K.

$$T_r = \frac{1}{2}\left\{ \frac{1}{2}(295+278) + \frac{1}{2}(270+287) \right\} = 282.5\text{K}$$

Gas properties:

$$M = 29 \text{ kg/kmol}$$
$$c_p = 1008 \text{ J/kg K}$$
$$\mu = 17.65 \times 10^{-6} \text{ kg/ms}$$
$$\text{Pr} = 0.69$$

For surface 6 of Table 8.4b,

$$b = 6.35 \text{ mm} = 6.35 \times 10^{-3} \text{ m}$$
$$D_h = 3.08 \text{ mm} = 3.08 \times 10^{-3} \text{ m}$$
$$t_w = 0.30 \text{ mm} = 0.30 \times 10^{-3} \text{ m}$$
$$t_f = 0.152 \text{ mm} = 0.152 \times 10^{-3} \text{ m}$$
$$L_f = 2.96 \text{ mm} = 2.96 \times 10^{-3} \text{ m}$$
$$A_f/A = 0.756$$
$$\beta' = 1204$$
$$\text{Scale factor} = 1$$

$k_w = 64$ W/m K (from Table A.1a)

a_0, a_1, etc.: see Table 8.4c, item 6

b_0, b_1, etc.: see Table 8.4c, item 6

Pressure drop data:

Hot

$P_1 = 0.986 \times 10^5$ Pa

$\Delta P = 330$ Pa

Coefficients from Fig. 8.21c

The average density of the hot stream is

$$\rho_m = \frac{P_m M}{\mathscr{R} T_m} = \frac{(98,600 - 165)(29)}{(8314)(0.5)(295 + 278)} = 1.198 \text{ kg/m}^3$$

$$N_{tu}^1 = 2N_{tu} = (2)(2.92) = 5.84$$

$$\text{Pr}^{2/3} = (0.69)^{2/3} = 0.78$$

Substituting in Eq. (8.80) and setting $\text{StPr}^{2/3}/f = 0.1$,

$$G^2 \simeq \frac{8\rho_m}{\text{Pr}^{2/3}} \left(\frac{\Delta P}{N_{tu}^1} \right) \left(\frac{\text{StPr}^{2/3}}{f} \right) = \frac{(8)(1.198)}{0.78} \left(\frac{330}{5.84} \right) (0.1) = 69.5$$

$$G \simeq 8.34 \text{ kg/m}^2 \text{ s}$$

The iterative calculation steps are:

1. $\text{Re} = \dfrac{GD_h}{\mu} = \dfrac{(8.34)(3.08 \times 10^{-3})}{17.65 \times 10^{-6}} = 1455$

2. $A_c = \dfrac{\dot{m}}{G} = \dfrac{0.5}{8.34} = 0.0600 \text{ m}^2$

   $$\sigma = \frac{b\beta' D_h}{8(b + t_w)} = \frac{(6.35 \times 10^{-3})(1204)(3.08 \times 10^{-3})}{8(6.35 \times 10^{-3} + 0.3 \times 10^{-3})} = 0.443$$

   $$A_{fr} = \frac{A_c}{\sigma} = \frac{0.0600}{0.443} = 0.135 \text{ m}^2$$

3. $f = -0.2340 + 245.09(1455)^{-1} - 18.574(1455)^{-0.5} + 3.090(1455)^{-0.2} = 0.168$

4. $\text{StPr}^{2/3} = -0.01066 + 7.803(1455)^{-1} - 0.5942(1455)^{-0.5} + 0.1343(1455)^{-0.2}$

   $$= 0.0104$$

   $$\text{St} = (0.0104/0.78) = 0.0133$$

5. $h_c = Gc_p\text{St} = (8.34)(1008)(0.0133) = 111.8 \text{ W/m}^2 \text{ K}$

6. $\beta^2 = \dfrac{2h_c}{k_w t_f} = \dfrac{(2)(111.8)}{(64)(0.152 \times 10^{-3})} = 2.299 \times 10^4; \quad \beta = 152$

   $\beta L_f = (152)(2.96 \times 10^{-3}) = 0.449; \quad \eta_f = \tanh(0.449)/(0.449) = 0.938$

7. $\dfrac{1}{U} = \dfrac{2}{h_c[\eta_f(A_f/A) + (1 - A_f/A)]} + \dfrac{t_w}{k(1 - A_f/A)}$

$\quad = \dfrac{2}{(111.8)[(0.938)(0.756) + (1 - 0.756)]} + \dfrac{0.3 \times 10^{-3}}{64(1 - 0.756)}$

$\quad U = 53.2 \ \text{W/m}^2 \ \text{K}$

8. $A = \dfrac{\dot{m}c_p N_{tu}}{U} = \dfrac{(0.5)(1008)(2.92)}{(53.2)} = 27.6 \ \text{m}^2$

9. $L = \dfrac{A D_h}{4 A_c} = \dfrac{(27.6)(3.08 \times 10^{-3})}{(4)(0.0600)} = 0.354 \ \text{m}$

10. $K_c = 1.18, \quad K_e = -0.06$

11. $\dfrac{\Delta P}{P_1} \simeq \dfrac{G^2}{2\rho_m P_1} \left[(1 - \sigma^2 + K_c) + \dfrac{f}{4}\dfrac{A}{A_c} - (1 - \sigma^2 - K_e) \right]$

$\quad = \dfrac{(8.34)^2}{2(1.198)(98,600)} \left\{ [1 - (0.443)^2 + 1.18] + \dfrac{0.168}{4}\dfrac{27.6}{0.0600} \right.$

$\quad\quad \left. - [1 - (0.443)^2 - (-0.06)] \right\}$

$\quad = 2.94 \times 10^{-4}(1.98 + 19.32 - 0.86)$

$\quad = 6.01 \times 10^{-3}$

$\quad \Delta P = (98,600)(6.01 \times 10^{-3}) = 592 \ \text{Pa}$

This pressure drop is higher than the required value of 330 Pa. For a second iteration, we take

$$G = \left(\dfrac{330}{592} \right)^{1/2} (8.34) = 6.23 \ \text{kg/m}^2 \ \text{s}$$

and obtain the following results:

1. $\text{Re} = 1086$

2. $A_c = 0.0803 \ \text{m}^2, \quad A_{fr} = 0.181 \ \text{m}^2$

3. $f = 0.193$

4. $\text{StPr}^{2/3} = 0.0117, \quad \text{St} = 0.0150$

5. $h_c = 94.1 \ \text{W/m}^2 \ \text{K}$

6. $\beta = 139, \quad \eta_f = 0.947$

7. $U = 45.2 \ \text{W/m}^2 \ \text{K}$

8. $A = 32.0 \ \text{m}^2$

9. $L = 0.307 \ \text{m}$

10. K_c, K_e: no change

11. $\Delta P/P_1 = 3.34 \times 10^{-3}$; $\Delta P = 329$ Pa

The error in this estimate of ΔP is negligible. The height of the exchanger is

$$H = \frac{A_{\text{fr}}}{L} \simeq \frac{0.181}{0.307} = 0.590 \text{ m}$$

Comments

1. HEX2 calculates ρ_1, ρ_2, and ρ_m for use in the complete pressure drop equation, Eq. (8.56): the acceleration pressure drop is not neglected.

2. All input values are saved by HEX2 and become default values for subsequent use of the program.

EXAMPLE 8.18   Economic Analysis of a Recuperator for an Air-Heating System

An economic analysis is required for the recuperator of Example 8.17. The interest rate for a loan over 20 years is 8%, the fixed cost of the exchanger is $300, and the cost of heat transfer surface per unit area is $10/m^2. The fixed cost for insurance and taxes is $20/yr, and the cost per unit area is 0.57$/m^2 yr. The cost of electricity is 0.08 $/kW h, and the heating system operates 16 hours per day for 190 days of the year. If the recuperator were not used, an electrical heater would have to make up the required heat: (i) Determine the benefit-cost differential for the exchanger sized in Example 8.17. (ii) Use HEX2 to map the benefit-cost differential as a function of cold-stream outlet temperature and hot stream pressure drop.

Solution

Given: Recuperator for an air-heating system.

Required: Economic analysis for a louvered-fin-surface cross-flow exchanger.

Assumptions: 1. A fan efficiency (including motor, etc.) of 65%.
                        2. Negligible salvage value.

(i) From Eq. (8.61), the annuity is

$$r = \frac{i}{1 - [1/(i+1)]^n} = \frac{0.08}{1 - (1/1.08)^{20}} = 0.102 \text{ yr}^{-1}$$

and, from Eq. (8.66), the total annual cost is

$$\begin{aligned}
\mathcal{C}_a &= (rC_{if} + \mathcal{C}_{if}) + (rc_i + c_t)A \\
&= [(0.102)(300) + 20] + [(0.102)(10) + 0.57](32.0) \\
&= 30.6 + 20 + 50.9 \\
&= \$101.5/\text{yr}
\end{aligned}$$

The operating cost is obtained from Eq. (8.67):

$$\mathscr{C}_p = \frac{e_p \tau}{\eta_p} \times \frac{\dot{m}\Delta P}{\rho} \simeq \frac{(0.08 \times 10^{-3})(16)(190)}{0.65} \times \frac{(2)(0.5)(330)}{1.198} = 103.1 \text{ \$/yr}$$

The annual value of the heat recovery is given by Eq.(8.69):

$$\mathscr{V} = e\dot{Q}\tau$$

$$= e\dot{m}c_p(T_{C,\text{out}} - T_{C,\text{in}})\tau$$

$$= (0.08 \times 10^{-3})(0.5)(1008)(287 - 270)(16)(190)$$

$$= \$2084/\text{yr}$$

The benefit-cost differential is then

$$\mathscr{V} - \mathscr{C} = 2084 - 103 - 102 = \$1879/\text{yr}$$

(ii) Using HEX2, the following table of $(\mathscr{V} - \mathscr{C})$ can be prepared.

				$T_{C,\text{out}}$ [K]			
		286		287		288	
ΔP	L	$\mathscr{V} - \mathscr{C}$	L	$\mathscr{V} - \mathscr{C}$	L	$\mathscr{V} - \mathscr{C}$
Pa	m	\$/yr	m	\$/yr	m	\$/yr
270	0.247	1791	0.299	1901	0.372	2003
330	0.258	1775	0.313	1885	0.391	1988
390	0.267	1757	0.324	1869	0.407	1973

Comments

The table indicates that an increase in $T_{C,\text{out}}$ and a decrease in ΔP will further improve the benefit-cost differential.

8.7 CLOSURE

A great variety of heat exchangers used in engineering practice were classified according to geometric flow configuration. For purposes of analysis, we chose to differentiate single-stream and two-stream steady-flow exchangers, although the single-stream exchanger can be viewed as a limit case of the two-stream exchanger when the capacity ratio $R_C \rightarrow 0$. The thermal performance of steady-flow exchangers can be obtained using either the logarithmic mean temperature difference (LMTD) approach, or the effectiveness-number of transfer units $(\varepsilon - N_{\text{tu}})$ approach. The emphasis was on the latter procedure in this text.

The design of heat exchangers requires that a specified heat transfer performance be achieved, subject to various constraints. The most important of these is often a pressure drop constraint, which, through the pumping power required, is related to operating costs. The capital cost of the exchanger depends on size and weight, and an economic analysis is used to trade off the competing objectives of low operating costs and low capital costs. Other design issues include fluid and exchanger material compatibility and ease of maintenance, but these issues are left to be dealt with by heat exchanger design handbooks.

The computer program HEX2 serves to introduce the student to computer-aided heat exchanger design. HEX2 sizes balanced cross-flow plate-fin heat exchangers to give a specified heat transfer performance, subject to a pressure drop constraint. In addition, HEX2 calculates the benefit-cost differential for the resulting design when used for heat recovery. Since HEX2 applies to a very limited class of exchangers and applications, it is not intended to be a general design tool. Rather, it was included here to give the student ideas about how to use computers to aid the design process.

REFERENCES

1. Bowman, R. A., Mueller, A. C, and Nagle, W. M., "Mean temperature difference in design," *Trans. ASME*, 62, 283–294 (1940).

2. Shah, R. K., and Mueller, A. C, "Heat exchanger basic thermal design methods," in Rohsenow, W. M., Harnett, J. P., and Ganić, E. N., eds., *Handbook of Heat Transfer Applications*, 2nd ed., Chap. 4, Part 1, McGraw-Hill, New York (1985).

3. Taborek, J., "Charts for mean temperature difference in industrial heat exchanger configurations," in Hewitt, G. E, coord., ed., *Hemisphere Handbook of Heat Exchanger Design*, Sec. 1.5, Hemisphere, New York (1990).

4. Turton, R., Ferguson, D., and Levenspiel, O., "Charts for the performance and design of heat exchangers," *Chemical Engineering*, 93, 81–88 (August 18, 1986).

5. Kays, W. M., and London, A. L., *Compact Heat Exchangers*, 3rd ed., Krieger, Melbourne, Fla. (1998).

6. Soland, J. G., Mack, W. M., Jr., and Rohsenow, W. M., "Peformance ranking of plate-fin heat exchanger surfaces," *J. Heat Transfer*, 100, 514–519 (1978).

7. Linnhoff, B., and Turner, J. A., "Heat-recovery networks: new insights yield big savings," *Chemical Engineering*, 88, 56–70 (November 2, 1981).

8. Linnhoff, B., Townsend, D. W., Boland, D., Hewitt, G. E, Thomas, B. E. A., Guy, R. H., and Marsland, R. H., *User Guide on Process Integration for the Efficient Use of Energy*, IChemE, Rugby (1982).

9. Joo, Y, Dieu, K., and Kim, C-J., "Air cooling of IC chip with novel microchannels monolithically formed on chip front surface," *Cooling and Thermal Design of Electronic Systems* (HTD 319, EEPD, 15, 117-121). ASME Int. Mechanical Engineering Congress and Exposition, San Francisco (1995).

10. Edwards, D. K., and Matavosian, R., "Thermoeconomically optimum counterflow heat exchanger effectiveness," *J. Heat Transfer*, 104, 191–193 (1982).

EXERCISES

8-1. An oil supply of 1 kg/s is to be heated from 15°C to 70°C. A 0.3 kg/s supply of water at 95°C is available. Is this supply sufficient?

8-2. A boiler is required to generate 3.0 kg/s of steam at a pressure of 0.13 MPa. Waste heat is available in the form of 40 kg/s of exhaust gas at 300°C. Is this energy supply sufficient? Take $c_p = 1180$ J/kg K for the exhaust gas.

8-3. A 1.8 kg/s supply of saturated steam at 0.11 MPa is available for heating water in a bottle washing operation at a soft drink bottling plant. What is the maximum flow rate of water that can be heated from 20°C to 70°C?

8-4. A shell-and-tube heat exchanger is used to heat fuel oil with hot engine exhaust. The mild steel tubes are $1^1/_2$ in nominal diameter and 14 gage. Design values of the inside and outside heat transfer coefficients are 40 and 100 W/m^2 K, respectively.

(i) Calculate the design value of the overall heat transfer coefficient based on tube outside area.

(ii) Using the fouling resistances given in Table 8.1, estimate the reduction in U expected during service due to fouling.

8-5. An air-cooled oil cooler has 1 cm–O.D., 1 mm-wall-thickness steel tubes, with 2 cm–O.D., 0.4 mm-thickness spiral fins at a pitch of 2 mm. If the inside and outside heat transfer coefficients are 200 and 25 W/m^2 K, respectively, determine the overall heat transfer coefficient-perimeter product. Take $k = 55$ W/m K for the steel.

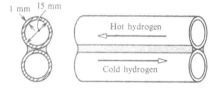

8-6. A twin-tube heat exchanger for a hydrogen cryogenic refrigeration system is made by welding two 15 mm–O.D., 1 mm-wall-thickness tubes as shown, and forming them into a coil that is inserted in an evacuated container. If there is balanced flow of hydrogen gas at 2.0×10^{-6} kg/s, and the tube material is stainless steel, estimate the overall heat transfer coefficient. Take $k = 15.0$ W/m K for the tube wall, and evaluate the hydrogen properties at 150 K and 2 torr.

8-7. A coaxial-tube heat exchanger has an inner 316 stainless steel tube of 3 cm outside diameter and 2 mm wall thickness. Helium at 0.22 MPa flows at 80 m/s through the inner tube. Water flows through the outer tube at 0.97 kg/s, which has an inside diameter of 5 cm. A change in service conditions requires the overall heat transfer coefficient to be increased. It is proposed to artificially roughen the inner surface of the inner tube by fitting an insert that gives parallel ribs 1 mm high at a 10 mm pitch. Determine the percentage increase in the overall heat transfer coefficient that will be achieved. Take the bulk temperatures of the helium and water as 600 and 310 K, respectively.

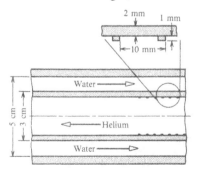

8-8. A 9 m-long stage of a multistage flash vaporization desalination plant operates at 27,150 Pa. The condenser tube bundle comprises 150 2 cm–O.D. titanium tubes with a 1 mm wall thickness, through which cooling sea water flows at 120 kg/s. If the cooling water enters the stage at 320 K, at what temperature does it exit? Take the outside heat transfer coefficient for film condensation to be 10,000 W/m² K. At what rate does the stage produce fresh water? Take $k = 18$ W/m K for the titanium.

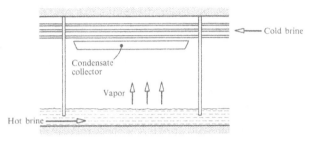

8-9. Hot brine from a geothermal well passes through 120 3 cm–O.D., 1 mm-wall-thickness, 16 m-long titanium alloy ($k = 14$ W/m K) tubes. Refrigerant-113 boils on the exterior of the tubes at 150°C, with an outside heat transfer coefficient of 20,000 W/m² K. The hot brine enters the tubes at 210°C, and the bulk velocity in each tube is 2 m/s. Calculate the following quantities.

   (i) Re_D, the Reynolds number inside a tube
  (ii) $h_{c,i}$, the inside heat transfer coefficient
 (iii) U, the overall heat transfer coefficient
 (iv) N_{tu}, the number of transfer units of the boiler
  (v) ε, the exchanger effectiveness
 (vi) $T_{H,out}$, the hot stream outlet temperature
(vii) $\dot{Q}$, the heat transferred in the boiler
(viii) $\dot{m}_C$, the R-113 vapor production rate

For the brine, take $k = 0.6$ W/m K, $\rho = 900$ kg/m³, $c_p = 4000$ J/kg K, and $v = 0.20 \times 10^{-6}$ m²/s; for R-113, take $h_{fg} = 0.2 \times 10^6$ J/kg.

8-10. Performance tests have been performed on a brine/water tubular heat exchanger for geothermal energy utilization. The tubes are 2 m long and have a 10.26 mm inside diameter. Brine at 350 K is supplied at 0.134 kg/s per tube. Silica is found to deposit at a rate of 6.6×10^{-7} g/cm² min. If the initial overall heat transfer coefficient is 5000 W/m² K, and the effect of scale roughness is to increase the inside heat transfer coefficient by 60%, determine the overall heat transfer coefficient after 1000 hours of operation. Take the density and thermal conductivity of silica scale as 2200 kg/m³ and 0.6 W/m K, respectively.

8-11. A power station operating on a mercury-steam binary cycle has mercury condensing at 530 K on the outside of 6 cm–O.D. steel tubes. Water boils inside the tubes at 500 K. If the overall heat transfer coefficient based on tube outside area is 20,000 W/m² K, how many tubes 3 m long are required for a turbine steam consumption of 25 kg/s?

8-12. Paralleling the analysis of the evaporator as a single-stream exchanger in Section 8.4, analyze a simple condenser to show that its performance is given by

$$\varepsilon = 1 - e^{-N_{tu}}; \quad \varepsilon = \frac{T_{C,\text{out}} - T_{C,\text{in}}}{T_{\text{sat}} - T_{C,\text{in}}}$$

where $N_{tu} = U\mathscr{P}L/\dot{m}_C c_{pC}$, and subscript C refers to the cold (coolant) stream.

8-13. Steam is condensed on a bundle of 400 tubes through which there is a flow of cold water. The tubes have an outside diameter of 2 cm, and the overall heat transfer coefficient based on outside area is 600 W/m² K. The water flow rate is 0.2 kg/s per tube, and it enters at 290 K. If the outlet temperature is 350 K when the condenser is operating at atmospheric pressure, how much steam is condensed on each tube, and how long is the tube bundle? The specific heat of water at 320 K is 4174 J/kg K, and the enthalpy of vaporization for steam at atmospheric pressure is 2.257×10^6 J/kg.

8-14. A 3 m-long stage of a multistage flash vaporization desalination plant operates at 3×10^4 Pa. The condenser tube bundle comprises 1300, 3 cm–O.D. titanium tubes through which cooling water flows at 60 kg/s.

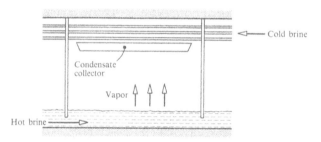

If the cooling water (sea water) enters the stage at 335 K, at what temperature does it exit? At what rate does the stage produce fresh water? Take the overall heat transfer coefficient based on the tube outside area as 1200 W/m² K and the specific heat of sea water as 4200 J/kg K.

8-15. A geothermal power plant uses isobutane as the secondary working fluid. After expanding through the turbine, isobutane vapor condenses in a shell-and-tube condenser at 325 K. The condenser coolant is water at 305 K supplied from a cooling tower at a rate of 500 kg/s. The condenser shell contains 4000 tubes of 25 mm O.D. and 2 mm wall thickness; the overall heat transfer coefficient based on tube outside area is estimated to be 450 W/m² K. If it is desired to have a condenser effectiveness of 80%, determine

   (i)   the outlet water temperature.
   (ii) the number of transfer units required.
   (iii) the length of the tube bundle.

Also, if the thermal efficiency of the power cycle is 30%, determine the power output of the turbine.

8-16. In a test of a shell-and-tube steam condenser, 140 kg/s of coolant water at 300 K is supplied to the tubes. When the pressure in the shell is maintained at 0.010 MPa, the condensate flow rate is measured to be 3.0 kg/s. Determine

  (i) the outlet water temperature.
  (ii) the heat exchanger effectiveness.
  (iii) the number of transfer units.
  (iv) the $U\mathscr{P}L$ product.

8-17. An air-cooled R-12 air-conditioning system condenser is required to condense 0.011 kg/s of saturated R-12 vapor at 320 K. The frontal area of the exchanger is 0.4 m^2, and air at 295 K enters the exchanger at 2 m/s.

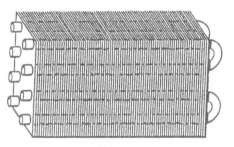

  (i) If the overall heat transfer coefficient is estimated to be 40 W/m^2 K, determine the heat transfer area required.
  (ii) Plot a graph of the required heat transfer area for an air inlet temperature in the range 290–305 K.
  (iii) If the overall heat transfer coefficient increases approximately as air velocity to the 0.6 power, plot a graph of the required heat transfer area for an air velocity range of 2–10 m/s.

Take $h_{fg} = 0.1237 \times 10^6$ J/kg for R-12; $\rho = 1.177$ kg/m^3 and $c_p = 1005$ J/kg K for air.

8-18. A single-pass shell-and-tube condenser is required to condense 1 kg/s of steam at 6224 Pa, by using cooling water at 295 K. If the allowable water temperature rise is 12 K, determine the length of a condenser containing 300 1 in, 18 gage brass tubes. Take the steam-side heat transfer coefficient as 6500 W/m^2 K, and allow for 0.2 mm-thick scale of conductivity 1.5 W/m K on the inside wall.

8-19. A one-shell-pass, two-tube-pass exchanger is to be used to condense 100 kg/s of steam at 4714 Pa using water at 290 K. If the expected overall heat transfer coefficient is 4000 W/m^2 K and the water temperature rise must not exceed 10 K, determine the heat transfer area required.

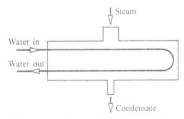

8-20. Saturated steam at 0.115 MPa is supplied to an air heater and condenses on 1 in, 18 gage horizontal brass tubes 0.7 m long. The tubes are staggered on 4 cm centers to give four rows, each with 15 tubes. Air enters the tubes at 1 atm, 285 K with a velocity of 2.0 m/s.

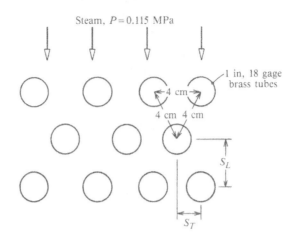

Determine the outlet air temperature, steam consumption rate, and air-side pressure drop.

8-21. A horizontal hot air duct in an attic has a 25 cm inside diameter and is 15 m long. The air velocity is 8 m/s. If the air enters at 35°C and the attic air temperature is 15 °C, what is the heat loss from the duct if it is uninsulated? (*Hint:* The duct can be modeled as a single-stream heat exchanger.)

8-22. A power plant stack is 100 m high and has an inside diameter of 6 m. The wall is made of concrete and varies in thickness linearly from 50 cm at the bottom to 25 cm at the top. Exhaust gas enters the base of the stack at a rate of 100 kg/s and at a temperature of 600 K. Air at 300 K and 1 bar blows at 5 m/s across the stack. Determine the exit temperature of the stack gas. (*Hint:* The stack can be modeled as a single-stream heat exchanger.)

8-23. Brine is to be evaporated in a multistage falling-film evaporator, and is fed to the top of a bundle of vertical 5 cm–O.D., 2 mm-wall-thickness stainless steel tubes 3 m high, at 0.010 kg/s per tube. Hot water at 340 K is fed to the bottom of the bundle and flows inside the tubes at 0.050 kg/s tube. If the saturation temperature in the stage is 330 K, what is the vapor production per tube?

8-24. Superheated steam at 130°C and 1 atm pressure enters a 5 cm–O.D., 2 mm-wall-thickness steel pipe at 0.0139 kg/s. The pipe is insulated with a 3 cm-thick layer of insulation ($k = 0.07$ W/m K) and is located in a factory where the ambient air temperature is maintained at 20°C. The heat transfer coefficient on the outside of the insulation accounting for both convection and radiation is estimated to be 9 W/m^2 K.

(i) Calculate the distance along the pipe to where steam commences to condense.
(ii) How much further does the steam flow before its bulk temperature falls to within 1°C of saturation?

(*Hint:* Once the steam starts to condense, the inner wall of the tube can be assumed to be at 100°C.)

8-25. A stream of 1 g/s of hydrogen gas at 1 MPa and 80°C is to be cooled. Available is a tank containing 100 kg of water in which is submerged a 4.0 m length of 15 mm–O.D., 2 mm-wall-thickness copper tube wound in a coil. The tank contains a stirrer driven by a small electric motor, and the walls of the tank are well insulated. If the initial temperature of the water is 20°C and the H_2 gas is passed through the coil, will the outlet temperature be less than 28°C after 2 hours of operation? (*Hint:* An analytical result can be obtained by modeling the coil as a single-stream heat exchanger, and the tank of water as a lumped thermal capacity system.)

8-26. A diesel power generation plant produces 3 kg/s of exhaust gas at 500°C. The gas is to be cooled to 200°C in a boiler that generates steam at 180 kPa. Determine the heat transfer area required and the rate of steam generation if the overall heat transfer coefficient is estimated to be 150 W/m^2 K. Take $c_p = 1090$ J/kg K for the exhaust gas.

8-27. Dale is a muscle car aficionado and has a 1970 Plymouth Barracuda with a 7.21 liter motor. Not satisfied with the power output, he has now acquired a supercharger to boost the inlet air pressure by 50%. The supercharger is connected to the carburetor inlet by a 7.6 cm–I.D. aluminum tube, 0.61 m long. Dale expects that this tube with suitable exterior fins can act as an intercooler and significantly reduce the temperature of the air entering the motor. Nominal conditions include an air temperature of 395 K at the outlet of the supercharger, an air flow rate of 0.5 kg/s, and air ducted into the engine compartment at 310 K. Dale hopes to reduce the air temperature from 395 K to below 340 K: give an opinion on the merits of his proposal.

8-28. Water flows at 1.2×10^{-3} kg/s through each of a number of parallel tubes bonded to the back face of a flat-plate solar collector. An approximate analysis of the collector indicates that heat will be collected at a rate of $1.3(90 - T)$ W/m per tube, where T is the local bulk temperature of the water, in degrees Celsius. It is desired to build a unit that can heat water from 15°C to 60°C.

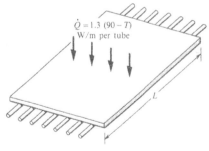

   (i) What is the required heat exchanger effectiveness?
   (ii) How many transfer units are required?
   (iii) How long should the unit be?

8-29. Vegetable oil is to be maintained at 50°C in a 3 m-diameter, 11 m-high tank. A water supply at 80°C is available, and a single helix coil of 8 mm–O.D. stainless steel tube has been proposed as the heat transfer surface. The tank is insulated with a 7 cm-thick medium-density fiberglass blanket, and the ambient temperature can go as low as 0°C. If the overall heat transfer coefficient from the water to the oil is estimated to be 160 W/m^2 K, determine the water flow rate and length of tube required for an 80% effective heat exchanger. Make reasonable assumptions for this preliminary design calculation.

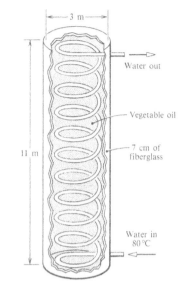

8-30. Exhaust gases exit a combustion test rig at 600°C and flow at 0.1 kg/s along a 5 m length of 4 cm–O.D., 2 mm-wall-thickness steel tube suspended from the roof of the laboratory. An estimate of the tube wall temperature is required for an ambient temperature of 22°C. Assume that the inside and wall resistances are negligible and that the heat loss from the outside is by natural convection and radiation ($\varepsilon = 0.9$ for the tube surface). (*Hint:* The governing differential equation requires numerical solution.)

8-31. A coaxial tube exchanger is operated under the following conditions.

	$\dot{m}$ kg/s	c_p J/kg K	T_{in} K	T_{out} K
Cold fluid	0.125	4200	313	368
Hot fluid	0.125	2100	483	

(i)   Determine T_{out} for the hot fluid.
(ii) Determine the exchanger effectiveness.
(iii) What is the ratio of required area for parallel versus counterflow operation?

8-32. A single-pass counterflow exchanger is required to cool 7000 kg/h of oil from 365 K to 330 K. Cooling water is available at 4000 kg/h and 290 K. If the overall heat transfer coefficient is 300 W/m^2 K, determine the surface area required. For the oil take $c_p = 2100$ J/kg K.

8-33. An economizer for a power plant is required to heat 8 kg/s of water from 340 K to 480 K. Flue gas is available at 25 kg/s and 800 K. Determine

(i) the outlet temperature of the flue gas.
(ii) the heat transfer area required for a counterflow exchanger if the overall heat transfer coefficient is 50 W/m^2 K.

Approximate the flue gas properties using those of air.

8-34. An aircraft oil cooler is to be designed to reduce the oil temperature from 390 K to 365 K. The oil (SAE 50) flow rate is 1.5 kg/s. If the overall heat transfer coefficient is 140 W/m^2 K and the entering air temperature is 310 K, find the necessary transfer area for

(i) counterflow.
(ii) parallel flow.

Assume balanced flow, $C_H = C_C$.

8-35. A two-shell-pass, four-tube-pass exchanger is available to cool 5 kg/s of liquid ammonia at 70°C, against 8 kg/s of water at 15°C. If the heat transfer area is 40 m^2 and the expected overall heat transfer coefficient is 2000 W/m^2 K when the ammonia is on the shell side, determine the outlet temperature of the ammonia.

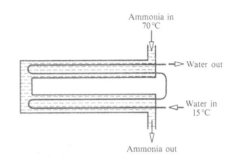

8-36. Interstage cooling is used to reduce the compression work required in a multistage compressor of a helium-cooled nuclear reactor. In one such cooler, 2.5 kg/s of helium at 2.5 bar is cooled from 400 K to 310 K by 5.0 kg/s of cooling water at 300 K. It is proposed to have a single pass of helium through 1/2 in, 18 gage copper tubes at a velocity of 30 m/s, with the water in counterflow on the shell side. If the shell-side heat transfer coefficient is 8000 W/m^2 K, determine the number of tubes required and the length of the exchanger. Also calculate the helium pressure drop.

8-37. Show that the effectiveness of a cross-flow exchanger with the C_{max} stream mixed and the C_{min} stream unmixed is given by

$$\varepsilon = \frac{1}{R_C}\left\{1 - \exp\left[R_C\left(e^{-N_{tu}} - 1\right)\right]\right\}$$

8-38. A diesel truck engine is turbocharged with 500 kg/h of air. The air is heated to 420 K by the compressor in the turbocharger. It is desired to cool the air to 392 K before it enters the engine. Water at 380 K is taken from the truck radiator at a rate of 1000 kg/h to provide cooling. An intercooler configured as a counterflow exchanger is used. The air-side heat transfer coefficient is 30 W/m^2 K, the water-side is 9000 W/m^2 K, and the air side is finned ($\eta_f = 1$) with a ratio of air to water area of 5 to 1. Determine the required effectiveness, the overall heat transfer coefficient based on water-side area, and the required water-side heat transfer area.

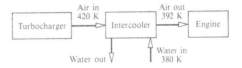

8-39. A heat exchanger is to be designed to cool 15 kg/s of a solution from 70°C to 40°C using 13 kg/s of water available at a temperature of 10°C. The specific heat of the solution is 3550 J/kg K. If the overall heat transfer coefficient may be taken to be 2000 W/m^2 K, calculate the required heat transfer areas for the following flow configurations:

   (i) Parallel flow
   (ii) Counterflow
   (iii) A shell-and-tube exchanger with the hot solution in the shell for one shell pass and four tube passes
   (iv) Same as (iii), but two shell passes

8-40. A 2.5 m-long coaxial-tube heat exchanger consists of a 6.38 mm–O.D., 4.52 mm–I.D. brass tube surrounded by a 9.63 mm–I.D. glass tube. The heat exchanger is tested in a laboratory with hot water inside the inner tube and cold water in the annulus. In a counterflow test the following data were recorded:

$$\dot{m}_C = 0.121 \text{ kg/s}; \ T_{C,in} = 23.5°C; \ T_{C,out} = 30.0°C$$
$$\dot{m}_H = 0.0562 \text{ kg/s}; \ T_{H,in} = 44.8°C; \ T_{H,out} = 31.3°C$$

Compare the effectiveness obtained in the test with the expected value.

8-41. A heat exchanger is to be designed for preheating secondary air supplied to an afterburner that reduces emissions from a gasoline engine. The exhaust gas flow rate varies from 20 to 150 kg/h, and the secondary air should be supplied at 10% of the exhaust mass flow rate. The exhaust gas leaves the afterburner at 850 K, the ambient air temperature is 290 K, and the secondary air should be heated to 780 K. A first design proposal is to construct a coaxial-tube counter flow exchanger by utilizing the exhaust pipe as the inner tube. If the exhaust pipe is 2 in schedule 10 steel pipe, and the proposed outer shell is $2^1/_2$ in schedule 10 pipe, determine the required exchanger length if the outer shell is well insulated. Comment on the feasibility of the design.

8-42. In cardiac surgery it is often necessary to achieve a certain level of hypothermia just prior to the operation, with restoration to normal body temperature immediately thereafter. For this purpose a heat exchanger may be installed in the arterial inlet tube leading from a heart-lung machine. Important constraints on the design include the following:

   (i) Blood flow must be upward to prevent trapping of gas bubbles.
   (ii) There should be minimal agitation of the blood to prevent blood trauma.
   (iii) Materials in contact with the blood must be inert, smooth, and easily sterilized.
   (iv) The coolant should be tap water.

A stainless steel shell-and-tube exchanger installed vertically is proposed. Water is on the shell side, and there are sufficient baffles to assume overall counterflow. The shell contains 24 $^1/_4$ in 20 gage tubes in a square array at a pitch of $^3/_4$ in. A blood flow of 2000 cc/min is to be cooled from 37.2°C to 27.5°C, with a cooling water supply of 6000 cc/min at 15°C. The baffles give a water velocity transverse to the tubes of 0.15 m/s. What length of exchanger is required? Take blood properties as follows: $\rho = 993$ kg/m^3, $c_p = 3850$ J/kg K, $k = 0.52$ W/m K, $\mu = 3.7 \times 10^{-3}$ kg/m s.

8-43. Use L'Hôpital's rule to derive Eq. (8.45) from Eq. (8.41).

8-44. Derive the $\varepsilon - N_{tu}$ relation for a counterflow exchanger, Eq. (8.41).

8-45. A cross-flow plate heat exchanger is to be designed for waste heat recovery from the exhaust stream in a copper smelting plant. A flow of 10 kg/s of exhaust gas enters the exchanger at 260°C, while 10 kg/s of air enters at 20°C. If the air is to be heated at 180°C and the overall heat transfer coefficient is estimated to be 50 W/m² K, determine the required heat transfer area for the following flow configurations:

  (i) Both fluids unmixed
  (ii) Both fluids mixed
  (iii) C_{max} mixed, C_{min} unmixed
  (iv) C_{max} unmixed, C_{min} mixed

The specific heat of the exhaust gas can be taken as 1220 J/kg K.

8-46. A counterflow heat exchanger for waste heat recovery consists of banks of finned heat-pipes, as shown in the sketch. The fins can be taken to have a fin effectiveness of unity, but the heatpipes have a thermal resistance that can be defined by $\dot{Q} = (T_e - T_c)/R_{hp}$. Derive the heat exchanger effectiveness for n discrete banks and balanced flow.

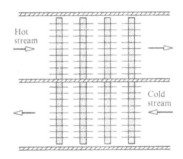

8-47. A counterflow heat exchanger is used to heat 6.0 kg/s of water from 35°C to 90°C by 14 kg/s of oil ($c_p = 2100$ J/kg K) supplied at 150°C. The design value of the overall heat transfer coefficient is 120 W/m² K. A second unit is to be built at another plant location; however, it has been proposed that the single exchanger be replaced by two smaller counterflow exchangers of equal heat transfer area. The two exchangers are to be connected in series on the water side and in parallel on the oil side. The oil flow is split equally between the two exchangers, and it may be assumed that the overall heat transfer coefficient for the smaller exchangers is also 120 W/m² K. Compare the transfer areas of the two arrangements.

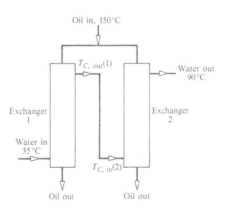

8-48. An oil is cooled in a coaxial-tube parallel-flow heat exchanger. A water flow of 0.1 kg/s enters the inner tube at 15°C and is heated to 40°C. The oil flows in the annulus and is cooled from 130°C to 55°C. It is proposed to cool the oil to a lower outlet temperature by increasing the length of the exchanger. Determine the following:

(i) The minimum temperature to which the oil may be cooled
(ii) The outlet oil temperature as a function of the fractional increase in exchanger length
(iii) The outlet temperature of each stream if the existing unit were switched to counterflow operation
(iv) The minimum temperature to which the oil may be cooled in counterflow operation
(v) The ratio of required length for counterflow to that for parallel flow as a function of outlet oil temperature

Neglect heat loss to the surroundings, and make any other reasonable assumptions.

8-49. Exhaust gases at 500°C from a 4-liter, 6-cylinder automotive engine, operating at 3000 rpm and 80% volumetric efficiency, are to be cooled in a coaxial-tube counterflow heat exchanger by 0.24 kg/s of cooling water supplied at 15°C. The exchanger is made from 15 m of 3 cm–O.D., 1 mm-wall-thickness 316 stainless steel tube inside a 4 cm–I.D. tube, with the gas flowing in the inside tube. Estimate the outlet temperature of the exhaust gas. Neglect heat losses to the surroundings.

8-50. A 6 kg/s flow of sulfuric acid ($c_p = 1480$ J/kg K) is to be cooled in a two-stage counterflow heat exchanger. The hot acid at 175°C is fed to a tank where it is stirred in contact with cooling coils; the continuous discharge from this tank at 89°C flows into a second stirred tank and leaves the second tank at 46°C. Cooling water at 20°C enters the cooling coil of the cold tank and leaves the cooling coil of the hot tank at 80°C. If the overall heat transfer coefficients are 52 and 38 W/m² K in the hot and cold tanks, respectively, determine the heat transfer surface areas required. Neglect heat losses to the surroundings, and assume that the acid temperature in each tank equals its outlet temperature.

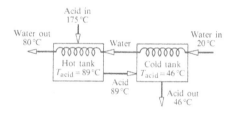

8-51. A one-shell-pass, two-tube-pass heat exchanger is used to cool a vegetable oil using river water. The design specifications are: $U = 340$ W/m² K, $T_{C,in} = 20°C$, $T_{C,out} = 70°C$, $T_{H,in} = 150°C$, and $T_{H,out} = 60°C$. After operating for one year, the performance is found to have deteriorated, and the oil is cooled only to 80°C. Fouling of the oil side is suspected; determine the apparent fouling resistance. Take $c_p = 1800$ J/kg K for the oil.

8-52. Air for a furnace is preheated from 20°C to 120°C in a preheater by flue gases available at 320°C. The furnace burns 8300 kg/h of coke (assumed to be 100%

carbon) with 10% excess air. Using an overall heat transfer coefficient of 20 W/m^2 K for comparison purposes, determine the heat transfer area required for the following configurations:

(i) Counterflow
(ii) Single-pass cross-flow, both streams mixed
(iii) Single-pass cross-flow, air mixed, gas unmixed

8-53. A counterflow heat exchanger is to be designed to recover waste heat from a geothermal brine. A 7.2 kg/s brine flow at 340 K is to be cooled to 308 K by a 9 kg/s water supply at 300 K. The unfouled overall heat transfer coefficient is estimated to be 1500 W/m^2 K, and a brine-side fouling resistance of 0.002 (W/m^2 K)$^{-1}$ should be allowed for. What is the required heat transfer area? The brine specific heat has been measured to be 3480 J/kg K.

8-54. Oil is heated in a 4 m-long coaxial-tube heat exchanger, with oil in laminar flow in the inside tube and water in turbulent flow in the outer jacket. The inner tube has a 2 cm outside diameter and 1.5 mm wall thickness, and the outer jacket has a 3 cm inside diameter. The flow rates of oil and water are 1.1 and 2 kg/s, respectively. The oil enters at 10°C, and the water enters at 90°C. Estimate the oil outlet temperature. Approximate the oil properties by those for SAE 50 oil, and comment on the effects of oil property variation along the exchanger.

8-55. A heat exchanger is used to preheat hydrogen fuel for a hydrogen combustor. The combustor burns 8.7 kg/s of fuel, and the exhaust gas flow rate is 55.0 kg/s. The exchanger is located in a 1 m-square duct and is a cross-flow unit with two passes of the hydrogen through 1 cm–O.D., 1 mm-wall-thickness Inconel-X tubes arranged in a square array with a pitch of 2 cm. The exhaust gases enter the exchanger at 0.55 MPa and 1400 K, while the hydrogen enters at 7.6 MPa and 20 K. If the required outlet temperature of the hy-

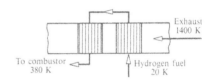

drogen is 380 K, determine the number of tubes per pass. Approximate the exhaust gas properties using those for air. (*Hint:* Use the LMTD approach and Fig. C.4*d* of Appendix C.)

8-56. An off-the-shelf counterflow heat exchanger with transfer surface area of 50 m^2 is required to cool 4 kg/s of oil from 90°C to 40°C. A water supply at 20°C is available. If the overall heat transfer coefficient is expected to be approximately 370 W/m^2 K, determine the required water flow rate and the water outlet temperature. Take $c_p = 2030$ J/kg K for the oil.

8-57. Exercise 8–56 is perhaps oversimplified, since the overall heat transfer coefficient will vary with water flow rate. For the heat transfer surface configuration under consideration, the following data for U as a function of water flow rate have been calculated.

Water flow rate, kg/s	1	1.5	2.0	2.5	3.0
U, W/m^2 K	330	358	379	396	411

(Notice that U does not vary strongly with water flow rate because the oil-side heat transfer resistance dominates.) Determine a new estimate of the required water flow rate and the water outlet temperature.

8-58. A coaxial-tube counterflow heat exchanger uses water to cool a hot-air stream containing corrosive vapor. Since gas-side heat transfer coefficients are typically much less than water-side coefficients, the outside of the 5 cm–O.D., 3 mm-wall-thickness stainless steel inner tube has 20 axial fins, as shown. The fins have a rectangular profile with a thickness of $2t = 3$ mm and a length of 1 cm. The outer tube has a 7 cm I. D. The water and air flow rates are both 0.2 kg/s, and the water and air inlet temperatures are 30°C and 90°C, respectively. If the exchanger is 6 m long, determine the outlet temperatures and the heat transfer rate. Use $k = 15$ W/m K for the stainless steel.

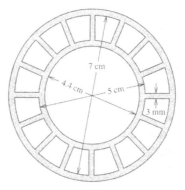

8-59. An off-the-shelf heat exchanger with transfer surface area of 50 m^2 is required to cool 4 kg/s of oil at 90°C. A water supply of 3 kg/s at 20°C is available. The overall heat transfer coefficient is expected to be approximately 370 W/m^2 K.

(i) Determine the oil outlet temperature for counterflow operation.
(ii) If it is desired to have an outlet temperature 3°C lower than the value obtained in part (i), determine the required change to the water flow rate. Take $c_p = 2030$ J/kg K for the oil.

8-60. A shell-and-tube heat exchanger is to have one pass on the shell side and eight passes on the tube side. It is required to cool 100 kg/s of a petroleum oil inside the tubes from 135°C to 50°C. Water is supplied to the shell at 160 kg/s and enters at 15°C. The exchanger has 200 steel tubes of 3 cm O.D. and 2 mm wall thickness. The inside and outside heat transfer coefficients are estimated to be 260 and 970 W/m^2 K, respectively, and a fouling resistance of 3.5×10^{-4}

$[W/m^2\ K]^{-1}$ should be included on the water side. Determine the required surface area using

(i) the LMTD method.
(ii) the $\varepsilon - N_{tu}$ method.

Take $c_p = 2200$ J/kg K for the oil.

8-61. A space station active temperature control system operating at part load circulates 10 kg/h of pressurized water. Part of the loop is a 10 m-long, straight, 2 cm–O.D., 1 mm-wall-thickness tube located in an unpressurized compartment of the station, where the surroundings are at a temperature of $-23.2°C$. If the water enters the length of tube at $96.8°C$, determine the outlet temperature. The tube wall emittance is 0.71.

8-62. A counterflow heat exchanger is to be designed to cool 0.195 kg/s of water from 5°C to 2°C against 30% aqueous ethylene glycol solution supplied at $-5°C$.

(i) Determine the overall heat transfer coefficient times area product required for an equal mass flow of glycol.
(ii) If a twin-tube heat exchanger (see Example 8.12) with 12 mm–O.D., 1 mm-wall-thickness copper tubes is to be used, determine the tube lengths required.

8-63. A counterflow heat exchanger is to be designed to cool 0.2 kg/s of water from 5°C to 2°C against 20% aqueous ethylene glycol solution supplied at 0°C.

(i) Determine the overall heat transfer coefficient times area product required for equal mass flows.
(ii) A double-tube configuration is to be used (see Fig. 8.1c). The glycol is flowing in a copper inner tube, and the outer annular cross-sectional area is to be chosen to give equal cold- and hot-side convective heat transfer resistances. Determine the heat exchanger lengths required for tube inner diameters in the range 5 mm–20 mm.

8-64. A counterflow waste-heat recuperator is designed to achieve an effectiveness of 0.6 when heating 35 kg/s of air with 20 kg/s of waste gas. By what factor must the transfer area be increased to raise the effectiveness of 0.7? Assume that the overall heat transfer coefficient remains the same, and take $c_p = 1000$ J/kg K for the air and 1100 J/kg K for the waste gas.

8-65. An interstage cooler for a multistage turbocompressor is required to cool 2 kg/s of nitrogen gas at 500 kPa from 165°C to 25°C. A counterflow exchanger is to be used with coolant water supplied at 20°C.

(i) If the water outlet temperature should not exceed 50°C, what is the minimum water flow required?
(ii) A shell-and-tube exchanger is to be used with a single tube pass and a single shell pass. If the gas flows through 2 cm–O.D., 2 mm-wall-thickness tubes at

a mass velocity of 100 kg/m^2 s, determine the number and length of tubes required. Neglect the tube wall and water-side thermal resistances.

8-66. An interstage cooler for an air compressor is required to cool 2 kg/s air at 400 kPa from 150°C to 30°C. A 2 kg/s water supply at 20°C is available as coolant. A shell-and-tube exchanger has been proposed, with the air flowing inside 2 cm–O.D., 1.5 mm-wall-thickness tubes at a mass velocity of 90 kg/m^2 s. An initial design based on a single-tube-pass and single-shell-pass counterflow unit gave an exchanger that was too long for the space available: alternative configurations are to be investigated.

   (i) Calculate the water outlet temperature, exchanger effectiveness, and the number of transfer units for the counterflow unit.
   (ii) Is it possible to use a two-tube-pass, one-shell-pass configuration? If not, what is the lowest air temperature that can be achieved with this configuration?
   (iii) Repeat (ii) for a two-tube-pass, two-shell-pass configuration.

8-67. An oil heater is to heat 5 kg/s of oil from 25°C to 50°C with condensing saturated steam at 570 kPa. A single-pass shell-and-tube configuration is to be used, with the oil inside 34 mm–O.D., 2 mm-wall-thickness tubes. Pressure-drop considerations require an oil mass velocity of 70 kg/m^2 s in the tubes. If the steam-side heat transfer coefficient is estimated to be 9000 W/m^2 K, determine the steam consumption rate, and the number and length of tubes required. Use properties of SAE 50 oil.

8-68. The exhaust gas from a diesel power generation plant is used to boil water in a single-tube-pass shell-and-tube heat exchanger. The exhaust flow rate is 5 kg/s; the gas enters the tubes at 110 kPa and 420°C, and is to leave at 210°C. Saturated water at 1170 kPa is supplied to the shell. The tubes are made of steel, and have a 30 mm O.D. and a 2 mm wall thickness. The mass velocity of the gas in the tubes should not exceed 17 kg/m^2 s to avoid an excessive pressure drop. Determine the number and length of tubes required. Ignore the water-side and tube-wall thermal resistances, but include a gas-side fouling resistance of 0.002 [W/m^2 K]$^{-1}$. Take $c_p = 1100$ J/kg K for the exhaust gas, but use air properties to estimate the convective heat transfer coefficient.

8-69. A one-shell-pass, two-tube-pass heat exchanger is required to cool 30,000 kg/h of oil from 150°C to 70°C against 40,000 kg/h of coolant water supplied at 20°C. If the overall heat transfer coefficient is estimated to be 280 W/m^2 K, determine the oil and water outlet temperatures by

   (i) the LMTD method.
   (ii) the $\varepsilon - N_{tu}$ method.

Use the properties of SAE 50 oil.

8-70. Tables 8.3a and 8.3b, items 9, give formulas for effectiveness and number of transfer units of a two-stream, shell-and-tube heat exchanger with n shell passes and $2n, 4n, \ldots$ tube passes. Show that if the capacity rate ratio $R_C = 1$, the formulas reduce to

$$\varepsilon = \frac{n\varepsilon_1}{1 + (n-1)\varepsilon_1}; \quad E = \frac{2^{1/2}n(1-\varepsilon)}{\varepsilon}$$

8-71. A shell-and-tube exchanger is to be used to cool 5 kg/s of oil used in a heat-treating process for quenching metal parts. The oil enters the exchanger shell at 80°C and must be cooled to 25°C. Coolant water is available at 15°C and a flow rate of 5 kg/s. An overall heat transfer coefficient of 250 W/m² K based on tube outside area is expected for clean surfaces if 30 mm–O.D., 2 mm-wall-thickness steel tubes are used.

   (i) Is it possible to use a single-shell-pass exchanger? If not, explain why not.
  (ii) Determine the heat transfer surface area for a two-shell-pass exchanger.
 (iii) If an oil-side fouling factor of 0.9×10^{-3} [W/m² K]$^{-1}$ is allowed for, what is the percentage increase in area required?

Take the oil specific heat as 2180 J/kg K.

8-72. A single-shell-pass, four-tube-pass heat exchanger is tested with water flowing at 24 kg/s in the tubes and oil flowing in the shell at 20 kg/s. The tube bundle has 80 steel tubes, 30 m long, with a 24 mm O.D. and a 2 mm wall thickness. In a test, the measured oil inlet and outlet temperatures are 85.8°C and 48.1°C when the water inlet temperature is maintained at 20°C. Estimate the outside heat transfer coefficient. Take the oil specific heat as 2320 J/kg K.

8-73. A cross-flow heat exchanger is used to heat air with the hot exhaust from a gas turbine. The exhaust flow rate is 0.2 kg/s and enters the exchanger at 900°C. Air is supplied at 1.2 kg/s and must be heated from 25°C to 125°C. Which of the four cross-flow configurations given as items 4 through 7 of Table 8.3b will give the minimum heat transfer area? Assume that $U = 250$ W/m² K for all cases, and use a specific heat of the exhaust that is 2% higher than that of air at the same temperature.

8-74. A two-tube-pass, single-shell-pass heat exchanger transfers heat from the collector loop to the storage tank in a solar heating system for a building in Denver. The maximum heat transfer rate to be accommodated is 24 kW (for a 50 m² collector at noon on a sunny day in fall or spring). Water at 1 kg/s from the storage tank enters the tubes at 50°C. The collector loop contains 50% ethylene glycol aqueous solution flowing at 0.66 kg/s and enters the shell at 67.5°C. The tube bundle consists of 50 tubes of 24 mm O.D. and 2 mm wall thickness, for which an overall heat transfer coefficient based on outside area of 835 W/m² K is estimated. Determine the length of tubes required. For the glycol solution use $c_p = 3690$ J/kg K.

8-75. A 2 kg/s stream of water is to be heated from 30°C to 80°C by 5 kg/s of exhaust gases available at 200°C. A finned-tube cross-flow exchanger is available for which a 1 m² cross-sectional area is provided for the gas flow. For the resulting gas velocity, an overall heat transfer coefficient of 160 W/m^2 K is estimated. If both streams can be considered to be unmixed, calculate the heat transfer area required

(i) using the LMTD method.
(ii) using the $\varepsilon - N_{tu}$ method.

The specific heat of the exhaust gases can be taken as 1100 J/kg K.

8-76. A steam condenser is required to condense 5.0 kg/s of steam at 22.7 kPa. Cooling water is available at 295 K. A shell-and-tube exchanger is to be used with a single pass of the water through 2 cm–O.D., 1.5 mm–wall-thickness tubes at a bulk velocity of 2 m/s. The overall heat transfer coefficient based on tube outside area is estimated to be 3800 W/m^2 K.

(i) What is the minimum cooling water flow rate that can be used?
(ii) Calculate the number of tubes and tube bundle length for water flow rates of 1.05, 1.10, 1.15, 1.2, and 1.25 times the minimum value.

8-77. Ambient air at $-10°$C for a building heating system is to be preheated against a balanced flow of warm air at 20°C ventilated from the building. The flow rates are 200 m^3/min, and it is desired to recover 80% of the energy of the discharge air. If the allowable pressure drop for the cold-side core is 500 Pa, specify suitable dimensions for a cross-flow exchanger using the louvered fin surface given as item 5 of Table 8.4c. Use a plate thickness of 0.3 mm, a scale factor of 2, and AISI 1010 steel. HEX2 should be used to perform the calculations.

8-78. When flow rates are low, heat conduction in the flow direction adversely affects heat exchanger performance, particularly if a high effectiveness is desired. The parameter that determines whether axial heat conduction is negligible can be determined by a scale analysis of the governing differential equations. Consider a coaxial-tube parallel-flow exchanger with balanced flow, equal hot- and cold- side heat transfer coefficients, and negligible tube-wall thermal resistance.

(i) By performing energy balances on the two streams and the tube wall, show that the governing differential equations are

$$\dot{m}c_p\frac{dT_H}{dx} = -h_c\mathscr{P}(T_H - T_w)$$

$$k_wA_w\frac{d^2T_w}{dx^2} = -hc\mathscr{P}(T_H - T_w) + h_c\mathscr{P}(T_w - T_C)$$

$$\dot{m}c_p\frac{dT_C}{dx} = h_c\mathscr{P}(T_w - T_C)$$

(ii) Following the procedure in Section 2.4.5, define appropriate dimensionless and normalized variables and hence show that axial conduction is negligible if $\lambda = k_wA_w/\dot{m}c_pL \ll 1$, where L is the exchanger length.

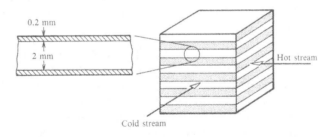

8-79. A low-pressure helium exchanger has a flow area of 0.016 m². A test is conducted in which 2×10^{-3} kg/s of helium is heated from 20 K to 380 K, and the inlet and outlet pressures are measured to be 820 and 110 Pa, respectively. Determine the viscous contribution to the pressure drop.

8-80. A 1 m–cube compact heat exchanger core has the plain fin heat transfer surface given as item 3 of Table 8.4. On the hot side, 0.9 kg/s of air enters at 360 K and 1.09×10^5 Pa, and it exits at 320 K. Calculate the pressure drop.

8-81. Rework Example 8.15 for parallel plates without fins. Assume that transverse mixing is negligible.

8-82. A flow of 50 m³/min of air at 30°C is to be cooled to 20°C against a balanced stream of air at 15°C.

   (i) How many transfer units are required for a cross-flow exchanger with both streams unmixed?
   (ii) Design an appropriate exchanger using a parallel-plate geometry with a gap width of 2 mm and a plate thickness of 0.2 mm. If the pressure drop for each stream should not exceed 40 Pa, specify the dimensions of the exchanger.

8-83. To cool a house in a hot but moderately dry climate, a flow of 40 m³/min of ambient air at 40°C is to be cooled against a balanced stream of 15°C air from an evaporative cooler.

   (i) How many transfer units are required for a counterflow exchanger to cool the air to 20°C?
   (ii) Design an appropriate exchanger using a parallel-plate geometry with a plate pitch of 2.5 mm. Specify the frontal area required for each stream and the

exchanger length if the pressure drop for each stream must not exceed 30 Pa. Assume that the plates are very thin.

8-84. A counterflow twin-tube heat exchanger is to be used for balanced flows of air at 3.0×10^{-3} kg/s. The hot stream enters at 360 K and must be cooled to 320 K; the cold stream enters at 290 K. If the average pressure in each stream is 1 atm and the allowable pressure drop for the hot stream is 104 Pa, determine suitable dimensions. Copper tubes may be assumed, giving a tube wall fin effectiveness of approximately unity.

8-85. To cool a small warehouse in Tucson, Arizona, a flow of 100 m³/min of ambient air at 35°C is to be cooled to 26°C against a balanced stream of 18°C air from an evaporative cooler. Specify the dimensions of an appropriate cross-flow heat exchanger. Use a simple parallel-plate matrix with a plate thickness of 0.2 mm and a gap width of 2.5 mm. The allowable pressure drop for each stream is 50 Pa.

8-86. Repeat Exercise 8–85 using the plain fin surface given as item 4 of Table 8.4. Use a scale factor of 3 and AISI 1010 steel, and increase the allowable pressure drop to 1000 Pa.

8-87. A single-pass cross-flow heat exchanger is to be used as a recuperator to recover energy from the flue gases of a combustor by preheating the combustion air. Inconel-X-750 tubes of outer and inner diameters 60 and 54 mm, respectively, are arranged in a staggered array, with both transverse and longitudinal pitches of 120 mm. The exhaust gas flows through the tube bank, and the combustion air is inside the tubes. The combustor burns 1.2 kg/s of a fuel with stoichiometric ratio of 15.2, and 10% excess air is supplied. The air enters the tubes at 310 K, and the flue gas enters the tube bank at 1100 K. The inside of the tubes can be assumed clean, but a fouling resistance of 0.01 (W/m² K)⁻¹ should be allowed for on the outside of the tubes. If the air is to be heated to 500 K and the pressure drop of the flue gas stream should not exceed 2000 Pa, suggest suitable dimensions for the tube bank. Approximate the flue gas properties by those for air. Use CONV to obtain $\bar{h}_c$ and ΔP.

8-88. One method of cooling IC chips is to provide channels on one side of the chip for a coolant flow. Various designs have been proposed, which use a wide range of channel sizes (see, for example, Exercises 8-89 and 8-90). To evaluate the cooling system as a heat exchanger, an appropriate model is required. Two models that are seen in the literature disagree on a key simplifying assumption. One assumes that the chip can be assumed isothermal: then the coolant channels are single-stream heat exchangers and the temperature difference decreases exponentially along the channel. The other assumes that the heat flux into the coolant is uniform along the channel: then both the chip and coolant temperatures vary linearly along the channel. If either of these assumptions is valid, the analysis is much simplified—otherwise it is necessary to consider the coupling of heat conduction in the chip and convection along the channel. It is satisfactory to assume that the heat is generated uniformly in the chip. The question then is whether conduction along the chip is large enough to give a nearly isothermal chip.

(i) Following the approach used in Exercise 8–78, formulate model equations that allow the effect of conduction to be evaluated.

(ii) Scale these equations and determine the criterion for when the chip can be assumed isothermal.

(iii) For the case where the heat flux into the coolant is uniform along the channel, show that the outlet coolant temperature is given by

$$T_{C,out} = \frac{T_{C,in} + N_{tu}T_{w,max}}{1 + N_{tu}}$$

where $N_{tu} = h_c \mathscr{P} L / \dot{m} c_p$.

8-89. One method of cooling IC chips is to etch deep microgrooves in silicon wafers bonded to the back side of the chip and to cover the open ends with a silicon cap. Consider a chip that is 1 cm square and has rectangular grooves 50 μm wide and 200 μm deep, at a pitch of 100 μm. Coolant water is available at 20°C. For chip power dissipation in the range 100–1000 W, determine the required water flow rates and pressure drops. The silicon thermal conductivity (148 W/m K) is high enough for the fin effectiveness to the channel walls to be close to unity. Take a maximum allowable chip temperature of 90°C.

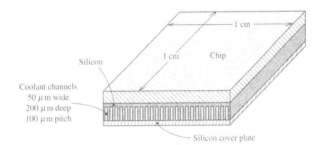

8-90. A novel method has been recently proposed for fabricating IC chip cooling channels, using microelectroplating techniques developed for MEMS technology [9]. Metallic microchannels 5–40 μm wide and 10–40 μm deep are built directly onto the front of the chip without etching or bonding, and can be incorporated directly into existing IC chip manufacturing procedures. Also, the thermal resistance between the heat-generating IC elements and the coolant is much smaller than for heat sinks bonded onto the back side of the chip (see Exercise 8-89). In one design, copper microchannels on a 1 cm-square chip with water as a coolant are proposed. The channels are 20 μm wide and 40 μm deep, and the wall thickness is 5 μm. Inlet water is at 20°C and the maximum chip temperature allowed is 90°C. Structural considerations limit the operating pressure to 2×10^5 Pa gage. What is the maximum power that can be dissipated by the cooling system?

8-91. Repeat Exercise 8-90 for air as coolant.

8-92. A heat exchanger consists of a stack of perforated plates of thickness t, hole diameter d, and open-area ratio ε_p. The stack is in a duct of cross-sectional area

A_f, through which fluid flows at a rate $\dot{m}$ kg/s. The wall of the duct is heated by a condensing vapor that maintains it at a constant temperature. For the parameter range of concern, the following assumptions can be made:

The flow through the holes is laminar and fully developed.

Entrance and exit contributions to the pressure drop across a plate are negligible.

Heat transfer on the front and back faces of the plate is negligible.

The fin effectiveness of the plate is unity—the plate is isothermal (see Example 2.7).

Constant-property, low-speed, continuum flow.

(i) Show that the core velocity equation of Kays and London, Eq. (8.58), becomes

$$\frac{V^2}{(P_1/\rho)} = \frac{(\Delta P/P_1)}{N_{tu}} \cdot \frac{\varepsilon_p^2 N_{tu_p}}{Eu_p}$$

where V is the superficial velocity (based on A_f).

(ii) Hence show that frontal area is inversely proportional to ε_p^2 and independent of hole diameter.

(iii) If the plate thickness is fixed, show that the number of plates required to give a specified N_{tu} is proportional to hole diameter squared and inversely proportional to open-area ratio.

(iv) Comment on the restrictions imposed by the assumption of isothermal plates.

8-93. A cross-flow plate-fin heat exchanger thermal-hydraulic design is described in Example 8.14. Refine the calculations by allowing for a plate thickness of 0.2 mm and a fin thickness of 0.1 mm if the material is AISI 1010 steel.

8-94. A flow of 40 m³/min of air at 60°C is to be cooled to 28.3°C against a balanced flow stream of air at 13°C.

(i) How many transfer units are required for a cross-flow exchanger with both streams unmixed?

(ii) Design an exchanger using square passages with 2 mm sides. Since the air streams contain corrosive vapors, a plastic with $k = 0.2$ W/m K is to be used for the plates and fins. The plate and fin thicknesses should both be 0.2 mm. Specify the dimensions of the exchanger if the pressure drop for each stream should not exceed 425 Pa.

8-95. In a batch process, 10,000 kg of 95°C wash water is dumped into a drain every hour in a plant that operates two eight-hour shifts per day, 310 days per year. At present, mains water at 18°C is heated by steam to 110°C to provide clean water to wash the next batch. The cost of the steam is estimated to be 0.7 cent/MJ of vaporization enthalpy. To reduce energy costs at the plant, it is proposed to pump the dirty wash water to an insulated holding tank for use to preheat the incoming water in a balanced-flow counterflow heat exchanger. The initial cost of the tank, pumps, and plumbing is estimated to be $200,000, and the heat exchanger can be acquired at $400/m^2 of heat transfer surface. Taxes and insurance are 15% of the exchanger cost, and capital can be loaned at 10% interest for 15 years. The exchanger can be designed to have an overall heat transfer coefficient of 1600 W/m^2 K. Should the proposal be adopted? If so, specify the optimal heat transfer surface area and the savings per year.

8-96. Rework the analysis in Section 8.6.4 for an unbalanced counterflow exchanger to obtain the relation between the optimal N_{tu} and the thermoeconomic parameter $\mathscr{E}$.

8-97. An oil supply of 0.31 kg/s must be heated from 15°C during 7200 hours per year of operation. A 0.75 kg/s supply of water at 150°C is available for preheating, if worthwhile. The energy gained by preheating will save 1 cent/MJ, but the heat exchanger will cost $250/m^2 yr to own and operate. If the exchanger design under consideration has an overall heat transfer coefficient of 170 W/m^2 K, should the preheater be installed? If so, how large should it be, what would it cost, and what would be the savings per year? Take $c_p = 2015$ J/kg K for the oil.

8-98. In optimizing the design of a heater, it is sometimes appropriate to prescribe $\dot{m}_C$, $T_{C,in}$, and $T_{C,out}$ for the cold stream and $T_{H,in}$ for the hot stream. The problem is to determine the optimal value of $\dot{m}_H$. We can write $T_{H,in} = T_0 + \dot{m}_{fuel}\Delta h_c/C_H$ if the hot stream is heated by burning a fuel, where Δh_c is the heat of combustion; then the annual cost of the fuel consumed is $\mathscr{C}_f = \dot{m}_{fuel}\tau\Delta h_c\,\nu$. The cost of the exchanger remains as specified by Eq. (8.66), and U is assumed to be independent of $\dot{m}_H$. As the exchanger is lengthened, the cost goes up, but since the effectiveness increases, the value of the fuel used is reduced. Determine the relation between effectiveness ε and thermoeconomic parameter $\mathscr{E}$ that maximizes the benefit-cost differential. Consider two cases:

(i) $C_H = C_{max}$.
(ii) $C_H = C_{min}$.

Make appropriate simplifying assumptions. (See also reference [10].)

8-99. In a laundry, 67°C dirty wash water is dumped into the drain, and 70°C clean water is required. Presently 15°C water is heated in an electric hot water heater, and the electricity costs 9 cents/kW h. The water is required at a rate of 5000 kg/h, 12 hours per day, 312 days/yr. To conserve energy it is proposed to install a counterflow heat exchanger to preheat the feed to the electric water heater. The installation will cost $20,000 plus $900 per square meter of heat exchanger surface. The interest rate to amortize the investment over 12 years is 10% per annum. Taxes and insurance are expected to have a fixed cost of $500 per annum plus $50/yr per square meter of heat exchanger surface. If the overall heat transfer coefficient is estimated to be 1000 W/m^2 K, determine the optimal heat transfer area of the exchanger and the corresponding net annual savings.

8-100. Referring to Examples 8.17 and 8.18, is there a combination of pressure drop and
outlet temperature that maximizes the benefit-cost differential? If there is,
comment on the feasibility of the design.

8-101. Referring to Examples 8.17 and 8.18, explore the effects of the characteristic
dimension of the louvered fin surface by varying the scale factor from 0.5 to 2.0.

8-102. Referring to Examples 8.17 and 8.18, explore the effect of core material by
considering the following materials:

    (i) Aluminum alloy, $k = 180$ W/m K
    (ii) Brass, $k = 111$ W/m K
    (iii) AISI 302 stainless steel, $k = 15$ W/m K

Take $c_i = \$10/m^2$ for the aluminum, and $\$15/m^2$ for the brass and stainless
steel.

8-103. Referring to Examples 8.17 and 8.18, explore the effect of type of heat transfer
surface by reworking the problem for the louvered fin surface given as item 5 of
Table 8.4. Notice that this surface has the same plate spacing as surface 6 but
fewer fins and a larger hydraulic diameter.

8-104. Referring to Examples 8.17 and 8.18, explore the effect of type of heat transfer
surface by reworking the problem for the plain fin surface given as item 3 of Table
8.4. Notice that this surface has approximately the same plate spacing as
surface 6.

8-105. Referring to Example 8.18, explore the effect of annual interest rate of capital
borrowed from the bank. Let i vary from 5 to 15% for a loan amortized over
20 years.

8-106. Referring to Example 8.18, explore the effect of the cost of electricity by varying
c_p from 0.05 to 0.15 \$/kW h.

A

PROPERTY DATA

Table A.1a   Solid metals: Melting point and thermal properties at 300 K

Metal (% composition)	T_{MP} K	ρ kg/m^3	c J/kg K	k W/m K	α^a m^2/s$\times 10^6$
Aluminum					
Pure	933	2702	903	237	97.1
Duralumin	775	2770	875	174	71.8
(4.4 Cu, 1.0 Mg, 0.75 Mn, 0.4 Si)					
Alloy 195, cast		2790	883	168	68.1
(4.5 Cu)					
Beryllium	1560	1850	1825	200	59.2
Bismuth	545	9780	122	7.9	6.59
Cadmium	594	8650	231	97	48.4
Copper					
Pure	1358	8933	385	401	117
Electrolytic tough pitch		8950	385	386	112
(Cu + Ag, 99.90 minimum)					
Commercial bronze	1293	8800	420	52	14.1
(10 Al)					
Brass	1188	8530	380	111	34.2
(30 Zn)					
German silver		8618	410	116	32.8
(15 Ni, 22 Zn)					
Constantan		8920	420	22.7	6.06
(40 Ni)					
Constantan		8860		23	
(45 Ni)					
Gold	1336	19300	129	317	127
Iron					
Pure	1810	7870	447	80.2	22.8
Armco		7870	447	72.7	20.7
(99.75 pure)					
Cast		7272	420	51	16.7
(4 C)					
Carbon steels					
AISI 1010		7830	434	64	18.8
(0.1 C, 0.4 Mn)					
AISI 1042, annealed		7840	460	50	13.9
(0.42 C, 0.64 Mn, 0.063 Ni, 0.13 Cu)					
AISI 4130, hardened and tempered		7840	460	43	11.9
(0.3 C, 0.5 Mn, 0.3 Si,					
0.95 Cr, 0.5 Mo)					
Stainless steels					
AISI 302 (18–8)		8055	480	15	3.88
(0.15 C, 2 Mn, 1 Si, 16–18 Cr,					
6–8 Ni)					
AISI 304	1670	7900	477	15	3.98
(0.08 C, 2 Mn, 1 Si, 18-20 Cr,					
8–10 Ni)					

(Continued)

Table A.1*a*   *(Concluded)*

Metal (% composition)	T_{MP} K	ρ kg/m^3	c J/kg K	k W/m K	α m^2/s$\times 10^6$
Stainless steels *(continued)*					
AISI 316		8238	468	13	3.37
(0.08 C, 2 Mn, 1 Si, 16–18 Cr,					
10–14 Ni, 2–3 Mo)					
AISI 410		7770	460	25	7.00
(0.15 C, 1 Mn, 1 Si, 11.5–13 Cr)					
Lead	601	11340	129	35.3	24.1
Magnesium					
Pure	923	1740	1024	156	87.6
Alloy A8					
(8 Al, 0.5 Zn)					
Molybdenum	2894	10240	251	138	53.6
Nickel					
Pure	1728	8900	444	91	23.0
Inconel-X-750	1665	8510	439	11.7	3.13
(15.5 Cr, 1 Nb, 2.5 Ti, 0.7 Al, 7 Fe)					
Nichrome	1672	8314	460	13	3.40
(20 Cr)					
Nimonic 75		8370	461	11.7	3.03
(20 Cr, 0.4 Ti)					
Hasteloy B		9240	381	12.2	3.47
(38 Mo, 5 Fe)					
Cupro-Nickel		8800	421	19.5	5.26
(50 Cu)					
Chromel-P		8730		17	
(10 Cr)					
Alumel		8600		48	
(2 Mn, 2 Al)					
Palladium	1827	12020	244	71.8	24.5
Platinum					
Pure	2045	21450	133	71.6	25.1
60 Pt-40 Rh	1800	16630	162	47	17.4
(40 Rh)					
Silicon	1685	2330	712	148	89.2
Silver	1235	10500	235	429	174
Tantalum	3269	16600	140	57.5	24.7
Tin	505	7310	227	66.6	40.1
Titanium					
Pure	1943	4500	522	21.9	9.32
Ti-6A1-4V		4420	610	5.8	2.15
Ti-2Al-2 Mn		4510	466	8.4	4.0
Tungsten	3690	19300	132	174	68.3
Zinc	693	7140	389	116	41.8
Zirconium					
Pure	2125	6570	278	22.7	12.4
Zircaloy-4		6560	285	14.2	7.60
(1.2-1.75 Sn, 0.18-0.24 Fe,					
0.07-0.13 Cr)					

[a]This table and subsequent ones are to read as $\alpha \times 10^6 = 97.1$, that is, $\alpha = 97.1 \times 10^{-6}$ m^2/s.

Table A.1b Solid metals: Temperature dependence of thermal conductivity k [W/m K] (see Table A.1a for metal compositions)

Metal	Temperature, K								
	200	300	400	500	600	800	1000	1200	1500
Aluminum									
Pure	237	237	240	236	231	218			
Duralumin	138	174	187	188					
Alloy 195, cast		168	174	180	185				
Copper									
Pure	413	401	393	386	379	366	352	339	
Commercial bronze	42	52	52	55					
Brass	74	111	134	143	146	150			
German silver		116	135	145	147				
Gold	323	317	311	304	298	284	270	255	
Iron									
Armco	81	73	66	59	53	42	32	29	31
Cast		51	44	39	36	27	23		
Carbon steels									
AISI 1010		64	59	54	49	39	31		
AISI 1042		52	50	48	45	37	29	26	30
AISI 4130		43	42	41	40	37	31	27	31
Stainless steels									
AISI 302		15	17	19	20	23	25		
AISI 304	13	15	17	18	20	23	25		
AISI 316		13	15	17	18	21	24		
AISI 410	25	25	26	27	27	29			
Lead	37	35	34	33	31				
Magnesium									
Pure	199	156	153	151	149	146			
Alloy A8			84						
Nickel									
Pure	105	91	80	72	66	68	72	76	83
Inconel-X-750	10.3	11.7	13.5	15.1	17.0	20.5	24.0	27.6	30.0
Nichrome		13	14	16	17	21			
Platinum	73	72	72	72	73	76	79	83	90
Silver	420	429	425	419	412	396	379	361	
Tantalum	58	58	58	59	59	59	60	61	62
Tin	73	67	62	60					
Titanium									
Pure	25	22	20	20	19	19	21	22	25
Ti-6Al-4V		5.8							
Tungsten	185	174	159	146	137	125	118	112	106
Zirconium									
Pure	25	23	22	21	21	21	23	26	29
Zircaloy-4	13.3	14.2	15.2	16.2	17.2	19.2	21.2	23.2	

Table A.1c Solid metals: Temperature dependence of specific heat capacity c [J/kg K] (see Table A.1a for metal compositions)

Metal	Temperature, K								
	200	300	400	500	600	800	1000	1200	1500
Aluminum									
Pure	798	903	949	996	1033	1146			
Duralumin		875							
Alloy 195, cast		883							
Copper									
Pure	356	385	397	412	417	433	451	480	
Commercial bronze	785	420	460	500					
Brass	360	380	395	410	425				
German silver		410							
Gold	124	129	131	133	135	140	145	155	
Iron									
Armco	384	447	490	530	574	680	975	609	634
Cast		420							
Carbon steels									
AISI 1010		434	487	520	559	685	1168		
AISI 1042		460	500	530	570	700	1430		
AISI 4130		460	500	530	570	690	840		
Stainless steels									
AISI 302		480	512	531	559	585	606		
AISI 304	402	477	515	539	557	582	611	640	682
AISI 316		468	504	528	550	576	602		
AISI 410		460							
Lead	125	129	132	136	142				
Magnesium									
Pure	934	1024	1074	1170	1170	1267			
Alloy A8		1000							
Nickel									
Pure	383	444	485	500	512	530	562	594	616
Inconel-X-750	372	439	473	490	510	546	626		
Nichrome		460	480	500	525	545			
Platinum	125	133	136	139	141	146	152	157	165
Silver	225	232	239	244	250	262	277	292	
Tantalum	133	140	144	145	146	149	152	155	160
Tin	215	227	243						
Titanium									
Pure	405	522	551	572	591	633	675	680	686
Ti-6Al-4V		610							
Tungsten	122	132	137	140	142	145	148	152	157
Zirconium									
Pure	264	278	300	312	322	342	362	344	344
Zircaloy-4		285	300	314	327	348	369		

Table A.2   Solid dielectrics: Thermal properties

Dielectric	T K	ρ kg/m³	c J/kg K	k W/m K	α m²/s × 10⁶
Aluminum oxide, Al₂O₃					
Sapphire	300	3970	765	46	15.2
Alumina	300	3970	765	36	11.9
	400		940	27	7.2
	600		1110	16	3.6
	1000		1225	7.6	1.6
	1500			5.4	
Carbon					
Diamond (type IIb)	300	3300	510	1300	772
ATJ-S graphite	300	1810	1300	98	42
	1000		1926	55	16
	2000		2139	38	9.8
	3000		2180	33	8.4
Pyrolytic graphite	300	2210	709	1950	1240
k parallel to layers	600		1406	892	287
	1000		1793	534	135
	2000		2043	262	58
k perpendicular to layers	300	2210	709	5.7	3.64
	600		1406	2.68	0.86
	1000		1793	1.60	0.40
	2000		2043	0.81	0.18
Graphite fiber epoxy	200	1400	640	8.7	9.7
(25% volume) composite	300		935	11.1	8.5
k parallel to fibers	400		1220	13.0	7.6
k perpendicular to fibers	200	1400	640	0.68	0.76
	300		935	0.87	0.66
	400		1220	1.1	0.64
Carbon-carbon weave	300	1860	810	110	73
	1000		1800	56	17
	2000		2140	36.5	9.2
	3000		2220	34.5	8.4
	4000		2260	34	8.1
	5000		2270	34	8.1
Ice	273	910	1930	2.22	1.26
Plastics					
Cellulose acetate	300	1300	1510	0.24	0.12
Neoprene rubber	300	1250	1930	0.19	0.079
Phenolic, filled	300	1760	1260	0.50	0.23
Polyamide (nylon)	300	1140	1670	0.24	0.13
Polyethylene (high density)	300	960	2090	0.33	0.16
Polypropylene	300	1170	1930	0.17	0.075
Polyvinylchloride	300	1714	1050	0.092	0.051
Teflon	300	2200	1050	0.35	0.15
	400			0.45	0.19

(*Continued*)

Table A.2 *(Concluded)*

Dielectric	T K	ρ kg/m^3	c J/kg K	k W/m K	α m^2/s$\times 10^6$
Silicon dioxide, SiO$_2$	200	2650		16.4	
Crystalline (quartz)	300		745	10.4	5.3
k parallel to c-axis	400		885	7.6	3.2
	600		1075	5.0	1.7
k perpendicular to c-axis	200	2650		9.5	
	300		745	6.2	3.1
	400		885	4.7	2.0
	600		1075	3.4	1.2
Polycrystalline (fused silica glass)	300	2220	745	1.38	0.83
	400		905	1.51	0.75
	600		1040	1.75	0.75
	800		1105	2.17	0.88
	1000		1155	2.87	1.11
	1200		1195	4.0	1.51
Titanium dioxide, TiO$_2$ (rutile)	300	4157	710	8.4	2.8
	600		880	5.0	1.4
	1200		945	3.3	0.84
Uranium oxide, UO$_2$	300	10,890	240	7.9	3.0
	500		265	6.0	2.1
	1000		305	3.9	1.2
	1500		325	2.6	0.73
	2000		355	2.3	0.59
	2500		405	2.5	0.57

Table A.3   Insulators and building materials: Thermal properties

	T K	ρ kg/m^3	c J/kg K	k W/m K	α m^2/s$\times 10^6$
Asbestos paper, laminated and corrugated	300	190		0.078	
4 ply	320			0.085	
	340			0.091	
	360			0.097	
	380			0.101	
8 ply	300	300		0.068	
	320			0.073	
	340			0.077	
	360			0.080	
	380			0.083	
Brick					
B&W K-28 insulating	600			0.03	
	1300			0.04	
Chrome	400	3010	835	2.3	0.92
	800			2.5	
	1200			2.0	
Fireclay	400	2645	960	0.9	0.35
	800			1.4	
	1200			1.7	
	1600			1.8	
Common	300	1920	835	0.72	0.45
Face	300	2083		1.3	
Concrete					
Stone 1-2-4 mix	300	2100	880	1.4	0.75
Cork	300	160	1680	0.043	0.16
Cotton	300	80	1300	0.06	0.58
Glass					
Fused silica	300	2220	745	1.38	0.83
Borosilicate (Pyrex)	300	2640	800	1.09	0.51
Soda-lime (25 Na$_2$O, 10 CaO)	300	2400	840	0.88	0.44
Cellular glass	240			0.048	
	260			0.051	
	280			0.054	
	300	145		0.058	
	320			0.063	
	340			0.067	
Fiberglass, paper-faced batt	300	16	835	0.046	3.4
	300	40		0.035	
	260			0.029	
	280			0.033	
	300	28		0.038	
	320			0.043	
	340			0.048	
	360			0.054	
	380			0.060	
	400			0.066	

(*Continued*)

Table A.3   *(Concluded)*

	T K	ρ kg/m^3	c J/kg K	k W/m K	α m^2/s$\times10^6$
Loose fill					
Cellulose, wood or paper pulp	290			0.038	
	300	45		0.039	
	310			0.042	
Vermiculite, expanded	240			0.058	
	260			0.061	
	280			0.064	
	300	122		0.069	
	320			0.074	
	240			0.052	
	260			0.056	
	280			0.059	
	300	80		0.063	
	320			0.068	
Magnesia	300	270		0.062	
(85 %)	350			0.068	
	400			0.073	
	450			0.078	
	500			0.082	
Paper	300	930	2500	0.13	0.056
Polystyrene, rigid	240			0.023	
	260			0.024	
	280			0.026	
	300	30–60	1210	0.028	0.4–0.8
	320			0.030	
Polyurethane, rigid foam	300	70		0.026	
Rubber					
Hard	270	1200	2010	0.15	0.062
Neoprene	300	1250	1930	0.19	0.079
Rigid foamed	260			0.028	
	280			0.030	
	300	70		0.032	
	320			0.034	
Snow	273	110		0.049	
		500		0.190	
Soil					
Dry	300	1500	1900	1.0	0.35
Wet	300	1900	2200	2.0	0.5
Woods					
Oak, parallel to grain	300	820	2400	0.35	0.18
perpendicular to grain	300	820	2400	0.21	0.11
White Pine, parallel to grain	300	500	2800	0.24	0.17
perpendicular to grain	300	500	2800	0.10	0.071
Wool, sheep	300	145		0.05	

Table A.4   Thermal conductivity of selected materials at cryogenic temperatures

Material	Temperature, K				
	5	10	30	100	200
Metals					
Aluminum (2024-T4)	3.5	7.7	21	50	72
Brass				71	94
Copper (OFHC)			950	430	400
Carbon steel (1020)	4.3	12	34	64	70
Stainless steel (303)	0.29	0.71	3.5	9.0	12
(304)	0.16	0.82	3.3	9.5	13
Titanium	0.15	0.80	2.0	4.5	6.5
Nonmetals					
Diamond (type 2A)	45	310	3300	10,000	
Glass, Pyrex				0.56	0.89
Glass, Phoenix		0.11	0.17	0.55	
Nylon-66	0.018	0.033	0.21		
Polyethylene	0.040	0.15	0.75		
Silicon	350	2000	4700	900	
Silicone rubber				0.18	0.21
Teflon			0.18	0.62	1.0
Vacuum grease					
(Dow-Corning silicone)	0.021				
Varnish (G.E. #7031,					
thermosetting)	0.065	0.075	0.15	0.25	0.36

Table A.5a Total hemispherical emittance at $T_s \simeq 300$ K, and solar absorptance[a]

Material and Surface Condition	Total Hemispherical Emittance	Solar Absorptance
Aluminum		
Foil, as received	0.05	
Foil, bright dipped	0.03	0.10
Vacuum-deposited on duPont Mylar	0.03	0.10
Alloy 6061, as received	0.04	0.37
Alloy 7075-T6, sandblasted with		
60 mesh silicon carbide grit	0.30	0.55
Weathered alloy 75S-T6	0.20	0.54
Aluminized silicone resin paint,		
Dow-Corning XP-310	0.20	0.27
Hard-anodized	0.80	0.23
Soft-anodized	0.76	0.55
Roofing	0.24	
Asbestos		
Board	0.93	
Cloth	0.87	
Slate	0.94	
Asphalt	0.88	
Brass		
Oxidized	0.60	
Polished	0.04	
Brick	0.90	0.63
Carbon		
Graphite, crushed on sodium silicate	0.88	0.96
Lampblack	0.92	
Chromium		
Bright plate	0.16	
Heated 50 hr at 870 K	0.18	0.78
Coal	0.78	
Concrete, rough	0.91	0.60
Copper		
Electroplated	0.03	0.47
Black oxidized in Ebanol C	0.16	0.91
Oxidized plate	0.76	
Earthenware		
Glazed	0.90	
Matte	0.93	
Frost, rime	0.99	
Glass		
Polished	0.87–0.92	
Pyrex	0.80	
Smooth	0.91	
Second-surface mirror	0.81	0.13
Gold		
On stainless steel	0.09	
On 3M tape Y8194	0.025	

(Continued)

Table A.5a   *(Continued)*

Material and Surface Condition	Total Hemispherical Emittance	Solar Absorptance
Granite	0.44	
Gravel	0.30	
Ice		
Crystal	0.96	
Smooth	0.97	
Inconel X		
Bright	0.21	0.90
Oxidized 4 hr at 1270 K	0.72	
Oxidized 10 hr at 980 K	0.79	
Iron		
ARMCO, bright	0.12	
ARMCO, oxidized	0.30	
Cast, oxidized	0.57	
Rusted	0.83	
Wrought, polished	0.29	
Wrought, dull	0.91	
Limestone	0.92	
Magnesium oxide	0.72	
Marble		
Polished	0.89	
Smooth	0.56	
White	0.92	
Mortar, lime	0.90	
Mylar, duPont film, aluminized on second surface		
6 μm thick	0.37	
25 μm thick	0.63	
75 μm thick	0.81	
Nickel		
Electroplated	0.03	0.22
Tabor solar absorber, electro-oxidized on copper		
110–30	0.05	0.85
125–30	0.11	0.85
Oak, planed	0.88	
Paints		
Black		
Parson's optical	0.92	0.97
Silicone high heat	0.90	0.94
Epoxy	0.87	0.95
Gloss	0.90	
Enamel, heated 1000 hr at 650 K	0.80	
Silver Chromatone	0.24	0.20
White		
Acrylic resin	0.90	0.26
Gloss	0.85	
Epoxy	0.85	0.25

(Continued)

Table A.5a *(Concluded)*

Material and Surface Condition	Total Hemispherical Emittance	Solar Absorptance
Paper		
Roofing	0.88	
White	0.86	
Plaster, rough	0.89	
Platinum-coated stainless steel	0.12	
Refractory		
Black	0.94	
White	0.90	
Rubber	0.88	
Sand	0.75	
Sandstone, red	0.59	
Silica		
Sintered, powdered, fused	0.82	0.08
Second-surface mirror		
Aluminized	0.81	0.14
Silvered	0.81	0.07
Silver		
Polished	0.02	
Plated on nickel on stainless steel	0.08	
Heated 300 hr at 650 K	0.15	
Slate	0.85	
Snow, fresh	0.82	0.13
Soil	0.94	
Spruce, sanded	0.80	
Stainless steel		
AISI 312, heated 300 hr at 530 K	0.26	
AISI 301, with armco black oxide	0.75	0.89
AISI 410, heated to 980 K	0.15	0.76
AISI 303, sandblasted heavily with 80 mesh		
aluminum oxide grit	0.42	0.85
Teflon	0.85	0.12
Titanium		
75 A	0.12	
75 A, oxidized 300 hr at 730 K in air	0.21	0.80
C-110 M, oxidized 100 hr at 700 K in air	0.06	0.52
C-110 M, oxidized 300 hr at 730 K in air	0.20	0.77
Tungsten, polished	0.03	
Water	0.90	0.98
White potassium zirconium silicate spacecraft coating	0.87	0.13
Zinc, blackened by Tabor solar collector		
electrochemical treatment, 120–20	0.14	0.89

[a]Since the solar spectrum outside the Earth's atmosphere is different from that at ground level, appropriate values of solar absorptance are a little different. The values in this table are for extraterrestrial conditions, except those for brick, concrete, snow, and water.

Table A.5*b*   Temperature variation of total hemispherical emittance for selected surfaces

Material and Surface Condition	Temperature, K						
	200	400	600	800	1000	1200	1400
Aluminum, polished foil	0.03	0.05	0.06	0.07			
Aluminum alloy 245T, polished	0.03	0.04	0.05	00.6	0.08		
Aluminum oxide, Al_2O_3		0.78	0.69	0.61	0.54	0.49	0.42
Cadmium sulphide, CdS	0.56	0.27	0.16				
Carbon		0.83	0.82	0.81	0.81	0.80	0.80
Copper, polished	0.02	0.03	0.03	0.04	0.04	0.05	0.05
Copper, oxidized at 1000 K			0.50	0.58	0.80		
Gold, polished foil	0.02	0.03	0.05	0.06	0.07	0.08	
Iron, polished	0.05	0.09	0.14	0.20	0.25	0.30	0.35
Iron oxides					0.82	0.82	0.80
Magnesium oxide, MgO	0.72	0.73	0.62	0.52	0.47	0.41	0.36
Molybdenum, polished	0.05	0.07	0.08	0.10	0.12	0.14	0.17
Monel metal, polished	0.14	0.15	0.17	0.19	0.22	0.26	0.34
Sodium chloride, NaCl	0.44	0.24					
Nickel, polished	0.08	0.10	0.11	0.12	0.15	0.17	0.21
Platinum, polished	0.05	0.07	0.09	0.11	0.13	0.15	0.17
Silver, polished	0.02	0.02	0.03	0.03	0.03	0.04	0.04
Silicon carbide, SiC	0.84	0.84	0.83	0.83	0.83	0.83	0.83
Silicon oxide, SiO_2, fused	0.72	0.79					
Tantalum, polished	0.01	0.03	0.05	0.08	0.11	0.13	0.16
Titanium alloy A110-AT, polished	0.14	0.17	0.21	0.23	0.26	0.28	0.29
Tungsten, polished		0.03	0.05	0.07	0.09	0.12	0.16
Zirconium oxide, ZrO_2	0.84	0.73	0.63	0.51	0.44	0.42	0.40
Zinc sulphide	0.56	0.30					

Table A.6a   Spectral and total absorptances of metals for normal incidence

Spectral absorptance, normal incidence, $1 < \lambda < 25\mu m$

$$\alpha(\lambda, T_s) \simeq A \left[\frac{(1 + \lambda^2/\lambda_{12}^2)^{1/2} - 1}{\lambda^2/2\lambda_{12}^2} \right]^{1/2} + \frac{B}{C + \lambda^2}$$

Total absorptance, normal incidence, $330 \text{ K} < T_e < 2200 \text{ K}$

$$\alpha(T_s, T_e) \simeq A \left[\frac{(1 + 0.078C_2^2/\lambda_{12}^2 T_e^2)^{1/2} - 1}{0.039C_2^2/\lambda_{12}^2 T_e^2} \right]^{1/2} + \frac{18.6(BT_e^2/C_2^2)}{1 + 18.6(CT_e^2/C_2^2)}$$

where $C_2 = \hbar c_0/\kappa_B = 14,389 \, \mu\text{m K}$

| | Parameter ($T_s \simeq 300$ K) | | | |
Metal	A	B	C	λ_{12}
Aluminum foil	0.0165	0.23	8.9	14
Cadmium, 99.99%, rolled plate	0.054	2.15	3.2	9
Chromium, polished electroplate	0.076	1.58	3.9	3
Columbium, 99.99%, rolled plate	0.15	0.29	$\simeq 0$	1
Copper, 99.99%, polished	0.018	0.077	3.2	45
Gold, 99.99%, polished	0.020	0.056	1.4	45
Indium, 99.99%, scraped	0.060	0.24	1.3	6
Inconel X, rolled plate	0.44	0.036	$\simeq 0$	1
Lead, 99.99%, scraped	0.16	0.39	1.1	4
Manganese, 99.99%, polished	0.19	4.8	11	8
Molybdenum, 99.99%	0.033	0.36	$\simeq 0$	7
Nickel, 99.99%, polished	0.029	0.83	2.4	5
Platinum, 99.99%, cold rolled	0.038	0.42	$\simeq 0$	4
Rhodium, polished electroplate	0.06	1.27	10	6
Silver, polished electroplate	0.011	0.16	11	70
Stainless steel, 303, lapped	$\simeq 0.71$	$\simeq 0$	$\simeq 0$	$\simeq 0.125$
Tin, 99.99%, rolled plate	0.052	0.56	0.8	7
Titanium, polished electroplate	0.13	2.9	8.3	12
Titanium, 99%, lapped	0.09	6.5	15	12
Tungsten, 99.99%, lapped	0.05	0.49	0.3	3
Vanadium, 99.99%, rolled plate	0.17	0.66	0.93	1
Zinc, 99.9%	0.036	0.26	$\simeq 0$	8
Zirconium, 99.99%, rolled plate	0.64	4	35	1

From *Advances in Thermophysical Properties at Extreme Temperatures and Pressures,* Am. Soc. Mech. Engrs., New York: 1965, pp. 189–199.

Table A.6b   Spectral absorptances at room temperature and an angle of incidence of 25° from the normal [for nonconductors $\alpha(25°) \simeq \alpha$ (hemispherical)]

	Wavelength, λ [μm]										
	0.3	0.35	0.4	0.45	0.5	0.6	0.7	0.8	1.0	1.5	2.0
Bright metals:											
Aluminum	0.05	0.05	0.07	0.07	0.08	0.11	0.12	0.14	0.08	0.04	0.035
Chromium	0.52	0.48	0.43	0.40	0.39	0.37	0.37	0.37	0.40	0.34	0.26
Copper					0.53	0.23	0.14	0.10	0.06	0.032	0.029
Gold	0.80	0.78	0.75	0.74	0.60	0.17	0.11	0.08	0.043	0.034	0.027
Stainless steel		0.61	0.57	0.54	0.53	0.49	0.46	0.44	0.34	0.28	0.25
Titanium	0.71	0.65	0.59	0.56	0.53	0.48	0.47	0.43	0.44	0.40	0.34
Paints and coatings:											
3M black velvet	0.97	→	→	→	→	→	→	→	→	→	0.96
Hard-anodized aluminum	0.95	0.94	0.93	→	→	→	→	→	0.92	0.90	0.85
Anodized titanium					0.53	0.48	0.48	0.48	0.50	0.50	0.52
White epoxy paint			0.60	0.12	0.10	0.15	0.21	0.09	0.08	0.30	0.43
Flame sprayed alumina	0.60	0.48	0.34	0.29	0.27	0.24	0.23	0.23	0.25	0.32	0.51
Aluminum paint			0.25	0.25	0.25	0.26	0.29	0.31	0.28	0.25	0.23

	Wavelength, λ [μm]										
	3	4	5	6	8	10	12	15	20	30	40
Bright metals:											
Aluminum	0.029	0.026	0.023	0.021	0.019	0.018	0.017	0.015	0.014	0.013	0.012
Chromium	0.19	0.145	0.110	0.088	0.078	0.065	0.059	0.05	0.047	0.036	0.030
Copper	0.029	0.022	0.021	0.020	0.020	0.018	0.018	0.018	0.018		
Gold	0.025	0.023	0.023	0.022	0.022	0.020	0.020	0.020	0.020		
Stainless steel	0.20	0.17	0.15	0.14	0.12	0.11	0.10	0.09	0.077	0.062	0.053
Titanium	0.29	0.24	0.22	0.20	0.17	0.15	0.14	0.13	0.11	0.09	0.08
Paints and coatings:											
3M black velvet	0.96	0.96	0.95	0.96	0.91	0.95	0.95	0.94	0.94	0.97	0.97
Hard-anodized aluminum	0.92	0.74	0.70	0.83	0.96	0.98	0.84	0.83	0.80	0.80	
Anodized titanium	0.89	0.76	0.76	0.82	0.83	0.90	0.91	0.88	0.85		
White epoxy paint	0.93	0.90	0.90	0.91	0.93	0.91	0.93	0.90	0.84	0.81	0.80
Flame sprayed alumina	0.73	0.53	0.62	0.88	0.98	0.98	0.74	0.79	0.75		
Aluminum paint	0.23	0.23	0.22	0.22	0.26	0.22	0.21	0.20	0.19		

Table A.7   Gases[a]: Thermal properties

Gas	T K	k W/m K	ρ^b kg/m^3	c_p J/kg K	$\mu \times 10^{6c}$ kg/m s	$\nu \times 10^{6c}$ m^2/s	Pr
Air	150	0.0158	2.355	1017	10.64	4.52	0.69
(82 K BP)	200	0.0197	1.767	1009	13.59	7.69	0.69
	250	0.0235	1.413	1009	16.14	11.42	0.69
	260	0.0242	1.360	1009	16.63	12.23	0.69
	270	0.0249	1.311	1009	17.12	13.06	0.69
	280	0.0255	1.265	1008	17.60	13.91	0.69
	290	0.0261	1.220	1007	18.02	14.77	0.69
	300	0.0267	1.177	1005	18.43	15.66	0.69
	310	0.0274	1.141	1005	18.87	16.54	0.69
	320	0.0281	1.106	1006	19.29	17.44	0.69
	330	0.0287	1.073	1006	19.71	18.37	0.69
	340	0.0294	1.042	1007	20.13	19.32	0.69
	350	0.0300	1.012	1007	20.54	20.30	0.69
	360	0.0306	0.983	1007	20.94	21.30	0.69
	370	0.0313	0.956	1008	21.34	22.32	0.69
	380	0.0319	0.931	1008	21.75	23.36	0.69
	390	0.0325	0.906	1009	22.12	24.42	0.69
	400	0.0331	0.883	1009	22.52	25.50	0.69
	500	0.0389	0.706	1017	26.33	37.30	0.69
	600	0.0447	0.589	1038	29.74	50.50	0.69
	700	0.0503	0.507	1065	33.03	65.15	0.70
	800	0.0559	0.442	1089	35.89	81.20	0.70
	900	0.0616	0.392	1111	38.65	98.60	0.70
	1000	0.0672	0.354	1130	41.52	117.3	0.70
	1500	0.0926	0.235	1202	53.82	229.0	0.70
	2000	0.1149	0.176	1244	64.77	368.0	0.70
Ammonia	250	0.0198	0.842	2200	8.20	9.70	0.91
(239.7 K BP)	300	0.0246	0.703	2200	10.1	14.30	0.90
	400	0.0364	0.520	2270	13.8	26.60	0.86
	500	0.0511	0.413	2420	17.6	42.50	0.83
Argon	150	0.0096	3.28	527	12.5	3.80	0.68
(87.3 K BP)	200	0.0125	2.45	525	16.3	6.65	0.68
	250	0.0151	1.95	523	19.7	10.11	0.68
	300	0.0176	1.622	521	22.9	14.1	0.68
	400	0.0223	1.217	520	28.6	23.5	0.67
	500	0.0265	0.973	520	33.7	34.6	0.66
	600	0.0302	0.811	520	38.4	47.3	0.66
	800	0.0369	0.608	520	46.6	76.6	0.66
	1000	0.0427	0.487	520	54.2	111.2	0.66
	1500	0.0551	0.324	520	70.6	218.0	0.67

(*Continued*)

Table A.7   *(Continued)*

Gas	T K	k W/m K	ρ^b kg/m³	c_p J/kg K	$\mu \times 10^{6c}$ kg/m s	$\nu \times 10^{6c}$ m²/s	Pr
Carbon dioxide	250	0.01435	2.15	782	12.8	5.97	0.70
(195 K subl.)	300	0.01810	1.788	844	15.2	8.50	0.71
	400	0.0259	1.341	937	19.6	14.6	0.71
	500	0.0333	1.073	1011	23.5	21.9	0.71
	600	0.0407	0.894	1074	27.1	30.3	0.71
	800	0.0544	0.671	1168	33.4	49.8	0.72
	1000	0.0665	0.537	1232	38.8	72.3	0.72
	1500	0.0945	0.358	1329	51.5	143.8	0.72
	2000	0.1176	0.268	1371	61.9	231.0	0.72
Refrigerant-22	240	0.00744	4.13	629	10.32	2.351	0.87
(232.2 K BP)	250	0.00808	4.22	655	10.78	2.558	0.87
	260	0.00871	4.05	687	11.25	2.775	0.89
	270	0.00932	3.90	726	11.72	3.003	0.91
	280	0.00993	3.76	773	12.21	3.245	0.95
	290	0.0105	3.63	829	12.74	3.505	1.00
	300	0.0111	3.51	897	13.31	3.790	1.07
	310	0.0117	3.40	978	13.96	4.106	1.16
	320	0.0123	3.29	1074	14.70	4.464	1.28
Refrigerant-134a	240	0.00886	5.18	762	9.49	1.832	0.82
(246.9 K BP)	250	0.00979	4.97	795	9.92	1.995	0.81
	260	0.0107	4.78	831	10.36	2.166	0.81
	270	0.0115	4.61	871	10.81	2.347	0.82
	280	0.0124	4.44	915	11.27	2.538	0.83
	290	0.0133	4.29	966	11.77	2.745	0.86
	300	0.0142	4.15	1025	12.31	2.970	0.89
	310	0.0153	4.01	1096	12.91	3.219	0.93
	320	0.0164	3.89	1186	13.60	3.500	0.98
	330	0.0177	3.77	1308	14.42	3.827	1.06
	340	0.0192	3.66	1492	15.47	4.230	1.20
	350	0.0209	3.55	1816	16.91	4.760	1.47
Helium	50	0.046	0.974	5200	6.46	6.63	0.73
(4.3 K BP)	100	0.072	0.487	5200	9.94	20.4	0.72
	150	0.096	0.325	5200	13.0	40.0	0.70
	200	0.116	0.244	5200	15.6	64.0	0.70
	250	0.133	0.195	5200	17.9	92.0	0.70
	300	0.149	0.1624	5200	20.1	124.0	0.70
	400	0.178	0.1218	5200	24.4	200.0	0.71
	500	0.205	0.0974	5200	28.2	290.0	0.72
	600	0.229	0.0812	5200	31.7	390.0	0.72
	800	0.273	0.0609	5200	37.8	620.0	0.72
	1000	0.313	0.0487	5200	43.3	890.0	0.72

(Continued)

Table A.7 *(Continued)*

Gas	T K	k W/m K	ρ^b kg/m^3	c_p J/kg K	$\mu \times 10^{6c}$ kg/m s	$\nu \times 10^{6c}$ m^2/s	Pr
Hydrogen	20	0.0158	1.219	10400	1.08	0.893	0.72
(20.3 K BP)	40	0.0302	0.6094	10300	2.06	3.38	0.70
	60	0.0451	0.4062	10660	2.87	7.06	0.68
	80	0.0621	0.3047	11790	3.57	11.7	0.68
	100	0.0805	0.2437	13320	4.21	17.3	0.70
	150	0.125	0.1625	16170	5.60	34.4	0.73
	200	0.158	0.1219	15910	6.81	55.8	0.68
	250	0.181	0.0975	15250	7.91	81.1	0.67
	300	0.198	0.0812	14780	8.93	109.9	0.67
	400	0.227	0.0609	14400	10.8	177.6	0.69
	500	0.259	0.0487	14350	12.6	258.1	0.70
	600	0.299	0.0406	14400	14.3	350.9	0.69
	800	0.385	0.0305	14530	17.4	572.5	0.66
	1000	0.423	0.0244	14760	20.5	841.2	0.72
	1500	0.587	0.0164	16000	25.6	1560	0.70
	2000	0.751	0.0123	17050	30.9	2510	0.70
Mercury	650	0.0100	3.761	104	64.08	17.04	0.67
(630 K BP)	700	0.0108	3.493	104	69.25	19.83	0.67
	800	0.0124	3.056	104	79.45	26.00	0.67
	900	0.0139	2.716	104	89.30	32.87	0.67
	1000	0.0154	2.445	104	98.67	40.36	0.67
	1200	0.0181	2.037	104	115.9	56.93	0.67
	1400	0.0206	1.746	104	132.1	75.68	0.67
	1600	0.0231	1.528	104	148.3	97.11	0.67
	1800	0.0258	1.358	104	165.1	121.5	0.67
	2000	0.0282	1.222	104	180.9	148.0	0.67
Nitrogen	150	0.0157	2.276	1050	10.3	4.53	0.69
(77.4 K BP)	200	0.0197	1.707	1045	13.1	7.65	0.69
	250	0.0234	1.366	1044	15.5	11.3	0.69
	300	0.0267	1.138	1043	17.7	15.5	0.69
	400	0.0326	0.854	1047	21.5	25.2	0.69
	500	0.0383	0.683	1057	25.1	36.7	0.69
	600	0.044	0.569	1075	28.3	49.7	0.69
	800	0.055	0.427	1123	34.2	80.0	0.70
	1000	0.066	0.341	1167	39.4	115.6	0.70
	1500	0.091	0.228	1244	51.5	226.0	0.70
	2000	0.114	0.171	1287	61.9	362.0	0.70

(Continued)

Table A.7 *(Concluded)*

Gas	T K	k W/m K	ρ^b kg/m^3	c_p J/kg K	$\mu \times 10^{6c}$ kg/m s	$\nu \times 10^{6c}$ m^2/s	Pr
Oxygen	150	0.0148	2.60	890	11.4	4.39	0.69
(90.2 K BP)	200	0.0192	1.949	900	14.7	7.55	0.69
	250	0.0234	1.559	910	17.8	11.4	0.69
	300	0.0274	1.299	920	20.6	15.8	0.69
	400	0.0348	0.975	945	25.4	26.1	0.69
	500	0.042	0.780	970	29.9	38.3	0.69
	600	0.049	0.650	1000	33.9	52.5	0.69
	800	0.062	0.487	1050	41.1	84.5	0.70
	1000	0.074	0.390	1085	47.6	122.0	0.70
	1500	0.101	0.260	1140	62.1	239	0.70
	2000	0.126	0.195	1180	74.9	384	0.70
Saturated steam	273.15	0.0182	0.0048	1850	7.94	1655	0.81
(not at 1 atm)	280	0.0186	0.0076	1850	8.29	1091	0.83
	290	0.0192	0.0142	1860	8.69	612	0.84
	300	0.0198	0.0255	1870	9.09	356.5	0.86
	310	0.0204	0.0436	1890	9.49	217.7	0.88
	320	0.0210	0.0715	1890	9.89	138.3	0.89
	330	0.0217	0.1135	1910	10.3	90.7	0.91
	340	0.0223	0.1741	1930	10.7	61.4	0.92
	350	0.0230	0.2600	1950	11.1	42.6	0.94
	360	0.0237	0.3783	1980	11.5	30.4	0.96
	370	0.0246	0.5375	2020	11.9	22.1	0.98
	373.15	0.0248	0.5977	2020	12.0	20.1	0.98
	380	0.0254	0.7479	2057	12.3	16.4	1.00
Superheated	400	0.0277	0.555	1900	14.0	25.2	0.96
steam	500	0.0365	0.441	1947	17.7	40.1	0.94
(373.2 K BP)	600	0.046	0.366	2003	21.4	58.5	0.93
	800	0.066	0.275	2130	28.1	102.3	0.91
	1000	0.088	0.220	2267	34.3	155.8	0.88
	1500	0.148	0.146	2594	49.1	336.0	0.86
	2000	0.206	0.109	2832	62.7	575.0	0.86

[a] At 1 atm pressure unless otherwise noted.
[b] Calculated using the ideal gas law.
[c] This table and subsequent ones are to be read as $\nu \times 10^6 = 4.52$, that is, $\nu = 4.52 \times 10^{-6}$ m^2/s.

Table A.8   Dielectric liquids: Thermal properties

Saturated Liquid (Melting point) (Boiling point) (Latent heat at BP)	T K	k W/m K	ρ kg/m^3	c_p J/kg K	$\mu \times 10^4$ kg/m s	$\nu \times 10^6$ m^2/s	Pr
Ammonia	220	0.547	705	4480	3.35	0.475	2.75
(195 K MP)	230	0.547	696	4480	2.82	0.405	2.31
(240 K BP)	240	0.547	683	4480	2.42	0.355	1.99
(1.37×10^6 J/kg)	250	0.547	670	4500	2.14	0.320	1.76
	260	0.544	657	4550	1.93	0.293	1.61
	270	0.540	642	4620	1.74	0.271	1.49
	280	0.533	631	4710	1.60	0.253	1.41
	290	0.522	616	4800	1.44	0.234	1.33
	300	0.510	602	4900	1.31	0.217	1.26
	310	0.496	587	4990	1.19	0.202	1.19
	320	0.481	572	5080	1.08	0.188	1.14
Carbon dioxide	220	0.080	1170	1850	1.39	0.119	3.22
(195 K subl.)	230	0.096	1130	1900	1.33	0.118	2.64
(0.57×10^6 J/kg)	240	0.110	1090	1950	1.28	0.117	2.27
	250	0.115	1045	2000	1.21	0.116	2.11
	260	0.113	1000	2100	1.14	0.114	2.11
	270	0.108	945	2400	1.04	0.111	2.33
	280	0.100	885	2850	0.925	0.105	2.64
	290	0.090	805	4500	0.657	0.094	3.78
	300	0.076	670	11000	0.549	0.082	7.95
Engine oil, unused	280	0.147	895	1810	21900	2450	27000
(SAE 50)	290	0.146	889	1850	10900	1230	13900
	300	0.145	883	1900	5030	570	6600
	310	0.144	877	1950	2500	285	3400
	320	0.143	871	1990	1370	157	1910
	330	0.142	865	2030	796	92	1140
	340	0.141	859	2070	515	60	760
	350	0.139	854	2120	350	41	530
	360	0.138	848	2160	255	30.1	400
	370	0.137	842	2200	189	22.5	300
	380	0.136	837	2250	147	17.6	245
	390	0.135	832	2290	112	13.5	191
	400	0.134	826	2330	88.4	10.7	154
	410	0.133	820	2380	71.3	8.7	128
	420	0.132	815	2420	57.9	7.1	106

(*Continued*)

Table A.8   *(Continued)*

Saturated Liquid (Melting point) (Boiling point) (Latent heat at BP)	T K	k W/m K	ρ kg/m^3	c_p J/kg K	$\mu \times 10^4$ kg/m s	$\nu \times 10^6$ m^2/s	Pr
Refrigerant-22 (CHClF$_2$)	220	0.120	1444	1090	3.54	0.245	3.22
(115.6 K MP)	230	0.115	1416	1098	3.26	0.230	3.11
(232.2 K BP)	240	0.111	1386	1108	2.97	0.214	2.97
(0.234 $\times$ 10^6 J/kg)	250	0.106	1356	1120	2.69	0.198	2.84
	260	0.102	1324	1136	2.42	0.183	2.70
	270	0.0974	1292	1158	2.17	0.168	2.58
	280	0.0932	1258	1185	1.95	0.155	2.48
	290	0.0891	1222	1220	1.75	0.143	2.40
	300	0.0850	1183	1263	1.57	0.133	2.33
	310	0.0810	1143	1315	1.41	0.123	2.29
	320	0.0770	1098	1378	1.27	0.116	2.27
Refrigerant-134a	210	0.123	1478	1219	7.45	0.504	7.42
(CF$_3$CH$_2$F)	220	0.118	1451	1226	6.21	0.428	6.45
(172 K MP)	230	0.113	1423	1239	5.21	0.366	5.69
(246.9 K BP)	240	0.109	1394	1255	4.62	0.331	5.33
(0.217 $\times$ 10^6 J/kg)	250	0.104	1364	1275	3.82	0.280	4.67
	260	0.0997	1339	1299	3.35	0.251	4.37
	270	0.0951	1302	1326	2.94	0.226	4.10
	280	0.0904	1270	1357	2.58	0.203	3.87
	290	0.0857	1235	1392	2.31	0.187	3.75
	300	0.0810	1199	1434	2.09	0.174	3.70
	310	0.0763	1160	1484	1.86	0.160	3.62
	320	0.0715	1118	1546	1.64	0.147	3.55
Nitrogen	70	0.151	841	2025	2.17	0.258	2.91
(63.3 K MP)	77.4	0.137	809	2060	1.62	0.200	2.43
(77.4 K BP)	80	0.132	796	2070	1.48	0.186	2.32
(0.200 $\times$ 10^6 J/kg)	90	0.114	746	2130	1.10	0.147	2.05
	100	0.097	689	2310	0.87	0.126	2.07
	110	0.080	620	2710	0.71	0.115	2.42
	120	0.063	525	4350	0.48	0.091	3.30
Oxygen	60	0.19	1280	1660	5.89	0.46	5.1
(55 K MP)	70	0.17	1220	1666	3.78	0.31	3.7
(90 K BP)	80	0.16	1190	1679	2.50	0.21	2.6
(0.213 $\times$ 10^6 J/kg)	90	0.15	1140	1694	1.60	0.14	1.8
	100	0.14	1110	1717	1.22	0.11	1.5

(Continued)

Table A.8 *(Continued)*

Saturated Liquid (Melting point) (Boiling point) (Latent heat at BP)	T K	k W/m K	ρ kg/m^3	c_p J/kg K	$\mu \times 10^4$ kg/m s	$v \times 10^6$ m^2/s	Pr
Therminol 60^a	230	0.132	1040	1380	6210	597	6490
(205 K MP)	250	0.131	1030	1460	686	66.6	765
(561 K 10% BP)	300	0.129	995	1640	63.8	6.41	81.1
	350	0.125	960	1820	21.5	2.24	31.3
	400	0.120	924	1990	10.8	1.17	17.9
	450	0.115	888	2160	6.62	0.745	12.4
	500	0.108	849	2320	4.59	0.541	9.86
	550	0.100	808	2470	3.47	0.429	8.57
Water	275	0.556	1000	4217	17.00	1.70	12.9
(273 K MP)	280	0.568	1000	4203	14.50	1.45	10.7
(373 K BP)	285	0.580	1000	4192	12.50	1.25	9.0
(2.26 x 10^6 J/kg)	290	0.591	999	4186	11.00	1.10	7.8
	295	0.602	998	4181	9.68	0.97	6.7
	300	0.611	996	4178	8.67	0.87	5.9
	310	0.628	993	4174	6.95	0.70	4.6
	320	0.641	989	4174	5.84	0.59	3.8
	330	0.652	985	4178	4.92	0.50	3.2
	340	0.661	980	4184	4.31	0.44	2.7
	350	0.669	973	4190	3.79	0.39	2.4
	360	0.676	967	4200	3.29	0.34	2.0
	370	0.680	960	4209	2.95	0.31	1.81
	373.15	0.681	958	4212	2.85	0.30	1.76
	380	0.683	953	4220	2.67	0.28	1.65
	390	0.684	945	4234	2.44	0.26	1.51
	400	0.685	937	4250	2.25	0.24	1.40
	420	0.684	919	4290	1.93	0.21	1.21
	440	0.679	899	4340	1.71	0.19	1.09
	460	0.670	879	4400	1.49	0.17	0.98
	480	0.657	857	4490	1.37	0.16	0.94
	500	0.638	837	4600	1.26	0.15	0.91
	520	0.607	820	4770	1.15	0.14	0.90
	540	0.577	806	5010	1.05	0.13	0.91
	560	0.547	796	5310	0.96	0.12	0.93
	580	0.516	787	5590	0.87	0.11	0.94

^{a}Registered trademark of Monsanto Chemical Company, St. Louis; also sold under the brand name "Santotherm."

Table A.9 Liquid metals: Thermal properties

Liquid metal (Melting point) (Boiling point) (Latent heat at BP)	T K	k W/m K	ρ kg/m^3	c_p J/kg K	$\mu \times 10^4$ kg/m s	$v \times 10^6$ m^2/s	Pr
Lead	650	16.7	10530	158	23.9	0.227	0.023
(601 K MP)	700	17.5	10470	156	21.1	0.202	0.019
(2020 K BP)	800	19.0	10350	155	17.2	0.166	0.014
(0.850 $\times$ 10^6 J/kg)	900	20.4	10230	155	14.9	0.146	0.011
Lithium	500	43.7	514	4340	5.31	1.033	0.053
(453 K MP)	600	46.1	503	4230	4.26	0.847	0.039
(1613 K BP)	700	48.4	493	4190	3.58	0.726	0.031
(19.5 $\times$ 10^6 J/kg)	800	50.7	483	4170	3.10	0.642	0.025
	900	55.9	473	4160	2.47	0.522	0.018
Mercury	300	8.4	13530	140	14.9	0.110	0.025
(234 K MP)	400	9.8	13280	140	11.3	0.085	0.016
(630 K BP)	500	11.0	13040	140	9.78	0.075	0.012
(0.292 $\times$ 10^6 J/kg)	600	12.1	12780	140	8.31	0.065	0.010
Potassium	400	45.5	814	800	4.9	0.60	0.0086
(337 K MP)	500	43.6	790	790	2.8	0.35	0.0050
(1032 K BP)	600	41.6	765	780	2.1	0.28	0.0040
(2.02 $\times$ 10^6 J/kg)	700	39.5	741	770	1.9	0.25	0.0036
	800	36.8	717	750	1.6	0.23	0.0034
	900	34.4	692	740	1.5	0.21	0.0031
Sodium	500	79.2	900	1335	4.2	0.47	0.0071
(371 K MP)	600	74.7	868	1310	3.1	0.36	0.0055
(1156 K BP)	700	70.1	840	1280	2.5	0.30	0.0046
(3.86 $\times$ 10^6 J/kg)	800	65.7	813	1260	2.2	0.27	0.0042
	900	62.1	792	1255	2.0	0.25	0.0040
	1000	59.3	772	1255	1.8	0.23	0.0038
	1100	56.7	753	1255	1.6	0.21	0.0035

Table A.10a Volume expansion coefficients for liquids

Liquid	T K	$\beta \times 10^3$ 1/K	Liquid	T K	$\beta \times 10^3$ 1/K
Ammonia	293	2.45	Hydrogen	20.3	15.1
Engine oil	273	0.70	Mercury	273	0.18
(SAE 50)	430	0.70		550	0.18
Ethylene glycol	273	0.65	Nitrogen	70	4.9
$C_2H_4(OH)_2$	373	0.65		77.4	5.7
Refrigerant-22	250	2.27		80	5.9
	260	2.41		90	7.2
	270	2.58		100	9.0
	280	2.78		110	12
	290	3.03		120	24
	300	3.35	Oxygen	89	2.0
	310	3.75	Sodium	371	0.27
	320	4.30	Therminol 60	230	0.79
	330	5.09		250	0.75
	340	6.34		300	0.70
	350	8.64		350	0.70
				400	0.76
Refrigerant-134a	230	2.00		450	0.84
	240	2.09		500	0.96
	250	2.20		550	1.1
	260	2.32			
	270	2.47			
	280	2.65			
	290	2.86			
	300	3.13			
	310	3.48			
	320	3.95			
	330	4.61			
	340	5.60			
	350	7.32			
Glycerin	280	0.47			
$C_3H_5(OH)_3$	300	0.48			
	320	0.50			

Table A.10*b*   Density and volume expansion coefficient of water

T K	ρ kg/m^3	$\beta \times 10^6$ 1/K
273.15	999.8679	−68.05
274.00	999.9190	−51.30
275.00	999.9628	−32.74
276.00	999.9896	−15.30
277.00	999.9999	1.16
278.00	999.9941	16.78
279.00	999.9727	31.69
280.00	999.9362	46.04
285.00	999.5417	114.1
290.00	998.8281	174.0
295.00	997.8332	227.5
300.00	996.5833	276.1
310.00	993.4103	361.9
320.00	989.12	436.7
330.00	984.25	504.0
340.00	979.43	566.0
350.00	973.71	624.4
360.00	967.12	697.9
370.00	960.61	728.7
373.15	957.85	750.1
380.00	953.29	788
390.00	945.17	841
400.00	937.21	896
450.00	890.47	1129
500.00	831.26	1432

Table A.11   Surface tensions in contact with air

Liquid	T K	$\sigma \times 10^3$ N/m	Liquid	T K	$\sigma \times 10^3$ N/m
Water	275	75.3	Oxygen (*continued*)	90	13.5
	280	74.8		100	11.1
	290	73.7	Potassium	400	110
	300	71.7		500	105
	310	70.0		600	97
	320	68.3		700	90
	330	66.6		800	83
	340	64.9		900	76
	350	63.2	Sodium	500	175
	360	61.4		700	160
	370	59.5		900	140
	373.15	58.9		1100	120
	380	57.6	Nitrogen	68	11.00
	390	55.6		70	10.53
	400	53.6		72	10.07
	420	49.4		74	9.62
	440	45.1		76	9.16
	460	40.7		77.4	8.85
	480	36.2		78	8.72
	500	31.6		80	8.27
	550	19.7		82	7.84
	600	8.4		84	7.42
	647.1	0.0		86	6.99
Ammonia	220	39		88	6.57
	240	34		90	6.16
	260	30	Refrigerant-22	220	20.2
	280	25		230	18.6
	300	20		240	16.9
	320	16		250	15.3
Carbon dioxide	248	9.1		260	13.7
	293	1.2		270	12.2
Hydrogen	15	2.8		280	10.7
	20	2.0		290	9.2
	25	1.1		300	7.8
Lead	600	470		310	6.4
	700	452		320	5.1
	800	437	Refrigerant-134a	210	21.5
	900	421		220	19.9
Lithium	500	390		230	18.3
	700	360		240	16.7
	900	335		250	15.1
Mercury	300	470		260	13.6
	400	450		270	12.1
	500	430		280	10.7
	600	400		290	9.28
	700	380		300	7.92
Oxygen	60	20.7		310	6.60
	70	18.3		320	5.33
	80	16.0		330	4.12

Table A.12a Thermodynamic properties of saturated steam

T K	$P \times 10^{-5}$ Pa	v m³/kg	ρ kg/m³	$h_{fg} \times 10^{-6}$ J/kg
273.15	0.00610	206.4	0.00484	2.501
274.00	0.00649	194.6	0.00514	2.499
275.00	0.00698	181.8	0.00550	2.496
276.00	0.00750	169.8	0.00589	2.494
277.00	0.00805	158.8	0.00630	2.492
278.00	0.00863	148.6	0.00673	2.490
279.00	0.00925	139.1	0.00719	2.488
280.00	0.00991	130.4	0.00767	2.486
281.00	0.01061	122.2	0.00818	2.484
282.00	0.01136	114.6	0.00873	2.482
283.00	0.01215	107.6	0.00929	2.479
284.00	0.01299	101.0	0.00990	2.476
285.00	0.01388	94.75	0.01055	2.473
286.00	0.01482	89.06	0.01123	2.471
287.00	0.01582	83.73	0.01194	2.468
288.00	0.01688	78.75	0.01270	2.466
289.00	0.01800	74.09	0.01350	2.463
290.00	0.01918	69.74	0.01434	2.461
291.00	0.02043	65.68	0.01523	2.459
292.00	0.02176	61.89	0.01616	2.456
293.00	0.02315	58.35	0.01714	2.454
294.00	0.02463	55.05	0.01817	2.451
295.00	0.02619	51.96	0.01925	2.449
296.00	0.02783	49.07	0.02038	2.447
297.00	0.02957	46.37	0.02157	2.444
298.00	0.03139	43.82	0.02282	2.442
299.00	0.03331	41.42	0.02414	2.439
300.00	0.03533	39.15	0.02554	2.437
301.00	0.03746	37.05	0.02700	2.434
302.00	0.03971	35.07	0.02851	2.432
303.00	0.04206	33.21	0.03011	2.430
304.00	0.04454	31.46	0.03179	2.427
305.00	0.04714	29.81	0.03355	2.425
306.00	0.04987	28.26	0.03539	2.423
307.00	0.05274	26.81	0.03730	2.421
308.00	0.05576	25.44	0.03931	2.418
309.00	0.05892	24.16	0.04139	2.416
310.00	0.06224	22.95	0.04357	2.414
311.00	0.06572	21.81	0.04585	2.412
312.00	0.06936	20.73	0.04824	2.409

(*Continued*)

Table A.12a   (*Continued*)

T	$P \times 10^{-5}$	v	ρ	$h_{fg} \times 10^{-6}$
K	Pa	m³/kg	kg/m³	J/kg
313.00	0.07318	19.72	0.05071	2.407
314.00	0.07717	18.75	0.05333	2.404
315.00	0.08135	17.83	0.05609	2.401
316.00	0.08573	16.97	0.05893	2.399
317.00	0.09031	16.16	0.06188	2.396
318.00	0.09511	15.39	0.06498	2.394
319.00	0.10012	14.66	0.06821	2.391
320.00	0.10535	13.98	0.07153	2.389
321.00	0.11082	13.33	0.07502	2.387
322.00	0.11652	12.72	0.07862	2.384
323.00	0.12247	12.14	0.08237	2.382
324.00	0.12868	11.59	0.08628	2.379
325.00	0.13514	11.06	0.09042	2.377
326.00	0.14191	10.56	0.09470	2.375
327.00	0.14896	10.09	0.09911	2.372
328.00	0.15630	9.644	0.1037	2.370
329.00	0.16395	9.219	0.1085	2.367
330.00	0.17192	8.817	0.1134	2.365
331.00	0.18021	8.434	0.1186	2.363
332.00	0.18885	8.072	0.1239	2.360
333.00	0.19783	7.727	0.1294	2.358
334.00	0.20718	7.400	0.1351	2.355
335.00	0.2169	7.090	0.1410	2.353
336.00	0.2270	6.794	0.1472	2.351
337.00	0.2375	6.512	0.1536	2.348
338.00	0.2484	6.244	0.1602	2.346
339.00	0.2597	5.987	0.1670	2.343
340.00	0.2715	5.741	0.1742	2.341
341.00	0.2837	5.509	0.1815	2.339
342.00	0.2964	5.288	0.1891	2.336
343.00	0.3096	5.077	0.1970	2.334
344.00	0.3233	4.876	0.2051	2.332
345.00	0.3375	4.684	0.2135	2.329
346.00	0.3521	4.500	0.2222	2.326
347.00	0.3673	4.325	0.2312	2.324
348.00	0.3831	4.158	0.2405	2.321
349.00	0.3994	3.999	0.2501	2.319
350.00	0.4164	3.847	0.2599	2.316
351.00	0.4339	3.701	0.2702	2.313
352.00	0.4520	3.562	0.2807	2.311

(*Continued*)

Table A.12a (*Concluded*)

T K	$P \times 10^{-5}$ Pa	v m^3/kg	ρ kg/m^3	$h_{fg} \times 10^{-6}$ J/kg
353.00	0.4708	3.429	0.2916	2.308
354.00	0.4902	3.301	0.3029	2.306
355.00	0.5103	3.179	0.3146	2.303
356.00	0.5310	3.062	0.3266	2.301
357.00	0.5525	2.951	0.3389	2.299
358.00	0.5747	2.844	0.3516	2.296
359.00	0.5976	2.742	0.3647	2.294
360.00	0.6213	2.644	0.3782	2.291
361.00	0.6457	2.550	0.3922	2.288
362.00	0.6710	2.460	0.4065	2.285
363.00	0.6970	2.373	0.4214	2.283
364.00	0.7240	2.291	0.4365	2.280
365.00	0.7518	2.212	0.4521	2.277
366.00	0.7804	2.136	0.4682	2.274
367.00	0.8100	2.063	0.4847	2.272
368.00	0.8405	1.993	0.5018	2.269
369.00	0.8719	1.925	0.5195	2.267
370.00	0.9044	1.861	0.5373	2.265
371.00	0.9377	1.798	0.5562	2.263
372.00	0.9722	1.738	0.5754	2.260
373.00	1.0076	1.681	0.5949	2.257
373.15	1.0133	1.673	0.5977	2.257
380.00	1.2875	1.337	0.7479	2.238
390.00	1.7952	0.9800	1.020	2.211
400.00	2.4563	0.7308	1.368	2.183
410.00	3.303	0.5535	1.807	2.154
420.00	4.371	0.4254	2.351	2.124
430.00	5.701	0.3311	3.020	2.093
440.00	7.335	0.2609	3.833	2.059
450.00	9.322	0.2082	4.803	2.025
460.00	11.708	0.1671	5.984	1.990
470.00	14.551	0.1353	7.391	1.953
480.00	17.908	0.1109	9.017	1.914
490.00	21.839	0.09172	10.90	1.872
500.00	26.401	0.07573	13.20	1.827
510.00	31.676	0.06374	15.69	1.779
520.00	37.726	0.05427	18.43	1.729
530.00	44.618	0.04639	21.56	1.676
540.00	52.420	0.03919	25.52	1.621
550.00	61.200	0.03175	31.50	1.563

Table A.12*b* Thermodynamic properties of saturated ammonia

T K	P kPa	v m^3/kg	$h_{fg} \times 10^{-6}$ J/kg
224	42.98	2.5055	1.414
226	48.27	2.2479	1.409
228	54.09	2.0212	1.404
230	60.48	1.8213	1.398
232	67.46	1.6445	1.392
234	75.10	1.4878	1.387
236	83.42	1.3487	1.381
238	92.50	1.2248	1.375
240	102.29	1.1143	1.369
242	112.96	1.0155	1.363
244	124.52	0.9271	1.358
246	137.05	0.8478	1.351
248	150.58	0.7765	1.345
250	165.07	0.7123	1.339
252	180.68	0.6544	1.333
254	197.50	0.6021	1.327
256	215.47	0.5548	1.320
258	234.80	0.5118	1.314
260	255.41	0.4728	1.307
262	277.46	0.4374	1.301
264	300.98	0.4051	1.294
266	326.04	0.3757	1.287
268	352.80	0.3487	1.280
270	381.12	0.3241	1.273
272	411.23	0.3016	1.267
274	443.24	0.2809	1.259
276	477.14	0.2619	1.252
278	512.97	0.2445	1.245
280	550.86	0.2285	1.238
282	590.87	0.2136	1.230
284	633.16	0.2000	1.223
286	677.70	0.1873	1.215
288	724.66	0.1756	1.207
290	774.15	0.1648	1.199
292	826.05	0.1548	1.191
294	880.67	0.1455	1.183
296	938.00	0.1369	1.175
298	998.03	0.1289	1.167
300	1061.35	0.1214	1.159
302	1127.39	0.1144	1.150

(*Continued*)

Table A.12b (*Concluded*)

T K	P kPa	v m³/kg	$h_{fg} \times 10^{-6}$ J/kg
304	1195.88	0.1079	1.141
306	1268.41	0.1019	1.133
308	1344.02	0.0962	1.124
310	1423.12	0.0909	1.115
312	1505.21	0.0860	1.106
314	1590.85	0.0814	1.097
316	1680.86	0.0770	1.087
318	1773.70	0.0729	1.078
320	1871.02	0.0690	1.068

Table A.12c Thermodynamic properties of saturated nitrogen

T K	P kPa	v m³/kg	$h_{fg} \times 10^{-6}$ J/kg
77.4	101.3	0.2209	0.1995
80	137.8	0.1656	0.1962
85	228.0	0.1028	0.1892
90	358.7	0.06681	0.1813
95	538.1	0.04486	0.1726
100	777.2	0.03123	0.1623
105	1085	0.02227	0.1509
110	1473	0.01613	0.1373
115	1954	0.01156	0.1201
120	2537	0.008101	0.0962
125	3236	0.004889	0.0495
126	3392	0.003216	0

Table A.12d Thermodynamic properties of saturated mercury

T K	P MPa	v m³/kg	$h_{fg} \times 10^{-6}$ J/kg
385	0.00007	229.1	0.2971
390	0.00009	182.0	0.2971
395	0.00011	146.0	0.2970
400	0.00014	117.6	0.2969
405	0.00018	94.76	0.2968
410	0.00022	77.00	0.2967
415	0.00027	63.15	0.2966
420	0.00033	51.84	0.2965
425	0.00041	42.72	0.2964
430	0.00049	35.46	0.2963
435	0.00060	29.51	0.2962
440	0.00073	24.68	0.2961
445	0.00088	20.74	0.2960
450	0.00105	17.53	0.2959
460	0.00150	12.68	0.2957
470	0.00208	9.223	0.2955
480	0.00287	6.863	0.2954
490	0.00390	5.135	0.2952
500	0.00524	3.902	0.2950
520	0.00913	2.331	0.2946
540	0.01527	1.459	0.2942
560	0.02460	0.9370	0.2938
580	0.03815	0.6210	0.2935
600	0.05749	0.4276	0.2931
620	0.08450	0.3010	0.2927
640	0.12117	0.2180	0.2923
660	0.16969	0.1602	0.2920
680	0.23281	0.1201	0.2916
700	0.31357	0.09179	0.2912
720	0.41544	0.07133	0.2908
740	0.54145	0.05629	0.2905
760	0.69616	0.04501	0.2901
780	0.88308	0.03645	0.2897
800	1.10570	0.02989	0.2893
820	1.36881	0.02477	0.2890
840	1.67803	0.02072	0.2886
860	2.03709	0.01750	0.2882
880	2.44991	0.01491	0.2878
900	2.92017	0.01282	0.2874
920	3.45295	0.01109	0.2871
940	4.05586	0.00967	0.2867
960	4.72617	0.00849	0.2863
980	5.47829	0.00749	0.2859
1000	6.30516	0.00665	0.2856

Table A.12e   Thermodynamic properties of saturated refrigerant-22 (chlorodifluoromethane)

T K	P MPa	v m^3/kg	$h_{fg} \times 10^{-6}$ J/kg
175	0.002393	6.9904	0.2686
180	0.003710	4.6346	0.2657
185	0.005596	3.1545	0.2627
190	0.008232	2.1993	0.2597
195	0.011836	1.5673	0.2568
200	0.016663	1.1395	0.2538
205	0.023007	0.8439	0.2508
210	0.031206	0.6355	0.2477
215	0.041634	0.4860	0.2447
220	0.054710	0.3770	0.2416
225	0.070890	0.2962	0.2385
230	0.090670	0.2355	0.2354
235	0.11458	0.1893	0.2321
240	0.14319	0.1537	0.2289
245	0.17709	0.1259	0.2255
250	0.21691	0.1041	0.2221
255	0.26332	0.08665	0.2185
260	0.31699	0.07266	0.2149
265	0.37864	0.06132	0.2111
270	0.44898	0.05206	0.2072
275	0.52878	0.04444	0.2032
280	0.61881	0.03812	0.1990
285	0.71986	0.03284	0.1947
290	0.83274	0.02841	0.1902
295	0.95828	0.02466	0.1854
300	1.0973	0.02148	0.1805
305	1.2508	0.01875	0.1753
310	1.4195	0.01641	0.1698
315	1.6045	0.01438	0.1640
320	1.8067	0.01262	0.1579
325	2.0271	0.01108	0.1514
330	2.2667	0.009731	0.1444
335	2.5268	0.008533	0.1368
340	2.8085	0.007464	0.1285
345	3.1132	0.006503	0.1193
350	3.4425	0.005626	0.1089
355	3.7983	0.004813	0.09678
360	4.1829	0.004034	0.08184
365	4.6000	0.003227	0.06127

Table A.12*f*   Thermodynamic properties of saturated refrigerant-134a (tetrafluoroethane)

T K	P MPa	v m³/kg	$h_{fg} \times 10^{-6}$ J/kg
245	0.0904	0.1804	0.2179
250	0.1134	0.1461	0.2147
255	0.1408	0.1193	0.2115
260	0.1733	0.09827	0.2082
265	0.2113	0.08153	0.2047
270	0.2555	0.06812	0.2011
275	0.3066	0.05728	0.1973
280	0.3653	0.04844	0.1934
285	0.4322	0.04119	0.1894
290	0.5080	0.03519	0.1851
295	0.5936	0.03019	0.1807
300	0.6897	0.02600	0.1761
305	0.7970	0.02246	0.1713
310	0.9165	0.01945	0.1662
315	1.0490	0.01687	0.1609
320	1.1954	0.01466	0.1554
325	1.3567	0.01274	0.1495
330	1.5339	0.01107	0.1433
335	1.7282	0.009611	0.1366
340	1.9410	0.008322	0.1294
345	2.1736	0.007177	0.1217
350	2.4277	0.006151	0.1131
355	2.7056	0.005220	0.1035

Table A.13a   Aqueous ethylene glycol solutions: Thermal properties

T K	20 (263.7 K)	Percent Glycol, by Mass (Freezing Point)			
		30 (257.6 K)	40 (248.7 K)	50 (237.6 K)	60 (213.2 K)
Thermal Conductivity, k [W/m K]					
250	-	-	0.456	0.425	0.400
260	-	0.488	0.456	0.423	0.397
270	0.513	0.488	0.456	0.422	0.394
280	0.519	0.491	0.456	0.419	0.393
290	0.522	0.493	0.456	0.418	0.389
300	0.525	0.494	0.456	0.418	0.388
310	0.531	0.495	0.456	0.417	0.385
320	0.538	0.497	0.456	0.416	0.381
Density, ρ [kg/m^3]					
250	-	-	1069.0	1084.0	1098.5
260	-	1051.6	1066.0	1080.5	1094.0
270	1034.1	1048.1	1061.5	1076.0	1089.0
280	1031.1	1044.1	1057.0	1071.0	1085.0
290	1026.6	1039.1	1052.1	1066.0	1079.0
300	1022.6	1035.6	1047.1	1060.5	1073.0
310	1018.6	1031.1	1043.1	1054.1	1066.5
320	1014.1	1025.6	1037.1	1048.1	1060.0
Specific Heat, c_p [J/kg K]					
250	-	-	-	3480	3260
260	-	-	3710	3490	3300
270	4050	3930	3720	3510	3330
280	4030	3920	3740	3550	3360
290	4020	3930	3750	3570	3400
300	4010	3940	3760	3600	3430
310	4010	3950	3780	3620	3460
320	4020	3960	3800	3650	3500
Dynamic Viscosity, μ [kg/m s $\times 10^3$]					
250	-	-	-	29.0	45.0
260	-	-	11.2	16.2	20.0
270	3.6	5.0	7.0	9.5	14.1
280	2.4	3.4	4.9	6.2	8.8
290	1.8	2.3	3.1	4.1	5.6
300	1.4	1.8	2.3	3.0	4.0
310	1.1	1.4	1.8	2.3	3.0
320	0.9	1.1	1.5	1.8	2.4

Table A.13*b*   Aqueous sodium chloride solutions: Thermal properties

| T | Percent NaCl, by Mass | | | | |
K	5	10	15	20	25
Freezing Point [K]					
	270.2	266.6	262.2	256.7	264.3
Thermal Conductivity, k [W/m K]					
260	-	-	-	0.434	-
270	-	0.504	0.478	0.449	0.420
280	0.542	0.516	0.490	0.464	0.438
290	0.560	0.534	0.508	0.481	0.454
300	0.575	0.550	0.525	0.498	0.470
Density, ρ [kg/m^3]					
260	-	-	-	1160	-
270	-	1076	1115	1156	1189
280	1035	1073	1111	1153	1184
290	1033	1070	1108	1150	1180
300	1032	1069	1107	1149	1178
Specific Heat, c_p [J/kg K]					
260	-	-	-	3365	-
270	-	3680	3515	3380	3280
280	3920	3705	3535	3395	3290
290	3930	3720	3555	3415	3300
300	3940	3730	3575	3425	3310
Dynamic Viscosity, μ [kg/m s$\times 10^3$]					
260	-	-	-	4.70	-
270	-	2.30	2.60	3.05	3.83
280	1.50	1.65	1.95	2.25	2.60
290	1.17	1.30	1.50	1.75	2.03
300	0.90	1.00	1.18	1.38	1.65

Table A.14*a*   Dimensions of commercial pipes [mm] (ASA standard)

Nominal Pipe Size (= I.D., in)		Schedule					
		5	10	40	80	160	XX Strong
$\frac{1}{4}$	O.D.	13.716	13.716	13.716	13.716		
	Wall	1.245	1.651	2.235	3.023		
	I.D.	11.227	10.414	9.246	7.671		
$\frac{3}{8}$	O.D.	17.145	17.145	17.145	17.145		
	Wall	1.245	1.651	2.311	3.200		
	I.D.	14.656	13.843	12.522	10.744		
$\frac{1}{2}$	O.D.	21.336	21.336	21.336	21.336	21.336	21.336
	Wall	1.651	2.108	2.769	3.734	4.750	7.468
	I.D.	18.034	17.120	15.799	13.868	11.836	6.401
$\frac{3}{4}$	O.D.	26.670	26.670	26.670	26.670	26.670	26.670
	Wall	1.651	2.108	2.870	3.912	5.534	7.823
	I.D.	23.368	22.454	20.930	18.847	15.596	11.024
1	O.D.	33.401	33.401	33.401	33.401	33.401	33.401
	Wall	1.651	2.769	3.378	4.547	6.350	9.093
	I.D.	30.099	27.864	26.645	24.308	20.701	15.215
$1\frac{1}{2}$	O.D.	48.260	48.260	48.260	48.260	48.260	48.260
	Wall	1.651	2.769	3.683	5.080	7.137	10.160
	I.D.	44.958	42.723	40.894	38.100	33.985	27.940
2	O.D.	60.325	60.325	60.325	60.325	60.325	60.325
	Wall	1.651	2.769	3.912	5.537	8.712	11.074
	I.D.	57.023	54.788	52.502	49.251	42.901	38.176
3	O.D.			88.900	88.900	88.900	88.900
	Wall			5.486	7.62	11.125	15.240
	I.D.			77.927	73.660	66.650	58.420
4	O.D.			114.300	114.300	114.300	114.300
	Wall			6.020	8.560	13.487	17.120
	I.D.			102.260	97.180	87.325	80.061
5	O.D.			141.300	141.300	141.300	141.300
	Wall			6.553	9.525	15.875	19.050
	I.D.			128.194	122.250	109.550	103.200
6	O.D.			168.275	168.275	168.275	168.275
	Wall			7.150	10.973	18.237	21.946
	I.D.			153.975	146.329	131.801	124.384
8	O.D.			219.075	219.075		
	Wall			8.179	12.700		
	I.D.			202.717	193.675		
10	O.D.			273.050	273.050		
	Wall			9.271	15.062		
	I.D.			254.508	242.926		
12	O.D.			323.850	323.850		
	Wall			10.312	17.450		
	I.D.			303.255	288.950		

Table A.14b   Dimensions of commercial tubes [mm] (ASTM standard)

Nominal Size (= O.D., in)	O.D.	Gage (BWG)	Wall	I.D.
$\frac{3}{16}$	4.775	20	0.889	2.997
$\frac{1}{4}$	6.350	22	0.711	4.928
		20	0.889	4.572
		18	1.245	3.861
$\frac{5}{16}$	7.950	20	0.889	6.172
		18	1.245	5.461
$\frac{3}{8}$	9.525	20	0.889	7.747
		18	1.245	7.036
		16	1.651	6.223
$\frac{1}{2}$	12.700	20	0.889	10.922
		18	1.245	10.211
		16	1.651	9.398
		14	2.108	8.484
$\frac{5}{8}$	15.875	18	1.245	13.386
		16	1.651	12.573
		14	2.108	11.659
$\frac{3}{4}$	19.050	20	0.889	17.272
		18	1.245	16.561
		16	1.651	15.748
		14	2.108	14.834
$\frac{7}{8}$	22.225	16	1.651	18.923
1	25.400	20	0.889	23.622
		18	1.245	22.911
		16	1.651	22.098
		14	2.108	21.184
		12	2.769	19.863
		10	3.404	18.593
$1\frac{1}{4}$	31.750	18	1.245	29.261
		16	1.651	28.448
		14	2.108	27.534
		12	2.769	26.213
		10	3.404	24.943
$1\frac{1}{2}$	38.100	18	1.245	35.611
		16	1.651	34.798
		14	2.108	33.884
		12	2.769	32.563
		10	3.404	31.293
		8	4.191	29.718
2	50.800	18	1.245	48.311
		16	1.651	47.498
		14	2.108	46.584
		12	2.769	45.263
		10	3.404	43.993
		8	4.191	42.418

Table A.14c Dimensions of seamless steel tubes for tubular heat exchangers [mm] (DIN 28 180)

Outside diameter	Wall thickness				
	1.2	1.6	2.0	2.6	3.2
Unalloyed and alloy steel tubes					
16	x	x	x		
20		x	x	x	
25		x	x	x	x
30		x	x	x	x
38			x	x	x
Austenitic stainless steel tubes					
16	x	x	x		
20	x	x	x	x	
25		x	x	x	x
30		x	x	x	x
38		x	x	x	x

Table A.14d Dimensions of wrought copper and copper alloy tubes for condensers and heat exchangers [mm] (DIN 1785-83)

Outside diameter	Wall thickness				
	0.75	1.0	1.25	1.5	2.0
8	x	x	x		
10	x	x	x		
11	x	x	x		
12	x	x	x		
14	x	x	x		
15	x	x	x		
16	x	x	x	x	
18		x	x	x	
19		x	x	x	x
20		x	x	x	x
22		x	x	x	x
23		x	x	x	x
24		x	x	x	x
25		x	x	x	x
28		x	x	x	x
30		x	x	x	x
32		x	x	x	x
35		x	x	x	x

Table A.14e Dimensions of seamless cold drawn stainless steel tubes [mm] (LN 9398)[a]

Outside diameter	Wall thickness								
	0.5	0.6	0.8	1.0	1.2	1.6	2.0	2.5	3.2
5	x		x						
6	x		x	x					
8	x		x	x	x				
10	x		x		x	x			
12			x		x		x		
14				x					
16			x	x	x		x	x	
18		x				x			
20			x	x	x	x		x	x
22			x						
25				x	x	x		x	
28				x					
32			x		x		x		
36									
40			x			x			
45			x			x	x		
50						x	x		

[a] Sizes in conformance with ISO 2964, R20.

Table A.14f Dimensions of seamless drawn wrought aluminum alloy tubes [mm] (LN 9223)[a]

Outside diameter	Wall thickness					
	0.8	1.0	1.2	1.6	2.0	2.5
12	x					
14						
16	x					
18						
20	x					
22						
25	x					
28						
32	x	x	x			
36						
40	x					
45		x	x			
50		x	x	x		
56						
63		x	x		x	
70						
80				x		x

[a] Sizes in conformance with ISO 2964, R20.

Table A.15   U.S. standard atmosphere

Altitude m	Temperature K	Pressure Pa	Density Ratio ρ/ρ_0	Gravity m/s^2	Mean Free Path m	Molecular Weight
0	288.150	1.01325+5	1.0000+0	9.8066	6.6328-8	28.964
200	286.850	9.89454+4	9.8094-1	9.8060	6.7617-8	28.964
400	285.550	9.66114+4	9.6216-1	9.8054	6.8936-8	28.964
600	284.250	9.43223+4	9.4366-1	9.8048	7.0288-8	28.964
800	282.951	9.20775+4	9.2543-1	9.8042	7.1672-8	28.964
1000	281.651	8.98762+4	9.0748-1	9.8036	7.3090-8	28.964
2000	275.154	7.95014+4	8.2168-1	9.8005	8.0723-8	28.964
3000	268.659	7.01211+4	7.4225-1	9.7974	8.9361-8	28.964
4000	262.166	6.16604+4	6.6885-1	9.7943	9.9166-8	28.964
5000	255.676	5.40482+4	6.0117-1	9.7912	1.1033-7	28.964
6000	249.187	4.72176+4	5.3887-1	9.7882	1.2309-7	28.964
7000	242.700	4.11052+4	4.8165-1	9.7851	1.3771-7	28.964
8000	236.215	3.56516+4	4.2921-1	9.7820	1.5453-7	28.964
9000	229.733	3.08007+4	3.8128-1	9.7789	1.7396-7	28.964
10,000	223.252	2.64999+4	3.3756-1	9.7759	1.9649-7	28.964
20,000	216.650	5.52930+3	7.2579-2	9.7452	9.1387-7	28.964
30,000	226.509	1.19703+3	1.5029-2	9.7147	4.4134-6	28.964
40,000	250.350	2.87143+2	3.2618-3	9.6844	2.0335-5	28.964
47,400	270.650	1.10220+2	1.1581-3	9.6620	5.7272-5	28.964
50,000	270.650	7.97790+1	8.3827-4	9.6542	7.9125-5	28.964
52,000	270.650	6.22283+1	6.5386-4	9.6481	1.0144-4	28.964
60,000	255.722	2.24606+1	2.4973-4	9.6241	2.6560-4	28.964
70,000	219.700	5.52047+0	7.1457-5	9.5941	9.2821-4	28.964
80,000	180.65	1.0366+0	1.632-5	9.564	4.065-3	28.964
90,000	180.65	1.6438-1	2.588-6	9.535	2.563-2	28.96
100,000	210.02	3.0075-2	4.060-7	9.505	1.629-1	28.88
105,000	233.90	1.4318-2	1.728-7	9.490	3.810-1	28.75
110,000	257.00	7.3544-3	8.024-8	9.476	8.150-1	28.56
115,000	303.78	4.1224-3	3.774-8	9.461	1.719+0	28.32
120,000	349.49	2.5217-3	1.988-8	9.447	3.233+0	28.07
125,000	442.35	1.6863-3	1.041-8	9.432	6.118+0	27.81
130,000	533.80	1.2214-3	6.195-9	9.417	1.019+1	27.58
135,000	624.30	9.3330-4	4.017-9	9.403	1.560+1	27.37
140,000	714.22	7.4104-4	2.770-9	9.388	2.248+1	27.20
145,000	803.74	6.0560-4	2.001-9	9.374	3.095+1	27.05
150,000	892.79	5.0617-4	1.498-9	9.360	4.114+1	26.92
155,000	957.94	4.2992-4	1.181-9	9.345	5.197+1	26.79
160,000	1022.23	3.6943-4	1.159-9	9.331	6.454+1	26.66

Table A.16   Selected physical constants

Universal gas constant

$\mathscr{R} = 8314.5$ J/kmol K

   $= 1.9872$ kcal/kmol K (thermochemical calorie)

Standard gravitational acceleration

$g = 9.8067$ m/s^2

Atomic mass unit

$u = 1.6605 \times 10^{-27}$ kg

Avogadro's number

$\mathscr{A} = 6.0221 \times 10^{26}$ molecules/kmol

Boltzmann constant

$\kappa_B = 1.3806 \times 10^{-23}$ J/K molecule

Planck's constant

$\hbar = 6.6261 \times 10^{-34}$ J s

Speed of light *in vacuo*

$c_0 = 2.9979 \times 10^8$ m/s

Stefan-Boltzmann constant

$\sigma = 2\pi^5 \kappa_B^4 / 15\hbar^3 c_0^2 = 5.6704 \times 10^{-8}$ W/m^2 K^4

Elementary electric charge

$e = 1.6022 \times 10^{-19}$ A s

UNITS, CONVERSION FACTORS, AND MATHEMATICS

Table B.1a   Base and supplementary SI units

Quantity	Name	Symbol
Length	meter	m
Mass	kilogram	kg
Time	second	s
Electric current	ampere	A
Thermodynamic temperature	kelvin	K
Amount of substance[a]	mole	mol
Luminous intensity	candela	cd
Plane angle	radian	rd
Solid angle	steradian	sr

[a] The mole is the gram mole of the cgs units system.

Table B.1b   Derived SI units

Quantity	Name	Symbol	Definition
Frequency	hertz	Hz	s^{-1}
Force	newton	N	$kg\ m/s^2$
Pressure, stress	pascal	Pa	N/m^2
Energy, work	joule	J	N m
Power	watt	W	J/s
Electric charge, quantity of electricity	coulomb	C	A s
Electric potential, electromotive force	volt	V	J/C
Electrical capacitance	farad	F	C/V
Electrical resistance	ohm	Ω	V/A
Electrical conductance	siemens	S	Ω^{-1}
Magnetic flux	weber	Wb	V s
Magnetic flux density	tesla	T	Wb/m^2
Inductance	henry	H	Wb/A
Luminous flux	lumen	lm	cd sr
Illuminance	lux	lx	lm/m^2
Celsius temperature	degree Celsius	°C	$1°C = 1\ K$[a]

[a] Celsius temperature $T - T_0$, where T is in kelvins and $T_0 = 273.15$ K. The unit *degree Celsius* is equal to the unit *kelvin*.

Table B.1c   Recognized non-SI units

Quantity	Name	Symbol	Definition
Time	minute	min	60 s
	hour	h	60 m
	day	d	24 h
Plane angle	degree	°	$(\pi/180)$ rad
	minute	′	$(1/60)°$
	second	″	$(1/60)'$
Volume	liter	l	10^{-3} m^3
Mass	ton (metric)	t	10^3 kg
Energy	electron volt	eV	1.60219×10^{-19} J
Mass of atoms	atomic mass unit	u	1.66057×10^{-27} kg
Length	astronomical unit	AU	149597.870×10^6 m
	parsec	pc	206,265 AU
Pressure	bar	bar	10^5 Pa

Adapted from NZS 6501:1092 "Units of Measurement," Standards Association of New Zealand, Wellington.

Table B.1d   Multiples of SI units

Factor	Prefix	Symbol
10^{18}	exa	E
10^{15}	peta	P
10^{12}	tera	T
10^9	giga	G
10^6	mega	M
10^3	kilo	k
10^2	hecto	h
10	deca	da
10^{-1}	deci	d
10^{-2}	centi	c
10^{-3}	milli	m
10^{-6}	micro	μ
10^{-9}	nano	n
10^{-12}	pico	p
10^{-15}	femto	f
10^{-18}	atto	a

Table B.2   Conversion factors

Temperature	0.555 K/°R $T[°R] = T[°F] + 459.67$	$T[K] = T[°C] + 273.15$
Length	0.3048 m/ft 2.54 cm/in	1609 m/mi
Velocity	0.3048 (m/s)/(ft/s) 0.4470 (m/s)/(mph)	0.2778 (m/s)/(km/h) 1.6093 (km/h)/(mph)
Volume	2.832×10^{-2} m³/ft³ 3.785×10^{-3} m³/gal 45 gal/bbl (oil)	10^{-3} m³/liter 4.545×10^{-3} m³/Imperial gal.
Mass	0.4536 kg/lb	14.59 kg/slug
Force	4.448 N/lb$_f$	10^{-5} N/dyne
Stress	47.88 (N/m²)/(lb$_f$/ft²) 6895 (N/m²)/psi	10^{-1} (N/m²)/(dyne/cm²)
Pressure	6895 Pa/psi 1.0133×10^5 Pa/atm 760 torr/atm	10^5 Pa/bar 133.3 Pa/atm
Energy, work	1055 J/Btu 4187 J/kcal^a 1.6021×10^{-19} J/ev	10^{-7} J/erg 1.356 J/ft lb$_f$
Power	0.2931 W/(Btu/hr)	0.7457 kW/hp
Heat flux	3.155 (W/m²)/(Btu/ft² hr)	4.187×10^4 (W/m²)/(cal/cm² s)
Heat transfer coefficient	5.678 (W/m² K)/(Btu/ft² hr °F)	4.187×10^4 (W/m² K)/(cal/cm² s °C)
Mass flux	1.3563×10^{-3} (kg/m² s)/(lb/ft² hr)	10^{-1} (kg/m² s)/(g/cm² s)
Mole flux	1.3563×10^{-3} (kmol/m² s)/(lb mole/ft² hr)	10^{-1} (kmol/m² s)/(g mole/cm² s)
Density	16.018 (kg/m³)/(lb/ft³) 515.3 (kg/m³)/(slug/ft³)	10^3 (kg/m³)/(g/cm³)
Enthalpy	2326 (J/kg)/(Btu/lb)	4187 (J/kg)/(cal/g)
Specific heat	4187 (J/kg K)/(Btu/lb °F)	4187 (J/kg K)/(cal/g °C)
Dynamic viscosity	47.88 (kg/m s)/(lb$_f$ s/ft²) 10^{-3} (kg/m s)/cp	10^{-1} (kg/m s)/poise 1 (kg/m s)/(N s/m²)
Diffusivity	2.581×10^{-5} (m²/s)/(ft²/hr)	10^{-4} (m²/s)/(cm²/s)
Thermal conductivity	1.731 (W/m K)/(Btu/hr ft °F)	418.7 (W/m K)/(cal/s cm °C)

^{a} I. T. calorie (International Steam Table calorie). Also in use is the thermochemical calorie, for which there are 4184 J/kcal.

Table B.3 Bessel functions

The differential equation

$$\frac{d^2y}{dx^2} + \frac{1}{x}\frac{dy}{dx} + \left(a^2 - \frac{n^2}{x^2}\right)y = 0$$

has the solution

$$y = AJ_n(ax) + BY_n(ax) \quad \text{for } n = 0, \text{ or an integer}$$

where J_n and Y_n are order n *Bessel functions* of the first and second kinds, respectively.
The differential equation

$$\frac{d^2y}{dx^2} + \frac{1}{x}\frac{dy}{dx} + \left(-a^2 - \frac{n^2}{x^2}\right)y = 0$$

has the solution

$$y = CI_n(ax) + DK_n(ax)$$

where I_n and K_n are order n *modified Bessel functions* of the first and second kinds, respectively.
Rules for differentiation are as follows:

$$\frac{d}{dx}[J_0(ax)] = -aJ_1(ax)$$

$$\frac{d}{dx}[Y_0(ax)] = -aY_1(ax)$$

$$\frac{d}{dx}[I_0(ax)] = aI_1(ax)$$

$$\frac{d}{dx}[K_0(ax)] = -aK_1(ax)$$

Table B.3a Bessel functions of the first and second kinds, orders 0 and 1

x	$J_0(x)$	$J_1(x)$	$Y_n(x)$	$Y_1(x)$
0.0	1.00000	0.00000	$-\infty$	$-\infty$
0.2	+0.99002	+0.09950	−1.0811	−3.3238
0.4	+0.96039	+0.19603	−0.60602	−1.7809
0.6	+0.91200	+0.28670	−0.30851	−1.2604
0.8	+0.84629	+0.36884	−0.08680	−0.97814
1.0	+0.76520	+0.44005	+0.08825	−0.78121
1.2	+0.67113	+0.49830	+0.22808	−0.62113
1.4	+0.56686	+0.54195	+0.33790	−0.47915
1.6	+0.45540	+0.56990	+0.42043	−0.34758
1.8	+0.33999	+0.58152	+0.47743	−0.22366
2.0	+0.22389	+0.57672	+0.51038	−0.10703
2.2	+0.11036	+0.55596	+0.52078	+0.00149
2.4	+0.00251	+0.52019	+0.51042	+0.10049
2.6	−0.09680	+0.47082	+0.48133	+0.18836
2.8	−0.18503	+0.40971	+0.43591	+0.26355
3.0	−0.26005	+0.33906	+0.37685	+0.32467
3.2	−0.32019	+0.26134	+0.30705	+0.37071
3.4	−0.36430	+0.17923	+0.22962	+0.40101
3.6	−0.39177	+0.09547	+0.14771	+0.41539
3.8	−0.40256	+0.01282	+0.06540	+0.41411
4.0	−0.39715	−0.06604	−0.01694	+0.39792
4.2	−0.37656	−0.13864	−0.09375	+0.36801
4.4	−0.34226	−0.20278	−0.16333	+0.32597
4.6	−0.29614	−0.25655	−0.22345	+0.27374
4.8	−0.24042	−0.29850	−0.27230	+0.21357
5.0	−0.17760	−0.32760	−0.30851	+0.14786
5.2	−0.11029	−0.34322	−0.33125	+0.07919
5.4	−0.04121	−0.34534	−0.34017	+0.01013
5.6	+0.02697	−0.33433	−0.33544	−0.05681
5.8	+0.09170	−0.31103	−0.31775	−0.11923
6.0	+0.15065	−0.27668	−0.28819	−0.17501
6.2	+0.20175	−0.23292	−0.24830	−0.22228
6.4	+0.24331	−0.18164	−0.19995	−0.25955
6.6	+0.27404	−0.12498	−0.14523	−0.28575
6.8	+0.29310	−0.06252	−0.08643	−0.30019
7.0	+0.30007	−0.00468	−0.02595	−0.30267
7.2	+0.29507	+0.05432	+0.03385	−0.29342
7.4	+0.27859	+0.10963	+0.09068	−0.27315
7.6	+0.25160	+0.15921	+0.14243	−0.24280
7.8	+0.25541	+0.20136	+0.18722	−0.20389

(*Continued*)

Table B.3*a*   *(Concluded)*

x	$J_0(x)$	$J_1(x)$	$Y_n(x)$	$Y_1(x)$
8.0	+0.17165	+0.23464	+0.22352	−0.15806
8.2	+0.12222	+0.25800	+0.25011	−0.10724
8.4	+0.06916	+0.27079	+0.26622	−0.05348
8.6	+0.01462	+0.27275	+0.27146	−0.00108
8.8	−0.03923	+0.26407	+0.26587	+0.05436
9.0	−0.09033	+0.24531	+0.24994	+0.10431
9.2	−0.13675	+0.21471	+0.22449	+0.14911
9.4	−0.17677	+0.18163	+0.19074	+0.18714
9.6	−0.20898	+0.13952	+0.15018	+0.21706
9.8	−0.23227	+0.09284	+0.10453	+0.23789
10.0	−0.24594	+0.04347	+0.05567	+0.24902

Table B.3*b*   Modified Bessel functions of the first and second kinds, orders 0 and 1

x	$e^{-x}I_0(x)$	$e^{-x}I_1(x)$	$e^x K_0(x)$	$e^x K_1(x)$
0.0	1.00000	0.00000		
0.2	0.82693	0.08228	2.14075	5.83338
0.4	0.69740	0.13676	1.66268	3.25867
0.6	0.59932	0.17216	1.41673	2.37392
0.8	0.52414	0.19449	1.25820	1.91793
1.0	0.46575	0.20791	1.14446	1.63615
1.2	0.41978	0.21525	1.05748	1.44289
1.4	0.38306	0.21850	0.98806	1.30105
1.6	0.35331	0.21901	0.93094	1.19186
1.8	0.32887	0.21772	0.88283	1.10480
2.0	0.30850	0.21526	0.84156	1.03347
2.2	0.29131	0.21208	0.80565	0.97377
2.4	0.27662	0.20848	0.77401	0.92291
2.6	0.26391	0.20465	0.74586	0.87896
2.8	0.25280	0.20073	0.72060	0.84053
3.0	0.24300	0.19682	0.69776	0.80656
3.2	0.23426	0.19297	0.67697	0.77628
3.4	0.22643	0.18922	0.65795	0.74907
3.6	0.21934	0.18560	0.64045	0.72446
3.8	0.21290	0.18210	0.62429	0.70206
4.0	0.20700	0.17875	0.60929	0.68157
4.2	0.20157	0.17553	0.59533	0.66274
4.4	0.19656	0.17245	0.58230	0.64535
4.6	0.19191	0.16949	0.57008	0.62924
4.8	0.18758	0.16667	0.55861	0.61425
5.0	0.18354	0.16397	0.54780	0.60027
5.2	0.17974	0.16138	0.53760	0.58718
5.4	0.17618	0.15890	0.52795	0.57490
5.6	0.17282	0.15652	0.51881	0.56335
5.8	0.16965	0.15424	0.51012	0.55246
6.0	0.16665	0.15205	0.50186	0.54217
6.2	0.16381	0.14994	0.49399	0.53243
6.4	0.16110	0.14792	0.48647	0.52318
6.6	0.15853	0.14597	0.47929	0.51440
6.8	0.15608	0.14409	0.47242	0.50604
7.0	0.15373	0.14228	0.46584	0.49807
7.2	0.15149	0.14054	0.45953	0.49046
7.4	0.14935	0.13886	0.45346	0.48318
7.6	0.14729	0.13723	0.44763	0.47622
7.8	0.14532	0.13566	0.44202	0.46955

(*Continued*)

Table B.3b   (*Concluded*)

x	$e^{-x}I_0(x)$	$e^{-x}I_1(x)$	$e^x K_0(x)$	$e^x K_1(x)$
8.0	0.14343	0.13414	0.43662	0.46314
8.2	0.14160	0.13267	0.43141	0.45699
8.4	0.13985	0.13124	0.42638	0.45108
8.6	0.13816	0.12986	0.42152	0.44539
8.8	0.13653	0.12852	0.41683	0.43991
9.0	0.13495	0.12722	0.41229	0.43462
9.2	0.13343	0.12596	0.40790	0.42952
9.4	0.13196	0.12473	0.40364	0.42459
9.6	0.13054	0.12354	0.39951	0.41983
9.8	0.12916	0.12238	0.39551	0.41522
10.0	0.12783	0.12126	0.39163	0.41076

Table B.4 The complementary error function

$$\text{erfc } \eta = 1 - \frac{2}{\pi^{1/2}} \int_0^{\eta} e^{-u^2} du$$

η	erfc η	η	erfc η	η	erfc η
0.00	1.0000	0.76	0.2825	1.52	0.03159
0.02	0.9774	0.78	0.2700	1.54	0.02941
0.04	0.9549	0.80	0.2579	1.56	0.02737
0.06	0.9324	0.82	0.2462	1.58	0.02545
0.08	0.9099	0.84	0.2349	1.60	0.02365
0.10	0.8875	0.86	0.2239	1.62	0.02196
0.12	0.8652	0.88	0.2133	1.64	0.02038
0.14	0.8431	0.90	0.2031	1.66	0.01890
0.16	0.8210	0.92	0.1932	1.68	0.01751
0.18	0.7991	0.94	0.1837	1.70	0.01621
0.20	0.7773	0.96	0.1746	1.72	0.01500
0.22	0.7557	0.98	0.1658	1.74	0.01387
0.24	0.7343	1.00	0.1573	1.76	0.01281
0.26	0.7131	1.02	0.1492	1.78	0.01183
0.28	0.6921	1.04	0.1413	1.80	0.01091
0.30	0.6714	1.06	0.1339	1.82	0.01006
0.32	0.6509	1.08	0.1267	1.84	0.00926
0.34	0.6306	1.10	0.1198	1.86	0.00853
0.36	0.6107	1.12	0.1132	1.88	0.00784
0.38	0.5910	1.14	0.1069	1.90	0.00721
0.40	0.5716	1.16	0.10090	1.92	0.00662
0.42	0.5525	1.18	0.09516	1.94	0.00608
0.44	0.5338	1.20	0.08969	1.96	0.00557
0.46	0.5153	1.22	0.08447	1.98	0.00511
0.48	0.4973	1.24	0.07950	2.00	0.00468
0.50	0.4795	1.26	0.07476	2.10	0.002980
0.52	0.4621	1.28	0.07027	2.20	0.001863
0.54	0.4451	1.30	0.06599	2.30	0.001143
0.56	0.4284	1.32	0.06194	2.40	0.000689
0.58	0.4121	1.34	0.05809	2.50	0.000407
0.60	0.3961	1.36	0.05444	2.60	0.000236
0.62	0.3806	1.38	0.05098	2.70	0.000134
0.64	0.3654	1.40	0.04772	2.80	0.000075
0.66	0.3506	1.42	0.04462	2.90	0.000041
0.68	0.3362	1.44	0.04170	3.00	0.000022
0.70	0.3222	1.46	0.03895	3.20	0.000006
0.72	0.3086	1.48	0.03635	3.40	0.000002
0.74	0.2953	1.50	0.03390	3.60	0.000000

For computer use: erfc $\eta = (a_1 t + a_2 t^2 + a_3 t^3) e^{-t^2}$; $t = (1 + p\eta)^{-1}$; $p = 0.47047$; $a_1 = 0.3480242$; $a_2 = -0.0958798$; $a_3 = 0.7478556$

C

CHARTS

Heat Exchangers

C.4a   LMTD correction factor for a heat exchanger with one shell pass and 2, 4, 6, ...tube passes

C.4b   LMTD correction factor for a cross-flow heat exchanger with both fluids unmixed

C.4c   LMTD correction factor for a cross-flow heat exchanger with both fluids mixed

C.4d   LMTD correction factor for a cross-flow heat exchanger with two tube passes (unmixed) and one shell pass (mixed)

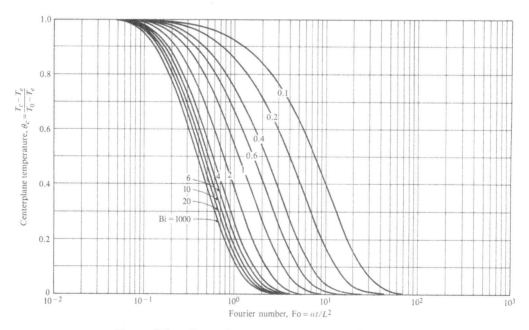

Figure C.1a   Centerplane temperature response for a convectively cooled slab; $Bi = h_c L/k$, where L is the slab half-width

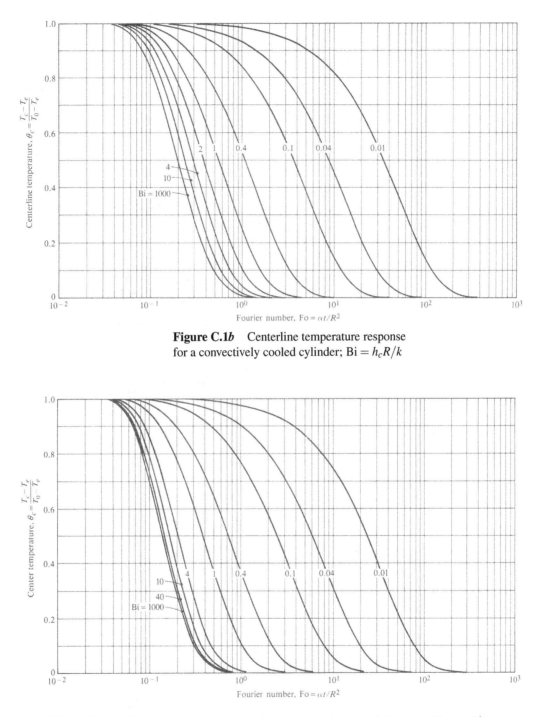

Figure C.1b   Centerline temperature response
for a convectively cooled cylinder; $Bi = h_c R/k$

Figure C.1c   Center temperature response for a convectively cooled sphere; $Bi = h_c R/k$

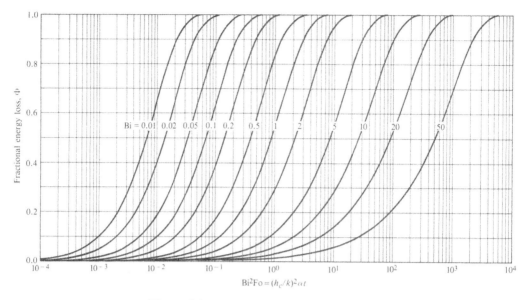

Figure C.2a Fractional energy loss for a convectively cooled slab; $Bi = h_c L/k$, where L is the slab half-width

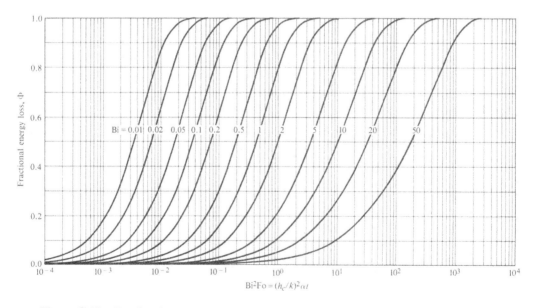

Figure C.2b Fractional energy loss for a convectively cooled cylinder; $Bi = h_c R/k$

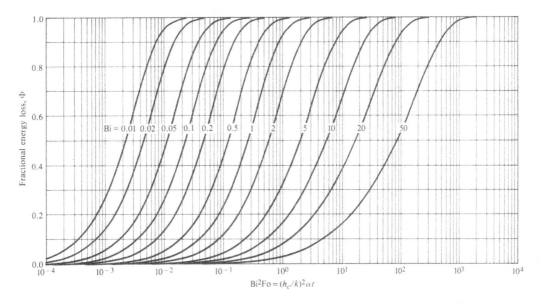

Figure C.2c Fractional energy loss for a convectively cooled sphere; $Bi = h_c R/k$

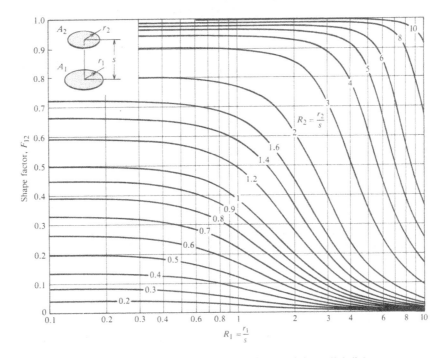

Figure C.3a Shape (view) factor for coaxial parallel disks

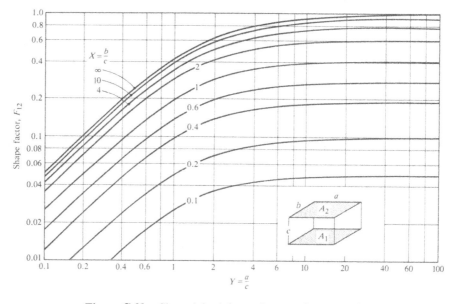

Figure C.3b   Shape (view) factor for opposite rectangles

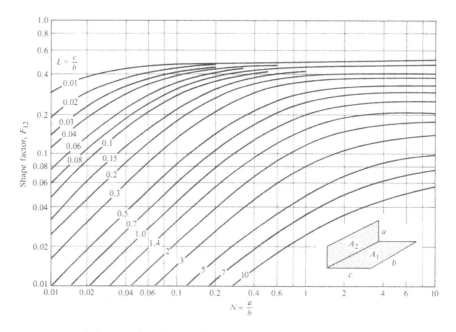

Figure C.3c   Shape (view) factor for adjacent rectangles

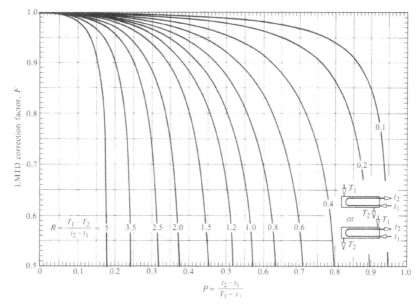

Figure C.4a   LMTD correction factor for a heat
exchanger with one shell pass and 2, 4, 6, ... tube passes

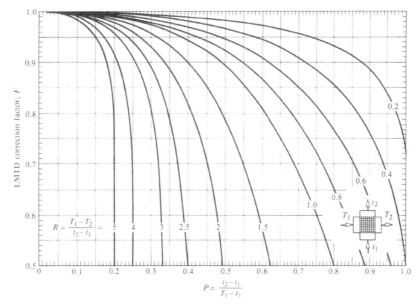

Figure C.4b   LMTD correction factor for a cross-flow
heat exchanger with both fluids unmixed

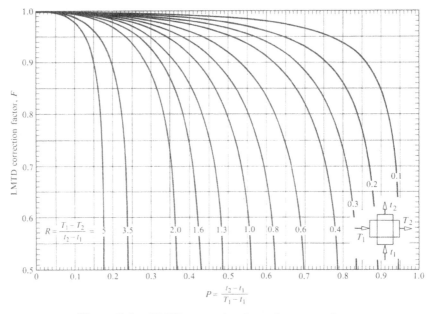

Figure C.4c LMTD correction factor for a cross-flow heat exchanger with both fluids mixed

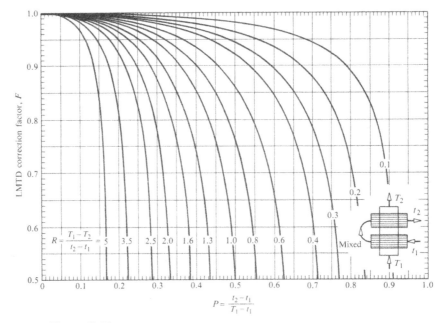

Figure C.4d LMTD correction factor for a cross-flow heat exchanger with two tube passes (unmixed) and one shell pass (mixed)

BIBLIOGRAPHY

GENERAL

There are many general heat transfer texts that can be consulted by the student seeking a different perspective on the material presented in *Basic Heat Transfer*. Following is a selection of books that the student might find useful.

Holman, J. P., *Heat Transfer*, 10th ed., McGraw-Hill, New York, N.Y. (2009).

Bergman, T. L., Lavine, A. S., Incropera, F. C, and DeWitt, D. P., *Fundamentals of Heat and Mass Transfer*, 7th ed., John Wiley & Sons, New York (2011).

Lienhard, J. H. V, and Lienhard, J. H. IV, *A Heat Transfer Textbook*, 4th ed., Dover Publications Inc., Mineola, N.Y. (2011).

Çengel, Y. A., and Ghajar, A. J., *Heat and Mass Transfer: Fundamentals and Applications*, 5th ed., McGraw-Hill, New York, N.Y. (2014).

Pitts, D. , and Sissom, L. E., *Schaum's Outline of Heat Transfer*, 2nd ed., McGraw-Hill, New York, N.Y. (2011).

Rathore, M. M., and Kapuno, R. A., *Engineering Heat Transfer*, 2nd ed., Jones & Bartley Learning, Mississauga, Ontario, Canada (2011).

Somewhat more advanced texts include:

Bird, R. B., Stewart, W. E., and Lightfoot, E. N., *Transport Phenomena*, revised 2nd ed., John Wiley & Sons, Hoboken, N.J. (2006).

Eckert, E. R. G., and Drake, R. M., Jr., *Analysis of Heat and Mass Transfer*, CRC Press, Boca Raton, F.L. (1986).

Kutateladze, S. S., *Fundamentals of Heat Transfer*, 2nd ed., Academic Press, New York, N.Y. (1963).

Rohsenow, W. M., and Choi, H. Y., *Heat, Mass, and Momentum Transfer*, Prentice-Hall, Englewood Cliffs, N.J. (1961).

Faghri, A., Zang, X., and Howell, J., *Advanced Heat and Mass Transfer*, Global Digital Press, Columbia, M.O. (2010).

References of historical relevance are:

Boelter, L. M. K., Cherry, V. H., Johnson, H. A., and Martinelli, R. C., *Heat Transfer Notes*, McGraw-Hill, New York, N.Y. (1965).
Jakob, M., *Heat Transfer*, 2 vols., John Wiley & Sons, New York (1949).
McAdams, W. H., *Heat Transmission*, 3rd ed., McGraw-Hill, New York (1954).

Useful reference works include:

Kreith, F. (Editor-in-Chief), *CRC Handbook of Thermal Engineering*, CRC Press, Boca Raton, F.L. (1999).
Rohsenow, W. M., Hartnett, J. P., and Ganic, E. N., eds., *Handbook of Heat Transfer Fundamentals*, 2nd ed., McGraw-Hill, New York (1985).

An introduction to temperature measurement is given by:

Beckwith, T. G., Marangoni, R. D., and Lienhard, J. H., V: *Mechanical Measurements*, 6th ed., Prentice-Hall, Upper Saddle River, N.J. (2006).

Introductory texts on the design of thermal systems are:

Boehm, R. F., *Design Analysis of Thermal Systems*, John Wiley & Sons, New York (1987).
Burmeister, L. C, *Elements of Thermal-Fluid System Design*, Prentice-Hall, Englewood Cliffs, N.J. (1998).
Jaluria, Y., *Design and Optimization of Thermal Systems*, 2nd ed., CRC Press, Boca Raton, F.L. (2007).
Janna, W.S., *Design of Thermal Systems*, 4th ed., Cengage Learning, Stamford, C.T. (2015).

CHAPTER 2

Conduction in nuclear fuel elements is dealt with in:

El-Wakil, M. M., *Nuclear Heat Transport*, 3rd printing, American Nuclear Society, LaGrange Park, 111. (1981), Chap. 13.

More detailed treatments of fins are given in:

Kraus, A. D., Aziz, A., and Welty, J. *Extended Surface Heat Transfer*, John Wiley & Sons, New York, N.Y. (2001).
Webb, R. L., and Kim, N., *Principles of Enhanced Heat Transfer*, 2nd ed., CRC Press, Boca Raton, F.L. (2005).

CHAPTER 3

A treatise on the mathematical analysis of heat conduction is:

Carslaw, H. S., and Jaeger, J. C, *Conduction of Heat in Solids*, 2nd ed., Oxford
    Science Publications, New York, N.Y. (1986).

Carslaw and Jaeger's book contains a large number of solutions for steady and
transient heat conduction problems.

Transform methods are emphasized in:

Hahn, D. W, and Özisik, M. N., *Heat Conduction*, 3rd ed., John Wiley & Sons, New
    York (2012).

A good elementary heat conduction text is

Myers, G. E., *Analytical Methods in Conduction Heat Transfer*, 2nd ed., AMCHT
    Publications, Madison, W.I. (1998).

A comprehensive compilation of conduction shape factors is given by:

Hahne, E., and Grigull, U., "Formfactor und Formwiderstand der Stationären
    Mehrdimensionalen Wärmeleitung," *Int. J. Heat Mass Transfer*, 18, 751–767
    (1975).

More advanced treatments of numerical methods for heat conduction may be
found in:

Patankar, S. V., *Numerical Heat Transfer and Fluid Flow*, Hemisphere, Washington,
    and McGraw-Hill, New York (1980).
Minkowycz, W. J., Sparrow, E. M., Schneider, G. E., and Pletcher, R. H., *Handbook
    of Numerical Heat Transfer*, John Wiley & Sons, New York (1988).
Reddy, J.N., and Gastling, D.K. *The Finite Element Method in Heat Transfer and
    Fluid Mechanics*, 3rd ed., CRC Press, Boca Raton, F.L. (2010).

CHAPTER 4

Useful compilations of correlation formulas for convective heat transfer may be
found in the following reference works.

Hewitt, G. F., coord, ed., *Hemisphere Handbook of Heat Exchanger Design, 2: Fluid
    Mechanics and Heat Transfer*, Hemisphere, New York (1990).
Kakaç, S., Shah, R. H., and Aung, W., *Handbook of Single Phase Convective Heat
    Transfer*, John Wiley & Sons, New York (1987).
Norris, R. H., et al., eds., *Heat Transfer and Fluid Flow Data Books*, General Electric
    Co., Schenectady, N.Y. (1943, with supplements to current date).
Rohsenow, W. M., Hartnett, J. P., and Ganic, E. N., eds., *Handbook of Heat Transfer
    Fundamentals*, 2nd ed., McGraw-Hill, New York (1985).
Webb, R. L., and Kim, N., *Principles of Enhanced Heat Transfer*, 2nd ed., CRC
    Press, Boca Raton, F.L. (2005).
Žukauskas, A., and Ulinskas, R., *Heat Transfer in Tube Banks in Crossflow*,
    Hemisphere, New York (1988).

CHAPTER 5

Advanced texts on convective heat transfer are:

Bejan, A., *Convection Heat Transfer*, 4th ed., John Wiley & Sons, New York (2013).
Burmeister, L. C, *Convection Heat Transfer*, 2nd ed., John Wiley & Sons, New York (1993).
Gebhart, B., Jaluria, Y., Mahajan, R. L., and Sammakia, B., *Buoyancy-Induced Flows and Transport*, CRC Press, Boca Raton, F.L. (1988).
Ghiaasiaan, S.M., *Convective Heat and Mass Transfer*, Cambridge University Press (2014).
Kays, W. M., Crawford, M. E., and Weigand, B. *Convective Heat and Mass Transfer*, 4th ed., McGraw-Hill, New York (2004).
Oosthuizen, P. H., and Naylor, D., *An Introduction to Convective Heat Transfer Analysis*, WCB/McGraw-Hill, New York, N.Y. (1999).

Useful supporting fluid mechanics texts are:

Schlichting, H., and Gersten, K. *Boundary Layer Theory*, 8th revised ed., Springer-Verlag, Berling, Germany (2000).
White, F. M., *Viscous Fluid Flow*, 3rd ed., McGraw-Hill, New York (2005).

CHAPTER 6

Recommended advanced texts on radiation heat transfer are:

Brewster, M. Q., Thermal Radiative Transfer and Properties, John Wiley & Sons, New York (1992).
Edwards, D. K., *Radiation Heat Transfer Notes*, Hemisphere, Washington, D.C. (1981).
Hottel, H. C, and Sarofim, A. F., *Radiative Transfer*, McGraw-Hill, New York (1967).
Modest, M. F., *Radiative Heat Transfer*, 3rd ed., Academic Press, Waltham, M.A. (2013).
Siegel, R., and Howell, J. R., and Mengüç, M. P. *Thermal Radiation Heat Transfer*, 5th ed., CRC Press, Boca Raton, F.L. (2010).

Solar radiation is treated in:

Duffie, J. A., and Beckman, W. A., *Solar Engineering of Thermal Processes*, 4th ed., John Wiley & Sons, New York (2013).
Goswami, D. Y., Kreith, F., and Kreider, J.F., *Principles of Solar Engineering*, 2nd ed., Taylor and Francis, New York, N.Y. (2000).
Kalogirou, S. A., *Solar Energy Engineering*, Academic Press, Waltham, M.A. (2009).

Design methods for low temperature solar collectors are given by:

Edwards, D. K., *Solar Collector Design*, Franklin Institute Press, Philadelphia (1978).

Atmospheric radiation is covered in:

Bohren, C. F., and Clothiaux, E. E., *Fundamentals of Atmospheric Radiation*, Wiley-VCH, Weinheim, Germany (2006).
Liou, K. N., *An Introduction to Atmospheric Radiation*, 2nd ed., Academic Press, Waltham, M.A. (2002).

CHAPTER 7

Reference works for condensation and boiling include:

Carey, V. P., *Liquid-Vapor Phase-Change Phenomena*, 2nd ed., CRC Press, Boca Raton, F.L. (2007).
Collier, J. G., and Thome, J. R., *Convective Boiling and Condensation*, 3rd ed., Oxford University Press, Oxford (1996).
Dwyer, O. E., *Boiling Liquid-Metal Heat Transfer*, American Nuclear Society, Hinsdale, I.L. (1976).
Hsu, Y. Y, and Graham, R. W., *Transport Processes in Boiling and Two-Phase Systems*, 2nd ed., American Nuclear Society (1986).
Kandilkar, S. G., *Handbook of Phase Change: Boiling and Condensation*, CRC Press, Boca Raton, F.L. (1999).
Stephan, K., *Heat Transfer in Boiling and Condensation*, Springer-Verlag, Berlin (1992).
Tong, L. S., and Tang, Y. S., *Boiling Heat Transfer and Two-Phase Flow*, 2nd ed., Taylor and Francis, Bristol, Pa. (1997).

Reference works for heatpipes include:

Reay, D. A., McGlen, R., and Kew, P., *Heat Pipes*, 6th ed., Elsevier, Waltham, M.A. (2014).
Faghri, M., *Heat Pipe Science and Technology*, Taylor and Francis, New York, N.Y. (1995).
Peterson, G. P., *Introduction to Heat Pipes*, John Wiley & Sons, New York (1994).
Terpstra, M., and van Veen, J. G., *Heat Pipes Construction and Application*, Elsevier Science Publishing, New York, N.Y. (2011).

CHAPTER 8

Useful reference books are:

Thulukkanam, K., *Heat Exchanger Design Handbook*, 2nd ed., CRC Press, Boca Raton, F.L. (2013).

Kakaç, S., Liu, H., and Pramuanjaroenkij, A., *Heat Exchangers: Selection, Rating and Thermal Design*, 3rd ed., CRC Press, Boca Raton, F.L. (2012).

Shah, R. K., and Sekulic, D. P., *Fundamentals of Heat Exchanger Design*, John Wiley & Sons, Hoboken, N.J. (2003).

Fraas, A. P., *Heat Exchanger Design*, 2nd ed., John Wiley & Sons, New York (1989).

Hausen, H., *Heat Transfer in Counterflow, Parallel Flow and Cross Flow*, McGraw-Hill, New York (1983).

Hesselgreaves, J. E., *Compact Heat Exchangers: Selection, Design and Operation*, Pergamon Press, Oxford, U.K. (2001).

Hewitt, G. F., coord. ed. *Hemisphere Handbook of Heat Exchanger Design*, Hemisphere, New York (1990).

Hewitt, G. F., Shires, G. L., and Bott, T. R., *Process Heat Transfer*, CRC Press, Boca Raton, Fla. (1994).

Idelchik, I. E., *Handbook of Hydraulic Resistance*, 3rd ed., Jaico Publishing House, New Delhi, India (2005).

Kakaç, S., and Liu, H., *Heat Exchangers*, CRC Press, Boca Raton, Fla. (1997).

Kays, W. M., and London, A. L., *Compact Heat Exchangers*, 3rd ed., Krieger, Melbourne, Fla. (1998).

Leong, K. C., Toh, K. C., and Leong, X. C., "Shell and Tube Heat Exchanger Design Software for Educational Applications," Int. J. Engng. Ed., 14, 217-224 (1998).

Shah, R. K., and Mueller, A. C, "Heat Exchangers," Chap. 4 in Rohsenow, W. M., Hartnett, J. P., and Ganic, E. N., eds., *Handbook of Heat Transfer Applications*, McGraw-Hill, New York (1985).

Standards of Tubular Exchangers Manufacturers Association, 9th ed., TEMA, Tarrytown, N.Y. (2007).

APPENDIX A

The most comprehensive compilation of thermophysical property data available is:

Touloukian, Y. S., and Ho, C. Y, eds., *Thermophysical Properties of Matter*, 13 vols., IFI/Plenum Press, New York (1970–1977).

Other useful sources include:

National Institute of Standards and Technology (NIST), *Thermophysical Properties of Fluid Systems*, webbook.nist.gov/fluid.

Sharqawy, M. H., Lienhard, J. H. V., and Zubair, S. M. *Thermophysical Properties of Seawater*, web.mit.edu/seawater.

American Society of Heating, Refrigeration and Air Conditioning Engineers, *ASHRAE Handbook of Fundamentals*, ASHRAE, Atlanta, Ga. (2013).

American Society of Metals, *Metals Handbook Desk Edition*, 2nd ed., Davis, J. R. (ed.), CRC Press, Boca Raton, F.L. (1998).

Hewitt, G. E, coord, ed., *Hemisphere Handbook of Heat Exchanger Design*, Hemisphere, New York (1990).

Irvine, T. F., Jr., and Lilley, P. E., *Steam and Gas Tables with Computer Equations*, Academic Press, Orlando, Fla. (1984).

Millat, J., Dymond, J. H., and Nieto de Castro, C. A., eds. *Transport Properties of Fluids*, Academic Press, Waltham, M.A. (2005).

Vargaftik, N. B., *Handbook of Thermal Conductivity of Liquids and Gases*, CRC Press, Boca Raton, F.L. (1993).

HEAT TRANSFER JOURNALS

The two most widely read heat and mass transfer journals are:

International Journal of Heat and Mass Transfer
Journal of Heat Transfer (Transactions of the American Society of Mechanical Engineers)

Other relevant journals include:

AIAA Journal of Thermophysics and Heat Transfer
AIChE Journal
Applied Thermal Engineering
Chemical Engineering Progress
Combustion and Flame
Combustion Science and Technology
Experimental Heat Transfer
Experimental Thermal and Fluid Science
Heat and Fluid Flow
Heat Transfer Engineering
Heat Transfer: Japanese Research
Heat Transfer: Recent Contents (titles)
Heat Transfer: Soviet Research
International Journal of Heat and Fluid Flow
Journal of Aerosol Science
Journal of Computational Physics
Journal of Enhanced Heat Transfer
Journal of Fluid Mechanics
Journal of Quantitative Spectroscopy and Radiative Transfer
Journal of Solar Energy Engineering
Letters in Heat and Mass Transfer
Nuclear Engineering and Design
Numerical Heat Transfer, Part A: Applications
Numerical Heat Transfer, Part B: Fundamentals
Physics of Fluids
Proceedings of the Combustion Institute
Progress in Energy and Combustion Science
Renewable Energy
Solar Energy
Wärme- und Stoffübertragung

NOMENCLATURE

A area, m^2; amplitude, m

A_c cross-sectional area; area for flow, m^2

A_{eff} effective area, m^2

A_f fin surface area, m^2

A_{fr} frontal area, m^2

A_p prime (unfinned) area, profile area of a straight fin, particle surface area, m^2

a surface area per unit volume (packed beds), m^{-1}

$\mathscr{A}$ Avogadro's number, molecules/kmol

BP   boiling point

Bi   Biot number, Eq. (1.40)

Bo   Boussinesq number, Eq. (4.31)

Br   Brinkman number, Eq. (4.24)

C flow thermal capacity (flow rate times specific heat), W/K; cost, \$; thermal capacity, J/K; electrical capacitance, F

C_D drag coefficient, Eq. (4.68)

C_f skin friction coefficient, Eq. (4.14)

C_{fb} constant in film boiling correlations

$C_{\max}$ constant in boiling peak heat flux correlations

$C_{\min}$ constant in boiling minimum heat flux correlations

C_{nb} constant in nucleate boiling correlation

$\mathscr{C}$ annual cost, \$/yr

c specific heat, J/kg K; average molecular speed, m/s; unit cost (e.g., per unit transfer area), \$/m^2; speed of light, m/s

c_0 speed of light in a vacuum, m/s

c_p specific heat at constant pressure, J/kg K; pumping power unit cost, \$/W h

c_v specific heat at constant volume, J/kg K

$\mathscr{c}$ unit annual cost (e.g., per unit transfer area), \$/m^2 yr

$\mathscr{c}_{pL}$ annual cost of pumping power per unit length, \$/m yr

D diameter, m

D_h	hydraulic diameter, m	h'_{fg}	enthalpy of vaporization plus sub-cooling correction, J/kg
E	energy, J; emissive power, W/m^2; voltage, V	h_{fs}	enthalpy of solidification, J/kg
$\mathscr{E}$	dimensionless thermoeconomic parameter, Eq. (8.98)	h_{sg}	enthalpy of sublimation, J/kg
e	elementary electric charge, A s	$\hbar$	Planck's constant, J s
Ec	Eckert number, Eq. (4.23)	I	intensity, W/m^2 sr; modified Bessel function of first kind; electrical current, A
Eu	Euler number, Eq. (4.17)		
F	force, N; LMTD correction factor; function	i	annual interest rate
F_{ij}	shape (view) factor	$\mathbf{i}$	unit vector in the x direction
Fo	Fourier number, Eq. (3.35), (3.90)	J	radiosity, W/m^2; Bessel function of first kind
$\mathscr{F}_{ij}$	transfer factor		
f	friction factor, Eq. (4.15) or (4.119)	Ja	Jakob number, Eq. (3.90)
G	irradiation, W/m^2; mass velocity, kg/m^2 s	$\mathbf{j}$	unit vector in the y direction
		K	modified Bessel function of second kind
G_0	solar constant, kW/m^2		
Gr	Grashof number, Eq. (4.27)	K_e, K_c	expansion and contraction coefficients, Eqs. (8.72) and (8.73)
g	gravitational acceleration, m/s^2		
$\mathbf{g}$	gravity vector, m/s^2	k	thermal conductivity, W/m K; absorptive index
H	elevation or height, m; total enthalpy, J/kg		
		k_s	equivalent sand grain roughness, m
h	heat transfer coefficient, W/m^2 K; enthalpy, J/kg; characteristic roughness height, m; metric coefficient	$\mathbf{k}$	unit vector in the z direction
		L	length, m; liquid stream superficial velocity, kg/m^2 s
h_b	boiling heat transfer coefficient, W/m^2 K	L_c	characteristic length defined by Eq. (7.63), $[\sigma/(\rho_l - \rho_v)g]^{1/2}$, m
h_c	convective heat transfer coefficient, W/m^2 K	L_{eff}	effective heatpipe length, m
		$\mathscr{L}$	characteristic length; effective beam length, m
h_i	interfacial conductance, W/m^2 K	$\mathscr{L}_m$	mean beam length, m
h_r	radiative heat transfer coefficient, W/m^2 K	$\mathscr{L}_m^0$	geometric mean beam length, m
		ℓ	Prandtl mixing length, m
h_{fg}	enthalpy of vaporization, J/kg	ℓ_t	transport mean free path, m
		M	molecular weight, kg/kmol

$\dot{m}$	mass flow rate, kg/s	R	radius, m; gas constant J/kg K; thermal resistance, K/W; electrical resistance, Ω
$\dot{m}''$	mass flow rate per unit area across a phase interface (mass transfer rate), kg/m^2 s		
		R_C	capacity ratio, Eq. (8.34)
MP	melting point	R_c	critical bubble radius, m
m	mass of a molecule, kg	R_f	fouling factor, [W/m^2 K]$^{-1}$
N	number of plates; number of tube rows transverse to flow	Ra	Rayleigh number, Eq. (4.29)
		Re	Reynolds number, Eqs. (4.13) and (7.18)
N_{tu}	number of transfer units, Eq. (8.33)		
N_{tu}^1	one-side number of transfer units, Eq. (8.59)	$\mathscr{R}$	universal gas constant, J/kmol K
		r	radial coordinate, m; recovery factor; annual charge (annunity), yr^{-1}
Nu	Nusselt number, Eq. (4.19)		
$\mathscr{N}$	molecule number density, molecules/m^3	r_p	pore radius of a wick, m
		S	conduction shape factor, m; surface area, m^2; contact area, m^2
n	refractive index; loan period, yr		
n	number fraction	S_p	pellet surface area, m^2
P	pressure, Pa	S'	surface area per unit width, m
P_E	equivalent broadening pressure ratio, Eq. (6.98)	S_L	longitudinal pitch, m
		S_T	transverse pitch, m
P_L	S_L/D	St	Stanton number, Eq. (4.21)
P_T	S_T/D	T	temperature, K or °C
Pe	Peclet number, Eq. (4.22)	t	time, s; thickness, m
Pr	Prandtl number, Eq. (4.18)	t_c	time constant, s
p	pitch, m; momentum, kg m/s	U	overall heat transfer coefficient, W/m^2 K; internal energy, J
$\mathscr{P}$	perimeter, m		
Q	thermal energy, J	u	specific internal energy, J/kg; velocity component in x direction, m/s; orthogonal curvilinear coordinate
$\dot{Q}$	net rate of heat transfer into a system, rate of heat flow, W		
		u_b	bulk velocity, m/s
$\dot{Q}_v$	internal heat source, W	V	velocity, m/s; volume, m^3
$\dot{Q}_v'''$	volumetric heat source per unit volume, W/m^3	v	specific volume, m^3/kg; velocity component in y direction, m/s
$\mathbf{q}$	heat flux vector, W/m^2	v_t	characteristic turbulence speed, m/s
q	heat flux, W/m^2		

$\mathcal{V}$ characteristic velocity, m/s; annual value, \$/yr

$\mathpzc{v}$ value of heat energy, \$/W h

v   velocity vector, m/s

W width of a surface, m; work done on a system, J; mass, kg

$\dot{W}$ rate of doing work, power, W

We   Weber number, Eq. (7.136)

w mass content of a system, kg; velocity component in z direction, m/s

x rectangular coordinate, m

Y Bessel function of second kind

y rectangular coordinate, m

z rectangular coordinate, elevation, m

GREEK SYMBOLS

α thermal diffusivity, m^2/s; absorptance

β thermal coefficient of volume expansion, K^{-1}; fin parameter; geometric mesh factor

β' heat transfer area per unit volume (plate-fin exchanger), m^{-1}

Γ flow rate per unit width, kg/m s

Δ finite increment; thermal boundary layer thickness, m

Δ_2 energy thickness, m, Eq. (5.62)

δ film thickness, m; hydrodynamic boundary layer thickness, m

δ_1 displacement thickness, m

δ_2 momentum thickness, m

δ_f equivalent stagnant film thickness, m

ε emittance; heat exchanger effectiveness; mass exchanger effectiveness

ε_M eddy diffusivity of momentum (eddy viscosity), m^2/s

ε_H eddy diffusivity of heat, m^2/s

ε_v void fraction

ζ dimensionless time

η dimensionless spatial coordinate

η_f fin efficiency, Eq. (2.42)

η_p pump efficiency

η_t total surface efficiency, Eq. (2.46)

θ angle, rad; contact angle, °; dimensionless temperature

κ absorption coefficient, m^{-1}; Darcy permeability, m^2; von Kármán's constant

κ_B Boltzmann constant, J/K

λ eigenvalue; wavelength, μm

μ dynamic viscosity, kg/m s

ν kinematic viscosity, m^2/s; wave-number, cm^{-1}

ν_f frequency, Hz (s^{-1})

ξ unheated starting length, m; dimensionless spatial coordinate; general spatial variable, m

ρ density, kg/m^3; reflectance

σ surface tension, N/m; Stefan-Boltzmann constant, W/m^2 K^4; core-to-frontal-area ratio; electrical conductivity, Ω^{-1} m^{-1}

τ shear stress, N/m^2; time period, s; transmittance; annual operating time, h

Φ fractional loss of energy, Eq. (3.73); arrangement factors for tube banks, Eqs. (4.116) and (4.117)

ϕ angle, rad

χ fin parameter, Eq. (2.42); tube bank pressure drop correction factor, Eq. (4.119)

Ψ Prandtl number function for natural convection, Eq. (4.84)

ψ tube bank pitch factor, Eq. (4.115)

Ω angular velocity, rad/s

ω solid angle, sr; angular velocity, rad/s

SUBSCRIPTS

$\odot$	related to the Sun
a	radiation-emitting gaseous chemical species; annual
aw	adiabatic wall
b	bulk or mixed mean value for a stream; blackbody; boiling point
C	cold side or stream; capillary
c	convection; condenser; centerline; critical; coolant
conv	convection
e	external; free-stream; evaporator
F	friction
f	fin; fixed; forced; friction
fr	frontal
G	gravity; gas stream
g	gas
H	hot stream; Hemholtz instability
i	inside; internal; interfacial; initial
L	liquid stream
LTC	lumped thermal capacity
l	liquid
lm	logarithmic mean value
M	momentum; matrix
MP	melting point
m	mean; m-surface
N	normal
n	natural
o	outside; reservoir
opt	optimal
p	pumping
r	radiation; reservoir; reference
rad	radiation
s	solar; salvage; s-surface (in a fluid, adjacent to an interface or wall)
sat	saturated
T	thermal; Taylor instability
t	turbulent; taxes
tr	transition
u	u-surface (in a condensed phase, adjacent to an interface)
v	vapor phase
w	at a solid wall; wall material
0	initial
∞	far away; far upstream
λ	spectral value (hemispherically averaged for surface radiation properties)

SUPERSCRIPTS

B	buoyancy
b	black
E	forced diffusion in an electric field
H	hot stream
i	internal
o	geometric value
oa	overall
0	reference state
$*$	reduced value; dimensionless; molar; limit of zero mass transfer
$+$	directed outward from a surface
$-$	directed toward a surface
$'$	fluctuating component; per unit length
$''$	per unit area
$'''$	per unit volume

OVERSCORES

$-$	average
$\cdot$	per unit time

INDEX

821